A Dictionary of
Astronomy

Ian Ridpath is a renowned author and broadcaster on astronomy and space and is also editor of *Norton's Star Atlas*, the world's most famous star atlas. He is author of a standard series of sky guides for amateur astronomers, illustrated by Wil Tirion. His other books include *Star Tales*, about the origin and mythology of the constellations.

(⊕) SEE WEB LINKS

Many entries in this dictionary have recommended web links. When you see the above symbol at the end of an entry go to the dictionary's web page at www.oup.com/uk/reference/resources/astronomy click on **Web links** in the Resources section and locate the entry in the alphabetical list, then click straight through to the relevant websites.

🖾 Links to over 100 images are also available on the dictionary's web page. When you see the symbol at the end of an entry, go to the web page (details above), click on **illustrations** in the Resources section and locate the entry in the alphabetical list.

Oxford Paperback Reference

The most authoritative and up-to-date reference books for both students and the general reader.

forthcoming

Many of these titles are also available online at
www.Oxfordreference.com

A Dictionary of
Astronomy

SECOND EDITION REVISED

IAN RIDPATH

OXFORD
UNIVERSITY PRESS

OXFORD
UNIVERSITY PRESS

Great Clarendon Street, Oxford OX2 6DP

Oxford University Press is a department of the University of Oxford.
It furthers the University's objective of excellence in research, scholarship,
and education by publishing worldwide in

Oxford New York

Auckland Cape Town Dar es Salaam Hong Kong Karachi
Kuala Lumpur Madrid Melbourne Mexico City Nairobi
New Delhi Shanghai Taipei Toronto

With offices in

Argentina Austria Brazil Chile Czech Republic France Greece
Guatemala Hungary Italy Japan Poland Portugal Singapore
South Korea Switzerland Thailand Turkey Ukraine Vietnam

Oxford is a registered trade mark of Oxford University Press
in the UK and in certain other countries

Published in the United States
by Oxford University Press Inc., New York

First published 1997
Revised edition 2003
Reprinted with updates and corrections 2004
Second edition 2007
Second edition revised 2012

British Library Cataloguing in Publication Data
Data available

Library of Congress Cataloging in Publication Data
Data available

Typeset by SPI Publisher Services, Pondicherry, India
Printed in Great Britain by
Clays Ltd, St Ives plc

ISBN 978-0-19-960905-5

10 9 8 7 6 5 4 3 2 1

List of Contributors

Editor

Ian Ridpath FRAS

Contributors

M. A. Barstow BA, PhD, CPhys, MInstP, FRAS
Neil Bone BSc
P. A. Charles BSc, PhD
C. J. Clarke BA, DPhil
R. J. Cohen MSc, BSc, PhD
Peter Coles MA, DPhil, FRAS
Storm Dunlop FRAS
M. G. Edmunds MA, PhD, FRAS
R. M. Green MA, PhD
D. H. P. Jones MA, BSc, PhD, FRAS

A. W. Jones PhD, CPhys, MInstP
C. Kitchin BA, BSc, PhD, FRAS
John W. Mason BSc, PhD, FRAS
Andrew Murray MA
J. B. Murray MA, MPhil, PhD
Gillian Pearce BSc, PhD, BM, BCh, FRAS
Kenneth J. H. Phillips BSc, PhD
Ian Ridpath FRAS
A. E. Roy BSc, PhD, FRSE, FRAS
Robin Scagell FRAS
John Woodruff FRAS

Additional contributors to the Second Edition

R. W. Argyle BSc, MSc, FRAS
D. S. Baskill Mphys, PhD
Edward R. Boyce BSc, PhD
Stephen Eales BA, PhD
J. R. C. Garry, BSc, MSc, PhD, FRAS
Ian D. Howarth BSc, PhD, FRAS
D. McNally BSc. MSc, PhD, FRAS
R. A. Marriott
Ian Morison BA, MSc, FRAS
J. D. Shanklin MA, FRAS
P. T. Wallace, BSc, FRAS
P. R. Young BA, MA, PhD, FRAS

Additional contributors to this edition

W. Jeffrey Hughes BSc, PhD, FRAS
Jon Loveday BSc, PhD, FRAS
Robert Connon Smith BSc, PhD, FRAS

Contents

Preface

Over 4000 entries in this *Dictionary of Astronomy* cover all aspects of the subject, from the smallest and nearest objects in the Solar System to the largest and most remote structures in the Universe. The terms and names it defines range from those in common use by amateur astronomers to those familiar only to professionals. Certain entries—notably those dealing with the main objects in the Solar System, and the principal entries for stars and galaxies—provide coverage in greater depth. Relevant concepts from physics are also defined.

Entries are ordered alphabetically on a letter-by-letter basis up to the first comma. This principle gives, for example, the sequence of headwords **diverging lens, D layer, D lines, dMe star, Dobsonian telescope;** but **Hubble, Edwin Powell, Hubble classification, Hubble constant**, Headwords which include a number are ordered as though the numbers were written out in words. For example, **47 Tucanae** will be found under F, and **61 Cygni** under S. The same principle applies to headwords in which a number follows a letter, such as **H I region** ('H one'), **H II region** ('H two'), and **H₂0 maser** ('H two 0'). **S0 galaxy** is ordered as if spelt 'S nought', with apologies to American users who would look under 'S zero'. Similarly, headwords containing a Greek letter are ordered as if the letter is spelt out: for example, **Hα** is treated as 'H alpha'.

Where several variants of a given term exist, our choice of headword for the main entry was strongly influenced by *The Astronomy Thesaurus*, compiled for the International Astronomical Union by Robyn and Robert Shobbrook. The present dictionary is the first to benefit from this valuable listing, which helps to standardize astronomical terminology.

Variants of a term are included in the dictionary with a cross-reference directing the reader to the main entry. For example, a reader looking up either **microwave background radiation** or **cosmic background radiation** will be referred to the main headword, which is cosmic microwave background. Cross-references within an entry are indicated by prefixing the term with an asterisk, thus: *cosmic background radiation. Other cross-references are printed in small capitals, for example '*see* BIG BANG'.

Different senses of the same headword are numbered 1,2,.... Where necessary, these numbers are appended to cross-references, as in *dispersion (1), for example.

Some terms do not have a full entry, but are defined in a different entry, in which they arc printed in italic. For example, a reader looking up **radiative recombination** or **signal-to-noise ratio** will find a cross-reference to, respectively, **free-bound transition** and **sensitivity**, under which the terms are defined.

As in all areas of science and technology, abbreviations are commonly encountered in modern astronomy, particularly in connection with the names of observatories, organizations, and telescopes. In this dictionary, such names are written out in full in the main headword, with a cross-reference from the abbreviated form: for example **Hubble Space Telescope**, cross-referenced at **HST, Infrared Astronomical Satellite**, cross-referenced at **IRAS**. Exceptions are made where an acronym has become the commonly accepted name, such as **MACHO, SIMBAD.** and **WIMP**.

We have tried to identify all those persons who are mentioned in the text by full name, nationality, and date(s). In a few cases we were unsuccessful in tracking down all these biographical details, and would be pleased to hear from anyone who can supply additional information. Seldom-used parts of personal names are enclosed in parentheses, especially if the forename by which someone is best known is not their first, for example (Alfred Charles) Bernard Lovell.

A book like this would not exist without its contributors. My grateful thanks go to those listed on p. vi, who wrote their entries skilfully and then had to endure what must at times have seemed a near-endless stream of editorial queries. My thanks go also to those staff members of observatories who provided information on their faculties. I owe a special debt to those who freely helped in tracking down biographical information, notably John Woodruff, Christof Plicht, and Thomas R. Williams. As ever, the resources of the Royal Astronomical Society proved invaluable, and the help of its librarian, Peter Hingley, is gratefully acknowledged.

Professor Archie Roy, who initiated the project, kept a paternal eye on it throughout. At Oxford University Press, Angus Phillips kept patience with an editing process that stretched out far longer than either of us had anticipated. John Woodruff's thorough copy-editing ensured a high degree of consistency and accuracy throughout.

IAN RIDPATH

Brentford, Middlesex
1997 June

Preface to the Second Edition revised

Over 50 entries have been added to this revised edition of the *Dictionary of Astronomy*. Many of these concern new space missions and observatories. Nearly 500 other entries have been revised and updated, including links to official websites for selected entries, notably those concerning various space missions, observatories and other institutions, and online catalogues. Distances for bright stars have been brought into line with the new reduction of the *Hipparcos Catalogue* released in 2008. Spellings of star names are now the same as those in the charts on the International Astronomical Union's website, rather than the less common variants used in the *Hipparcos Catalogue* adopted previously.

IAN RIDPATH

Brentford, Middlesex
2011 January

Å Symbol for *angstrom.

AAO Abbr. for *Australian Astronomical Observatory.

AAS Abbr. for *American Astronomical Society.

AAT Abbr. for *Anglo-Australian Telescope.

AAVSO Abbr. for *American Association of Variable Star Observers.

A band A broad *Fraunhofer line in the Sun's spectrum at around 759 nm, due to absorption by oxygen in the Earth's atmosphere. Because the oxygen is in molecular form, the A band is actually a group of close, regularly spaced lines over the range 759–768 nm, unresolved at low resolution.

Abell Catalogue A catalogue of 2712 rich clusters of galaxies published in 1958 by the American astronomer George Ogden Abell (1927–83) from inspection of the *Palomar Observatory Sky Survey photographs. The catalogue had well-defined criteria for selection of the clusters (*see* ABELL CLUSTER). A later extension to the southern sky (published 1989) was based on photographs taken with the *United Kingdom Schmidt Telescope in Australia.

( ) SEE WEB LINKS
• Detailed description and full catalogue downloadable from the CDS.

Abell cluster A cluster of galaxies listed in the *Abell Catalogue. To appear in the catalogue, a cluster must satisfy selection criteria which include containing more than 50 galaxies and having a dense concentration (richness). The clusters are classified as regular (R) or irregular (I) in appearance, ranked in increasing richness from 1 to 5, and increasing distance from 1 to 6. The approximate frequency of Abell clusters is one per 2.4×10^5 cubic megaparsecs.

Abell radius A radius of about 2 megaparsecs within which at least 50 galaxies of a particular range of brightness must be found if the cluster is to qualify as an *Abell cluster.

aberration, constant of *See* ANNUAL ABERRATION.

aberration, optical An imperfection or error in the image produced by a lens, mirror, or optical system. There are six types of aberration: *chromatic aberration, *spherical aberration, coma (*see* COMA, OPTICAL), *field curvature, *distortion, and *astigmatism. Chromatic aberration is not present in images formed by mirrors. All can be corrected to a greater or lesser extent by suitable optical design.

aberration of starlight The small apparent difference between the observed direction of a star and its true direction (see diagram). It is due to the combined effect of the observer's motion across the path of incoming starlight and the finite velocity of light. The actual amount of displacement and its direction depend on the observer's speed and direction of motion. Aberration of starlight resulting from the Earth's orbital motion is termed *annual aberration; the much smaller effect resulting from the Earth's rotation is *diurnal aberration. *Planetary aberration is a combined result of the observer's motion and the time taken for light to travel from a body in the Solar System to the observer.

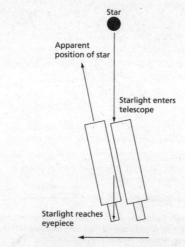

Aberration of starlight: The Earth's orbital motion alters the apparent direction of incoming starlight. In a similar effect, vertically falling raindrops appear to be travelling at an angle as seen from a moving vehicle.

ablation The wearing away of the outer layers of a body by melting, erosion, vaporization, or some other process due to aerodynamic effects as the body moves at high speed through a planetary atmosphere. Ablation can affect natural bodies such as meteoroids, or artificial objects such as spacecraft. Ablation of a spacecraft's protective heat shield prevents overheating of the spacecraft's interior during atmospheric entry.

ablation age The period of time since the outer glassy layers of a tektite solidified following ablation during its re-entry into the Earth's atmosphere. The ablation ages of known tektites vary from about 600000 to 35 million years.

absolute magnitude (*M*) 1. The brightness that a star would have if it were at a distance of 10 parsecs in perfectly clear space without *interstellar absorption. Absolute magnitude is usually deduced from the *visual magnitude measured through a V filter, and is then written M_V. If defined for another wavelength it carries a different subscript (U, B, etc.). When radiation at all wavelengths is included it becomes the absolute *bolometric magnitude, M_{bol}. The Sun has an absolute magnitude of +4.8. Most other stars range between −9 (supergiants) and +19 (red dwarfs). 2. The brightness that a Solar System body would have if it were at a distance of 1 AU from both the Sun and Earth and fully illuminated by the Sun (i.e. with a *phase angle of 0°).

absolute temperature Former name for *thermodynamic temperature.

absolute zero The zero point on the *thermodynamic temperature scale, equal to −273.15° Celsius or −459.67° Fahrenheit. It is often stated that all motion of atoms and molecules ceases at absolute zero, but in fact a small amount of energy (the **zero-point energy**) still remains. Absolute zero is the coldest temperature theoretically possible, but it can never be attained in practice.

absorption The transfer of energy from a photon to an atom or molecule. If the energy of the photon is just sufficient to raise an electron from one energy level to another, the result is an absorption line at a particular wavelength; this is called a *bound–bound transition. Absorption at a wide range of wavelengths is called **continuous absorption**, and occurs when the absorbed photons have energies greater than the minimum needed to eject an electron completely; this is called a *bound–free transition, or *photoionization. Absorption may occur whenever radiation traverses a gas, for example the outer layers of a star, the Earth's atmosphere, a nebula, or interstellar space (*see* INTERSTELLAR ABSORPTION).

absorption coefficient A measure of the decrease in intensity of radiation as it passes through a medium. It is the reciprocal of the distance required to reduce the radiation to $1/e$ of its original value (where e is a constant equal to 2.718).

absorption edge A limiting feature in a series of absorption lines from a single element (e.g. hydrogen), representing the wavelength at which the element becomes ionized. In the *Balmer series of hydrogen, the lines become closer together as the series limit of 364.6nm is reached, forming an edge at that wavelength (*see* BALMER LIMIT). Similarly, in the *Lyman series the lines become closer towards the *Lyman limit of 91.2nm.

absorption line A dark feature in the spectrum of a star, formed by cooler gas in the star's outer layers (the *photosphere) that absorbs radiation emitted by hotter gas below. The *Fraunhofer lines in the solar spectrum are the best-known examples. Each line is a unique signature of the element or molecule that forms it, which enables the chemical composition of the star to be determined. *See also* BAND.

absorption nebula *See* DARK NEBULA.

absorption spectrum A spectrum that consists primarily of dark *absorption lines, created when light from a hot source passes through cooler material. The spectra of normal, cool stars such as the Sun fall into this category.

abundance *See* ELEMENTS, ABUNDANCE OF.

Acamar The star Theta Eridani, magnitude 2.9. It is a double star, consisting of an A5 subgiant and an A1 dwarf, magnitudes 3.2 and 4.3, 161 l.y. away.

acceleration of free fall (*g*) The acceleration experienced by an object falling freely in a gravitational field, also known as the **acceleration due to gravity**. Its mean value at the Earth's surface is 9.807 m/s²; this varies slightly with latitude, because the Earth is not perfectly spherical. On any body the acceleration due to gravity can be found from the formula $g = GM/R^2$, where M is the mass of the body, R is its radius, and G is the universal *gravitational constant.

accretion The process by which the mass of a body increases by the accumulation of matter, in the form of either gas or small solid bodies which collide with and adhere to the body. The bodies in the Solar System are thought to have grown by accretion; some stars are surrounded by an *accretion disk.

accretion disk A structure that forms around a compact object (e.g. a white dwarf, neutron star, or black hole) when matter flows towards it. Accretion disks are found in interacting binary stars, and are assumed to exist in *active galactic nuclei and *quasars. In a binary, mass lost from the secondary star can form a disk of gas around the compact object. The disk may have a *hot spot (1) where the stream of material hits its outer edge. Material is fed from the inner edge of the disk through a *boundary layer* (which may radiate as much energy as the disk itself) on to the compact object. When the compact object has an extremely strong magnetic field, as in an *AM Herculis star, the material may form an **accretion column** over each magnetic pole, rather than a disk. The gravitational energy that is released can cause high ultraviolet or X-ray luminosities, and may accelerate jets of material from the disk to very high speeds.

ACE Abbr. for *Advanced Composition Explorer.

Achernar The star Alpha Eridani, a B3 dwarf of magnitude 0.45, ninth-brightest in the sky, 139 l.y. away. Its name comes from Arabic and means 'river's end'.

Achilles Asteroid 588, the first *Trojan asteroid to be discovered, by Max *Wolf in 1906. It is a member of the group of Trojans at the L_4 *Lagrangian point 60° ahead of Jupiter. Achilles is a D-class asteroid of diameter 135 km. Its orbit has a semimajor axis of 5.195 AU, period 11.84 years, perihelion 4.43 AU, aphelion 5.96 AU, and inclination 10°.3.

achondrite A class of stony meteorite usually (though not always) lacking the tiny, rounded inclusions known as chondrules found in *chondrites. Achondrites make up about 9% of all meteorite falls. They consist principally of one or more of the minerals plagioclase, pyroxene, and olivine. The main distinction between achondrites and chondrites is that the achondrites have different abundances of calcium and similar elements, and almost no metal or sulphide. Achondrites are thought to have crystallized from a magma in the same way as terrestrial rocks. The achondrites are divided into five main classes. The two main calcium-rich classes (containing more than 5% calcium) are the *pigeonite–plagioclase achondrites* (*eucrites) and the *plagioclase–hypersthene achondrites* (*howardites). There are three main calcium-poor classes (usually less than 1% calcium): the *hypersthene achondrites* (*diogenites), the *olivine–pigeonite achondrites* (*ureilites), and the *enstatite achondrites* (*aubrites). The eucrites, howardites, and diogenites are often collectively referred to as the *basaltic achondrites. The *pigeonite–maskelynite achondrites* (*shergottites), *augite–olivine achondrites* (*nakhlites), and *olivine achondrites* (*chassignites) comprise the rare *SNC meteorites, which are thought to come from Mars. There is also a very rare class of augite achondrite, the angrites, named after the Angra dos Reis meteorite, which fell in Brazil in 1869.

achromatic Describing a lens consisting of two or more optical components (*elements*), intended to correct for *chromatic aberration. Commonly used as the objective of small refractors, the achromatic lens (or *achromat*) was invented in 1729 by the English optician Chester Moor Hall (1703–71) and first manufactured commercially by J. *Dollond in 1758. It has one element of *crown glass and another of *flint glass. The *dispersion (1) of the crown glass compensates for the chromatic error of the flint glass, while still leaving some refractive power. The two-element design is termed an **achromatic doublet**. It is practically impossible to correct all wavelengths of light, however, and most lenses adopt a compromise, bringing two particular wavelengths to a common focus, thus reducing the false colour. A lens that corrects for more than two wavelengths is termed an *apochromatic lens.

achromatism Freedom from false colour (*chromatic aberration) in an optical system. In reality, no optical system containing lenses can ever be completely free from false colour, but the aim is to reduce the amount of false colour to acceptable proportions. A mirror is completely achromatic.

A-class asteroid A rare class of asteroid that has both a moderately high albedo (0.13–0.35) and an extremely reddish spectrum at wavelengths shorter than 0.7 μm. Strong absorption in the near-infrared is interpreted as indicating the presence of the mineral olivine. Members of this class include (246) Asporina, diameter 60 km, and (446) Aeternitas, diameter 45 km.

acronical Referring to the rising or setting of a celestial object at or shortly after sunset. A planet's rising is acronical when it is at *opposition.

Acrux *See* ALPHA CRUCIS.

active galactic nucleus (AGN) The central region of a galaxy which emits far more radiation than can be produced by stars alone. A galaxy with such a nucleus is known as an **active galaxy**. The first active galaxies were discovered in 1943 by C. K. *Seyfert, who noticed that some spirals have an exceptionally bright core; these are now termed *Seyfert galaxies. *Quasars, discovered in the 1960s, appear to be a more luminous version of Seyferts in which the nucleus is so bright that the surrounding galaxy can hardly be seen at all. About 10% of active galaxies have the additional characteristic of being strong radio sources, or 'radio-loud'. The radio emission from such *radio galaxies is often from regions well outside the galaxy itself known as *lobes, although in some cases there are radio *jets pointing back towards the centre of the galaxy, indicating that it is the nucleus which is the ultimate source of the energy.

The widely accepted theory of the origin of the energy in active galaxies is that they contain a central black hole with a mass of up to 10^9 solar masses. Three main pieces of evidence lead to this conclusion. First, active galactic nuclei are often highly variable, meaning that the radiation is from a region not much bigger than the Solar System. Second, the spectral lines of an active galactic nucleus are often very broad, up to 10000 km/s, indicating the presence of something very massive in the centre. Third, there is no credible alternative.

The radiation is produced not by the black hole itself but by gas in a surrounding *accretion disk. The viscosity in the disk causes gas gradually to spiral towards the black hole. As it does so, gravitational energy is converted into heat, and the hot gas then emits large amounts of radiation. Although this theory is widely accepted, there are still many unanswered questions, including how the jets seen in radio galaxies are produced.

In addition to Seyfert galaxies, quasars, and radio galaxies, the range of active galaxies now includes *blazars and *BL Lac objects, plus many subspecies. However, it seems likely that at heart most active galactic nuclei are physically similar. Their diversity in outward appearance has two main causes. One is differences in the energy output of the 'central engine' itself—quasars, for example, are particularly powerful. The second factor is the angle from which the AGN is viewed.

Surrounding the black hole and accretion disk is thought to be a ring of molecular gas and dust. Within the ring there are tiny dense clouds of gas which are heated by the radiation from the accretion disk. It is these clouds which are the source of the broad spectral lines seen from quasars and some types of Seyfert galaxy (Seyfert 1s). The clouds farther from the ring are also heated by the radiation from the accretion disk, and these clouds produce the narrow spectral lines seen from the other main class of Seyfert galaxies (Seyfert 2s). Radio jets (where they exist) are emitted along the axis of the accretion disk.

This model can explain many of the apparent differences between active galactic nuclei. If we view the AGN along the axis of the ring, we see the radiation emitted by the accretion disk and also broad spectral lines. However, if we view the AGN through the ring, this radiation and the broad spectral lines are hidden by the dust in the ring, and we can see only the narrow spectral lines produced by the clouds farther out. The appearance of a 'radio-loud' AGN will also depend on the viewing angle. Viewed from close to the axis of the ring, a radio-loud AGN may be classified as a blazar and the radio jet will often appear to be moving faster than the speed of light (*superluminal velocity); viewed from close to the plane of the ring, the AGN will be classified as a radio galaxy, with not many signs of activity in the nucleus, but with huge lobes of radio-emitting plasma on either side of the galaxy.

It now seems quite likely that most galaxies have passed through an AGN phase. Astronomers have recently discovered that most nearby galaxies contain *supermassive black holes, including one of a few million solar masses in the centre of our Galaxy. Observations show that there were many more active galactic nuclei in the past than there are today, and it seems likely that the inactive nearby galaxies once contained an AGN, but for some reason in these galaxies the central engine has now run out of fuel. Possibly all the gas in the vicinity of the black hole has now been consumed.

active optics A system that compensates for the deforming effects of gravity on a telescope's mirrors, maintaining their surface accuracy and alignment. The image of a guide star is examined as the telescope tracks it across the sky. Actuators behind the mirrors control movable supports to preserve the mirrors' shape and alignment. The first large telescope to employ active optics was the *New Technology Telescope at the European Southern Observatory.

active prominence A solar prominence with very rapid motion (up to 2000 km/s), often associated with a flare. Active prominences are located at low latitudes on the Sun, where sunspots and active regions are usually found. The main categories include *loop prominences, *coronal rain, *surge prominences, *sprays, and *arch filament systems.

active region An area on the Sun where magnetic fields emerge through the photosphere into the chromosphere and corona. Active regions on the photosphere include *sunspots and *faculae. Their counterparts higher in the chromosphere are *plages. Also in the chromosphere are dark *fibrilles and *filaments. Active regions in the corona are areas of enhanced density and temperature sometimes called *coronal condensations. Other examples of active regions are areas on the photosphere where sunspots have faded, and *X-ray bright points in the corona. *Flares occur in active regions. Active regions are allocated numbers by the US government's National Oceanic and Atmospheric Administration (NOAA) in order of appearance. The present numbering system started on 1972 January 5. Four-digit numbers are assigned up to 9999, after which the numbering returns to 0000, as happened on 2002 June 14; the numbers are prefixed by AR for Active Region. Active regions that exist for more than one solar rotation are given a new number each time they reappear.

Adams, John Couch (1819–92) English mathematical astronomer. In 1843 he began to calculate the orbit of a new planet whose gravitational effects would explain why Uranus did not follow its predicted path, but he did not complete his calculations or publish them. In 1846 the new planet, subsequently named Neptune, was sighted by J. G. *Galle as a result of independent calculations by U. J. J. *Le Verrier. Adams and Le Verrier were eventually both credited with predicting Neptune's existence, although it is now clear that the credit belongs mainly to Le Verrier. Adams' later work included calculation of the Moon's *secular acceleration, and of the orbital elements of the Leonid meteor swarm.

Adams, Walter Sydney (1876–1956) American spectroscopist, born in Syria. He was the first to detect systematic differences in the spectra of giant and dwarf stars. From 1914, with the German astronomer Arnold Kohlschütter (1883–1969), Adams developed methods of establishing the surface temperature, luminosity, and distance of stars from their spectra. In 1918 he showed that the density of the white dwarf Sirius B is $50000 g/cm^3$. Adams also carried out spectrographic studies of the atmospheres of Mars and Venus.

Adams Ring The outermost of Neptune's rings, named after J. C. *Adams. It lies 62930km from Neptune's centre, and is about 20km wide. It contains four denser sections, or ring arcs, named Courage, Liberté, Egalité, and Fraternité.

adaptive optics An optical design that can rapidly counteract the effects of atmospheric seeing on an image. This may be done by deforming a mirror in the light path of a telescope to keep a star's image as point-like as possible. The system may use as a reference a real star, or an artificial star produced by shining a laser up through the layers of air that are causing the bad seeing. Any extended objects in the field, such as galaxies, will also be sharpened. This technique can increase the resolution of a ground-based telescope by a factor of 40.

Adhara The star Epsilon Canis Majoris, magnitude 1.5. It is a B2 supergiant, 405 l.y. away, with a companion of magnitude 7.4.

adiabatic process A change or process in which no heat enters or leaves a system, as occurs for example in an expanding or contracting gas cloud. An adiabatic change is usually accompanied by a rise or fall in the temperature of the system; ionization of atoms or dissociation of molecules may also occur. *See also* ISOTHERMAL PROCESS.

Adonis Asteroid 2101, the second of the *Apollo group to be discovered, by the Belgian astronomer Eugène Joseph Delporte (1882–1955) in 1936, when it passed within 0.015 AU (2.2 million km) of the Earth. It was not seen again until 1977. Adonis is about 1km in diameter. Its orbit has a semimajor axis of 1.875 AU, period 2.57 years, perihelion 0.44 AU, aphelion 3.31 AU, and inclination 1°.3.

Adrastea The second-closest satellite of Jupiter, distance 128980km, orbital period 0.298 days; also known as Jupiter XV. Adrastea is $20 \times 16 \times 14$km in size. It was discovered in 1979 with the Voyager 2 spacecraft. It lies less than one Jovian radius above the planet's cloud tops and just within the outer rim of Jupiter's main ring. Material knocked off Adrastea by micrometeorite impacts is thought to contribute to the ring.

ADS **1**. Abbr. for *Aitken Double Star Catalogue.
2. Abbr. for *Astrophysics Data System.

Advanced Composition Explorer (ACE) A NASA spacecraft to study the composition of the *solar wind and cosmic rays, launched 1997 August. It is stationed in a halo orbit around the L_1*Lagrangian point 1.5 million km sunwards of the Earth. ACE carries nine instruments to sample low-energy particles of solar origin and high-energy particles from the Galaxy. From its position upstream in the solar wind it provides advance warning of geomagnetic storms.

(((∰))) SEE WEB LINKS
• Official mission website.

advance of perihelion A gradual turning of the major axis of an orbit in the same direction as the body moves along the orbit; also known as **apsidal motion**. An advance of perihelion means that the *longitude of perihelion of the orbit increases. In the Solar System, the effect is caused mostly by the gravitational attractions of the planets. A similar effect, the **advance of pericentre**, is seen in the orbits of binary stars, caused by the oblateness of the stars themselves. In the 19th century, a small unexplained advance of Mercury's perihelion of some $43''$ per century was attributed to an undiscovered planet within the orbit of Mercury (*see* VULCAN). This effect is now known to be caused by the curvature of space near the Sun, as predicted by the general theory of relativity.

Advanced Technology Solar Telescope A 4-m reflector for solar studies being built at an altitude of 3055 m on Haleakala, Hawaii, by a consortium of over 20 US institutions led by the *National Solar Observatory. Using *adaptive optics it will be able to resolve features on the Sun as small as 30 km across at wavelengths from the near ultraviolet to the infrared (0.3–28 μm). When opened in 2014 or later it will be the largest solar telescope in the world.

(((∰))) SEE WEB LINKS
• Official telescope website.

advection Transfer by a moving fluid, generally in a direction perpendicular to a gravitational field. The term is applied to both the shifting of planetary atmospheric gases by horizontal motion, and also to the resultant transfer of heat, for example from low to high latitudes. More recently, advection has also come to mean the transfer of heat vertically within a planetary body, for example by hot, molten material rising through the lithosphere of a planet.

aeon A period of 10^9 years (i.e. a billion years); US *eon*.

aerial Another name for an *antenna.

aerobraking The technique of using the atmospheric drag of a planet to modify the orbit of a spacecraft, thereby saving propellant. It was first used at another planet by the US Venus probe Magellan in 1993. Aerobraking lowered the high point of Magellan's orbit above Venus from 8500 to 600 km and shortened its orbital period from 195 to 94 min. Aerobraking can also be used to help space probes enter orbit around a planet, as with the Mars Global Surveyor craft in 1997.

aerolite An alternative name for a *stony meteorite, now largely obsolete.

aeronomy The study of processes and phenomena in the Earth's upper atmosphere, from the *mesopause to its uppermost limit which, for most practical purposes, can be taken to lie at an altitude of around 300 km. Among the processes studied are *airglow, reactions in the *chemosphere, and the formation of *noctilucent clouds and the *ionosphere.

aerosol A small particle, either solid or liquid, suspended in an atmosphere. Aerosols cause *atmospheric extinction. Some properties of the particles, such as their mean size and oblateness, may be inferred from the way in which an aerosol layer scatters light.

Ae star A star of spectral type A which exhibits emission lines in its spectrum (hence the suffix 'e'). Usually these are sharp emission lines of hydrogen, and are superimposed on an otherwise normal spectrum. These lines arise in a surrounding expanding shell or disk of material. Ae stars are young stars still in the process of formation. *See also* HERBIG AE/BE STAR.

aether *See* ETHER.

afocal Describing an optical system in which an image is transferred without bringing it to a focus. Afocal photography, for example, involves pointing a camera, focused on infinity, into the eyepiece of a telescope whose image also appears at infinity. When an image is transferred as a beam of parallel light rays, the beam is termed an **afocal beam**.

afterglow The electromagnetic radiation often observed after a *gamma-ray burst (GRB). As a shockwave of gas from a GRB ploughs into the surrounding gas and dust, kinetic energy is converted into radiation, creating the afterglow. The emission is *synchrotron radiation from electrons accelerated in magnetic fields within the shock and can be observed at X-ray, optical, and radio wavelengths in succession as the shockwave expands and cools. The afterglow appears first hours to weeks after the initial GRB is detected, and later may be detected at longer wavelengths. It can last about a week as seen at X-ray wavelengths, for several weeks at optical wavelengths, and up to a year at radio wavelengths.

AGB star Abbr. for *asymptotic giant branch star.

Agena Alternative name for the star *Hadar or Beta Centauri.

agglutinate A small object consisting of impact glass and fragments of minerals or rocks, all welded together into an **aggregate**. Agglutinates are produced by the impact of micrometeorites into the lunar regolith or other planetary surface.

AGILE An Italian satellite for high-energy astrophysics, launched in 2007 April to observe at energies of 30 MeV–50 GeV and 10–40 keV; the name is an acronym of Astro-rivelatore Gamma a Immagini LEggero. AGILE carries three main instruments: a Gamma-Ray Imaging Detector (GRID) with an exceptionally large field of view, covering the energy range 30 MeV–50 GeV across one-fifth of the entire sky simultaneously; a hard X-ray coded-mask imaging detector, Super-AGILE, observing at 10–40 keV; and a caesium iodide Mini-Calorimeter (MC), a non-imaging detector for the 0.25–200 MeV range, providing spectral and timing information of transient events. The satellite's targets include gamma-ray bursts, AGNs, solar flares, and pulsars.

(()) SEE WEB LINKS
• Official mission website.

AGK Abbr. for *Astronomischen Gesellschaft Katalog*, a series of catalogues of star positions. AGK1 covered most of the sky, observed by meridian circles around the world; it was published between 1890 and 1954. AGK2 (1951–8) was a repetition of AGK1, with additional stars, from the north pole to −2° declination observed photographically from Hamburg and Bonn in about 1930. It contains 181581 stars, mostly to about magnitude 9 plus some fainter stars. AGK2A (published 1943) contains positions of 13747 reference stars for AGK2, mostly of 8th and 9th magnitudes, derived from meridian observations made from Germany and Pulkovo, Russia. AGK3 (published 1975) was a reobservation of AGK2; the prefix AGK was retained even though the catalogue was not published under the auspices of the Astronomischen Gesellschaft. AGK3R contains the positions of 21499 reference stars for AGK3, observed in an international programme from about a dozen observatories. *See also* SOUTHERN REFERENCE STARS.

(()) SEE WEB LINKS
• Detailed description and full catalogue downloadable from the CDS.

AGN Abbr. for *active galactic nucleus.

Ahnighito meteorite Another name for the *Cape York meteorite.

airglow A weak background emission of light over the whole sky, resulting mainly from the excitation of atmospheric oxygen at an altitude of around 100 km by solar radiation; also known as *nightglow*. Airglow emissions vary with the time of day. Green oxygen emission at a wavelength of 557.7 nm predominates at night. Sodium and red oxygen emissions are prominent in the twilight airglow. Airglow is also produced during daylight, and is 1000 times as intense as at night.

airmass The pathlength of starlight through the Earth's atmosphere, expressed in relation to the value at the zenith. Airmass is 1 at the zenith and approximately 2 at an altitude of 60°. It is approximately the secant of the zenith distance, but rises more slowly as the star approaches the horizon, becoming 1% less than the secant at 17° altitude.

air shower Another name for a *cosmic-ray shower.

Airy, George Biddell (1801–92) English astronomical administrator and geophysicist. During his tenure as the seventh Astronomer Royal, the Royal Observatory at Greenwich became a model of efficiency for positional astronomy. He belittled pure research, however, which made Greenwich a late starter in the fields of spectroscopy and astrophysics. The transit telescope he installed at Greenwich in 1851 now defines the position of 0° longitude on Earth. Airy's only significant astronomical discovery was of irregularities in the orbits of Venus and the Earth. In 1854, by making gravity measurements at the top and bottom of a mineshaft, he estimated the Earth's mass.

Airy disk The central spurious disk of the image of a star formed by a telescope. Because of *diffraction, even with perfect optics a star's image is never point-like, but consists of a central disk, the Airy disk, surrounded by several fine *diffraction rings. All telescopes of given size have the same size of Airy disk, which gets smaller with increasing aperture. The size of the Airy disk is given approximately in radians by 1.22λ times the f/number, where λ is the wavelength of the light. The size of the Airy disk limits the resolving power of a telescope, although in apertures larger than about 100 mm the Airy disk is often smaller than the false disk caused by *seeing. It is named after G. B. *Airy.

Aitken, Robert Grant (1864–1951) American astronomer. At Lick Observatory he carried out a vast survey of double stars, initially with William Joseph Hussey (1862–1926), discovering over 3100 new pairs. In 1923 he was the first to see the companion of *Mira. His *New General Catalogue of Double Stars* was published in 1932 (*see* AITKEN DOUBLE STAR CATALOGUE).

Aitken Double Star Catalogue (ADS) Popular name for the *New General Catalogue of Double Stars Within 120° of the North Pole* by R. G. *Aitken, published in 1932, containing measurements of 17 180 double stars. It succeeded the *General Catalogue of Double Stars* published in 1906 by the American observer Sherburne Wesley Burnham (1838–1921). Double stars are often referred to by their ADS number as listed in Aitken's catalogue.

Al Velorum star A pulsating variable star of short period (0.04–0.2 days) that closely resembles a *Delta Scuti star, although with a greater amplitude (0.3–1.2 mag. or more) and somewhat higher luminosity. The Al Velorum stars may be the older of the two groups. They are also called *high-amplitude Delta Scuti stars*.

AJ Abbr. for *Astronomical Journal.*

Akari A Japanese infrared astronomy satellite, known before its launch in 2006 February as ASTRO-F; the name means 'light' in Japanese. Akari carries a 0.69-m telescope equipped with two main instruments: the Far-Infrared Surveyor (FIS) to make an all-sky survey in the wavelength range 50–180 μm; and the InfraRed Camera (IRC) for targeted observations in the near and medium infrared range, 1.7–25.6 μm. Akari's observations extend and improve those of the *Infrared Astronomical Satellite (IRAS).

al- For personal names beginning 'al-', *see* the next element of the name.

Albategnius Latinized name of al-*Battānī.

reflected by the surface [handwritten annotation]

al-Battānī *See* BATTĀNĪ.

albedo The fraction of the total light or other radiation falling on a non-luminous body, such as a planet, or on a planetary surface feature, that is reflected from it. In general, the albedo is equal to the amount of light reflected divided by the amount of light received. Albedo values range from 0.0 (0%) for a totally black surface that absorbs all incident light, to 1.0 (100%) for a perfect reflector. Planets or planetary satellites with dense atmospheres have much higher albedos than those with transparent or no atmospheres. The albedo may vary over the surface, so for practical purposes an average albedo is specified. Natural surfaces reflect different amounts of light in different directions, and albedo can be expressed in several ways according to whether the measurement is made in one direction or averaged over all directions. *See* BOND ALBEDO; GEOMETRICAL ALBEDO; HEMISPHERICAL ALBEDO.

albedo feature A feature on a planet that is markedly darker or brighter than its surroundings. Albedo features do not always correspond to a topographic or geological feature, as differences in albedo can arise from variations in surface composition as well as topography. For example, one of the darkest features on Mars seen through a telescope is *Syrtis Major, but spacecraft have found no distinctive differences in topography or type of terrain between it and its surroundings. In general, though, albedo features are related to different terrain types, as on the Moon where the dark maria are plains of solidified lava and the bright spots are relatively recent impact craters.

Albireo The star Beta Cygni, one of the best-known double stars in the sky. It consists of a K3 bright giant of magnitude 3.1 and a B9.5 dwarf of magnitude 5.1. The stars show contrasting colours of orange and blue-green. The primary is also a close visual binary with a period of just over 200 years. Albireo lies 434 l.y. away.

Alcor The star 80 Ursae Majoris, an A5 dwarf of magnitude 4.0, distance 82 l.y. It forms a naked-eye double with *Mizar.

Alcyone The star Eta Tauri, magnitude 2.9, the brightest member of the *Pleiades star cluster. It is a B7 giant 403 l.y. away.

Aldebaran The star Alpha Tauri. It is a K5 giant that varies irregularly by about 0.1 mag. from its average brightness of magnitude 0.87. It appears to be a member of the *Hyades star cluster but is in fact only 67 l.y. away, less than half the cluster's distance.

Alderamin The star Alpha Cephei, magnitude 2.4. It is an A7 dwarf 49 l.y. away.

Alexandra family A small family of asteroids at a mean distance of 2.6–2.7 AU from the Sun, with inclinations of 11–12°. The family is unusual in that its members are of varying composition (classes C, G, and T). The family is named after the C-class (54) Alexandra, 166 km in diameter, discovered in 1858 by the German astronomer Hermann Mayer Salomon Goldschmidt (1802–66). Alexandra's orbit has a semimajor axis of 2.713 AU, period 4.47 years, perihelion 2.18 AU, aphelion 3.25 AU, and inclination 11°.8.

ALEXIS A small US Department of Energy satellite carrying six telescopes to map the soft X-ray and extreme ultraviolet background over the whole sky. ALEXIS (Array of Low-Energy X-ray Imaging Sensors) was launched in 1993 April. The mission ended in 2005 April.

(⊕) SEE WEB LINKS
• Official mission website.

Alfvén, Hannes Olof Gösta (1908–95) Swedish physicist. In the 1930s he developed a theory of sunspot formation based on the idea that under certain conditions a magnetic field can be 'frozen in' to a plasma. In 1942 he proposed that waves (now called *Alfvén waves) can propagate through a plasma under conditions similar to those found in the Sun's atmosphere. His work inaugurated the study of *magnetohydrodynamics, for which he was awarded a share of the 1970 Nobel Prize in Physics.

Alfvén surface The surface of the region surrounding a neutron star within which ionized gas is pulled around by the star's magnetic field as it spins. It is named after H. O. G. *Alfvén.

Alfvén wave A transverse wave that occurs in a region containing a magnetic field and a plasma. The ionized and therefore highly conducting material of the plasma is said to be 'frozen in' to the magnetic field and is forced to take part in its wave motion. At least part of the energy heating the Sun's corona is thought to be provided by Alfvén waves propagating from the outer layers of the Sun. The waves are named after H. O. G. *Alfvén. *See also* MAGNETOHYDRODYNAMICS.

Algenib The star Gamma Pegasi, a B2 subgiant of magnitude 2.8, lying 392 l.y. away. Algenib is also an alternative name for the star *Mirfak (Alpha Persei).

Algieba The star Gamma Leonis. It is a binary consisting of a K1 giant of magnitude 2.6 and a G7 giant of magnitude 3.5 (combined magnitude 2.0). The stars orbit each other with a period of 510 years, and lie 130 l.y. away.

Algol The star Beta Persei, the first eclipsing binary to be discovered. Algol was found to be variable in 1669 by the Italian astronomer and mathematician Geminiano Montanari (1633–87), but the period was first determined by J. *Goodricke in 1782–3. Algol varies from magnitude 2.1 to 3.4 in a period of 2.8673043 days, although the period has varied slightly. The variable period and the emission lines that are sometimes detectable in its spectrum are evidence for *mass transfer, indicating that the system is a *semidetached binary. The eclipsing pair, which consists of a B8 dwarf and a K2 subgiant, is accompanied by a third component (Algol C), magnitude 4.7, having an orbital period of 1.862 years. The system is a faint X-ray source and also emits radio bursts. Algol lies 90 l.y. away.

Algol star A type of eclipsing binary with periods of constant (or almost constant) brightness between well-defined eclipses; abbr. EA. This feature is an indication that the system is a *detached binary or *semidetached binary. The secondary minimum may be invisible. Periods range from 0.2 to 10000 days, and the amplitudes may reach several magnitudes. If *mass transfer occurs, the material is accreted directly on to the hot star, not via an *accretion disk. Mass transfer via an accretion disk does occur in the *W Serpentis stars*, which have wider separations and may be pre-Algol stars. The terms *W Serpentis star* and *Serpentid* are sometimes used to include both groups of stars.

Algonquin Radio Observatory A radio observatory in Algonquin Park near Lake Traverse, Ontario, site of a 45.7-m antenna opened in 1966. It was operated by the National Research Council of Canada until 1991 and then jointly by the Geodetic Survey Division of Natural Resources Canada, Ottawa, and the Space Geodynamics Laboratory of York University, Toronto. Since 2008 it has been operated by Thoth Technology Inc. of Kettleby, Ontario. It is now used primarily for geodetic and astrometric Very Long Baseline Interferometry (VLBI).

(⊕) SEE WEB LINKS
• Official observatory website.

Alhena The star Gamma Geminorum, magnitude 1.9. It is a subgiant of type A1 lying 109 l.y. away.

aliasing A phenomenon in which a digitized signal is observed to contain spurious low-frequency components. These occur if the original signal is sampled at a rate insufficient to record the highest frequencies present (*undersampling*).

alidade A simple sighting device for measuring altitudes. It consists of a bar pivoted so as to swing in a vertical plane and be aligned with a celestial object. The object's altitude can then be read off from a scale. Alidades were often incorporated in ancient position-measuring instruments, such as *astrolabes.

Alioth The star Epsilon Ursae Majoris. It is an A0 subgiant of magnitude 1.8 with strong lines of chromium in its spectrum. It is a variable of the Alpha^2 Canum Venaticorum type, fluctuating by a few tenths of a magnitude with a period of 5.1 days. It lies 83 l.y. away.

Alkaid The star Eta Ursae Majoris, also known as Benetnasch. It is a B3 dwarf of magnitude 1.9, lying 104 l.y. away.

Allegheny Observatory The observatory of the University of Pittsburgh, at an altitude of 380 m in Pittsburgh, Pennsylvania, founded in 1860 but relocated on its present site, Riverview Park, in 1912. Its main instrument is the 0.76-m Thaw refractor, in operation since 1914 but fitted with a new objective lens in 1985. Other instruments include the 0.74-m Keeler astrometric reflector, opened in 1906 but given new optics in 1992.

(((•))) SEE WEB LINKS
• Official observatory website.

Allende meteorite A meteorite of *carbonaceous chondrite type, which fell near Pueblito de Allende, northern Mexico, on 1969 February 8, scattering thousands of fragments over an area of 48 × 7 km. The parent body probably weighed over 30 tonnes. More than 2 tonnes of CV3-type material were collected, the largest piece weighing 110 kg. The formation age of this material is 4.6 billion years, making it some of the oldest primordial planetary material to be recovered.

Allen Telescope Array (ATA) A radio telescope array at *Hat Creek Radio Observatory, California, designed to search for extraterrestrial intelligence while simultaneously carrying out astronomical observations. The ATA is jointly owned and operated by the SETI Institute of Mountain View, California, and the University of California, Berkeley. When completed it will consist of 350 antennas, each of 6.1 m diameter, which can be combined electronically to form 16 different beams, each having the sensitivity of a single 114-m dish and the resolution of a 700-m dish. The ATA observes at wavelengths from 3 cm to 60 cm simultaneously. Its first 42 antennas began operation in 2007. Two intermediate stages, with 98 and 206 antennas, will enter service before the full array is completed. The ATA takes its name from the Paul G. Allen Foundation, which provided funding.

(((•))) SEE WEB LINKS
• Official telescope website.

all-sky camera A camera with a field of view that includes all, or nearly all, of the sky on one frame. The images from such cameras are circular, and have the zenith at the centre and the horizon around the edge. They are used in particular for meteor and fireball 'patrols'. A simple design of all-sky camera consists of a convex mirror which reflects the image of the sky to a camera above it, pointing downward. More advanced designs use ultra-wide-angle lenses.

ALMA Abbr. for *Atacama Large Millimeter Array.

Almach The star Gamma Andromedae; also spelled Almaak and Alamak. It is a double star consisting of a K3 bright giant and a B9 dwarf, magnitudes 2.3 and 4.8 (combined magnitude 2.1). The fainter star is itself a close double with an orbital period of 64 years. Almach is 393 l.y. away.

Almagest A compendium of astronomical and mathematical knowledge written by *Ptolemy in about AD 150. It incorporates the star catalogue compiled by *Hipparchus, upon whose work Ptolemy may have drawn for other parts of the book. It is the most complete surviving treatise on ancient astronomy, and contains descriptions of the 48 Greek constellations on which our present-day constellation system is based. Its original Greek title was *Syntaxis*; *Almagest*, the name it acquired when translated into Arabic in about AD 820, means 'the greatest'.

almanac A publication, usually issued yearly, listing predicted dates and times of forthcoming celestial phenomena and positions of celestial objects, along with other

information of interest to astronomers, navigators, and surveyors. Examples are *The
Astronomical Almanac and *The *Nautical Almanac*.

almucantar A small circle on the celestial sphere that is parallel to the horizon. All objects
on an almucantar are at the same altitude at a given time.

Alnair The star Alpha Gruis, magnitude 1.7. It is a B7 dwarf, 101 l.y. away.

Alnath An alternative name for the star *Elnath.

Alnilam The star Epsilon Orionis, magnitude 1.7, the middle star of the belt of Orion. It is a
B0 supergiant, estimated to be about 2000 l.y. away.

Alnitak The star Zeta Orionis, one of the three stars of the belt of Orion. It is a supergiant
of type O9.5, magnitude 1.9, with a close companion of magnitude 3.9, a B0 giant that orbits
it every 1500 years; the combined magnitude of the pair is 1.7. Alnitak is 740 l.y. away.

α Symbol for *right ascension.

Alpha Capricornid meteors A meteor shower of generally low activity (maximum
ZHR 10) between July 3 and August 15. There appear to be several maxima, the principal
one around July 30. Alpha Capricornid meteors are often long, slow, and bright. At maximum,
the radiant lies at RA 20h 28m, dec. −10°.

Alpha Centauri The closest star to the Sun, also known as Rigil Kentaurus. It is actually a
triple system, consisting of a bright binary with a period of 80 years and a faint red dwarf
2° away called *Proxima Centauri. The binary consists of a G2 dwarf of magnitude −0.01 and
a K1 dwarf of magnitude 1.3. To the naked eye they appear as a single star of magnitude
−0.28, the third-brightest in the sky. The binary pair lie 4.32 l.y. from the Sun, 0.09 l.y.
farther than Proxima Centauri.

Alpha Crucis The brightest star in the constellation Crux, also called Acrux. It is a double star
consisting of a B0.5 subgiant, magnitude 1.3, and a B1 dwarf, magnitude 1.7. To the naked
eye they appear as a single star of magnitude 0.8. Alpha Crucis is 322 l.y. distant.

Alpha Cygnid meteors A putative meteor shower with ill-defined activity limits,
emanating throughout July and August from an apparently stationary radiant at RA 21h 00m,
dec. +48° near Deneb. Observed rates seldom exceed 1–3 meteors/h, and there are doubts
as to the shower's authenticity.

Alpha Cygni star A type of supergiant pulsating variable star exhibiting *non-radial
pulsation; abbr. ACYG. The spectral types are Be-Ae Ia, and the optical amplitude is
approximately 0.1 mag. Multiple pulsation frequencies are superimposed, giving rise to
light-curves that often appear highly irregular. The periods range from a few days to several
weeks.

alpha particle (α-particle) The nucleus of a helium-4 atom. It consists of two protons
and two neutrons and hence is positively charged. Alpha particles are emitted by the
nuclei of atoms in a process of radioactive decay known as **alpha decay**.

Alphard The star Alpha Hydrae, magnitude 2.0. It is a K3 giant, 180 l.y. away.

Alpha Scorpiid meteors A meteor shower, active from April 20 to May 19, producing
rather low observed rates (maximum ZHR 10). Its parent meteor stream lies close to the ecliptic,
and has been split into at least two branches by planetary perturbations. Maxima occur on
April 28 from a radiant at RA 16h 32m, dec. −24°, and on May 13 from RA 16h 04m, dec. −24°.
It is not recognized as a genuine shower by some authorities.

Alpha² Canum Venaticorum star A type of main-sequence, extrinsic variable star;
abbr. ACV. Stellar rotation produces brightness variations of 0.01–0.1 mag., accompanied by
changes in the strengths of the spectral lines and the magnetic fields. The rotational periods
range from 0.5 days to more than 160 days. The spectra (B8p–A7p) exhibit abnormally

strong lines of silicon, strontium, chromium, and rare-earth elements. Stars in the ACVO subtype undergo *non-radial pulsation (periods 0.004–0.01 days), with small (0.01-mag.) variations superimposed on the fluctuations caused by the rotation.

Alphecca The star Alpha Coronae Borealis, also spelt Alphekka and also known as Gemma. It is an eclipsing binary of the Algol type, varying between magnitudes 2.2 and 2.3 with a period of 17.4 days. The two stars are dwarfs of spectral types A0 and G5. Alphecca lies 75 l.y. away.

Alpheratz The star Alpha Andromedae, magnitude 2.1. It is a B9 subgiant 97 l.y. away. An alternative name is Sirrah.

Alrescha The star Alpha Piscium, also spelt Alrischa. It is a close binary consisting of two A-type stars of peculiar spectra and unknown luminosity class, magnitudes 4.2 and 5.2, with an orbital period of about 930 years. Together they appear as a star of magnitude 3.8. They lie 151 l.y. away.

Alshain The star Beta Aquilae. It is a G8 subgiant of magnitude 3.7, lying 45 l.y. away.

al-Ṣūfī *See* ṢŪFĪ.

Altair The star Alpha Aquilae, magnitude 0.76, the twelfth-brightest star in the sky. It is an A7 dwarf 16.7 l.y. away. Altair forms one corner of the so-called Summer Triangle of stars, completed by Deneb and Vega.

altazimuth mounting A method of mounting a telescope so that it can pivot up and down (in altitude) around one axis and horizontally (in azimuth) around the other axis. Following an object across the sky thus usually requires simultaneous movements around each axis. Moreover, as objects cross the sky their orientation in the field of view changes. The *equatorial mounting was therefore long preferred for large telescopes, but with the advent of readily available computer control, which can easily compensate for the varying movements and field rotation, the altazimuth mounting has been adopted for large telescopes as it simplifies construction.

altitude (*h* or *a*) The angular distance of a celestial object above or below the observer's horizon. Altitude is 0° at the horizon and 90° at the zenith. *See also* ZENITH DISTANCE.

al-Ṭūsī *See* ṬŪSĪ.

aluminizing The process of depositing a reflective aluminium coating on a mirror, the successor to *silvering. The mirror is placed in a vacuum chamber and the aluminium is vaporized by heating. With no air molecules to collide with, the aluminium atoms travel directly to the mirror, coating it with a thin, even layer (about 100 nm thick). Subsequent layers of other materials, such as silicon dioxide, may then be deposited in a similar way to protect the coating. Large telescopes have aluminizing chambers housed within the observatory building so that the mirror can be re-aluminized on site whenever necessary. The reflectivity of a fresh aluminium coating with overcoating is 89%, declining by a few per cent a year as a result of oxidation of the surface.

Amalthea The third-closest satellite of Jupiter, distance 181 200 km, orbital period 0.498 days; also known as Jupiter V. Its axial rotation period is the same as its orbital period. Amalthea was discovered in 1892 by E. E. *Barnard. It is noticeably elongated, measuring 250 × 146 × 128 km. Its surface is heavily scarred by impact craters, the largest 90 km wide. The surface has a very low albedo, 0.09, and is distinctly reddish, possibly because of sulphur from Io's volcanoes. Spacecraft measurements yield a mean density of 0.85 g/cm^3, indicating a porous structure rich in water ice.

Ambartsumian, Viktor Amazaspovich (1908–96) Armenian astrophysicist. His major work was on the origin and evolution of stars, for which he was the first to take proper account of their physical properties. He discovered and named stellar *associations, and did important early work on radio galaxies and *active galactic nuclei. He also studied mass ejection from novae and planetary nebulae. Ambartsumian was instrumental in the founding of *Byurakan Astrophysical Observatory.

ambipolar diffusion The process by which ions and neutral atoms or molecules in interstellar gas travel at different speeds in the presence of the interstellar magnetic field. A gas cloud that is collapsing to form stars can detach itself from the interstellar magnetic field through this process, so avoiding the build-up of magnetic pressure that would otherwise halt the collapse. Provided that the collapse is slow, the neutral atoms and molecules can slip past the magnetic field lines in the gas. In this way the bulk of the cloud collapses but some of the magnetic flux leaks out.

AM Canum Venaticorum star A rare form of cataclysmic binary that appears to consist of a pair of white dwarfs. The few known are extremely hydrogen-deficient and have practically pure helium spectra. There are no outbursts; the variations are like those of a *W Ursae Majoris star or *ellipsoidal variable, accompanied by rapid *flickering. There may be mass transfer to an *accretion disk around the primary, and the system may represent an extinct nova. The type star has the shortest period (18 min) of any known eclipsing binary.

American Association of Variable Star Observers (AAVSO) An amateur society founded in 1911, with headquarters in Cambridge, Massachusetts. Observers around the world submit over 400 000 variable star estimates to the AAVSO each year. There are over 18 million archival observations available online. AAVSO activities also include solar observing.

(((●))) **SEE WEB LINKS**
• Official website.

American Astronomical Society (AAS) An organization founded in 1899 for the promotion of astronomy and related branches of science, with its headquarters in Washington, DC. It publishes the *Astronomical Journal, *Astrophysical Journal, and a quarterly *Bulletin* (BAAS).

(((●))) **SEE WEB LINKS**
• Official website.

Ames Research Center A NASA scientific and engineering establishment in Mountain View, California. It was founded in 1939 for aeronautical research and became part of NASA in 1958. As well as aeronautics, Ames specializes in space life sciences and various branches of astronomy including Solar System exploration and infrared astronomy.

(((●))) **SEE WEB LINKS**
• Official website.

AM Herculis star A type of cataclysmic binary whose light exhibits linear and, more significantly, extremely strong circular *polarization (hence the alternative name *polar*); abbr. AM. The polarization varies smoothly over the orbital period. Such systems consist of a K–M dwarf and a strongly magnetic compact object, accretion taking place down an **accretion column** over the latter's magnetic poles. The compact star's rotation is synchronized with the orbital period. Optical variations are in the form of *flickering, with amplitudes up to 4–5 mag. The type star also exhibits X-ray variability.

Amor group A group of asteroids that cross the orbit of Mars, but not that of Earth; also known as *Earth-approaching* (or *Earth-grazing*) *asteroids*. Their orbits have perihelia between 1.017 AU (Earth's aphelion distance) and 1.3 AU (the perihelion of Mars). Close approaches to Mars and Earth can turn Amors into Earth-crossers (*see* APOLLO GROUP) temporarily, and vice versa (*see also* NEAR-EARTH ASTEROID). The group is named after (1221) Amor, diameter 1 km, discovered in 1932 by the Belgian astronomer Eugène Joseph Delporte (1882–1955). Amor's orbit has a semimajor axis of 1.921 AU, period 2.66 years, perihelion 1.09 AU, aphelion 2.76 AU, and inclination 11°.9. Amor group members show a broad variety of compositional types, evidently having originated from several sources. Asteroids may be perturbed into Amor-type orbits either by Jupiter, from near the 3 : 1 and 5 : 2 Kirkwood gaps in the main *asteroid belt, or by Mars, from near the inner edge of the main belt. The two largest members of the Amor group are the S-class (1036) Ganymed, diameter 32 km, and *Eros.

amplitude **1**. The overall range of brightness of a variable star: the difference between its maximum and minimum magnitudes.
2. In physics, the maximum difference of a cyclical quantity from its mean value, i.e. half the peak-to-peak value.

Am star A star of spectral type A whose spectrum additionally contains very strong metallic lines (hence the suffix 'm') more typical of type F. Am stars are usually members of close binary systems. They rotate more slowly than normal A stars, thereby allowing some elements to sink and some to rise in their atmospheres, and this produces the observed abundance anomalies. Sirius is an Am star.

anaemic spiral galaxy A type of spiral galaxy with characteristics intermediate between normal gas-rich spirals and gas-poor lenticular galaxies. They are seen most frequently in rich clusters of galaxies. Examples are NGC 4941 and 4866 in the Virgo Cluster, and NGC 4921 in the Coma Cluster.

analemma A curve that depicts the changing altitude of the Sun during the year at noon, and also the difference between apparent solar time and mean solar time. The analemma is shaped like a thin figure of eight, its height corresponding to the Sun's altitude and its width to the time difference. It was originally a refinement to sundials that allowed corrections to be made for the *equation of time.

Ananke A retrograde satellite of Jupiter, 21 048 000 km from the planet; also known as Jupiter XII. Ananke orbits Jupiter in 624.1 days at an inclination of almost 149° to Jupiter's equator. It is about 20 km in diameter, and was discovered in 1951 by the American astronomer Seth Barnes Nicholson (1891–1963). Ananke is the largest member of the so-called Ananke retrograde irregular group of Jovian satellites which all have similar orbital eccentricity, inclination, and semimajor axis: Euporie, Orthosie, Euanthe, Thyone, Mneme, Harpalyke, Hermippe, Praxidike, Thelxinoe, and Iocaste.

anastigmatic Describing a lens or optical system that has zero *spherical aberration, coma (*see* COMA, OPTICAL), and *astigmatism. Certain designs of *Maksutov and *Schmidt–Cassegrain telescope are **anastigmats**, as are all photographic lenses. Anastigmats usually have at least three optical elements, and often many more.

Anaxagoras (*c.*500–*c.*428 BC) Greek philosopher, born in modern Turkey. In his cosmogony a vortex developed in the primordial Universe, causing dense, wet, dark, and cold matter to fall inwards and form the Earth, while rarefied, dry, light, and hot matter was forced outwards. The Sun, Moon, and stars were torn from the Earth by friction. His claim that the Sun is a red-hot stone allegedly led to his prosecution for impiety and banishment from Athens. Anaxagoras seems to have known the true cause of eclipses—that they are caused by the blocking of light from the Sun.

Anaximander (*c.*610–*c.*540 BC) Greek philosopher, born in modern Turkey. He developed a cosmogony in which universes come into existence from an ageless and eternal reservoir, into which they are eventually reabsorbed. In our universe rotation has put heavy material at the centre (the Earth) and fire at the periphery (the stars). The significance of this world-view is its all-embracing principle governing everything, which can be regarded as the first exposition of a universal law. Anaximander has wrongly been credited with discovering the equinoxes and the obliquity of the ecliptic.

Andromeda (And) (*gen.* **Andromedae**) A constellation of the northern sky, representing Princess Andromeda of Greek mythology. Its brightest stars are Alpha Andromedae (*Alpheratz) and Beta Andromedae (*Mirach). It contains the notable double star Gamma Andromedae (*Almach) and the 9th-magnitude planetary nebula NGC 7662. The constellation's best-known feature is the *Andromeda Galaxy.

Andromeda Galaxy The spiral galaxy nearest to our own, and the largest member of the *Local Group; also called M31 and NGC 224. It lies 2.5 million l.y. away and is visible to the naked eye as an elongated patch of light of total magnitude 3.4 in the constellation Andromeda. The Andromeda Galaxy has two arms and is classified as an Sb spiral. Its total mass is over 400 billion solar masses, slightly more massive than our own Galaxy. On long-exposure photographs it can be traced across over 4° of sky, corresponding to a diameter of about 150000 l.y. It has around 600 globular clusters, over three times as many as our own Galaxy. There are two close dwarf elliptical companion galaxies, both of 8th magnitude: M32 (NGC 221), and NGC 205 (sometimes known as M110), plus more than ten dwarf spheroidal systems. 📷

Andromedid meteors A meteor shower seen in 1872, 1885, 1899, and 1904; also known as the *Bielids*. Activity resulted from the Earth's passage through debris from Comet *Biela, which is presumed to have disintegrated in the mid-19th century. The Earth's approach to the node of the comet's orbit on 1872 November 27 was marked by a meteor storm of around 6000 meteors/h. The 1885 return was even more spectacular, with rates of perhaps 75000 meteors/h. Gravitational perturbations have pulled the meteor stream orbit away from that of the Earth, so that few, if any, Andromedids are now seen.

Anglo-Australian Observatory (AAO) Former name of the *Australian Astronomical Observatory.

Anglo-Australian Telescope (AAT) A 3.9-m reflector opened in 1974 at the Anglo-Australian Observatory (now the *Australian Astronomical Observatory), New South Wales. In 1995 it was fitted with a corrector lens at the prime focus which gives it a two-degree field of view (2dF); an array of fibre optics at the prime focus allows simultaneous spectroscopy of up to 400 stars or galaxies over this wide field of view.

(🌐) SEE WEB LINKS
• Official telescope website.

Ångström, Anders Jonas (1814–74) Swedish physicist and astronomer. He noted that the dark *Fraunhofer lines in the Sun's spectrum had the appearance of a reversed emission spectrum. In 1861 he began an intensive study of the solar spectrum, confirming the presence of hydrogen in the Sun. In 1868 he published an atlas of the solar spectrum and measurements of the wavelengths of over a thousand spectral lines. These measurements were expressed in units of 10^{-10}m, a length subsequently named the angstrom unit in his honour.

angstrom (Å) A unit of length equal to 10^{-10}m. It was used mainly for specifying the wavelength of radiation, but has been replaced by the *nanometre (nm). One angstrom is 0.1 nm. It is named after A. J. Ångström.

angular acceleration The rate of change in the *angular velocity of a spinning body, or in the angular velocity of a celestial object in its orbit about another body.

angular diameter The apparent size of an object, such as the diameter of a planet, expressed in degrees, minutes, or seconds of arc.

angular distance *See* ANGULAR SEPARATION.

angular momentum The momentum a body has by virtue of its rotation. A body such as a planet has two types of angular momentum: one type results from its motion in orbit around the Sun, and the other from its spin on its own axis. Angular momentum in orbit is given by the body's mass multiplied by its orbital *angular velocity and by its distance from the Sun. Angular momentum of spin depends on the mass of the body's individual parts and their distances from its centre (the *moment of inertia*) multiplied by its angular velocity of spin. As a consequence of *conservation of angular momentum*, a body spins faster as it gets smaller, such as when a gas cloud shrinks to become a star.

angular resolution The distance, in angular measure, between two close objects that can just be separated by an optical system, such as a telescope. *See also* RESOLVING POWER.

angular separation The apparent distance between two objects, such as two stars, expressed in degrees, minutes, or seconds of arc.

angular velocity (ω) The rate of rotation of a body, either about its own axis or in its orbit about another body. For example, the Earth rotates with a certain angular velocity about its polar axis, but it also has angular velocity in its revolution about the Sun.

anisotropy A characteristic of a substance or body in which physical properties are different in different directions. In astronomy, the temperature of the cosmic microwave background is observed to be anisotropic on a large angular scale as a result of the Sun's motion through space (*dipole anisotropy*), and on a small scale as a result of fluctuations in the density of the early Universe. *See also* ISOTROPY.

Ankaa The star Alpha Phoenicis, magnitude 2.4. It is a K0 giant 85 l.y. away.

annealing In optical manufacture, the process of allowing glass or ceramic to cool slowly at a controlled rate, to avoid stress patterns within the material. A large mirror blank, for example, is cast from molten material. If cooling is too rapid, temperature variations will arise and some areas will solidify more rapidly than others, creating internal stresses that can cause the mirror blank to shatter when it is being ground to shape. Annealing of a large telescope mirror can take several months.

annual aberration The small displacement in position of a star's image during the year due to the motion of the Earth around the Sun. Annual aberration was discovered by J. *Bradley in 1728 from observations of the changes in distance from the zenith of the star Gamma Draconis. The ratio of the Earth's mean velocity to the speed of light gives the *constant of aberration*, 20″.5. This is the maximum amount by which a star can appear to be displaced from its mean position. During the course of a year, the star appears to move around its mean position in a shape that ranges from a circle for a star at the ecliptic pole, via a progressively flattened ellipse, to a straight line for a star on the ecliptic. *See also* E-TERMS.

annual equation A periodic disturbance in the celestial longitude of the Moon, resulting from the changing gravitational pull of the Sun around the Earth's elliptical orbit; also known as **annual inequality**. It has an amplitude of 11 minutes of arc and a period of one *anomalistic year.

annual inequality Another name for annual equation.

annual parallax The maximum apparent difference in position of a star during the course of a year due to the changing position of the Earth in its orbit around the Sun; also known as *heliocentric parallax*. The amount of displacement is equal to the angular separation between the Earth and Sun as seen from the star. *See also* TRIGONOMETRIC PARALLAX.

annual variation The yearly changes in the right ascension and declination of a star due to annual precession and proper motion. These quantities are tabulated in star catalogues.

annular eclipse A solar eclipse occurring when the Moon is near apogee and its apparent diameter is smaller than that of the Sun, so that a ring, or **annulus**, of the Sun's disk remains visible at mid-eclipse along the narrow central ground track on the Earth. Annular eclipses can last up to 12m 30s.

anomalistic month The mean interval of time between successive passages of the Moon through its closest point to Earth (its perigee), equal to 27.55464 days. The Moon's orbit is not fixed in space but can be regarded as a slowly rotating ellipse, the rotation due principally to perturbation by the Sun. Hence the perigee point advances slowly around the orbit, completing one circuit in 8.85 years. As a consequence, the anomalistic month is about 5.5 hours longer than the *sidereal month.

anomalistic year The interval of time between successive passages of the Earth through the perihelion of its orbit, equal to 365.2596 days. Due to perturbations by the other planets, the Earth's orbit is not a fixed ellipse but one that is slowly advancing. As a result the anomalistic year is about 5 min longer than the Earth's orbital period (the *sidereal year).

anomalous Cepheid A star which has a period characteristic of comparatively long-period RR Lyrae variables of type RRAB, but which is considerably more luminous; abbr. BLBOO from the prototype BL Boötis.

anomalous iron An *iron meteorite with a chemistry and structure unlike those of any of the primary groups into which other iron meteorites are divided.

anomaly An angle that describes a body's position at a given time in an elliptical orbit. There are three types of anomaly: *eccentric anomaly, *mean anomaly, and *true anomaly. All are measured in the direction of the body's motion and take as their starting-point the body's periapsis (i.e. its closest point to the object it orbits).

anorthosite A medium- to coarse-grained igneous rock, grey to white in colour. It is composed almost entirely of calcium-rich plagioclase feldspar (95%), one of the most common rock-forming minerals, together with minor amounts of pyroxene (4%), olivine, and iron oxides. Anorthosite is thought to comprise a significant fraction of the lunar crust, and is present at all the Apollo and Luna landing sites. Samples of anorthosites 4.4–4.5 billion years old were among the oldest lunar rocks returned by spacecraft.

ANS Abbr. for *Astronomical Netherlands Satellite.

ansae The extensions of Saturn's rings either side of the planet; sing. *ansa*. The word is Latin for 'handles', since the rings project from the planet like handles.

antapex *See* SOLAR APEX.

Antares The star Alpha Scorpii, an M1.5 supergiant over 10 000 times as luminous as the Sun and with a diameter about 400 times the Sun's. It is a semiregular variable, ranging between magnitudes 0.9 and 1.2 with a period of about 5 years. Antares lies 554 l.y. away. It has a B2 dwarf companion of magnitude 5.4 with an orbital period of about 1200 years. The name Antares is Greek, and is usually translated as 'rival of Mars', because of its strong red colour, but sometimes as 'like Mars'.

antenna A device used in radio astronomy to detect radio waves; also known as an **aerial**. In an antenna, radio waves induce an oscillating electric current in a conductor. The simplest antenna is a *dipole antenna, which is a simple metal rod. More complex antennas, such as the *Yagi antenna, may consist of a dipole or folded dipole (the *driven element*) with other conducting rods (*parasitic elements*) which direct or reflect the waves on to the dipole and so increase the antenna's *directivity and *gain. The most powerful radio telescopes, especially at high frequencies, use parabolic (dish) antennas to collect waves and bring them to a focus. Large numbers of antennas may be combined in an *array* to achieve high resolution. Antennas may also be used to transmit radio waves.

Antennae A pair of interacting galaxies, NGC 4038 and 4039, lying 60 million l.y. away in the constellation Corvus. They have two long, faint, curved, tails extending for 360 000 l.y. (over 20′ on the sky), which give them their name. The tails are probably composed of stars pulled out by gravitational tidal interactions during the collision of a spiral and a lenticular galaxy.

antenna pattern A graph of the sensitivity or gain of a radio telescope in various directions. A typical antenna pattern shows several sensitivity maxima (*lobes*) of which the strongest (the *main lobe* or *main beam*) defines the direction in which the telescope is most sensitive. An antenna pattern with a narrow main lobe of circular cross-section and negligible *side lobes is known as a *pencil beam*, while a pattern with a wide, flat main lobe is known as *fan beam*. *See also* POLAR DIAGRAM.

antenna temperature A measure of signal strength in radio astronomy. It is defined as the temperature of a black-body enclosure which, if completely surrounding a radio telescope, would produce the same signal power as the source under observation. Antenna temperature is a property of the source, not of the antenna itself. Where a uniform extended source fills the telescope beam, the antenna temperature is equal to the *brightness temperature of the source.

anthelion The point in the sky directly opposite the Sun.

anthelion radiant A broad, weak source of meteors (usually fewer than five per hour) located on the ecliptic some 15° east of the *anthelion point. Meteors from this radiant are caused by particles of interplanetary dust moving faster than the Earth which enter the atmosphere as they overtake us. Activity from the anthelion radiant was formerly attributed to a sequence of individual minor showers, but is now recognized to be part of a general background that is active throughout the year and that has no specific parent body. The only exception is the *Taurid meteor shower of October and November, which is associated with Comet *Encke.

anthropic principle The proposition that the presence of human life is connected with the properties of the Universe. There are various forms of the anthropic principle. The least controversial is the *weak anthropic principle*, according to which human life occupies a special place in the Universe because it can evolve only where and when the conditions are suitable for it. A more speculative version, the *strong anthropic principle*, asserts that the laws of physics must have those properties that allow life to evolve. In recent years, the weak anthropic principle has been used to explain many properties of the Universe, such as the values of the fundamental constants and the number of dimensions of space, on the grounds that if they did not have these values human life would never have evolved. However, the validity of arguments of this kind is very uncertain unless, as some speculative theories suggest, there are actually many universes, each having different values for the fundamental constants and possibly different numbers of dimensions (*see* MULTIVERSE). If that is true, some of these universes would contain no life at all, some would contain very different life from ours, and our Universe would obviously contain our kind of life.

anticentre *See* GALACTIC ANTICENTRE.

anti-dwarf nova Another name for a *VY Sculptoris star.

antimatter Matter composed of *antiparticles*: subatomic particles that have identical *rest mass to corresponding particles of ordinary matter but opposite charge, and are opposites in other fundamental properties. For example, the antiparticle of the electron is the positron, which has a positive charge equal to the electron's negative charge; the antiproton has a negative charge equal to the proton's positive charge. When matter and antimatter meet, they annihilate each other, releasing energy. The Universe seems to be almost entirely in the form of matter rather than antimatter; why this should be so is presumably related to events shortly after the Big Bang.

antitail A protuberance from a comet that appears to point towards the Sun, often looking like a spike. It is actually part of the comet's dust tail, and consists of larger (millimetre-sized) particles lagging behind the comet in its orbit. The antitail does not actually point sunwards, but results from a perspective effect when the comet is seen from a certain angle. Antitails are uncommon. Where present, they are usually most prominent when the Earth passes through, or close to, the plane of the comet's orbit. The cometary dust, which lies in a thin sheet, is then seen edge-on.

Antlia (Ant) (*gen.* **Antliae**) An unremarkable constellation of the southern sky, representing an air pump. Its brightest star, Alpha Antliae, is magnitude 4.3.

Antoniadi, Eugène Michael (1870–1944) French astronomer, born in Turkey. He specialized in the planets, of which he was one of the most outstanding observers of his time. His highly detailed drawings show, for example, spots on Saturn's disk and 'spokes' in its rings, the South Tropical Disturbance on Jupiter, and surface details on Mercury. Antoniadi's chief interest, though, was Mars. He was sceptical about the Martian canals, and was able to demonstrate that they were illusory. He devised the *Antoniadi scale for the quality of seeing.

Antoniadi scale A scale of *seeing conditions, used by amateur astronomers, that describes the effect of atmospheric motions on an image, usually that of the Moon or a planet. Its categories are:
  I perfect, without a quiver
  II slight undulations, but with calm periods lasting several seconds
  III moderate, with some greater air movements
  IV poor, with the image in constant movement
  V very bad, making observations very difficult
It was devised by E. M. *Antoniadi.

Antu The first 8.2-m Unit Telescope (UT1) of the European Southern Observatory's *Very Large Telescope in Chile, opened in 1998. Its name means 'the Sun' in the local Mapudungun language.

ap-, apo- Prefixes referring to the point in an elliptical orbit that lies farthest from the object being orbited, as in *aphelion and *apogee.

Apache Point Observatory An observatory at an altitude of 2788m in the Sacramento Mountains near Sunspot, New Mexico. It houses a 3.5-m reflector opened in 1994; a 2.5-m reflector built for the *Sloan Digital Sky Survey which began operation in 1998; and a 1-m reflector owned by New Mexico State University. The observatory is owned by the Astrophysical Research Consortium (ARC), whose members are the University of Chicago, the University of Colorado, Johns Hopkins University, Princeton University, the University of Virginia, the University of Washington, and New Mexico State University; the last-named also operates the observatory for the Consortium.

((()) SEE WEB LINKS
• Official observatory website.

apastron The point in an elliptical orbit around a star that is farthest from the centre of the star.

aperture The diameter of the main lens or mirror of a telescope, or, in radio astronomy, of the collecting dish. For *Schmidt cameras, however, it is the diameter of the *corrector plate. *See also* UNFILLED APERTURE.

aperture efficiency A measure of the efficiency with which a telescope can collect radio waves. It is expressed as the *effective aperture divided by the geometrical area. For a simple parabolic antenna, aperture efficiencies rarely exceed 65%, due to the difficulties of designing a primary feed to be equally sensitive to waves reflected from all points of the dish.

aperture ratio The ratio of the aperture (diameter) of a lens or mirror to its focal length, usually expressed as a true ratio, e.g. 1 : 8. It is more common, however, to refer to the *focal ratio, which is the inverse of the aperture ratio and is expressed in the form $f/8$ (*see* F/NUMBER).

aperture synthesis A technique used in radio astronomy to achieve high angular resolution. It uses an array of telescopes to simulate a single telescope of large aperture (see the diagram). In principle, an aperture of any size can be synthesized by moving the two elements of an *interferometer to occupy all possible positions in the desired aperture. In practice, all synthesis telescopes make use of the fact that

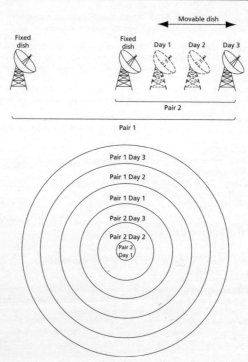

aperture synthesis: An aperture synthesis telescope consists of an array of dishes connected in pairs. As the Earth rotates, each pair traces out one ring of a much larger dish, whose diameter is equal to the maximum separation of the pairs. Some of the dishes may be movable, to provide a wider range of separations

over a period of 12 hours the Earth's rotation will move the elements to sweep out half a ring of the synthesized aperture (*supersynthesis*, or *Earth-rotation synthesis*); the other half of the ring can be derived from the observations of the first half. The elements need then be moved only to sweep out successive rings. In practice, some aperture-synthesis telescopes employ several movable dishes to reduce observing time, while in others the dishes remain fixed (an *unfilled aperture*). Aperture synthesis requires complex data-reduction techniques and powerful computers. Examples of aperture synthesis telescopes are the *Multi-Element Radio-Linked Interferometer Network (MERLIN), the *Ryle Telescope, the *Very Large Array, and the Westerbork Synthesis Radio Telescope (*see* WESTERBORK RADIO OBSERVATORY).

APEX Abbr. for *Atacama Pathfinder Experiment.

apex *See* SOLAR APEX.

aphelion The point in an elliptical orbit around the Sun that is farthest from the Sun's centre.

Aphrodite Terra A large upland area on Venus, near the equator. It is about 17 500 km long, and has about the same area as Africa. It includes four smaller highland areas: Ovda Regio and Thetis Regio in its western half, and Atla Regio and Ulfrun Regio to the east. There are many troughs in Aphrodite, including *Diana Chasma near the centre, and volcanoes such as Maat Mons, 8.4 km high.

ApJ Abbr. for *Astrophysical Journal.*

aplanatic Describing a lens or optical system that is free from both *spherical aberration and coma (*see* COMA, OPTICAL). Examples of aplanatic systems include some types of achromatic lens and the *Ritchey–Chrétien telescope.

apoapsis The point in an elliptical orbit that lies farthest from the centre of the object being orbited.

apocentre The point in an elliptical orbit that lies farthest from the centre of mass of the orbiting system, such as a binary star or a planet and satellite.

apochromatic Describing a lens or optical system with very little *chromatic aberration. In an apochromatic lens (also known as an **apochromat**) the chromatic aberration is completely cancelled out at three or more wavelengths, compared with two for a typical achromatic lens. Apochromats usually have at least three optical *elements, made from different types of glass. *Fluorite has become a popular material for apochromats, giving virtually perfect colour correction with only two elements, but at considerable cost.

apodization A technique in which the normal performance of an instrument is deliberately degraded in such a way that the instrument's performance is actually improved for one specific application. It may be applied to optical and radio telescopes (where it is called *tapering the antenna*), and is particularly important in *Fourier transform spectrometers where it leads to the elimination of spurious spectral features at the expense of a slight reduction in spectral resolution. A common application is the use of an **apodizing screen** in the observation of a close visual binary where the image of a faint companion star falls on one of the diffraction rings surrounding the Airy disk. A circular screen which gradually becomes opaque towards the edge reduces the brightness of the rings; a screen with a square aperture produces a cross-shaped diffraction pattern which can be rotated clear of the companion star.

apogee The point in an elliptical orbit around the Earth that is farthest from the Earth's centre. 📷

apohele An asteroid whose orbit lies entirely within that of the Earth (an aphelion of 0.983 astronomical units or less). The first proposed member of the group was 1998 DK_{36}, but its orbit is poorly determined. The first confirmed apohele was 2003 CP_{20}, now known as (163693) Atira, aphelion 0.98 AU, and the second was 2004 JG_6, aphelion 0.97 AU.

Apollo group A group of asteroids that cross the Earth's orbit, but whose average distances from the Sun are greater than that of Earth; also known as *Earth-crossing asteroids*. Their perihelion distances are 1.017 AU (Earth's aphelion) or less. Most Apollo asteroids are small (up to 5 km diameter) and highly irregular in shape. They are named after (1862) Apollo, the first of the group to be discovered, by the German astronomer Karl Reinmuth (1892–1979) in 1932. Apollo is a Q-class asteroid of diameter 1.5 km. It came within 0.07 AU (10.5 million km) of Earth in 1932, but was then lost until 1973. Apollo's orbit has a semimajor axis of 1.470 AU, period 1.78 years, perihelion 0.65 AU, aphelion 2.30 AU, and inclination $6°.4$. Apollo can approach to within 0.028 AU (4.2 million km) of Earth's orbit and can also make close approaches to Venus and Mars. Some Apollo asteroids such as *Phaethon and *Icarus approach closer to the Sun than Mercury. The largest Apollo asteroid is (1866) Sisyphus, diameter 8 km. Other notable named Apollos include *Hephaistos, *Toro, and *Toutatis. The Apollo asteroids were once thought to be extinct cometary nuclei, but in fact Jupiter may perturb asteroids from near the 3:1 Kirkwood gap in the main asteroid belt into Apollo-type orbits. *See also* NEAR-EARTH ASTEROID.

Apollonius of Perga (c.262–c.190BC) Greek mathematician, born in modern Turkey. He showed that the ellipse, parabola, and hyperbola are all curves formed by a plane intersecting a cone in different ways, i.e. that they are *conic sections*. The orbital path of an unperturbed body moving in a gravitational field follows one of these three curves, as would come to be appreciated by later astronomers such as E. *Halley, who translated Apollonius' book *Conics*. Apollonius also originated the mathematical concept of motion based on epicycles and deferents, later taken up by *Hipparchus and *Ptolemy to explain planetary motion.

Apollo project The American space programme to land humans on the Moon. The
three-seater Apollo spacecraft were launched to the Moon by Saturn V rockets. In orbit
around the Moon, two astronauts entered the Lunar Module and descended to the
surface. After four manned test flights, the first manned landing was made on 1969 July 20
by Neil Alden Armstrong (1930–) and Edwin Eugene 'Buzz' Aldrin (1930–) during the
Apollo 11 mission. On subsequent missions, astronauts spent longer on the lunar surface
and were able to explore greater distances by means of an electrically powered lunar rover.
A total of 381 kg of rock samples was brought back by the Apollo astronauts. (*See* Table 1,
Appendix.)

apparent diameter The size of a celestial object or feature as seen by an observer,
expressed in angular measure; the same thing as angular diameter.

apparent magnitude (*m*) The brightness of a celestial object as measured by the
observer. Sirius, the brightest star, has an apparent magnitude of −1.44, and the faintest
stars visible to the naked eye under the most favourable conditions have magnitudes of
about +6.5. Stars of magnitude +23 are regularly measured at modern professional
observatories, and as faint as +30 with the Hubble Space Telescope. If *m* carries no
subscript it is assumed to be the *visual magnitude. The notation m_{bol} denotes the
apparent *bolometric magnitude.

apparent noon The time at which the Sun crosses the observer's meridian, and
hence reaches its greatest altitude above the horizon. Due to the *equation of time,
apparent noon can differ by some minutes from mean noon as recorded by a clock
keeping local mean solar time.

apparent place The predicted coordinates of a star as it would be seen from the
centre of the Earth, referred to the true equator and equinox at a specific date. It includes
displacements from the heliocentric direction given in a star catalogue due to
precession, nutation, aberration, proper motion, annual parallax, and gravitational
deflection of light.

apparent retrogression The temporary reversal in the direction of movement of
an outer planet on the celestial sphere as it is overtaken by the Earth; also known as *retrograde
motion*.

apparent sidereal time Time as given by direct observation of the stars; strictly, the hour
angle of the true equinox. The vernal equinox (the First Point of Aries) is not a completely
fixed sidereal direction but varies due to general precession and nutation. Nutation is a
periodic variation and imposes irregularities of about a second on *sidereal time. The
principal period of this variation is 18.6 years, which is the period of regression of nodes
of the Moon's mean orbit around the ecliptic. Apparent sidereal time corrected for this
effect gives mean sidereal time.

apparent solar time Time as given by the daily movement of the Sun across the sky;
strictly, the *hour angle of the Sun plus 12 hours, which is added so that the solar day
begins at midnight. Apparent solar time is the time shown on a sundial. The Sun's hour
angle increases due to the Earth's rotation, but slightly more slowly than that of the stars
because the Sun moves against the star background as the Earth orbits it. However, this
movement is not entirely uniform because the Earth's orbit is elliptical, and consequently
apparent solar time can be anything up to a quarter of an hour ahead of or behind *mean
solar time. The difference between apparent solar time and mean solar time is known as
the *equation of time.

apparition The period of time during which an object in the Solar System is
visible from Earth, such as an evening apparition of Venus or the apparition of a periodic
comet. The term is not usually applied to objects that are more regularly visible, such as
the Sun or stars.

appulse The apparent close approach of two celestial bodies when one moves into the same line of sight as the other, such as when a planet appears to pass another planet or a star.

apse, apsis *See* APSIDES.

apsidal motion A progressive change in the orientation of the major axis (i.e. the *line of apsides*) of an orbit; also known as *advance of perihelion. There are several possible causes, including perturbations by other bodies; mass transfer between stars in a binary system; the elliptical shape of stars in a close binary; and the curvature of space around a massive body.

apsides The two points in an elliptical orbit that lie closest to or farthest from the centre of the body being orbited; sing. *apsis* or *apse*. The closest point is the *periapsis* (or *periapse*) and the farthest point the *apoapsis* (or *apoapse*). The straight line joining these two points is the *line of apsides*, and is the same as the major axis of the orbit. In the case of the Earth's orbit around the Sun, the apsides are perihelion and aphelion.

Ap star A peculiar star (hence the suffix 'p') of spectral type A that has much stronger absorption lines than normal of elements such as manganese, mercury, silicon, chromium, strontium, and europium; also known as a *peculiar A star*. Such stars have unusually strong magnetic fields, thousands of times stronger than the Sun's typical surface field. This leads to certain elements being localized in spots, similar to sunspots. As the star rotates the spectral features therefore vary. Hotter and cooler versions of Ap stars are, respectively, Bp and Fp stars.

Apus (Aps) (*gen.* **Apodis**) An insignificant constellation near the south celestial pole, representing a bird of paradise. Its brightest star is Alpha Apodis, magnitude 3.8.

Aquarid meteors Any of several meteor showers that are active from Aquarius during July and August, principally the *Delta Aquarids and *Iota Aquarids. Meteors emanating from this part of the sky that cannot be identified with a specific radiant are simply termed 'Aquarids' to distinguish them from sporadic meteors.

Aquarius (Aqr) (*gen.* **Aquarii**) A constellation of the zodiac, popularly known as the water-carrier, through which the Sun passes from the third week of February to the second week of March. Its brightest stars are Alpha Aquarii (Sadalmelik) and Beta Aquarii (Sadalsuud), both of magnitude 2.9. It contains the 7th-magnitude globular cluster M2, the *Helix Nebula, and the *Saturn Nebula. Three meteor showers radiate from Aquarius each year, the *Delta Aquarids, the *Eta Aquarids, and the *Iota Aquarids.

aqueous alteration A process in which changes to minerals occur through reactions with water. The effects of aqueous alteration are apparent in certain meteorites (particularly carbonaceous chondrites) and in the surface layers of some asteroids, notably Ceres.

Aquila (Aql) (*gen.* **Aquilae**) A constellation on the celestial equator, representing an eagle. It is marked by the bright star *Altair (Alpha Aquilae), flanked by Beta Aquilae (*Alshain) and Gamma Aquilae (*Tarazed). Eta Aquilae is one of the brightest Cepheid variables, ranging from magnitude 3.5 to 4.4 with a period of 7.2 days.

Ara (Ara) (*gen.* **Arae**) A small constellation of the southern sky, representing an altar. Its brightest stars are Alpha and Beta Arae, both magnitude 2.8. Ara contains the open cluster NGC 6193 and the globular cluster NGC 6397.

arachnoid A cobweb-shaped landform on Venus showing radiating lineaments centred on a volcanic dome and surrounded by concentric ridges or fractures. Arachnoids are up to 150km in diameter and occur mostly in two lowland regions of Venus, Sedna Planitia and Ganiki Planitia. Other sparse groupings are known.

Arago, (Dominique) François (Jean) French scientist and statesman. With the
French physicist Augustin Jean Fresnel (1788–1827), he established the wave
theory of light and studied its *polarization. In 1811 he invented the *polariscope* for
measuring the degree of polarization of light. This instrument allows light from solid
and fluid surfaces to be distinguished, and Arago used it in several astronomical studies.
One was an extensive study of the total solar eclipse of 1842, in which he determined
that the Sun's limb is gaseous, and examined polarized light from the chromosphere
and corona.

Arago ring A ring of Neptune, fourth in order from the planet, named after F. *Arago.
It lies 57 600 km from Neptune's centre and is less than 100 km wide.

arc A part of a circle. Angular measure is given in units of arc. *See* DEGREE OF ARC; MINUTE
OF ARC; SECOND OF ARC; RADIAN.

archaeoastronomy The study of the astronomical significance of archaeological remains
and artefacts, particularly large stone structures (megaliths). Structures such as Stonehenge
in England and the Great Pyramid in Egypt have components which are oriented to align
with the positions of celestial bodies at significant times of the year. This suggests that they
were constructed at least partly for astronomical purposes, although further interpretation
remains largely speculative.

arch filament system Narrow arches or loops above the Sun's photosphere, visible in
hydrogen light as darker absorption features against the solar disk, in a similar manner to
normal *filaments; when seen at the limb, though, they appear bright. Arch filament systems
are a characteristic feature of developing *active regions on the Sun and connect areas of
opposite magnetic polarity (usually sunspots) in the innermost part of young *bipolar groups.
Individual arches have typical lengths of 30 000 km and lifetimes of some tens of minutes,
although the entire arch filament system can last for several days. Their absorption of
hydrogen light implies the arches are cool and dense, and so they are believed to be
photospheric plasma that has risen into the corona.

arc minute (arcmin) *See* MINUTE OF ARC.

arc second (arcsec) *See* SECOND OF ARC.

Arcturus The star Alpha Boötis, magnitude −0.05, the brightest star north of the celestial
equator and the fourth-brightest star of all. It is a K1.5 giant, 37 l.y. away.

arcus An arc-shaped feature on a planetary body; pl.*arcus*. The term is not a geological one,
but is used in the nomenclature of individual features on planets and satellites. An example
is Hotei Arcus on Titan.

area photometer A specialized photometer used to measure simultaneously the light
at each point of an extended object such as a galaxy or nebula; also known as an *imaging
photometer*. Most area photometers rely on CCD detectors or infrared arrays.

Arecibo Observatory The site of the world's largest radio astronomy dish, 305 m in
diameter, opened in 1963 and upgraded in 1974 when the original wire-mesh surface was
replaced by solid panels. The observatory is situated 12 km south of Arecibo, Puerto Rico, and is
operated by the National Astronomy and Ionosphere Center (NAIC), which has its
headquarters at Cornell University, New York State. The dish is suspended in a natural
hollow in the ground, and scans a strip of sky overhead as the Earth rotates. Radio sources
can be tracked within 20° of the zenith by using movable feeds suspended above the stationary
reflector, giving coverage in declination from about 38° north to 1° south. The Arecibo dish
is used for atmospheric studies as well as astronomy.

(( )) SEE WEB LINKS
• Official observatory website.

Arend–Roland, Comet (C/1956 R1) A long-period comet discovered in 1956 November by the Belgian astronomers Silvain Arend (1902–92) and Georges Roland (1922–91); formerly designated 1957 III. It reached perihelion, 0.32 AU, on 1957 April 8. Closest approach to Earth, 0.57 AU, was on April 21. At peak brightness, the comet reached magnitude −1, with a tail of 25–30°. It also had a prominent *antitail. Comet Arend–Roland has a hyperbolic orbit (eccentricity 1.0002) of inclination 119°.9.

(((⊕))) SEE WEB LINKS
• Information page at Cometography website.

areo- Prefix referring to Mars; for example, areography is the study of the surface of Mars.

Argelander, Friedrich Wilhelm August (1799–1875) German astronomer, born in modern Lithuania. From a study of the proper motions of several hundred stars he confirmed F. W. *Herschel's calculated position of the Sun's *apex. His main achievement, begun in 1852, was a survey of all stars down to 9th magnitude in the northern hemisphere. The resulting star atlas and catalogue were published in 1859–62 as the *Bonner Durchmusterung. He subsequently initiated the first *AGK catalogue. Argelander also introduced the subdivision of magnitudes into tenths, the step method for estimating stellar magnitudes, and the practice of assigning capital roman letters to variable stars.

Argelander step method A visual method of estimating the magnitude of a variable star, described by F. W. A. *Argelander. It involves comparing the variable with a comparison star, assigning a step value that reflects the ease with which the brightness of the variable can be distinguished from that of the comparison. Estimates of the form 'A(3)V, V(1)B' result, and the magnitude of the variable (V) can be calculated from the known magnitudes of the comparisons (A and B), as with the *fractional method. *See also* POGSON STEP METHOD.

Argo Navis One of the 48 constellations known to the Greeks, representing the ship of the Argonauts, but broken up in the 18th century by N. L. de *Lacaille into *Carina, *Puppis, and *Vela.

argon–potassium method *See* POTASSIUM–ARGON METHOD.

argument of perihelion (ω) An angle that defines the direction of the major axis of an orbit around the Sun. It is the angle between the ascending node and the perihelion of an orbit, measured in the plane of the orbit and in the direction of orbital motion. *See also* ELEMENTS, ORBITAL.

Argyre Planitia A lowland plain within the Argyre impact basin in the southern hemisphere of Mars, visible telescopically from Earth as a pale circular patch, centred at −50° lat., 44° W long. Argyre is the second-largest impact basin on Mars, about 800 km wide and over 4 km deeper than its surroundings.

Ariel The fourth-largest satellite of Uranus, 1158 km in average diameter; also known as Uranus I. Ariel orbits Uranus in 2.52 days at a distance of 190 950 km, keeping the same face towards the planet. It was discovered in 1851 by W. *Lassell. Ariel is an icy body, with a surface scored by giant steep-sided troughs, termed *chasmata*, between lightly cratered regions. The longest trough, Kachina Chasmata, is over 600 km long, and the largest crater, Yangoor, is 78 km wide. The erosion of some craters and the presence of smooth-floored material in the chasmata is evidence of volcanic or tectonic resurfacing by viscous ice flows.

Ariel satellites A series of UK scientific satellites launched by NASA between 1962 and 1979. Ariel 1, the first UK satellite, studied solar X-rays and the Earth's outer atmosphere. The next three studied the atmosphere and radio astronomy. The longest-lived mission was the X-ray astronomy satellite Ariel 5, which operated from 1974 to 1980. Ariel 6, the last UK satellite, was primarily concerned with cosmic rays but also carried X-ray instruments.

⊕ SEE WEB LINKS
• Information page at Goddard Space Flight Center

Aries (Ari) (Arietis) A constellation of the zodiac, representing a ram. The Sun lies in Aries for the last 10 days of April until mid-May. Its brightest star is Alpha Arietis (*Hamal). Gamma Arietis is a pair of white stars of magnitudes 4.6 and 4.7.

Aries, first point of *See* FIRST POINT OF ARIES.

Aristarchos Telescope A 2.28-m reflector at an altitude of 2340 m at Helmos Observatory on Mt Helmos, Greece, 130 km west of Athens, opened in 2004. It is owned and operated by the National Observatory of Athens.

⊕ SEE WEB LINKS
• Official telescope website.

Aristarchus of Samos (*c.*320–*c.*250 BC) Greek mathematician and astronomer. He attempted to calculate the sizes and distances of the Sun and Moon, establishing that the Sun is much larger than the Earth and much farther off than the Moon. To Aristarchus is due the first heliocentric theory. He placed the Sun at the centre of the Earth's orbit, and the fixed stars on a sphere a great distance from the Sun. The theory attracted little support mainly because *Aristotle's geocentric theory was held in high esteem, but also because the idea of a moving Earth was disliked.

Aristotle (384–322 BC) Greek philosopher. He based his geocentric model of the Universe on the system of concentric spheres proposed by *Eudoxus (as modified by *Callippus), increasing the number of spheres to 49 to account for the movement of all celestial bodies. (This was later modified by *Ptolemy, who replaced spheres by epicycles.) The outermost sphere, which carried the fixed stars, controlled the motion of the others and was itself controlled by a supernatural agency. Aristotle's world-view of a Universe in which everything had its natural place, with a changing Earth surrounded by the eternal, perfect, and incorruptible heavens, was not seriously challenged for almost two thousand years. Aristotle demonstrated that the Earth was spherical, from the shadow it cast during a lunar eclipse, and calculated its size, obtaining a result that was 50% greater than the true value.

Arizona meteor crater *See* METEOR CRATER.

Arizona Radio Observatory (ARO) A division of *Steward Observatory founded in 2002 which owns and operates a 12-m millimetre-wave dish on Kitt Peak, Arizona, and the 10-m Submillimeter Telescope (SMT) on Mount Graham, Arizona. The 12-m dish, opened in 1967 and upgraded in 1983, was originally owned by the National Radio Astronomy Observatory but was transferred to ARO in 2002, while the SMT, formerly known as the Heinrich Hertz Telescope, was opened in 1994 and transferred to ARO in 2002.

⊕ SEE WEB LINKS
• Official observatory website.

AR Lacertae star A close binary containing two evolved stars neither of which has yet overflowed its *Roche lobe so that the pair are still a *detached system; abbr. AR. They belong to the *RS Canum Venaticorum class of binary. The rotational period of the stars is locked to the orbital period, resulting in enhanced activity from both components. Enhanced versions of solar-like flares excite the coronae of the stars, producing light at extreme ultraviolet and X-ray wavelengths, while the interaction of electrons and magnetic fields gives rise to radio emission. AR Lacertae itself is also an eclipsing binary.

armillary sphere An instrument used since ancient times to demonstrate and observe the movements of the heavens. It consists of a number of rings (**armillaries**) representing celestial great circles such as the meridian, horizon, celestial equator, and ecliptic, arranged to form a skeletal celestial sphere. Armillary spheres can still be found in the form of equatorial sundials, in which the ring representing the celestial equator is marked with hours. The name comes from the Latin word **armilla**, meaning 'ring' or 'bracelet'.

arm population Those young stars concentrated in the spiral arms of galaxies; they are *extreme Population I stars. Their youth is indicated by their high content of heavy elements, by their proximity to the gas and dust from which they formed, and by the presence among them of massive stars which have only a short lifetime. They are found in loose open clusters or associations.

Arneb The star Alpha Leporis. It is an F0 supergiant about 2200 l.y. away, of magnitude 2.6.

array 1. A group of telescopes or antennas (*elements*) arranged to operate as a single instrument. *See* INTERFEROMETER, PHASED ARRAY.
   2. A regular arrangement of detectors such as CCDs or infrared detectors placed in the focal plane of a telescope, either in a line (a *one-dimensional array*) or in a plane (a *two-dimensional array*), and designed to record an image of a source.

artificial satellite A spacecraft sent into orbit around the Earth. The first artificial Earth satellite was Sputnik 1, launched by the former Soviet Union on 1957 October 4. A spacecraft can become an artificial satellite of the Moon or a planet if it goes into orbit around it.

ASCA A Japanese X-ray astronomy satellite, known as Astro-D before its launch in 1993 February; also known in Japanese as Asuka. The satellite had four grazing-incidence telescopes each covering the energy range 0.5–12keV (0.10–2.5nm). ASCA (Advanced Satellite for Cosmology and Astrophysics) was the first X-ray mission to use *CCD spectrometers. It operated until 2000 July and re-entered in 2001 March.

(((⊕))) SEE WEB LINKS
• Information page at Goddard Space Flight Center.

ascending node (☊) The point in an orbit at which a body moves from south to north across a reference plane, such as the plane of the ecliptic or of the celestial equator. The *longitude of the ascending node is one of the elements of an object's orbit. *See* ELEMENTS, ORBITAL; NODE.

Asclepius Asteroid 4581, a member of the *Apollo group, discovered in 1989 by the American astronomers Henry Edward Holt (1929–) and Norman Gene Thomas (1930–). It can approach to within 0.004 AU (600000km) of Earth's orbit. Asclepius is about 400m in diameter. Its orbit has a semimajor axis of 1.023 AU, period 1.03 years, perihelion 0.66 AU, aphelion 1.39 AU, and inclination $4°.9$.

ashen light A phenomenon in which the unlit portion of Venus's globe glows faintly when the phase is a thin crescent. It may be caused by refraction of sunlight in Venus's very dense atmosphere or by atmospheric electrical phenomena. The ashen light is fainter than *earthshine on the Moon, and can be seen only when Venus is very close to inferior conjunction, but even then it is rare.

ASKAP Abbr. for *Australian Square Kilometre Array Pathfinder.

ASP Abbr. for *Astronomical Society of the Pacific.

aspect The position of a body in the Solar System relative to the Sun, as seen from Earth. The main aspects are *conjunction, *greatest elongation, *opposition, and *quadrature.

aspheric Describing a surface of a lens or mirror that has been given an optical figure which is not part of a sphere. The most common aspheric surface is the *paraboloid, widely used for telescope mirrors. Another common aspheric surface is that used for the corrector plates of *Schmidt–Cassegrain telescopes. In objective lenses most surfaces are spheroidal, one surface being aspheric to correct for aberrations.

association, stellar A very loose concentration of young stars (perhaps 100 in a region several hundred light years across), conspicuous only because it consists of distinctive stars. An *OB association consists predominantly of the most massive stars, of spectral types O and B; an *R association consists of medium-mass stars, surrounded by reflection nebulae; and *T associations are the birthplaces of the lowest-mass stars, in the form of T Tauri stars. All three types of association can be found together, and often have a star cluster within them. Like open clusters, associations are born from nebulae in the spiral arms of our Galaxy, but are so scattered that they disperse in 10 million years or so.

Association of Universities for Research in Astronomy (AURA) A consortium of US universities founded in 1957 to provide observing facilities for astronomers. AURA operates the *National Optical Astronomy Observatory, the *National Solar Observatory, the *Space Telescope Science Institute, and the *Gemini Observatory. Its headquarters are in Washington, DC. It has 37 US member institutions plus international affiliates in Australia, Canada, the Canary Islands, Chile, Germany, and Japan.

(⊕) SEE WEB LINKS
• Official website.

A star A star of spectral type A, whose spectrum is dominated by absorption lines of hydrogen (the *Balmer series); in fact, their hydrogen absorption lines are the strongest of any normal stars. They appear blue-white in colour. A-type stars on the main sequence have temperatures in the range 7200–9500 K, are 7–50 times more luminous than the Sun, and are of 1.5–3 solar masses. Sirius, the brightest star in the sky, is of spectral type A1, and Vega is type A0. A-type supergiants, such as Deneb, are more massive stars (up to 16 solar masses) evolving off the main sequence, with temperatures up to 9700 K, and luminosities over 35 000 times the Sun's. There are a number of peculiar groups among stars of this spectral type, particularly the *Ae stars, *Am stars, and *Ap stars. Also, two of the principal types of pulsating variable (*RR Lyrae stars and *Delta Scuti stars) are evolved stars whose surface temperatures are in the A-star range.

ASTE Abbr. for *Atacama Submillimeter Telescope Experiment.

asterism A distinctive pattern of stars that forms part of one or more constellations. For example, the familiar shape of the Plough or Big Dipper is an asterism within the constellation Ursa Major, while the *Square of Pegasus and the *False Cross are asterisms formed from stars of more than one constellation. The term is also used to describe a group of stars that appear to form a cluster, even though they are not related.

asteroid Any of the many small rocky or metallic objects in the Solar System, mostly lying in a zone (the *asteroid belt) between the orbits of Mars and Jupiter; also known as a *minor planet*. They range in diameter from almost 1000 km for *Ceres (the first asteroid discovered, in 1801) down to less than 10 m for the smallest so far detected. The total mass of all asteroids is 4×10^{21} kg, about one-twentieth the mass of the Moon. At the end of 2010 over half a million asteroids were known and the total was growing by about 10% per year.

When an asteroid is discovered it is given a temporary designation, consisting of the year of discovery followed by two letters; the first indicates the half-month during which the asteroid was discovered, and the second order of discovery within that half-month. Only when an accurate orbit has been determined is it assigned a permanent number, and the discoverer then has the right to name it. Increasing numbers are being discovered in dedicated searches such as the *Catalina Sky Survey, *Lincoln Near-Earth Asteroid Research, *Pan-STARRS, and *Spacewatch. There are thought to be between 1–2 million larger than 1 km, mostly in the main belt, although only a small percentage of these are currently known.

The orbits of most asteroids have higher eccentricities and inclinations than those of the major planets. Within the main asteroid belt, orbital eccentricities average about 0.15, and inclinations about 10°; occasionally they exceed 0.5 and 30°, respectively, more typical of the orbits of short-period comets. Indeed, some objects classified as asteroids may be defunct cometary nuclei. Rotation periods of asteroids range from a few hours to several weeks, but are typically 6–24 hours. The larger asteroids are roughly spherical, but those smaller than 150 km are commonly elongated or irregular. Radar studies of a few asteroids have revealed that some may be dumbbell-shaped or possibly double; these include *Castalia and *Toutatis. A few asteroids have small moons, the first of which was photographed in orbit around *Ida by the Galileo space probe.

Some main-belt asteroids form *groups* with similar orbital characteristics (semimajor axis, orbital eccentricity, and inclination), for example the *Cybele, *Hilda, *Hungaria, and *Phocaea groups. Where the group seems to have originated from the break-up of a single parent body, it is called a *Hirayama family. A small percentage of asteroids orbit outside the main asteroid belt. Members of the *Amor group cross the orbit of Mars, while *Apollo and *Aten group asteroids cross that of Earth; these three groups are collectively termed *near-Earth asteroids. Farther out, the *Trojan asteroids orbit at Jupiter's distance, while beyond Neptune is the *Kuiper Belt.

Asteroids are divided into various classes according to their reflectance spectra, which reveal differences in composition. The proportion of different asteroid classes changes markedly with increasing distance from the Sun. *S-class (silicaceous) asteroids predominate in the inner main belt (at less than 2.4 AU). *C-class (carbonaceous) asteroids are more prevalent in the middle and outer regions of the belt, with a peak near 3 AU. The dark asteroids near the outer edge of the main belt have a reddish tinge, and may be richer in organic components; these are the *P-class asteroids. Still farther out, many of the Trojan asteroids are even redder; they are termed *D-class asteroids. There is an apparent concentration of *M-class (metallic) asteroids in the middle of the belt, at 2.5–3.0 AU.

Asteroids are thought to have formed through the accretion of metre-sized bodies, but were prevented from aggregating into a planet by the gravitational effect of Jupiter, which had already formed. In addition, some *planetesimals left over from the formation of Jupiter may have been scattered into the asteroid belt. The largest asteroids were heated by the decay of radioactive isotopes within them. They melted and became differentiated, acquiring a metallic core, overlain by a mantle and crust. Subsequent collisions led to fragmentation, and almost every asteroid is probably a fragment of a once-larger body. In addition, most meteorites are believed to be pieces of asteroids.

(((( ))) SEE WEB LINKS

- Small-body Database Browser at NASA's Jet Propulsion Laboratory. Find orbital and physical data for any known asteroid.

asteroid belt The region of the Solar System between the orbits of Mars and Jupiter in which most asteroids are found; also known as the *main belt*. It extends from 2.1 to 3.3 AU from the Sun, corresponding to orbital periods of 3.0–6.0 years. Asteroids are not uniformly distributed throughout the belt. There are sparsely populated zones, known as *Kirkwood gaps, as well as concentrations of asteroids with similar orbital elements, forming groups and *Hirayama families. The proportion of various asteroid classes changes markedly through the belt (*see* ASTEROID). 📷

asteroid family See HIRAYAMA FAMILY.

asteroseismology The measurement and interpretation of oscillations or pulsations of stellar photospheres. It is the extension of *helioseismology to other stars. Asteroseismology is often used to refer to the study of weak, solar-like oscillations in stars, but can also include the larger amplitude variations of, for example, rapidly oscillating *Ap stars and *Beta Cephei stars. Solar-like oscillations have been identified in several dwarf and subgiant stars, with over 30 oscillation modes detected in Alpha Centauri A. In most cases the disk of the star cannot be resolved so it is global oscillations that are measured, which probe the deeper layers of the star. Two techniques are used to measure stellar oscillations: photometry and Doppler spectroscopy. The former is difficult for detecting solar-like oscillations as the brightness variations are typically a few millionths of a magnitude, at the limit of what is measurable from ground-based observatories. The perfect seeing available in space has, however, allowed such tiny oscillations to be measured by satellites such as *Microvariability and Oscillations of Stars (MOST), *COROT, and *Kepler. Solar-like oscillations have been detected at ground-based observatories using *radial-velocity spectrometers that measure the Doppler motions on the stars' surfaces. Observations are made at several sites around the globe to provide continuous coverage for long periods.

asthenosphere A weak layer within a planetary body, beneath the rigid outer layer known as the *lithosphere. It is slightly weaker than the lithosphere and deformable over long periods, allowing continental drift to take place on the Earth. On other planets and satellites the asthenosphere allows the topography to smooth out slowly, highlands becoming lower and deep basins shallower. The depth of the asthenosphere varies according to the size, density, composition, and thermal structure of the planet or satellite. On Earth, the top of the asthenosphere is around 100km below the surface, but on the Moon it lies 800km below the surface, more than half-way to the centre.

astigmatism A defect of a lens or an optical system in which the focal length along one diameter differs from that along another. Typically, as the focus is approached, the image of a star appears first as a short straight line; then, at the best focus, as a small circle (the *least circle of confusion* or *focal circle*); then, once the best focus is passed, as a line perpendicular to the first. Astigmatism may be caused by distortion of an optical component, often as a result of mechanical stress; or it may be inherent in an optical design, in which case its effects increase with distance from a system's optical axis. It can occur in the human eye as well as in optical instruments.

Astraea Asteroid 5, the fifth asteroid to be discovered, by the German amateur astronomer Karl Ludwig Hencke (1793–1866) in 1845. Astraea is an S-class asteroid of diameter 120km. Its orbit has a semimajor axis of 2.577 AU, period 4.14 years, perihelion 2.09 AU, aphelion 3.07 AU, and inclination 5°.4.

astration The cycle in which interstellar material forms into stars, is enriched with heavy elements as a result of nuclear reactions, and is then returned to interstellar space via stellar winds, planetary nebulae, or supernovae.

astrobiology The scientific study of the possibility of life elsewhere in the Universe; also known as *exobiology*. Astrobiology embraces branches of biology (for example microbiology, biochemistry, and ecology) in addition to astronomy. Considerations include assessment of the likely conditions required for the origin and development of life, and the means of detecting such life. Investigations such as the study of life in hostile environments on Earth, the search for water on Mars and sampling of the soil, the study of organic molecules in meteorites and comets, the synthesis of prebiotic compounds, and the various SETI programmes based on radio astronomy can be considered part of astrobiology.

astrobleme An eroded impact crater on Earth, identifiable by its geological structure and highly shocked rocks; the name means 'star wound'. About 180 terrestrial impact structures are known. Evidence of an impact origin includes signs of high-pressure shock waves, such as *shatter cones, shock-induced minerals (e.g. *coesite and *stishovite), and microscopic linear patterns in quartz called *shock lamellae*. The largest are Vredefort in South Africa (diameter 300km), Sudbury in Ontario, Canada (250km), *Chicxulub in Mexico (170km), Popigai in Russia (100km), and Manicouagan in Quebec, Canada (100km). Virtually all are on land, but Montagnais (diameter 45km) is on the continental shelf off Nova Scotia, and Chicxulub extends into the Gulf of Mexico from the Yucatán Peninsula. The oldest, Suavjärvi (in Russia) and Vredefort, are just over 2 billion years old, but the majority (about 60%) are less than 250 Myr old.

SEE WEB LINKS
• Database of terrestrial impact structures.

astrochemistry The study of chemical reactions that occur naturally in space. Molecules are able to form in space at very low temperatures (e.g. 20 K) and at pressures far lower than are achievable in a laboratory on Earth. Many chemical species that are unstable on Earth exist in space, and are detectable by their spectral lines at radio, infrared, optical, and ultraviolet wavelengths. There are two broad types of formation process: chemistry in gas clouds (including *photodissociation) and chemistry on the surfaces of dust grains. Many astrochemical reactions involve only gases. In some cases atoms or molecules in the gas may become ionized by the passage of *cosmic rays, and the subsequent reactions between ions and molecules are more rapid as a result. In the other main process, dust particles act as catalysts, providing a surface on which atoms, ions, and molecules can stick and then react. The dust also acts as a shield and prevents starlight from breaking up the molecules again. Heating of the dust by newly formed stars may release complex molecules from the grains back into the gas cloud and drive a new series of chemical reactions. *See also* INTERSTELLAR MOLECULE.

astrodynamics The branch of *celestial mechanics that deals with the motions of artificial satellites and space probes. Astrodynamics involves planning and controlling the trajectories of spacecraft, by techniques such as *gravity assist.

astrograph A telescope specifically designed for photographing comparatively wide areas of sky. Traditionally, an astrograph is a refractor with a lens corrected to give its best images at blue wavelengths, to which early photographic emulsions were sensitive. The astrograph is mounted together with a visual refractor of similar focal length for accurate tracking on a guide star. Today, *Schmidt cameras are used instead, typically having fields of view of 6°. *See also* NORMAL ASTROGRAPH.

Astrographic Catalogue A series of star catalogues covering the whole sky, obtained from measurements of photographs taken with *normal astrographs at many observatories between about 1890 and 1950. It was published in 21 sections, each covering a zone of declination measured from a particular observatory, between 1902 and 1963. The general limiting magnitude is about 11 (photographic), but some observatories measured fainter stars. In 1997 the US Naval Observatory published the *AC 2000* catalogue, containing 4 621 836 stars measured as part of the original Astrographic Catalogue project but with their positions converted to the same reference frame as the modern *Hipparcos Catalogue*. A revised version, *AC 2000.2*, was released in 2001, incorporating photometry for 2.4 million stars from the *Tycho-2 catalogue*.

((()) SEE WEB LINKS
• Detailed description and full catalogue downloadable from the CDS.

astrolabe An ancient device for measuring the altitudes of stars, like a simple sextant. A basic astrolabe consists of a disk hung vertically, with a sight (the *alidade*) which is pivoted so that it may be pointed at a chosen star. The local time can then be read off the face (the *tablet*) of the astrolabe. Different tablets can be used for different latitudes. Sophisticated modern versions are used for high-precision measurements of star positions; *see* DANJON ASTROLABE; PRISMATIC ASTROLABE.

astrology The supposed influence of the relative positions of the planets on people's personalities and events in their lives. In its modern form astrology is a pseudoscience, but in ancient times astrology and astronomy were intertwined. Often, the motive for keeping observational records was astrological. Ancient Chinese records of celestial events, from which the fortunes of entire dynasties were divined, are now of great value in the study of historical eclipses, novae, and comets.

astrometric binary A binary star in which the presence of an invisible or unresolved component is inferred from irregularities in the proper motion of the visible component. Similarly, a known binary star may prove to be an astrometric multiple star. Some astrometric binaries have subsequently been resolved by *speckle interferometry. Components calculated to have extremely low mass may be either faint red dwarfs, brown dwarfs, or possibly even planets.

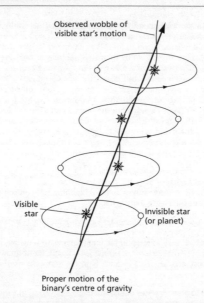

Observed wobble of
visible star's motion

Visible
star

Invisible star
(or planet)

Proper motion of the
binary's centre of gravity

astrometric binary: The presence of an unseen companion causes a visible star to wobble from side to side as it
moves through space.

astrometry The measurement of positions, parallaxes, and proper motions on the sky.
It can be divided broadly into two categories: global and small-field astrometry.
Global astrometry is concerned with mapping and cataloguing positions and motions over
large areas of sky. It was traditionally based on optical observations made with meridian
instruments and astrolabes. Currently, optical interferometers are being developed to
achieve greater accuracy. A stellar reference frame of bright stars is derived from such
observations, and interpolation to fainter stars is achieved by photographic surveys. The
accuracy of all ground-based optical astrometry is limited by thermal and mechanical
instabilities in the telescopes, but also mainly by uncertainties in the amount of atmospheric
refraction. The *Hipparcos satellite was launched to overcome these limitations. Global
astrometry at radio wavelengths is carried out by interferometers, with both short and very
long baselines.
   In *small-field astrometry*, relative positions are measured within the area observable with
long-focus telescopes, by means of photographic plates or, more recently, CCDs. Its main aims
are to measure relative proper motions and trigonometric parallaxes of stars, to discover
*astrometric binaries, and to identify faint optical counterparts of objects detected at other
wavelengths.

astronavigation The technique of position-finding by reference to celestial bodies. An
observer uses a sextant to observe the altitude of a celestial body, such as the Sun, the Moon, a
planet, or a bright star. From several observations of the same object at different times, or of
different objects at the same time, the observer's position can be calculated. Traditional
astronavigation on Earth has been largely superseded by methods using artificial satellites or
radio beacons.

Astronomer Royal An honorary title held by a prominent British astronomer, created
in 1675 by King Charles II when the Royal Observatory was founded. Until 1971 the

Astronomer Royal was also Director of the Royal Greenwich Observatory, but the post was
subsequently made separate.

Astronomers Royal

Name	Held office
John *Flamsteed (1646–1719)	1675–1719
Edmond *Halley (1656–1742)	1720–42
James *Bradley (1693–1762)	1742–62
Nathaniel Bliss (1700–1764)	1762–4
Nevil *Maskelyne (1732–1811)	1765–1811
John *Pond (1767–1836)	1811–35
George B. *Airy (1801–92)	1835–81
William H. M. *Christie (1845–1923)	1881–1910
Frank W. *Dyson (1868–1939)	1910–33
Harold Spencer *Jones (1890–1960)	1933–55
Richard *Woolley (1906–86)	1956–71
Martin *Ryle (1918–84)	1972–82
Francis Graham-Smith (1923–)	1982–90
Arnold Whittaker Wolfendale (1927–)	1991–4
Martin John Rees (1942–)	1995–

Astronomer Royal for Scotland A title created in 1834, originally given to
the director for the Royal Observatory at Edinburgh but since 1995 a separate honorary
position.

Astronomical Almanac, The A yearly publication containing tables of the Sun,
Moon, and planets and their satellites, and other astronomical data. It is compiled
and published jointly by the US Naval Observatory and HM Nautical Almanac Office.
It began in Britain in 1766 under the title *The Nautical Almanac and Astronomical Ephemeris*,
but from 1914 two separate versions were issued, one for astronomers and one for
navigators (*see* NAUTICAL ALMANAC, THE). In 1960 the name of the astronomical version
was abridged to *The Astronomical Ephemeris*. The current title was adopted in 1980 when
it merged with *The American Ephemeris and Nautical Almanac*, published in the US since
1852 (for the year 1855).

(⊕) SEE WEB LINKS
- UK official site. Some free online content.
- US official site. Some free online content.

Astronomical Journal (AJ) A publication containing original research and observations,
founded in 1849 by B. A. *Gould and now issued monthly by the *American Astronomical
Society.

(⊕) SEE WEB LINKS
- Official website with contents and abstracts of recent issues and free online access to
older issues.

Astronomical Netherlands Satellite (ANS) A Dutch satellite for ultraviolet and X-ray
astronomy, launched in 1974 August. It carried a 0.22-m telescope for broad-band photometry
at ultraviolet wavelengths. Two X-ray devices, consisting of reflecting collectors with

proportional counter detectors, covered the energy range 2–40 keV (0.03–0.62 nm). ANS operated until 1976.

 SEE WEB LINKS
• Information page at Goddard Space Flight Center.

Astronomical Society of the Pacific (ASP) An international scientific and educational organization based in San Francisco, USA, founded in 1889. Its monthly *Publications* contain technical papers, and it also publishes a bimonthly popular magazine, *Mercury*.

 SEE WEB LINKS
• Official website.

astronomical triangle A triangle on the celestial sphere in which the three sides are formed by the arcs of great circles. Such a triangle is called a *spherical triangle* in spherical trigonometry (the geometry of figures drawn on a sphere).

astronomical twilight The period before sunrise and after sunset when the centre of the Sun's disk is between 12° and 18° below the horizon. It is regarded as the time during which 6th-magnitude stars are visible at the zenith in a clear sky. In summer, astronomical twilight can last all night at latitudes higher than 48½° because the Sun is never more than 18° below the horizon. *See also* TWILIGHT.

astronomical unit (AU) A unit of length, formerly the mean distance of the Earth from the Sun, but now defined more technically, as below. In the original definition of the AU by C. F. *Gauss, the Earth's mean distance from the Sun (i.e. the semimajor axis of the Earth's orbit) was given by the precise form of Kepler's third law as found by Newton:

$$n^2 a^3 = k^2 (m + m_E),$$

where n is the Earth's mean motion (in radians per day), a is the semimajor axis of the Earth's orbit (in AU), m and m_E are the masses of the Sun and Earth respectively (in solar masses), and k is the *Gaussian gravitational constant. The astronomical unit is now defined as the distance from the Sun of a massless particle moving in a circular orbit around the Sun with an orbital period of one *Gaussian year* of $2\pi/k$ ephemeris days. The Earth's mean distance from the Sun is 1.000 000 031 AU, where 1 AU = 149 597 870 km.

astronomy The study of the space beyond the Earth and its contents; phenomena in the Earth's upper atmosphere that have their origin in space, such as aurorae and meteors, are also included. Before the application of the telescope to astronomy early in the 17th century, astronomy was concerned purely with measuring the positions and movements of celestial bodies, a branch now known as *astrometry. The realization in the 17th century that *gravitation governed the movements of celestial bodies led to mathematical methods for calculating orbits, now known as *celestial mechanics. In the 19th century the development of spectroscopy, by which the composition of bodies could be determined by analysing their light, marked the birth of *astrophysics. The recognition in the 20th century that galaxies exist outside our own and the discovery of the expansion of the Universe gave rise to modern *cosmology, which deals with the origin and evolution of the Universe.

***Astronomy and Astrophysics* (A&A)** A publication founded in 1969 by the merger of several existing European journals. It is a joint enterprise of scientific organizations in various European and South American countries. It is published four times monthly and contains papers on all aspects of astronomy and astrophysics.

 SEE WEB LINKS
• Official website with contents and abstracts of recent issues and free online access to older issues.

astrophotography The application of photography in astronomy. Images can build up for long periods on a photographic emulsion or *charge-coupled device (CCD), revealing stars and other objects invisible to the eye, and are recorded permanently, allowing their positions and brightnesses to be measured accurately.

Astrophotography began in earnest in 1883, when A. A. *Common photographed the Orion Nebula with a 36-inch (0.91-m) reflector, recording stars which could not be seen visually with the same instrument. Astronomers were quick to exploit the new medium, and for the next hundred years it was the primary means of making optical observations. Its application to *spectroscopy was equally important.

In astrophotography, a photographic emulsion or CCD is placed at the focus of a telescope, instead of an eyepiece. The emulsion may be on a glass plate or film. A shutter is required for making the exposure, which can last many minutes or even hours. During the exposure the telescope must be guided accurately on the object being photographed (*see* GUIDER).

Conventional emulsions suffer from *reciprocity failure, which limits effective exposure times to less than 30 min. Emulsions specifically designed for astrophotography are available with reduced reciprocity failure. Alternatively, conventional emulsions can be *hypersensitized to improve their long-exposure performance.

Colour photography is possible using conventional colour film, but the best results with long exposures come from *three-colour photography*. Separate exposures are made on hypersensitized black-and-white emulsions, each filtered to record only red, green, or blue light. The separate exposures are then combined in the darkroom to give a full-colour picture.

Photographic emulsions have now been almost entirely replaced for professional purposes by CCDs, which have a greater and more linear sensitivity to light.

Astrophysical Journal (ApJ) A publication founded in 1895 by G. E. *Hale and James Edward Keeler (1857–1900), now issued by the *American Astronomical Society three times a month. It publishes research papers in astronomy and, especially, astrophysics. Published with it is the *Astrophysical Journal Letters*, founded in 1958, which contains shorter and more topical papers. The *Astrophysical Journal Supplement* series, founded in 1954, appears monthly and carries longer and more technical papers.

((⊕)) SEE WEB LINKS

• Official website with contents and abstracts of recent issues and free online access to older issues.

astrophysics The study of the physical nature of the Universe and the objects in it, notably stars, galaxies, and the composition of the space between them. Astrophysics originated in the application of spectroscopy to the study of starlight in the 19th century. It complements the traditional branches of astronomy, *astrometry and *celestial mechanics, which are concerned with the positions and motions of objects.

Observational astrophysics interprets the electromagnetic and gravitational radiation emitted by celestial objects. Theoretical astrophysics attempts to explain the processes involved, which can lead to new understanding of the behaviour of matter under conditions not encountered on Earth. For example, nuclear physics had to develop before energy generation inside stars could be understood, and study of objects such as white dwarfs and neutron stars has helped confirm predictions about the behaviour of matter under extreme compression and intense gravitational fields.

Astrophysics can also probe the extremely tenuous gas between the stars, where complex molecules are formed and high-energy particles called cosmic rays move close to the speed of light. It also addresses ultimate questions about the origin of the Universe, the conditions shortly after its creation, and the origin of the chemical elements. Closer to home, astrophysics deals with matters such as the environment of the planets and the effects of the solar wind, which could explain short-term changes in our weather and long-term changes in climate. Many areas of physics are involved in such studies, including spectroscopy, plasma physics, atomic physics, and relativity.

In recent years, astrophysical advances have resulted from observations by satellites in space, which allow astronomers to study the Universe at all wavelengths from radio waves to gamma rays.

Astrophysics Data System (ADS) An online database of journals and books hosted at the Harvard–Smithsonian Center for Astrophysics in Cambridge, Massachusetts, and funded by NASA. Publications in the ADS cover the fields of astronomy, astrophysics, physics, and geophysics. The ADS provides open access not only to published papers but also to preprints of forthcoming papers. The ADS first went online in 1994.

(⊕) SEE WEB LINKS
• Official ADS website, allowing search of their digital library.

ASTROSAT An Indian satellite for X-ray and ultraviolet astronomy, scheduled for launch in 2012. It will carry an Ultraviolet Imaging Telescope (UVIT) with two 0.38-m mirrors for imaging at far ultraviolet (130–180 nm), near ultraviolet (180–300 nm), and optical wavelengths; three Large-Area Xenon Proportional Counters (LAXPC) for timing and low-resolution spectrometry at X-ray energies 3–80 keV; a coded-mask camera with Cadmium-Zinc-Telluride detector array (CZT) for higher spectral resolution in the range 10–150 keV; a Soft X-ray Imaging Telescope (SXT) for energies 0.3–8 keV; and a Scanning X-ray Sky Monitor (SSM) for monitoring X-ray transients and other variable sources at 2–10 keV.

(⊕) SEE WEB LINKS
• Official mission website.

astrum A star-shaped feature consisting of a pattern of cracks radiating from a focus on the surface of Venus; pl.*astra*. The term is not a geological one, and has not yet been officially allotted to any feature.

Asuka An alternative name for *ASCA. Asuka is an ancient Japanese word meaning 'flying bird'.

asymptotic giant branch star (AGB star) A star that occupies a strip in the Hertzsprung–Russell diagram that is almost parallel to, and just above, the giant branch. Stars evolve from the *horizontal branch to the asymptotic giant branch when they have exhausted the helium in their core and are burning it in a shell (*see* HELIUM SHELL FLASH). At the end of the AGB stage, stars with between about 1 and 8 solar masses undergo extensive mass loss through a vigorous stellar wind, exposing the core and leaving the star with a surrounding envelope of gas and dust. A star in such a stage is known as a *post-AGB star*, or a *protoplanetary nebula.

Atacama Cosmology Telescope (ACT) A 6-m millimetre-wave telescope at an altitude of 5190 m on Cerro Toco in the Atacama desert of northern Chile designed to map the cosmic microwave background at wavelengths of 2, 1.4, and 1.1 mm. It was opened in 2007 and is operated by an international consortium of observatories and institutions led by Princeton University. The ACT is similar to the *South Pole Telescope.

(⊕) SEE WEB LINKS
• Official telescope website.

Atacama Large Millimeter Array (ALMA) An aperture-synthesis array for millimetre- and submillimetre-wavelength astronomy being jointly built by the European Southern Observatory, the US National Radio Astronomy Observatory, and the National Astronomical Observatory of Japan, at an altitude of 5000 m on the Chajnantor plateau of the Atacama desert in northern Chile. When it comes into full operation in 2013, ALMA will consist of 66 dishes operating at wavelengths of 0.3 to 3.6 mm. The ALMA main array consists of fifty 12-m dishes that can be arranged with baselines up to 16 km, while the Atacama Compact Array (ACA) consists of four 12-m dishes and twelve 7-m ones in an area 250 m wide for imaging extended sources. 📷

(⊕) SEE WEB LINKS
• Official telescope website.
• Information page at NRAO.

Atacama Pathfinder Experiment (APEX) A 12-m antenna for submillimetre astronomy, jointly owned by the Max Planck Institut für Radioastronomie, *Onsala Space

Observatory, and the *European Southern Observatory (ESO), sited at an altitude of 5100 m
on the Chajnantor plateau, Chile. It works at wavelengths between 0.2 and 1.5 mm
and began operation in 2005. APEX is a forerunner of the *Atacama Large Millimeter Array
at the same site.

(⊕) SEE WEB LINKS
• Official telescope website.

Atacama Submillimeter Telescope Experiment (ASTE) A Japanese 10-m dish
for submillimetre astronomy at an altitude of 4860 m at Pampa La Bola in the Atacama
desert, northern Chile. It observes at wavelengths from 0.4 to 1.1 mm and began operation
in 2004.

(⊕) SEE WEB LINKS
• Official telescope website.

ataxite A class of iron meteorite which is nickel-rich, containing more than 12%
nickel. The ataxites contain discrete spindles of kamacite (a nickel-poor iron–nickel
alloy), fringed by taenite (up to 50% nickel), forming an octahedral arrangement
(*see* OCTAHEDRITE). A mixture of kamacite and taenite, called plessite, fills the spaces
between the spindles. Ataxite meteorites do not show an obvious *Widmanstätten pattern,
nor can the fine striations known as *Neumann lines be seen. The *Hoba West meteorite
is an ataxite.

Aten group A group of asteroids that cross the Earth's orbit, and whose average distances
from the Sun are less than that of Earth; along with the more abundant *Apollo group, they are
termed *Earth-crossing asteroids*. Their aphelia are greater than 0.983 AU (Earth's perihelion
distance) and their orbital periods are shorter than 1 year. The group is named after (2062)
Aten, the first of the group to be discovered, by the American astronomer Eleanor Kay Helin,
née Francis (1932–2009), in 1976. It is an S-class asteroid of diameter about 1 km. Aten's orbit
has a semimajor axis of 0.967 AU, period 0.95 years, perihelion 0.79 AU, aphelion 1.14 AU, and
inclination $18°.9$. Other major members of the group include (3753) Cruithne, diameter 3 km;
*Hathor; and *Ra-Shalom.

Atira asteroid An alternative name for an *apohele asteroid.

Atlas The second-closest satellite of Saturn, distance 137 670 km, orbital period 0.602 days;
also known as Saturn XV. It was discovered in 1980 on images from the Voyager 1 spacecraft.
It is $41 \times 35 \times 18$ km in size, and orbits at the outer edge of Saturn's A Ring, where its
gravitational field affects the ring structure.

atmosphere The gaseous envelope surrounding an astronomical body. Several
planets (including the Earth) retain considerable atmospheres because of their strong
gravity. Motions of the gas within planetary atmospheres in response to heating, coupled
with rotational forces, generate weather systems. The planetary satellites Titan and Triton
also possess atmospheres. Pluto has a 'seasonal' atmosphere, which forms when the
planet is close to perihelion, condensing out at aphelion. *See also* SOLAR ATMOSPHERE;
STELLAR ATMOSPHERE.

atmospheric extinction The loss of starlight in passing through the Earth's
atmosphere. Most of the loss arises from *Rayleigh scattering by molecules of nitrogen and
oxygen. At certain wavelengths there is *selective absorption from molecules of oxygen,
ozone, and water vapour. Particles of dust and industrial pollutants can also contribute to
extinction by *Mie scattering. Atmospheric extinction is proportional to the *airmass and
the atmospheric pressure. At sea level and in a perfectly clear sky a star 80° from the
zenith appears 1 mag. fainter than it would at the zenith. This in turn is 0.3 mag. fainter than
if no atmosphere were present. These figures apply only to visual observations; in blue
light, the extinction is greater, while in red light it is less. This is why the Sun and Moon
in particular appear red when near the horizon.

a

Atmospheric Extinction	
zenith distance °	Extinction (relative to zenith)_mag.
85	1.75
80	1.00
75	0.65
70	0.45
60	0.25
50	0.10
40	0.05

atmospheric refraction The displacement in apparent direction of a celestial object caused by the refraction of its light in passing through the Earth's atmosphere. Refraction increases the observed altitude of an object. For an object on the horizon, the amount is about 34′ (i.e. just over half a degree). According to a simple atmospheric model, the refraction at zenith distances up to about 45° is proportional to the tangent of the zenith distance, but the exact figure depends on the atmospheric pressure, temperature, and humidity. *See also* HORIZONTAL REFRACTION.

atmospheric scattering *See* ATMOSPHERIC EXTINCTION; MIE SCATTERING; RAYLEIGH SCATTERING.

atmospheric window A range of wavelengths to which the Earth's atmosphere is relatively transparent, so that astronomical observations can be made from the ground. The major windows are in the visible, infrared, and radio parts of the spectrum. The visible window extends from about 0.3 to 0.9 μm. There are *infrared windows at several wavelengths between about 1.25 and 30 μm, and beyond 300 μm. The *radio window extends from about 1 cm to 30 m.

ATNF Abbr. for *Australia Telescope National Facility.

atom The smallest part of a chemical element that can take part in a chemical reaction. An atom is composed of a *nucleus* made up of protons and neutrons, surrounded by electrons orbiting at different *energy levels. The number of protons in the nucleus of an element's atom is referred to as the **atomic number** of the element or atom; the total number of protons and neutrons is the *mass number*. In a *neutral atom* the number of protons is equal to the number of electrons.

atomic clock A device that generates a signal at a precisely known frequency which is locked on to some fundamental atomic resonance. Caesium-based frequency standards are the most basic since the *second is defined in terms of properties of the caesium atom (*see* ATOMIC TIME). Atomic clocks achieve an accuracy of about one part in 10^{14}, equivalent to a second in 3 million years. Slightly greater accuracy can be achieved, over short periods of time at least, by hydrogen maser devices which provide a signal at the frequency corresponding to the 21-cm radio line.

atomic time The time-scale used for all precise timekeeping, including civil time. It is based upon atomic frequencies and is the most accurate and consistent available today. The fundamental unit is the SI *second, which is defined in terms of a particular spectral line of the caesium-133 atom. The frequency of this microwave line is adopted as precisely 9 192 631 770 Hz.

The SI second is the basis for *International Atomic Time* (TAI). TAI was officially introduced by international agreement in January 1972, but had been available since 1955. When TAI was officially introduced, the length of the SI second was the same as that of a second in

*Ephemeris Time, the time-scale then used in astronomy. The two time-scales differed by a fixed amount (ET = TAI + 32.184s). Strictly speaking, however, the two definitions of the second were conceptually different. This difference was removed in 1984 when Terrestrial Dynamical Time (now known simply as *Terrestrial Time) was introduced to replace Ephemeris Time. Terrestrial Time has the SI second as its basic unit and differs from TAI only by the constant offset mentioned above.

TAI is the basis not just of astronomical time-scales but also of civil timekeeping. Broadcast time signals use *Coordinated Universal Time (UTC), which is TAI with an offset of a whole number of seconds. This offset requires occasional adjustment by the insertion of *leap seconds*.

Atria The star Alpha Trianguli Australis, a K2 giant of magnitude 1.9, lying 391 l.y. away.

attenuation 1. The reduction in strength of an electromagnetic wave by absorption or scattering in the medium through which it passes; for example, the extinction of starlight by interstellar dust.

2. The deliberate reduction of the signal strength in a radio receiver to prevent following stages of amplification being overloaded.

AU Symbol for *astronomical unit.

aubrite A class of calcium-poor achondrite meteorite; also known as *enstatite achondrites*. Aubrites consist almost entirely of the silicate mineral enstatite. They resemble the *enstatite chondrites, to which they are probably related. Aubrites have large grain sizes (sometimes exceeding several centimetres), which suggests that they formed within cooling magmas; alternatively, they may have originated through processes occurring in the solar nebula. Almost all known aubrites are *breccias. They are named after a meteorite that fell at Aubres, France, on 1836 September 14.

Auger shower Another term for a *cosmic-ray shower, named after the French physicist Pierre Victor Auger (1899–1993).

augmentation The increase in the apparent diameter of a celestial body as seen from the surface of the Earth compared with the apparent diameter it would have if seen from the centre of the Earth.

AURA Abbr. for *Association of Universities for Research in Astronomy.

Auriga (Aur) (*gen.*** Aurigae)** A constellation of the northern sky, representing a charioteer. Its brightest star is *Capella (Alpha Aurigae). Auriga contains three 6th-magnitude open clusters, M36, M37, and M38. *Epsilon Aurigae is a long-period eclipsing binary. Zeta Aurigae is another eclipsing binary, varying between magnitudes 3.7 and 4.0 with a period of 972 days; it is the prototype of the *Zeta Aurigae stars.

Aurora A European Space Agency programme, with Canadian involvement, for the long-term exploration of the Solar System, notably Mars, the Moon, and the asteroids. The ultimate goal is an international mission to land humans on Mars in the 2030s. The first ESA mission in the Aurora programme is *ExoMars, an orbiter and lander to be launched in 2016. That will be followed by surface rovers and missions to bring back samples from Mars. Development of a manned lunar base is also likely to be a stepping stone to Mars.

 SEE WEB LINKS

• ESA mission website.

aurora An emission of light from the Earth's high atmosphere, caused principally by oxygen atoms or nitrogen molecules that are excited by electrons accelerated within the magnetosphere. The visible aurora is dominated by the green (557.7 nm wavelength) and red (630 nm) emissions of oxygen, occurring respectively at altitudes of 100 km and above about 400 km, and red (661–686 nm) nitrogen emissions at about 95 km. Violet purple (391.4nm) nitrogen emissions are sometimes seen in the sunlit uppermost parts of aurorae at altitudes of 1000km.

Popularly known as the *northern lights* (or *southern lights* in the southern hemisphere), the aurora takes a number of characteristic forms. These may range from a *glow*, low over the northern horizon (from which the **aurora borealis**—'northern dawn'—takes its name), through *arcs* and *bands*, which may be homogeneous, or may show vertical rays. Isolated *rays* and *patches* of auroral light may also be seen. Most spectacular of all is the *corona*, a perspective effect whereby rays appear to converge on a single region of the sky almost overhead during a particularly intense storm. During strong activity, the rays and other structures move, causing a 'curtain' effect, and there may often be rapid changes in brightness. The aurora is seen from the southern hemisphere as the **aurora australis**, a mirror-image of activity present at the same time over the opposite hemisphere of the Earth.

Auroral activity is present more or less continuously around the high-latitude *auroral ovals. Observers at lower latitudes, such as in the British Isles, southern United States, or Australasia, see auroral activity only when the magnetosphere is disturbed by violent solar events. Mid-latitude aurorae are usually triggered by solar flares or *coronal mass ejections, and are most likely around times of high sunspot activity. Aurorae also occur on Jupiter, Saturn, Uranus, and Neptune. 📷

auroral oval A high-latitude ring of more or less permanent aurora, girdling the geomagnetic pole at a distance of 2000–2500km under quiet geomagnetic conditions. There are two ovals, one in either hemisphere, each the mirror-image of the other. They are displaced such that the dayside edge is closer to the pole than that on the night side. Under disturbed geomagnetic conditions, the auroral ovals brighten, broaden, and expand towards the equator, particularly on the night side, so the aurora becomes visible at lower latitudes. The auroral ovals remain relatively fixed in space above the rotating Earth.

auroral substorm *See* SUBSTORM.

Australian Astronomical Observatory An observatory at an altitude of 1150 m on Siding Spring Mountain near Coonabarabran, New South Wales, founded in 1973 and owned and operated by the Australian Department of Innovation, Industry, Science and Research. Its headquarters are in Epping, NSW. Its instruments are the 3.9-m Anglo-Australian Telescope and the 1.2-m United Kingdom Schmidt Telescope. It was originally known as the Anglo-Australian Observatory and was jointly owned by the UK and Australia, but came under full Australian ownership in 2010.

((⊕)) SEE WEB LINKS
• Official observatory website.

Australian Square Kilometre Array Pathfinder (ASKAP) An aperture synthesis radio telescope consisting of 36 dishes, each of 12 m diameter with baselines up to 6 km, being built at the *Murchison Radio-astronomy Observatory in Western Australia. It is due for completion in 2013 and will become part of the *Australia Telescope National Facility. As well as being a major instrument in its own right, ASKAP will serve as a testbed for the *Square Kilometre Array.

((⊕)) SEE WEB LINKS
• Official telescope website.

Australia Telescope National Facility (ATNF) The collective name for a group of eight radio antennas, located at three observatories in New South Wales, Australia, owned and operated by the Commonwealth Scientific and Industrial Research Organisation (CSIRO). The antennas can be used individually, or in various combinations as a long-baseline interferometer. The heart of the ATNF is the Compact Array at the *Paul Wild Observatory, Narrabri, consisting of six dishes, 22m in diameter, in a line 6km long. The largest individual instrument is the 64-m dish at *Parkes Observatory, 300km to the south of the Compact Array. Between these two sites lies another 22-m dish at Mopra Observatory, near the town of

Coonabarabran. The Australia Telescope began operation in 1988. Its headquarters are in Epping, NSW.

 SEE WEB LINKS
• Official telescope website.

australite A tektite from the largest strewn field of tektites in the world, which covers the whole of southern Australia, including Tasmania. Estimates of the *ablation age of australites are in the range 600 000–750 000 years. The 1-km Mount Darwin crater in western Tasmania has a similar age (730 000 years), but it seems too small to be their source.

autocorrelation A mathematical operation used in signal processing, especially as the first stage in obtaining the spectrum of a radio source. The **autocorrelation function** of a signal is a measure of the degree to which successive segments of the signal resemble each other. *See* AUTOCORRELATOR.

autocorrelator A digital device used in radio astronomy to make a spectrum of a radio source.

autoguider A *guider that causes a telescope to follow a guide star automatically by feeding control signals to the telescope's drive system. It may use a photomultiplier which views the star through the centre of a rotating mask, or a *charge-coupled device (CCD) with software which monitors the position of the guide star. Some CCDs use a system known as *track and accumulate* in which software compensates for drift of the image when superimposing several short exposures to give the effect of one long one. The autoguider may be attached to either a *guide telescope or an *off-axis guider.

autoionization A process by which an excited electron within an atom or molecule imparts sufficient energy to another electron in the same atom or molecule for it to escape. It is the inverse of *dielectronic recombination. *See also* BOUND–FREE TRANSITION.

autumnal equinox *See* EQUINOX.

averted vision A technique used to see faint objects visually. It involves looking to one side of an object's position, rather than directly at it, so that light falls on to the outer part of the retina, which is more sensitive than the centre of vision. To avoid the risk of the object falling on the eye's blind spot, it should be kept on the side of the field of view nearest the nose.

Avior A name used by navigators for the star Epsilon Carinae, magnitude 1.9. It is a K3 giant, 605 l.y. away.

axion A hypothetical elementary particle which has been proposed as a candidate for *dark matter in the Universe. Axions are expected to have been produced abundantly in the early phases of the hot Big Bang. Although the axion is predicted to have a very tiny mass (about 10^{-11} of the mass of an electron) and might therefore be expected to be a form of *hot dark matter, it interacts so weakly with radiation that it is, in fact, a plausible candidate for *cold dark matter.

axis, optical *See* OPTICAL AXIS.

axis, rotation The imaginary line around which a body spins. The axis of rotation joins a body's north and south poles.

azimuth (*A*) The direction to a celestial object measured in degrees, clockwise from north around the observer's horizon. Azimuth is 0° for an object due north, 90° due east, 180° due south, and 270° due west.

Azophi Latinized name of al-*Ṣūfī.

B

BAA Abbr. for *British Astronomical Association.

Baade, (Wilhelm Heinrich) Walter (1893–1960) German-American astronomer. In 1944 he identified two different *populations of stars in the Galaxy, and found that both contained Cepheid variables. Moreover, each population of Cepheids had its own *period–luminosity law. In 1952 he began to search for Cepheids in the Andromeda Galaxy, M31, but found none, even though they should have been luminous enough to be visible; he reasoned that M31 was over twice as distant as had been thought, which made the Universe both larger and older than had been believed. Baade collaborated with R. L. B. *Minkowski on identifying radio sources, and with Minkowski and F. *Zwicky on supernovae. He also discovered ten asteroids, including *Hidalgo and *Icarus.

Baade's Window A small area of sky in the constellation Sagittarius that is relatively free from obscuring dust, and through which an optical telescope can see to the central region of our Galaxy and beyond. The window is at galactic longitude 0°.9 and latitude −3°.9, near the globular cluster NGC 6522. W. *Baade used this window to observe RR Lyrae variable stars in the central bulge of our Galaxy, and determine their distance and hence the size of our Galaxy.

Baade–Wesselink method A method of determining the distances of certain types of pulsating variable star, particularly Cepheids and RR Lyraes. It is jointly named after W. *Baade, who first proposed it in 1926, and the Dutch astronomer Adriaan Jan Wesselink (1909–95), who refined it in 1946. In the original method, the variation in angular diameter of a Cepheid as it pulsates is inferred by means of model atmosphere calculations from the measured changes in its brightness. Spectroscopy is used to measure the corresponding changes in radial velocity, providing the actual distance over which the star's surface has moved. By dividing the measures of angular and linear diameter, the distance to the star is obtained. More recently, it has proved possible to measure the angular diameter of the pulsating star directly using optical interferometers, thus allowing a more accurate measurement of the star's distance. This newer approach is known as the *geometric Baade–Wesselink method.*

Babcock, Harold Delos (1882–1968) American solar astronomer and physicist, father of H. W. *Babcock. He studied the *Zeeman effect in the Sun's spectrum, and in 1928 published the wavelengths of over 20000 solar spectral lines. In 1952 Babcock and his son developed the solar magnetograph, with which they made the first-ever measurements of magnetic fields on the Sun's surface.

Babcock, Horace Welcome (1912–2003) American astronomer, son of H. D. *Babcock. After the Second World War he collaborated with his father in solar research. In 1946 he discovered the first *magnetic star, 78 Virginis. Babcock also studied the rotation of the Andromeda Galaxy and invented the concept of *adaptive optics.

background noise An unwanted, randomly fluctuating signal that may mask the signal from a radio source. Background noise has many origins: the random motions of electrons in the receiving apparatus (*thermal noise*), radiation from our Galaxy (*synchrotron radiation*), and the *cosmic microwave radiation from the Big Bang are three major contributors. Noise extending uniformly over a broad band of frequencies is termed *white noise*.

background radiation Electromagnetic radiation arriving at a telescope detector or receiver that does not come from the source under observation. In radio astronomy, background radiation comes from the Milky Way and the *cosmic microwave background from the Universe as a whole. In infrared astronomy, background radiation from the atmosphere and from the telescope itself can be considerable, but may be reduced by careful design of the telescope and instruments.

backscattering The phenomenon that occurs when radiation is scattered predominantly backwards along its original path. *Rayleigh scattering and *Thomson scattering generate equal amounts of forward and backscattering. Particular cases of backscattering are the rainbow, and the *heiligenschein* from dew on grass which appears as a halo around the head of the observer's shadow. *See also* FORWARD SCATTERING.

Bailey type A subdivision of the *RR-Lyrae type of variable stars, described by the American astronomer Solon Irving Bailey (1854–1931), based upon the shape and amplitude of the light-curve. In order of decreasing amplitude, the original types were designated a, b, and c, with mean periods of 0.48, 0.58, and 0.32 days, respectively. There is a continuous transition between types a and b, both of which have asymmetric light-curves, and so they are now regarded as one type, RRAB. Light-curves of type RRC are distinct, and frequently appear almost sinusoidal. Since Bailey's time, another subclass has emerged known as the double-mode RR Lyraes, abbreviated as RR(B) or RRd; these stars pulsate simultaneously in the fundamental and first overtone radial modes.

Baily, Francis (1774–1844) English stockbroker and astronomer. In 1826 he produced a catalogue containing highly accurate positions for 2881 stars. His published account of an annular eclipse he observed from Scotland in 1836 described 'a row of lucid points, like a string of bright beads . . . running along the lunar disc with beautiful coruscations of light', and he listed other observers who had previously seen the same thing. This phenomenon has since become known as *Baily's beads. He is also noted for his editing of historical star catalogues including those of Ptolemy, Tycho Brahe, Hevelius, and Flamsteed.

Baily's beads A phenomenon seen immediately before or after totality in a solar eclipse, when sunlight shines through valleys along the Moon's limb. The appearance is like a curved string of pearls. It is named after F. *Baily, who described the effect following an annular eclipse in 1836. One bead can on occasion appear very much brighter than the others, giving the effect known as a *diamond ring*.

Baker–Schmidt telescope A modified form of *Schmidt camera, designed by the American optician James Gilbert Baker (1914–2005), that uses a secondary convex mirror within the telescope to produce a flat field free from *astigmatism and coma (*see* COMA, OPTICAL), within a comparatively short tube. The photographic plate, also located within the instrument, faces outwards rather than towards the main mirror, as in a conventional Schmidt.

Baldwin effect The inverse relationship between the total absolute luminosity of a *quasar and the equivalent width of its various emission lines, including Lyman-alpha. The origin of this effect is not well understood. It may result from the orientation of the quasar, but it might also be caused by different degrees of ionization in quasars of different luminosity. It is named after the American astronomer Jack Allen Baldwin (1945–).

Balmer limit A break in the spectrum of a star or galaxy at 364.6nm; also known as the **Balmer jump**. It arises at the termination of the *Balmer series of hydrogen, at which point the atom becomes ionized. The size of this jump (i.e. the level of the continuum on either side of the break) is an important indicator of the physical conditions in the emitting region, particularly the pressure.

Balmer lines *See* BALMER SERIES.

Balmer series A sequence of absorption or emission lines in the visible part of the spectrum, due to hydrogen; also known as **Balmer lines**. In order of decreasing wavelength they are Hα (656.3nm, red), Hβ (486.1nm, blue-green), Hγ (434.0nm, blue), and so on, becoming closer together as they approach 364.6nm (in the ultraviolet), the **Balmer limit**. Balmer absorption lines are caused by jumps of electrons from the second *energy level to higher levels, and emission lines when the electrons drop back to the second energy level. They are named after the Swiss mathematician Johann Jakob Balmer (1825–98). *See also* HYDROGEN SPECTRUM.

band A broad absorption feature in a spectrum due to molecules or very closely spaced absorption lines of a single element; for example the b band in the Sun's spectrum, which is due to many lines of magnesium.

bandpass filter A filter designed to transmit a range of frequencies (the *bandwidth*) between two cutoff frequencies and to reject frequencies outside that range.

bandwidth 1. The range of frequencies occupied by a signal.
2. The range of frequencies that can be transmitted by a filter, amplifier, or other device. *See also* COHERENCE BANDWIDTH.

bar A unit of pressure, often used to express atmospheric pressures. One bar is equal to 10^5 pascals, approximately the Earth's atmospheric pressure at sea level. The *millibar* (100 Pa) is also frequently used.

barium star A red-giant star of spectral type G or K in which heavier elements such as barium appear in the spectrum with unusually high abundance; also known as a *Ba II star* or a *heavy-metal star*. Barium stars are similar to *CH stars, but are more metal-rich and have insufficient carbon to be considered as *carbon stars. About 1% of G and K giants are barium stars. It seems probable that all barium stars are in binary systems with white-dwarf companions and long orbital periods (hundreds of days). Barium stars may therefore be the result of accretion of part of the matter ejected by an initially more massive companion when it was in a red-giant phase before becoming a white dwarf.

Barlow lens A diverging lens, usually a doublet, placed in front of the focal point of the telescope to multiply the magnification of an eyepiece. A typical Barlow lens effectively doubles the focal length of the telescope and hence doubles the power of any eyepiece used with it. The value can be varied somewhat by altering the distance between the Barlow and the eyepiece. It was invented in 1834 by the English physicist and mathematician Peter Barlow (1776–1862).

Barnard, Edward Emerson (1857–1923) American astronomer. He discovered 16 comets, including in 1892 the first comet to be discovered photographically. Also in 1892 he found Amalthea, the fifth satellite of Jupiter. He pioneered the photography of Milky Way starfields, publishing a catalogue of star-poor regions now known to be *dark nebulae. His catalogue numbers (prefixed by B) are still in use. In 1916 he discovered *Barnard's Star.

Barnard's Galaxy Popular name for the dwarf irregular galaxy NGC 6822, a member of our Local Group about 1.6 million l.y. away in Sagittarius. It is about 10 arcminutes in diameter and has an integrated magnitude of 8.3, but is not prominent visually. It was discovered by E. E. *Barnard in 1884.

Barnard's Loop A large, faint emission nebula about 1500 l.y. away in Orion, discovered by E. E. *Barnard. It is a semicircular arc, 14° across, centred approximately on the Orion Nebula. Powerful young stars from the Orion OB1 association ionize the loop, which is thought to have been formed by a supernova explosion 3 million years ago. It is expanding at 10–20 km/s.

Barnard's Star The second-closest star to the Sun, 5.9 l.y. away in Ophiuchus. It is an M4 dwarf of apparent magnitude 9.5, with a luminosity 2000 times less than the Sun's. It has the largest proper motion of any star known, $10''.36$ per year, as was discovered in 1916 by E. E. *Barnard, after whom the star is named.

barn-door mount Another name for a *Scotch mount.

Barnes–Evans relationship A correlation between the surface brightness of a star at visual wavelengths and the star's (V – R) colour as measured in *Johnson photometry. The correlation is independent of the star's *luminosity class and nearly independent of *interstellar absorption. The relationship allows the angular diameter of any star to be found from its V magnitude and (V – R) colour and is used in measuring the distance to pulsating variables by the *Baade–Wesselink method and also the distances of novae and eclipsing binaries. It is named after the American astronomer Thomas Grady Barnes III (1944–) and the Welsh astronomer David Stanley Evans (1916–2004), who first published it in 1976.

barred spiral galaxy A type of disk galaxy with spiral arms extending from an almost rectangular or cigar-shaped bar of stars across its central regions, which can account for up to one-third of the galaxy's total light output; Hubble type SB. The central bars are typically between 2.5 and 5 times as long as they are wide. The masses of barred spirals range from about 10^9 to 5×10^{11} solar masses, and their diameters from about 10000 to over 300000 l.y. Almost half of all disk galaxies, including our own, contain a noticeable bar; similar but much less prominent structures may exist undetected in most disk galaxies. The bars often show sharp, straight dust lanes on the edge leading the rotation. There is often a cluster of bright nebulae (*H II regions) at the outer ends of the bar, where the spiral arms usually start. Many barred galaxies show a narrow ring of stars near the outer end of the bar. The bars probably differ in width and thickness as well as length, and there may be considerable streaming motion or flow of interstellar gas along them. It is not yet clear whether bars are a permanent structure in a galaxy, or represent a transient and perhaps recurrent feature. ◙

barrel distortion An optical defect in which the magnification of a lens is slightly greater close to its optical axis than at the edges. The image of a square object then has convex sides, hence the name.

Barringer Crater *See* METEOR CRATER.

barycentre The *centre of mass of a system of massive bodies. The system orbits around its barycentre. In a system of two bodies of equal mass, the barycentre lies midway between them. If the masses are unequal, the barycentre will lie closer to the greater mass. If one mass is much greater than the other, the barycentre may actually lie within the body of the larger mass, as in the case of the barycentre of the Earth–Moon system which lies about 1600 km below the surface of the Earth. The barycentre of the Solar System lies just outside the Sun's surface, so that the Sun actually performs a complicated orbit about that point.

Barycentric Celestial Reference System (BCRS) A non-rotating set of spacetime coordinates centred on the *barycentre of the Solar System which takes into account distortions in spacetime caused by the masses of the Sun and planets.

barycentric coordinates A system of coordinates with their origin at the centre of mass of the Solar System. The reference plane is usually the ecliptic. *See also* HELIOCENTRIC COORDINATES.

Barycentric Coordinate Time (TCB) The time-scale used in the *Barycentric Celestial Reference System, the unit of which is the SI second. TCB differs from *Terrestrial Time (TT) because of the movement of the Earth in the Barycentric Celestial Reference System and the Earth's immersion in the gravitational potential of the Sun and other Solar System bodies. These effects cause an overall rate change of about 0.5 s per year on which are superimposed smaller periodic variations. The time-scale *Barycentric Dynamical Time (TDB) is a convenient linear transformation of TCB that keeps in step with TT on the average.

Barycentric Dynamical Time (TDB) A time-scale originally intended to be used for calculating the positions of planets and other Solar System bodies using methods consistent with general relativity. Compared with *Terrestrial Time, TDB exhibits small periodic variations, caused by the effects on terrestrial clocks of the changing speed and gravitational potential experienced as the Earth orbits the Sun.

baryon Any member of a class of subatomic particles that includes protons and neutrons, as well as numerous others such as the Σ^+ (sigma-plus) and Ω^- (omega-minus) particles. Together with the *mesons, baryons form one of the two major classes of subatomic particles, the *hadrons. Baryons are composite particles, made up from three *quarks, and have a *spin of ½. All isolated baryons other than the proton are unstable and decay down to a proton plus other particles. *See also* NON-BARYONIC MATTER.

baryon star A star composed principally of *baryons. In practice the term is a synonym for a *neutron star, since the electrical repulsion of protons would disrupt a pure proton star.

basalt A dark-coloured, fine-grained igneous rock low in silica, and composed essentially of calcium-rich plagioclase feldspar (over 50%) and pyroxene. Basalt is the most common extrusive igneous rock, solidified from molten magma or volcanic fragments erupted on to the surface of the Earth, Moon, and other planets. The lunar *mare regions are covered with basaltic rocks.

basaltic achondrite A collective name by which the most abundant achondritic meteorites, the *eucrites, *howardites, and *diogenites, are often known. Together with the *mesosiderites (which are apparently related to the howardites), these meteorite types are also called the *eucrite association*. The compositional similarities between eucrites, diogenites, and howardites suggest that they originated in a single asteroid. Vesta is a prime candidate for this parent body, since it shows a eucritic reflectance spectrum on one hemisphere, and a diogenitic spectrum on the other.

baseline The distance between any two aerials in an *interferometer. For a given observing frequency, the resolution of an interferometer improves in proportion to the maximum baseline in the array.

basin, impact A large impact crater within which one or more concentric inner rings of mountains are present. There are three morphological types: *peak-ring basins*, in which there is a single inner ring; *central peak basins*, in which both an inner ring and a *central peak are present; and *multi-ringed basins*, in which there is more than one inner ring. The rings are thought to form when the central region of the crater floor rebounds upwards and then collapses again, producing the first inner ring. In larger craters the central region may continue to oscillate up and down, forming additional rings. As many as six concentric rings have been noted in some basins. The *Orientale Basin on the Moon is a beautifully preserved multi-ringed basin 930km in diameter. The Moon's *Imbrium Basin, nearly 1300km across, is larger, but much of it has been covered by younger lavas. The *South Pole Aitken Basin, 2400 km wide, is the largest lunar impact structure of all. The largest basin in the Solar System is *Valhalla on Callisto, 3000km across. The Hellas basin on Mars (*see* HELLAS PLANITIA) is 2200km in diameter, but this has been heavily eroded, and its floor is covered in sediments. The size above which inner rings start to form varies from planet to planet, and also between different regions of the same planet. The smallest basins on the Moon are about 140km in diameter, but on Mercury basins as small as 100km are found.

al-Battānī, Muḥammad ibn Jābir (858–929) Arab astronomer, born in modern Turkey; also known by the Latinized name Albategnius. He was one of the first Arab astronomers to grasp the importance of accurate observations. He produced a set of tables, including a catalogue of star positions more accurate than those in Ptolemy's *Almagest*, that was to influence medieval European astronomers. Al-Battānī refined the values of the precession of the equinoxes, the obliquity of the ecliptic, and the length of the tropical year, and found that over the course of the year the Earth–Sun distance varies.

Ba II star Another name for a *barium star.

Bautz–Morgan class A category in a classification scheme for clusters of galaxies based on the contrast between the brightest member and the typical bright galaxies in the cluster. Class I clusters are dominated by an extremely large, luminous supergiant galaxy. In Class III, no member appears to stand out against the bright galaxy population. Class II is intermediate. The classification is named after the American astronomers Laura Patricia Bautz (1940–) and W. W. *Morgan, who published it in 1970.

Bayer, Johann (1572–1625) German lawyer and astronomer. In his star atlas *Uranometria* (1603) he allocated Greek letters to the main stars in each constellation, usually in order of brightness, and this system of Bayer letters has persisted. He also depicted twelve new southern constellations introduced by Dutch navigators, making *Uranometria* the first atlas to cover the entire sky.

Bayer letters The system of identifying stars in a constellation by Greek letters according to their approximate order of brightness. They were introduced in 1603 by J. *Bayer in his star atlas *Uranometria*.

B band A broad *Fraunhofer line in the Sun's spectrum at around 687 nm, due to absorption by molecular oxygen in the Earth's atmosphere. The B band is actually a group of close, regularly spaced lines spanning about 2 nm, unresolved at low resolution.

B-class asteroid A subclass of the *C-class asteroids. Its members are distinguished by a moderately low albedo (0.04–0.08), but higher than that of a typical C-class asteroid. The second-largest main-belt asteroid, *Pallas, is now included in this class. Other members include (379) Huenna, diameter 92 km, and (431) Nephele, diameter 95 km.

BCRS Abbr. for *Barycentric Celestial Reference System.

BD Abbr. for *Bonner Durchmusterung*.

beam 1. A stream of radiation or particles confined to a narrow range of directions.
  2. An imaginary stream of radiation originating at the focus of a telescope (or central point of an array of telescopes) and illuminating the volume of space from which the telescope or array can receive radiation. The term is commonly used to refer to the patch of sky to which the telescope is sensitive. Optical and infrared telescopes generally have a simple circular beam, while radio telescopes (especially interferometers) may have beams of a complex shape. *See also* FAN BEAM; PENCIL BEAM.

beaming The channelling of radiation or particles into a beam. Jets are emitted in beams along the rotation axis of an accretion disk around protostars and at the centres of active galaxies. The radio emission from pulsars appears to be emitted along the object's magnetic axes. Other examples of beaming include secondary cosmic-ray showers, which are observed mostly around the zenith due to the increased atmospheric absorption at large zenith angles. *See also* RELATIVISTIC BEAMING.

beamwidth A measure of the angular width of a radio telescope beam. *Full-width at half-maximum* (FWHM) or the *half-power beamwidth* (HPBW) is the width of the beam at a point where the power received from a point source is half its peak value. As a rule, the smaller the beamwidth, the better the resolution of the telescope.

beat Cepheid A Cepheid variable that is oscillating in both the fundamental and first overtone *pulsation modes; abbr. CEP(B). The periods interact to produce a longer, *beat period*. Such stars, also known as *double-mode Cepheids*, account for nearly half of all Cepheids with periods of 2–4 days.

Becklin–Neugebauer Object (BN Object) A powerful source of infrared radiation discovered by the American astronomers Eric Edward Becklin (1940–) and Gerry Neugebauer (1932–) in 1966 behind the Orion Nebula. It is thought to be a massive young star 10 000 times as luminous as the Sun. The star is embedded in a dense cocoon of gas and dust that completely obscures it from optical telescopes.

Becrux *See* BETA CRUCIS.

bediasite A type of tektite found in Texas, USA. The name derives from the Bedias tribe, who inhabited the area where the tektites were found. Bediasites have *ablation ages of 33–35 million years. This is very similar to the derived ages of the *georgiaites and the Martha's Vineyard tektite, and together they make up the oldest group of tektites.

Beehive Cluster Alternative name for the open cluster *Praesepe.

Belinda The ninth-closest satellite of Uranus, distance 75 260 km, orbital period 0.624 days; also known as Uranus XIV. Belinda is elongated in shape, with a mean diameter of 66 km, and was discovered in 1986 with the Voyager 2 spacecraft.

Bellatrix The star Gamma Orionis, magnitude 1.6. It is a B2 giant, 252 l.y. away.

Bell Burnell, (Susan) Jocelyn (1943–) English astronomer. She was the first to notice a signal from what proved to be a *pulsar (subsequently designated CP 1919), recorded in 1967 August during a survey of radio galaxies. Bell, A. *Hewish, and colleagues began detailed observations, and Bell found a second pulsar the following December.

Belt of Orion A line of three bright stars, *Alnilam, *Alnitak, and *Mintaka, in the constellation Orion, representing the hunter's belt.

Benetnasch An alternative name for the star *Alkaid (Eta Ursae Majoris).

Bennett, Comet (C/1969 Y1) A long-period comet discovered on 1969 December 28 by the South African amateur John ('Jack') Caister Bennett (1914–90); formerly designated 1970 II. Comet Bennett reached perihelion, 0.54 AU, on 1970 March 20, and was prominent throughout 1970 April. Its maximum magnitude was 0, and its tail reached 20°. Spiral jets were seen emerging from the nucleus. Rapid changes in the gas tail were caused by turbulent conditions in the solar wind close to sunspot maximum. Its orbit has an eccentricity of 0.996, inclination 90°.0, and a period about 1700 years.

 SEE WEB LINKS
• Information page at Cometography website.

bent-pillar mounting A form of *German mounting for telescopes that features a pillar bent through an angle equal to the observer's latitude. This bent pillar carries the polar axis. It is also referred to as a *knee mounting* or *knee pedestal* because of its shape. The design was developed by the German firm of Carl Zeiss for long-focus refractors.

BepiColombo An ESA–Japanese mission to the planet Mercury, planned for launch in 2014, arriving at Mercury in 2020. It will consist of two orbiters, the Mercury Planetary Orbiter (MPO) to study Mercury's surface and the Mercury Magnetospheric Orbiter (MMO) to study its magnetosphere and space environment; the magnetosphere orbiter will be provided by Japan. The mission is named after the Italian mathematician and engineer Giuseppe ('Bepi') Colombo (1920–84), who first explained Mercury's unique rotation.

 SEE WEB LINKS
• ESA mission website.

BeppoSAX An Italian X-ray astronomy mission with participation from the Netherlands and ESA, launched in 1996 April. It is named after the Italian astronomer Giuseppe ('Beppo') Paolo Stanislao Occhialini (1907–93); SAX is the acronym for '*Satellite per Astronomia a raggi X*'. The satellite carried a range of instruments for broad-band timing and spectroscopy over the energy range 0.1–300 keV. BeppoSAX made spectroscopic studies of X-ray sources and looked for *gamma-ray bursts (GRB), observing the first GRB X-ray *afterglow. It ceased operation in 2002 April and re-entered the atmosphere a year later.

 SEE WEB LINKS
• Official mission website.

Berkeley–Illinois–Maryland Association Array (BIMA Array) *See* COMBINED ARRAY FOR RESEARCH IN MILLIMETER-WAVE ASTRONOMY.

Bessel, Friedrich Wilhelm (1784–1846) German mathematician and astronomer. After compiling a *fundamental catalogue of 3000 stars (1818), he began an extensive programme of measuring star positions and proper motions. By 1833 he had highly accurate positions for 50000 stars, which he combined with refinements of previous observations into a catalogue of 63000 stars, marking the beginning of modern astrometry. In 1838 he announced the first reliable determination of a star's parallax (61 Cygni), just 6% above the present-day value. In 1844 he deduced that oscillations in the proper motions of Sirius and Procyon indicated the existence of unseen companion stars, subsequently detected visually.

Besselian day numbers A set of five quantities that are tabulated at daily intervals in *The Astronomical Almanac*. When combined with a set of star constants, they allow the apparent right ascension and declination of a star to be calculated from its mean place. The latter, together with the star constants, can be obtained from a star catalogue. In this way the effects of general precession, nutation, and annual aberration are included in an efficient manner, avoiding more detailed calculation. They are named after the German astronomer F. W. *Bessel, who first introduced them. *See also* DAY NUMBER.

Besselian elements 1. Values used in computing an occultation of a star or planet by the Moon. They include the Universal Time (UT) of conjunction in right ascension between the Moon and the star (or planet); the *Greenwich hour angle of the star (or planet) at that time; and values of particular geometrical properties of the Moon, the star (or planet), and the observer's location.
　　2. Eight values used in computing the circumstances of a solar eclipse. They describe the position and direction of the axis of the Moon's shadow, and the size of the shadow relative to the Earth. Besselian elements are named after F. W. *Bessel, who devised the method of computation.

Besselian epoch An obsolete method of specifying the date of astronomical events. It was the standard before 1984, but has since been replaced by the simpler *Julian epoch. The prefix 'B' is now added to Besselian epoch dates to avoid confusion with the Julian epoch. The Besselian epoch was expressed as the year plus a decimal part, e.g. 1975.2406. The time-scale used was *Ephemeris Time, and the unit of time was the *Besselian year. The standard epochs of 1900.0 and 1950.0 used earlier this century were Besselian epochs, differing by 50 tropical years. In fact,

　　B1950.0 = 1950 January 0.923d ET = 1949 December 31 22h 09m ET.

Besselian year The interval of time required for the right ascension of the fictitious *mean sun to increase by 24h. It is defined as beginning when the mean sun's longitude is precisely 280° (or RA 8h 40m); this value was chosen as it corresponds roughly to January 1. The Besselian year is virtually identical to the *tropical year and is 365.2422 days long. It was the basic unit of the now-superseded *Besselian epoch.

Be star A star of spectral type B, on or near the main sequence, which exhibits emission lines of hydrogen (hence the 'e' suffix); Be stars are sometimes referred to as 'classical' to distinguish them from Herbig Be stars (*see* HERBIG AE/BE STAR) and *B[e] stars. The emission is usually variable, as is the brightness of the star. This class, which represents about 20% of the B-star population, consists of rapidly rotating B stars which have thrown off an equatorial disk or ring. Excitation of the gas in this disk by the star's ultraviolet light causes the emission. Equatorial rotation velocities are 250–450 km/s, significantly higher than for normal B stars. Bright Be stars include *Achernar and Gamma Cassiopeiae. *See also* GAMMA CASSIOPEIAE STAR; SHELL STAR.

B[e] star A star of spectral type B which exhibits *forbidden lines in emission in its spectrum, and showing an infrared excess. Most known B[e] stars are giants or supergiants. The infrared excess is attributed to dust, possibly in a circumstellar disk.

Beta Canis Majoris star *See* BETA CEPHEI STAR.

Beta Cephei star A type of pulsating variable, of early spectral type (O8–B6), lying just above the main sequence; abbr. BCEP. They are also known as **Beta Canis Majoris stars**. Magnitude variations are 0.01–0.3, with periods of 0.1–0.6 days. There is no generally accepted explanation for the variations, but both *radial pulsation and *non-radial pulsation appear to be involved. A short-period subgroup (BCEPS) has periods one-tenth those of the normally observed ones. *See also* 53 PERSEI STAR.

Beta Crucis The second-brightest star in the constellation Crux, also known as Becrux or Mimosa. It is a B0.5 giant 279 l.y. away and a variable of the Beta Cephei type, ranging between magnitudes 1.2 and 1.3 with a period of 5.7 hours.

beta decay Radioactive decay in which an atomic nucleus spontaneously decays into a daughter nucleus, releasing two subatomic particles. Either a neutron turns into a proton, releasing an electron plus an antineutrino; or a proton turns into a neutron, releasing a positron plus a neutrino. The resultant nucleus has the same mass number as the original nucleus (i.e. the same total number of protons and neutrons), but the atomic number differs by one. The electrons or positrons emitted are known as **beta particles**.

Beta Lyrae star A type of eclipsing binary with a light-curve that shows continuous variation throughout the orbital period, which is normally 1 day or more; abbr. EB. A secondary minimum is always observed and is normally much shallower than the primary minimum, which usually has an amplitude of less than 2 mag. Beta Lyrae systems were once thought to be *contact binaries with approximately ellipsoidal components. However, most appear to be *semidetached binaries in which there is *mass transfer towards the smaller component. In Beta Lyrae itself, a bright giant secondary is rapidly losing mass to what appears to be a main-sequence primary, which is hidden by a thick accretion disk. Energy from the infall of material excites high-temperature emission from a shell of material that surrounds both components and is being lost from the system.

beta particle (β-particle) A particle that is emitted in *beta decay: either an electron or its antiparticle, the positron.

Beta Persei star An obsolescent term for an Algol-type eclipsing binary. *See* ALGOL STAR.

Beta Pictoris An A3 dwarf star of magnitude 3.9, lying 63 l.y. away. In 1983 the Infrared Astronomical Satellite (IRAS) discovered that Beta Pictoris is surrounded by a disk of dust and gas larger than our Solar System, from which planets are forming. The disk contains at least one planet which has been photographed orbiting the star.

Beta Regio A broad, rifted upland area on Venus centred at +25° lat., 283° long. It is about 2800km across, and is cut by a broad north–south valley, Devana Chasma, which extends for 4600km southwards into the neighbouring upland of Phoebe Regio. Two huge volcanoes, Rhea Mons and Theia Mons, both over 200km in diameter and more than 5km high, are situated in the northern and southern half of Beta Regio, respectively; Theia, which lies on Devana Chasma, has a large caldera over 50km across with radiating lava flows. Beta Regio showed up on the first radar maps of Venus made from Earth, and the Venera 9 craft landed on its north-eastern flank.

Beta Taurid meteors A daytime meteor shower following a similar orbit to Comet *Encke, occurring between June 5 and July 18. Peak activity, detected by radar, occurs around the end of June. The meteors cannot be seen visually because the radiant lies close to the Sun. The *Tunguska event of 1908 has been attributed to a large meteoroid from the Beta Taurid stream.

Betelgeuse The star Alpha Orionis, the tenth-brightest star in the sky. It is an M2 supergiant and a semiregular variable, fluctuating between magnitudes 0.0 and 1.3 in a period of several years, with an average magnitude of 0.5. It lies around 500 l.y. away, much closer than the stars in the Orion Association, and has a luminosity 10000 times the Sun's. Betelgeuse is extremely large, about 500 times the diameter of the Sun. Its variations in brightness occur as it swells and contracts in size. 📷

Bethe, Hans Albrecht (1906–2005) German-American physicist. In 1938, following a suggestion by C. F. von *Weizsäcker, he and the American physicist Charles Louis Critchfield (1910–94) worked out the details of the *proton–proton chain, and argued that it would account for energy production in stars less massive than the Sun. The following year, independently of Weizsäcker, he suggested that the *carbon–nitrogen cycle, in which hydrogen is converted into helium, is the means by which the Sun and more massive stars generate their energy. For this work he received the 1967 Nobel Prize in Physics. With the German-born theoretical physicist Walter Heinrich Heitler (1904–81) he developed the cascade theory of *cosmic rays.

Bethe–Weizsäcker cycle Another name for the *carbon–nitrogen cycle.

Bianca The third-closest satellite of Uranus, distance 59170km, orbital period 0.435 days; also known as Uranus VIII. Bianca is 42km in diameter and was discovered in 1986 on images taken by the Voyager 2 spacecraft.

Bianchi cosmology The study of universes which are homogeneous but not isotropic. Standard cosmological models assume that the Universe, on cosmological scales, is both homogeneous (looks the same at every place) and isotropic (looks the same in every direction)—that is, it conforms to the *cosmological principle. If the assumption is relaxed, so that the former condition holds but not the latter, then the allowed solutions of the equations of general relativity are called **Bianchi models**, after the Italian mathematician Luigi Bianchi (1856–1928).

biconcave lens A lens with two concave surfaces, so that it is thinner in the middle than at the edges. These lenses are always *diverging lenses, and objects seen through them look smaller.

biconvex lens A lens with two convex surfaces, so that it bulges in the middle. A simple magnifying glass is a typical example. A biconvex lens is always a *converging lens.

Biela, Comet 3D/ A now-disintegrated periodic comet discovered in 1772 by the French amateur astronomer Jacques Laibats-Montaigne (1716–88), rediscovered in 1805 by J. L. *Pons, and rediscovered again in 1826 by the Austrian amateur Wilhelm von Biela (1782–1856), who calculated the orbit and showed that all three objects were identical. At the return of 1845/6, the comet's nucleus was observed to have split into two. The double comet returned in 1852, but nothing was seen of it thereafter. Confirmation that it had disintegrated came in 1872 November when its debris produced a storm of *Andromedid meteors. Comet Biela had an orbital period of 6.6 years, perihelion 0.86 AU, eccentricity 0.76, and inclination $12°.6$.

(((🌐))) **SEE WEB LINKS**
• Information page at Cometography website.

Bielid meteors *See* ANDROMEDID METEORS.

Big Bang theory The most widely accepted theory of the origin and evolution of the Universe. According to the Big Bang theory, the Universe originated from an initial state of high temperature and density and has been expanding ever since. The best current measurements place the occurrence of the Big Bang at 13.73 billion years ago, ±0.1 billion years; in other words, this is the age of the Universe.

The theory of general relativity predicts the existence of a *singularity at the very beginning, where the temperature and density were infinite. Most cosmologists interpret this singularity as meaning that general relativity breaks down at the *Planck era under the extreme physical conditions of the very early Universe, and that the very beginning must be addressed using a theory of *quantum cosmology. With our present knowledge of high-energy particle physics, we can run the clock back through the *lepton era and *hadron era to about a millionth of a second after the Big Bang, when the temperature was 10^{13} K. Using more speculative theory, cosmologists have tried to push the model to within 10^{-35} s of the singularity, when the temperature was 10^{28} K.

The Big Bang theory accounts for the expansion of the Universe; the existence of the *cosmic microwave background; and the abundances of light nuclei such as helium, helium-3, deuterium, and lithium-7, which are predicted to have been formed about 1 second after the Big Bang when the temperature was 10^{10} K. The cosmic microwave background provides the most direct evidence that the Universe went through a hot, dense phase. In the Big Bang theory, the microwave background is accounted for by the fact that, for the first million years or so (i.e. before the *decoupling of matter and radiation), the Universe was filled with plasma that was opaque to radiation and therefore in thermal equilibrium with it. This phase is usually called the *primordial fireball. When the Universe expanded and cooled to about 3000 K, at a time known as the *recombination epoch, it became transparent to radiation, which we now observe, much cooled and diluted, as thermal microwave radiation.

The discovery of the microwave background in 1965 resolved a long-standing battle between the Big Bang and its then rival, the *steady-state theory, which cannot explain the black-body form of the microwave background. Ironically, the term Big Bang was initially intended to be derogatory and was coined by F. *Hoyle, one of the strongest advocates of the steady state.

Big Bang Chronology		
Era	Time after Big Bang	Temperature
Planck era	0 to 10^{-43} s	? to 10^{34} K
radiation era[a]	10^{-43} s to 30000 years	10^{34} to 10^4 K
matter era[b]	30000 years to present	10^4 to 3K

[a] The time from about 10^{-6} or 10^{-5} s to about 1s or so is subdivided into the hadron and lepton eras.
[b] Includes the *recombination epoch, which took place about 300000 years after the Big Bang, at a temperature of about 3000K.

Big Bear Solar Observatory An observatory at an altitude of 2070m in the San Bernardino Mountains of southwestern California, opened in 1969. It is owned by the California Institute of Technology (Caltech), and has been operated since 1997 by the New Jersey Institute of Technology. The observatory benefits from particularly steady seeing, being sited on an artificial island in Big Bear Lake. The world's largest solar telescope, the 1.6-m New Solar Telescope (NST), was opened in 2009, replacing the 0.65-m vacuum reflector in operation since 1973.

((()) SEE WEB LINKS
• Official observatory website.

Big Crunch The putative end-state of a closed *Friedmann universe (i.e. one in which the density exceeds the *critical density). Such a universe expands from an initial Big Bang, reaches a maximum radius, and then collapses to a Big Crunch, where the density of matter becomes infinite. After a Big Crunch, there could be another phase of expansion and collapse, leading to an oscillating universe.

Big Dipper Popular name for the asterism formed by the seven main stars of Ursa Major (Alpha, Beta, Gamma, Delta, Epsilon, Zeta, and Eta Ursae Majoris), resembling a saucepan or ladle. An alternative name is the Plough.

billitonite A type of tektite found on Billiton Island, Indonesia, one of the many sites where tektites have been recovered in Southeast Asia. Billitonites are found in Quaternary gravels and tuffs, and so are less than a million years old. This agrees with the average *ablation age of the Southeast Asian tektites, around 600000 years.

BIMA Array Abbr. for Berkeley–Illinois–Maryland Association Array, a former interferometer for observations at millimetre wavelengths at *Hat Creek Radio Observatory, California, now part of the *Combined Array for Research in Millimeter-wave Astronomy.

binary galaxy A pair of galaxies in orbit around each other. True binary galaxies are very difficult to distinguish from chance superpositions of two galaxies in the line of sight. The statistical study of the orbits of binary pairs is valuable in trying to estimate the total masses of particular types of galaxy, although the extremely long orbital periods prevent the accurate determination of masses of individual galaxies.

binary pulsar A pulsar in orbit with another star. The existence of a companion star is revealed by a cyclic change in the pulse period as the two stars orbit each other. Over 135 binary pulsars were known by the end of 2010, with orbital periods from an hour and a half to several years, and pulsation periods from 1.6ms (*millisecond pulsars) to over 1s. The first known binary pulsar, PSR 1913+16, was discovered in 1974. It consists of a pulsar that pulses 17 times a second, in a highly eccentric orbit of period 7.75 hours around a second neutron star from which pulses are not observed. Each star is of 1.4 solar masses, close to the *Chandrasekhar limit, and the orbital period is gradually shortening due to loss of energy through *gravitational radiation. Other notable binary pulsars include PSR 1957+20, sometimes called the **Black Widow Pulsar**, in which intense radiation from the pulsar is evaporating its small companion star. The first example in which both members are detectable pulsars (i.e. a double pulsar), PSR J0737−3039AB, was discovered in 2003; in this system a pulsar with a period of 23 ms is orbited every 2.4 hours by a pulsar with a period of 2.8 s. A number of binary pulsars are now known to be *recycled pulsars which have been spun up to high rotational speeds by accretion of gas from the companion.

binary star A pair of stars bound together by their mutual gravitation, and orbiting their common centre of mass. Such a system is distinct from an *optical double, which is not gravitationally bound.

A *visual binary is resolved visually or photographically, whereas an *astrometric binary is detectable by irregularities in the proper motion of the one visible star or the *photocentre (2). Direct evidence for a companion is provided by the eclipses in an *eclipsing binary, and by the Doppler shift of spectral lines in a *spectroscopic binary.

The orbital periods of binaries range from minutes to thousands of years. In a *close binary*, the separation is comparable to the diameter of the stars. Such systems are subdivided according to the amount by which each component fills its *Roche lobe, giving rise to *detached, *semidetached, and *contact binaries. The last two categories include *interacting binaries, in which *mass transfer occurs. Many binaries are also variable stars, the most notable being the various forms of *cataclysmic binary, Type I *supernovae, and certain variable X-ray sources.

binding energy The energy released when protons and neutrons bind together to form an atomic nucleus, or the energy required to break up that nucleus. The binding energy per proton or neutron is greatest for atoms with mass numbers in the range 50–65 (the *iron peak). Energy is released when lighter atoms combine to form heavier ones (as in *nucleosynthesis inside stars), provided the products are no heavier than iron. Conversely, when heavy atoms break down to lighter ones (e.g. in fission reactors) energy is also released, provided the products remain heavier than iron. The binding energy is the equivalent of the *mass defect.

binoculars A pair of low-power telescopes mounted side by side so that both eyes can be used simultaneously. Conventional *prismatic binoculars* use a pair of *Porro prisms to fold each light path, thus shortening the distance between the objective lenses and the eyepieces. These prisms also erect the image. An alternative to Porro prisms are *roof prisms, which do the same thing but more compactly since the light emerges along the same axis as it enters. Binoculars are specified by a pair of numbers, such as 7 × 50. The first number refers to the magnification, and the second to the aperture in millimetres. Sizes commonly used for astronomy are 7 × 50, 10 × 50, and 11 × 80. The aperture divided by the magnification gives the size, in millimetres, of the *exit pupil. Amateur astronomers use binoculars for observing comets and large deep-sky objects such as star clusters, nebulae, and galaxies. They are also useful for locating stars somewhat fainter than can be seen by the naked eye, and for observing the brighter variable stars.

Biot, Jean-Baptiste (1774–1862) French physicist. In 1803 he investigated the meteorite fall at *l'Aigle in France. The Swiss physicist Marc Auguste Pictet-Turretin (1752–1825) had suggested that meteorites fell from the sky, and E. F. F. *Chladni had speculated that they were the remains of a shattered planet, but the idea was generally dismissed. By questioning witnesses of the fall and analysing specimens of meteorites and of the ground where they lay, Biot was able to demonstrate that meteorites have a cosmic origin. This was the first properly scientific study of meteorites.

bipolar group A sunspot pair or group in which the leading and following spots (in the sense of solar rotation) have opposite magnetic polarities. The polarities obey *Hale's law* (named after G. E. *Hale), which states that the leading and following spots have opposite polarities on either side of the equator. These polarities are reversed in each successive sunspot cycle. On the Mount Wilson sunspot classification system, bipolar groups are denoted by the Greek letter β.

bipolar nebula A gas cloud with two main lobes which lie symmetrically on either side of a central star. This bipolar shape arises from the ejection of material by the star in opposing directions. In some cases the outflowing material escapes along the rotation axis of a dense disk of material which surrounds the star, and which may completely obscure the star at optical wavelengths. Bipolar nebulae can be produced by outflow from very young or old stars.

bipolar outflow The flow of material in two opposing directions away from a central source, usually a star. Most young stars go through a violent phase of ejecting material. In some, jets are seen moving away from the star at speeds of many hundreds of kilometres per second. The ejected material sweeps up the surrounding molecular gas into two moving lobes of gas, forming a bipolar molecular outflow. Old stars that have evolved off the main sequence also eject material, and this may produce a bipolar outflow which we observe as a bipolar planetary nebula.

birefringent filter A type of *interference filter used for studying the Sun's corona; also known as a *Lyot filter* after its inventor, B. *Lyot. It uses a stack of double-refracting (i.e. *birefringent*) quartz crystals alternated with polarizing materials. Light passing through the crystals is separated into two beams, mutually polarized at right angles to each other. The crystals also rotate the direction of polarization of the beams, the extent depending on the wavelength of the light. By blocking successively narrow wavelength ranges using the polarizing material it is possible to achieve a filter which transmits a band as narrow as 0.4 nm, using a filter some 40 mm thick. These filters are costly and are rarely encountered.

Birmingham Solar Oscillations Network (BiSON) A network of six remotely controlled solar observatories monitoring solar oscillations (*see* HELIOSEISMOLOGY), jointly owned and operated by the University of Birmingham and Sheffield Hallam University, England. BiSON stations are sited at *Teide Observatory, Tenerife (opened 1975); *Mount Wilson Observatory, California (opened 1992 but previously on Haleakala, Hawaii, 1981–91); Carnarvon, Western Australia (opened 1985); the *South African Astronomical Observatory (opened 1990); *Las Campanas Observatory, Chile (opened 1991); and the *Paul Wild Observatory, New South Wales (opened 1992).

 SEE WEB LINKS
• Official BiSON website.

BiSON Abbr. for *Birmingham Solar Oscillations Network.

black body An imaginary object that is a perfect absorber of radiation (and also a perfect emitter) at all wavelengths. The wavelength at which a black body emits the peak of its radiation depends solely on its temperature.

black-body radiation The thermal radiation that would be emitted by a *black body. Black-body radiation has a characteristic spectrum, described by *Planck's law, in which the peak of the emission moves to shorter wavelengths as the temperature of the black body rises (*see* WIEN'S DISPLACEMENT LAW). The *cosmic background radiation is black-body radiation, and stars often act as black-body radiators in the optical part of the spectrum.

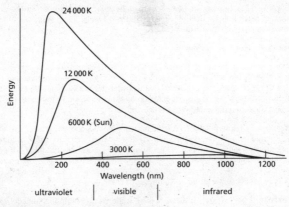

black-body radiation: The higher the temperature of a black body, the more the peak of the emitted radiation is displaced towards shorter wavelengths. Hot stars have a peak in the ultraviolet, while cool stars peak in the infrared. Solar-type stars have a peak in the visible part of the spectrum

black drop A phenomenon seen during transits of Venus and Mercury across the Sun. The black drop is a dark area that appears to link the limb of the Sun to the following limb of the transiting planet for a short time after second contact, and its preceding limb for a short time before third contact. It results from a combination of atmospheric seeing effects, diffraction in the telescope, and the limb darkening of the Sun. The phenomenon means that accurate timing of transits is impossible.

black dwarf A *degenerate star that has cooled until it is no longer visible.

Black Eye Galaxy The galaxy M64 (NGC 4826) in Coma Berenices. It is a 9th-magnitude Sb-type spiral with a dark lane of dust silhouetted against its nucleus, which gives the 'black eye' effect. Its distance is about 20 million l.y.

black hole An object with such a strong gravitational field that its escape velocity exceeds the velocity of light. One way in which black holes are believed to form is when massive stars collapse at the end of their lives. A collapsing object becomes a black hole when its radius has shrunk to a critical size, known as the *Schwarzschild radius, and light can no longer escape from it. The surface having this critical radius is referred to as the *event horizon, and marks the boundary inside which all information is trapped. Hence events within the black hole cannot be observed from outside. Theory indicates that both space and time become distorted inside the event horizon and that an object collapses to a single point, a *singularity, at the centre of a black hole. Black holes may have any mass. *Supermassive black holes (10^5 solar masses) may exist at the centres of active galaxies. At the other extreme, *mini black holes* of radii 10^{-10} m and masses similar to that of an asteroid may have been formed in the extreme conditions following the Big Bang.

No black hole has ever been observed directly. However, an *accretion disk may form around a black hole when matter falls towards it from a nearby companion star or other source. Energy predominantly at X-ray wavelengths is produced as matter in the accretion disk loses momentum and spirals in; these X-rays can be detected by satellites in orbit. Several black-hole candidates have been located in our Galaxy, most famously *Cygnus X-1. In addition, there is good evidence for a supermassive black hole of 3–4 million solar masses at the centre of our Galaxy.

There are several theoretically possible forms of black hole. A non-rotating black hole without electrical charge is known (after K. *Schwarzschild) as a *Schwarzschild black hole*. A non-rotating black hole with electrical charge is termed a *Reissner–Nordström black hole* after the German physicist Hans Jacob Reissner (1874–1967) and the Finn Gunnar

Nordström (1881–1923). In practice, black holes are likely to be rotating and uncharged, a form known as a *Kerr black hole. Black holes are not entirely black; theory suggests that they can emit energy in the form of *Hawking radiation.

Black Widow Pulsar *See* BINARY PULSAR.

blazar A class of extragalactic, violently variable objects that includes *BL Lacertae objects and *optically violently variable* (OVV) *quasars*, from which the name is contracted. They are thought to be the high-speed jet of plasma and radiation from an active galactic nucleus viewed nearly end-on. The OVV quasars have broad emission lines in their spectra, but otherwise show all the characteristics of BL Lac objects.

Blaze Star Popular name for the recurrent nova T Coronae Borealis. Two major outbursts have been observed, in 1866 and 1946, when the star reached magnitude 2.0, rapidly fading to about magnitude 9 within a month, and subsequently to its quiescent magnitude of 10.8. Other, smaller outbursts have been recorded at ultraviolet wavelengths. The star is a *spectroscopic binary, one component being an M giant, the other a massive white dwarf. The usually quoted orbital period of 227.6 days is in some doubt.

Blazhko effect A periodic modulation of the amplitude and period of certain *RR Lyrae stars, and other related pulsating variables. **Blazhko periods** range between about 10 days and over 530 days, with a marked peak at 20–40 days. The favoured explanation involves *double-mode pulsation, but in certain cases it appears that the star is an *oblique rotator. The Blazhko periods may themselves be variable, as in RR Lyrae itself, where a further 4-year cycle seems to occur. The effect is named after the Russian astronomer Sergei Nicolaevitch Blazhko (1870–1956).

BL Herculis star A subtype of short-period *W Virginis stars; abbr. CWB. They have periods of less than 8 days and show a characteristic bump on the descending portion of the light-curve.

blink comparator An instrument in which a pair of images can be viewed in rapid succession to detect any differences between them; also known as a **blink microscope**. The images are usually of the same part of the sky, taken at different times. Any object which has moved or changed brightness will appear to jump or blink as the operator switches between the two images. Blink comparators are used to search for asteroids, variable stars, and novae. Images taken using CCD cameras are compared using software which displays them alternately on a computer monitor.

Blinking Planetary The planetary nebula NGC 6826 in Cygnus. It is an 8th-magnitude nebula with a 10th-magnitude central star. Alternately looking at the central star and then away causes the star and nebula to appear to blink on and off. It is about 3200 l.y. away.

BL Lacertae object (BL Lac object) An apparently star-like object with a near-featureless spectrum; also known as a *Lacertid*. BL Lac objects show considerable brightness variations— often by several magnitudes over days or weeks—with high (and variable) polarization, and are sometimes associated with a compact radio source. The prototype was originally classified as a peculiar 14th-magnitude variable star in the constellation of Lacerta, hence the variable-star designation. The extragalactic nature of these objects was not established until absorption lines from low-mass giant stars, characteristic of the spectra of elliptical galaxies, were identified in the spectra of the faint nebulosity surrounding some of them. From these lines a redshift could be obtained and their distances deduced. About 1500 are known. They are now thought to be high-speed jets of plasma and radiation from an active galactic nucleus, viewed nearly end-on.

bloomed lens A lens with a coating that reduces reflections from its surface. When light passes between air and glass or back again, about 4–5% is reflected. By coating, or *blooming*, the lens with a thin film of a material such as magnesium fluoride, the reflection is reduced to 1–2%. A single coating affects light of a particular wavelength. To improve transmission across the visible spectrum, lenses are often multi-coated with several layers of different materials. The term originated from the similar appearance of a bloomed lens to the bloom on fruit such as plums.

blooming 1. The coating applied to a *bloomed lens.
   2. A defect obtained with images taken using CCDs (*see* CHARGE-COUPLED DEVICE), in which bright images have long bars on either side. This is caused when pixels become saturated with charge, which then spills over into adjacent pixels in the column.

blue clearing An occasion when the surface markings of Mars can be seen clearly through Earth-based telescopes fitted with a blue filter. Normally the surface features of Mars are at best very faint and at worst invisible through a blue filter, and the atmosphere and white clouds appear much brighter. The reality of blue clearings has been questioned.

blue compact dwarf galaxy A small galaxy with strong emission lines, or strong continuum emission from young stars, in the blue region of the spectrum. The emission is usually interpreted as evidence of a recent localized burst of star formation. *See also* COMPACT GALAXY.

blue giant A massive star that has exhausted the hydrogen fuel in its core and left the main sequence. Blue giants have a surface temperature of around 30000K and a luminosity some 10000 times that of the Sun. As they grow older they expand and cool, eventually becoming red giants.

blue Moon 1. A phenomenon caused by high-altitude dust in the Earth's atmosphere, which scatters red light preferentially, making the Moon appear blue (*see* MIE SCATTERING). Blue Moons were seen in 1883 after the eruption of the volcano Krakatoa, and in 1950 because of forest fires in Canada. The phenomenon is rare, hence the expression 'once in a blue Moon'.
   2. In modern US usage, the second full Moon in a calendar month. This occurs every 2–3 years on average.

Blue Planetary The 8th-magnitude planetary nebula NGC 3918 in Centaurus. The name was given it by J. F. W. *Herschel, who described it as like Uranus but larger. It is about 2600 l.y. away.

blue populous cluster *See* LARGE MAGELLANIC CLOUD.

blueshift A *Doppler shift of light towards the blue end of the spectrum, caused when the emitting source is approaching the observer. *See also* RADIAL VELOCITY; REDSHIFT.

blue straggler A blue main-sequence star, usually found in a globular or open cluster that appears to be slower in its evolution than other stars of similar mass and luminosity in the same cluster. On the *Hertzsprung–Russell diagram, blue stragglers are found on the main sequence above the *turnoff point of the cluster. They are thought to form either through mass transfer in a binary system or through merger of two stars. The latter process occurs either through direct collision of two single stars or through the merger of the two stars in a binary.

B magnitude The magnitude of a star in blue light on the *Johnson photometry system. The B filter is centred on a wavelength of 436nm and has a bandwidth of 89nm. It is the photoelectric equivalent of the older *photographic magnitude, but modified to be free of ultraviolet light. There are also B filters in *Geneva photometry, *Walraven photometry, and the *six-colour system, but the context should always make it clear when these are intended.

BN Object Abbr. for *Becklin–Neugebauer Object.

Bode, Johann Elert (1747-1826) German mathematician and astronomer. In 1772 he publicized a formula, now known as *Bode's law, which yielded the approximate distances of the six known planets; from this he predicted the existence between Mars and Jupiter of an undiscovered planet. His major publication was *Uranographia* (1801), a comprehensive atlas of the entire sky showing over 17000 stars and nebulae. For fifty years he oversaw the publication of astronomical data in the Berlin Academy's yearbook.

Bode's law A numerical sequence announced by J. E. *Bode in 1772 which matches the distances from the Sun of the six planets then known. It is also known as the Titius–Bode law, as it was first pointed out by the German mathematician Johann Daniel Titius (1729–96) in 1766. It is formed from the sequence 0, 3, 6, 12, 24, 48, 96, 192 by adding 4 to each number. The planets were seen to fit this sequence quite well—as did Uranus, discovered in 1781. However, Neptune and Pluto do not conform to the 'law'. Bode's law stimulated the search for a planet orbiting between Mars and Jupiter that led to the discovery of the first asteroids. It is often said that the law has no theoretical basis, but it does show how orbital *resonance can lead to *commensurability.

Bode's law								
Planet	Mercury	Venus	Earth	Mars	Ceres	Jupiter	Saturn	Uranus
Bode's law distance	4	7	10	16	28	52	100	196
Actual distance (10^{-1} AU)	3.9	7.2	10	15.2	28	52	95	192

Bok, Bartholomeus ('Bart') Jan (1906–83) Dutch-American astronomer. His lifelong work was the study of the Milky Way, much of it in collaboration with his wife, Priscilla Bok, née Fairfield (1896–1975). In particular he investigated its structure, its distribution of stars, interstellar matter, and star-forming regions. In the 1930s he discovered the objects now called Bok *globules, and demonstrated that stellar *associations are made up of young stars. In the early 1950s, with J. H. *Oort and others, he pioneered the mapping of the Galaxy at radio wavelengths.

Bok globule *See* GLOBULE.

bolide A fireball accompanied by one or more audible explosions, often associated with meteorite falls. The initial explosion is the sonic boom of the leading mass; smaller detonations follow if there are several major fragments. Many shock waves may be set up if considerable fragmentation occurs, causing a rumbling noise which follows the first major fragmentation.

bolometer A detector sensitive to radiation of all wavelengths. It has a blackened surface so that the incoming radiation raises its temperature, which is measured from a change in electrical resistance. While the bolometer can be used at any wavelength, it is best suited to measuring infrared wavelengths longer than 6μm.

bolometric correction The difference between the *visual and *bolometric magnitudes of an object. Two zero-points are in use, which differ by 0.07mag. One defines the Sun to have zero bolometric correction. The other has its zero-point set so that bolometric corrections for all stars are positive; this is because other stars emit more energy than the Sun at non-visual wavelengths, either in the ultraviolet for hotter stars or the infrared for cooler stars. Confusingly, some authorities define bolometric correction as bolometric magnitude minus visual magnitude, which makes all values negative.

bolometric magnitude (m_{bol}) The apparent magnitude a star would have if its energy output could be measured at all wavelengths in the absence of the Earth's atmosphere. The name arises because *bolometers are sensitive to radiation of all wavelengths. In practice, the bolometer is the best detector only for wavelengths longer than 6μm, and other detectors are used for shorter wavelengths. Between 0.3 and 23μm the star is compared with a reference source on the ground using only those wavelength windows where the Earth's atmosphere is relatively transparent. Even within these windows it is important to correct for *atmospheric extinction. Between these windows the star's flux is estimated with the help of a model atmosphere. Outside the range 0.3–23μm, the flux is measured from satellites.

Boltzmann constant (k or k_B) A constant that relates the kinetic energy of a particle in a gas to the temperature of that gas. It has the value $1.3806488 \times 10^{-23}$ joules per kelvin. The particles can be molecules, atoms, ions, or electrons. The Boltzmann constant relates pressure, p, and temperature, T, by the equation $p = nkT$, where n is the number of particles per unit volume. In astrophysics, this equation is of importance in understanding the interiors and surface layers of stars, and the atmosphere of planets. The constant is named after the Austrian physicist Ludwig Edward Boltzmann (1844–1906).

Bond, George Phillips (1825–65) American astronomer, son of W. C. *Bond. He worked closely with his father in studies of the Solar System and in developing astrophotography. In 1850 they took the first good photographs of the Moon, and followed this with the first photograph of a star (Vega). Bond himself studied the comparative brightnesses of the Sun, the Moon, and Jupiter (the *Bond albedo is named after him), and produced a notable report on Donati's Comet of 1858.

Bond, William Cranch (1789–1859) American astronomer, father of G. P. *Bond. He founded the Harvard College Observatory, originally sited in his house but transferred to Harvard in 1838. He was succeeded as its director by his son. The Bonds frequently worked together, discovering Saturn's satellite Hyperion in 1848, and Saturn's 'crêpe ring' (the C Ring) in 1850. They also collaborated in the development of astrophotography.

Bond albedo The fraction of the total light or other radiation falling on a non-luminous spherical body that is reflected from the body in all directions; also called *spherical albedo*. It is calculated over all wavelengths and its value therefore depends on the spectrum of the incident radiation. The Bond albedo determines the energy balance of a body such as a planet. It is named after G. P. *Bond.

Bondi, Hermann (1919–2005) British cosmologist and mathematician, born in Austria. In 1948, with T. *Gold and F. *Hoyle, he proposed the *steady-state theory of the Universe in which matter is continuously created. Although abandoned in favour of the Big Bang theory, it stimulated much important astrophysical research, particularly Hoyle's work on *nucleosynthesis. Bondi also studied relativity, showing in 1962 that the existence of gravitational waves follows from the theory of general relativity.

***Bonner Durchmusterung* (BD)** A catalogue of 324 198 stars down to 9th magnitude from the north celestial pole to declination 2° south. It was compiled at Bonn, Germany, by F. W. A. *Argelander; most of the observations were made by Eduard Schönfeld (1828–91) and (Carl Nicolaus) Adalbert Krüger (1832–96). It was published in three volumes in 1859–62, with an associated atlas in 1863. The catalogue was extended to 23° south in 1886 by Schönfeld, adding 133 659 stars, with an atlas in 1887. Many stars are still known by their BD numbers as given in these catalogues. *See also* CÓRDOBA DURCHMUSTERUNG.

(⊕) SEE WEB LINKS
• Detailed description and full catalogue downloadable from the CDS.

Boötes (Boo) (*gen.* Boötis) A constellation of the northern sky, representing a herdsman. It contains *Arcturus (Alpha Boötis), the brightest star north of the celestial equator. Epsilon Boötis (*Izar or Pulcherrima) is a double star with orange and blue components. The *Quadrantid meteor shower radiates every January from northern Boötes.

boson A particle such as a *photon, a *meson, an atomic nucleus of even mass number (e.g. the commonest type of helium nucleus), or the hypothetical *graviton which has a zero or integer value of *spin. They do not obey the Pauli exclusion principle. Bosons are named after the Indian physicist Satyendra Nath Bose (1894–1974). *See also* FERMION.

Boss General Catalogue (GC) Popular name for the five-volume *General Catalogue of 33,342 Stars* compiled in 1936–7 by the American astronomer Benjamin Boss (1880–1970). This catalogue contains positions and proper motions of all stars brighter than

7th magnitude over the whole sky, plus thousands of fainter stars for which accurate proper motions could be determined. It was a successor to the *Preliminary General Catalogue* of 6188 stars published in 1910 by his father, Lewis Boss (1846–1912), who also initiated the larger catalogue.

(⊕) SEE WEB LINKS
• Detailed description and full catalogue downloadable from the CDS.

boundary layer In a planetary atmosphere, the level of the atmosphere in direct contact with the planet's surface, in which friction between the surface and the air plays a significant role in determining atmospheric movements. On Earth, the boundary layer lies at the base of the *troposphere and varies in depth from a few hundred metres under stable conditions to 1–2 km when convection is strong.

bound–bound transition A change to the energy of an electron within an atom, or more rarely within a molecule, in which the electron remains attached (bound) to the atom or molecule both before and after the change. When the energy is increased, a photon is absorbed; when the energy is reduced, a photon is emitted. Bound–bound transitions produce the emission and absorption lines found in stellar spectra.

bound–free transition A change to the energy of an electron within an atom or a molecule in which the electron gains sufficient energy to escape. The electron goes from being bound to being free, and leaves behind an ion; hence this is another name for *ionization. The energy for the change may come from a photon, resulting in the absorption bands known as *ionization edges* in stellar spectra, or from collisions with other atoms or particles (*collisional ionization*). If the energy comes from another excited electron within the atom, the process is known as *autoionization*.

Bouwers telescope A design of telescope identical to the *Maksutov telescope, developed by the Dutch optician Albert Bouwers (1893–1972). His publication of the design in 1940, during World War II, predated that of the *Maksutov telescope, but gained little publicity because of the German occupation of Holland.

Bowen, Ira Sprague (1898–1973) American astrophysicist. In 1927 he explained the origin of strong green lines in the spectra of planetary nebulae. They are *forbidden lines produced by a process now called *Bowen fluorescence—in this case, transitions between atomic states in doubly ionized oxygen (O III) and not, as W. *Huggins had earlier speculated, by an unknown element termed 'nebulium'. This led to the correct identification of lines in the solar spectrum that had been similarly attributed to a hypothetical 'coronium', and thence to advances in the spectroscopic study of the compositions, temperatures, and densities of the Sun, stars, and nebulae.

Bowen fluorescence A mechanism that gives rise to certain strong emission lines from ionized atoms of oxygen, carbon, and nitrogen in diffuse nebulae. Extremely hot stars and accretion disks (at temperatures of 30000 K or more) produce copious amounts of extreme ultraviolet radiation at 30.4 nm from singly ionized helium atoms. These photons excite the ions of C III and N III in surrounding gas because the ions have a transition very close to this wavelength. These excited ions then return to the ground state by emitting a series of photons, including a group of lines in the 464–465 nm (blue) region. Bowen fluorescence also accounts for the exceptionally strong O III lines in the spectra of some *planetary nebulae. The mechanism was identified by I. S. *Bowen.

bow shock An abrupt boundary between regions of a fluid (gas or liquid) which are travelling at relative rates greater than the local speed of sound, such as that formed by the *solar wind when it encounters a planet, comet, or other body. It is similar to the wave that forms at the bow of a ship. An aircraft travelling faster than sound generates a bow shock and a wake whose passage is heard as a sonic boom. Similar situations arise in space when flowing gas meets a fluid or a solid obstacle at supersonic speeds. A fast wind from a young star meeting dense condensations of gas forms bow shocks around them.

Bp star A star of spectral type B which is peculiar in that it shows a deficiency relative to normal B stars of helium in its spectrum. Bp stars have strong magnetic fields, and are a hotter extension of *Ap stars to about spectral type B8.

Brackett series A sequence of absorption or emission lines in the near-infrared part of the spectrum, due to hydrogen. The Brackett series of lines arises from electron jumps between the fourth energy level and higher levels. The lines have wavelengths from 4.05 μm (Brackett-α) towards shorter wavelengths, the spacing between the lines diminishing as they converge on the series limit at 1.46 μm. They are named after the American physicist Frederick Sumner Brackett (1896–1972). *See also* HYDROGEN SPECTRUM.

Bradley, James (1693–1762) English astronomer. In collaboration with Samuel Molyneux (1689–1728) he attempted to detect stellar parallax, but instead discovered the annual *aberration of starlight. This discovery, announced in 1728, was the first direct observational verification that the Earth orbits the Sun. From the amount of aberration he calculated the speed of light to within 2% of the true value. In 1742 Bradley became the third Astronomer Royal. He subsequently discovered the *nutation of the Earth's axis, announced in 1748.

Bragg crystal spectrometer An instrument that uses a crystal to split X-rays according to energy, just as a prism splits visible light into the colours of the spectrum. It is named after the British physicist William Henry Bragg (1862–1942). The device has the highest resolution of any X-ray spectrometer, but suffers from low efficiency and can operate only in a very narrow band. It has been used to best effect in studies of the Sun, but has also been flown on X-ray satellites such as the *Einstein Observatory.

Brahe, Tycho (1546–1601) Danish astronomer. He was the most accomplished observer of the pre-telescope era, expert in constructing instruments for making accurate naked-eye positional measurements. He first gained fame through his report (*De nova stella*, 1573) of the 1572 supernova in Cassiopeia. In 1576 he constructed Uraniborg, an observatory on the island of Hven in the Baltic (a second observatory, Stjerneborg, was built in about 1584). He calculated that the comet seen in 1577 had a highly elongated orbit, which would pass through several of the 'spheres' on which the planets were supposedly carried, and this led him to doubt the reality of *Aristotle's planetary model. However, he rejected the heliocentric system proposed by *Copernicus. In the *Tychonic system, although the planets orbit the Sun, the Sun itself (and the Moon) revolve around a stationary Earth. Tycho made major contributions to the study of the Moon's orbit. In 1597 he moved to Prague, and employed J. *Kepler as his assistant. Kepler later made use of Tycho's observations when deriving his laws of planetary motion.

Brans–Dicke theory An alternative to Einstein's theory of general relativity which attempts to incorporate *Mach's principle. Among other things, this theory predicts that the value of Newton's *gravitational constant, *G*, should change with time. It was originated by the American physicists Carl Henry Brans (1935–) and R. H. *Dicke.

breccia A complex type of rock made of shattered, crushed, and (in some cases) once-melted fragments of rock, all cemented together by a finer-grained material (the *matrix*). It is a common result of impact processes on planetary surfaces. A *genomict* is a breccia in which the components originated in distinct but genetically related rocks. A *monomict* is a breccia in which all the components originated in the same type of rock. A *polymict* is a breccia in which the components originated in two or more rocks of differing compositions.

bremsstrahlung A class of radiation mechanisms in which a charged particle (normally an electron) is decelerated and loses energy in the form of electromagnetic radiation. The deceleration may be either by electric fields (thermal bremsstrahlung; *see* FREE–FREE TRANSITION) or by magnetic fields (*magnetobremsstrahlung). The name is German for 'braking radiation'.

bridge An apparent structure of stars or gas linking one galaxy with another. In some cases this may be the result of physical (tidal) interaction between the two galaxies, as in the case of the *Antennae. In other cases the feature may be an observational artefact, or a chance, line-of-sight superposition of luminous material in the galaxies that is unrelated to any interaction. A combination of both artefact and line-of-sight effects probably accounts for the apparent bridge between the quasar Markarian 205 and the spiral galaxy NGC 4319, which have very different velocities of recession (21 000 and 1700 km/s respectively). A chain of galaxies linking two clusters or groups of galaxies is also sometimes known as a bridge.

bright nebula A luminous cloud of interstellar gas and dust. The term includes *emission nebulae, in which the gas glows with its own light, and *reflection nebulae, in which the gas and dust reflect light from nearby stars.

brightness temperature (T_B or T_b) The apparent temperature of an astronomical source calculated on the assumption that the source is a black-body emitter. For a thin plasma, the brightness temperature is always less than the true temperature of the source. For non-thermal sources, the brightness temperature may bear no resemblance to the true temperature. Maser and synchrotron sources can have very high brightness temperatures.

Bright Star Catalogue A catalogue of stars brighter than about magnitude 6.5, published by Yale University Observatory and hence also known as the *Yale Bright Star Catalogue*. It contains positions, magnitudes, spectral types, and other data for the 9096 stars in the *Harvard Revised Photometry. The first edition of the catalogue appeared in 1930 and the fourth in 1982; a fifth edition was issued in electronic format in 1991. A *Supplement* containing another 2603 stars down to magnitude 7.1 was published in 1983.

• Detailed description and full catalogue downloadable from the CDS.

• Catalogue supplement, downloadable from the CDS.

British Astronomical Association (BAA) An organization of amateur astronomers based in London, founded in 1890. It publishes a bimonthly *Journal* and an annual *Handbook*.

SEE WEB LINKS

• Official BAA website.

broad-band photometry A general term used to describe photometric measurements through filters with a bandwidth in the range 30–100 nm. Typical examples are *Johnson photometry, *Kron–Cousins RI photometry, *RGU photometry, and the *six-colour system. The advantage of broad-band photometry over intermediate-band or narrow-band photometry is that the wider filters pass more light so that fainter stars can be observed.

Brocchi's Cluster An alternative name for the star cluster popularly known as the *Coathanger. It is named after the American amateur astronomer Dalmiro Francis Brocchi (1871–1955).

Brooks 2, Comet 16P/ *See* JUPITER COMET FAMILY.

Brown, Ernest William (1866–1938) English mathematician and celestial mechanician who worked mostly in the USA. His major work was on lunar theory, in which he built upon the work of G. W. *Hill. The resulting tables of the Moon's motion, completed in 1919, were based on constants derived from analyses by P. H. *Cowell of 150 years of observations made from the Royal Observatory at Greenwich, and were accurate to $1''$. Brown also worked on gravitational perturbations of other Solar System bodies.

brown dwarf An object that, because of its low mass (less than 0.08 solar masses), never becomes hot enough to begin hydrogen fusion in its core although it can produce some nuclear energy for a few million years from the fusion of deuterium, a heavy isotope of hydrogen, which requires lower temperatures than normal hydrogen fusion; hence it is considered to be not a star, but a *substellar object*. Brown dwarfs have surface temperatures of about 2000 K and cooler. They thus have a very low luminosity and are difficult to detect. Brown dwarfs may be at least as numerous as visible stars but because of their low masses are not thought to be a significant component of the galactic *dark matter. The first brown dwarf to be identified with certainty was a companion to the nearby red dwarf Gliese 229, photographed by the Hubble Space

Telescope in 1995. Brown dwarfs are now assigned the spectral types L and T (*see* L DWARF; T DWARF). An object below about 0.01 solar masses (about 10 Jupiter masses) is regarded as a planet.

B star A star of spectral type B, whose spectrum contains predominantly absorption lines of hydrogen and neutral helium. B-type stars are hot and appear blue in colour, emitting strongly in the ultraviolet. Temperatures for B-type stars on the main sequence are 10500–30000K. Main-sequence luminosities are 100–52000 times the Sun's, and masses are 3–18 solar masses. B-type supergiants are stars evolving from the top of the main sequence. As such, they have masses up to 25 solar masses and luminosities as high as 260000 times the Sun's. Regulus and Spica are B-type dwarfs, and Rigel is a B-type supergiant. Together with O stars they are the major constituents of *OB associations, which delineate the spiral arms of galaxies. Being massive, they have lifetimes of only a few tens of millions of years. *See also* BE STAR; BP STAR; BW STAR.

Bubble Nebula The diffuse nebula NGC 7635 in Cassiopeia. It consists of a faint spherical nebula around a 7th-magnitude star. The bubble has been generated in a cloud of hydrogen by the stellar wind from a hot star within the cloud.

Budrosa family A *Hirayama family of asteroids at a mean distance of 2.9 AU from the Sun. The orbits of family members are inclined at 6–8° to the plane of the Solar System. The family is small, and its members are of varying composition. The largest member, (349) Dembowska, of diameter 140km, is of the rare R class, and appears to be achondritic in composition. The M-class (338) Budrosa itself, diameter 63km, was discovered in 1892 by the French astronomer Auguste Honoré Charlois (1864–1910). Budrosa's orbit has a semimajor axis of 2.911 AU, period 4.97 years, perihelion 2.85 AU, aphelion 2.97 AU, and inclination 6°.0.

Bug Nebula The double-lobed planetary nebula NGC 6302 in Scorpius. It is an example of a *bipolar nebula, and lies about 6500 l.y. away.

bump Cepheid A subtype of the Delta Cephei variable stars; abbr. CEP(B). Bump Cepheids exhibit a 'bump' on the descending branch of the light-curve, and multiple periodicity, with two or more *pulsation modes. The fundamental period is 2–7 days.

Burbidge, Geoffrey Ronald (1925–2010) and **(Eleanor) Margaret,** née Peachey (1919–) English astrophysicists, married in 1948, who worked mostly in the USA. They collaborated with W. A. *Fowler and F. *Hoyle in the publication of a detailed theory (dubbed the B^2FH theory) of *nucleosynthesis in stars. The Burbidges also investigated quasars and active galactic nuclei, and were the first to suggest that quasars are more energetic versions of *Seyfert galaxies. In 1970 Geoffrey Burbidge raised the problem of the *missing mass, having found that only 25% of the mass of certain elliptical galaxies was accounted for by their luminous components.

burst A period of sudden intense emission, usually of X-rays or gamma rays, having a rapid rise and decay. Observed burst durations can be as short as a few hundredths of a second. The source of the emission is known as a *burster. See also* GAMMA-RAY BURST; X-RAY BURST.

Butcher–Oemler effect A property of most clusters of galaxies with redshifts of about 0.4, which show a much higher proportion of late-type spiral and irregular galaxies (disk galaxies) than do similar clusters closer to us. This effect implies a strong evolution of galaxies in the recent past. Many of the disk galaxies must have disappeared from view in the nearby clusters, probably as a result of mergers or by losing their gas. The effect was reported in 1978 by the American astronomers Harvey Raymond Butcher (1947–) and Augustus Oemler, Jr (1954–).

Butler matrix A circuit used to combine the outputs from an array of radio antennas to provide a number of simultaneous beams across the sky. It is a form of *phased array and is named after the American electronics engineer Jesse Lorenzo Butler (1923–).

Butterfly Cluster The 4th-magnitude open star cluster M6 (NGC 6405) in Scorpius, consisting of several dozen stars arranged in the shape of a butterfly. It is about ¼° wide and lies 2000 l.y. away.

butterfly diagram A graph on which the latitudes of sunspots are plotted against time. It shows how spots migrate from higher latitudes (30–40° north or south) towards the equator (latitude 5° or so) throughout each sunspot cycle, in accordance with *Spörer's law. The shape of the distributions, when plotted for both northern and southern hemispheres, resembles the wings of a butterfly. 📷

((🌐)) SEE WEB LINKS

• Regularly updated version of the diagram from NASA's Marshall Space Flight Center.

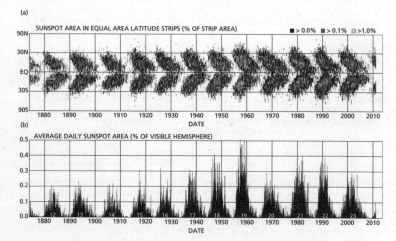

butterfly diagram: (a) Butterfly diagram for sunspots from 1874, and (b) corresponding plot of the total area covered by sunspots, measured as a percentage of the visible hemisphere

Butterfly Nebula A name applied to the diffuse nebula IC 2220 in Carina, and used as an alternative designation for the *Little Dumbbell and the *Bug Nebula.

BV photometry Measurement of a star's brightness at blue and yellow-green (visual) wavelengths; a subset of *Johnson photometry. UBV photometry is much more frequently used, but the U (ultraviolet) band is omitted in certain circumstances, for example for red stars where there is insufficient flux in U, for short-period variable stars to speed the observations, or when observing with a refractor, since an objective lens absorbs most of the U light. The B–V colour index is a good indication of the temperature of a star; in conjunction with the spectral type, it can be used to measure interstellar absorption.

Bw star An unusual star of spectral type B, but with weak helium lines (hence the suffix 'w') that make classification difficult; also known as a *helium-weak* or *helium-poor* star. Bw stars are found in the spectral range B2–7, and may also show enhanced lines of silicon or of mercury and manganese. Many are variable. They appear to be related to *Ap and *Bp stars.

BY Draconis star A type of *rotating variable star; abbr. BY. The variations are caused by non-uniform surface brightness and show a quasi-periodicity related to the rotation period. The latter ranges from a few hours to about 120 days, and the amplitude from a few thousandths to 0.5 mag. BY Draconis stars are K–M dwarfs. The non-uniform brightness is caused by *starspots and chromospheric activity, similar to that on the Sun. A BY Draconis star may simultaneously be a *flare star.

Byurakan Astrophysical Observatory The observatory of the Armenian Academy of Sciences, founded in 1946 by V. A. *Ambartsumian. It is located at Mount Aragats near Yerevan, Armenia, at an altitude of 1500m. Its main instruments are a 2.6-m reflector opened in 1976, and a 1-m Schmidt opened in 1960.

 SEE WEB LINKS

• Official observatory website.

Caelum (Cae) (*gen.* **Caeli**) A small and faint constellation of the southern sky, representing a sculptor's chisel. Its brightest star, Alpha Caeli, is magnitude 4.4.

Calar Alto Observatory The German–Spanish Astronomical Centre at 2170 m altitude on Calar Alto in the Sierra de los Filabres, Andalusia, Spain. It was founded in 1972 and is operated jointly by the Max-Planck-Institut für Astronomie, Heidelberg, and the Instituto de Astrofísica de Andalucía (CSIC), Granada. Its largest instrument is a 3.5-m reflector, opened in 1984. There are also a 2.2-m reflector, opened in 1979, and a 1.23-m reflector, opened in 1975. A 0.8-m Schmidt was moved to Calar Alto in 1980 from Hamburg, where it had been used since 1955. Since 1978 Madrid Observatory has also operated a 1.5-m reflector at the site.

(⊕) SEE WEB LINKS
• Official observatory website.

caldera A large volcanic depression, usually several kilometres in diameter, formed primarily by the collapse of a magma chamber roof or, less frequently, by a volcanic explosion. A caldera often contains smaller craters, cones, and other types of volcanic vent. Small collapse craters less than 1 km in diameter are known as *collapse pits*. Calderas are often non-circular in shape. Many of the small number of volcanic craters on the Moon appear to be calderas, some of them at the heads of *sinuous rilles. There are also huge calderas at the summits of many of the giant volcanoes of Mars. *Olympus Mons has a complex of summit calderas more than 60 km in diameter, apparently the result of several collapse events. Similar features are found on the volcanoes of Venus. ◙

calendar A scheme for organizing periods of time—days, weeks, months, years—and numbering or naming them. Many different calendars are in use, but only the *Gregorian calendar has world-wide recognition. Calendars usually attempt to recognize astronomical cycles, notably the phases of the Moon and the length of the year, but they do not necessarily closely follow them. For example, the months in the *Gregorian calendar have no connection with the actual phases of the Moon and are no more than arbitrary subdivisions of the year. The Gregorian calendar is a solar calendar designed to relate the dates as precisely as possible to the annual seasonal cycle. The year normally has 365 days, but periodically an additional day is inserted (or *intercalated*) to maintain the correspondence. Thus the vernal equinox will always occur on or within one day of March 21.

The Jewish calendar, on the other hand, can be described as *lunisolar*. Its months are tied closely to the lunar cycle and are either 29 or 30 days in length. There are normally 12 such months in a year, but so that close correspondence with the seasons is not quickly lost, an additional month is intercalated in a leap year. The Jewish calendar is based on the 19-year *Metonic cycle, which contains 235 months. Seven of the years of this cycle are therefore leap years.

The Islamic calendar is wholly lunar and the year always consists of 12 lunar months. Odd-numbered months have 30 days and even-numbered ones 29 days, so that the normal year has 354 days. Hence the Islamic year regresses through the seasons in a period of about 33 years, but this period has no significance in the Islamic calendar. There is, however, a calendrical sequence of 30 years, eleven of which are made leap years by adding one extra day to the first month. A very close correspondence is thus maintained with the astronomical lunar cycle, accurate to one day in 30000 years. *See also* JULIAN CALENDAR.

calendar year The number of full days that is treated as one year for civil or religious purposes. This number must vary from year to year in any calendar to keep it in step with astronomical cycles such as the orbital periods of the Earth or Moon, since a fraction of a day is involved in each case. Years are counted from some epoch. In the Gregorian calendar this epoch is the birth of Christ, estimated (probably incorrectly) to have occurred in AD 1. Years before this are designated BC, but there was no calendar year zero. For astronomical computing purposes, the calendar year 1BC is designated 0, 2BC is −1, and so on.

Caliban A retrograde satellite of Uranus, sixteenth in order from the planet at a distance of 7170000km; also known as Uranus XVI. It orbits every 579.6 days at an inclination of 140° to Uranus' equator. Caliban has a diameter of 72km and was discovered in 1997.

California Nebula The diffuse nebula NGC 1499 in Perseus. In shape, it resembles the US state of California, having a length of 2½° and a maximum width of ⅔°. It is faint and best seen on long-exposure photographs. It is illuminated by the hot 07 star Xi Persei, magnitude 4.0.

Callippus (c.370–c.300BC) Greek astronomer and mathematician. He modified *Eudoxus' scheme of Earth-centred spheres, adding extra ones for the Sun, Moon, and some of the planets, bringing the total to 34. Callippus' model was further refined by *Aristotle. Callippus calculated accurate lengths for the seasons, as measured between solstice and equinox.

Callisto The second-largest satellite of Jupiter, and the outermost of the four Galilean satellites; also known as Jupiter IV. It orbits 1883000km from Jupiter's centre in 16.689 days, keeping the same face turned towards the planet. Callisto is 4821km in diameter but, at magnitude 5.7, is the faintest of the Galilean satellites, because of its low albedo of 0.19. Callisto's density is 1.83g/cm^3. It has an icy surface uniformly covered with impact craters (the largest, Heimdall and Lofn, are 210 and 200km wide respectively), with little sign of the tectonic resurfacing activity seen on the other Galilean satellites. It has several multi-ringed basins, the two largest being *Valhalla (3000km across, the largest in the Solar System) and Asgard (1400km across), which are surrounded by numerous concentric rings.

Caloris Basin The largest impact basin on Mercury, 1300km in diameter. Much of its interior is filled by smooth plains, known collectively as Caloris Planitia. There are many wrinkle ridges and fractures, some of them radial and others concentric, the latter suggesting two or three inner rings. Caloris was probably formed about 4 billion years ago, during the period of late heavy bombardment that affected all the planets.

Caltech Submillimeter Observatory (CSO) An observatory with a 10.4-m dish for millimetre and submillimetre astronomy, opened in 1987 at an altitude of 4070m on Mauna Kea, Hawaii. It can work in conjunction with the nearby *James Clerk Maxwell Telescope to produce an interferometer with a baseline of 158m. It is owned and operated by the California Institute of Technology (Caltech).

(()) SEE WEB LINKS
• Official telescope website.

Calypso A satellite of Saturn, occupying the trailing *Lagrangian point (L_5) in the orbit of *Tethys; also known as Saturn XIV. Its distance is 294710km, and its orbital period 1.888 days. Calypso is 30 × 23 × 14km in size. It was discovered in 1980 on images from the Voyager 1 spacecraft.

Cambridge Optical Aperture Synthesis Telescope (COAST) An optical interferometer at Cambridge, England, that uses the technique of *aperture synthesis, originally developed by radio astronomers, to produce high-resolution images by combining the light from five 0.4-m mirrors with baselines up to 100m. It began operation in 1995.

(()) SEE WEB LINKS
• Official telescope website.

Camelopardalis (Cam) (*gen.* Camelopardalis) A faint constellation of the northern sky, representing a giraffe. Its brightest star is Beta Camelopardalis, magnitude 4.0.

Campbell, William Wallace (1862–1938) American astronomer and mathematician. In 1892 he studied the spectral changes of Nova Aurigae. In 1896 he began a long series of measurements of stellar radial velocities, publishing a catalogue of 3000 of them in 1928. This work improved knowledge of the Sun's motion in the Galaxy and of the Galaxy's rotation. Campbell led several expeditions to solar eclipses. At the eclipse of 1922 he detected the deflection of starlight by the Sun, as predicted by Einstein's theory of general relativity, confirming and improving the previous results of F. W. *Dyson and A. S. *Eddington.

Canada–France–Hawaii Telescope A 3.6-m reflector at an altitude of 4200 m on Mauna Kea, Hawaii, opened in 1979. It is jointly owned and operated by Canada, France, and the University of Hawaii through the Canada–France–Hawaii Telescope Corporation, which has its headquarters in Kamuela, Hawaii.

(⊕) SEE WEB LINKS
• Official telescope website.

canale A sinuous channel on the surface of Venus; pl.*canali*. Canali are found on low-lying plains and have near-constant cross-sections and widths. They are typically a few hundred kilometres long and a few kilometres wide, Baltis Vallis being the longest at 6800 km. The smoothness and great length of canali point to the flow of very fluid lava as their cause.

canals, Martian Supposed dark, straight lines on the surface of Mars seen through telescopes from Earth. They were first seen by G. V. *Schiaparelli in 1877, and were reported by others until the 1960s when close-up pictures of Mars from the *Mariner probes demonstrated them to be illusions. Early in the 20th century, P. *Lowell and others considered the canals to be evidence for an advanced Martian civilization.

Cancer (Cnc) (*gen.* **Cancri**) The faintest constellation of the zodiac, in which the Sun lies for three weeks from late July until mid-August. Cancer represents a crab. Its brightest star, Beta Cancri, is magnitude 3.5; Zeta Cancri is a double star of magnitudes 5.1 and 6.2. The main feature of the constellation is the open star cluster M44, *Praesepe. There is also a smaller 7th-magnitude star cluster, M67, ½° wide.

Cancer, Tropic of *See* TROPIC OF CANCER.

Canes Venatici (CVn) (*gen.* **Canum Venaticorum**) A constellation of the northern sky, representing a pair of hunting dogs. It contains the double star Alpha Canum Venaticorum (*Cor Caroli) and the variable Y Canum Venaticorum (*La Superba). Also in the constellation is the 6th-magnitude globular star cluster M3, and M51, the *Whirlpool Galaxy.

Canis Major (CMa) (*gen.* **Canis Majoris**) A constellation of the southern sky, popularly known as the Great Dog. It is distinguished by the brightest star in the sky, *Sirius (Alpha Canis Majoris). Other bright stars in Canis Major are Beta (*Mirzam), Delta (*Wezen), and Epsilon (*Adhara). Canis Major contains a notable open star cluster, M41, magnitude 4.5.

Canis Minor (CMi) (*gen.* **Canis Minoris**) A constellation of the equatorial region of the sky, popularly known as the Little Dog. It contains the bright star *Procyon (Alpha Canis Minoris) but little else of note.

Cannon, Annie Jump (1863–1941) American astronomer. In 1896 she began work under E. C. *Pickering at the Harvard College Observatory, where she undertook the classification of stellar spectra. She refined W. P. *Fleming's alphabetical sequence (A, B, C, etc.), dropping some categories and rearranging to yield a continuous series of spectral types: from hot O and B stars, through A, F, and G, to cool K and M stars. Cannon's classifications of the spectra of over 225 000 stars were published in the *Henry Draper Catalogue*.

Canopus The star Alpha Carinae, magnitude −0.62, the second-brightest star in the sky. It is an F0 bright giant, and lies 309 l.y. away.

cantaloupe terrain A type of terrain on Neptune's satellite Triton, consisting of near-circular shallow dimples up to about 50 km across, separated by series of overlapping ridges. The origin of such terrain is uncertain, but the dimples could be the product of melting and folding of the icy surface. It is so named because of its resemblance to the skin of a cantaloupe melon.

Capella The star Alpha Aurigae, magnitude 0.08, sixth-brightest in the sky. It is actually a spectroscopic binary, consisting of two G-type giants orbiting each other with a period of 104 days. Capella lies 43 l.y. away.

Cape Photographic Durchmusterung (CPD) The first star catalogue made by measuring photographs of the sky. The photographs were taken at the Cape Observatory, South Africa, by D. *Gill in 1885–90, and star positions on them were measured in the Netherlands by J. C. *Kapteyn. The result was a three-volume work published in 1896–1900, containing 454875 stars down to 10th magnitude from −18° to the south pole.

SEE WEB LINKS
• Detailed description and full catalogue downloadable from the CDS.

Cape photometry Photometric observations made at the South African Astronomical Observatory, mostly from its site in Cape Town (formerly the Royal Observatory, Cape of Good Hope). These observations were intended to extend the UBV system of *Johnson photometry to the southern hemisphere. Some U magnitudes were measured with refractors, in which the objective lens seriously reduced the amount of ultraviolet light transmitted; these observations are differentiated by the symbol U_c.

Cape RI photometry *See* KRON–COUSINS RI PHOTOMETRY.

Cape York meteorite The most massive iron meteorite, 30.9 tonnes, currently exhibited in a museum, on show at the Hayden Planetarium, New York. It is a medium *octahedrite. It was found at Cape York, Greenland, in 1897, and was nicknamed The Tent ('Ahnighito' in the local language). It was brought to New York with two other large iron meteorites found at the same time and on the same site, one called The Woman, weighing about 3 tonnes, the other called The Dog, weighing about 400 kg. All three are owned by the American Museum of Natural History.

Caph The star Beta Cassiopeiae, an F2 giant of magnitude 2.3 lying 55 l.y. away. It is a pulsating variable of the Delta Scuti type, but its variations are less than 0.1 mag.

Capricorn, Tropic of *See* TROPIC OF CAPRICORN.

Capricornid meteors A meteor shower active between July 5 and August 20. The shower is generally quite weak (maximum ZHR 5), with possible maxima on July 8, July 15, and July 26. The radiant on July 15 lies at RA 21 h 00 m, dec. −15°. Capricornid meteors are slow, yellowish, and sometimes bright. They are not recognized as a genuine shower by some authorities.

Capricornus (Cap) (*gen.* Capricorni) The smallest constellation of the zodiac, through which the Sun passes from late January to mid-February. Capricornus represents a goat with a fish's tail. Its brightest star is Delta Capricorni (Deneb Algedi), magnitude 2.9. Alpha Capricorni (Algedi or Giedi) is a wide but unrelated double of magnitudes 3.6 and 4.2.

captured rotation *See* SYNCHRONOUS ROTATION.

Carafe Galaxy An unusually shaped galaxy in Caelum, which forms a group with the elliptical galaxy NGC 1595 and the peculiar galaxy NGC 1598. The Carafe seems to have been distorted by gravitational interaction with the nearby NGC 1595, and is classified as a Seyfert galaxy with a ring. It has no NGC number of its own.

carbonaceous chondrite The most primitive of the three main classes of chondrite meteorite, with compositions that most nearly resemble that of the Sun, except for the most *volatile elements. They probably date from early in the formation of the Solar System. Carbonaceous chondrites are divided into four subgroups on the basis of composition: CI, CM, CO, and CV. The CI subgroup are the most primitive, having the lowest density, the highest content of volatiles and carbon, and compositions most closely resembling that of the Sun; they contain no chondrules. Type CM contain less than 15% chondrules with average diameter around 0.3mm. Type CO contain 35–40% chondrules, with typical diameters of 0.2–0.3mm. Type CV also contain 35–45% chondrules, but with average diameters of about 1mm. Each subgroup may be further divided into various subtypes on the basis of their texture and mineralogy (e.g. CI1, CM2, CO3, and CV3). All groups contain complex organic molecules.

carbon burning The set of nuclear reactions within stars by which carbon is converted into neon and sodium. Carbon burning occurs only in stars of about 10 solar masses or greater, and is the next set of reactions in the core after helium burning. Once carbon is exhausted in the core it may continue to burn in a surrounding shell.

carbon cycle *See* CARBON–NITROGEN CYCLE.

carbon flash The explosive release of energy, analogous to that in the *helium flash, in a star that begins to burn carbon in a degenerate core (*see* DEGENERACY). The carbon flash occurs only in stars of 1–6 solar masses, since less massive ones never undergo a carbon-burning stage, while in more massive ones the stage occurs non-explosively in a non-degenerate core.

carbon–nitrogen cycle (CN cycle) A chain of nuclear reactions in stars that converts hydrogen into helium, with the associated release of nuclear energy. Carbon, nitrogen, and oxygen act as catalysts to speed the six-stage reaction. It is also known variously as the **carbon cycle**, the *carbon-nitrogen-oxygen cycle* (CNO cycle), and the *Bethe-Weizsäcker cycle*. The cycle becomes increasingly important with rising temperature and dominates the *proton–proton reaction at temperatures above 18 million K, as in the cores of stars of over 2 solar masses.

carbon star A cool red-giant star in an advanced stage of evolution, displaying strong carbon features in the form of CN, CH, and C_2 (Swan) bands in its spectrum; also known as spectral type C. In carbon stars, the abundance of carbon is greater than that of oxygen. The additional presence of lithium indicates that these elements have been produced by nuclear reactions in the star's core and are now being transported by convection to its surface. Since carbon can be produced only by the *triple-alpha process at a very high temperature, these stars must be highly evolved. These rare but luminous objects include the former types R (K-type giants with temperatures of 4000–5000K) and N (M-type giants which are cooler, about 3000K) which were introduced in the *Harvard classification. The N-type carbon stars can be up to 10 times as luminous as the R type. Many are either semiregular variables, such as Y Canum Venaticorum (*La Superba), U Hydrae, and UU Aurigae, or Mira stars, for example R Leporis (*Hind's Crimson Star) and V Coronae Borealis. Visually they appear deep orange or red. At maximum, about mag. 4.8, U Hydrae and TX Piscium are the brightest carbon stars.

Carina (Car) (*gen.* **Carinae**) A southern constellation representing the keel of the ship *Argo Navis. It contains the second-brightest star in the sky, *Canopus (Alpha Carinae). Another bright star is Beta Carinae (*Miaplacidus). *Eta Carinae is a unique variable star embedded in the nebula NGC 3372. Major star clusters include NGC 2516 and NGC 3532, and also IC 2602, sometimes called the Southern Pleiades.

Carina Arm A spiral arm in our Galaxy seen in the constellations Vela, Carina, Crux, and Centaurus. It may be a continuation of the *Sagittarius arm, rather than a completely separate feature.

Carina–Sagittarius Arm A possible large-scale spiral arm in our Galaxy, a combination of the Carina and Sagittarius Arms.

CARMA Abbr. for *Combined Array for Research in Millimeter-wave Astronomy.

Carme A retrograde satellite of Jupiter, 23280000km from the planet's centre; also known as Jupiter XI. It orbits in 726.3 days at an inclination of 165° to Jupiter's equator. Carme is about 30 km in diameter, and was discovered in 1938 by the American astronomer Seth Barnes Nicholson (1891–1963). It is the largest member of the so-called Carme retrograde irregular group of Jovian satellites, which have comparable orbital inclinations and eccentricities. Members of this group include Arche, Pasithee, Chaldene, Kale, Isonoe, Aitne, Erinome, Taygete, Kalyke, Eukelade, and Kallichore.

Carnegie Observatories The astronomy department of the Carnegie Institution of Washington, with headquarters in Pasadena, California. It operates the *Las Campanas Observatory in Chile, which includes the twin *Magellan Telescopes.

((⊕)) SEE WEB LINKS
• Official observatory website.

Carpo A small satellite of Jupiter, also known as Jupiter XLVI. It orbits 17056000 km from the planet's centre every 455.1 days at an inclination of 55°. Carpo has a diameter of about 4 km and was discovered in 2003 from ground-based observations. It gives its name to the Carpo prograde group of Jovian satellites, of which it is so far the sole member.

Carrington, Richard Christopher (1826–75) English astronomer. In 1853 he established a private observatory where, concurrently with night-time observations for his *Catalogue of 3735 Circumpolar Stars* (published in 1857), he began a daytime programme of measurements of the heliographic positions of sunspots, which he continued until 1861. The results, published in 1863, included a calculation of the position of the Sun's axis of rotation to unprecedented accuracy, and measures of sunspot distribution and rotation period as functions of heliographic latitude. He originated the system of *Carrington rotations.

Carrington heliographic coordinates One of two *heliographic coordinate systems used for identifying the position of features on the Sun's surface. In the Carrington system, lines of longitude rotate with the Sun. A prime meridian, analogous to the Greenwich meridian on Earth, was defined to coincide with the central meridian of the Sun (as seen from Earth) at a specific time on 1853 November 9 when R. C. *Carrington began his observations. A new Carrington rotation begins each time the prime meridian crosses the central meridian. The rotation period of the coordinate system varies throughout the year because of the changing distance of the Earth from the Sun, with an average value of 27.2753 days (the mean synodic period). This corresponds to a sidereal rotation period of 25.38 days. The Carrington longitude of the central meridian is 360° at the beginning of the Carrington rotation, and decreases to zero at the end. The longitude of a feature on the Sun (such as a sunspot) remains approximately constant in the Carrington system, in contrast to the *Stonyhurst system.

Carrington rotation A system for identifying individual rotations of the Sun. On this system the Sun is assigned a sidereal rotation period of 25.38 days. The synodic period of a Carrington rotation varies slightly during the year because of the changing speed of the Earth in its orbit, but averages 27.2753 days. Carrington rotation no. 1 began on 1853 November 9. The system was originated by R. C. *Carrington.

Carte du Ciel A series of star charts obtained from photographic plates taken with *normal astrographs. These plates went to fainter magnitudes than those used for the associated *Astrographic Catalogue*. The charts show stars down to approximately photographic magnitude 14.

Cartesian coordinates A coordinate system in which the position of a point or object is given with reference to either two or three planes at right angles to each other. For a two-dimensional system, the planes are represented by horizontal and vertical axes labelled x and y respectively. For a three-dimensional system, as used for example when describing the position of a planet, a third axis is introduced, labelled z. The system was introduced by the French mathematician René Descartes (1596–1650).

Cartwheel Galaxy A ring-shaped galaxy 500 million l.y. away in Sculptor, produced when a dwarf galaxy passed through a larger spiral galaxy. The Cartwheel's rim is 170000 l.y. in diameter and is composed of gas and young stars. The central hub and faint radiating spokes are composed of old stars.

Cassegrain telescope A reflecting telescope in which a convex secondary mirror reflects light to a focus through a hole in the primary mirror; it is an adaptation of the *Gregorian telescope. The convex secondary has a hyperboloidal cross-section, while the primary mirror is paraboloidal. The primary mirror has a central hole through which the light is brought to the

Cassegrain focus. (The term **Cassegrain focus** is also applied to any focus similarly located behind a mirror, such as in a *Ritchey–Chrétien telescope.) A Cassegrain reflector is more compact than a Newtonian of the same focal length. It is also easier to mount heavy or bulky equipment at the Cassegrain focus than at the prime focus or Newtonian focus. The design was devised in the 17th century by Laurent Cassegrain (*c.*1629–1693), a French priest and teacher. *See also* NEWTONIAN–CASSEGRAIN TELESCOPE; SCHMIDT–CASSEGRAIN TELESCOPE.

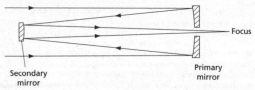

Focus

Secondary
mirror

Primary
mirror

Cassegrain telescope

Cassini French family, originally from Italy, which produced four generations of astronomers, cartographers, and geodesists. The most important astronomically were Giovanni Domenico *Cassini (sometimes referred to as Cassini I) and his son Jacques *Cassini (Cassini II). César François Cassini (Cassini III), also known as Cesare Francesco Cassini (1714–84), son of Jacques, began a topographic map of France that was completed by his own son, Jacques Dominique Cassini (Cassini IV) (1748–1845). These four Cassinis were successively directors of the Paris Observatory.

Cassini, Giovanni Domenico (1625–1712) Italian-born French astronomer who moved to Paris in 1669; also known as Jean Dominique Cassini. He measured the rotation periods of Mars and Jupiter, discovered four of Saturn's satellites, and in 1675 observed the division in Saturn's rings now named after him. In 1672, with the help of observations by his countryman Jean Richer (1630–96), he measured the distance to Mars by triangulation, and was thus able to refine the dimensions of the Solar System, obtaining a value for the astronomical unit that was short by only 7%. Cassini drew up improved tables for Jupiter's satellites, which played a part in O. C. *Römer's determination of the velocity of light.

Cassini, Jacques (1677–1756) French astronomer and geodesist, son of G. D. *Cassini; also known as Giacomo Cassini. His accurate determination of the length of an arc of a meridian passing through France, later extended by his son César François, showed that the Earth is an oblate spheroid (i.e. flattened at the poles). He compiled the first tables of the motion of Saturn's satellites and in 1738 made one of the first definite determinations of a star's proper motion, that of Arcturus.

Cassini Division The widest and most prominent gap in the rings of Saturn, about 4700km wide, separating the A and B Rings, discovered in 1675 by G. D. *Cassini. Its inner edge is 117500 km from Saturn's centre. Images from the Voyager and Cassini probes show that it is not empty, but that several thin rings lie within it. The division could be accounted for by satellites 10–30km in diameter clearing away ring particles, but no such satellites have been detected.

Cassini–Huygens A spacecraft to Saturn launched in 1997 October, jointly built by NASA, ESA, and the Italian Space Agency. It consisted of two parts: the Cassini orbiter, and the Huygens lander to study Saturn's largest satellite, Titan. Cassini–Huygens went into orbit around Saturn in 2004 July, having undergone two gravity-assist flybys of Venus and one each of the Earth and Jupiter *en route*; it made observations of Jupiter during its encounter in 2000 December. Huygens landed on Titan in 2005 January, measuring atmospheric conditions during its descent and transmitting from the surface for over an hour. The Cassini orbiter is studying Saturn's atmosphere, satellites, and rings.

🌐 **SEE WEB LINKS**
- Official mission website.
- ESA mission website.

Cassiopeia (Cas) (*gen.* **Cassiopeiae**) A constellation of the northern sky, representing Queen Cassiopeia of Greek mythology. Its brightest stars are Alpha (*Schedar) and Beta Cassiopeiae (*Caph). Gamma Cassiopeiae is a *shell star that can vary unpredictably between about magnitudes 3.0 and 1.6. M52 is a prominent open cluster, near which lies the strong radio source *Cassiopeia A. The supernova now known as *Tycho's Star flared up near Kappa Cassiopeiae in 1572.

Cassiopeia A A prominent radio source in the constellation Cassiopeia, and the brightest in the sky apart from the Sun. It has a ring-like structure, thought to be the expanding shell of a supernova remnant 10000 l.y. away. Faint traces of optical nebulosity are visible. Measurements of the rate of expansion indicate that the supernova must have occurred around 1660–80, but there are no recorded sightings of it, with the possible exception of a 6th-mag. star recorded near this position by J. *Flamsteed in 1680. An envelope of dust ejected by the star before it exploded may have dimmed the light from the supernova by several magnitudes, accounting for the lack of visual observations at the time. 📷

Castalia Asteroid 4769, a member of the *Apollo group, discovered in 1989 by the American astronomer Eleanor Kay Helin, née Francis (1932–2009). Radar observations show that it consists of two kilometre-sized lobes, apparently in contact, giving it a maximum length of about 1.8km. Castalia can approach to within 0.023 AU (3.4 million km) of Earth. Its orbit has a semimajor axis of 1.063 AU, period 1.10 years, perihelion 0.55 AU, aphelion 1.58 AU, and inclination 8°.9.

Castor The star Alpha Geminorum, which appears to the naked eye of magnitude 1.6 but which is actually a complex multiple star. A small telescope divides Castor into two stars of magnitudes 1.9 and 2.9, both A-type dwarfs, which orbit each other with a period of about 450 years. Each is a spectroscopic binary. There is a third star, known as Castor C or YY Geminorum, an eclipsing pair of M1 dwarfs that ranges from magnitude 9.2 to 9.6 in a period of 0.8 days. The six-star system lies 51 l.y. away.

cataclysmic binary A *semidetached binary that undergoes outbursts. The term is usually restricted to *cataclysmic variables of the *dwarf nova, *nova, and *nova-like types, excluding any binary that is a supernova, symbiotic star, or X-ray binary. In a cataclysmic binary a cool main-sequence star, usually in the spectral type range G8 to M8, loses mass via *Roche lobe overflow onto its white dwarf companion. If the white dwarf is non-magnetic, the overflowing material forms an *accretion disk around the white dwarf; the disk is cool at the outer edge, and becomes progressively hotter towards the inner edge, from which material falls preferentially onto the equatorial regions of the white dwarf. By contrast, if the white dwarf possesses a magnetic field the accretion flow is more or less dominated by the magnetic field and falls onto the white dwarf near its magnetic poles. When the magnetic field is sufficiently strong, no disk can form; accretion then occurs directly from the Roche lobe of the cool dwarf along magnetic field lines and onto the poles through an *accretion column*. Such stars are known as *AM Herculis stars after the type star, or *polars, because of the strong polarization in the light from them. For fields of intermediate strength, a disk may form but with a hole in the middle; accretion then occurs from the entire inner edge of the disk along field lines onto the magnetic poles, forming an *accretion curtain*. In this case, the star is usually called an *intermediate polar.

cataclysmic variable 1. A variable star that exhibits sudden outbursts generally arising either from the release of gravitational energy through accretion or from thermonuclear processes. The latter may occur on the surface or in the interior. The group includes many different types of variable, most of which are close binaries undergoing *mass transfer. All *supernovae, including those believed to consist of single stars, also come into this category. *See* DWARF NOVA; NOVA; NOVA-LIKE VARIABLE; SUPERNOVA; SYMBIOTIC STAR.

2. Another name for a *cataclysmic binary.

catadioptric system An optical system in which mirrors and lenses bring light to a focus. The best-known catadioptric designs are the *Schmidt–Cassegrain and *Maksutov telescopes, in both of which the incoming beam passes through a transparent *corrector plate. Alternative designs, such as the *Sampson catadioptric*, may use smaller lenses in the converging beam to correct some of the aberrations that would otherwise be present.

Catalina Sky Survey A search for asteroids and comets, particularly those passing close to Earth, undertaken with the 0.68-m Catalina Schmidt telescope of *Steward Observatory on Mt Bigelow, Arizona. The Catalina Sky Survey began in 1992 under the name Bigelow Sky Survey, but changed to the current name in 1998. The Catalina Schmidt telescope was upgraded from 0.4-m to 0.68-m aperture in 2003. Follow-up observations, and searches for fainter objects, are undertaken by the 1.5-m telescope at nearby Mount Lemmon Observatory. Southern sky coverage is provided by a cooperating survey at Siding Spring Observatory, Australia.

(⊕) SEE WEB LINKS
• Official project website.

catalogue equinox The zero point of right ascension to which the positions of stars in a particular catalogue are referred.

catena A line of craters, usually overlapping, also known as a crater chain; pl.*catenae*. The name is not a geological term, but is used in the nomenclature of individual features on planets and satellites. Examples are Tithoniae Catena on Mars and Gipul Catena on Callisto.

catoptric system An optical system that uses only mirrors to focus light. A conventional Newtonian reflecting telescope is an example.

Cat's Eye Nebula The 9th-magnitude planetary nebula NGC 6543, which lies 3000 l.y. away in Draco. The complex shape of the gas loops thrown off by the central star 1000 years or so ago gives it the appearance of a cat's eye. 📷

cavus A hollow or irregular depression on a planetary surface; pl.*cavi*. The name is not a geological term, but is used in the nomenclature of individual features, for example the Sisyphi Cavi on Mars.

CCAT Abbr. for *Cerro Chajnantor Atacama Telescope.

CCD Abbr. for *charge-coupled device.

CCD spectrometer A spectrometer that uses a *charge-coupled device (CCD) to measure the energy of incident photons. CCD spectrometers can be sensitive to a wide range of wavelengths, from infrared and optical to ultraviolet and X-rays. Each incident photon generates electrons in the CCD, the number of which is proportional to the energy of the original photon.

C-class asteroid The most common class of asteroid in the outer main asteroid belt. C-class asteroids have low albedos (0.03–0.08) and flat reflectance spectra at wavelengths longer than $0.4\,\mu m$. The 'C' is for carbonaceous, since they are believed to have surfaces that are similar in composition to *carbonaceous chondrite meteorites. Their spectra show an ultraviolet absorption feature at wavelengths shorter than $0.4\,\mu m$, thought to be due to *water of hydration. Examples include *Hygiea and *Davida. Classes B, F, and G are subclasses of C class.

cD galaxy A supergiant elliptical galaxy found at the centre of some rich clusters of galaxies. They are an extreme type of *D galaxy. Their surface brightness falls off much more slowly with radius than for most elliptical galaxies, and they show an extensive faint halo of stars. They are among the most luminous of all galaxies, with luminosities up to about 2×10^{12} Suns, and they may have double or multiple nuclei. The evolution of cD galaxies may involve accretion of gas from the cluster and *galaxy cannibalism.

CDM Abbr. for *cold dark matter.

CDS Abbr. for *Centre de Données astronomiques de Strasbourg.

celestial axis The line joining the north and south celestial poles, about which the celestial sphere rotates.

celestial coordinates Any system for locating the positions of objects on the celestial sphere. Various systems are defined, according to the chosen point of observation (the *origin*) and the plane of reference they use. *Geocentric coordinates give the position of an object as it would be seen from the centre of the Earth, the most commonly used point of origin. *Heliocentric coordinates, sometimes used for objects in the Solar System, give the position as would be seen from the centre of the Sun; a variant system, barycentric coordinates, gives the position as would be seen from the centre of mass of the Solar System. *Topocentric coordinates give positions as seen from a specific point on Earth. There are several different reference planes for celestial coordinates. *Equatorial coordinates, the most common system, use the celestial equator as the reference plane; their origin can be geocentric or topocentric. *Ecliptic coordinates take the ecliptic as their reference plane; they can be geocentric or heliocentric in origin. *Horizontal coordinates give positions with respect to an observer's horizon, and are topocentric. *Galactic coordinates are given with reference to the plane of the Galaxy. *See also* RECTANGULAR COORDINATES; SPHERICAL COORDINATES.

celestial equator The great circle on the celestial sphere that lies directly above the equator of the Earth. Any point on the celestial equator is equidistant from the north and south celestial poles. *Declination is zero on the celestial equator.

Celestial Intermediate Reference System (CIRS) A geocentric coordinate system based upon the present position of the celestial pole but with right ascensions reckoned from a point called the Celestial Intermediate Origin (CIO) instead of the equinox. The CIO is a point on the celestial equator at present very close to the prime meridian of the *Geocentric Celestial Reference System. CIRS directions are the modern equivalent of classical *apparent places. To express the direction of a star as a CIRS position, its catalogue coordinates in the *International Celestial Reference System are corrected for proper motion and parallax, for the gravitational deflection of light by the Sun, and for stellar aberration, and are then rotated to take account of precession and nutation. Calculating hour angles from CIRS right ascensions requires knowing the *Earth rotation angle, which can be thought of as a simplified sidereal time, free of precession and nutation effects. The difference between CIRS right ascensions and classical equinox-based right ascensions is termed the *equation of the origins*.

celestial latitude (β) A coordinate that gives the angular position of a celestial body north or south of the ecliptic as seen from the Earth; also known as *ecliptic latitude*. It is measured in degrees from 0° at the ecliptic to 90° at the ecliptic poles.

celestial longitude (λ) A coordinate that gives the angular position of a celestial body around the plane of the ecliptic as seen from the Earth; also known as *ecliptic longitude*. It is measured in degrees anticlockwise from 0° to 360° along the ecliptic, starting at the vernal equinox.

celestial mechanics The branch of astronomy that deals with the motions of bodies in orbit, such as planets, natural and artificial satellites, space probes, comets, and binary or multiple stars. The forces involved are gravitation and, for some artificial satellites, atmospheric drag and *radiation pressure from sunlight. Although in most cases the orbit of a body is nominally elliptical and subject to *Kepler's laws, the attraction of other bodies and the presence of other forces will cause small departures, known as *perturbations, from the elliptical form. *See also* N-BODY PROBLEM; THREE-BODY PROBLEM.

celestial meridian *See* MERIDIAN, CELESTIAL.

celestial pole Either of the two points about which the celestial sphere appears to rotate each day. The celestial poles lie on the celestial sphere directly above the Earth's geographical poles, and are 90° from the celestial equator. Because of *precession, the celestial poles describe a circle around the *ecliptic poles every 25 800 years.

celestial sphere An imaginary sphere of indefinite size, used as a background for assigning positional coordinates to celestial objects (see diagram). The sphere may be centred on the Earth, the observer, or any other point which acts as the origin of the chosen system of coordinates. As seen from Earth, the celestial sphere appears to rotate about the axis joining the *celestial poles every 23h 56m 04s (a *sidereal day), actually a result of the axial spin of the Earth. Two important circles on the celestial sphere are the *celestial equator and the *ecliptic. The angle between them is about 23°.4, and is known as the *obliquity of the ecliptic. The celestial equator and the ecliptic intersect at the vernal and autumnal *equinoxes. The positions of the celestial poles, and hence of the celestial equator, are gradually moving on the celestial sphere because of a slow wobbling of the Earth in space termed *precession. *See also* CELESTIAL COORDINATES.

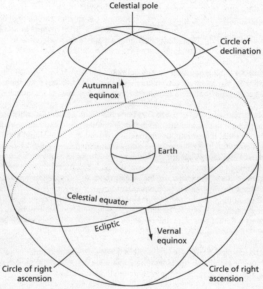

celestial sphere

Centaur group A group of unusual bodies in the outer Solar System, possibly giant icy *planetesimals or cometary nuclei. Group members have orbits which lie roughly between those of Jupiter and Neptune, in most cases crossing the orbits of one or more of the outer planets, with periods in the range 20–200 years. The first to be discovered were *Chiron, in 1977, and *Pholus, in 1992. Centaurs have diameters in the range 50–300 km, but there may be many smaller members still undiscovered. They are thought to be derived from the *Kuiper Belt and are now usually classified with the so-called *scattered-disk objects* (SDOs) which have eccentric orbits extending well beyond Neptune. Centaurs and SDOs could be captured into *Jupiter comet family or *Halley-family cometary orbits.

Centaurus (Cen) (*gen.* Centauri) A large and prominent southern constellation, representing a centaur. Its brightest star, *Alpha Centauri (Rigil Kentaurus or Toliman), is actually a triple star, one member of which, *Proxima Centauri, is the closest star to the Sun. Beta Centauri (*Hadar or Agena) is another bright star. *Omega Centauri is the most prominent globular cluster in the sky, and NGC 5128 is a peculiar radio galaxy known as *Centaurus A. NGC 3918 is a planetary nebula known as the *Blue Planetary.

Centaurus A A strong radio and X-ray source in the constellation Centaurus, identified with the 7th-magnitude giant elliptical galaxy NGC 5128. Centaurus A is a classical radio galaxy with two pairs of radio-emitting lobes, the largest spanning 1.5 million l.y., and an X-ray-emitting jet over 15000 l.y. long. At 11 million l.y. away, it is the nearest radio galaxy to the Sun. Although the parent galaxy is classified as elliptical, it has an uncharacteristic band of dust across it and is believed to be the result of a merger between an elliptical and a spiral galaxy.

Center for Astrophysics (CfA) *See* HARVARD–SMITHSONIAN CENTER FOR ASTROPHYSICS.

Central Bureau for Astronomical Telegrams An office of the International Astronomical Union that announces discoveries of objects and transient events such as comets, asteroids, novae, and supernovae which require immediate observation. The first telegram bureau was established by the Astronomische Gesellschaft at Kiel, Germany, in 1884, and moved to Copenhagen in 1914. Its work was adopted by the IAU when that body was founded in 1919. Since 1965 the Bureau has been based at the *Smithsonian Astrophysical Observatory. Nowadays the bureau sends out its announcements by electronic mail.

((⊕)) SEE WEB LINKS
• Official website.

central meridian An imaginary straight line joining the north and south poles of a planet's disk as seen from Earth. It is used as a reference for observers making longitude measurements of features on a planet. As the planet rotates, the time when a feature crosses, or *transits*, the central meridian is noted; its longitude can then be derived from published tables. Such observations are known as **central meridian transit** timings.

central peak A hill at or near the centre of a crater floor. Central peaks normally occur within impact craters of 15–120 km diameter, although this varies between different planets and satellites, and different terrain types. They are thought to be caused by rebound of the crater bottom immediately after formation. Central peaks can also occur within volcanic craters, but are comparatively rare.

Centre de Données astronomiques de Strasbourg (CDS) The world's leading astronomical data centre, based at the Observatoire de Strasbourg, France; the name is French for Strasbourg Astronomical Data Center. It was founded in 1972 as the Centre de Données Stellaires (Centre for Stellar Data), and was renamed in 1983 when it was expanded to cover all objects outside the Solar System, not just stars. CDS collects data from ground-based and space observatories, and makes them available in electronic form. CDS operates the *SIMBAD database.

((⊕)) SEE WEB LINKS
• Official website with searchable content.

centre of gravity The point in a body at which all external forces can be taken to act. In a uniform gravitational field this is the same as the body's *centre of mass. However, in a non-uniform gravitational field, such as that produced by the gravitational pulls of a number of external moving bodies, the body cannot be said to have a true centre of gravity.

centre of inertia Another name for *centre of mass.

centre of mass The point in a body or system of bodies which acts as if all the mass were concentrated there; also known as **centre of inertia**. The gravitational force between two massive bodies acts along the line joining their centres of mass. For a sphere of uniform density, or a sphere whose density varies radially from the centre, the centre of mass is at the centre of the sphere. In a uniform gravitational field, a body's centre coincides with its *centre of gravity.

centrifugal force An apparent force pulling an orbiting body away from the centre of its orbit. In a circular orbit, centrifugal force appears to be equal and opposite to *centripetal force, which is a real force. Centrifugal force is a consequence of the body's inertia, which would make the body continue in a straight line if the centripetal force acting on it were to cease.

centripetal force A force that tries to pull an orbiting body towards the centre of its orbit. This force is actually the same as the gravitational force. Gravity attracts the body, causing its path to curve away from the straight line that it would otherwise follow.

Cepheid instability strip The region on the Hertzsprung–Russell diagram occupied by several types of pulsating variable star, including Cepheid variables, classical Cepheids, Delta Scuti stars, RR Lyrae stars, and W Virginis stars. It runs diagonally upwards and to the right from the main sequence at about absolute magnitude 2, and includes extremely luminous supergiants at the very top; here it tends to merge with the broader region occupied by RV Tauri, red semiregular, and long-period (Mira) variables. The RR Lyrae variables are located where the instability strip intersects the horizontal branch in globular clusters, while classical Cepheids and W Virginis stars lie at the intersection of the strip with the supergiant branch in metal-rich and metal-poor stars, respectively. Hot, massive stars enter the strip as they evolve off the main sequence. During this phase of the stars' lives, their expanding outer layers become unstable and they pulsate regularly. Stars can move across the instability strip and back again several times as they continue to evolve.

Cepheid variable One of an important group of yellow giant or supergiant *pulsating variables, named after the prototype, Delta Cephei. This general term is commonly applied to more than one stellar type, in particular to the *classical Cepheids (sometimes known as *Delta Cephei stars*), and the less numerous *W Virginis stars.
The significance of Cepheid variables became apparent when H. S. *Leavitt discovered that their period was directly related to their absolute magnitude. The resulting *period–luminosity relationship is used to determine distances. Subsequent work established the existence of two distinct types, with essentially parallel period–luminosity relationships. The classical Cepheids are Population I objects with absolute magnitudes 0.7–2 mag. brighter than the Population II W Virginis stars, as well as larger masses and greater metallicity. Both types undergo *radial pulsations in the *fundamental mode. At maximum size, Cepheids are typically 7–15% larger than at minimum size.
In the past, other distinct types were regarded as forms of Cepheids, notably those called *dwarf Cepheids* (*AI Velorum and *Delta Scuti stars), and *short-period Cepheids* (*RR Lyrae stars). *See also* BEAT CEPHEID; BUMP CEPHEID; DOUBLE-MODE VARIABLE.

Cepheus (Cep) (*gen.* **Cephei**) A constellation of the north polar region of the sky, representing King Cepheus of Greek mythology. Its brightest star is Alpha Cephei (*Alderamin), but its best-known star is the double Delta Cephei, the brighter component of which is the prototype *Cepheid variable, ranging from magnitude 3.5 to 4.4 in a period of 5.37 days. Beta Cephei (Alfirk) is a double of unrelated stars, magnitudes 3.2 and 7.9, the brighter of which is slightly variable (*see* BETA CEPHEID VARIABLE). Mu Cephei is a variable red supergiant known as the *Garnet Star.

Cerenkov counter A device for detecting high-energy charged particles via the *Cerenkov radiation they generate. This radiation, which appears as a predominantly bluish light, can be detected using photomultiplier tubes which convert the light into electrical pulses.

Cerenkov radiation The light emitted by charged particles such as protons or electrons when they pass through a transparent medium (e.g. the Earth's atmosphere, glass, or certain plastics) at a speed greater than the speed of light in that medium. The effect is the electromagnetic equivalent of a sonic boom. The radiation can occur at any wavelength, but increases in intensity with frequency and so is strongest in the blue and ultraviolet. The high-energy particles involved may be the secondary products of gamma rays or cosmic rays. For example, when gamma rays with energies of 10^{12} eV pass through the Earth's atmosphere they generate secondary electrons, and ground-based optical telescopes can detect the flash of blue light they emit. Cerenkov radiation is named after the Russian physicist Pavel Alexeyevich Cerenkov (1904–90).

Ceres The first asteroid to be discovered, by G. *Piazzi on 1801 January 1, and hence given the number 1. Ceres is by far the largest asteroid, an oblate spheroid 975 × 909 km, mean diameter 952 km, mass 9×10^{20} kg (about 1% the mass of the Moon) and mean density about 2.0 g/cm^3.

Its rotation period is 9.075 hours. Ceres contains roughly one-third the mass of the entire asteroid belt. It is a member of the new class of *dwarf planets, along with Eris and Pluto. Its mean magnitude at opposition is 7.4; only Vesta can ever become brighter. Its orbit has a semimajor axis of 2.765 AU, period 4.60 years, perihelion 2.55 AU, aphelion 2.98 AU, and inclination 10°.6. Ceres is of *G class, with a reflectance spectrum that implies a composition resembling the *carbonaceous chondrite meteorites. Ceres is the largest member of a small family of asteroids with a mean distance of 2.76–2.80 AU from the Sun and orbital inclinations of 9–11°. Other Ceres family members include (39) Laetitia, (264) Libussa, (374) Burgundia (all S-class), and (446) Aeternitas (A-class).

Cerro Chajnantor Atacama Telescope (CCAT) A 25-m telescope for millimetre and submillimetre astronomy being built by a consortium of US, Canadian, and German institutes at an altitude of 5600 m near the summit of Cerro Chajnantor, Chile. It is expected to be in operation in 2017 or later and will act as a wide-field survey instrument to complement the observations of the *Atacama Large Millimeter Array.

(( )) SEE WEB LINKS
• Official telescope website.

Cerro Pachón A mountain in Chile which is the site of the 8.1-m Gemini South telescope (*see* GEMINI OBSERVATORY) and also the 4.1-m *Southern Astrophysical Research (SOAR) Telescope. Cerro Pachón lies 10 km southeast of *Cerro Tololo Inter-American Observatory. It has also been chosen as the site for the proposed *Large Synoptic Survey Telescope.

Cerro Paranal The site in northern Chile of the *Paranal Observatory.

Cerro Tololo Inter-American Observatory (CTIO) An observatory at an altitude of 2215 m on Cerro Tololo mountain, Chile, 55 km southeast of La Serena, where its headquarters are. Founded in 1963, it is part of the US *National Optical Astronomy Observatory. Cerro Tololo's main instrument is the 4-m Blanco Telescope, opened in 1974, named after the observatory's first director, the American astronomer Victor Manuel Blanco (1918–2011). This telescope is the southern twin of the 4-m instrument at *Kitt Peak National Observatory. Other instruments include a 1.5-m reflector, opened in 1968; a 1.3-m reflector originally used for the *Two-Micron All-Sky Survey, opened in 1997 but transferred to CTIO ownership in 2001; a 0.9-m reflector, opened in 1967; Yale University's 1-m reflector, installed here in 1973; the University of Michigan's 0.6-m Curtis Schmidt, originally opened in 1950 but moved here in 1967; and the 0.6-m Wisconsin H-Alpha Mapper (WHAM) telescope, originally opened on Kitt Peak in 1996 but moved to CTIO in 2009. The 1.5-m, 1.3-m, 1-m, and 0.9-m have been operated since 2003 by the Small and Moderate Aperture Research Telescope System (SMARTS) consortium of US institutions. The Curtis Schmidt is now used for monitoring orbital debris and is known as the Michigan Orbital Debris Survey Telescope (MODEST). In 2006 the University of North Carolina at Chapel Hill began operation of the PROMPT array (Panchromatic Robotic Optical Monitoring and Polarimetry Telescopes) consisting of six 0.4-m telescopes to study gamma-ray bursts. CTIO also operates the 4.1-m *Southern Astrophysical Research (SOAR) Telescope on Cerro Pachón, a peak 10 km to the southeast of Cerro Tololo.

(( )) SEE WEB LINKS
• Official observatory website.

Cetus (Cet) (*gen.* Ceti) The fourth-largest constellation, straddling the celestial equator. Cetus represents a sea-monster or whale. Its brightest star is Beta Ceti (*Deneb Kaitos). Alpha Ceti is known as *Menkar. The most celebrated star in the constellation is the red-giant variable *Mira (Omicron Ceti). Tau Ceti, magnitude 3.5, is a star similar to the Sun, lying 11.9 l.y. away. M77, a 9th-magnitude spiral, is the brightest *Seyfert galaxy.

CfA Abbr. for *Harvard–Smithsonian Center for Astrophysics.

CGRO Abbr. for *Compton Gamma Ray Observatory.

Chamaeleon (Cha) (*gen.* **Chamaeleontis**) A small, faint constellation near the south celestial pole, representing a chameleon. Its brightest stars, Alpha and Gamma Chamaeleontis, are both magnitude 4.1.

Chandler period The approximate period of the wandering of the Earth's geographical poles over the surface. Although this motion is irregular, a basic period of 428 days may be recognized (*see* CHANDLER WOBBLE).

Chandler wobble The irregular movement of the Earth's geographical poles over the surface, anticlockwise for the north pole and clockwise for the south. The effect is very small, the displacement of the pole being less than 10m, but it produces slight but detectable changes in latitude and longitude of the order of 0″.3. This polar motion is a displacement of the poles with respect to the Earth, not the star background, and hence is quite distinct from *precession, which is a variation in the direction of the Earth's axis in space. It is named after the American astronomer Seth Carlo Chandler (1846–1913).

Chandrasekhar, Subrahmanyan (1910–95) Indian-American astrophysicist, born in modern Pakistan. He was the first to identify *white dwarf stars as end-products of stellar evolution, and developed a theory which used relativistic effects to account for their *degeneracy pressure. He calculated an upper mass limit (the *Chandrasekhar limit) beyond which a star would enter a more dramatic final phase, setting the scene for the modern theory of *black holes. He studied how stars transfer energy by radiation in their atmospheres, publishing his findings in *Radiative Transfer* (1950). Chandrasekhar shared the 1983 Nobel Prize in Physics with W. A. *Fowler.

Chandrasekhar limit The maximum possible mass of a *degenerate star, above which it will be unable to support itself against the inward pull of its own gravity. For a star with no hydrogen content the limit is 1.44 solar masses, which is thus the maximum possible mass for a white dwarf. A degenerate star with a mass greater than this limit would collapse under gravity to become either a neutron star or a black hole. It is named after S. *Chandrasekhar.

Chandrasekhar–Schönberg limit The maximum mass of a star's helium core that can support the outer parts of the star against gravitational collapse, once the hydrogen at its centre has been exhausted. The limit is about 10–15% of the total mass of the star. If the mass of helium in the core exceeds this limit, the central parts collapse while the outer part expands rapidly to become a red giant. Calculations suggest that this happens only in massive stars. The limit is named after S. *Chandrasekhar and the Brazilian astrophysicist Mario Schönberg (1914–90).

Chandra X-ray Observatory A NASA X-ray astronomy satellite, one of the four *Great Observatories, launched in 1999 July. It uses a mirror *grazing-incidence telescope with a collecting area equivalent to that of a conventional 0.4-m optical telescope to observe X-rays of 0.1–10keV (0.12–12nm). Chandra provides sharper images (0″.5) and more detailed X-ray spectra than any previous X-ray mission. It is named after S. *Chandrasekhar.

(⊕) SEE WEB LINKS
• Official mission website.

Chandrayaan-1 An Indian Moon probe, with European and NASA involvement, launched 2008 October by the Indian Space Research Organization into a circular polar orbit around the Moon; the name Chandrayaan is Hindi for 'Moon Craft'. Its instruments included the Terrain Mapping Camera (TMC) to generate a 3-D map of the Moon; the Hyper Spectral Imager (HySI) for mineral mapping; the Lunar Laser Ranging Instrument (LLRI) for topographical mapping; the Miniature Synthetic Aperture Radar (Mini-SAR) to search for water ice at the Moon's poles; and a set of X-ray spectrometers for mapping specific elements on the surface. In 2008 November the Moon Impact Probe (MIP) was released and hit the lunar surface near the Moon's south pole. Contact was lost with Chandrayaan-1 in 2009 August.

(⊕) SEE WEB LINKS
• Official mission website.

Chang'e A series of Moon probes launched by the China National Space Administration, named after a Chinese Moon goddess. Chang'e 1 was a lunar orbiter, launched 2007 October. It made a three-dimensional map of the surface with a stereo camera system, studying the surface composition with a gamma/X-ray spectrometer, mapping the thickness of the lunar regolith with a microwave radiometer, and collecting data on the solar wind. The mission ended in 2009 March when Chang'e 1 was commanded to hit the Moon. Chang'e 2 was launched in 2010 October carrying an improved camera with which it photographed the Moon from a 100-km polar orbit. One of its aims was to help with the selection of a launch site for Chang'e 3, a lunar rover planned for launch in 2013. Later members of the series may bring back lunar samples.

chaos A distinctive area of broken terrain on a planetary surface. The name is not a geological term, but is used in the nomenclature of individual features, for example Iani Chaos on Mars.

chaotic orbit An orbit that can change in a largely unpredictable manner, or one where a tiny change in the position and/or the velocity of the orbiting body will produce major changes in the orbit. For example, when a small body such as a comet or asteroid makes a close encounter with a massive planet such as Jupiter, its post-encounter path will differ by millions of kilometres if the fly-by distance is altered by only a few hundred kilometres. The outcomes of such encounters are essentially unpredictable. The orbit of the object *Chiron, which is perturbed by Saturn and Uranus, is also chaotic.

chaotic terrain Collapsed and jumbled terrain on a planetary surface, first seen on Mars by the Mariner 6 and 7 probes in 1969. It is thought to result from the freezing and fracture of a partially melted ice-rich terrain. An example is Hydraotes Chaos, south of Chryse on Mars.

CHARA Array An optical and infrared interferometer consisting of six 1-m telescopes in a Y-shape with baselines up to 330 m, situated at *Mount Wilson Observatory. It is operated by the Center for High Angular Resolution Astronomy (hence CHARA) of Georgia State University. It made its first observations in 1999 and became fully operational in 2004.

(())) SEE WEB LINKS
• Official telescope website.

charge A fundamental property of elementary particles that causes them to attract or repel other particles by means of electric force. Charge is either positive or negative. Like charges repel (e.g. positive–positive) and unlike charges attract. Protons have a positive charge; electrons have a negative charge. Atoms that consist of equal numbers of protons and electrons are electrically neutral. If there is an excess of electrons the atom is negatively charged; an excess of protons makes it positively charged. *See also* ION; IONIZATION.

charge-coupled device (CCD) A silicon chip containing an array of light-sensitive diodes, used for capturing images. The photodiodes, arranged in an array of rows and columns, become charged when light falls on them. The amount of charge depends on the amount of light, which may be built up over time. These charges are read out column by column to provide an analogue signal of the image on the array, which is then converted to digital form for display and storage on a computer. CCDs are widely used both by professional and amateur astronomers as they are more sensitive to light than a photographic emulsion, give an output in almost constant proportion to the amount of light falling on them (i.e. a *linear response*), and have no *reciprocity failure. The image can be displayed almost immediately after the end of the exposure, and image processing can be used to enhance it. However, the detector area is much smaller than a photographic plate or film, which can be of any size required and provides much finer resolution.

Charon The largest satellite of Pluto; also known as Pluto I. It orbits 19 570 km from Pluto's centre in 6.387 days. Its orbital period and its axial rotation are the same as the rotation period of Pluto itself, so Charon not only keeps one face permanently turned towards Pluto but also hangs stationary over one point on its surface. Charon is 1210 km in diameter, over half the size of Pluto. Its mass is approximately one-tenth that of Pluto, and the *barycentre of the system lies above Pluto's surface. Charon's density is about $1.7\,\text{g/cm}^3$, suggesting a composition of rock and ice. Water ice has been detected on its surface. Charon was discovered in 1978 by the American astronomer James Walter Christy (1938–).

chasma A canyon or deep linear trough on a planetary surface; pl.*chasmata*. The name is not a geological term, but is used in the nomenclature of individual features, for example Tithonium Chasma on Mars, or Ithaca Chasma on Tethys.

chassignite A very rare type of achondrite meteorite, named after the 4-kg meteorite that fell at Chassigny, France, in 1815, the only known fall of this type. Chassignite (also known as *olivine achondrite*) is an olivine-rich rock that resembles lunar and terrestrial dunites in texture and general mineral composition, but contains olivine of higher iron content. The Chassigny meteorite contains 92% olivine. It has a formation age of 1.3 billion years and an *exposure age of about 12 million years. Chassignite belongs to the class of *SNC meteorites, which are thought to come from Mars.

chemical element *See* ELEMENT, CHEMICAL.

chemosphere A region of the upper atmosphere between altitudes of 40 and 80 km in which chemical processes driven by sunlight are significant. The chemosphere overlaps the upper *stratosphere and the *mesosphere. Chemical processes taking place at this level include the formation of ozone by the action of sunlight on oxygen, and various reactions involving oxides of nitrogen. The term is not in regular use by meteorologists.

Chicxulub An eroded impact crater approximately 170 km in diameter near the city of Mérida in the Yucatán peninsula of Mexico. It was formed 65 million years ago by the impact of an asteroid or comet of estimated diameter 10 km. The environmental effects caused by the ejecta from the impact are thought to have been responsible for the extinction of the dinosaurs. The crater is now buried under 1 km of limestone sediments and was discovered in the 1990s by geophysicists exploring for oil.

Chiron Asteroid 2060, an unusual object in the outer Solar System discovered in 1977 by the American astronomer Charles Thomas Kowal (1940–); also designated 95P/Chiron. Chiron lies well beyond the main asteroid belt, moving from within the orbit of Saturn out to the orbit of Uranus. It was the first of the *Centaur group to be discovered. Chiron is estimated to be about 200 km in diameter and has a rather dark, rocky or dusty surface. In 1989 it began to develop a faint coma like a comet, as a result of which it was given a periodic comet designation. Its orbit has a semimajor axis of 13.70 AU, period 50.7 years, perihelion 8.51 AU, aphelion 18.88 AU, and inclination 6°.9. Its rotation period is 5.92 hours.

((()) SEE WEB LINKS
• Information page at Cometography website.

Chladni, Ernst Florens Friedrich (1756–1827) German physicist of Hungarian descent. In 1794 he speculated that meteorites are fragments of a shattered planet. Their extraterrestrial origin was confirmed by J.-B. *Biot in 1803. Chladni suggested in 1819 that there could be a connection between meteors and comets.

chondrite A type of stony meteorite, so named because most (though not all) contain small spherules termed *chondrules. Chondrites are the most abundant class of meteorite, accounting for around 82% of all falls. They are largely composed of iron- and magnesium-bearing silicate minerals. Except for the most *volatile elements, their chemical composition closely resembles that of the Sun, and presumably the solar nebula. Chondrites are divided into three main classes primarily on the basis of their chemistry. *Enstatite or E-chondrites contain the most *refractory elements and are highly reduced. *Carbonaceous or C-chondrites contain about 3% carbon; these have the highest proportions of volatile substances, and are the most oxidized. *Ordinary chondrites, the most common type, are intermediate in their abundance of volatile elements and oxidation state. Each of the three main classes of chondrite is divided into several subgroups on the basis of composition, and each subgroup may be further subdivided into several types on the basis of their texture and mineralogy.

chondrule A small spheroidal object, typically about 0.2–3.8mm in size, although 0.5–1.5 mm is more usual, found in all chondritic meteorites except the CI subgroup of carbonaceous chondrites. The composition of chondrules varies widely, and may consist of one or more of any number of silicate minerals, or glass. Chondrules apparently existed independently before their incorporation in meteorites, but their precise origin is still unclear. Whatever their origin, chondrules are clearly some of the most primitive solid material in the Solar System.

chopper A mechanical device designed to switch the beam of a radio or infrared telescope rapidly between two position on the sky in order to subtract background radiation. At one position the beam is centred on the source under observation, and at the other it is looking at empty sky. Chop frequencies are typically several times a second. *See also* CHOPPING SECONDARY.

chopping secondary A secondary mirror of a radio or infrared telescope mounted in such a way that it can be vibrated so as to switch the telescope beam rapidly between two positions on the sky in a similar manner to a *chopper.

Christie, William Henry Mahoney (1845–1922) English astronomer. As the eighth Astronomer Royal he expanded the activities of the Royal Observatory at Greenwich into the fields of astrophotography and spectroscopy, and introduced new and improved telescopes. With E. W. *Maunder he began the Greenwich series of daily solar photographs.

chromatic aberration False colour in a refracting optical system, arising because light of different wavelengths is refracted by different amounts. The most common example is the coloured fringes that appear round images produced by a simple lens. Red light is refracted less than blue light, so the red focus lies farther from a lens than the blue focus. An image which is in focus in yellow light will therefore have a red fringe. The longer the focal ratio of a lens, the less chromatic aberration it will show. Early telescopes were made with very long focal ratios for this reason. The *achromatic lens was invented to overcome chromatic aberration.

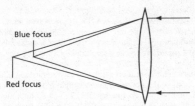

chromatic aberration: A lens brings blue light to a shorter focus than red light

chromosphere A region of a star's atmosphere above its photosphere. The Sun's chromosphere extends from the *temperature minimum, some 500km above the base of the photosphere, outwards by 9000km where it merges with the *corona. For the first 1500km the chromosphere is more or less continuous, but above this it breaks up into jagged *spicules. The temperature of the chromosphere rises from 4400K at 500km to about 6000K at 1000–2000km. There is a rapid rise to coronal temperatures at heights of about 2500km (the *transition region), the exact height depending on the strength of the local magnetic field. At the top of the chromosphere, the density is a millionth of its value at the bottom. Immediately before and after a total solar eclipse, the chromosphere is visible either as a crescent or a diamond ring, with red coloration due to *Hα emission from which it takes its name (meaning 'colour sphere'). Outside eclipses it is visible in Hα and calcium K-line *filtergrams, and from space in ultraviolet emission lines. The presence of chromospheres in nearby cool dwarf stars is deduced from similar emissions.

chromospheric network A global pattern in the Sun's chromosphere, formed by bright and dark *mottles visible in *Hα light or the calcium H and K lines. The network marks regions of strong magnetic field, and coincides with boundaries of the *supergranulation on the photosphere below. Coarse bright mottles outline the network. Fine mottles, both dark and bright, are arranged in clusters, with their bases rooted in coarse mottles. These small mottles are probably *spicules seen against the Sun's disk. The network extends into the *transition region above the chromosphere, but fans out into the corona and can no longer be detected there.

Chryse Planitia A large, roughly circular depression on Mars about 1700 km wide, perhaps an ancient impact basin, centred at +27° lat., 40° W long. Its floor is over 2.5 km below the so-called Martian datum height (the zero elevation contour, at which the average Martian atmospheric pressure is 6.1 mb). Many ancient channels that are thought to have once contained water run into Chryse. Radar sounding of Chryse Planitia shows signs of water ice 1 km below the surface, hence it is a promising place to look for signs of Martian life. Chryse Planitia was the landing site for Viking 1 in 1976 and for Mars Pathfinder in 1997.

CH star A red-giant *carbon star with especially strong bands in its spectrum due to the molecule CH; other carbonaceous molecules such as CN (cyanogen) and C_2 are also strong. CH stars are members of Population II, and are found in the galactic halo and in globular clusters.

Circinus (Cir) (*gen.* Circini) A small and insignificant constellation of the southern sky, representing a pair of draughtsman's compasses. Its brightest star, Alpha Circini, is magnitude 3.2.

Circlet An asterism in Pisces, consisting of a ring formed by seven stars of 4th and 5th magnitude: Gamma, Kappa, Lambda, TX (or 19), Iota, Theta, and 7 Piscium. The Circlet lies immediately south of the Square of Pegasus.

circular velocity The velocity of an object moving in a circular orbit about a planet or other body. If the masses of the orbiting object and the central body are m and M, and r is the radius of the orbit, then the circular velocity V is given by $V = \sqrt{[G(M + m)/r]}$, where G is the gravitational constant.

circumpolar A celestial object that does not rise or set as seen from a given location, but circles around the celestial pole. To be circumpolar, an object must have a polar distance that is less than the observer's latitude. At the Earth's pole all objects above the horizon are circumpolar, but at the equator no object is circumpolar.

circumstellar maser A maser found in the cool dusty envelope of a red giant star that is undergoing mass loss. Masers of different molecules lie in different zones around the star. Stars with oxygen-rich envelopes have circumstellar OH, H_2O, and SiO masers, while stars with carbon-rich envelopes have HCN masers. The *OH–IR sources have oxygen-rich envelopes with a characteristic twin-peaked OH maser at 1612 MHz and strong infrared emission from the dusty envelope.

circumstellar matter A general term for material, usually gas and dust, surrounding a star. After a star's birth, the surrounding cloud absorbs the visible light, re-emitting it as infrared radiation. Most young stars go through a violent phase, ejecting material at speeds of hundreds of kilometres per second. When this ejected material collides with the surrounding gas, bipolar molecular outflows (*see* BIPOLAR OUTFLOW) and *Herbig–Haro objects can result. If the star is hot its ultraviolet radiation ionizes the gas, producing an *emission nebula. These combined effects destroy the cloud, so that the star becomes visible. For most of the star's life a tenuous *stellar wind blows. At the end of a star's life it ejects mass copiously. Depending on the mass of the star and the stage in its evolution, this may result in a cool circumstellar envelope, a *planetary nebula, or even a *supernova remnant.

Cirrus Nebula Alternative name for the *Veil Nebula.

CIRS Abbr. for *Celestial Intermediate Reference System.

cislunar Between the Earth and the Moon.

civil time Time as defined for everyday purposes by civil governments. It depends on the longitude of the country or region (*see* ZONE TIME), but ultimately it is derived from *Coordinated Universal Time.

civil twilight The period before sunrise and after sunset when the centre of the Sun's disk is less than 6° below the horizon. It is regarded as the time during which outdoor activities such as sports can still take place without the need for artificial illumination. *See also* TWILIGHT.

civil year *See* CALENDAR YEAR; GREGORIAN CALENDAR.

Clairaut, Alexis Claude (1713–65) French mathematician and physicist. In 1736 he took part in an expedition to Lapland to measure the length of a degree of latitude; the result proved that the Earth is an oblate spheroid. The next year he provided a rigorous theoretical explanation for the aberration of starlight, which J. *Bradley had been unable to do. He used his own solution of the *three-body problem to derive an accurate prediction for the date of perihelion passage of Halley's Comet on its return in 1758/9.

Clark, Alvan (1804–87) American optician. His company at Cambridge, Massachusetts, made the lenses for the world's two largest refractors, the 36-inch (0.91-m) at *Lick Observatory and the 40-inch (1.01-m) of *Yerkes Observatory. In 1862 his son Alvan Graham Clark (1832–97) discovered the first white dwarf, the companion of Sirius, while testing an 18½-inch (0.47-m) Clark refractor.

classical Cepheid A Population I Cepheid variable, sometimes known as a *Delta Cephei star*; abbr. DCEP. In contrast to the superficially similar W Virginis stars of Population II, classical Cepheids are massive (5–15 solar masses), young bright giants or supergiants found exclusively in the disk population of galaxies, where they are often members of open clusters. The periods are 1–135 days, with amplitudes of 0.5–2 mag. The stars exhibit well-defined *period–luminosity and *period–luminosity–colour relations that are the basis of a widely used method of determining extragalactic distances.

clast A fragment of a rock or mineral that is included within another rock. *Breccias are clastic rocks composed of angular, broken rock fragments embedded within a finer-grained matrix.

clathrate A structure in which molecules of one substance (the *guest*) are physically trapped in cavities within the crystal lattice of another (the *host*), without specific chemical bonding between them. The noble gases and some hydrocarbons form clathrates with water, occupying cavities in an open ice structure; they are called **clathrate hydrates**. Cometary nuclei may contain ice in the form of clathrate hydrates.

CLEAN A technique widely used in radio astronomy to enhance radio maps. It consists of a series of steps to identify and remove the effect of distortion caused by unwanted *side lobes. The distortion (i.e. the side-lobe pattern) can usually be calculated or measured for a particular telescope or interferometer. The technique is particularly useful for producing good maps from *aperture synthesis observations in which only a fraction of the total aperture has been synthesized. The raw, or dirty, maps contain numerous false features caused by the side lobes. The CLEAN technique recognizes true features and separates them from the side-lobe pattern.

Clementine A space probe built by the US Naval Research Laboratory to test sensors originally developed by the Ballistic Missile Defense Organization for detecting and tracking missiles. It was launched in 1994 January and put into orbit around the Moon, where it mapped the surface at various wavelengths with a set of four cameras. It was then put into geocentric orbit, but an intended rendezvous with asteroid *Geographos was abandoned when the spacecraft accidentally used up its control gas.

(⊕) SEE WEB LINKS
• Official mission website.

Clerk Maxwell, James *See* MAXWELL, JAMES CLERK.

C line A *Fraunhofer line in the Sun's spectrum at 656.3 nm, due to absorption by hydrogen. It is better known as the Hα line in the *Balmer series.

clock drive *See* DRIVE.

clock star A bright star in the equatorial region of the sky with precisely known right ascension, for determining the error of clocks used to time meridian transits.

closed universe A universe which is finite in size, has a finite lifetime, and in which space is positively curved. A *Friedmann universe with density greater than the *critical density is an example. *See also* CURVATURE OF SPACETIME.

Clown Face Nebula Alternative name for the *Eskimo Nebula, NGC 2392, a planetary nebula in Gemini.

Cluster A fleet of four ESA satellites to investigate the interaction between the solar wind and the Earth's magnetic field. The first two, named Salsa and Samba, were launched in 2000 July, followed in August by Rumba and Tango. The four spacecraft are on highly elliptical polar orbits, extending out to 119 000 km from Earth, taking them in and out of the Earth's magnetosphere. Separation between the spacecraft ranges from a few hundred to 20 000 km, enabling them to measure features in the magnetosphere at various scales.

(((⊕))) SEE WEB LINKS
• ESA mission website.

cluster, star *See* GLOBULAR CLUSTER; OPEN CLUSTER; STAR CLUSTER.

cluster of galaxies An aggregation of galaxies, which may or may not be bound together by gravity. For example, our Galaxy, the Milky Way, is a member of the *Local Group, a rather small cluster of which the only other large member is the Andromeda Galaxy. At the other extreme are the *Abell clusters which contain many hundreds or even thousands of galaxies in a region just a few million light years across; prominent nearby examples are the *Virgo and *Coma Clusters. Between these two extremes, galaxies appear to be clustered in systems of varying density. The densest Abell clusters are held together by their own self-gravity. Such *rich clusters* are filled with X-ray emitting gas, at temperatures of up to 10^8 K, and tend to have giant elliptical galaxies at their centre. The less rich and more extended systems may not be bound by gravity. *See also* LARGE-SCALE STRUCTURE.

cluster variable *See* RR LYRAE STAR.

CMB Abbr. for *cosmic microwave background.

CME Abbr. for *coronal mass ejection.

CM relation Abbr. for *colour–magnitude relation.

CN band An absorption feature in a star's spectrum due to the cyanogen molecule, CN, which is commonly seen in cool, late-type stars, especially in red giants.

CN cycle Abbr. for *carbon–nitrogen cycle.

CNO cycle Abbr. for carbon–nitrogen–oxygen cycle (*see* CARBON–NITROGEN CYCLE).

CN star A giant star of spectral type G or K that has abnormally strong cyanogen (CN) absorption bands in its spectrum.

Coalsack A conspicuous dark nebula in the southern constellation Crux, easily visible to the naked eye as a dark patch against the brighter background of the Milky Way. It lies about 600 l.y. away and covers nearly $7° \times 5°$ of sky. Its true diameter is about 60 l.y., and its mass is about 3500 solar masses. An open cluster, the *Jewel Box, is visible next to the Coalsack, but is actually far more distant. The *Northern Coalsack* is another name for part of the *Great Rift. 📷

coaltitude Another name for *zenith distance.

COAST Abbr. for *Cambridge Optical Aperture Synthesis Telescope.

Coathanger A star cluster in the constellation Vulpecula, also known as *Brocchi's Cluster or Collinder 399. It consists of a group of ten stars of 5th, 6th, and 7th magnitudes that form a shape like a coathanger. The stars are not actually related, but lie in the same line of sight by chance.

COBE Abbr. for *Cosmic Background Explorer.

Cocoon Nebula The diffuse nebula IC 5146 in Cygnus, a complex of light and dark nebulosity surrounding a sparse cluster with a 10th-magnitude star.

cocoon star A star surrounded by a dense cloud of gas and dust which absorbs some of the star's radiant energy and re-emits it at infrared wavelengths. In extreme cases the star may be completely obscured optically, leaving only an infrared source. *OH–IR sources are examples of cocoon stars.

CoD Abbr. for *Córdoba Durchmusterung.

coded mask A method of recording an image of the sky at high X-ray and gamma-ray energies where reflecting telescopes are unable to operate. A patterned mask similar to a crossword grid is used to cast a shadow of the celestial X-ray or gamma-ray emission on to a detector. From the shadow, an image can be reconstructed in a computer by the process of *deconvolution.

coelostat A mirror system that reflects light from a chosen part of the sky into a fixed instrument, such as a telescope, with no rotation of the field of view. A single mirror, driven to turn once in 48 hours, will do this for a single declination only. To observe objects at a wide range of declinations, another mirror is needed to feed the light from the driven mirror into the telescope. This second mirror's position and tilt are altered to point at different declinations. Such an arrangement is useful with large or delicate apparatus which can therefore be mounted in a fixed position. *See also* HELIOSTAT; SIDEROSTAT.

coesite A rare mineral produced when the shock wave from a meteorite passes through rock containing quartz. It is a form of silica produced at temperatures of 450–800°C and a pressure of 38000 bar. Coesite-bearing sandstone fragments found in the vicinity of large craters are evidence that the crater is of meteoritic origin. It is named after the American chemist Loring Coes Jr (1915–78) who first synthesized it in 1953. *See also* STISHOVITE.

coherence A measure of the extent to which electromagnetic waves are in step with one another. On an astrophysical scale, *maser sources are the most notable emitters of coherent radiation.

coherence bandwidth The range of frequencies over which electromagnetic waves from a given source remain in step with one another (i.e. are *coherent*). Waves with a narrow coherence bandwidth will remain in step with one another for a longer time and over a longer distance than waves with a broad coherence bandwidth.

colatitude The latitude of a body subtracted from 90° (i.e. the *complement* of the latitude).

cold camera A camera specially constructed to cool the film or plate during an exposure, in order to reduce *reciprocity failure; also known as a *cooled camera*. Temperatures down to −78°C are achieved by keeping dry ice (solid carbon dioxide) in contact with the back plate of the camera, against which the film is pressed. A thick window is needed to insulate the film from the atmosphere. Alternatively, the insulation may be provided by a chamber with an optical window, either evacuated or filled with dry nitrogen gas. Some types of *CCD camera are used with nitrogen gas to increase sensitivity and reduce noise.

cold dark matter (CDM) A particular type of *non-baryonic matter that, according to some theories, is created in the early stages of the Big Bang and survives to the present time in sufficient numbers to contribute significantly to the present density of the Universe. The term 'cold' signifies that these particles move at speeds much less than that of light, usually because they are heavy. There are many possible candidates for cold dark matter, such as *axions, *photinos, and primordial (low-mass) black holes. Evidence for the existence of cold dark matter has come from recent observations that show the overall amount of *dark matter in the Universe is roughly 50 times greater than the amount of matter contained in stars. Other observations show that roughly 80% of this dark matter is non-baryonic.

collapsar An obsolete term for a very massive star that has collapsed to form a black hole or a neutron star.

colles Small hills or knobs on a planetary surface (used only in the plural form; sing. *collis*). The name is not a geological term, but is used in the nomenclature of individual features, for example Oxia Colles on Mars.

collimation The act of making a beam of light parallel by a suitable arrangement of lenses or mirrors (a **collimator**), or of bringing the components of a system, particularly a telescope, into correct alignment. In high-energy astrophysics (X-ray and gamma-ray astronomy), collimation is often used to restrict the field of view of non-imaging instruments to improve the accuracy of source location and reduce source confusion; the collimator is then made of a material which is opaque to the radiation in question, such as glass or stainless steel.

collimation error The error in alignment between the *optical axis of a telescope and the declination axis of its equatorial mount; the two should be exactly perpendicular if the telescope is to track stars precisely.

collimator A device, usually a lens, for producing a parallel beam of light. In a *spectroscope, the collimating lens collects light from the slit, which is at its focal point, and passes it as a parallel beam on to the full width of the prism or grating. Alternatively, a collimator may produce a parallel beam of light for testing an instrument. *See also* COLLIMATION.

Colombo Gap A division about 100 km wide near the inner edge of the C ring of Saturn, 77 800 km from the planet's centre. The narrow and non-circular Colombo (or Titan) Ringlet lies within it. It was discovered on images from Voyager 2 in 1981 and is named after the Italian astronomer Giuseppe Colombo (1920–84).

colorimetry An obsolete term for the measurement of *colour index.

colour A shortened form of the term *colour index.

colour excess (*E*) The difference, due to *interstellar absorption, between the observed colour of a star and its *intrinsic colour index. Interstellar absorption decreases towards longer wavelengths, so its effect is always to make a star appear redder; colour excesses are therefore always positive. The symbol for the colour excess in B − V is $E_{(B - V)}$, and similarly for other colours.

colour index The difference between the apparent magnitudes of a star at two different wavelengths, as measured through filters of different colours, e.g. B (blue) and V (yellow-green). B − V and U − B are typical colour indices. In *Johnson photometry and *Kron–Cousins RI photometry, all colour indices are made zero for a star of spectral classification A0V (e.g. Vega). Generally, hotter stars have negative colour indices, and cooler stars positive ones. Nowadays colour index is usually shortened to 'colour'.

colour–luminosity relation *See* COLOUR–MAGNITUDE RELATION.

colour–magnitude relation The correlation between *colour index and *absolute magnitude for stars on the lower main sequence; also known as **colour–luminosity relation**. If the absolute magnitudes of stars intrinsically fainter than the Sun are plotted against colour index, there appears a tight relationship which defines the *main sequence. The slope of the relationship depends on the colour index used. If B − V is used for the colour then the relationship is much steeper than for V − I.

colour temperature The temperature of a *black body which has the same *colour index as a star. Because stars do not radiate as perfect black bodies, the colour temperature differs according to which colour index is used, for example B − V or U − B.

Columba (Col) (gen. Columbae) A constellation of the southern sky, representing a dove. Its brightest star is Alpha Columbae (Phact), magnitude 2.7.

column density The amount of material in an imaginary cylinder (usually of cross-sectional area $1 \, cm^2$) between an observer and an astronomical object. Column densities in our Galaxy are usually in the range 10^{19}–10^{23} atoms/cm^2.

colure A great circle that passes through the celestial poles and cuts the ecliptic at either the equinox points (the *equinoctial colure*) or at the solstice points (the *solstitial colure*). The colures are *hour circles.

coma, cometary The envelope of gas and dust that surrounds the solid nucleus of an active comet (*see* NUCLEUS, COMETARY). The coma often appears as a teardrop, being largely shaped by the solar wind flowing around the comet. Near perihelion, the coma may be 100000km wide. A coma does not usually form until the comet is within 3–4 AU of the Sun. Coma production has, however, been recorded from *Chiron at over 11 AU.

coma, optical A defect of an optical component or system in which star images become increasingly fan-shaped away from the optical axis; the shape resembles the coma of a comet, hence the name. In *positive coma*, light passing through the outer zones of the optical system forms a disk rather than a point image. The size and position of the disk depend on the distance of each zone from the centre of the lens, so the images from different zones overlap to form a fan pointing towards the centre of the field. With *negative coma*, the fan points away from the field's centre. The effect is usually combined with other aberrations, such as *astigmatism and*chromatic aberration, which distorts the fan shape.

Coma Berenices (Com) (gen. Comae Berenices) A faint but distinctive constellation of the northern sky, representing the hair of the Egyptian Queen Berenice. It is notable for a scattering of faint stars that make up the *Coma Star Cluster. The constellation's brightest star is Beta Comae Berenices, magnitude 4.2. Coma Berenices contains a number of galaxies which are members of the *Virgo Cluster, as well as a considerably more distant cluster of galaxies, the *Coma Cluster. Its two most famous galaxies, M64 (the *Black Eye Galaxy) and NGC 4565, a 10th-magnitude spiral seen edge-on, are both about 20 million l.y. away, closer to us than either cluster.

Coma Cluster A large, regular cluster of more than 3000 galaxies, about 280 million l.y. away in Coma Berenices; also known as Abell 1656. It covers more than 4° of sky and has a true diameter of at least 20 million l.y. Most of the bright members are ellipticals or lenticulars. Two very bright giant ellipticals, NGC 4874 and NGC 4889, lie either side of the centre; these galaxies have about 10^{11} solar luminosities. The total mass of the cluster is estimated at 2×10^{15} solar masses.

Coma Star Cluster The open cluster Melotte 111 in Coma Berenices consisting of about 50 stars, the brightest of 4th and 5th magnitudes, scattered over several degrees of sky. The cluster is centred about 285 l.y. away.

Coma–Virgo Cluster *See* VIRGO CLUSTER.

Combined Array for Research in Millimeter-wave Astronomy (CARMA)

A millimetre-wave radio telescope created by combining the antennae of the former Owens Valley Radio Observatory's millimetre array and the Berkeley–Illinois–Maryland Association (BIMA) Array on a new site at an altitude of 2440m at Cedar Flat, in the Inyo Mountains of eastern California. The CARMA array consists of six antennae of 10.4 m (from the Owens Valley array) and nine of 6.1 m (from the BIMA array), distributed in a pseudo–random pattern giving baselines up to 2 km. It came into full operation in 2007. In 2008 these were joined by eight 3.5-m antennas belonging to the University of Chicago, completing a 23-element interferometer.

(((⊕))) **SEE WEB LINKS**

• Official telescope website.

combined magnitude The total brightness of two (or more) celestial objects, such as a close pair of stars that appear as one to the naked eye. Since the magnitude scale is logarithmic, the combined magnitude m is not simply the sum of the individual magnitudes m_1 and m_2, but must be found from the formula

$$m = m_1 - 2.5 \log \{1 + \text{antilog}[-0.4 (m_2 - m_1)]\}$$

For example, the combined magnitude of two stars each of magnitude 1 is 0.25.

comes The fainter component of a double star; pl.*comites*. The word is Latin for 'companion'.

comet A small body, composed of ice and dust, in orbit around the Sun. The name derives from the Greek *kometes*, meaning 'long-haired'. Comets are thought to exist in vast numbers in the *Oort Cloud and*Kuiper Belt, beyond the planets. From there they can be perturbed by the gravitational influence of passing stars into new orbits that bring them into the inner Solar System, where they become visible from Earth. When a comet is far from the Sun its nucleus is frozen solid and shines only by reflecting sunlight. As the nucleus nears the Sun it heats up and releases gas and dust, forming first a coma and, in some cases, a tail (*see* COMA, COMETARY; NUCLEUS, COMETARY; TAIL, COMETARY). The gas becomes ionized and emits light. Whereas the nucleus may be only 1 km or so across, the coma can extend for 10^5 km or more from the nucleus and the tail for 10^8 km}. Around the visible coma is an even larger cloud of hydrogen, detectable at ultraviolet wavelengths. Despite their size, a comet's coma and tail are of such low density that background stars can be seen through them. The mass of a typical comet is perhaps 10^{14} kg.

Each year over 200 comets are seen with telescopes and space satellites; only a few ever become bright enough to be visible with the naked eye. Most are new *long-period comets appearing for the first time, with orbital periods of over 200 years. The remainder are *periodic comets, either new discoveries or known objects following predicted orbits. The most famous of these, and the brightest, is *Halley's Comet. At the end of 2010 over 3000 comets were known, of which over 90% are long-period comets. During their passage through the inner Solar System comets can have their orbits altered by the gravitational influence of the planets, notably Jupiter. One spectacular example was Comet *Shoemaker–Levy 9, which hit Jupiter in 1994.

Some comets are discovered by amateur astronomers conducting deliberate searches, but most are found on images taken by professional astronomers; recently, over a hundred comets a year passing close to the Sun have been found on images taken by the *Solar and Heliospheric Observatory (SOHO) (*see* KREUTZ SUNGRAZER; SUNSKIRTER). Comets are named after their discoverers (now usually restricted to two names), or the spacecraft or survey which found them, and are also assigned a designation based on when they were discovered. According to a convention introduced in 1995, comets are identified by the year and a letter indicating the half-month in which they were discovered, plus the order of discovery in that half-month (e.g. C/1999 D3 would be the third comet discovered during the second half of 1999 February). The names of periodic comets are preceded by P/ and a number indicating the order in which their periodicity was established (e.g. 1P/Halley, 2P/Encke). Comets that are defunct—either observed to have disintegrated or simply disappeared—are given the prefix D/ (e.g. 3D/Biela, D/1993 F2 Shoemaker–Levy). Comets for which there are insufficient observations to calculate an orbit are given the prefix X/.

In recent years, comets have been extensively investigated by space probes. The first photographs of a cometary nucleus were taken in 1986 when the Russian *Vega probes and the European Space Agency's *Giotto flew past Halley's Comet. In 2005 NASA's *Deep Impact probe collided with a cometary nucleus and in 2006 *Stardust brought the first cometary samples back to Earth. In 2014 the European probe *Rosetta is due to go into orbit around the nucleus of comet Churymov-Gerasimenko.

Comets are believed to be icy planetesimals left over from the formation of the outer planets. The total population of the Oort Cloud and Kuiper Belt may be 10^{12} objects, with a combined mass greater than the Earth. The main component of cometary ice is frozen water, plus some methane (CH_4), carbon monoxide (CO), and carbon dioxide (CO_2). Several other carbon-containing molecules have also been detected, including formaldehyde (H_2CO), hydrogen cyanide (HCN), and methyl cyanide (CH_3CN). These same molecules are also found in interstellar nebulae, similar to the nebula from which the Solar System formed. Small (less than 1 mm) dust particles released from comets around perihelion contribute to the inner Solar System's *zodiacal dust cloud. Larger dust particles, of millimetre and centimetre size, from periodic comets give rise to meteor streams. 📷

cometary globule A small, dark nebula with a bright-rimmed head and a long diffuse tail, resembling a comet in appearance. The bright rim is the result of ionization. The tail may be up to 10 l.y. long, and it often points either towards or away from a powerful young star with which the nebula is associated.

comet family A group of comets which possess similar orbital characteristics, either because of a common origin or as a result of planetary perturbations. Members of some families, for example the *Jupiter comet family, do not have a common origin, and may have been captured from a wide range of original orbits. It was once believed that the gravitational influences of Saturn, Uranus, and Neptune also produced comet families, but the apparent association of comets with these planets is now thought to result from *resonances with Jupiter's orbital period. Other families, for example the *Kreutz sungrazers, clearly have a common origin from the break-up of a precursor body.

comet-seeker A low-power optical instrument with a small focal ratio, offering high-contrast and wide-field views. Large-aperture binoculars or telescopes with these characteristics are ideal for sweeping the sky for comets. Reflectors with suitable optics for comet-hunting and observing are sometimes called *rich-field telescopes.

commensurability The circumstance in which the ratio of the orbital periods of two bodies, such as satellites or planets, is an exact fraction such as one-half or two-thirds. In the Solar System there are many near-commensurabilities in orbital periods. For example, the orbital periods of Jupiter and Saturn have a ratio of nearly two-fifths, while the Saturnian satellite pairs Dione and Enceladus, and Mimas and Tethys, have ratios of almost one-half. Such situations give rise to *resonances. In some cases the commensurabilities provide stability, while in others they lead to instability, as with the *Kirkwood gaps in the asteroid belt where asteroids tend to avoid orbits in which they would have periods commensurable with that of Jupiter.

Common, Andrew Ainslie (1841–1903) English telescope-maker and astrophotographer. He made several large reflecting telescopes, including the 36-inch (0.91-m) Crossley Reflector later acquired by Lick Observatory. From 1880 he used another 36-inch instrument for photography, obtaining the first good pictures of the comet now designated C/1881 K1, and the first photograph of the Orion Nebula to record its filamentary structure. This, and other time exposures of subjects such as Jupiter and Saturn, were made possible by the accurate drive mechanism that he installed on the telescope.

common envelope binary A binary star in which two stellar cores are immersed in an extremely large, common envelope of gas. All close binaries that contain white dwarfs, neutron stars, or black holes have gone through such a process. These range from the relatively common *cataclysmic variables that feature a white dwarf to the more exotic objects such as X-ray binaries containing neutron stars or presumed black holes. In addition, some merged objects such as the *FK Comae Berenices stars could be products of the same process. The envelope arises when one member of the binary becomes a red giant and fills its *Roche lobe. Mass transfer leaves the core of the giant and the relatively denser companion deep inside a common envelope. Friction then causes the cores to spiral towards each other. The angular momentum and energy they lose in the process is transferred to the envelope, which then spins up and sheds mass in response. The phase comes to an end either when the whole envelope has been lost, leaving an exposed compact binary, or the two cores merge within the rapidly spinning envelope.

common proper motion (c.p.m.) The circumstance in which two or more stars move together through space; that is, they have the same *proper motion. This is an indication that they may form a true binary or multiple star, even if no orbital motion can be detected.

compact galaxy A type of galaxy which can only just be distinguished from stars on Schmidt camera sky survey plates. They have apparent diameters of 2–5″ and a region of high surface brightness that can be due either to a bright nucleus or a region of active star formation. Some 2000 such objects were catalogued by F. *Zwicky from Palomar Observatory Sky Survey plates.

compact object The small and dense end-product of stellar evolution–a white dwarf, neutron star, or black hole.

compact source In radio astronomy, a class of extragalactic sources of small angular size displaying no extended structure and tending to be brightest at short wavelengths. *See also* EXTENDED SOURCE (2).

comparator A device for comparing two images of the same part of the sky taken at different times, to search for objects which have changed or moved. The main types are the *blink comparator and *stereo comparator.

comparison spectrum A reference spectrum containing lines of accurately known wavelengths which can be used to calibrate astronomical spectra. A comparison spectrum of emission lines is usually produced in an arc-lamp attached to the spectrograph, and is recorded immediately before or after the spectrum of the object under study. Alternatively, absorption lines can be superimposed on the spectrum during the exposure using an *iodine cell.

composite-spectrum binary A *spectroscopic binary in which the components are of dissimilar spectral types. *See also* DOUBLE-LINED BINARY; SINGLE-LINED BINARY; SYMBIOTIC STAR.

Compton effect The change in wavelength and direction of an X-ray or gamma-ray photon when it collides with a particle, usually an electron; also known as **Compton scattering**. Some of the photon's energy is transferred to the particle and the photon is reradiated at a longer wavelength. The effect is used to detect gamma rays by converting the photon to one of lower energy, which can be more easily detected by, for example, a *proportional counter. The effect was discovered in 1923 by the American physicist Arthur Holly Compton (1892–1962). *See also* INVERSE COMPTON EFFECT.

Compton Gamma Ray Observatory (CGRO) A NASA satellite launched in 1991 April to observe gamma rays from celestial objects. CGRO was the second of the series of *Great Observatories. It was deorbited in 2000 June. The satellite carried four instruments covering the energy range 10 keV to 30 GeV (4×10^{-8} to 0.124 nm). The BATSE (Burst and Transient Source Experiment) instrument studied gamma-ray bursts, while OSSE (Oriented Scintillation Spectrometer Experiment), Comptel (Imaging Compton Telescope), and EGRET (Energetic Gamma Ray Experiment Telescope) recorded images of the gamma-ray sky in the 0.1–10 MeV (1.2×10^{-4} to 0.012 nm), 1–30 MeV (4×10^{-5} to 1.24×10^{-3} nm), and 0.02–30 GeV (4×10^{-8} to

conic section

6×10^{-5}nm) ranges respectively. The satellite was named after the American physicist Arthur Holly Compton (1892–1962).

 SEE WEB LINKS
• Official mission website.

Compton–Getting effect A small deviation from uniformity in the distribution of cosmic rays across the sky, caused by our Solar System's motion around the Galaxy at 200km/s. More cosmic rays should be observed in the direction of travel, but the effect is below the sensitivity of current instruments. It is named after the American physicists Arthur Holly Compton (1892–1962) and Ivan Alexander Getting (1912–2003), who discussed it in 1935.

Compton scattering Another name for the *Compton effect.

Compton upscattering Another name for *inverse Compton scattering.

Concordia family A small but well-defined *Hirayama family of asteroids in the main asteroid belt, mean distance 2.70–2.75AU from the Sun. Its members have orbital inclinations of 4–6°. There are only small differences between the C-class spectra of the three largest family members, (58) Concordia, (128) Nemesis, and (210) Isabella, but another allocated member, (340) Eduarda, has an S-class spectrum and may not be a true family member. Concordia itself, 93km in diameter, was discovered in 1860 by the German astronomer (Karl Theodor) Robert Luther (1822–1900). Concordia's orbit has a semimajor axis of 2.700AU, period 4.44 years, perihelion 2.58AU, aphelion 2.82AU, and inclination 5°.1.

Cone Nebula An elongated dark nebula that is part of the nebulosity surrounding the open cluster NGC 2264 in Monoceros.

confusion A phenomenon in which the number of sources in a given region of sky is so great that it is difficult to resolve individual objects. A sky survey is said to be **confusion-limited** if the brightness of the faintest sources that can be reliably resolved is determined not by the characteristics of the instrument but by confusion from fainter sources.

conic section A figure obtained by slicing a cone. There are four different types of conic section. If the cone is cut perpendicular to its axis, the resulting figure is a circle. If the cut is not perpendicular to the axis but still produces a closed curve, the curve is an ellipse. If the cone is cut parallel to one of its sloping sides, the resulting curve is a parabola, which is not closed. If the angle of cut is tilted still further, the open figure obtained is a hyperbola. An ellipse has an *eccentricity less than 1; a circle is a special case of an ellipse, where the eccentricity is 0. A parabola has an eccentricity of exactly 1. A hyperbola has an eccentricity greater than 1. The orbits of celestial bodies are conic sections.

| Circle | Ellipse | Parabola | Hyperbola |

conic section a

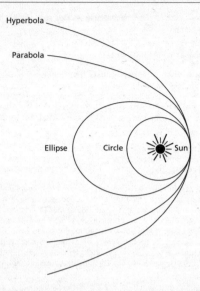

conic section b

conjunction The occasion when two Solar System bodies have the same *celestial longitude as seen from Earth; the bodies may be a planet and the Sun, two planets, or the Moon and a planet. Superior planets (i.e. those with orbits outside that of the Earth) are at conjunction when they lie directly behind the Sun as seen from Earth. Mercury and Venus, which are closer to the Sun, have two conjunctions; *inferior conjunction*, when they lie between the Earth and Sun; and *superior conjunction*, when they lie on the far side of the Sun.

constant of aberration *See* ANNUAL ABERRATION.

constellation Any of the 88 areas into which the celestial sphere is divided for the purposes of identifying objects, as adopted by the International Astronomical Union in 1922 (see Table 3, Appendix). In 1930 the IAU went on to adopt official constellation boundaries, defined by lines of right ascension and declination for the epoch 1875. This epoch was chosen because it had already been used by B. A. *Gould to draw boundaries for the southern constellations. The IAU boundaries, defined by the Belgian astronomer Eugène Joseph Delporte (1882–1955), were published in *Délimitation Scientifique des Constellations* (1930) and the related *Atlas Céleste* (1930).

The brightest stars in a constellation are identified by a Greek letter (the system of *Bayer letters) or by a number (its *Flamsteed number). When referring to stars identified in this way, the genitive case of the constellation's name is always used, as in Alpha Orionis, 61 Cygni, or Zeta Ursae Majoris. Three-letter abbreviations of constellation names, as laid down by the IAU, are also frequently encountered (e.g. Ori, Cyg, or UMa).

The constellations officially recognized today are based on a group of 48 Greek figures listed by *Ptolemy in the 2nd century AD, with subsequent additions by various others. At the end of the 16th century two Dutch navigators, Pieter Dirkszoon Keyser (c.1540–96) and Frederick de Houtman (1571–1627), created twelve new constellations in the far southern part of the sky, below the horizon to Greek astronomers. The Dutch celestial cartographer Petrus Plancius (1552–1622), the Latinized form of Pieter Platevoet, added three more constellations in the spaces between those known to the Greeks, and he separated the stars of Crux from Centaurus.

The northern constellations that we know today were completed by J. *Hevelius in 1687, who introduced several new figures, seven of which are still recognized today. In the 1750s the southern sky was filled out with fourteen new constellations by N. L. de *Lacaille, who also split the large Greek figure Argo Navis into three parts.

 SEE WEB LINKS

• International Astronomical Union official page on the constellations.

constructive interference *See* INTERFERENCE.

contact binary A close binary star in which both components fill their *Roche lobes. When both components have just reached the bounding *equipotential surface they are sometimes described as forming a *double contact binary*. More often, both components exceed the equipotential surface and share a common convective envelope. This configuration is sometimes described as an *overcontact binary. Further expansion would carry the envelope to the equipotential surface that contains the outer (L_2) *Lagrangian point, through which material would then be lost from the system.

continuous creation A hypothetical process which is a necessary part of the *steady-state theory of cosmology. If the Universe is expanding, then its density would decrease with time unless matter were somehow created to fill the increased volume. Although such a creation process has never been observed, the rate of creation required is so small (about 1 atom of hydrogen per cubic metre over the age of the Universe) that it is difficult to rule out by direct observations.

continuous spectrum A featureless, unbroken emission spectrum across a range of wavelengths, normally produced by an incandescent solid or a hot gas. Astronomical examples of continuous, or near-continuous, spectra include those of *BL Lacertae objects and white dwarfs of type DC.

continuum A featureless spectrum, or region of the spectrum between absorption or emission lines. Measurement of the strengths of such lines requires accurate determination of the strength of the continuum on either side of the line.

convection, stellar The transport of energy inside stars by currents of hot gas. It is not fully understood, the main uncertainty in the current theory being the so-called *mixing ratio, which expresses the size of the convective cells.

convective envelope The outer region of a star in which heat is transported mainly by convection. Only stars with masses up to and a little beyond that of the Sun have a convective envelope, in which the gas is cool enough to be only partially ionized. In such a region, as rising gas cools, its electrons and ions recombine, releasing heat which continues to drive the gas upwards. Only 1% of the Sun's mass is in its convective envelope, but in stars of half a solar mass the figure is over 40%.

convective equilibrium The state attained by a gas in which the rate of energy generation (such as by nuclear reactions in stars) is exactly balanced by the rate at which energy is transported outwards by convection, so that the gas remains at a constant temperature. *See also* RADIATIVE EQUILIBRIUM.

convective overshoot *See* OVERSHOOTING.

convective zone The region within a star in which energy is transported mainly by convection currents of gas. Such zones occur in the cores of massive stars and the surface layers (*convective envelopes) of low-mass stars. Stars of less than 0.4 solar masses are entirely convective.

convergent point The point in the sky towards which the proper motions of stars in a *moving cluster appear to converge.

converging lens A lens that is thicker at the centre than at the edges, so that parallel light passing through it converges to a focus; also known as a *positive lens*. A magnifying glass is an example. A converging lens produces a *real image*—one that can be seen on a screen. *See also* DIVERGING LENS.

convolution A mathematical operation used in signal processing in which one set of data is *smoothed* (averaged) by multiplying it by another. It is often used to represent the blurring of an astronomical image due to the performance limitations of the telescope or the distorting effects of the atmosphere. *See also* DECONVOLUTION.

Cooke, Thomas (1807–68) English optician. His largest telescope was the 25-inch (0.64-m) Newall refractor, at the time the largest in the world, made in 1868 for Robert Stirling Newall (1812–89), a wealthy English amateur astronomer. In 1890 Newall gave it to Cambridge University, who in 1957 donated it to the National Observatory of Athens, Greece. In 1922 the Cooke firm merged with an instrument company set up by Edward Troughton (1753–1835) and William Simms (1793–1860).

cooled camera *See* COLD CAMERA.

cooling flow A mechanism thought to occur in clusters of galaxies in which hot gas, lost by the galaxies through supernova explosions and stellar winds, cools by emitting X-rays and sinks on to a massive elliptical galaxy at or near the cluster's centre. As the gas cools it may form into low-mass stars, enhancing the mass of the giant elliptical.

coorbital Describing two or more bodies, such as satellites of a planet, which share the same or a similar orbit. There are two possible circumstances in which this can happen. In one, the bodies can be separated in the same orbit by 60° of longitude, so that they never come close, as with Saturn's satellite Tethys and its two smaller coorbital satellites Calypso and Telesto. Here, the smaller satellites lie at the *Lagrangian points in the orbit of Tethys. However, some coorbital bodies are in adjacent orbits with marginally different periods and hence can pass each other. This happens with Saturn's satellites Janus and Epimetheus, whose orbits are just 50km apart. These two satellites actually swap orbits when the inner satellite, moving faster than the outer one, overtakes the outer one and changes places with it.

Coordinated Universal Time (UTC) The time-scale that is provided by broadcast time signals, and is the basis for all civil timekeeping. The natural choice for this standard time would be *Universal Time (UT), but that contains unpredictable irregularities caused by changes in the rotation rate of the Earth. Consequently UTC is linked to *International Atomic Time (TAI), but an offset is introduced to keep it always within 0.9s of UT. This is done by introducing an occasional *leap second. The time-scale now widely known as GMT is in fact UTC.

coordinate systems *See* CELESTIAL COORDINATES.

coordinate time The coordinate representing a global measure of time in a relativistic theory of gravity, most usually the *general theory of relativity. This coordinate amounts to a universal way of labelling events. Each event has four coordinates: three in space and one in time. The coordinate time is useful because of its global definition, but it does not correspond to physical time. That will be different for different observers, who will measure their own *proper time. Strictly, the coordinate time is not uniquely defined as it depends on the choice of coordinates. However, the term is used in practice to mean the time coordinate in a standard coordinate system.

Copernican system A model of the Solar System proposed by N. *Copernicus in which the Sun lay at the centre with the planets orbiting around it; the stars lay at a vast distance beyond the planets. The model retained the circular orbits and epicycles of the *Ptolemaic system, but incorporated Copernicus's own observations. It also contained elements from variants of the Ptolemaic system proposed by the Arab astronomers al-*Ṭūsī and Ibn al-Shāṭir (1304–75), which Copernicus apparently knew about. In Copernicus's model the motion of the sky results from the Earth's axial rotation. The Copernican system reproduced planetary motion no better than the Ptolemaic system did, because the concept of elliptical orbits had not yet been introduced, and until that advance by J. *Kepler it found little acceptance. However, it was significant in removing the Earth from the centre of the Universe.

Copernicus, Nicolaus (1473–1543) Polish churchman and astronomer of German parentage (in Polish, Mikołaj Kopernigk). While a student he had come to believe that the Earth revolves around the Sun and not vice versa. Such a view was held to be heretical, the Church regarding the geocentric world-view of *Aristotle and *Ptolemy as consistent with its doctrines. Copernicus set down his basic ideas around 1510 in the *Commentariolus*, which was circulated anonymously. In 1512–29 he made the observations that he needed to support his theory, while carrying out ecclesiastic and local administrative duties. In 1539 the Austrian astronomer and mathematician Georg Joachim von Lauchen (1514–74), known as Rheticus, became a pupil of Copernicus and began to spread his ideas. The *Copernican system was published openly in 1543 in the book *De revolutionibus orbium coelestium*. However, the reality of a heliocentric Solar System came to be accepted only after the work of *Galileo and J. *Kepler.

Copernicus satellite A US astronomy satellite, also known as OAO-3, which carried a 0.81-m telescope for ultraviolet spectroscopy of hot stars at wavelengths of 70–300nm. Three small X-ray detectors were also carried. Copernicus was launched in 1972 August and operated until 1980. It studied interstellar matter, mainly by observing absorption features from interstellar gas superimposed on stellar spectra. Its name commemorated the 500th anniversary of the birth of Nicolaus Copernicus in 1973.

(⊕) SEE WEB LINKS
• Information page at Goddard Space Flight Center.

Coprates A dark, elongated albedo feature on Mars, visible in Earth-based telescopes, centred at $-15°$ lat., 65° W long. It corresponds to the *Valles Marineris canyon system, the eastern main valley of which is named Coprates Chasma. The feature was previously known as Agathodaemon.

Cor Caroli The star Alpha Canum Venaticorum, 115 l.y. away. It is a wide double star. The brighter component is an A0 bright giant of magnitude 2.9, the prototype of the Alpha^2 Canum Venaticorum variables (magnetic stars with variable spectra) and fluctuates by about 0.1 mag. in a period of 5.5 days. The companion is an F0 dwarf, magnitude 5.6.

Cordelia The innermost known satellite of Uranus, distance 49 750 km, orbital period 0.335 days; also known as Uranus VI. Cordelia is 26 km in diameter, and was discovered in 1986 on images from Voyager 2. Cordelia and *Ophelia are *shepherd moons for the Epsilon ring of Uranus.

Córdoba Durchmusterung (CoD) A catalogue of 613 953 stars down to 10th magnitude, from $-22°$ to the south pole, produced at the Córdoba Observatory, Argentina. It was the southern equivalent of the *Bonner Durchmusterung*. The first four volumes, by John Macon Thome (1843–1908), were published 1892–1914, and the fifth, by Charles Dillon Perrine (1867–1951), in 1932.

(⊕) SEE WEB LINKS
• Detailed description and full catalogue downloadable from the CDS.

core, planetary The densest, central part of a planetary body, differing markedly in composition from the layer above it. The core of the Earth is that part of the planet's interior over 2900 km or so below the surface. It consists of mainly iron–nickel alloy, and is further divided into a liquid *outer core* and a solid *inner core*.

core, stellar The central region of a star, in which hydrogen burning occurs while the star is on the main sequence. After the star has left the main sequence, the core is the site of further nuclear burning of other fuels such as helium and oxygen. About 30% of the mass of a star of 10 solar masses is in its core. In such massive stars the cores are entirely convective, ensuring that the region is well mixed and of uniform composition. Stars of less than 0.4 solar masses are convective in both their cores and their envelopes. Stars of 0.4–1.2 solar masses have radiative cores.

core collapse 1. The collapse of a star's core, as in the formation of a neutron star. The core collapses when it has evolved to the point where it consists entirely of iron, which cannot be burnt in nuclear reactions and hence produces no energy to support itself against gravity. The core then contracts, releasing gravitational potential energy which tears the nuclei of iron atoms into their constituent protons and neutrons. As the density rises, protons combine with electrons to form neutrons. Collapse is halted when the pressure from the *degeneracy of the neutron gas balances the inward pull of gravity. The entire process takes less than a second.

2. The collapse of the central part of a star cluster after energy is transferred to its outer parts through encounters between its constituent stars. The age of the cluster at which this should theoretically occur is, however, less than the true age of many globular clusters. The effects of encounters involving binary stars are thought to be important in averting core collapse.

Coriolis force An apparent force deflecting the motion of an object or a fluid moving over the surface of a rotating body such as a planet or star. For example, in the Earth's northern hemisphere objects are deflected to the right of their direction of travel, and in the southern hemisphere to the left. The effect applies to winds, water, and rocket launches. It is named after the French physicist Gustave Gaspard de Coriolis (1792–1843).

corona 1. An extremely hot (about 2 million K), highly ionized gas surrounding the Sun. Certain other stars also have coronae. The Sun's corona is visible at total eclipses as a white region extending out to several solar radii, displaying streamers, plumes, and bubbles or loops. The radiation of the *white-light corona has components due to emission lines (the *E corona), and scattering from electrons (the *K corona) and dust particles (the *F corona). The corona's outer extension is the *solar wind.

X-ray images of the Sun's corona show complex loop structures near sunspot groups, and near smaller *X-ray bright points. The X-ray emission, plus emission lines due to highly ionized atoms (*coronal lines), shows that the temperature is about 2 million K; even higher temperatures of 4 million K or more are found in *active regions. Magnetic fields, with a strength of about 10^{-3} tesla, govern the corona's shape. The magnetic fields form closed loops in active regions and much of the *quiet corona* (i.e. non-active regions), but in *coronal holes the magnetic field lines are open and stretch out into space, not returning to the Sun.

The energy that heats the corona is thought to come from motions in and below the solar photosphere. High-frequency photospheric motions create waves in the Sun's magnetic field that travel outwards into the corona. Conversion of the wave energy into heat can account for the high temperatures of the coronal gas. One method of dissipating the wave energy is if the wave resonates with the coronal gas. An alternative heating mechanism requires low-frequency photospheric motions to tangle up the magnetic lines of force in the corona. Eventually the stresses in the magnetic field are released through reconnection of field lines. The energy released in such events is expected to vary widely, from *flares to so-called *microflares* and even smaller *nanoflares*.

The appearance of the corona changes during the solar cycle. At solar maximum it consists of many active-region loops and streamers around the disk, but at solar minimum it is dominated by large coronal holes at each pole and a sheet-like structure near the equator.

Main-sequence stars cooler than spectral type F0 often have coronae with active regions, as indicated by their X-ray emission. This is particularly true of the M-type dwarf *flare stars. Coronae are also present in some interacting binary systems like the *RS Canum Venaticorum stars.

2. A large circular or elongated feature on a planetary surface, surrounded by concentric ridges; pl.*coronae*. The name, which means 'crown' or 'circle', is not a geological term, but is used in the nomenclature of individual features, for example Nightingale Corona on Venus, or Arden Corona on Uranus's satellite Miranda.

3. A region of tenuous, very hot gas extending out of the galactic plane in spiral galaxies such as the Milky Way; also called *galactic corona*. ▣

Corona Australis (CrA) (*gen.* **Coronae Australis**) A small southern constellation, popularly known as the Southern Crown. It is distinguished by an arc of faint stars, the brightest of which, Alpha and Beta Coronae Australis, are of magnitude 4.1.

Corona Borealis (CrB) (*gen.* **Coronae Borealis**) A small but distinctive constellation of the northern sky, popularly known as the Northern Crown. It features a prominent arc of stars, the brightest of which is Alpha Coronae Borealis (*Alphecca or Gemma). It contains two remarkable variables: R Coronae Borealis, the prototype of a group that undergoes sudden dips in light output (*see* R CORONAE BOREALIS STAR); and T Coronae Borealis, a recurrent nova known as the *Blaze Star.

coronagraph An instrument that enables the Sun's corona to be observed outside the times of total eclipses, invented by B. *Lyot in 1930. It is a form of refracting telescope with an objective lens kept free of dust and static charge, sometimes by means of a thin layer of oil. An *occulting disk is placed at the prime focus to form an artificial eclipse. A lens directly behind the occulting disk forms an image of the objective on a diaphragm, thereby removing most of the stray light from the objective. A third lens behind the diaphragm actually forms the image of the corona on film or a detector. Only observatories at high altitude with exceptionally clear atmospheric conditions are suitable sites for coronagraphs. Even then, only the inner part of the *E corona can be observed, although the *K corona can be imaged using polarization analysers. Spacecraft coronagraphs can observe the corona out to several solar radii, using electronic imaging instead of photography. The LASCO (Large Angle Spectroscopic Coronagraph) on the SOHO spacecraft observes the corona out to 30 solar radii.

coronal bright point One of the three key features of the Sun when seen in X-rays, the others being *active regions and *coronal holes. Unlike active regions, coronal bright points are distributed at all latitudes on the solar disk. Bright points have a central core around 10000 km wide and mostly occur above areas of opposite magnetic polarity on the photosphere; when the regions of opposite polarity encounter each other and cancel out, energy is released that heats the gas above the photosphere to 1–2 million K. Bright points also occur when newly emerged magnetic field reacts with the pre-existing magnetic field in the corona, again with the release of magnetic energy to heat the gas. Coronal bright points have typical lifetimes of a day. They are often referred to as *X-ray bright points* or *EUV bright points*, according to the wavelength at which they are observed.

coronal condensation A brighter area of the solar *corona in which the density and temperature of the gas is higher than its surroundings. Such areas are visible at the Sun's limb above sunspot groups and consist of loop structures which outline the corona's magnetic field. The term *active region is now preferred.

coronal hole A region in the Sun's corona of very low density, about 100 times less than that of coronal active regions. Coronal holes show up as apparent voids in X-ray images or, at the limb, by an absence of white-light emission in coronagraph images. A large coronal hole is always present at each of the Sun's poles. Low-latitude holes appear shortly before solar minimum, increasing in size over several months and sometimes coalescing with one of the polar holes. Other holes may contract and disappear. The magnetic field in coronal holes is in the form of open field lines stretching out into space, along which plasma flows to produce the high-speed streams in the *solar wind. 📷

coronal lines Emission lines in the spectrum of the Sun's corona caused by highly ionized atoms. Only a small number of coronal lines are emitted at visible wavelengths, and these can be studied from the ground only during eclipses or with *coronagraphs. The visible coronal lines are *forbidden lines (caused by transitions of very low probability in highly ionized atoms). Examples include the so-called red and green coronal lines at wavelengths of 637.5 and 530.3 nm due to Fe^{9+} (iron lacking 9 electrons) and Fe^{13+} (iron lacking 13 electrons), respectively, which were once thought to be caused by an unknown element, dubbed 'coronium'. Many more coronal lines are found at ultraviolet and X-ray wavelengths and can be detected only from space. These ultraviolet and X-ray lines have been observed not only in the Sun but also in a wide range of other stars with the Extreme Ultraviolet Explorer, the Chandra X-Ray Observatory, and XMM-Newton.

coronal loop A structure in the Sun's corona seen in X-ray, ultraviolet, or white-light images, consisting of an arch extending upwards from the photosphere. Loops come in a wide range of sizes, from *transequatorial loops* up to 1 solar radius in length (700000 km) that join active regions in opposite hemispheres, down to *quiet region loops* at around 5000 km long. The two ends of the loop, called the *footpoints*, are located in regions of the photosphere of opposite magnetic polarity to each other. This and other evidence suggests that coronal loops are magnetic flux tubes filled with hot plasma.

coronal mass ejection (CME) The large-scale ejection of matter from the Sun's corona at speeds of 10–3000km/s. The mass thrown off in such an event is about 10^{13}kg. Spacecraft coronagraphs show that a typical CME consists of a bright leading edge of gas forming a loop or, more probably, a bubble in the corona, ahead of a dark cavity. In *Hα light, an erupting prominence moves outwards within this dark cavity. A pair of legs, where the bright rim originally connected with the Sun, may persist for a day or more. Sometimes a flare seems to be triggered shortly after a CME, but the connection between CMEs and flares is unclear. A CME produces a disturbance in the solar wind preceded by a shock wave. Interplanetary space probes encountering such disturbances have recorded increased wind speeds and densities, and a rapidly varying magnetic field. When these interplanetary disturbances reach the Earth, they give rise to *geomagnetic storms. Their frequency varies with the sunspot cycle. At solar minimum about one CME occurs a week, rising to an average of two or three per day at solar maximum. ◙

coronal plume A faint feature in the Sun's corona that extends radially out from *coronal holes at the Sun's poles; also referred to as a *polar plume*. Coronal plumes can be seen at the Sun's limb in visible light with coronagraphs and during eclipses, as well as in ultraviolet emission lines formed at temperatures around 1 million degrees. The bases of the plumes can be identified on the Sun's disk at ultraviolet wavelengths as areas of bright emission in coronal holes. Plumes have been traced out to at least fifteen solar radii on images from the *SOHO spacecraft. Plumes are usually denser and cooler than their surroundings, and are most easily identified during solar minimum when coronal holes are clearly seen at the poles of the Sun. There is as yet no conclusive evidence for plumes in equatorial coronal holes.

coronal rain Globules of gas seen in *Hα light descending along curved paths to the chromosphere. They are the final stage in the development of *loop prominences, which are produced by a large flare on the Sun's limb. Several hours after the flare has occurred, the prominence breaks up into fragments which descend along the outline of the now-invisible loop as coronal rain.

COROT A joint French–ESA satellite, launched in 2006 December to observe up to 200000 stars with a 0.27-m telescope in search of dips in brightness caused by the transit of planets larger than the Earth. It is also examining 150 stars more carefully to detect even smaller brightness changes caused by oscillations of their surfaces, from which information can be deduced about their internal structure (a technique known as *asteroseismology). COROT stands for COnvection ROtation and planetary Transits.

((⊕)) SEE WEB LINKS
- ESA mission website.
- Official mission website.

co-rotation Where one body orbits another, the circumstance in which one or both bodies have an axial rotation period the same as the orbiting body's revolution period. The co-rotating body therefore keeps one face permanently turned towards the other, as in the case of the Moon facing the Earth. In the case of Pluto and its satellite Charon, both bodies are in co-rotation and so each keeps the same face turned to the other. *See also* SPIN–ORBIT COUPLING; SYNCHRONOUS ROTATION.

corpuscular radiation 1. Any particles (i.e. protons, electrons, neutrons, atomic nuclei, ions, and atoms) emitted at high speed from a star or other object, as for example in the *solar wind. Material can also be lost in the form of jets from young stars and *active galactic nuclei.
    2. Another name for *cosmic rays.

corrector plate A thin glass plate that corrects for various aberrations of an optical system. The front element of a *Schmidt–Cassegrain or *Maksutov telescope, which corrects for spherical aberration, are examples which cover the full aperture of the telescope.

correlation detection A method by which a signal of known form may be extracted from a noisy background. It is useful where the form of the signal is known, as in a line in a spectrum or stellar image, and the best estimates of the position and strength of the feature are required. It is also used to measure the arrival times of pulses from pulsars; a template of the average pulse profile is correlated with the input waveform to find the best-fit arrival time.

correlation function A mathematical function used in the statistical description of galaxy clustering. It measures the excess probability of finding a galaxy at a given distance from a particular galaxy, compared with what one would expect if galaxies were distributed randomly throughout space. A positive value of the function indicates that there are more pairs of galaxies with a given separation than there would be in a random distribution, while a negative value indicates that galaxies tend to avoid each other. Observations show that galaxies are highly clustered on scales up to several tens of millions of light years.

correlation receiver A radio receiver used in a simple radio interferometer in which signals from two antennas are multiplied together. As a source passes across the sky, the output from the receiver rises and falls as the signals from the two antennas pass alternately into and out of phase, producing a characteristic fringe pattern.

correlator A device used in radio interferometry to multiply the signals received by each pair of antennas in an array. The correlator is the first stage in the processing of the radio signals to create a map of the radio source.

Corvus (Crv) (gen. Corvi) A small constellation of the southern sky, representing a crow. Its brightest star is Gamma Corvi, magnitude 2.6. Delta Corvi is a double star, magnitudes 3.0 and 9.2.

COS-B A European gamma-ray satellite, launched in 1975 August and operated until 1982. It mapped the gamma-ray sky in the energy range 70–5000 MeV (2.5×10^{-7} to 2×10^{-5} nm) with a spark-chamber detector. It also detected several individual sources, including the Vela and Crab Pulsars and the quasar 3C 273.

(⊕) SEE WEB LINKS
• ESA mission website.

cosmic abundance *See* ELEMENTS, ABUNDANCE OF.

Cosmic Background Explorer (COBE) A NASA satellite launched in 1989 November to map the *cosmic microwave background (CMB). It carried three sets of instruments: a group of three Differential Microwave Radiometers (DMRs) to look for irregularities in the CMB; the Far-Infrared Absolute Spectrometer (FIRAS) to measure the spectrum of the CMB; and the Diffuse Infrared Background Experiment (DIRBE) to search for infrared radiation from forming galaxies. COBE's most important results were the discovery of irregularities in the CMB on the level of 1 part in 10^5, and the confirmation that the spectrum of the CMB is a black body with a temperature of 2.73 K. Its observations ceased at the end of 1993.

(⊕) SEE WEB LINKS
• Official mission website.

cosmic background radiation (CBR) *See* COSMIC MICROWAVE BACKGROUND.

cosmic censorship *See* NAKED SINGULARITY.

cosmic dust *See* DUST; GRAINS, INTERPLANETARY; GRAINS, INTERSTELLAR.

cosmic microwave background (CMB) A faint, diffuse glow coming from all directions in the sky, most intense at a wavelength of around 1 mm; also known as the **microwave background radiation** or **cosmic background radiation**. Its existence was predicted from the Big Bang theory, and it was discovered in 1965 by A. A. *Penzias and R. W. *Wilson. The CMB originated when the Universe was 300 000 years old and consisted of a plasma at a temperature of approximately 3000 K. The expansion of the Universe since then has redshifted the CMB photons to an apparent temperature of only about 3 K. Measurements by the *Cosmic Background Explorer satellite (COBE) and more recently the *Wilkinson Microwave Anisotropy Probe (WMAP) have shown that the CMB has a black-body spectrum with a temperature of 2.725K. COBE and WMAP have detected random irregularities in temperature of only a few millionths of a degree. The CMB is regarded as one of the three most important pieces of evidence for the Big Bang, the others being the recession of the galaxies and the cosmic abundance of helium. ◼

cosmic rays High-energy charged particles present in space with velocities near the speed of light. Cosmic rays consist mostly of protons (90%), the remainder being *alpha particles (helium nuclei, 8%) plus a few electrons and heavier nuclei. Cosmic rays have energies ranging from 10^8 to 10^{20} eV and probably beyond. Higher-energy cosmic rays (above about 300 MeV) come from outside the Solar System and are thought to be accelerated in interstellar shock waves created by supernovae; those with the highest energies of all are accelerated many times. The entry of galactic cosmic rays into the inner Solar System is modulated by the solar wind and the interplanetary magnetic field. The number of cosmic rays hitting Earth is largest at solar minimum. Lower-energy cosmic rays are generated intermittently by shock waves near the Sun (*solar energetic particles*) and at the *termination shock where the solar wind slows as it encounters the local interstellar medium.

A cosmic ray from space is known as a *primary cosmic ray. When such particles collide with atoms in the atmosphere they are absorbed, causing both nuclear and chemical reactions. Higher-energy cosmic rays produce secondary particles (*secondary cosmic rays), particularly neutrons, that can be detected on the ground. Cosmic rays of very high energy produce *cosmic-ray showers.

Cosmic rays were first detected during a balloon flight in 1912 by V. F. *Hess, and the term was coined in 1925 by the American physicist Robert Andrews Millikan (1868–1953).

cosmic-ray shower The cascade of secondary particles and photons produced when a *primary cosmic ray enters the Earth's atmosphere and collides with air molecules; also known as an *air shower* or an *Auger shower*. The secondary cosmic rays are initially *pions. Neutral pions decay (disintegrate) into gamma rays, which in turn generate electrons and positrons by *pair production. Deceleration of the electrons and positrons by the atmosphere produces more photons by *bremsstrahlung radiation. The charged pions decay into muons which themselves decay into electrons and neutrinos. The primary cosmic ray and many of the secondaries may undergo further collisions, yielding more particles. The resulting shower can cover an area several kilometres wide at ground level, depending on the energy of the particle that caused it. For example, the shower generated from a single cosmic ray of energy 10^{19}eV will cover 10 square kilometres and contain 10 billion particles at ground level.

cosmic scale factor A mathematical quantity that describes the changing separation of two points as the Universe expands. It can be thought of as a time-dependent magnification factor. In the expanding Universe, separations between points increase uniformly so that a regular grid at some particular time looks like a blown-up version of the same grid at an earlier time. Because the symmetry is preserved, one needs to know only the factor by which the grid has been expanded in order to establish the past grid from the later one. Likewise, one needs to know only the scale factor to obtain a picture of the past physical conditions from present data.

cosmic string A hypothetical one-dimensional (line-like) defect in the structure of *spacetime produced in some models of the early Universe. If produced in the framework of a *grand unified theory, such a string would be about 10^{-31}m thick, and have a mass of about 10^7 solar masses per light year. The strong gravitational effect of cosmic strings might have assisted in the formation of galaxies and of *large-scale structure in the Universe. There is, however, no evidence yet for their existence. *See also* COSMIC TEXTURE; DOMAIN WALL; MONOPOLE.

cosmic texture A hypothetical twisting of the fabric of *spacetime produced in some models of the early Universe. It is the three-dimensional analogue of the *magnetic monopole, *cosmic string, and *domain wall, but is much harder to visualize. Textures, if they exist, may be responsible for the origin of *large-scale structure in the Universe.

cosmic year The time taken for the Sun to orbit once around the centre of our Galaxy; equivalent to about 220 million years. Also known as a *galactic year*.

cosmochemistry *See* ASTROCHEMISTRY.

cosmogony The study of the origin and development of particular objects and systems in the Universe. It is sometimes used for the formation of galaxies and stars, but is more usually taken to refer specifically to the origin of the Solar System.

cosmological constant (Λ) A mathematical term introduced by A. *Einstein into the equations of general relativity to obtain a solution corresponding to a *static universe. This term describes a kind of pressure or (if it has the opposite sign) tension exerted by space itself, which can cause the Universe to expand or contract even in the absence of any matter. When the expansion of the Universe was discovered, Einstein came to regard his introduction of the cosmological constant as his 'greatest blunder'. Recently, however, observations of distant supernovae and other observations have shown that the expansion of the Universe appears to be accelerating (*see* COSMOLOGY). Because gravity is an attractive force that would make the expansion decelerate, something like a cosmological constant must exist to counteract and even override it. Astronomers often attribute the observed acceleration to a force called *dark energy. However, the true nature of dark energy remains one of the biggest questions in astronomy.

cosmological distance scale The calibration of measurements of relative distances between extragalactic objects. It is difficult to measure the distances of galaxies directly because individual calibration stars, such as Cepheid variables, cannot be seen at great distances. On the other hand, the relative distances of galaxies can be measured by means of correlations such as the *Tully–Fisher or *Faber–Jackson relations. These will show that one galaxy is, for example, twice as far away as another, but they cannot give exact distances unless the true distance to at least one galaxy is known. Once the distance scale is established, astronomers can determine the *Hubble constant with accuracy.

cosmological principle The proposal that we are in no special place and that the Universe looks much the same to all observers, wherever they are. In technical terms, this means that the Universe must be homogeneous and isotropic. This principle is not strictly true as the Universe is clearly irregular on small scales. There is good evidence, however, that on very large scales the Universe is quite smooth and regular. A consequence of the cosmological principle is that the *spacetime geometry of the Universe must be described by the *Robertson–Walker metric, which drastically restricts the set of cosmological models compatible with general relativity. *See also* PERFECT COSMOLOGICAL PRINCIPLE.

cosmology The study of the structure and evolution of the Universe. *Observational cosmologists* try to measure the properties of the Universe, such as its chemical composition, density, and rate of expansion. *Theoretical cosmologists* try to explain these properties using the laws of physics. In practice, there are very few observational cosmologists who do not try to understand their observations using theory, although there are many theoretical cosmologists who are not observers. In its broadest sense, cosmology also has philosophical and theological aspects.

Observations show that the Universe is *isotropic* and *homogeneous*—that is, it looks roughly the same in every direction, and an observer at any place in the Universe would see roughly the same numbers of galaxies and clusters of galaxies. The assumption that the Universe is homogeneous and isotropic is known as the *cosmological principle and is the starting point for most theoretical models. These models are usually based on the *general theory of relativity, the theory of gravitation devised by A. *Einstein.

Of the four forces of nature, gravity is the only one which operates over large scales. Because gravity is an attractive force, the Universe cannot be at rest—if the Universe were at rest even for an instant, the gravitational attraction between galaxies would immediately make it start to contract. Einstein believed the Universe was static, so he added an extra term, the *cosmological constant, to his equations to counter the effect of gravity. When the expansion of the Universe was discovered, he came to regard the introduction of this term as a mistake.

Mathematical models of the expansion which are based on the cosmological principle and the general theory of relativity (without the cosmological constant) are called *Friedmann universes.

The Friedmann equations lead to the startling conclusion that at some time in the past the Universe was infinitesimally small and infinitely dense. Evidence that it was also infinitely hot—i.e. that the Universe started in a hot *Big Bang—came from two lines of observation. The first was the discovery in 1965 of the *cosmic microwave background. The second was the measurement of the abundance of helium in the Universe. Although most elements are manufactured by nuclear fusion in stars, there is far too much helium around for it to be made in this way. In the Big Bang theory, the helium and a few other light elements were produced by nuclear fusion in the first three minutes after the Big Bang, and the cosmic background radiation was emitted by the Universe when it was in this early hot phase. There are currently no other plausible explanations of the helium abundance or the cosmic microwave background, and as a result the Big Bang theory is accepted by most cosmologists.

For almost seven decades, the aim of most observational cosmologists was to answer a single question: which one of the three possible Friedmann universes do we live in? The answer appeared to be governed by the average density of the Universe. If the average density were greater than a certain value, called the *critical density, the gravitational attraction between galaxies would be strong enough that the Universe would eventually collapse and go through a reverse Big Bang—the Big Crunch. If the average density were less than the critical density, the Universe would expand for ever. If the average density were equal to the critical density, the Universe would eventually stop expanding but only after an infinite amount of time.

Various approaches to this question were tried, but the methods had potential flaws and the answers they gave did not agree. In the late 1990s, however, a new set of observational programmes started to give a consistent set of answers to this and many other questions about the Universe. These programmes included detailed studies of the cosmic background radiation which allowed astronomers to estimate, among other things, the *curvature of spacetime; direct measurements of the amount of matter in the Universe; and studies of how the brightness of distant supernovae depend on redshift. All of these gave consistent answers, and the universe they describe—which may well be the one we live in—is consequently known as the **concordance universe**.

These results, confirmed and refined by NASA's *Wilkinson Microwave Anisotropy Probe (WMAP), indicate that the Universe started in a hot Big Bang 13.7 billion years ago. The curvature of spacetime is zero. The Universe is dominated by *dark matter rather than by the luminous stuff in stars and galaxies, but only about 20% of this dark matter consists of the protons and neutrons that make up our everyday world. The average density of the Universe is only about 30% of the critical density, and so it will expand for ever. Finally, and most surprisingly, the Universe appears to contain a mysterious force which is pushing the galaxies apart and acts rather like Einstein's cosmological constant. This force has been named *dark energy and is causing the Universe's expansion to accelerate. If these results are correct, they provide answers to some important questions about the Universe. However, there are two new questions for which convincing answers do not yet exist: we do not know with any certainty what the dark matter consists of, and we have even less idea what the dark energy is.

cosmos Another name for the *Universe.

Cosmos satellites An ongoing series of Earth satellites, initiated by the former Soviet Union and continued by Russia; the name is also spelt Kosmos. Cosmos satellites were the successors to the Sputnik series. Scientific satellites, military satellites, and failures of other missions have all been given the Cosmos name. The first Cosmos was launched in 1962 March.

COSPAR An organization established by the International Council of Scientific Unions in 1958 to continue the cooperative research programmes in space science begun during the *International Geophysical Year and to exchange the resulting data. COSPAR has its headquarters in Paris, and holds scientific assemblies every two years. Its name is an acronym of Committee on Space Research.

 SEE WEB LINKS
• Official website.

coudé focus A focal point in an equatorially mounted telescope that lies along the polar axis (see diagram). In a Cassegrain telescope, a single flat mirror in the centre of the tube reflects the converging light beam down the polar axis. Other variants, for refractors, involve two large flat mirrors of similar size to the objective, one to direct light along the polar axis and the other to view the chosen part of the sky. **Coudé** is French for 'elbow', referring to the bending of the light path through 90°. The coudé focus is stationary, so large or heavy equipment can be stationed there, often in a separate **coudé room**. It is often used for spectroscopy.

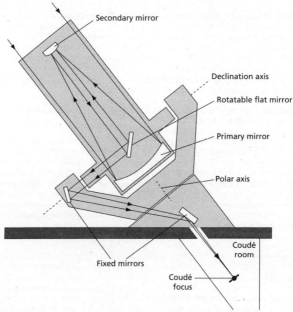

coudé focus

Couder telescope A *Ritchey–Chrétien reflecting telescope with an *aspheric concave primary mirror and a concave ellipsoidal secondary. It has a short focal ratio and wide field of view free from coma, spherical aberration, and astigmatism, but the field of view is curved and located within the instrument, making it unsuitable for visual work. The design is rarely encountered. It is named after the French optician André (Joseph Alexandre) Couder (1897–1979).

counterglow Another name for the *gegenschein.

Cowell, Philip Herbert (1870–1949) English celestial mechanician, born in India. He was concerned in particular with the motion of the nodes of the Moon's orbit and its *secular acceleration. **Cowell's method** for calculating the motions of mutually gravitationally interacting bodies was used by him and A. C. D. *Crommelin to pinpoint the return of Halley's Comet in 1910.

c.p.m. Abbr. for *common proper motion.

Crab Nebula A supernova remnant approximately 6500 l.y. away in the constellation Taurus; also known as M1 or NGC 1952, and as the radio source Taurus A. The Crab Nebula is the remains of a star that was seen to explode as a supernova of Type II in AD1054, reaching a maximum apparent magnitude of −6. In telescopes it appears as an elliptical nebulosity of 8th magnitude. Its true dimensions are 11 × 7.5 l.y. Optically, the nebula has two components: an outer region of reddish, twisted filaments of hydrogen gas; and an inner, whitish core that shows no spectral features. The light from the core is *synchrotron radiation, caused by high-speed electrons from the *Crab Pulsar. It is highly polarized and is continuous at all wavelengths, from gamma rays to radio waves. The expansion velocity of the outer filaments is about 1000km/s. There is evidence that the expansion is accelerating, driven by the radiation from the pulsar. ◙

Crab Pulsar The pulsar at the heart of the Crab Nebula; also known as PSR 0531 + 21 (formerly NP 0532). The Crab Pulsar has a period of 33.08ms and is slowing by 36.4ns per day. The rotational energy it loses is turned into synchrotron emission in the surrounding nebula. Its spin-down age is consistent with its birth in the supernova of AD1054 that gave rise to the Crab Nebula, making it one of the youngest pulsars known. The Crab Pulsar was discovered at radio wavelengths in 1968. In 1969 it became the first pulsar seen to be flashing optically. Pulses at the same frequency have since been detected at X-ray and gamma-ray wavelengths. Visually it appears as a star of 16th magnitude.

Crater (Crt) (*gen.* **Crateris**) A constellation of the southern sky representing a cup or bowl. Its brightest star, Delta Crateris, is of magnitude 3.6.

crater A bowl-shaped depression on a planetary surface. Craters are usually of either impact or volcanic origin; however, some craters on the satellites of the outer planets may be the result of *ice volcanism* in which internal heating melts and vaporizes the ice to produce gas explosions and water flows. Volcanic craters tend to be less circular than impact ones, and may be formed either by explosion or collapse. Volcanic craters of purely explosive origin are normally under 1km in diameter and are usually found at the top of a cone of ejected debris; collapse craters can be more than 100km across, and may have no outer slope at all. *Impact craters have been found on all bodies in the Solar System photographed by space probes except Jupiter's volcanic satellite Io, but volcanic craters do not seem to exist on the smaller Solar System bodies.

crater chain A series of craters approximately in line. They may be overlapping, touching, or separate from one another. Crater chains are usually of either secondary impact or volcanic origin. Secondary impact craters are normally oriented more or less radially around a much larger impact crater. Volcanic crater chains occur when several vents are active along the same volcanic fissure. They may be formed either mainly by collapse, when a series of calderas or collapse pits forms in a line, or by explosion, in which explosive activity occurs at several places along the fissure. Some crater chains on Mars appear to have formed by collapse along a fissure without volcanic activity being involved.

crater counting The counting of impact craters of various size ranges within a given area or geological unit of a planetary surface, in order to determine its age. The older a given feature, the more impacts will have struck it. On the Moon, for example, the densely cratered highlands are clearly much older than the sparsely cratered maria.

crêpe ring An old name for Saturn's C ring, which arose from the fact that it is partially transparent. Its inner edge lies 74500km from Saturn's centre (14200km above the cloud tops), and it extends outwards for 17500km to the start of the B ring. Many finer rings are visible within the C ring along with broader, well-defined bands.

crepuscular rays An atmospheric phenomenon resulting from the scattering of sunlight by dust particles in the *troposphere. Crepuscular rays are seen in twilight near sunset or sunrise, when the shadows cast by distant clouds give rise to a fan-like effect, the rays appearing to diverge from the Sun as a result of perspective. A similar effect is also seen when the Sun is behind dense convective clouds and the cloud edge casts shadows over sunlit areas away from the observer. An inverse effect is seen when bright sunbeams fall through small gaps in a sheet of cloud.

crescent The phase of a Moon or inferior planet when it is less than half illuminated as seen by the observer.

Cressida The fourth-closest satellite of Uranus, distance 61 770 km, orbital period 0.464 days; also known as Uranus IX. Cressida is 62 km in diameter, and was discovered in 1986 on images from the Voyager 2 spacecraft.

Crimean Astrophysical Observatory (CrAO) An observatory at an altitude of 600 m at Nauchny, Ukraine, owned and operated by the Ministry of Science and Technology of the Ukraine. Its main instrument is the Shajn 2.6-m reflector, opened in 1961 and named after Grigorii Abramovich Shajn (1892–1956), a former director. There are also two 1.25-m telescopes (opened in 1955 and 1981), a 1-m solar tower telescope (1955), and a gamma-ray telescope (1989). The CrAO also operates several smaller optical instruments on Mount Koshka, near Simeis, at an altitude of 346 m. A 22-m radio telescope, opened in 1966, is located near the foot of Mount Koshka. The observatory originated at Simeis in 1912 as an outstation of Pulkovo Observatory, but was destroyed in World War II. It was subsequently rebuilt, when it received its current name and the main site was moved to Nauchny. Until 1991 the CrAO was operated by the Academy of Sciences of the USSR.

(⊕) SEE WEB LINKS
• Official observatory website.

critical density The mean density of matter that is required for gravity to halt the expansion of the Universe, equivalent to about 10^{-29} g/cm^3. If the *cosmological constant is zero, a universe with a density below the critical density will expand for ever, whereas a universe with a density greater than the critical density will eventually collapse. A universe with exactly the critical density is described by the *Einstein–de Sitter model, which lies on the dividing line between these two extremes. The average density of material that can be directly observed in our Universe is less than 1% of the critical value. However, various observations suggest that most of the matter in the Universe is *dark matter, which cannot be seen directly but which can be detected through its gravitational effect. Even when the dark matter is added in, however, the average density of the Universe is still only about 30% of the critical density. It therefore seems likely that the Universe will expand for ever, especially because recent observations imply that the cosmological constant is not zero, which will cause the expansion to accelerate.

Crommelin, Andrew Claude de la Cherois (1865–1939) Irish astronomer of French descent. He specialized in calculating cometary orbits, often working with P. H. *Cowell; together they predicted the date of return of Halley's Comet in 1910, and calculated its previous apparitions back to the 3rd century BC. In 1929 he discovered that three comets (from 1818, 1873, and 1928) were one and the same; it was subsequently renamed Comet 27P/Crommelin, and returned in 1956 as he predicted.

cross-axis mounting A form of *English mounting in which the telescope is mounted on one side of the polar axis, and is balanced by a counterweight on the other side.

crossing time The time required for a star or galaxy to move from one side of a cluster to the other, if it is travelling at the typical random speed of objects in the cluster. This is the shortest time-scale on which significant dynamical events can occur, such as changes in the shape of the cluster and exchange of energy between cluster members. For a globular cluster the crossing time is approximately 100 000 years; for a rich cluster of galaxies it is about a billion years.

cross-wire micrometer A simple form of micrometer, consisting of two non-parallel wires, webs, or bars, usually in the form of a cross, in the field of view of an eyepiece. The eyepiece is oriented so that the wires cross the field diagonally, and timings are made as stars cross the wires as the Earth rotates. The interval between successive crossings of the wires depends on a star's position in the field of view. A star's position can be calculated by comparison with those of stars of known position.

crown glass A basic variety of optical glass, containing a high percentage of potassium and calcium. It has a *refractive index of about 1.5, and on average has a lower dispersion of light than the other main basic optical glass, *flint glass. Crown glass has a lower density than flint glass but is more durable.

crust The outermost layer of a planetary body, different in composition from the *mantle beneath. The Earth's crust, which forms the upper part of the lithosphere, varies in thickness between about 10km under the oceans to about 40km under the continents.

Crux (Cru) (gen. Crucis) The smallest constellation of all, better known as the Southern Cross. Its distinctive cross-shape is marked by four stars: *Alpha Crucis (Acrux); *Beta Crucis (Becrux or Mimosa); *Gamma Crucis (Gacrux); and the faintest of the four, Delta Crucis, magnitude 2.8. Mu Crucis is a wide double of magnitudes 4.0 and 5.2. NGC 4755 is a sparkling open cluster known as the *Jewel Box or the Kappa Crucis Cluster. Crux lies in a rich part of the Milky Way, part of which is blotted out by the dark *Coalsack nebula.

cryostat A device designed to maintain an infrared detector or the low-noise amplifier of a radio telescope at a very low temperature in order to reduce thermal noise. In infrared telescopes, cryostats can be evacuated vessels known as *dewars* cooled by liquid nitrogen (down to 77K) or liquid helium (down to 4K), and usually contain optical components, filters, and a preamplifier as well as the detector. In radio astronomy they are usually evacuated chambers cooled to about 12 K by helium-gas refrigerators.

cryovolcanism Rapid melting of ice to form a flow of liquid or gas, which then escapes onto the surface of a planetary body. Formations such as lava tubes, ejecta blankets, and plumes that are seen in terrestrial volcanoes all have cryovolcanic equivalents. In the outer Solar System water ice is a common cryovolcanic material, for example in the eruptions seen on Saturn's moon *Enceladus. On colder moons cryovolcanic processes are driven by other ices, such as nitrogen, which melt or vaporize at temperatures below 0°C. The nitrogen geysers of Triton are an example of cryovolcanism involving a substance other than water.

C star The spectral type assigned to *carbon stars in the *Morgan–Keenan classification, with subdivisions C0–C9. In the previous Harvard system these stars were classified as types R and N; those types are now classified respectively as C0–C4 and C5–C9. The C0–C4 subtypes follow a temperature sequence, but the differences in C5–C9 spectra are more closely linked to composition, particularly the carbon-oxygen ratio. Luminosity classes are difficult to assign to C stars, but an alternative designator is often used, based on the strength of the C_2 (Swan) band at 474nm, and written as a subscript in the range 1–5; for example, Y Canum Venaticorum has a spectral type $C5_4$.

CTIO Abbr. for *Cerro Tololo Inter-American Observatory.

Cubewano An object in the *Kuiper Belt with a mean distance from the Sun greater than about 41 AU and an orbit of low eccentricity; also known as a **main Kuiper Belt object** or **classical Kuiper Belt object.** Unlike the *Plutinos, Cubewanos are not in a resonance with Neptune. Most known objects in the Kuiper Belt fall into this category. The first such object to be discovered, 1992 QB_1, now numbered (15760), was the first known member of the Kuiper Belt; it has a perihelion of 40.88 AU, aphelion 47.50 AU, inclination 2°.2, and period 294 years. The name Cubewano comes from the pronunciation of 'QB1'.

culmination The moment at which a celestial object lies on an observer's meridian (the north–south line in the sky); also known as a *transit* or a *meridian passage. Upper culmination* and *lower culmination* refer to the instants of maximum and minimum altitude respectively for a circumpolar object. At upper culmination (also known as **culmination above pole**) it has an hour angle of 0h; at lower culmination (or **culmination below pole**) it is passing between the pole and the horizon, and has an hour angle of 12h. Lower culmination for non-circumpolar objects occurs below the horizon and is thus unobservable. When used without qualification, 'culmination' means upper culmination.

Curiosity A NASA Mars rover planned for launch at the end of 2011, landing in 2012 August. The six-wheeled Curiosity is larger and faster than any previous Mars rover, able to travel up to 90 m/h. The Mars Descent Imager (MARDI) will photograph the landing site as the probe descends. On the surface, the Mast Camera (Mastcam) will take colour stereo pictures of the landscape. It shares the lander's mast with the Chemistry and Camera spectrometer (ChemCam), which will fire laser pulses at rocks and analyse the vaporized gases. Curiosity has a robotic arm to place instruments against rocks. The Alpha Particle X-ray Spectrometer (APXS) will analyse rock composition, while a close-up camera called the Mars Hand Lens Imager (MAHLI) will provide microscopic views of individual rocks and dust. The arm will also collect samples for analysis by the Chemistry and Mineralogy instrument (CheMin) and the Sample Analysis at Mars (SAM) on board the rover, looking for the chemicals associated with life. The Rover Environmental Monitoring Station (REMS) is an onboard weather station.

(((⊕))) SEE WEB LINKS
• Official mission website.

curvature of field The condition in which an optical system forms an image not on a *focal plane, but on a curved *focal surface* (usually a concave sphere) which intersects the optical axis and is perpendicular to it at the focal point. If an eyepiece produces a curved field, the image cannot be focused across the whole field of view. Schmidt cameras, which have a curved field, require the photographic film or plate to be curved to match the focal surface.

curvature of space *See* CURVATURE OF SPACETIME.

curvature of spacetime A property of *spacetime in which the familiar laws of geometry no longer apply in regions where gravitational fields are strong. In general relativity the geometry of spacetime is intimately connected with the distribution of matter. In a space of only two dimensions, such as a flat rubber sheet, Euclidean geometry applies so that the sum of the internal angles of a triangle on the sheet is 180°. If a massive object is placed on the rubber sheet, the sheet will distort and the paths of objects moving on the sheet will become curved. This is, in essence, what happens in general relativity.

The Universe as a whole may have positive, negative, or zero curvature. A universe with *positive curvature* would curve back on itself like the surface of a sphere so that one could in principle travel out into space and eventually end back at the same place. Such a universe is a **closed universe,** having finite size. A universe with *negative curvature*, however, would be an *open universe*, infinite in size.

The curvature of space can be measured with geometric tests. On the surface of a sphere, which has positive curvature, the sum of the angles in a triangle is greater than 180°; on the surface of a saddle, which has negative curvature, the sum of the angles is less than 180°. Observations of the *cosmic background radiation provide such a geometric test, with the surprising result that the Universe has zero curvature (*see* COSMOLOGY). In other words, it is spatially flat (Euclidean) and infinite in both space and time. One possible explanation is provided by the theory of the *inflationary universe, which proposes that the Universe expanded extremely quickly shortly after the Big Bang. If inflationary theory is true, our entire observable Universe, over 10 billion light years in size, may be only a speck within the greater Universe—and this may explain why we fail to see any curvature. Returning to the analogy of the surface of a sphere, if the triangle on the sphere is small compared with the size of the sphere, the sum of the angles in a triangle will not differ significantly from 180°. If the theory of inflation is correct, the Universe may actually have positive or negative curvature, but we see too small a part of it to detect the curvature.

curvature radiation Another name for *synchrotron radiation.

curve of growth A method for determining the temperature and chemical abundances of stellar atmospheres. From the different profiles of weak and strong absorption lines of a given element, a diagram can be constructed showing how the equivalent linewidths increase (or 'grow') from weak to strong. The shape of this diagram is related to the total abundance of the element. The technique has now largely been superseded by computer-modelling of stellar atmospheres.

cusp The region near the pointed 'horns' of the crescent Moon, Venus, or Mercury, as seen from Earth. The term may also be used of other bodies seen as a crescent from space probes.

cusp cap A bright region at the cusps of a planet or satellite in crescent phase. The term is usually used in relation to Venus, where bright cusps at one or both poles have often been observed, sometimes bordered by a dark **cusp collar**. The term 'cap' is used by analogy with the polar ice caps of the Earth and Mars.

Cybele group A group of asteroids, often called the Cybeles, in the outer part of the main asteroid belt at a mean distance of 3.4 AU from the Sun, between the 2:1 and 5:3 Jovian resonances. The Cybeles, in common with the *Trojan asteroids, the *Hilda group, and the asteroid *Thule, are thought to be remnants of the primordial population at their respective distances. The group is named after the P-class asteroid (65) Cybele, diameter about 240 km, discovered in 1861 by the German astronomer Ernst Wilhelm Lebrecht Tempel (1821–89). Cybele's orbit has a semimajor axis of 3.430 AU, period 6.35 years, perihelion 3.06 AU, aphelion 3.80 AU, and inclination 3°.6.

cyclotron maser A maser source based on *cyclotron radiation, observed in the magnetospheres of certain planets and in active areas of the Sun's atmosphere.

cyclotron radiation Electromagnetic radiation emitted by a charged particle circling in a magnetic field substantially below the speed of light. It is named after the type of particle accelerator in which it was first observed. The radiation is circularly polarized and appears at a single frequency, the *gyrofrequency*, which is independent of the velocity of the particle. Cyclotron radiation has been detected from the magnetospheres of the outer planets. *See also* GYROSYNCHROTRON RADIATION.

Cygnus (Cyg) (*gen.* **Cygni**) A prominent constellation of the northern sky, representing a swan. It is sometimes known as the Northern Cross because of the shape formed by its main stars, the brightest of which is *Deneb (Alpha Cygni). *Albireo (Beta Cygni) is a famous double star; another notable double is *61 Cygni. P Cygni is a 5th-magnitude variable blue supergiant. M39 is a 5th-magnitude open cluster, and NGC 6826 is a planetary nebula known as the *Blinking Planetary. Cygnus lies in the Milky Way and is replete with fascinating objects: the radio source *Cygnus A; the X-ray source *Cygnus X-1; the brightest dwarf nova, *SS Cygni; the *Veil Nebula, part of the *Cygnus Loop; the *North America Nebula; and the *Great Rift.

Cygnus A A strong radio source approximately 10^9 l.y. away in the constellation Cygnus, identified with a giant elliptical galaxy of 15th magnitude. Cygnus A is the brightest extragalactic radio source. It is a classical *radio galaxy with prominent radio-emitting lobes on either side of the central source, fed by faint, narrow jets. The optical parent galaxy, first identified in 1952, has an unusual structure which was originally interpreted as two spiral galaxies in collision. Although this model was subsequently rejected, it remains possible that giant elliptical galaxies such as Cygnus A may have arisen from the merger of smaller galaxies.

Cygnus Loop A large supernova remnant in the constellation Cygnus at a distance of 1400 l.y. The loop is 3° across and expanding at over 100 km/s. It is about 5000 years old. Radio, infrared, and X-ray maps show a complete loop structure. However, optical telescopes show only fragments of the loop, notably the *Veil Nebula.

Cygnus Rift Another name for the *Great Rift.

Cygnus X-1 A strong source of X-rays 8000 l.y. away in Cygnus. The X-rays originate from the inner region of an *accretion disk, which is formed as material flows from the 9th-magnitude blue supergiant HDE 226868 towards a compact companion. This unseen companion is calculated to have a mass of 8–16 solar masses, considerably greater than the mass limit for a neutron star, leading to the conclusion that it must be a black hole. Gas is drawn from the supergiant star by the strong gravitational pull of the black hole. As the gas falls through the accretion disk, the gas heats up to millions of degrees, hot enough to emit X-ray radiation, which can be observed by space-borne X-ray observatories. 📷

Cynthian Referring to the Moon. The term comes from an identification of the Moon goddess with Artemis, twin sister of Apollo, supposedly born on Mount Cynthus on the island of Delos.

Cytherean Referring to the planet Venus. The term comes from Cythera, the mythical birthplace of Aphrodite (the Greek equivalent of Venus).

Dactyl The satellite of asteroid *Ida, discovered by the Galileo probe in 1993, the first confirmed satellite of an asteroid. Dactyl has dimensions of $1.6 \times 1.4 \times 1.2$ km and an estimated density of $2.2–3.0$ gm/cm^3. It was about 100 km from Ida's centre when imaged.

Dall–Kirkham telescope A variant of the *Cassegrain telescope, with a concave ellipsoidal primary mirror and a convex spheroidal secondary mirror. This combination produces an image free from *astigmatism, although with greater coma (*see* COMA, OPTICAL) off-axis. It is particularly suited to observations where good resolving power is more important than wide field of view, such as planetary observation. It was invented independently by the English amateur telescope-maker Horace Edward Stafford Dall (1901–86) and the American Alan R. Kirkham (1909–68).

Damocles Asteroid 5335, a body with an unusual orbit that takes it from near the orbit of Mars out beyond the orbit of Uranus, discovered in 1991 by the Scottish-born Australian astronomer Robert Houston McNaught (1956–). On the basis of its orbit, which has a period of 40.7 years and a high inclination, it would probably have been classed as a comet had it shown a trace of a coma or hint of a tail. However, Damocles appears to be a bare rocky body, although it could be an extinct *Halley-family comet. Its diameter is estimated at 12 km. Although Damocles' present orbit does not cross that of the Earth, in a few tens of thousands of years from now it may well do so due to planetary perturbations. Currently its semimajor axis is 11.84 AU, perihelion 1.59 AU, aphelion 22.09 AU, and inclination 61°.7. It was last at perihelion in 1990.

Damocloid An object having the appearance of an asteroid, but with a similar orbit to that of the *Halley family of comets. Damocloids are thought to be the nuclei of former comets now rendered inactive by degassing or mantling of the surface. They are named after their prototype, *Damocles. Such an object could impact the Earth with little notice on account of its low albedo (only a few per cent) and high approach velocity.

Danjon, André (1890–1967) French astronomer. From 1921 he built a variety of special photometers for studying lunar eclipses (for which he established the *Danjon scale) and variable stars, and calculated the first really accurate values for the albedos of the Moon, Venus, and Mercury. He improved the design of the prismatic astrolabe for measuring star positions (*see* DANJON ASTROLABE).

Danjon astrolabe A *prismatic astrolabe with a micrometer, designed in 1938 by the French astronomer A. *Danjon. It eliminates the observer's *personal equation in measuring the instant at which the direct and reflected images of a star coincide; hence it is also known as an *impersonal astrolabe*. The main prism is equilateral so that observations are made at a constant zenith distance of 30°. Danjon astrolabes have been used since the 1950s for measuring time and latitude and in observations for compiling star catalogues.

Danjon scale A scale for roughly estimating the darkness of a total lunar eclipse, originated by A. *Danjon. The five-point scale runs from 0 (extremely dark, Moon invisible) to 4 (extremely bright, eclipse having little effect on Moon's visibility). The darkness of a lunar eclipse is largely determined by the opacity of the Earth's atmosphere, which is affected by clouds and volcanic dust.

dark adaptation The increased sensitivity of the eye to light in dark conditions. It results from chemical changes within the eye, and takes at least 20min to complete. Dark adaptation is reversed when the eye is again exposed to bright light, completely in the case of blue or violet light, but only partially in the case of red light. It is for this reason that visual observers use dim red lights for examining charts, etc.

dark energy A force proposed to account for the apparent acceleration in the expansion of the Universe, as deduced from observations of supernovae in distant galaxies. One proposed explanation for the acceleration of the expansion is a *cosmological constant, but if such a constant exists its physical meaning is completely unknown. Astronomers have generally preferred to explain the acceleration as a consequence of the fields which fill space, such as the electromagnetic field. According to quantum mechanics, every field should have a minimum energy called the *vacuum energy*, and this vacuum energy could account for the observed acceleration. The problem with this explanation is that the predicted effect of the vacuum energy on the expansion of the Universe is roughly 10^{120} times greater than the acceleration actually measured. An alternative explanation for dark energy is a different type of field known as *quintessence. The cause of the cosmological acceleration remains an unsolved question. *See also* COSMOLOGY.

dark galaxy A region of space containing as much hydrogen gas and *dark matter as a large galaxy but in which no stars have formed; hence such objects are invisible to optical telescopes, but they can be detected at radio wavelengths from their hydrogen emission. The first dark galaxy, in the Virgo Cluster, was found in 2004. Theorists estimate that there could be up to 100 times as many dark galaxies as normal ones throughout the Universe.

dark matter Material whose presence can be inferred from its effects on the motions of stars and galaxies, but which cannot be seen directly because it emits little or no radiation; also known as *missing mass*. Evidence for dark matter in spiral galaxies is provided by their *rotation curves. The existence of dark matter in rich clusters of galaxies can be deduced from the motions of the member galaxies (*see* VIRIAL THEOREM). Additional evidence for the existence of dark matter comes from observations of distant supernovae and of the *cosmic microwave background. The different observational techniques all yield a similar value for the amount of dark matter, which is about 50 times greater than the total amount of matter contained in stars (*see* COSMOLOGY). Models of how elements such as helium were formed during the first few minutes after the *Big Bang imply that only about 20% of the dark matter consists of so-called *baryonic matter*, the protons and neutrons that make up stars, galaxies, and our everyday world. The composition of the remaining 80% is unknown, although various possibilities, including hypothetical particles such as *axions and *photinos, have been suggested. *See also* COLD DARK MATTER; HOT DARK MATTER; MACHO; NON-BARYONIC MATTER; WIMP

dark nebula A cloud of interstellar gas and dust that absorbs light from behind, so that it appears black against a brighter background. The absorbed light heats the dust particles, which then reradiate some of the energy as infrared radiation. Some of the background light is not absorbed but is scattered, or redirected. The *Horsehead Nebula in Orion is a well-known dark nebula; another is the *Coalsack in the southern Milky Way.

daughter isotope An isotope produced by the radioactive decay of the nuclei of another isotope (the *parent isotope). For example, lead-206 is a daughter isotope of uranium-238, which has a half-life of 4.5 billion years. The amount of any daughter isotope (or its decay products) in a sample increases with time, while that of its parent decreases; hence the relative proportions of the two can be used to derive the age of the sample (*see* RADIOACTIVE AGE DATING).

Davida Asteroid 511, discovered in 1903 by the American astronomer Raymond Smith Dugan (1878–1940). It is the sixth-largest main-belt asteroid, dimensions $360 \times 294 \times 254$ km, mean diameter 300km. It is of C class, with a rotation period of 5.13 hours. Davida's orbit has a semimajor axis of 3.166 AU, period 5.63 years, perihelion 2.58 AU, aphelion 3.76 AU, and inclination 15°.9.

David Dunlap Observatory (DDO) The former astronomical observatory of the University of Toronto, Canada, at Richmond Hill, Ontario, founded in 1935. Its main instrument is a 1.88-m reflector, opened in 1935, the largest optical telescope in Canada. In 2008 the Observatory closed for research purposes and is now operated as a public education facility by the Royal Astronomical Society of Canada.

(⊕) SEE WEB LINKS

• Official observatory website.

Davis, Raymond, Jr (1914–2006) American physical chemist. After World War II, he began to study methods for detecting neutrinos emitted, first, from nuclear reactors and later, in the 1960s, from the Sun (*see* SOLAR NEUTRINO UNIT), utilizing reactions of the neutrinos with the isotope chlorine-37 to produce argon-37. This work earned him a share of the 2002 Nobel Prize in Physics. Davis also used the decay of chlorine-36 to argon-36 to estimate the time that meteorites, including *Lost City, had spent in space.

Dawes, William Rutter (1799–1868) English physician and amateur astronomer. He specialized in making micrometric measurements of double stars. He made a special study of telescopic resolving power, and in 1867 he drew up a table showing what aperture of telescope is necessary to resolve two stars of a given separation (the *Dawes limit). In 1850 he observed Saturn's 'crêpe ring' (the C Ring) independently of W. C. and G. P. *Bond, but 10 days after they discovered it.

Dawes limit The maximum *resolving power that can be obtained in practice from a telescope of a given aperture. It was established by W. R. *Dawes on the basis of tests of the visibility of double stars through various apertures. It states that a pair of sixth-magnitude stars can just be seen as double by a 1-inch (25.4-mm) telescope if they are separated by 4″.56. The table gives the resolving power of various instruments based on the Dawes limit.

Dawes Limit

Aperture mm	Resolving power ″
25	4.63
50	2.32
75	1.54
100	1.16
125	0.93
150	0.77
175	0.66
200	0.58
250	0.46
300	0.39
400	0.29
500	0.23

Dawn A NASA space probe to visit the asteroids Vesta and Ceres, launched in 2007 September. Dawn used ion propulsion to travel to Vesta, arriving there in 2011 July and going into orbit around it for a year. Dawn carries two German-made cameras to photograph the asteroid's surface, as well as two spectrometers to study the mineral composition and a magnetometer to search for a magnetic field. In 2012 July Dawn will depart for Ceres, reaching there in 2015 February to repeat its studies.

(⊕) SEE WEB LINKS

• Official mission website.

day The rotation period of the Earth relative to some external point. Several types of day are defined according to the reference point chosen. For most purposes the day is defined as 86400s (i.e. 24 hours exactly), which corresponds to the *mean solar day, but the mean solar day depends on the rotation of the Earth, which is neither completely uniform nor completely predictable. The *apparent solar day*, the interval from one meridian transit of the Sun (apparent noon) to the next, has a considerable annual variation of about 30s due to the changing value of the *equation of time. This is averaged out to give the mean solar day. The *mean sidereal day*, on the other hand, is the Earth's period of axial rotation relative to the stars, with a minor correction of 0.0084s to allow for precession. It is the interval between successive transits of the mean *equinox, and is nearly 4min shorter than the mean solar day. Due to the Earth's orbital motion around the Sun there is precisely one more sidereal day in the year than the number of solar days.

Daylight Saving Time Another name for *Summer Time.

day number 1. Daily entries in *The Astronomical Almanac* of various quantities that assist in calculating the apparent places of stars from their mean places given in catalogues. The quantities tabulated include nutation, the obliquity of the ecliptic, and *Besselian day numbers. *Second-order day numbers* are used in high-precision calculations, particularly for stars at high declinations.

   2. The Julian Day Number, which is the whole-number part of the *Julian Date.

D-class asteroid A class of asteroid whose members are rare in the main belt but which are increasingly encountered beyond 3.3 AU from the Sun, a distance at which an asteroid has an orbital period exactly half that of Jupiter (i.e. a 2 : 1 resonance). D-class asteroids have a very low albedo (0.02–0.05) and a generally featureless reflectance spectrum. They are very red at longer wavelengths, possibly due to carbon-rich material. Members of this class include (1256) Normannia (a member of the *Hilda group), diameter 69km, and many *Trojan asteroids, including (911) Agamemnon, diameter 167km.

DCT Abbr. for *Discovery Channel Telescope.

DDO classification A classification of the luminosity of spiral and irregular galaxies, based on the fact that their luminosity is related to the appearance of their spiral arms, the brightest systems having the best-developed structure. Its name derives from the *David Dunlap Observatory, where it was developed by the Dutch-born Canadian astronomer Sidney van den Bergh (1929–). The DDO scheme retains the Sa, Sb, Sc, Irr notation of the *Hubble classification, but adds luminosity classes similar to those applied to stars, ranging from I (supergiants), via II (bright giants), III (giants), and IV (subgiants), to V (dwarfs). Class I corresponds roughly to an absolute blue magnitude of -20.5 (i.e. 2×10^{10} solar luminosities) and class V to blue magnitude -14 (10^8 solar luminosities). No Sa or Sb spirals are fainter than class III.

DDO photometry A system of *intermediate-band photometry developed at the *David Dunlap Observatory to study G and K giant stars and dwarfs of differing chemical composition. Seven filters are used, at wavelengths of 346, 381, 417, 426, 452, 489, and 513nm.

dec. Abbr. for *declination.

deceleration parameter (q) A figure that describes the slowing-down of the expansion of the Universe. In a *Friedmann universe, the deceleration parameter is simply half the *density parameter, Ω. A value of q greater than 0.5 indicates that the expansion is decelerating quickly enough for the Universe eventually to collapse. A value less than 0.5 indicates that the expansion will continue for ever. In models with a *cosmological constant, q can even be negative, indicating an accelerated expansion, as in the *inflationary universe.

de Chéseaux, Comet (C/1743 X1) A long-period comet discovered in late 1743 by several observers including the Swiss astronomer (Jean) Philippe Loys de Chéseaux (1718–51), who computed its orbit and described its multiple tail. It reached a peak magnitude of about -7 late in 1744 February, when it was visible in daylight. After perihelion, 0.22 AU, on 1744

March 1, the comet displayed a fan-shaped tail consisting of six or seven rays. The tail was visible in the dawn sky before the comet's head had risen, and extended up to 90°. The comet's orbit is parabolic, with an eccentricity of 1.0 and inclination 47°.1.

declination (dec., δ) A coordinate on the celestial sphere, the equivalent of latitude on Earth. Declination is measured in degrees north or south of the celestial equator, from 0° at the celestial equator to +90° at the north celestial pole and −90° at the south celestial pole. It is an *equatorial coordinate.

declination axis The axis of an *equatorial mounting around which a telescope can be moved in declination. The declination axis is at right angles to the *polar axis.

deconvolution Any computational method used in signal processing to correct in part for the broadening or blurring effect of an instrument or the atmosphere. Deconvolution methods are used extensively in aperture synthesis to reduce the distorting effects of the telescope beam pattern. See CLEAN; MAXIMUM-ENTROPY METHOD.

decoupling The stage early in the history of the Universe when, according to the Big Bang theory, particles of matter ceased to interact with radiation. Decoupling happened at different times, and therefore at different temperatures, for different particles. Neutrinos, for example, decoupled from the background radiation at a temperature of about 10^{10}K (about 1s after the Big Bang) while ordinary matter decoupled at a temperature of a few thousand degrees K (after about 300000 years). After matter and radiation decoupled, the background radiation propagated freely through the expanding Universe.

dedispersion A technique used in radio astronomy to compensate for the smearing effect of interstellar *dispersion (2). The radio band being observed is divided into many narrow channels, each of which is measured separately. Provided the dispersion measure of the source is known, appropriate time delays can be introduced between the channels to remove the dispersive delay across the band. The signals from the channels are then recombined to form a dedispersed signal. Dedispersion is widely used for the observation of pulsars, whose narrow pulses are readily smeared out by interstellar dispersion.

Deep Impact A NASA spacecraft, launched in 2005 January, that rendezvoused with Comet 9P/Tempel 1 in 2005 July. It released a 370-kg impactor which hit the comet's nucleus, producing a crater and ejecting dust and ice that was studied by the main flyby craft and ground-based telescopes, improving knowledge of the structure and composition of the nucleus. The probe was subsequently put on course to Comet 103P/Hartley which it flew past in 2010 October in the *EPOXI mission.

(((●))) SEE WEB LINKS
• Official mission website.

deep-sky object An object beyond the Solar System. The term is not usually applied to individual stars, but to objects of the kind found in the *Messier Catalogue and *New General Catalogue, i.e. star clusters, nebulae, and galaxies.

Deep Space Network (DSN) A world-wide network of radio telescopes for tracking and communicating with space probes, owned by NASA and operated by the *Jet Propulsion Laboratory. The network consists of three deep-space communications complexes, located on three continents to allow continuous coverage as the Earth rotates. The complexes are at Goldstone in the Mojave Desert of California, 72km north of Barstow; Robledo de Chavela, 60 km west of Madrid, Spain; and the Tidbinbilla Nature Reserve, 40km southwest of Canberra, Australia. Each complex has one 70-m and one 26-m antenna. In addition, the Goldstone complex has four 34-m antennae, Madrid has three, and Australia has two. The 70-m dishes were enlarged from 64m in 1983–8, before Voyager 2's encounter with Neptune. The smaller dishes can be used for tracking some Earth-orbiting scientific satellites. The dishes are also used for radio and radar astronomy observations. NASA plans to replace the 70-m dishes

with a new generation of 34-m antennas by 2025. The first site to be upgraded will be
Canberra, where up to three 34-m antennas will come into operation by 2018.

(⊕) SEE WEB LINKS
• Official website.

Deep Space 1 (DS1) A small NASA space probe, the first mission of the *New Millennium
Program, launched in 1998 October. It flew past the asteroid (9969) Braille in 1999 July and
comet 19P/Borrelly in 2001 September, imaging them both.

(⊕) SEE WEB LINKS
• Official mission website.

defect of illumination That part of the disk of a satellite or planet that is not illuminated
as seen from Earth, usually expressed in angular measure (e.g. seconds of arc). The defect of
illumination decreases as the phase increases, and vice versa.

deferent In the *Ptolemaic system of planetary motion, a hypothetical circle around the
Earth, although the Earth was not necessarily at the exact centre. Along the circumference of the
deferent moved the centre of a smaller circle, the *epicycle. The deferent was effectively the
orbit of a planet. *See also* EQUANT.

deflection of light The bending of light by the gravitational field of the Sun. At the Sun's
limb, the deflection amounts to $1''.75$ radially away from the Sun. This effect is corrected for
when reducing star positions from mean place to apparent place.

degeneracy A state of matter attained when atomic particles are packed together as tightly
as is physically possible, at densities of several thousand tonnes per cubic metre. Particles
which are very close together are forbidden by the *Pauli exclusion principle* to have the same
energy, and, as a result, the particles repel each other. This causes **degeneracy pressure**
which, unlike thermal pressure, depends only on density and not on temperature. It provides
the main support against gravity in white dwarfs (*electron degeneracy*) and neutron stars
(*neutron degeneracy*). Degenerate matter is found also in the cores of low-mass stars that
have exhausted their central hydrogen, in brown dwarfs, and in the central regions of the
giant planets.

degenerate star A star that has collapsed to a high density so that *degeneracy pressure
is its main support against further collapse. Collapse to a degenerate state occurs either after
nuclear fuel is exhausted, as in white dwarfs and neutron stars, or, as in brown dwarfs,
which have insufficient mass to raise the core temperature high enough to ignite hydrogen.

degree of arc (°) A unit of angular measure, equivalent to 1/360 of a circle.

degrees of freedom The number of independent variables in a moving system such as a
group of orbiting bodies. In the *n-body problem, the number of degrees of freedom rises with
the number of gravitating bodies involved, greatly increasing the complexity of the problem.

Deimos The outer satellite of Mars, distance 23460 km from the planet's centre, orbital period
1.262 days. Deimos is irregular in shape, $16 \times 12 \times 10$ km across its three axes. It spins on its
shortest axis in the same time that it takes for one orbit, and so keeps the same face towards
Mars. Deimos was discovered in 1877 by A. *Hall. Its surface is covered with impact
craters, but it is smoother than Phobos as many of its craters appear to have been infilled.
Deimos may be a captured C-class asteroid.

delay line A length of cable inserted between an antenna and a receiver to introduce a
precise time delay into the signal. Delay lines have many applications in radio astronomy,
especially where signals from different elements of an interferometer have travelled different
distances and need to be brought back into phase. Delay lines have now been largely
supplanted by the technique of digitally sampling the data and by using **digital delay lines**
which store the data in memory for an appropriate time.

Delphinus (**Del**) (*gen.* **Delphini**) A small but distinctive constellation of the equatorial region of the sky, representing a dolphin. Its two brightest stars are Alpha Delphini (Sualocin) and Beta Delphini (Rotanev), magnitudes 3.8 and 3.6 respectively; the names of these stars written backwards spell Nicolaus Venator, the Latinized form of Niccolò Cacciatore (1780–1841), an Italian astronomer who named them after himself. Gamma Delphini is an attractive double of yellow stars, magnitudes 4.3 and 5.1.

δ Symbol for *declination.

Delta Aquarid meteors A meteor shower producing moderate activity between July 15 and August 20. The meteor stream's orbit lies close to the ecliptic, and has been split into two major components. At maximum, on August 6, the Northern Delta Aquarids appear from a radiant at RA 23h 04m, dec. +02°. The Southern Delta Aquarids peak on July 29 from a radiant at RA 22h 36m, dec. −17°. The southern component is the more active, with peak ZHR around 25, compared with 10 for the northern component.

Delta Cephei star *See* CLASSICAL CEPHEID.

Delta Delphini star A giant star of spectral type late A to early F, with weak Ca II lines. The type star is a spectroscopic binary with an orbital period of 40.58 days in which both components are Delta Scuti stars; the period of the primary is 0.158 day, and that of the secondary 0.134 day.

Delta Scuti star A type of pulsating variable with a short period (0.01–0.2 day) and small amplitude (0.003–0.9 mag.); abbr. DSCT. The pulsation is driven primarily by instabilities in the hydrogen convection zone, and multiple *pulsation modes occur simultaneously. These Population I stars have spectral types from A0 to F5 and lie at the lower end of the *Cepheid instability strip, either on the main sequence or among the subgiants and giants. In older literature, they were often described by the misleading terms *dwarf Cepheid*, *RRs variable*, or *ultra-short-period variable*. *See also* AI VELORUM STAR; SX PHOENICIS STAR.

delta T (ΔT) In astronomical time scales, the difference between *Terrestrial Time (TT) and *Universal Time (UT). Currently this difference is nearly one minute and is increasing at the rate of about 2s every three years. It arises because the Earth's rate of axial rotation is not precisely uniform, but shows certain irregularities and is gradually slowing down. Universal Time is tied to the rotation of the Earth through its link with sidereal time, while Terrestrial Time is defined by the orbital dynamics of bodies in the Solar System.

Demon Star Popular name for the variable star *Algol, so named because it lies in the head of Medusa, the Gorgon, in the constellation Perseus.

Deneb The star Alpha Cygni, magnitude 1.25. It is an A2 supergiant with a luminosity 50000 times the Sun's. Its distance is about 1400 l.y., making it the most distant of all the first-magnitude stars. It forms one corner of the so-called *Summer Triangle of stars.

Deneb Kaitos The star Beta Ceti, magnitude 2.0. It is a G9 giant, 96 l.y. away. An alternative name is Diphda.

Denebola The star Beta Leonis, magnitude 2.1. It is an A3 dwarf, 36 l.y. away.

Denning, William Frederick (1848–1931) English amateur astronomer. His chief contributions were to the study of meteors, for which he calculated the heights and velocities of over 1200, and published a catalogue of meteor shower radiants. He deduced that the temperature of the atmosphere at about 100km was higher than had been believed. Denning also discovered nebulae, comets, and novae, including Nova Aquilae in 1918.

density parameter (Ω) The ratio of the actual *mean density of matter in the Universe to the *critical density required for the Universe to collapse. In the absence of a *cosmological constant, a universe with Ω less than 1 will expand for ever, whereas a universe with Ω greater than 1 will eventually collapse. Current observations imply that the density parameter is about 0.3. The Universe will therefore expand for ever. The observations also imply that the

Universe contains something like a cosmological constant, which is causing this expansion to accelerate (*see* COSMOLOGY).

density wave A wave-like concentration of the density of interstellar material and stars that propagates through the disk of a spiral galaxy, giving rise to the spiral structure. The compression of the interstellar matter as it flows into the spiral arms causes star formation, and the bright, new-born stars outline the arms. The generation of the spiral wave may be aided by the existence of a central bar in the galaxy, or by tidal interaction with a companion galaxy.

descending node (☊) The point in an orbit at which a body moves from north to south across a reference plane, such as the plane of the ecliptic or of the celestial equator. *See also* NODE.

Desdemona The fifth-closest satellite of Uranus, distance 62660 km, orbital period 0.474 days; also known as Uranus X. Desdemona is 54 km in diameter, and was discovered in 1986 on images from the Voyager 2 spacecraft.

de Sitter, Willem (1872–1934) Dutch mathematician and astronomer. He was an early supporter of the theory of relativity, assessing its implications for astronomy. From it he derived what is now called the *de Sitter universe, the first theoretical model of an expanding Universe. In other work he refined the orbits and masses of Jupiter's Galilean satellites, and showed the rotation of the Earth to be gradually slowing.

de Sitter universe A model of an expanding universe in which there is no matter or radiation but the expansion is driven by a *cosmological constant. It was proposed in 1917 by W. *de Sitter. Although this model is physically unrealistic, it introduced the idea that the real Universe might be expanding. An expansion phase which is very similar to that in the de Sitter model also plays an important role in modern theories of the *inflationary universe.

Despina The third-closest satellite of Neptune, orbiting every 0.335 days at a distance of 52 526 km, slightly closer than the inner edge of the planet's Le Verrier ring; also known as Neptune V. Despina is 148 km in diameter, and was discovered in 1989 on images from the Voyager 2 spacecraft.

destructive interference *See* INTERFERENCE.

detached binary A close binary star in which neither star fills its *Roche lobe. It may be an *eclipsing binary of the *Algol type, a *reflection variable, or an *ellipsoidal variable. In general, no mass transfer or mass loss can occur.

deuterium An isotope of hydrogen, the nucleus of which is composed of one proton and one neutron. Deuterium is expected to have been produced in the Big Bang as a by-product of the nuclear reactions that produce helium. This makes it potentially important as a test of the Big Bang model, because deuterium cannot easily be made in stars, and any significant quantity of deuterium observed today is therefore presumably of primordial origin.

de Vaucouleurs, Gérard Henri (1918–95) French-American astronomer. In 1953, from the distribution galaxies brighter than mag. 12.5, he inferred the existence of the *local supercluster. Also in the 1950s he studied the structure and rotation of the Magellanic Clouds. With his first wife Antoinette de Vaucouleurs, née Piétra (1921–87), he published the *Reference Catalogue of Bright Galaxies* (three editions, 1964–91), which collated different types of data from many sources. This work was linked with his exploration of the *cosmological distance scale. His use of many different 'standard candles' led to a value of the *Hubble constant of around 100 km/s/Mpc, higher than the 50 km/s/Mpc then being proposed by A. *Sandage and others. He also studied the planets, notably Mars.

dewar Another name for a *cryostat in which liquid nitrogen or helium is used to cool the detectors.

dew cap An extension to the tube of a *refracting telescope to prevent dew from forming on the objective lens, or to the tube of a *catadioptric telescope to prevent dew forming on the corrector plate. It works by reducing cooling of the objective by shielding it from cold air currents. The dew cap should be between two and three times longer than the telescope's aperture, with a diameter which does not restrict the field of view at low powers. If it is too long, however, it may impair the *seeing because of air currents circulating within it.

D galaxy An extremely bright giant elliptical galaxy near the centre of a cluster of galaxies, in which it will be the *first-ranked* (i.e. brightest) member. They are assigned the letter D on *Morgan's classification, which signifies that they are dustless. The more extreme supergiant examples at the centres of rich clusters are classified as *cD galaxies. D galaxies are often radio sources, and are surrounded by extended envelopes of faint stars.

diagonal A flat mirror or prism that reflects a light beam through 90°, for example the secondary mirror in a Newtonian reflector. A *star diagonal*, consisting of a short tube into which the eyepiece fits, is used in refractors to give a more comfortable viewing position, while a *Sun diagonal* is similar but has a non-silvered mirror to cut down the light intensity. Both of these have the drawback that they give a laterally inverted mirror image.

diamond ring An effect seen immediately before or after totality in a solar eclipse. The diamond ring is produced by sunlight shining through a valley on the lunar limb, causing a single dazzlingly bright point of light.

Diana Chasma A deep trough in the middle of Aphrodite Terra on Venus at $-15°$ lat., $155°$E long. It is one of several deep troughs in the region, and contains some of the lowest elevations on Venus, over 1 km deep in places. It is 100 km wide and over 900 km long, and was discovered by radar from *Arecibo Observatory.

dichotomy The moment when a planetary disk appears exactly half-illuminated by the Sun, as seen from Earth. The term is used exclusively for the planets Venus and Mercury; the same phase for the Moon is known as *first quarter* when it occurs between new Moon and full Moon, and *last quarter* when it occurs between full Moon and new Moon.

dichroic mirror A partially reflecting mirror that also transmits some light. Such mirrors are produced by the deposition of thin films of materials which reflect particular colours and transmit others by interference. A typical dichroic mirror may transmit blue light and reflect yellow light, for example. *Nebula filters are produced by the same process.

Dicke, Robert Henry (1916–97) American physicist and astronomer. In 1961 he suggested that the gravitational constant varies with time (*see* BRANS–DICKE THEORY). In 1964, with the Canadian-born American physicist Phillip James Edwin Peebles (1935–) and others, he began to develop a *hot Big Bang theory, independently of G. *Gamow. The theory predicted the existence of the *cosmic microwave background, discovered shortly after by A. A. *Penzias and R. W. *Wilson. He also invented the *Dicke radiometer and *Dicke switch, and in 1957 set out what has become known as the weak *anthropic principle.

Dicke radiometer A radio receiver designed to measure weak signals in the presence of noise; also known as a **Dicke receiver**. The input to the receiver is rapidly switched (by a *Dicke switch) between the antenna and a reference noise source. It is useful where accurate measurements of absolute flux are required, and has been used to measure the very weak signal from the cosmic microwave background. It is named after R. H. *Dicke.

Dicke switch A solid-state device designed to switch between the two inputs to a Dicke radiometer. The switching rate is normally around 10–1000 times per second.

dielectronic recombination A process by which an ion recombines with an electron in nebulae and hot interstellar gas, at temperatures above 10000 K. The ion captures the electron into a particular energy level, but the electron's energy is transferred to another electron in a lower orbit, which is then excited to a higher level. As the electron returns to its lower level, it radiates away this energy, and the ion is said to have *recombined*. It is termed dielectronic recombination because two electrons take part in the process, and is the opposite of *autoionization.

differential rotation The rotation of a non-solid body in which different parts turn at different speeds. Stars and planetary atmospheres rotate differentially. For example, the equatorial regions of the Sun rotate about 25% faster than the polar regions, and the rotation of Jupiter is divided into System I (the equatorial region) and System II (the rest of the planet). Disk-like systems such as the rings of Saturn, accretion disks around stars, and disk galaxies also rotate differentially because they are composed of individual parts, each with a separate orbit.

differentiation A geological process by which a magma separates into layers of igneous rock of differing density and composition within a body such as a planet or asteroid. On the planetary scale, differentiation typically produces a core, a mantle, and a crust.

diffraction The slight bending of light around the edge of an obstacle in its path. It is a consequence of the wave nature of light. Because of diffraction, a star image consists of an *Airy disk surrounded by several *diffraction rings, produced by diffraction at the edge of the telescope lens or mirror. In a reflecting telescope which has a secondary mirror supported by one or more arms, diffraction around the arms causes the images of stars to have spikes. Diffraction also occurs at other wavelengths, such as radio. A point radio source occulted by the Moon, for example, disappears or reappears with brightness oscillations, rather than abruptly, as a *diffraction pattern sweeps across the observatory. The size of the diffraction pattern increases with the wavelength of the radiation involved.

diffraction grating A surface on which are ruled very fine and evenly spaced straight grooves (typically 100–1000 per mm) which break light into a spectrum by diffraction; finer spacings give greater dispersion of the spectrum, although the resolution is ultimately limited by the size of the grating. There are two main types, *transmission gratings and *reflection gratings, both of which are used for astronomy. Unlike prisms, diffraction gratings produce several sets, or *orders*, of spectra which are arranged symmetrically either side of the main spectrum, becoming progressively fainter and more dispersed with distance. To minimize light loss, the surface grooves are normally cut asymmetrically, with a sawtooth cross-section, so that most of the diffracted light is concentrated into a single spectral order; such a grating is said to be *blazed*.

diffraction-limited A definition of quality in an optical system, indicating that it meets the *Rayleigh criterion. This is met when there is no more than a quarter of a wavelength of light between the wavefronts reaching the focal point from all parts of the aperture. However, the practical resolving power of a telescope of a given aperture is better than this theoretical limit, so diffraction-limited optics are not necessarily perfect.

diffraction pattern Any pattern produced by an optical system as the result of *diffraction. The spikes on a star's image produced by the arms supporting the secondary mirror of a telescope are one example; the *Airy disk and *diffraction rings which appear in the image of a star seen through a telescope are another.

diffraction rings The faint rings that appear around the *Airy disk of a star image as seen through a telescope; they are a result of *diffraction. The Airy disk itself is in effect ring zero, with the first ring having as little as 1.7% of its intensity. Diffraction rings are most noticeable in instruments of small aperture. Their size decreases with increasing aperture.

diffuse interstellar bands Relatively broad absorption features in the spectra of distant sources that are caused by material in interstellar space. A variety of complex molecules have been suggested as the cause of these features, but none have been identified as responsible. Over 200 such bands are known, the strongest being at 443.0 and 617.7nm.

diffuse nebula A cloud of interstellar gas and dust which glows because of the effect on it of ultraviolet radiation from nearby stars. The term *H II region is now preferred. The description 'diffuse' dates from earlier times when nebulae were defined according to their optical appearance. A diffuse nebula was one which retained its fuzzy appearance under magnification through a large telescope, as opposed to one which could be resolved into stars.

diogenite A class of calcium-poor achondrite meteorite; also known as *hypersthene achondrites*. They are named after the Greek philosopher Diogenes of Apollonia (*fl.* 5th cent. BC), who recognized the cosmic origin of meteorites. Diogenites are composed almost entirely of the mineral bronzite (orthopyroxene), with lesser amounts of the other minerals present in ordinary chondrites. The diogenites probably formed by partial melting of a parent body with a chondritic composition, followed by slow cooling within the crust of that parent body. *See also* BASALTIC ACHONDRITE.

Dione The fourth-largest satellite of Saturn, 1123 km in diameter; also known as Saturn IV. Dione orbits 377 650 km from Saturn in 2.737 days, sharing its orbit with the much smaller *Helene. Its axial rotation is the same as its orbital period. Dione was discovered in 1684 by G. D. *Cassini. The two hemispheres of Dione show different crater densities, the trailing face being more heavily cratered. The trailing hemisphere also displays extensive bright complexes of scarps and fissures which give an overall appearance of wispy streaks many hundreds of kilometres in length. Dione's density is $1.5 \, \text{g/cm}^3$, suggesting that it consists of ice with a rocky core. 📷

dioptric system An optical system that uses only lenses to focus light. A refracting telescope is an example.

Diphda An alternative name for the star *Deneb Kaitos.

dipole 1. A system of two equal and opposite closely spaced electric charges (*electric dipole*) or magnetic poles (*magnetic dipole*).
2. A *dipole antenna.
3. Any system having a bipolar structure about an axis of symmetry (e.g. a **dipole field**, or **dipole anisotropy**).

dipole antenna An antenna consisting of a straight metal rod or wire broken in the centre to make two terminals. The terminals are connected to a *feed* which carries the signal to the receiver. The maximum sensitivity is in the direction perpendicular to the rods. *See* FOLDED DIPOLE; FULL-WAVE DIPOLE; HALF-WAVE DIPOLE.

Dirac cosmology A cosmological theory built around the so-called large numbers hypothesis, which relates the fundamental constants of subatomic physics to large-scale properties of the Universe such as its age and mean density. It is due to the British mathematical physicist Paul Adrien Maurice Dirac (1902–84). Dirac's theory is not widely accepted, but it introduced ideas related to the *anthropic principle.

directivity The ratio of the maximum sensitivity of an antenna to its average sensitivity. Directivity is proportional to the effective aperture of the antenna and inversely proportional to the square of the wavelength. Where there are no losses in the antenna, the directivity is equal to the gain over an isotropic antenna.

direct motion The movement of a body such as a planet from west to east on the celestial sphere; or the movement of a body in its orbit in an anticlockwise sense as seen from above the Sun's north pole; or the rotation of a body on its axis in an anticlockwise sense as seen from above the Sun's north pole. Direct motion, also known as *prograde motion*, is the normal direction of orbital motion and axial rotation of bodies in the Solar System; the opposite direction is *retrograde. Objects with direct motion have an orbital or axial inclination less than $90°$.

dirty snowball A model of a comet's nucleus, proposed by F. L. *Whipple in 1949, and confirmed by the *Giotto space probe to Halley's Comet in 1986. According to this model, cometary nuclei consist of a relatively solid conglomerate of frozen ices and dust, rather than a loose 'flying sandbank' as suggested by an alternative model.

disconnection event A separation in the gas tail of a comet, resulting from changes in the strength and direction of the local interplanetary magnetic field (IMF) carried by the solar wind. The existing gas tail separates from the coma and is swept away downwind, to be replaced by a new tail which develops in a slightly different direction governed by the prevailing IMF. The new gas tail can, in turn, be subject to disconnection. Several such events were recorded in the tail of Halley's Comet at its 1910 return, which coincided with a period of high solar activity.

Discovery Channel Telescope (DCT) A 4.2-m telescope owned and operated by Lowell Observatory, due to be opened in 2011. It will be used in the search for near-Earth objects and extrasolar planets, and for studying the Kuiper Belt, as well as for educational and broadcasting purposes. The DCT is sited at an altitude of 2360 m on a peak called Happy Jack in the Coconino National Forest, about 65 km southeast of Flagstaff, Arizona. It will be able to switch from an ultra-wide-field (prime focus) mode for surveys to a longer focal length for spectroscopy and detailed imaging.

((∰)) SEE WEB LINKS
• Official telescope website.

Discovery Program A NASA series of relatively simple, low-cost planetary missions capable of being developed and built within three years. The first was *NEAR Shoemaker. Others include *Dawn, *Deep Impact, *Genesis, *Kepler, *Lunar Prospector, *Mars Pathfinder, *Messenger, and *Stardust.

((∰)) SEE WEB LINKS
• Official Discovery Program website.

dish antenna Another name for a *parabolic antenna.

disk galaxy A type of galaxy whose main structure is a thin disk of stars in approximately circular orbits about the centre, and whose light output typically falls off exponentially with radius. The term applies to all types of galaxy other than ellipticals, dwarf spheroidals, and some peculiar galaxies. The disk in *lenticular galaxies (type S0) contains very little interstellar material, while the disks of spiral and irregular galaxies contain considerable amounts of gas and dust in addition to the stars.

disk population Those stars which, like the Sun, lie in a flattened disk in the Galaxy, and move in nearly circular orbits around its centre. They are Population I stars of all ages up to the age of the disk, but in general are younger than stars in the *galactic halo. Disk stars in the solar neighbourhood orbit the Galaxy at similar speeds to the Sun, and therefore have low velocities relative to it. *See also* HALO POPULATION.

disparition brusque *See* FILAMENT.

dispersion 1. The splitting of light into its constituent colours by a prism or diffraction grating. *High dispersion* means that the spectrum of an object is spread out more than in *low dispersion*, and thus more details can be seen; the degree of dispersion thus governs spectral resolution. A high-resolution spectrograph will produce a typical dispersion of 1–2 nm/mm.
2. A phenomenon in which electromagnetic waves of different frequencies travel at different speeds in certain media, such as ionized gases (plasmas). Interstellar matter is largely ionized hydrogen and is therefore dispersive. The time delay of a wave introduced by dispersion depends on the electron density along the line of sight, and is inversely proportional to the square of the observing frequency. Dispersion smears out rapid fluctuations in the source, a matter of great importance in the observation of pulsars. *See also* DEDISPERSION.

dispersion measure A measure of the amount of dispersion suffered by radio waves from an astronomical object. It depends on the electron density along the line of sight and is conventionally measured in parsecs per cubic centimetre. Dispersion measure is readily obtained from accurate timing of pulsars and is a key method of estimating the electron density in interstellar space.

distance modulus The difference between the apparent magnitude of an object, m, corrected for *interstellar absorption, and its absolute magnitude, M. This difference is directly linked to the distance in parsecs, r, by the formula

$m - M = 5 \log (r/10)$.

Put another way, if the distance modulus is known, the distance in parsecs can be found from

$r = 10 \times 1.585^{(m - M)}$.

Distance moduli are used for finding the distances to objects too far away to show a measurable parallax (i.e. beyond a few hundred light years). The distances of extragalactic objects are often referred to in terms of their distance modulus, rather than in parsecs or light years.

distortion Variation of magnification across the field of view of a lens. It can be either *pincushion or *barrel distortion.

diurnal Happening daily, or during the course of a day.

diurnal aberration The aberrational displacement of a star's position due to the velocity of the observer on the rotating Earth. It is much smaller than *annual aberration, amounting to only $0''.3$ on the equator.

diurnal inequality A variation in the observed movement of a celestial object caused by the observer's change in position as the Earth rotates. During the course of a night, the positions of the planets against the stellar background are apparently shifted because the observer is carried around by the rotating Earth.

diurnal libration An effect which allows observers on Earth to see almost 1° farther around the east and west limbs of the Moon's Earth-facing hemisphere. It results from the fact that the observer's position relative to the Moon changes between moonrise and moonset (see the diagram). The amount of diurnal libration depends on the observer's position on Earth, and can be up to $1°02'$ for an observer at the equator.

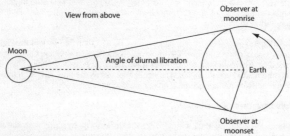

diurnal libration: A slight change in an observer's position relative to the Moon during the course of a day

diurnal motion Daily motion, such as the diurnal rotation of the celestial sphere.

diurnal parallax An alternative name for *geocentric parallax.

diverging lens A lens that is thicker at the edges than at the middle, so that parallel light entering it is diverged away from a point; also known as a *negative lens*. The view through a diverging lens is upright but appears smaller than in reality. It cannot be used by itself to give a real image that can be captured on a screen, but instead gives a *virtual image*—one that can be seen only by looking through the lens. *See also* CONVERGING LENS.

D layer The lowest region of Earth's *ionosphere, between 80 and 90 km in altitude. The D layer principally absorbs, rather than reflects, radio waves, particularly around sunspot maximum. *Sudden ionospheric disturbance events result from enhancements of the dayside D layer by short-wave radiation from solar flares.

D lines Two prominent lines close together in the yellow part of the spectrum at wavelengths of 589.0 nm (D_2) and 589.6 nm (D_1), produced by neutral atoms of sodium. They were designated as feature D in the Sun's spectrum by J. von *Fraunhofer. The sodium D lines are intrinsically strong in relatively cool stars, including the Sun, and are also present in the spectra of very distant stars due to absorption by sodium atoms in interstellar space.

dMe star An M-type red-dwarf star whose spectrum displays emission lines (hence the suffix 'e') due to hydrogen (Hα) and calcium (H and K lines). This activity indicates enhanced coronal emission (a dMe star is usually also an X-ray source) as a result of the powerful magnetic fields generated by the deep convection zone in the star.

Dobsonian telescope A Newtonian telescope on a particular type of altazimuth mounting. The base is a flat board, on which rotates an open-topped box; the altitude axis of the telescope rests in V-shaped cutouts in the sides of the box (see the diagram). An important feature of the design is the use of Teflon pads to provide low-friction bearing surfaces. The design is named after the American amateur astronomer John Lowry Dobson (1915–).

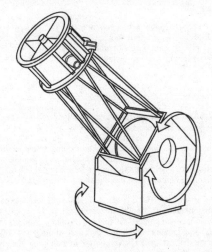

Dobsonian telescope

Dog Star Popular name for *Sirius, the most brilliant star in the sky, which lies in the constellation Canis Major, the Great Dog.

Dollond, John (1706–61) English optician. In 1753 he invented the *heliometer, and four years later discovered that an *achromatic lens could be made by using two different types of glass, crown glass and flint glass. Although the principle had been discovered earlier, it was not widely known and Dollond received the credit for his independent invention. In 1765 his eldest son, Peter Dollond (1730–1820), developed the achromatic triplet lens by placing convex lenses of crown glass either side of a biconcave lens of flint glass.

domain wall A hypothetical two-dimensional defect in the structure of *spacetime, produced in some models of the early Universe. In essence, it is a wall-like structure in which energy is trapped. A theory of particle physics that predicts large numbers of domain walls would also predict a highly inhomogeneous Universe, contrary to observations. *See also* COSMIC STRING; COSMIC TEXTURE; MAGNETIC MONOPOLE.

dome 1. A steep-sided extrusion of viscous lava forming a bulbous mass over a volcanic vent; also known as a *volcanic dome*. An example is the dome that formed within the Mount St Helens crater, Washington state, after the dramatic eruption in 1980; other examples occur on Venus.
   2. A low, circular raised area above a lunar mare, often with a pit in the centre; also known as a *lunar dome*. Slopes of lunar domes are usually only a few degrees. They are thought to be volcanic vents, in many cases the source of the fluid lavas that formed the maria, and are

thus similar to shield volcanoes on Earth. Examples of lunar domes are found near the crater Hortensius.

Dominion Astrophysical Observatory (DAO) An optical observatory at an altitude of 230 m in Victoria, British Columbia, Canada, founded in 1916 and operated by the Herzberg Institute of Astrophysics. Its main instrument is the 72-inch (1.85-m) Plaskett reflector, named after the Observatory's first director, J. S. *Plaskett, opened in 1918 and given a new mirror in 1974. Other instruments include the 1.22-m McKellar reflector, opened in 1962 and given a new mirror in 1985. The DAO also houses the Canadian Astronomy Data Centre.

(⊕) SEE WEB LINKS
• Official observatory website.

Dominion Radio Astrophysical Observatory (DRAO) A radio observatory at Penticton, British Columbia, Canada, founded in 1960. It is operated by the Herzberg Institute of Astrophysics. Its main instruments are a Synthesis Telescope, opened in 1972, currently consisting of seven 9-m radio dishes on a baseline 600 m long, and a 25.6-m dish opened in 1960.

(⊕) SEE WEB LINKS
• Official observatory website.

Donati, Comet (C/1858 L1) A long-period comet discovered on 1858 June 2 by the Italian astronomer Giovanni Battista Donati (1826–73); formerly designated 1858 VI. In 1858 September it reached magnitude −1. Perihelion, 0.58 AU, was on 1858 September 30, with closest approach to Earth (0.5 AU) on October 9. Around this time the comet showed a prominent curved dust tail resembling a scimitar, for which it is famous. At best, this dust tail stretched for 60°, with two thin, straight ion tails at a tangent to it. The nucleus threw off shells of material as it rotated. The comet's orbit has a period of nearly 2000 years, eccentricity 0.996, and inclination 117°.0.

Doppler, Christian Johann (1803–53) Austrian physicist. In 1842 he described the *Doppler shift for sound waves, and gave the mathematical equation for the variation of pitch with the velocity of the source. He realized that the effect would occur with all wave motions, including light. However, he incorrectly believed that the colours of stars indicated their motions towards or away from the Earth.

Doppler broadening The widening of spectral lines caused by large-scale motions of the emitting or absorbing gas. For example, the Doppler shift in the approaching and receding hemispheres of a rotating star will cause the absorption lines from that star to be broadened.

Doppler effect *See* DOPPLER SHIFT.

dopplergram An image of the Sun that shows the line-of-sight movement of gas. Such images are derived from the Doppler shift of the gas measured in a particular spectral line either by a spectrometer or with narrow-bandpass filters. The latter method is used for *helioseismology measurements.

Doppler shift The change in wavelength of electromagnetic radiation as a result of relative movement between the source and the observer. If the source is moving towards the observer, the wavelengths become shorter and the spectral lines are shifted towards the blue end of the spectrum (a *blueshift*). If the source is receding, the wavelengths become longer and spectral lines are moved towards the red end of the spectrum (a *redshift). The effect is named after C. J. *Doppler.

Dorado (Dor) (*gen.* **Doradus)** A constellation of the southern sky, representing a dolphinfish, also known as mahi-mahi. Its brightest star is Alpha Doradus, magnitude 3.3. Beta Doradus is one of the brightest Cepheid variables, ranging between magnitudes 3.5 and 4.1 with a period of 9.83 days. Doradus contains most of the *Large Magellanic Cloud, including the *Tarantula Nebula and the site of *Supernova 1987A.

dorsum A ridge on a planetary surface; pl.*dorsa*. The name is not a geological term, but is used in the nomenclature of individual features, for example Schiaparelli Dorsum on Mercury.

Double Cluster A pair of 4th-magnitude open star clusters in Perseus, also known as NGC 869 and 884, or h and χ (Chi) Persei, each about ½° across. They lie about 7500 l.y. away in the Perseus spiral arm of our Galaxy, and are members of the Perseus OB1 association. NGC 869 is the richer, with around 200 stars against 150 in NGC 884. Both clusters are only a few million years old.

Double Double The quadruple star Epsilon Lyrae. With binoculars, or even sharp eyesight, the star appears as a wide double of magnitudes 4.6 and 4.7 which form a genuine binary. Moderate-sized telescopes reveal that each star is itself double. The pair known as Epsilon^1 consists of an A4 dwarf and an F1 dwarf of magnitudes 5.0 and 6.1, orbiting each other in a period of about 1725 years; the Epsilon^2 pair consists of an A8 dwarf and an F0 dwarf, magnitudes 5.2 and 5.5, orbital period about 720 years. The four stars lie about 160 l.y. away.

double-lined binary A spectroscopic binary in which both components have similar brightnesses; sometimes also called an *SB2 system*. During certain phases of the orbit the spectral lines appear doubled because of the components' differing radial velocities. The effect is particularly noticeable if the spectral types of the two components are similar. *See also* COMPOSITE-SPECTRUM BINARY; SINGLE-LINED BINARY.

double-mode variable A pulsating variable whose complex variations consist of superimposed oscillations at the *fundamental mode and an additional overtone. The first overtone is found in a *beat Cepheid, and other types of variable. The *Blazhko effect in certain RR Lyrae stars appears to be explained by the presence of the third overtone.

double pulsar Two pulsars in orbit around each other, as distinct from a binary in which only one member is a pulsar (*see* BINARY PULSAR). The first, and so far only, example of a double pulsar, J0737–3039AB, was discovered in 2003 in the constellation Puppis. Each of the pulsars has a mass about 1.3 times that of the Sun, and their orbital period is 2.4 hours. The theory of general relativity predicts that such a binary should lose energy through the emission of *gravitational waves, which makes the double pulsar an excellent natural laboratory for testing relativity. As deduced from the change in orbital period, the separation of the pulsars is reducing by about 7mm per day, in exact agreement with theory. It is expected that the stars will eventually merge in approximately 85 million years. In addition, the *Shapiro delay in the travel time of the pulse from the more distant pulsar as it passes through the curved space around the nearer one is exactly as predicted.

double quasar A quasar whose image has been split into two by the *gravitational lens effect of a massive galaxy or cluster of galaxies along the line of sight. The first example found, 0957+561 A and B, is a quasar at a redshift of 1.41 whose twin images lie 6″ apart. In this case the lensing is caused by a cluster at a redshift of 0.36, one of whose galaxies is visible 1″ away from one of the quasar images. The discovery of this double quasar in 1979 was the first confirmation (apart from measurements at eclipses of small deflections of light by the Sun) of the existence of gravitational lenses. More complex gravitational lens effects produce more than two images.

double star Two stars that appear close to one another on the sky. Such pairs may be divided into two classes: *optical doubles*, where the components are not gravitationally bound; and *physical doubles*, in which the stars are orbiting their common barycentre. The term 'double star' is often restricted to the former group, and the term *binary star is used for the latter. In fact, optical doubles are relatively uncommon, and the majority of doubles are actually true binary systems.

Double Star (mission) A pair of satellites launched and operated by the China National Space Administration to study the Earth's magnetosphere. The satellites were called TC-1 and TC-2 (Tan-Ce is Chinese for 'Explorer'). TC-1, also known as the equatorial satellite, was launched in 2003 December into a highly elliptical orbit inclined at 28°.5 to the equator to study the Earth's *magnetotail. TC-2, known as the polar satellite, was launched into a polar orbit in 2004 July to study events over the magnetic poles, including the development of aurorae.

The Double Star satellites worked in conjunction with ESA's *Cluster mission. TC-1 re-entered in 2007 October.

(()) SEE WEB LINKS
● ESA mission website.

doublet 1. Two closely spaced lines in a spectrum that are due to the same element, for example the sodium doublet at 589.0 and 589.6nm (the sodium *D lines). A doublet results from electrons jumping from a lower level to one of two closely spaced upper levels.

2. A lens consisting of two components, usually making an *achromatic lens. These usually have one component of *crown glass and one of *flint glass.

DQ Herculis star See INTERMEDIATE POLAR.

Draco (Dra) (*gen.* **Draconis**) The eighth-largest constellation, representing a dragon curled around the north celestial pole. Its brightest star is Gamma Draconis (*Eltanin). Alpha Draconis is known as *Thuban. Draco contains several interesting double stars. They include Mu Draconis, a close pair each of magnitude 5.7 which orbit in a period of 670 years, and Nu Draconis, a binocular pair, both magnitude 4.9. NGC 6543 is a 9th-magnitude planetary nebula.

draconic month The mean interval of time between successive passages of the Moon through the ascending node of its orbit, the point at which it passes from south to north of the ecliptic; also known as a *nodical month*. The Moon's orbit is inclined at about 5° to the ecliptic but it is not fixed in space. It has a slow variation which causes the two nodes to regress around the ecliptic every 18.61 years. Consequently the draconic month (27.2122 days) is about 2.5 hours shorter than the *sidereal month.

Draconid meteors Another name for the *Giacobinid meteors.

Drake equation A formula for calculating the number of civilizations in our Galaxy from which we might detect artificial transmissions. The equation, proposed in 1961 by the American radio astronomer Frank Donald Drake (1930–), multiplies together the following factors: R^*, the rate of formation of stars in our Galaxy with suitable environments for the development of intelligent life; f_p, the fraction of those stars that have planetary systems; n_e, the number of planets in each system that are suitable for life; f_l, the fraction of such planets on which life actually arises; f_i, the fraction of life-bearing planets on which intelligent life emerges; f_c, the fraction of civilizations that release detectable signs of their existence into space; and L, the average lifetime of such civilizations. The solution to the equation depends on the values assigned to each factor, many of which are not well determined. Possible results for the number of communicative civilizations in the Galaxy at present range from almost zero to many thousands.

Draper, Henry (1837–82) American doctor and pioneer astrophotographer, son of J. W. *Draper. From 1863, using telescopes and spectrographs of his own construction, he took 1500 high-quality lunar photographs, and obtained the best contemporary photographs of the Sun's spectrum. In 1872 he recorded the first spectrum of another star, Vega, and identified in it prominent hydrogen lines. In 1879 he switched to dry photographic plates and obtained excellent spectra of stars, the Moon, Mars, Jupiter, the comet now designated C/1881 K1, and the Orion Nebula. After his death, his widow Mary Anna Draper, née Palmer (1839–1914), donated the funds that made possible the *Henry Draper Catalogue.

Draper, John William (1811–82) English scientist, father of H. *Draper; he moved to the USA in 1832. Draper was a pioneer photographer. In the winter of 1839/40 he obtained the first daguerreotype (early form of photograph) of the Moon, in which the maria were distinctly visible. In 1843 he took what is believed to be the first photograph of the Sun's infrared spectrum, recording three Fraunhofer lines, and he photographed the Sun's ultraviolet spectrum at about the same time.

drawtube The tube that carries the eyepiece of a telescope.

Dreyer, Johan Ludvig Emil (1852-1926) Danish astronomer, known also as John Louis Emil Dreyer. He moved to Ireland in 1874 to work, originally at Lord *Rosse's observatory, and later compiled the *New General Catalogue* (NGC) and its supplements, the two *Index Catalogues* (IC). Also a historian of astronomy, Dreyer edited the scientific papers of William Herschel and wrote a biography of his childhood hero, Tycho Brahe.

drifting sub-pulse A *sub-pulse seen to occupy a different position in successive pulse profiles recorded from a pulsar and giving the impression of drifting. The origin of sub-pulses is not understood.

drift scan A technique for surveying the sky in which the telescope is fixed and the rotation of the Earth carries celestial objects through its beam. Many radio interferometers designed for survey work are not physically steerable and rely on drift scanning to cover large areas of sky. Drift scanning has the advantage of a uniform, precise scan rate, and consequent accurate positioning, but it is relatively inflexible. *See also* RASTER SCAN.

drive A means of turning the mounting of a telescope to counteract the motion of a celestial object across the sky as the Earth rotates; also called a *clock drive*. An electric or a clockwork motor is used to drive the telescope or, in older telescopes, a slowly falling weight. Electronics can control the rate of the motor and provide speed corrections to compensate for mechanical errors in the telescope mounting and the effects of atmospheric refraction.

D star The spectral classification of white dwarf stars. The original sequence used was DB, DA, DF, DG, and DK, in order of declining effective temperature from 100000 to 4000K (i.e. the second letter follows the normal sequence of spectral types, B to K). White dwarfs with a continuous (featureless) spectrum were classified as type DC. Subsequently, types DO (extremely hot, showing only ionized helium lines) and DQ (containing atomic or molecular carbon in addition to their mainly helium atmospheres) were added, and type DZ was introduced to combine the former types DF, DG, and DK. Also, a temperature index was added from 0 (hottest) to 9 (coolest). In the fully developed scheme, a DC9 white dwarf, for example, is very cool and contains no detectable absorption lines. The majority of white dwarfs are type DA (as is Sirius B), showing only broad absorption lines of hydrogen, their breadth resulting from the star's enormous surface gravity. Most of the rest are type DB, which show only neutral helium lines. The metal lines in type DZ are thought to be caused by material accreted by the white dwarf from interstellar space.

Dubhe The star Alpha Ursae Majoris, magnitude 1.8. It is a K0 giant lying 123 l.y. away. It is a close double star, with an A8 dwarf companion of magnitude 4.8 that orbits it in a period of 45 years. It is one of the two Pointers that indicate the direction of the Pole Star, Polaris.

Dumbbell Nebula An 8th-magnitude planetary nebula in Vulpecula, also known as M27 and NGC 6853. It is the most prominent planetary nebula for small telescopes. The Dumbbell Nebula is so named because of its double-lobed appearance, like an hourglass, in large apertures or on photographs. The Dumbbell lies about 1400 l.y. away. 📷

dust Small particles of solid matter found in space: in the Solar System (interplanetary dust and cometary dust), around stars (circumstellar dust), and between the stars (interstellar dust). The individual particles are usually called *dust grains*, and are of size 10nm and upwards. Dust comprises about 1% by mass of interstellar matter. It extinguishes and reddens starlight, and can also be detected by its absorption and emission of infrared radiation and by its polarizing effect on starlight. The exact composition of interstellar dust is uncertain: silicates, graphite, and carbides are among the components identified through infrared absorption measurements. Dust may account for the *diffuse interstellar bands in the optical spectra of stars. Dust is produced in the cool outer envelopes of red-giant stars, in novae, and in supernovae. Dust is also present in the Solar System as interplanetary dust and cometary dust. Some dust is driven out of the Solar System by the radiation pressure of the Sun's light, while new dust is generated by asteroid collisions and the break-up of comets. *See also* GRAINS, INTERPLANETARY; GRAINS, INTERSTELLAR.

dust tail *See* TAIL, COMETARY.

duty cycle The fraction (typically 5 %) of a pulsar's period that is occupied by pulsed emission.

dwarf Cepheid A now-obsolete name once applied to *AI Velorum stars and *Delta Scuti stars.

dwarf galaxy A small, low-luminosity galaxy. Dwarf ellipticals (dE) and dwarf irregulars (dIrr) are probably just fainter examples of this general type of galaxy but *dwarf spheroidals are a separate type. Dwarfs in general are the most common galaxies in the Universe. There are no dwarf spirals, since no spirals are known with an absolute blue magnitude below −17 (about 10^9 solar luminosities).

dwarf nova A form of cataclysmic binary, more specifically known as a *U Geminorum star, that undergoes repeated outbursts. These occur when thermal instabilities trigger a sudden brightening of the accretion disk around the white-dwarf primary. The individual maxima have a general resemblance to the light-curve of a nova, although with a smaller amplitude (2–6 mag.), and a shorter duration (a few days). The intervals between outbursts are far shorter than those in a *recurrent nova, and although irregular and unpredictable, show a mean cycle length for each individual star. Accretion on to the primary will probably eventually produce a nova outburst.

dwarf planet A new category of planet introduced in 2006 by the International Astronomical Union. As defined by the IAU, a dwarf planet has mass sufficient for its own gravity to have rendered it spherical or near-spherical, yet too small to have removed material around its orbit by accretion or gravitational disturbance. The first five recognized dwarf planets are *Ceres, *Eris, *Pluto, Haumea, and Makemake. Dozens more may exist beyond Neptune.

dwarf spheroidal galaxy The most common type of galaxy in the Universe, although not recognized until 1938 (the Sculptor dwarf was the first example discovered) because of their low luminosity and low surface brightness; symbol dSph. Dwarf spheroidals are similar to ellipticals in that they contain little or no interstellar material, but they are not so centrally concentrated, have low density, low surface brightness, and a decrease of light output with radius which implies that they are more closely related to small *disk galaxies. Dwarf spheroidals have diameters up to about 10 000 l.y. and masses up to about 10^7 Suns. Half the galaxies in the *Local Group are dwarf spheroidals, mostly clustered around M31 and our own Galaxy.

dwarf star A star on the main sequence of the Hertzsprung–Russell diagram. Dwarf stars are of *luminosity class V. Most stars are of this type, as is the Sun. They range in mass from about 0.1 to 100 solar masses. The term is used because stars on the main sequence are smaller than those of the same mass which have evolved into giants. White dwarfs, black dwarfs, and brown dwarfs are not dwarfs in this sense of being main-sequence stars.

dwarf variable An obsolescent term for any low-luminosity variable star. It was generally used for various forms of *irregular variable (particularly for *T Tauri stars), *flare star, *flash star, or *dwarf nova.

Dwingeloo galaxy A galaxy found in the Dwingeloo Obscured Galaxy Survey (DOGS), a radio sky survey using the 25-m radio telescope of the *Dwingeloo Radio Observatory, Netherlands. The survey looks for 21-cm radiation from galaxies obscured from optical view by gas and dust in the plane of our own Galaxy. Dwingeloo 1, the first such galaxy found, in 1994, is a barred spiral about 10 million l.y. away, and is probably a member of a group associated with the *Maffei Galaxies.

Dwingeloo Radio Observatory A radio observatory at Dwingeloo, about 60 km southwest of Groningen, in the Netherlands, owned by the Netherlands Institute for Radio Astronomy (ASTRON), which has its headquarters here. It contains a single 25-m dish, opened in 1956, now used mostly by amateur astronomers and for educational purposes.

dynamical equinox The intersection of the celestial equator with the ecliptic, as determined from observations of the Sun or other members of the Solar System. The offset of the dynamical equinox from the catalogue equinox is derived by measuring the positions of Solar System objects relative to the catalogue positions of stars.

dynamical friction The force experienced by a mass such as a star cluster or dwarf galaxy travelling through an extended distribution of stars. The force arises because the gravitational effect of the mass as it moves causes a slight enhancement of the density of stars behind it, and the consequent increased gravitational influence acts as a drag on the mass. Dynamical friction is believed to be important in the merging of galaxies, and in the decay of the orbits of globular clusters around galaxies.

dynamical parallax An estimate of the parallax of a visual binary for which the period and semimajor axis have been obtained. The semimajor axis in astronomical units is derived from Kepler's third law by assuming a value for the total mass of the system based on the stars' spectral types.

dynamical time The time-scale that is used in calculating orbital motions within the Solar System. The underlying physical law governing such motions is the law of gravitation. In Newtonian gravitation, time plays a fundamental role; it is absolute and universal, and the problem is to work out the positions of bodies for each instant of time. The time-scale that was used until 1984 was called *Ephemeris Time. Since then, however, relativistic effects have been included. According to the theory of relativity, time is different for each observer. It has therefore been necessary to introduce two distinct dynamical time-scales, namely *Terrestrial Time (originally Terrestrial Dynamical Time) and *Barycentric Dynamical Time. There are only very slight periodic differences between the two, of the order of a thousandth of a second.

dynamo The activity in an electrically conducting fluid that generates an electrical current and a magnetic field. Dynamo action in the plasma in the outer layers of the Sun, and in the core or mantle of the Earth and other planets, is thought to produce the magnetic fields of these bodies.

Dyson, Frank Watson (1868–1939) English astronomer. He became the ninth Astronomer Royal in 1910. His main concern was with the accurate measurement of star positions and proper motions, which revealed the distribution of stars in our part of the Galaxy. From observations at several eclipses he identified elements in the solar chromosphere from lines in its spectrum; this success motivated the Greenwich eclipse expedition of 1919 which confirmed Einstein's prediction that a gravitational field bends starlight. Dyson introduced the broadcasting of radio time signals from Greenwich in the 1920s.

Eagle Nebula A diffuse nebula surrounding the star cluster M16 (NGC 6611) in Serpens, also known as IC 4703. It is about ½° wide. Dark columns of gas and dust, termed **elephant trunks**, protrude into the brighter parts of the nebula. New stars are being born from denser pockets within the elephant trunks, a process photographed in 1995 by the Hubble Space Telescope. The Eagle Nebula lies 8500 l.y. away. ▣

early-type galaxy An elliptical or lenticular galaxy—one without spiral arms. The name comes from the position of these galaxies in the *tuning-fork diagram of galaxy shapes. For similar reasons, an Sa spiral galaxy may be referred to as an early-type spiral, as opposed to a late-type Sc or Sd spiral. *See also* HUBBLE CLASSIFICATION.

early-type star Any massive, hot star of spectral type O, B, or A. The designation 'early' derives from an old, mistaken idea that stars evolved from being hot and young to cool and old. The term is also used to refer to the hotter end of any given spectral class: for example, a K1 star is earlier than a K5 star. *See also* LATE-TYPE STAR.

Earth (⊕, ♁) The third planet from the Sun. At perihelion each January it is 147 099 590 km from the Sun, as against 152 096 150 km at aphelion in July. Seen from far off in space it has a strong blue coloration, caused by its atmosphere. The Earth is slightly ellipsoidal in shape (equatorial diameter 12 756 km, polar diameter 12 714 km). Its age is 4.57×10^9 years.

Earth			
Physical data			
Diameter (equatorial)	Oblateness	Inclination of equator to orbit	Axial rotation period (sidereal)
12 756 km	0.0034	23°.44	23.934 hours
Mean density	Mass	Mean albedo (geometric)	Escape velocity
5.52 g/cm³	5.974×10^{24} kg	0.37	11.19 km/s

Orbital data				
Mean distance from Sun				
10^6 km	AU	Eccentricity of orbit	Inclination of orbit to ecliptic	Orbital period (sidereal)
149.598	1.0	0.017	0°.0	365.256 days

Averaged over a year, the Earth's atmosphere has a composition (by volume) of 78% nitrogen, 21% oxygen, and 0.9% argon, plus carbon dioxide, hydrogen, and other gases in much smaller quantities. Water vapour is also present in variable quantities. White clouds of condensed water vapour may obscure a quarter of the Earth's surface at a time, belts of cloud being common around the equator and at temperate and polar latitudes. The pressure of the atmosphere at sea level varies around 1000 mbar. The average atmospheric temperature at the surface is 15°C, but ranges from −50°C average in winter in Siberia up to +40°C in the Sahara in summer.

Liquid water covers 71% of the Earth's surface. Volcanism and impact cratering occur; more than 500 volcanoes are presently known to have been active on land over the course of recorded human history. Impact cratering was an important process early in the Earth's history, but the atmosphere now protects the surface from all but the largest meteoroids. Over 170 impact crater sites have been found on the Earth, but most are old and heavily eroded. The dominant geological process on the Earth's surface is erosion and deposition by water or ice. Liquid water is also responsible for the development of life, which has itself played an important role in transforming the landscape.

The Earth's outer layer is the *lithosphere*, topped by the crust, which together vary in thickness between 70 km in parts of the oceans to 150 km in the thickest parts of the continents. Below this is the *mantle*, which stretches down to a depth of about 2900 km, where the iron–nickel core begins. This core has allowed the development of a magnetic field, which attains a strength of about 3×10^{-5} tesla near the equator. Convection within the mantle, coupled with the thin crust, has given rise to plate tectonics and continental drift, creating vast ranges of mountains and the ocean deeps. The Earth has one natural satellite, the Moon. 📷

Earth coorbital asteroid An asteroid moving in a similar orbit to that of the Earth. Such asteroids have an orbital period of one year and hence are said to be in 1:1 mean motion resonance with the Earth. Most of the objects in this category probably come from the population of *near-Earth asteroids, but some may be ejecta from past impacts on the Moon and a few may be man-made space debris.

Earth-crossing asteroid Another name for an asteroid in the *Apollo or *Aten group. *See also* NEAR-EARTH ASTEROID.

Earth-grazing asteroid Another name for an *Amor group asteroid; also known as an **Earth-approaching asteroid**. *See also* NEAR-EARTH ASTEROID.

earthlight Another name for *earthshine.

Earth rotation angle (ERA) (θ) The modern equivalent of Greenwich Sidereal Time (GST), referred to a point called the Celestial Intermediate Origin (CIO) instead of the equinox. The CIO is a point defined not by its position but by its motion, and at present lies very close to the prime meridian of the *Geocentric Celestial Reference System (within 0.1 arcsec throughout the 21st century). Unlike GST, which is a complicated function of both UT1 and Terrestrial Time and includes precession and nutation terms, ERA is a simple linear function of UT1 alone.

earthshine A faint illumination of the dark part of the Moon when it is a thin crescent, caused by sunlight reflected from the Earth; also known as **earthlight**. The effect is popularly termed 'the old Moon in the new Moon's arms'.

eccentric anomaly (*E*) The angle between the periapsis of an orbit and a given point on a circle around the orbit, as seen from the centre of the orbit (see diagram). The point concerned is found by drawing a line perpendicular to the major axis through the actual position of the orbiting body until it cuts the circle around the orbit, as in the diagram. The diameter of the circle is equal to the major axis of the orbit, and its centre is also the orbit's centre. The angle of eccentric anomaly is measured in the direction of orbital motion. *See also* KEPLER'S EQUATION.

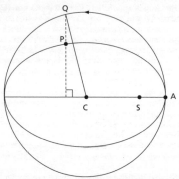

eccentric anomaly: The eccentric anomaly of a planet's position is the angle ACQ, where A is the perihelion, C is the centre of a circle drawn around the planet's elliptical orbit, and Q is a position on that circle found by drawing a line perpendicular to the major axis of the orbit through the actual position of the planet, P. The Sun is at S.

eccentricity (e) A measure of the shape of an orbit, or how far it diverges from a circle. If $e = 0$, the orbit is a perfect circle. If e is less than 1 the orbit is an *ellipse, and the nearer e is to 1, the more elongated the ellipse. If $e = 1$ exactly, the orbit is a *parabola, and if e is greater than 1 the orbit is a *hyperbola. The orbits of the planets and major satellites are ellipses of small eccentricity. Eccentricity is one of the elements of an orbit (*see* ELEMENTS, ORBITAL).

echelle grating A *diffraction grating with rulings spaced relatively widely (typically 30–100 per millimetre). Because of the small number of rulings, the grating produces numerous overlapping spectra that must be separated before they reach the detector. Echelle gratings are easier to manufacture than gratings with a higher density of rulings, but require a more complex spectrograph design (*see* ECHELLE SPECTROGRAPH).

echelle spectrograph A spectrograph that uses an *echelle grating to achieve very high spectral resolution and wide wavelength coverage. The overlapping spectra produced by the grating must be separated before they reach the detector. The separation is usually done by a prism or a *grism (sometimes called a *cross-disperser*), which places the sorted spectra one above the other. In combination with sensitive detectors such as CCDs, echelle spectrographs are becoming increasingly important in astronomy for high-resolution spectroscopy.

E-class asteroid A rare class of asteroid whose members have a featureless reflectance spectrum, flat to slightly reddish over the wavelength range 0.3–1.1 µm. E-class asteroids are distinguished from the spectrally identical classes M and P by their higher albedos (0.25–0.60). The E is for enstatite, since they are believed to have surfaces that are similar in composition to enstatite achondrite meteorites (*see* AUBRITE). Members of this class include (44) Nysa, diameter 70 km, and (214) Aschera, diameter 23 km.

eclipse The entry of one celestial body into the shadow of another; strictly speaking, an eclipse of the Sun is an occultation, not an eclipse. In any one year, the maximum total of solar and lunar eclipses visible from Earth is seven. The minimum number is two, both of which must be solar. Eclipses frequently come in pairs, with a lunar eclipse preceded or followed within a fortnight or so by a solar eclipse at the opposite node of the Moon's orbit. As seen from any one place, lunar eclipses are twice as common as solar eclipses.

Solar eclipses occur when the new Moon lies close to the node of its orbit, and at the same longitude as the Sun. The Moon obscures at least part of the Sun's disk as seen from a comparatively limited *ground track* where the cone of shadow cast by the Moon falls on the Earth's surface. Along the centre of this track a *total eclipse* may be seen; eclipses occurring around lunar apogee may be *annular eclipses. A *partial eclipse* is seen either side of the central path of totality or annularity, which sweeps eastwards at around 3200 km/h. The maximum diameter of the Moon's umbral cone at the Earth is 270 km, but the actual width of the shadow path on the Earth's surface can become much greater than this at high latitudes, where the umbral cone strikes the surface at an oblique angle. At a total eclipse, the Moon moves across the Sun's disk for an hour or so, until the Sun is completely covered and its corona becomes visible. By coincidence, the Sun and Moon have much the same apparent angular diameter (about 0°.5). Total solar eclipses are rare at any one place on the Earth, so astronomers usually have to travel long distances to see them. A solar eclipse can last up to 3 hours from *first contact* to *fourth contact*; totality has a theoretical maximum duration of 7 m 32 s, but is usually much shorter. A *lunar eclipse* is visible from Earth wherever the Moon is above the observer's horizon, and occurs as the full Moon passes through the Earth's shadow. A *total lunar eclipse* occurs when the Moon passes through the dark central umbra of the Earth's shadow. *Partial lunar eclipses* are also seen, as are *penumbral eclipses, which are often barely noticeable. The Earth's shadow is much broader than the Moon itself, so that lunar eclipses may last for up to 4 hours from *first contact* to *fourth contact*; totality lasts up to 1 h 47 m.

Planetary satellites are also eclipsed by the shadows of their primaries; those of the Galilean satellites of Jupiter are readily observable.

( ) SEE WEB LINKS

• NASA eclipse website.

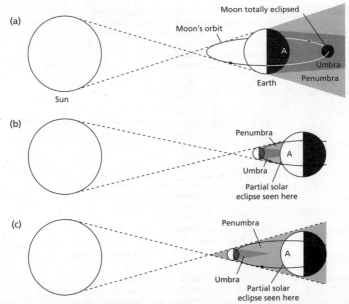

eclipse: (a) In a total lunar eclipse the Moon, viewed from point A, is faintly illuminated by sunlight refracted through the Earth's atmosphere. (b) An observer at A sees the Sun's disk completely obscured by the Moon in a total solar eclipse, but the solar corona becomes visible. (c) In an annular eclipse of the Sun, an observer at A sees the Moon framed by the outline of the Sun. In (b) and (c), observers within the Sun's penumbra see a partial solar eclipse.

eclipse season The time during which the Sun is close enough to one of the nodes of the Moon's orbit for an eclipse to take place. The eclipse season lasts about 37 days for solar eclipses, and about 24 days for lunar eclipses. Eclipse seasons recur every 173.31 days; two eclipse seasons equal one *eclipse year.

eclipse year The interval of time between successive passages of the Sun through a node of the Moon's orbit, when eclipses can occur. It lasts 346.62 days, the length of two *eclipse seasons. An eclipse year is shorter than a sidereal year because the nodes of the Moon's orbit regress westwards around the ecliptic by about 19° per year. There are almost exactly 19 eclipse years in a *saros.

eclipsing binary A binary star whose orbital plane is sufficiently close to our line of sight for partial or total eclipses of at least one of the components to occur. Regardless of the masses, the star with the higher surface luminosity is called the *primary*, and *primary minimum occurs when it is eclipsed by the fainter *secondary*. On the basis of the light-curve, eclipsing binaries are divided into three types: *Algol stars, *Beta Lyrae stars, and *W Ursae Majoris stars. Eclipses may also occur in various *cataclysmic variables, such as dwarf novae, novae, symbiotic stars, and related types.

ecliptic The apparent path of the Sun against the star background over the course of a year. The movement of the Sun along the ecliptic is actually a result of the Earth's movement in its orbit around the Sun. Therefore the ecliptic is actually the plane of the Earth's orbit projected on to the celestial sphere. Because of the Earth's axial tilt, the ecliptic is inclined at about 23°.4 to the celestial equator, an angle known as the *obliquity of the ecliptic. The ecliptic crosses the celestial equator at the *equinoxes. It takes its name from the fact that eclipses occur when the Moon is near the plane of the ecliptic.

ecliptic coordinates A system of coordinates that specifies the position of an object in the Solar System relative to the plane of the Earth's orbit, the ecliptic. Ecliptic coordinates of objects as they would be seen from the centre of the Earth (i.e. *geocentric ecliptic coordinates*) are given in terms of *celestial latitude and *celestial longitude (also known as ecliptic latitude and longitude). Ecliptic coordinates of objects as they would be seen from the centre of the Sun (i.e. *heliocentric ecliptic coordinates*) are given in terms of *heliocentric latitude and *heliocentric longitude.

ecliptic latitude Another name for *celestial latitude.

ecliptic limits The greatest angular distance that the Sun can be from the *nodes of the Moon's orbit and still cause an eclipse. The actual value varies with the changing distances, and hence apparent sizes, of the Sun and Moon. Maximum values (the *major ecliptic limits*) are attained when the Earth is at perihelion and the Moon is at perigee, while minimum values (the *minor ecliptic limits*) occur at aphelion and apogee. For a partial solar eclipse, the *solar ecliptic limit* (i.e. the greatest possible distance of the Sun from the node of the Moon's orbit at new Moon) has a maximum value of 18°.4; for a total solar eclipse it is 11°.8. For a partial lunar eclipse, the *lunar ecliptic limit* (i.e. greatest possible distance of the Sun from the Moon's node at full Moon) has a maximum value of 12°.2; for a total lunar eclipse, it is 5°.9.

ecliptic longitude Another name for *celestial longitude.

ecliptic pole Either of the two points on the celestial sphere that lie 90° north and south of the plane of the ecliptic. The north ecliptic pole lies in Draco, and the south ecliptic pole in Dorado. The ecliptic poles lie at the centre of the circle of precession swept out by the celestial poles every 25 800 years.

E corona That part of the light from the Sun's *corona that is due to emission lines from hot gas (the 'E' stands for emission). These emission lines include the so-called *forbidden lines of highly ionized atoms of iron, calcium, and some other elements. The E corona is much fainter than the *K and *F coronae.

Eddington, Arthur Stanley (1882–1944) English astrophysicist. From 1917 he studied stellar structure, applying discoveries in atomic physics to the understanding of radiation pressure, and using A. *Einstein's discovery of mass–energy equivalence to explain energy generation in stars. He obtained observational proof that gravity bends light, as predicted by the general theory of relativity, when he measured slight apparent changes in the positions of stars seen near the Sun during the total solar eclipse of 1919; the accuracy of his results has since been questioned, but their announcement influenced the acceptance of general relativity. In 1924 he derived the *mass–luminosity relation. His work was summarized in *The Internal Constitution of the Stars* (1926).

Eddington limit The theoretical upper limit to the luminosity of a star of given mass, at which the outward force of radiation on the stellar surface just balances the inward force of gravity. Stars with a greater luminosity would be blown apart by their own radiation. The Eddington limit for the Sun is 30 000 times its actual luminosity. The maximum mass of a star set by the Eddington limit is about 120 solar masses. It is named after A. S. *Eddington.

Edgeworth–Kuiper Belt An alternative name for the *Kuiper Belt.

E-ELT Abbr. for *European Extremely Large Telescope.

effective aperture 1. In radio astronomy, the ratio between the power delivered to the receiver by an antenna and the flux density of the incoming radiation; also known as **effective area**. The effective aperture of a parabolic antenna is generally smaller than its geometric aperture, because of imperfections in the reflecting surface and the inefficiency of the feed in collecting radio waves reflected from all parts of the dish. An effective aperture can be defined even if no geometric aperture exists, as for example in an interferometer. **2**. The diameter of a single mirror that would have the same light-gathering power as a telescope mirror composed of several smaller individual reflectors.

effective area 1. In radio astronomy, an alternative term for *effective aperture (1). **2**. The unrestricted photon-collecting area of an optical, infrared, or X-ray telescope after obstructions in the optical path have been taken into account.

effective focal length The *focal length at which an optical system appears to be working. For example, when the eyepiece of a telescope is used to project an image on to film, the image is magnified as if it were produced by a telescope of several times the telescope's actual focal length. The magnification is given by $(d/F) - 1$, where d is the distance between the film plane and the *field stop within the eyepiece, and F is the eyepiece's focal length. The focal length is increased by this factor.

effective temperature (T_{eff} or T_e) The temperature of a *black body that radiates the same total energy per unit area as a given object, such as a star. It is the most useful measure of a star's surface temperature. The effective temperature of the Sun, for example, is 5800 K. However, in other circumstances it may be misleading, such as for a planetary nebula which may have a *kinetic temperature (obtained from the mean energies of the particles) of 10 000 K, but an effective temperature of only 50 K because of the low density of particles. *See also* COLOUR TEMPERATURE.

Effelsberg Radio Observatory The observatory of the Max-Planck-Institut für Radioastronomie, in the Eifel mountains 40 km southwest of Bonn, Germany. It is the site of a 100-m fully steerable dish, opened in 1972, which was the largest fully steerable radio dish until the opening of the 100 × 110-m *Green Bank Telescope in 2000. The outer part of the Effelsberg dish was resurfaced in 1998 and a new secondary with active optics was added in 2006.

(⊕) SEE WEB LINKS
• Official observatory website.

Egg Nebula A *protoplanetary nebula (1) in Cygnus, also known as CRL 2688. It was discovered at infrared wavelengths in 1974 by the US Air Force Cambridge Research Laboratories; a photograph of it taken by the Hubble Space Telescope in 1995 illustrates how a red giant star ejects matter at the end of its life. The Egg Nebula is approximately 3000 l.y. away.

Einstein, Albert (1879–1955) German-Swiss-American theoretical physicist. His theories of relativity helped to shape 20th-century science and had profound implications for astronomy. The *special theory of relativity (published in 1905) arose out of the failure to detect the *ether, and built on the work of the Dutch physicist Hendrik Antoon Lorentz (1853–1928) and the Irish physicist George Francis Fitzgerald (1851–1901). It yields the relation $E = mc^2$ between mass and energy, which was the key to understanding energy generation in stars. The *general theory of relativity (announced 1915, published in expanded form 1916), which encompasses gravitation, assumes great importance in very large-scale systems, and rapidly had an impact on cosmology. Astronomy has furnished observational evidence to support these theories. Einstein produced no subsequent work of great significance, searching unsuccessfully for a theory that would link relativity with electromagnetic forces (a so-called *grand unified theory). His Nobel Prize in Physics in 1921 was awarded primarily for his work on the photoelectric effect.

Einstein coefficient One of three quantities, known as *transition probabilities, used to describe the rate at which photons are absorbed or emitted from atoms or ions. The Einstein coefficients are widely used in the understanding of stellar spectra.

Einstein Cross An example of a *gravitational lens effect in which four images of a background object are formed. The first such example to be detected, the quasar G2237+0305, displays four distinct images arranged symmetrically around the image of a foreground galaxy, which is acting as the lens. The lensed quasar is about 8 billion l.y. away, while the galaxy lies 500 million l.y. from us. This object is sometimes also known as the Huchra Lens after its discoverer, the American astronomer John Peter Huchra (1948–2010). Similar objects subsequently discovered are now also termed Einstein crosses.

Einstein–de Sitter universe A type of universe in which the mean density of matter is precisely matched to the critical density. Such a model will not actually collapse, but will expand for ever with a continually decreasing expansion rate. This model lies on the dividing line between a closed *Friedmann universe (which collapses) and an open Friedmann universe (which does not). This model has the mathematical virtue of simplicity, in that it is spatially flat (*see* CURVATURE OF SPACETIME). It is named after A. *Einstein and W. *de Sitter.

Einstein Observatory Popular name for the NASA satellite HEAO-2, the second *High Energy Astronomical Observatory, launched in 1978 November. It was the first X-ray astronomy mission that could take images of the X-ray sky, for which it used grazing-incidence mirrors with a collecting area of approximately 200 cm². Einstein had four instruments, each covering the energy range 0.1–4 keV (0.3–12 nm) for X-ray imaging and spectroscopy. It operated until 1981 April.

((())) SEE WEB LINKS
• Information page at Goddard Space Flight Center.

Einstein ring A circular image of a distant source produced by a *gravitational lens. Theoretically, such an image can be produced when a point mass lies exactly on the line of sight to a distant galaxy or quasar. In practice, near-circular arcs have been detected in images of rich clusters of galaxies, and the radio emission from the quasar MG1654+1348 appears to have been lensed in this way.

Einstein shift Another name for *gravitational redshift.

ejecta Material thrown out from an explosive event, such as a crater-forming impact or a volcanic eruption. An **ejecta blanket** is the area immediately outside the rim of an impact crater, where the ejecta has completely covered the underlying terrain.

Elara A satellite of Jupiter, thirteenth in order of distance from the planet; also known as Jupiter VII. It orbits in 259.1 days at a distance of 11 716 000 km, close to *Lysithea's orbit. It is 80 km in diameter, and was discovered in 1905 by the American astronomer Charles Dillon Perrine (1867–1951). It is a member of the *Himalia group of Jovian satellites.

E layer A component of the Earth's *ionosphere, at an altitude of 110 km; also known as the *Heaviside layer* after the English physicist Oliver Heaviside (1850–1925). It is used as a reflective surface by radio operators. Electron densities in the E layer show diurnal variation, being greater during daytime. Absorption of radio waves by the denser daytime E layer can be a problem. Localized patches of *sporadic E* ionization which commonly appear at the same atmospheric level during the summer months can also disrupt short-wave radio communication.

electromagnetic radiation Energy arising from the acceleration of electrically charged entities (e.g. electrons). Electromagnetic radiation can be considered to be composed of waves or particles, since it displays properties of both; this is referred to as the *wave-particle duality*. Electromagnetic waves are composed of oscillating electric and magnetic fields which lie at right angles to each other and to the direction of travel. They propagate through a vacuum at the speed of light, c; the speed is slower when travelling through a medium such as air, water, or glass. The waves have a wavelength, λ, and a frequency, f or v, which are linked by the equation $c = f\lambda$. Electromagnetic radiation may also be regarded as being composed of a stream of particles of zero mass called photons. The energy, E, of a photon is related to frequency by Planck's formula $E = hf$, where h is the *Planck constant. Hence the higher the frequency, the greater the energy of the radiation.

electromagnetic spectrum The complete range of electromagnetic radiation. From longest to shortest wavelengths, the electromagnetic spectrum consists of radio waves (10^5–10^{-3} m), infrared waves (10^{-3}–10^{-6} m), visible light (4–7×10^{-7} m), ultraviolet waves (10^{-7}–10^{-9} m), X-rays (10^{-9}–10^{-11} m), and gamma rays (10^{-11}–10^{-14} m). The speed of electromagnetic radiation, c, is constant in a vacuum (*see* LIGHT, VELOCITY OF). Since c is equal to wavelength multiplied by frequency (expressed in appropriate units), for c to remain constant the frequency must become greater as the wavelength becomes shorter. For example, radio waves have long wavelength and low frequency, but gamma rays have high frequency and short wavelength. So both wavelength and frequency can take on a wide range of values. The visible region of the electromagnetic spectrum (i.e. light), detectable by the human eye, is subdivided into red (at the long-wavelength end), orange, yellow, green, blue, indigo, and violet (the short-wavelength end).

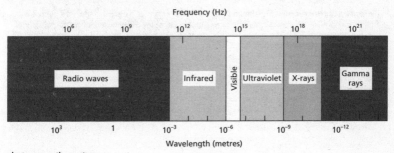

Frequency (Hz)

| 10^6 | 10^9 | 10^{12} | 10^{15} | 10^{18} | 10^{21} |

Radio waves | Infrared | Visible | Ultraviolet | X-rays | Gamma rays

| 10^3 | 1 | 10^{-3} | 10^{-6} | 10^{-9} | 10^{-12} |

Wavelength (metres)

electromagnetic spectrum

electron An elementary particle carrying a negative charge. Electrons are found in *energy levels* orbiting the nucleus of an atom. In a neutral atom, the number of electrons is equal to the number of protons. When detached from an atom they are called *free electrons*.

Electrons have a charge of 1.602×10^{-19} coulomb and a mass of 9.109×10^{-31} kg. The antiparticle of the electron is the *positron*.

electron degeneracy The state of *degeneracy attained when the density of matter is so high that electrons cannot be packed any closer together. Electron degeneracy supports white dwarf stars against further collapse. The only other type of degeneracy in astronomical objects is the *neutron degeneracy found in neutron stars.

electron density The number of free electrons per unit volume of space. Typical values range from less than 10^{-4} electrons/cm³ in intergalactic space to 0.03 electrons/cm³ in the disk of our Galaxy, 10^4 electrons/cm³ in H II regions, and 10^8 electrons/cm³ in stellar winds. Electron density in the Galaxy can be mapped from the *dispersion measure of pulsars. Conversely, for a part of the Galaxy in which the electron density is well-known, the distance of a radio source may be derived from its dispersion measure. Electron density is directly related to the amount of ionized hydrogen, and is a useful probe of conditions in interstellar space.

electron scattering opacity The main source of resistance to the flow of radiant energy in very hot gases, i.e. those with temperatures above 10^6 K. This opacity is due to free electrons, which do not absorb radiation but scatter it, thus increasing the time the radiation takes to pass through the gas. It is the main source of opacity in stars with masses greater than the Sun's. In less massive stars, the *Kramers opacity is more important.

electron temperature The temperature of the free electrons in a plasma, determined from their mean kinetic energy; also known as *kinetic temperature*. Inside stars, this measure of temperature is likely to give close agreement with other methods, but in rarefied gases found in nebulae and the outer atmospheres of stars it is often quite different. For example, the electron temperature of the solar corona ranges from 500 000 to 2 million K, although its *effective temperature is only about 50–100 K.

electronvolt (eV) A unit of energy used in atomic and nuclear physics. It is defined as the energy acquired by an electron in falling through a potential difference of one volt. One electronvolt is equal to 1.602×10^{-19} joule. Electronvolts are used as a measure of the energy of cosmic rays and high-energy photons. X-rays and gamma rays can have energies of 100 000 eV or more. By comparison, optical photons have energies of 2–3 eV. The *rest mass of atomic particles can also be expressed in terms of electronvolts, since mass and energy are equivalent. Hence the rest mass of an electron is about 500 keV and that of protons and neutrons about 1000 MeV. Electronvolts can be converted to wavelength (λ) by the formula $\lambda = 1239.8$nm/eV.

element, chemical A substance that consists of atoms of the same type, and cannot be broken down into chemically simpler substances. There are 92 naturally occurring elements, although others have been synthesized bringing the known total to more than 100.
An element is characterized by its *atomic number*, which is the total number of protons in the nucleus of each of its atoms. Hydrogen has an atomic number of 1, helium 2, and so on. The complete list of elements, arranged to show similarities in their properties, is known as the *periodic table*. An element can have different *isotopes, which are atoms with different numbers of neutrons in their nuclei. For example, deuterium (1 proton, 1 neutron) and tritium (1 proton, 2 neutrons) are isotopes of hydrogen. In astronomy, all elements heavier than helium are often known collectively as *metals*.

element, optical An individual optical unit within an optical system. A standard *achromatic lens has two optical elements, for example.

elementary particle A fundamental constituent of matter; also known as a *subatomic particle*. Elementary particles are divided into two main classes: *hadrons, which themselves consist of units called *quarks; and *leptons, which are not composed of quarks and seem to have no internal structure. Hadrons which consist of three quarks are also known as *baryons; examples are the proton and the neutron. Hadrons which consist of two quarks are known as *mesons. Leptons include the electron, muon, and neutrino. Elementary particles have properties of *charge, *spin, and *rest mass. They can be classified by the interactions they take part in. Hadrons participate in strong interactions, weak interactions, and, if they carry charge, electromagnetic interactions. Leptons do not participate in strong interactions. About 200 different elementary particles are thought to exist.

elements, abundance of The relative amount of each element or isotope in a given object such as a planet, star, or galaxy. *Cosmic abundance* refers to the proportions in the Universe as a whole. The amount of an element may be expressed either as the number of atoms, or by mass. The standard cosmic abundance is based on that of the Solar System, which has been determined from the Sun's spectrum and from analysis of meteoric debris. The cosmic abundance is dominated by hydrogen, with about 25% helium by mass and all remaining elements amounting to less than 2%.

Cosmic Abundance of The Most Common Elements

Element	Percentage abundance by mass	Percentage abundance by number of atoms
Hydrogen	73.5	92.1
Helium	24.9	7.8
Oxygen	0.73	0.061
Carbon	0.29	0.030
Iron	0.16	0.004
Neon	0.12	0.008
Nitrogen	0.10	0.008
Silicon	0.07	0.003
Magnesium	0.05	0.002
Sulphur	0.04	0.001
Argon	0.02	0.001
Nickel	0.01	0.0002

elements, orbital Six quantities that describe the size, shape, and orientation of an orbit. They can be used to calculate the position of a body in its orbit at any given time. For planets orbiting the Sun, the ecliptic and the vernal equinox (the first point of Aries) are used as a reference plane and reference direction respectively (see the diagram). The orbital elements are then: the *longitude of the ascending node, Ω; the *inclination, i; the *argument of perihelion, ω; the *semimajor axis, a; the *eccentricity, e; and a final number that gives the position of the planet in the orbit at a given time (or *epoch*). This can be the time of *perihelion passage, T (or τ), the *longitude at the epoch, L, or the mean anomaly at the epoch, M. For a comet with a highly eccentric orbit, the semimajor axis is usually replaced by the perihelion distance, q.

If the orbiting body is the Moon, the orbital elements are modified; the argument of perihelion is replaced by *argument of perigee*, and time of perihelion passage by *time of perigee passage*. Furthermore, if the orbiting body is an artificial Earth satellite, the ecliptic is replaced as reference plane by the equator, and the longitude of the ascending node is replaced by the *right ascension of the ascending node, α.

For any other body with satellites, such as Jupiter or Saturn, the reference plane is either the planet's equator if the satellite is near the planet, or the planet's orbital plane if the satellite is strongly perturbed by the Sun.

When an orbit is subject to significant perturbations, its orbital elements are given for a specified time; these are known as *osculating elements.

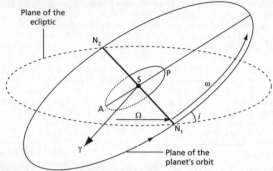

elements, orbital: The elements of a planet's orbit around the Sun.

A	*aphelion*
P	*perihelion*
N_1	*ascending node*
N_2	*descending node*
S	*Sun*
♈	*vernal equinox*
Ω	*longitude of the ascending node*
i	*inclination*
ω	*argument of perihelion*

E line A *Fraunhofer line at 527 nm wavelength in the spectrum of the Sun and other cool stars, primarily due to absorption by neutral iron atoms.

Ellerman bomb A small, very bright point in solar active regions, visible in *filtergrams tuned to the *line wings of the Hα line. Ellerman bombs occur at the *footpoints* of arch filaments, often in the early stages of the development of major active regions. An alternative name is *moustache*, from the appearance of the Hα line, which has emission wings extending either side of a dark core. It is named after the American astronomer Ferdinand Ellerman (1869–1940).

ellipse A closed curve like a flattened circle. The orbits of most celestial bodies are ellipses; the body being orbited lies at one *focus of the ellipse. The longest diameter of an ellipse is termed its *major axis*, and the shortest diameter is the *minor axis*. Half the major axis is the *semimajor axis*, usually denoted by the letter a; the *semiminor axis* (half the minor axis) is usually denoted by b. The distance between the two foci of an ellipse divided by the length of the major axis is the ellipse's *eccentricity, e. The eccentricity defines the shape of the ellipse, the ellipse being a circle when $e = 0$. When e is almost 1, the ellipse is long and narrow. For example, Halley's Comet follows an orbit for which $e = 0.967$.

ellipsoid A body or surface in which every cross-section passing through the centre is a circle or ellipse. The Moon's shape conforms closely to an ellipsoid with three unequal axes, the shortest being the rotation axis and the longest aligned towards the Earth. Strictly speaking, the Earth is also ellipsoidal in shape since the equator is slightly elliptical, although the difference between its longest and shortest axes is less than 1 km. *See also* SPHEROID.

ellipsoidal variable A type of small-amplitude variable (range below 0.2 mag.), consisting of a close, but not interacting, binary system; abbr. ELL. The two components are tidally distorted into ellipsoidal shapes. The inclination of the orbital plane to the line of sight is such that eclipses do not occur, but the orbital motion causes fluctuations in the visible area of the components and thus in the apparent magnitude of the system.

elliptic aberration *See* E-TERMS.

elliptical galaxy A type of galaxy with a smooth, featureless circular or elliptical appearance, no spiral arms, and little or no interstellar gas or dust; symbol E. Elliptical galaxies range from about 10^7 solar masses and a few thousand light years in diameter for *dwarf ellipticals* to over 10^{12} solar masses and over 100 000 l.y. in diameter for *giant ellipticals*. Elliptical galaxies are classified from E0 to E7 according to their apparent shape, E0 appearing circular from our viewpoint and E7 the most elliptical. The degree of ellipticity is calculated from the ratios of the major (*a*) and minor (*b*) axes, using the formula $10(a - b)/a$. Dwarf ellipticals are given the designation dE, while giant and supergiant ellipticals are known as *D galaxies and *cD galaxies respectively; such massive examples are usually found at the centres of clusters.

The stars in elliptical galaxies are mostly old (Population II), although some ellipticals also contain intermediate-age stars that formed more recently. The light from elliptical galaxies falls off in a characteristic way from the centre to the edge, except where the galaxy has been disturbed in some way, for example by tidal forces from a passing galaxy or the addition of a faint envelope by *galaxy cannibalism, as occurs in D and cD galaxies. Relative to their luminosity, elliptical galaxies have the greatest number of globular star clusters of any type of galaxy; a large elliptical may possess several thousand of them. A high proportion of the bright galaxies in rich clusters of galaxies are ellipticals (40%), whereas the general proportion of ellipticals outside rich clusters is much lower, around 10%.

The intrinsic shape of elliptical galaxies can be spheroidal (cigar- or discus-shaped) or truly ellipsoidal (triaxial) with different dimensions along all three axes. Some ellipticals appear to be rotating sufficiently quickly to explain their flattened shape, but many (particularly the large ellipticals) show very little rotation. The details of the origin of elliptical galaxies remain controversial. They could either be old systems which formed rapidly and then quickly used up or lost their interstellar gas, or they could result from the merger of spiral galaxies.

ellipticity Another term for *oblateness.

Elnath The star Beta Tauri, magnitude 1.65, a B7 giant 134 l.y. away. The name is also spelt Alnath.

elongation **1.** The angle between the Sun and a body in the Solar System, as seen from Earth. Elongations are measured from 0° to 180° east or west of the Sun, in the plane passing through the Sun, Earth, and body. When a body is in *conjunction* with the Sun, its elongation is 0°; at *opposition* its elongation is 180°; and when at *quadrature* its elongation is 90° east or west. *Greatest elongation* is the maximum angular separation of Mercury or Venus from the Sun, either west or east.

2. The angle between a planet and a satellite, as seen from Earth. Elongations are measured from 0° east or west of the planet, in the plane passing through the planet, Earth, and satellite.

Eltanin The star Gamma Draconis, magnitude 2.2. It is a K5 giant lying 154 l.y. away. An alternative version of the name is Etamin.

Elysium Planitia An extensive plain on Mars some 3000 km across, centred at approximately +2° lat., 205°W long. Elysium Mons, a volcano 11 km high, lies in eastern Elysium, and Apollinaris Patera, a smaller volcano, lies in the southeast, where it is bordered by the cratered southern uplands of Mars.

e-MERLIN See MULTI-ELEMENT RADIO-LINKED INTERFEROMETER NETWORK.

emersion The reappearance of a star following *occultation by the Moon; or, the emergence of an object from shadow at the end of an eclipse.

emission The release of a photon from an atom or ion when an electron jumps to a lower energy level. Emission from an astronomical object indicates that there must be a source of energy within it. A hot gas radiates because the motion of the atoms causes collisions which excite electrons in the atoms to higher energy levels, or even causes them to break free of the atoms, producing ions. When the electrons return to lower energy levels, they emit the energy difference in the form of photons of specific wavelengths, giving rise to emission lines. If the electron is free (unbound) and recombines with an atom or ion, then the emission can occur at a continuous range of wavelengths, producing a *continuous spectrum. *See also* BOUND–BOUND TRANSITION; FREE–BOUND TRANSITION

emission line A bright line in a spectrum at a particular wavelength, given out by hot or excited atoms. Emission lines can appear superimposed on a normal absorption spectrum, as caused by hot gas around a star, or they can appear on their own, as in the spectrum of a nebula excited by radiation from a nearby hot star. They can be used to determine the temperature, pressure, and chemical composition of the emitting gas.

emission nebula A luminous cloud of gas and dust in space which shines with its own light. The light can be generated in several ways. Usually the gas glows because it is exposed to a source of ultraviolet radiation; examples are H II regions and planetary nebulae, which are ionized by central stars. Gas may also glow because it has become ionized in a violent collision with another gas cloud, as in *Herbig–Haro objects. Finally, some of the light from supernova remnants such as the Crab Nebula is produced by the process of *synchrotron radiation, in which charged particles spiral around the interstellar magnetic field.

emission spectrum A spectrum that is dominated by, or consists solely of, bright emission lines. An emission-line spectrum indicates the presence of hot gas and a source of energy, as in a very hot star (*see* EMISSION). Examples of objects with emission-line spectra include Wolf–Rayet stars, planetary nebulae, and quasars.

emissivity (ϵ) A measure of an object's ability to radiate electromagnetic radiation compared with that of a *black body at the same temperature. A black body, which is a perfect emitter, has an emissivity of 1, whereas a perfect reflector has an emissivity of 0.

emulsion, photographic *See* PHOTOGRAPHIC EMULSION.

Enceladus The eleventh-closest satellite of Saturn, distance 238 200 km, orbital period 1.370 days; also known as Saturn II. Its axial rotation period is the same as its orbital period. Enceladus was discovered in 1789 by F. W. *Herschel. It is 513 × 503 × 497 km in diameter. Impact craters cover part of its surface, but over half the surface consists of smooth plains with few craters. The smooth plains are separated from the cratered areas by long sinuous ridges. There are also grooves, folds, faults, and other signs of large-scale fracturing and past geological activity on the surface. Some form of *cryovolcanic process is still underway—plumes of ice particles and water vapour were observed rising from its southern polar region by the Cassini spacecraft. Enceladus has an albedo of nearly 1.0, which means that it reflects nearly all the sunlight that falls on it. Hence it can absorb very little heat, and the surface temperature is −201°C. Enceladus lies within Saturn's diffuse E Ring, and contributes material to the ring.

Encke, Comet 2P/ The comet of shortest known period, 3.3 years, which has been observed at more returns than any other. It also has the smallest aphelion distance of any normal comet, 4.1 AU, and can be followed around its entire orbit. It was first seen in 1786 by the French astronomer Pierre François André Méchain (1744–1804), and again by C. L. *Herschel in 1795. It was rediscovered twice by J. L. *Pons, in 1805 and 1818. The common identity of all four comets was established in 1819 by J. F. *Encke, after whom it is now named. Comet Encke has probably occupied its current orbit for several thousand years, losing much of its gas and dust in that time. At most returns it is comparatively faint. The comet is the progenitor of a complex dust stream which produces the *Taurid and *Beta Taurid meteor showers. It has been suggested that Encke is the largest surviving fragment of a single large object that broke up long ago. Its orbit has a perihelion of 0.34 AU, eccentricity 0.85, and inclination 11°.8. The nucleus has an estimated diameter of about 5 km.

(⊕) SEE WEB LINKS
• Information page at Cometography website.

Encke, Johann Franz (1791–1865) German mathematician and astronomer. In 1819 he calculated the orbit of a comet (subsequently named Comet *Encke) that had been discovered by J. L. *Pons, and found that its period was less than 4 years, still the shortest known. In 1837 he became the first to observe the gap in Saturn's A Ring, since named the *Encke Division. He calculated the Sun's distance from observations of the transits of Venus in 1761 and 1769.

Encke Gap A division in Saturn's rings about 325 km wide, centred at 133 570 km from the planet's centre, near the outer edge of the A Ring. It was the first division to be discovered after Cassini's Division, and is named after J. F. *Encke. The small satellite *Pan orbits within it.

energy level The discrete energy state of an electron in an atom or molecule. Electrons can occupy only specific energy levels and jump from one energy level to another. The lowest energy level is called the *ground state of the atom or molecule. Within an atom, energy levels correspond to the distance from the nucleus. The electron requires energy to jump from a state of lower energy (an energy level close to the nucleus) to a higher energy level (further from the nucleus); the atom is then said to be *excited*. Energy levels are defined by a *quantum number*, symbol n. For the ground state, $n = 1$. Higher energy levels are numbered 2, 3, etc., outwards from the nucleus. When an electron jumps from a higher energy level to a lower energy level it emits a *photon, the energy of which is equal to the difference between the two energy levels.

English mounting A form of *equatorial mounting in which the telescope is carried within a rectangular cradle or yoke, which is itself supported on two columns or piers aligned north–south; also known as a *yoke mounting*. The yoke forms the *polar axis, and the telescope pivots within this yoke around the *declination axis. In variants known as *modified English mounting* or *cross-axis mounting*, the yoke is replaced by a single beam, the telescope being on one side and a counterweight on the other. Instruments with English mountings include the 100-inch (2.5-m) Hooker Telescope at Mount Wilson Observatory.

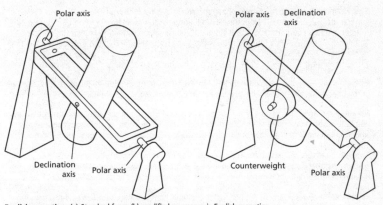

English mounting: (a) Standard form; (b) modified or cross-axis English mounting.

Enif The star Epsilon Pegasi, magnitude 2.4. It is a K2 supergiant 690 l.y. away. It reportedly flared up to magnitude 0.7 for one night in September 1972. It has a wide companion of magnitude 8.4, which is not related.

Ensisheim meteorite The oldest meteorite whose fall can be dated with certainty. A fireball was observed over Battenheim on 1492 November 16, and after a violent explosion a 127-kg stone meteorite fell on the village of Ensisheim in Alsace. It is an *ordinary chondrite of type LL6.

enstatite chondrite The rarest and, in general, the least primitive of the three main classes of chondrite meteorite. They are much less oxidized than the ordinary chondrites. Iron in the metallic state constitutes 15–25% of their mass, and the principal silicate is a nearly iron-free pyroxene known as enstatite. The enstatite chondrites are divided into two subgroups according to whether the ratio of iron to silicon in the meteorite is high (the EH group) or low (the EL group). The EH and EL groups are probably derived from separate parent bodies. Most enstatite chondrites are breccias. They are chemically similar to the *aubrites (enstatite achondrites), and may have originated from the same parent bodies.

entrainment A phenomenon in which the medium through which a jet propagates is drawn into the jet itself. One example is the presence of neutrons in the solar wind. Since the neutrons cannot have been accelerated by the processes thought to produce the proton and electron components of the solar wind, they must have been entrained into the cosmic rays from the outer layers of the Sun. Jets from radio galaxies may also contain material entrained from the galaxy.

entropy (S) A measure of disorder in a system; the higher the entropy, the greater the disorder. In a closed system an increase in entropy is accompanied by a decrease in energy availability. The Universe itself can be regarded as a closed system; therefore its entropy is increasing and its available energy is decreasing. *See also* HEAT DEATH OF THE UNIVERSE.

envelope A cloud of gas and dust surrounding a star or other astronomical object. Stellar envelopes are of many types. Hot young stars generate hot glowing envelopes, either by ionizing the surrounding gas or by ejecting hot material. Evolved stars shed their outer layers and generate cool circumstellar envelopes which are rich in dust and molecules. When the hot core of the old star is exposed, the envelope is ionized and may be detected as an emission nebula—as a *planetary nebula, for example.

eon *See* AEON.

Eos family A *Hirayama family of asteroids in the outer region of the main asteroid belt at a mean distance of 3.02 AU from the Sun. Most Eos family members have surface properties roughly intermediate between S and C types, leaning towards the S type especially in albedo. The family is named after (221) Eos, an S-class asteroid, diameter 104 km, discovered in 1882 by the Austrian astronomer Johann Palisa (1848–1925). Its orbit has a semimajor axis of 3.014 AU, period 5.23 years, perihelion 2.70 AU, aphelion 3.33 AU, and inclination 10°.9.

epact A term used in ecclesiastical lunar tables to calculate the date of Easter. It gives the age of the Moon (i.e. the number of days since New Moon) at the start of each year.

Ep galaxy An elliptical galaxy showing some peculiarity in appearance, such as a dust patch with a few blue supergiant stars, filamentary gas with strong optical emission lines, or an uncharacteristic fall-off in light with distance from the centre. For example, M87, the giant elliptical galaxy in the Virgo Cluster, is peculiar because it has a long, narrow jet extending from its nucleus. M32, a dwarf elliptical companion of the Andromeda Galaxy, is classified as peculiar because its outer regions appear to have been removed in gravitational encounters with its massive neighbour. The nearest radio galaxy, *Centaurus A, is also an Ep galaxy.

ephemeris A table giving the predicted positions of a celestial object at given times; pl.*ephemerides*. Collections of ephemerides for various objects are published in yearly almanacs, such as The *Astronomical Almanac*.

Ephemeris Time (ET) The time-scale that was used in calculating orbits within the Solar System from 1960 to 1984. Its fundamental unit was the *ephemeris second* which was defined by the statement that the *tropical year at the epoch 1900.0 was precisely 31 556 925.9747 ephemeris seconds. Ephemeris Time was in many ways inconvenient, and in 1984 it was replaced by Terrestrial Dynamical Time (now known as *Terrestrial Time) which has the SI second as its fundamental unit.

epicycle In the *Ptolemaic system of planetary motion, a hypothetical small circle whose centre moved along the circumference of a bigger circle, the *deferent (see diagram). Combinations of epicycles could be used to reproduce the observed motions of the planets, without abandoning the Greek dogma that only circular motions were allowable in the heavens.

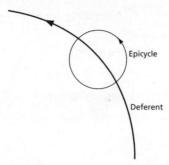

epicycle

Epimetheus A satellite of Saturn, in virtually the same orbit as *Janus between Saturn's F and G Rings, at a distance of 151 410 km; also known as Saturn XI. Its orbital period is 0.694 days, the same as its axial rotation period, and its size is 130 × 114 × 106 km. Epimetheus and Janus are *coorbital and regularly exchange orbits when one catches up with the other. Epimetheus was discovered in 1980 on images from Voyager 1.

epoch The date and time at which an astronomical observation was made, or the date for which positions of celestial objects and orbital elements are calculated. Because of *precession and *nutation, celestial coordinates change with time, so that positions of celestial objects must be referred to a given date. The standard epoch now commonly used in ephemerides and star catalogues is 2000 January 1, 12 h (written as 2000.0).

EPOXI The extended mission of the *Deep Impact spacecraft, which flew past Comet Hartley 2 (103P/Hartley) in 2010 November, photographing its nucleus. There was no spare impactor so this time there was no impact on the nucleus. EPOXI is a combination of two acronyms: EPOCH, Extrasolar Planet Observation and Characterization, which observed eight stars known to have planets for signs of transits; and DIXI, the Deep Impact Extended Investigation, which concentrated on the comet.

((⊕)) SEE WEB LINKS
• Official mission website.

e-process The so-called *equilibrium process* by which silicon is converted to heavier elements such as iron and nickel through a set of nuclear reactions in certain massive stars. The process is a complex one in which the equilibrium between the rates of certain reactions is important.

Epsilon Aurigae An eclipsing binary with a very long orbital period (9892 days) and many unusual characteristics. A visible star, now thought to be an F0 giant, is eclipsed for approximately 610 days by a larger, unseen object. This dark companion is thought to consist of a B5 main-sequence star surrounded by a disk of dust and gas with a radius of about 4 AU seen virtually edge-on. During eclipses, the apparent magnitude falls from 3.0 to 3.8. Epsilon Aurigae is about 2000 l.y. away. The last eclipse was in 2009–11; the next is due to start in 2036.

equant In the *Ptolemaic system of planetary motion, a point to one side of the Earth with respect to which an *epicycle would have uniform angular motion.

equation 150

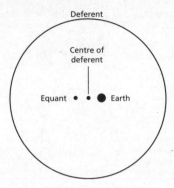

equant

equation A mathematical relationship between a number of variables and constants. Examples in astronomy are *Kepler's equation, the *equation of the centre, and the *equation of time. The term also appears in *personal equation, a correction for the observer's personal error when measuring or timing something.

equation of light The correction required to account for the time that light takes to cross the orbit of the Earth. Light takes almost exactly 499s to travel from the Sun to the Earth, so that the time recorded for an event, such as an occultation, could vary by over 16½ minutes depending on where the Earth lies in its orbit. For events outside the Solar System, it is common to record timings as though they had been made from the Sun (*heliocentric time*). Within the Solar System, such as when timing eclipses of a planet's satellites, not only the position of the Earth needs to be taken into account but also that of the planet, and those of the satellites.

equation of the centre The *true anomaly minus the *mean anomaly of a body moving in an elliptical orbit. In other words, it is the angular difference between the actual position of a body in its orbit and the position at which it would lie if it moved at constant angular velocity.

equation of the equinoxes The difference between apparent sidereal time and mean sidereal time. It is also known as *nutation in right ascension*.

equation of the origins The difference between the *Earth rotation angle and the Greenwich Sidereal Time (GST) of a celestial object. Put another way, it is the angular distance between the equinox and the Celestial Intermediate Origin (CIO), a point on the celestial equator that lies very close to the intersection with the prime meridian of the *Geocentric Celestial Reference System. Whereas the path followed by the CIO is always at right angles to the equator, the equinox drifts along the equator because of precession and nutation.

equation of time The difference between *apparent solar time and *mean solar time; technically, the difference between the *hour angle of the true Sun and that of the fictitious *mean sun. Whereas the mean sun by definition proceeds uniformly, the true Sun may be some minutes ahead of or behind its expected position because of the eccentricity of the Earth's orbit. Moreover, the true Sun moves along the ecliptic, not the celestial equator, which is an additional contributor to the equation of time. The difference reaches a maximum in early November, when apparent solar time is over 16 min ahead of mean solar time, and in mid-February, when apparent solar time is over 14 min behind mean solar time. The equation of time is zero on four occasions during the year: on April 15, June 14, September 1, and December 25.

equator The intersection of the surface of a body with a plane perpendicular to the body's axis of rotation. The plane of the equator passes through the body's centre and divides the body into northern and southern hemispheres. All points on the equator are equidistant from the body's poles of rotation. *See also* CELESTIAL EQUATOR; GALACTIC EQUATOR.

equatorial bulge The increased equatorial diameter of a rapidly rotating planet, caused by centrifugal force. The size of the effect depends on the body's density and its rotation period. The Earth, which rotates in 24 hours, has a slight equatorial bulge, its equatorial diameter being about 0.3% greater than its polar diameter. Saturn, which has a rotation period of just over 10 hours and a mean density less than that of water, has the greatest equatorial bulge of all the planets, with an equatorial diameter nearly 11% greater than its polar diameter. *See also* OBLATENESS.

equatorial coordinates A system of coordinates that specifies the position of a celestial object relative to the celestial equator. The equatorial coordinate system is most commonly used for giving positions on the celestial sphere. The coordinates normally used are *right ascension, the equivalent of longitude on Earth, and *declination, the equivalent of terrestrial latitude. However, *hour angle and *polar distance may be used instead.

equatorial horizontal parallax The difference between the position of an object at 0° declination when on the horizon of an observer at the equator, and the position in which the object would be seen from the centre of the Earth. It is thus the parallactic displacement corresponding to a distance equal to the equatorial radius of the Earth. *See also* HORIZONTAL PARALLAX.

equatorial mounting A telescope mounting with one axis (the *polar axis) parallel to the Earth's axis and the other (the *declination axis) at right angles to it. The telescope is attached to the declination axis. The Earth's rotation is counteracted simply by rotating the mount around the polar axis. From the mid-19th century until recently almost all large telescopes were on equatorial mountings. Some very large instruments now have computer-controlled *altazimuth mountings. Designs of equatorial mounting include the *English, *fork, *German, *horseshoe, and *Springfield mountings.

equinoctial colure The *hour circle that passes through the celestial poles and the vernal and autumnal equinoxes.

equinox Either of the two points at which the Sun's apparent yearly path (the ecliptic) intersects the celestial equator; or the dates on which this occurs—on March 20 (*vernal equinox*) and September 22 or 23 (*autumnal equinox*). These points are also known as the *first point of Aries and the *first point of Libra. When the term 'equinox' is used without qualification, the vernal (spring) equinox is meant. Around the time of the equinoxes, night and day are equal in length the world over. The equinox is not a fixed point, but moves because of *precession and *nutation. If only precession is taken into account, the resulting point is known as the *mean equinox of date*. If, however, nutation is also included, the point is known as the *true equinox*.

equipotential surface A surface surrounding a body or group of bodies over which the strength of the gravitational field is constant. For a spherical body such as a star or planet, the equipotential surface is a sphere centred on the body. For a body of oblate shape, or two or more bodies such as a binary star, the equipotential surfaces are of complicated shapes. The *Roche lobes of a binary star are the equipotential surfaces about the two stars which meet at the inner *Lagrangian point, which lies on the line joining the stars' centres.

equivalence principle The principle that the effects of a gravitational field cannot be distinguished from those of uniform acceleration. In a small region within a gravitational field, the laws of physics may be regarded as the same as in an accelerated frame in the absence of gravity. The principle arises from general relativity and is another way of stating that gravitational mass equals inertial mass.

equivalent focal length The *focal length of an optical system expressed as if it consisted of a single optical element. It is used to describe the focal length of an enclosed optical system, such as an eyepiece, which has no position from which the focal distance can be measured.

equivalent width A measure of the strength of a line in a spectrum. An emission line is an increase in the intensity of a spectrum, while an absorption line is a dip. The equivalent width is the width of the spectral continuum that has the same area as the line. Values are negative for emission lines, positive for absorption lines.

Equuleus (Equ) (*gen.* **Equulei)** The second-smallest constellation in the sky, representing a foal. It contains little of note. Alpha Equulei (Kitalpha), magnitude 3.9, is its brightest star.

ERA Abbr. for *Earth rotation angle.

Eratosthenes (*c.*276–*c.*194 BC) Greek writer on many subjects, born in modern Libya. He made the first known calculation of the Earth's circumference that was based on a scientifically sound method. According to tradition, he did this by measuring the Sun's altitude as seen from Alexandria at the summer solstice, when he knew the Sun to be overhead at Syene, some 1000 km to the south. However, it is possible that he may actually have used a simpler method, which is to measure the amount by which the angle from the zenith to the horizon exceeded 90° as seen from a high point, such as a lighthouse. The effect of atmospheric refraction would give a result about 15% in excess of the true value, which is what he obtained. Eratosthenes also measured the *obliquity of the ecliptic.

Erfle eyepiece A design of wide-field eyepiece consisting of three lenses, two or three of which may be *doublets, that gives a field of view of 65–70°. Although the field of view is slightly curved and the images at the edge of the field suffer from *astigmatism and*chromatic aberration, the Erfle is cheaper to manufacture than many other wide-field designs. It was the first wide-field eyepiece, and was designed in 1917 by the German optician Heinrich Valentin Erfle (1884–1923). The *König eyepiece* is an adaptation of the Erfle, with shorter focal length, named after its inventor, the German optician Albert König (1871–1946).

ergosphere The region just outside the *event horizon of a rotating black hole (i.e. a *Kerr black hole), within which an observer is forced to rotate with the black hole, and cannot remain stationary with respect to the rest of the Universe. The outer boundary of the ergosphere is the *static limit. In a non-rotating *Schwarzschild black hole, the event horizon and stationary limit coincide. The ergosphere derives its name from the fact that energy can be extracted from the black hole in this region via the *Penrose process (the prefix 'ergo' means 'work').

Eridanus (Eri) (*gen.* **Eridani)** An extensive constellation, the sixth-largest, representing a river running from the celestial equator deep into the southern sky. Its brightest star is *Achernar (Alpha Eridani). Epsilon Eridani, magnitude 3.7, is a K2 dwarf only 10.5 l.y. away. Omicron-2 Eridani, also known as 40 Eridani, magnitude 4.4, is accompanied by a magnitude 9.5 white dwarf, the most easily seen white dwarf in the sky. Theta Eridani (*Acamar) is an attractive double star, as is 32 Eridani, magnitudes 4.8 and 6.1.

Eris The first *trans-Neptunian object found with a diameter larger than that of Pluto, and the largest of the new class of *dwarf planets; also known as minor planet (136199). Eris is about 2400 km in diameter and has a mean density of 2.3 g/cm^3 and a mass 1.27 times that of Pluto. It has a satellite, Dysnomia, about 0.1 its diameter, which orbits it every 15.8 days. It was discovered in 2005 by the American astronomers Michael Edwards Brown (1965–), Chadwick Aaron Trujillo (1973–), and David Lincoln Rabinowitz (1960–). Its orbit has a perihelion of 38.54 AU, aphelion 97.56 AU, inclination 43°.9, and period 561.35 years, placing it among the *scattered-disk objects.

Eros Asteroid 433, discovered in 1898 by the German astronomer Gustav Witt (1866–1946) and independently on the same day by the Frenchman Auguste Charlois (1864–1910). Eros was the first known asteroid with an orbit that brings it within the orbit of Mars; it is a member of the *Amor group. The *NEAR Shoemaker space probe went into orbit around Eros in 2000 February, photographing its ancient, cratered surface. Eros is highly elongated, 34 × 11 × 11 km, with a mass of 6.7 × 10^{15} kg and a density of 2.7. It is of S class, with a rotation period of 5.27 hours. Its orbit has a semimajor axis of 1.458 AU, period 1.76 years, perihelion 1.13 AU, aphelion 1.78 AU, and inclination 10°.8.

eruptive binary A frequently used synonym for any *cataclysmic variable that is also a binary system. In fact, the outburst mechanisms in an *eruptive variable differ from those in a cataclysmic variable.

eruptive prominence A solar prominence that has previously been quiescent but suddenly starts to ascend at a few hundred kilometres per second, eventually disappearing from view. Eruptive prominences are often visible at the Sun's limb in association with *coronal mass ejections. On the Sun's disk, the equivalent phenomenon is a disappearing *filament.

eruptive variable A variable star that exhibits sudden changes in brightness, in the form of outbursts, flares, or fades. The variations are caused by activity in the star's chromosphere or corona, and may be accompanied by mass loss through the ejection of shells or as an enhanced stellar wind. *See* FLARE STAR; FLASH STAR; FU ORIONIS STAR; GAMMA CASSIOPEIAE STAR; IRREGULAR VARIABLE; NEBULAR VARIABLE; R CORONAE BOREALIS STAR; RS CANUM VENATICORUM STAR; S DORADUS STAR; T TAURI STAR; UV CETI STAR.

ER Ursae Majoris star A type of *SU Ursae Majoris star with a very short interval (20–50 days) between *superoutbursts*, i.e. outbursts that are about 2 magnitudes brighter and five times longer than normal outbursts. Such stars typically spend a third to half of their time in superoutburst, with a rapid succession of normal outbursts (typically about one every 4 days) in between. The mass transfer rate is nearly ten times that in a typical SU Ursae Majoris star.

ESA Abbr. for *European Space Agency.

escape velocity The minimum velocity required to escape from a gravitational field. The escaping object can be anything from a gas molecule to a spacecraft. A body's escape velocity is given by $\sqrt{(2GM/R)}$, where G is the *gravitational constant, M is the mass of the body, and R is the distance of the escaping object from the body's centre. An object moving at less than escape velocity enters an elliptical orbit; at exactly escape velocity it follows a *parabolic trajectory; and if the object exceeds the escape velocity it moves away on a *hyperbolic trajectory.

Eskimo Nebula The 8th-magnitude planetary nebula NGC 2392 in Gemini; an alternative title is the Clown Face Nebula. The nebula has a surrounding fringe, like an Eskimo's hood, which gives it its name. It is about 3000 l.y. away.

ESO Abbr. for *European Southern Observatory.

ET Abbr. for *Ephemeris Time.

Eta Aquarid meteors A prominent meteor shower, particularly for observers in the southern hemisphere, produced by debris from *Halley's Comet. Activity is seen between April 24 and May 20, with a broad maximum centred on May 5, when the radiant lies at RA 22 h 20 m, dec. −01°. Peak ZHR values may approach 50. Eta Aquarid meteors are very swift (geocentric velocity 67 km/s), and a high proportion leave persistent trains. Bright Eta Aquarids are quite common, and often appear yellow.

Eta Carinae A *hypergiant irregular variable, which became the second brightest star in the sky (magnitude −0.8) in 1843. Its current magnitude is around 5.5, but it shows fluctuations over periods of decades. It is a *luminous blue variable of absolute magnitude −10, and is officially classified as an *S Doradus star. It lies inside a cluster of massive stars. It is now thought to be a massive binary with an orbital period of 5.5 years and an eccentricity of about 0.75. The only spectrum visible is that of the surrounding *Homunculus Nebula. Eta Carinae is an intense infrared source and its extreme mass loss (about 0.1 solar masses per year) involves energies close to those of some supernovae. It lies about 7500 l.y. away.

Eta Carinae Nebula The bright diffuse nebula NGC 3372 in Carina, about 2° across, containing numerous stars and clusters. It is divided into two parts by a dark V-shaped lane. At its heart is the variable star *Eta Carinae, which is concealed by the smaller *Homunculus Nebula. The Eta Carinae Nebula contains the Keyhole Nebula, a smaller dark nebula silhouetted against the brighter background near Eta Carinae. Its distance is about 7500 l.y.

etalon Two parallel glass plates with reflecting surfaces between which light is reflected numerous times before exiting. The effect of interference produces an outgoing beam that contains only a limited range of wavelengths, as in an *interference filter. Etalons in which the gap between the plates can be varied to select different wavelengths are used in a *Fabry–Perot interferometer.

Etamin An alternative name for the star *Eltanin.

E-terms A component of *annual aberration that depends on the eccentricity of the Earth's orbit; also known as **elliptic aberration**. It amounts to only $0''.34$ at maximum. Until 1984 the catalogue positions of stars were not corrected for the E-terms, but since then the E-terms have been taken into account when defining a star's *mean place.

ether A hypothetical medium once thought to permeate all space, through which electromagnetic radiation supposedly travelled; formerly spelt *aether*. On the basis of this supposition, the Earth should move with respect to the ether, and it was predicted that the speed of light would vary when measured in different directions. Experiments in the 19th century (e.g. the *Michelson–Morley experiment) failed to detect any such variation in speed. The ether is now regarded as unnecessary, since it is recognized that electromagnetic radiation can propagate through empty space.

eucrite A class of calcium-rich achondrite meteorite; also known as *pigeonite–plagioclase achondrites*. Two subclasses are recognized: the *non-cumulate eucrites*, and the less-common *cumulate eucrites*. In their mineralogy, texture, and composition they closely resemble terrestrial and lunar basalts. The non-cumulate eucrites are essentially basalts that originated as surface lava flows, while the cumulate eucrites (and closely related *diogenites) are intrusive rocks formed at shallow depth. Their parent body may have been the asteroid Vesta. Their name comes from the Greek *eukritos*, meaning 'easily discerned'. *See also* BASALTIC ACHONDRITE.

Eudoxus of Cnidus (*c*.400–*c*.350 BC) Greek mathematician and astronomer, born in modern Turkey. He developed a model of planetary motion in which the Sun, Moon, and planets were carried around the Earth on a series of 27 Earth-centred spheres, with axes at different angles and rotating at different speeds. This theory of 'heavenly spheres', as modified first by *Callippus and ultimately by *Ptolemy, remained the orthodox view of the Universe for two thousand years. Eudoxus is also reputed to have introduced the constellation system from Egypt.

Euler, Leonhard (1707–83) Swiss mathematician. He refined the mathematical theory of the Moon's motion, and calculated improved orbits for Jupiter and Saturn. His lunar theory helped the German cartographer and astronomer Johan Tobias Mayer (1723–62) to produce highly accurate lunar tables; its full merits would come to be appreciated by G. W. *Hill, over a century later. Euler's work on optical systems strongly influenced the technical development of telescopes. He also worked on planetary perturbations, the orbits of comets, and the tides.

Eunomia family A small family of asteroids in the central region of the main asteroid belt, at a mean distance of 2.6–2.7 AU and with a mean orbital inclination of about 12°. The largest member is the S-class (15) Eunomia, diameter 255 km, while the two next-largest members, (85) Io and (141) Lumen, are of C class. Eunomia was discovered in 1851 by the Italian astronomer Annibale de Gasparis (1819–92). Its orbit has a semimajor axis of 2.642 AU, period 4.30 years, perihelion 2.14 AU, aphelion 3.14 AU, and inclination 11°.7.

Europa The fourth-largest satellite of Jupiter and the second of the four Galilean satellites from the planet; also known as Jupiter II. Europa orbits in 3.551 days at a distance of 671 000 km. Its period of axial rotation is the same as its orbital period. Its diameter is 3122 km, making it only slightly smaller than our Moon. Europa's density of $3.01\,g/cm^3$ indicates that it is mainly composed of silicate rock, mixed with at least 5% water. It has a bright, icy surface of albedo 0.67, dominated by networks of dark, linear cracks, some over 1000 km long. Many of these are curved and are termed *flexus*; they are thought to arise from tidal stresses. Impact craters on Europa are few and far between, indicating that the surface is young and perhaps being resurfaced by icy flows from below the surface. Indirect evidence suggests the widespread presence of liquid water below the icy crust, but neither the depth nor the extent of this subsurface sea are known. 📷

European Extremely Large Telescope (E-ELT) A proposed optical and infrared telescope to be built by the European Southern Observatory. It will be built on Cerro Armazones at an altitude of 3060m in the central Atacama Desert, Chile, some 130 km south of the town of Antofagasta and about 20 km from Cerro Paranal, home of ESO's Very Large Telescope. Its main mirror will have an aperture of 42 m and consist of 984 hexagonal segments, each 1.45 m across. A sophisticated *adaptive optics system will be used to sharpen the images. The E-ELT is scheduled for completion in 2018.

(()) SEE WEB LINKS
• Official telescope website.

European Northern Observatory The name by which the *Teide Observatory on Tenerife and the *Roque de los Muchachos Observatory on La Palma in the Canary Islands are jointly known.

(()) SEE WEB LINKS
• Official observatory website.

European Southern Observatory (ESO) An organization founded in 1962 by a consortium of European countries to establish observing facilities in the southern hemisphere. It currently has fourteen member nations: Austria, Belgium, the Czech Republic, Denmark, Finland, France, Germany, Italy, the Netherlands, Portugal, Spain, Sweden, Switzerland, and the UK. Its headquarters are at Garching, near Munich, Germany. ESO owns and operates two observing sites in Chile: *La Silla Observatory and *Paranal Observatory, the latter being the site of ESO's *Very Large Telescope. ESO is also participating in the *Atacama Large Millimeter Array (ALMA) project and is building the 42-m *European Extremely Large Telescope.

(()) SEE WEB LINKS
• Official observatory website.

European Space Agency (ESA) An organization of European countries for cooperation in space research and technology, with its headquarters in Paris. ESA establishments include the European Space Research and Technology Centre (ESTEC) at Noordwijk, Netherlands; the European Space Operations Centre (ESOC), Darmstadt, Germany, which controls European space missions; ESRIN, the centre for data from remote-sensing satellites (formerly the European Space Research Institute), at Frascati, Italy; and the European Astronaut Centre at Cologne, Germany. The ESA launch site is at Kourou, French Guiana, on the Atlantic coast of South America. ESA came into being in 1975 as successor to the European Space Research Organization (ESRO) and the European Launcher Development Organization (ELDO). Its 18 member nations are: Austria, Belgium, the Czech Republic, Denmark, Finland, France, Germany, Greece, Ireland, Italy, Luxembourg, the Netherlands, Norway, Portugal, Spain, Sweden, Switzerland, and the UK. Canada, Hungary, Romania, Poland, Estonia, and Slovenia are Cooperating States.

(()) SEE WEB LINKS
• Official ESA website.

European VLBI Network (EVN) A consortium of radio observatories established in Europe in 1980 but now expanded worldwide, giving baselines of over 9000 km. It is the most sensitive VLBI array in the world and currently consists of over 20 individual radio telescopes in a dozen countries: China (Sheshan 25-m and Urumqi 25-m), Finland (Metsähovi 14-m), Germany (Effelsberg 100-m and Wettzell 20-m), Italy (Medicina 32-m, Noto 32-m, and Sardinia 64-m), Netherlands (Westerbork array), Poland (Torun 32-m), Russia (32-m dishes at Svetloe, Zelenchukskaya, and Badary), South Africa (Hartebeesthoek 26-m), Spain (Yebes 40-m and 14-m), Sweden (Onsala 25-m and 20-m), UK (Jodrell Bank 76-m and 25-m, and Cambridge 32-m), and the USA (Arecibo 305-m).

(((⊕))) SEE WEB LINKS
• Official EVLBI website.

EUV Abbr. for *extreme ultraviolet.

EUVE Abbr. for *Extreme Ultraviolet Explorer.

eV Symbol for *electronvolt.

evection A periodic disturbance in the Moon's position that results from changes in the eccentricity of its orbit caused by the Sun's attraction. It amounts to a maximum of 76′ in longitude and has a period of 31.8 days.

evening star Popular name for the planet Venus when it appears after sunset in the evening twilight sky.

event horizon The surface of a black hole. For a non-rotating black hole, it is a spherical boundary at the black hole's *Schwarzschild radius where the escape velocity becomes equal to the speed of light, so no events occurring within it can be seen from outside. However, the effects of the black hole's powerful gravitational field can still be felt outside the event horizon. For a rotating black hole, the event horizon is elliptical (*see* KERR BLACK HOLE).

Evershed effect An outward flow of gas in the penumbra of a sunspot, starting from the boundary with the umbra and moving radially out through the penumbra, and sometimes a little beyond it. The effect can be observed spectroscopically, particularly for spots near the solar limb, but may also be visible as outward-moving facular points, forming a so-called *moat*. The maximum outflow velocity is about 2 km/s. There is an *inverse Evershed flow* at higher altitudes, in which the flow (up to 20 km/s) is directed both inwards and downwards towards the sunspot. The effect is named after the English astronomer John Evershed (1864–1956), who discovered it in 1909.

EVN Abbr. for *European VLBI Network.

evolved star A star that has exhausted the hydrogen fuel in its core, and has evolved off the main sequence. Depending on its mass, an evolved star may be burning other nuclear fuels in its core and hydrogen in a thin shell (as in giants), or it may consist of spent nuclear fuel (as in neutron stars and white dwarfs).

excitation A process by which an electron bound to an atom acquires sufficient energy to transfer from a lower energy level to a higher energy level. The atom is then *excited*. Excitation can occur in two ways. In *collisional excitation* a particle such as an electron collides with an atom and transfers some of its energy to the atom. If the energy corresponds exactly to the difference between two energy levels, the atom becomes excited. In *radiative excitation* a photon of radiation is absorbed by an atom. The photon's energy must exactly equal the difference between the two energy levels. In both cases, when the electrons return to lower energy levels they emit photons.

excitation temperature The temperature of a gas or plasma obtained from the relative numbers of atoms or ions in the ground state and in excited states.

exit cone The range of directions within which light can escape from the gravitational field of an object, centred on the vertical to the object's surface. The concept is of significance only for an object collapsing to become a black hole. Then, rays of light sent out at a large angle to the vertical are bent back to the object's surface by its strong gravitational field. As the object collapses, the cone angle will decrease, closing up entirely as the object reaches its *event horizon.

exit pupil The smallest cross-section through the beam of light from an eyepiece through which all of the light from the eyepiece passes. The observer's eye should be at the exit pupil to see the full and brightest field of view. The diameter of the exit pupil is the *focal length of the eyepiece divided by the *focal ratio of the telescope. Thus an 18-mm eyepiece will have an exit pupil of 3 mm on an $f/6$ telescope. A simple way of finding the diameter of the exit pupil for binoculars is to divide the aperture by the magnification. For 7×50 binoculars, the size of the exit pupil will therefore be just over 7 mm. If the exit pupil is larger than the diameter of the dark-adapted eye (about 7 mm on average), some light will not enter the eye and will be wasted.

EX Lupi star A subclass of eruptive *T Tauri stars, also known as **EXors** or **subfuors**. EX Lupi stars exhibit optical outbursts of 1–4 mag., each lasting some 10–100 days and separated by several months. They are K or M dwarfs, and at maximum their spectrum is dominated by emission lines similar to those of classical T Tauri stars. As in the case of *FU Orionis stars, the eruptions of EX Lupi stars are attributed to an enhanced rate of accretion from a circumstellar disk onto the star's surface.

exobiology Another name for *astrobiology.

ExoMars A joint ESA–NASA programme to send two missions to Mars. The first, planned for launch in 2016, will consist of the Trace Gas Orbiter to detect atmospheric trace gases such as methane, plus a simple lander known as the Entry, Descent, and Landing Demonstrator. The second mission, planned for launch in 2018, would consist of two rovers, one European and the other American, delivered to the same site. ExoMars is the first part of the *Aurora programme.

(⊕) SEE WEB LINKS
• ESA mission website.

exoplanet Another name for an *extrasolar planet.

EXor See EX LUPI STAR.

Exosat The first ESA mission entirely devoted to X-ray astronomy, launched in 1983 May; the name is short for European X-ray Observatory Satellite. Exosat carried three instruments: an imaging telescope which operated at 0.05–2 keV (0.6–25 nm); a proportional counter array; and a gas scintillation proportional counter array; together they spanned 1–50 keV (0.025–1.24 nm). Exosat was placed into a highly elliptical orbit with a period of 96 hours, allowing extended observations of variable X-ray sources. It operated until 1986.

(⊕) SEE WEB LINKS
• ESA mission website.

exosphere The outermost part of the Earth's atmosphere, lying at an altitude above 500 km, and petering out at greater heights into interplanetary space. At such levels the atmospheric particle densities are extremely low; the density at 1000 km altitude is estimated to be only 7.3×10^5 particles/cm^3 (compared with 2.5×10^{19} particles/cm^3 at sea level). Under such conditions, collisions between particles are rare.

expanding universe Any model universe in which the space between widely separated objects is expanding. In the real Universe, neighbouring objects such as close pairs of galaxies do not move apart because their mutual gravitational attraction exceeds the effect of the cosmological expansion. However, the distance between two widely separated galaxies, or clusters of galaxies, will increase as the Universe expands.

Explorer An ongoing series of US scientific spacecraft. Explorer 1, launched on 1958 January 31, was the first successful US satellite; it discovered the Earth's *Van Allen Belts. Later Explorers have included the *International Ultraviolet Explorer and the *Cosmic Background Explorer. A concurrent *Small Explorer Program* (SMEX), consisting of simpler, cheaper satellites, began in 1992 with the *Solar, Anomalous, and Magnetospheric Particle Explorer (SAMPEX), later joined by the Medium-class Explorer (MIDEX) series, the first of which was the *Far Ultraviolet Spectroscopic Explorer (FUSE).

(((∰))) SEE WEB LINKS
• Official Explorer program website.

exposure age The period of time during which a meteoroid has been exposed to cosmic radiation in space; also known as *cosmic-ray exposure age*, or *radiation age*. This is usually the time between a meteoroid leaving its parent body (such as an asteroid) and arriving on Earth. Bombardment of the meteoroid by cosmic rays in space produces radioactive isotopes (such as helium-3, neon-21, and argon-38), or phenomena such as *fission tracks, whose abundances can be used to estimate the exposure age. Typical exposure ages range from a few million years to a few hundred million years.

extended source 1. A source of angular size greater than the resolution of the instrument used to observe it; therefore, an extended source is said to be *resolved*. To the human eye, the Sun and Moon are extended sources, but the stars and planets are not. *See also* POINT SOURCE.
2. In radio astronomy, a class of sources of large angular extent displaying extended structure such as jets and lobes, and tending to be brightest at long wavelengths. *See also* COMPACT SOURCE.

extinction, atmospheric *See* ATMOSPHERIC EXTINCTION.

extinction, interstellar The dimming of starlight caused by a combination of *interstellar absorption and *scattering.

extragalactic nebula An obsolete term for a *galaxy.

extrasolar planet A planet orbiting a star other than the Sun; also known as an *exoplanet*. Planets of other stars are too faint to be seen directly with existing instruments, so indirect methods of detection are necessary. The first extrasolar planet, found in 1992, orbits a pulsar, PSR 1257+12. A cyclic change in the timing of the pulsar's radio emissions was interpreted as motion of the pulsar about its centre of mass with an orbiting planet. A similar approach, known as the radial velocity method, has been used to find other extrasolar planets. The technique looks for a cyclical Doppler shift in light from a star that would occur as it orbited its common centre of mass with one or more planets. Using this technique, astronomers in 1995 detected the first known extrasolar planet around an ordinary star, 51 Pegasi, 50 l.y. away. The planet has a mass about half that of Jupiter and an orbital period of 4.2 days. The first multi-planet system, also discovered by the radial velocity method, was found in 1999 around Upsilon Andromedae. An alternative approach, the transit method, is to look for dips in the brightness of a star as an orbiting planet moves in front of it, although this is restricted to planets whose orbital plane lies close to our line of sight. The first transit of an extrasolar planet across the face of a star, HDE 209458 in Pegasus, was observed in 1999. A third technique, which has also borne fruit, looks for a spike on the light curve of a gravitational *microlensing event that would be caused by a planet accompanying the lensing star.

Masses of the first 500 known extrasolar planets range from 25 Jupiters (on the verge of being a *brown dwarf) to a few times that of the Earth. Periods range from less than a day to several centuries, the longest being Fomalhaut b, nearly 900 years, although 96% are under 10 years. Orbital radii are 0.01 AU to over 500 AU from their parent stars, although only a few are greater than 25 AU. Planet-finding spacecraft such as *COROT and *Kepler are using the transit technique to detect smaller drops in light, and hence smaller planets, than is possible from the ground. 📷

(((⊕))) SEE WEB LINKS

• Extrasolar planets encyclopedia, regularly updated.

• NASA website on exoplanets, regularly updated.

extreme Population I star A star belonging to the youngest stellar population, such as a *T Tauri star, a star newly arrived on the *zero-age main sequence, or a massive OB star with its associated *H II region. Such stars have high metal abundances (similar to that of the Sun or greater). They are found in localized regions of the galactic disk, notably the spiral arms, where star formation has occurred very recently.

extreme ultraviolet (EUV) A region of the spectrum between the ultraviolet and X-ray regions, covering the wavelengths 10–100 nm. The first EUV sources were detected by a telescope on the Apollo–Soyuz mission in 1975. The first survey of the EUV sky was carried out by *Rosat in 1990/91. The *Extreme Ultraviolet Explorer continued astronomy in this waveband.

Extreme Ultraviolet Explorer (EUVE) A NASA satellite to study the sky at extreme ultraviolet wavelengths, launched in 1992 June. First it made a six-month survey of the sky using three *grazing-incidence telescopes, covering the wavelengths 6–70 nm. A fourth telescope pointing at right angles to the others surveyed a limited area of sky with deeper exposure. This telescope included a spectrometer that was used as the primary instrument after the initial sky survey. It ceased operation in 2001 January.

(((⊕))) SEE WEB LINKS

• Information page at Goddard Space Flight Center.

extrinsic variable A variable star whose variations arise for some external reason, such as rotation, orbital motion, or obscuration, as distinct from intrinsic processes such as pulsation. *See* ALPHA² CANUM VENATICORUM STAR; BY DRACONIS STAR; ECLIPSING BINARY; ELLIPSOIDAL VARIABLE; FK COMAE BERENICES STAR; INTRINSIC VARIABLE; MAGNETIC VARIABLE; ROTATING VARIABLE; SX ARIETIS STAR.

eye lens The lens in an eyepiece that is closest to the observer's eye.

eyepiece A lens or combination of lenses used to magnify the image formed by a telescope; also known as an *ocular*. The simplest form of eyepiece is a single converging lens of short focal length, but this has severe aberrations except at the very centre of its field of view. In practice, therefore, eyepieces usually have at least two elements. A *field lens*, facing the objective or mirror, gathers light over a wider field than a single lens, while the **eye lens**, through which the observer looks, provides the magnification. This combination gives a good field of view while keeping aberrations under control. A *field stop* (a diaphragm located so that it is in focus as seen through the eye lens) provides a hard edge to the field of view. Many additional elements can be included to improve performance. Popular designs include the *Erfle, *Huygenian, *Kellner, *monocentric, *Nagler, *orthoscopic, *Plössl, and *Ramsden eyepieces.

eye relief The distance between the rear lens (the **eye lens**) of an eyepiece and the *exit pupil. If this distance is too short, the eye will be uncomfortably close to the eye lens; if it is too long, it may be hard to keep the eye in the best position to view the image. In general, eye relief gets shorter as the focal length of the eyepiece decreases, but it is possible to design eyepieces of short focal length with good eye relief. Good eye relief is useful for spectacle wearers.

Faber–Jackson relation The observed relationship between the luminosity of an elliptical galaxy and the random speed of stars near its centre. The relation has been useful in estimating the relative distances of galaxies, since the absolute magnitude of a galaxy can be determined from stellar motions within it, deduced from its spectrum. The relation is named after the American astronomers Sandra Moore Faber (1944–) and Robert Earl Jackson (1949–), who published it in 1976.

Fabricius, David (1564–1617) German astronomer and priest, originally named Goldschmidt; father of J. *Fabricius. He was one of the first telescopic observers. Kepler used observations by Fabricius and Tycho Brahe in calculating an elliptical orbit for Mars. Fabricius noted the fading of the star *Mira.

Fabricius, Johann(es) (1587–1615) German astronomer and physician, son of D. *Fabricius. With his father, he used the method of projection to study the Sun telescopically. He studied sunspots, of which he was one of a number of independent discoverers along with C. *Scheiner, *Galileo, and the Englishman Thomas Harriot (1560–1621). By observing the change in position of sunspots, Fabricius found that the Sun rotates.

Fabry lens A small lens placed within a photometer to provide an image of the telescope's objective lens on the *photocathode. It is placed behind a mask in the *focal plane of the objective, which isolates the object chosen for measurement. By imaging the objective lens rather than the object, small movements of the image caused by air currents do not cause it to move across the photocathode, which may vary in sensitivity across its surface. The use of such a lens was first suggested by the French physicist (Marie Paul Auguste) Charles Fabry (1867–1945).

Fabry–Perot interferometer An optical instrument used for high-resolution spectroscopy of extended objects such as galaxies and nebulae. Light from an object is passed through an *etalon, which transmits only a narrow range of wavelengths. The spacing (or gap) between the plates of the etalon can be adjusted in steps, making it possible to scan the spectral region of interest, for example a particular group of spectral lines, and producing an image of the object at each chosen wavelength. In combination with sensitive detectors such as CCDs, Fabry–Perot interferometers can achieve a typical resolution of 0.03nm. The design was first constructed by the French physicists (Marie Paul Auguste) Charles Fabry (1867–1945) and (Jean-Baptiste Gaspard Gustav) Alfred Pérot (1863–1925) in the late 19th century.

facula 1. A brighter and hotter patch on the Sun's photosphere, visible in white light and best seen near the solar limb against the background of limb darkening. Faculae often appear shortly before a sunspot group forms, and remain visible for several days or weeks after the spots have vanished. High-latitude (polar) faculae also occur, well away from sunspots; these, unlike sunspot faculae, are most numerous on the rising part of the sunspot cycle. Faculae are slightly hotter (by about 300K) than the surrounding photosphere. They are locations of strong magnetic fields (0.1 tesla), and coincide with bright patches in the chromosphere (*plages) and the chromospheric network. Normally they can be resolved into small **facular bright points**, around 150km wide, lasting 20min or so. There is often a connection between faculae and structure within large sunspots, in particular *light-bridges*, which are ridges of bright material crossing a sunspot.

2. A bright spot on a planetary surface; pl.*faculae*. The name is not a geological term, but is used in the nomenclature of individual features, for example Memphis Facula on Ganymede.

fall A meteorite that has actually been seen to fall and has been recovered at the place of impact. In many cases the meteorite is recovered immediately or very soon after falling, and is therefore relatively free of terrestrial contamination and weathering. Sometimes, however, the meteorite is not recovered until months or years after its observed fall. Only about 4% of known meteorites were seen to fall. *See also* FIND.

falling star Popular US name for a *meteor.

False Cross A cross-shape in the southern sky formed by four stars, Iota and Epsilon Carinae and Kappa and Delta Velorum, sometimes mistaken for the true Southern Cross, the constellation *Crux.

Fanaroff–Riley class A category in a classification of extragalactic radio sources, typically radio galaxies or quasars, based on the separation between the brightest regions in their radio-emitting lobes. In Class I (FRI) sources, the bright regions are separated by less than half the overall extent of the radio lobes. Hence these regions are closer to the nucleus than in Class II (FRII) sources, where they are separated by more than half the total length of the source. Class II sources are in general considerably more powerful than Class I. The scheme is named after the South African astronomer Bernard Lewis Fanaroff (1947–   ) and the British astronomer Julia Margaret Riley (1947–   ).

fan beam An antenna pattern characterized by a wide, flat main lobe in the shape of a fan. Fan beams are produced by certain types of radio interferometer, notably the *Mills cross.

Faraday effect The rotation of the plane of polarization of an electromagnetic wave as it passes through a region containing free electrons and a magnetic field; also known as **Faraday rotation**. The amount of rotation, in radians, is given by $RM\lambda^2$, where RM is the *rotation measure* of the source and λ is the wavelength. Observation of the Faraday effect in pulsars is the most important means of determining the magnetic field of the Galaxy. It is named after the English physicist Michael Faraday (1791–1867).

far infrared The longer wavelengths of the infrared part of the electromagnetic spectrum. Although ill-defined, the far infrared is generally taken to cover wavelengths of about 35–300 μm, which are blocked by the Earth's atmosphere. The region between 300 μm and 1 mm, which is accessible from high mountain observatories, is now often referred to as the *submillimetre band*.

farrum A pancake-shaped feature on a planetary surface, especially Venus; pl.*farra*. The name is not a geological term, but is used in the nomenclature of individual features, for example Carmenta Farra on Venus.

far ultraviolet The wavelength range from 91.2 nm (the *Lyman limit) to approximately 200 nm; atmospheric absorption of radiation at these wavelengths prevents it from reaching the ground. Space-borne telescopes operating in this region have included the *International Ultraviolet Explorer, the *Far Ultraviolet Spectroscopic Explorer, and the *Hubble Space Telescope. At its short-wavelength end, the far ultraviolet overlaps with the *extreme ultraviolet.

Far Ultraviolet Spectroscopic Explorer (FUSE) A NASA ultraviolet astronomy satellite, launched in 1999 June. It performed high-resolution spectroscopy in the wavelength range 90–120 nm using four mirrors each 0.39 by 0.35 m. It was many thousands of times more sensitive than the *Copernicus satellite, the only previous mission to cover this region. Observations ceased in 2007 July.

(() SEE WEB LINKS

• Official mission website.

Fast Auroral Snapshot Explorer (FAST) A NASA satellite for studying the processes that cause aurorae, launched in 1996 August.

 SEE WEB LINKS
• Official mission website.

fast nova A type of nova that declines by 3 magnitudes from maximum in 100 days or less; abbr. NA. An alternative classification is based on the time taken to decline by 2 magnitudes from maximum: very fast, less than 10 days; fast, 11–25 days; moderately fast, 26–80 days; slow, 81–150 days; very slow, 151–250 days.

Faulkes Telescopes A pair of 2-m reflectors, of identical design to the *Liverpool Telescope, for use by students in the UK, Hawaii, and Australia over the Internet. FT-North is sited at an altitude of 3050 m at the University of Hawaii's observatory on Haleakala mountain, Maui island, and opened in 2004. FT-South, which came into operation in 2007, is located at *Siding Spring Observatory in New South Wales, Australia. Since 2005 they have been part of the Las Cumbres Observatory Global Telescope Network.

SEE WEB LINKS
• Official telescopes website.

fault A fracture in a planetary surface, on either side of which there has been relative movement parallel to the sides of the fracture. In a *normal fault*, the relative movement is vertical. In a *transcurrent* or *strike–slip fault*, the relative movement is horizontal. In a *thrust fault*, the fault plane is at an angle to the horizontal, and the upper side rides over the lower.

F band A *Fraunhofer line in the Sun's spectrum at 486.1 nm, due to absorption by hydrogen. It is also known as the Hβ line in the *Balmer series.

F-class asteroid A subclass of the *C-class asteroids. Its members have low albedos (0.03–0.07), and a featureless reflectance spectrum, flat to slightly bluish over the wavelength range 0.3–1.1 μm. F-class asteroids are distinguished from class C by having a weak to non-existent ultraviolet absorption feature. Members of this class include (142) Polana, diameter 55 km, and (213) Lilaea, diameter 83 km.

F corona The outer part of the Sun's corona, which is illuminated by sunlight scattered or reflected from solid dust particles. The same phenomenon also produces the *zodiacal light much farther from the Sun. The dust grains are a few micrometres across, and are located in a disk extending outwards from about 1 solar radius (750 000 km) from the Sun's surface. Unlike the electrons responsible for the *K corona, the dust grains are relatively slow-moving. Thus the light scattered from them has the same spectrum as the photosphere, including its Fraunhofer lines (hence the letter 'F'). The F corona is the brightest part of the corona beyond 1.5 solar radii from the Sun's surface.

feed 1. The point at which a cable or waveguide is connected to an antenna.
 2. A small antenna placed at the focus of a radio telescope to collect incoming radio waves. It is commonly either a dipole antenna or a horn antenna (a **feedhorn**).

Fermi Gamma-ray Space Telescope An international gamma-ray observatory launched in 2008 June, a successor to the *Compton Gamma Ray Observatory; the partners are NASA, the US Department of Energy, France, Germany, Japan, Italy, and Sweden. Fermi carries two instruments: the Large Area Telescope (LAT), which maps the gamma-ray sky at energies from 20 MeV to more than 300 GeV; and the GLAST Burst Monitor (GBM), which detects gamma-ray bursts at energies from 8 keV to 30 MeV. The GBM's field of view covers the entire sky not obscured by the Earth.

SEE WEB LINKS
• Official mission website.
• NASA mission page.

fermion A type of elementary particle such as an electron, neutron, or proton having a + ½ or −½ value of *spin. A maximum of two electrons can therefore occupy the ground state, provided they have opposite spins. Hence elements other than hydrogen and helium must have some of their electrons in higher levels than the ground state. Fermions obey the Pauli exclusion principle. They are named after the Italian–American physicist Enrico Fermi (1901–54). *See also* BOSON.

fibrille A feature in the low chromosphere of the Sun; also spelt **fibril**. Fibrilles are fine dark lines arranged in a near-spiral pattern around sunspots as seen in *Hα light. They are particularly prominent around large, mature spots, from which they extend outwards for about 10 000 km, forming a *superpenumbra*. The whole pattern lasts as long as the active region containing the sunspots, but individual fibrilles live for only 20 min or so. Fibrilles are thought to delineate the magnetic field in the chromosphere around sunspots.

field curvature *See* CURVATURE OF FIELD.

field equations Equations that describe the properties of fields, such as electromagnetic fields (Maxwell's equations). However, the term is usually taken to refer to A. Einstein's field equations, which describe the curvature of spacetime produced by a gravitational field.

field flattener A glass plate or plates, located between the objective and the *focal plane of a telescope, optically figured to correct for aberrations such as coma (*see* COMA, OPTICAL) or *curvature of field. On large telescopes, a field corrector is particularly important at the *prime focus to give a wide, coma-free field.

field lens The lens of an *eyepiece that is farthest from the observer's eye.

field of view The full angular extent of an image. It may refer to the actual dimensions of the sky being viewed, usually about 1° or less, or the apparent size of the *field stop as seen through an eyepiece, which can be between about 20° and 90°, depending on the design of the eyepiece.

field star A star that is visible in the same field of view as a star cluster, but is not a member of that cluster, being either closer to us or more distant. Similarly, a **field galaxy** lies in the same line of sight as a group of galaxies, but is not a member of the group.

field stop A circular aperture in an *eyepiece which limits the field of view.

53 Persei star A subgroup of *Beta Cephei stars in which *non-radial pulsation is particularly significant. In this group, multiple periods (0.15–2 days) are often present, but all tend to be highly unstable. The type star, 53 Per (also known as V469 Per), has a range of 4.81–4.86 mag. and a period of 0.304 days.

figure The shape given to an optical surface during manufacture. The process of producing or altering the shape is known as **figuring**.

filament 1. A long 'tongue' of relatively cool material (10 000 K) suspended in the much hotter solar corona (2 million K). Filaments appear dark when seen silhouetted against the Sun's disk in *Hα light, but at the limb they appear as *prominences. Quiescent filaments (the equivalent of *quiescent prominences at the limb) may show gradual changes, but portions of the filament may move more quickly, at speeds of a few kilometres per second. Loop filaments (the disk equivalent of *loop prominences) are sometimes seen near very large flares. The equivalent of an *eruptive prominence is the disappearing filament, sometimes called a *disparition brusque*, from the French meaning 'sudden disappearance'. Disappearing or *winking* filaments may occur as a result of a *Moreton wave. An *arch filament system is a set of cool loops associated with emerging active regions.
  2. A chain-like cluster or supercluster of galaxies. The *large-scale structure of galaxy distribution in the Universe is dominated by such features, which may be tens of millions of light years long.

filamentary nebula A group of elongated clouds of gas and dust with a fine thread-like structure as seen from Earth. Many filamentary structures may actually be sheets or parts of shells seen edge-on, rather than threads. The best-known filamentary nebulae, such as the *Veil Nebula, are supernova remnants. Although these filaments have temperatures of 10000K they are actually the coolest parts of the remnant, other parts of which remain at over 1 million K.

filar micrometer A device for measuring the separations of two objects as seen through a telescope, such as the components of a double star. It consists of two fixed fine wires or webs in the field of view of an eyepiece, crossing at right angles, plus a third movable web, parallel to one of the fixed webs. The position of the movable web can be measured precisely against a scale. Many filar micrometers can also be rotated against an angular scale; such a device is known as a *position micrometer*. Both separation and *position angle can then be measured.

filigree A very fine structure on the Sun's photosphere, consisting of tiny bright dots arranged along the dark lanes separating granules. Sometimes there is a clustering of such dots to form *crinkles*. The smallest dots are about 150km across and last for less than 30min. The filigree structures are hotter (by a few hundred degrees K) than the surrounding photosphere in which they are located, and mark sites of strong (0.1 tesla) magnetic field. They are visible in spectroheliograms made in weak Fraunhofer lines or the *line wings of the Hα line.

filled aperture The collecting area of a radio or optical telescope that is entirely occupied by a reflecting surface or receivers, as distinct from an *interferometer, which consists of individual elements with spacing between. *See also* UNFILLED APERTURE.

filled-centre supernova remnant Another name for a *plerion.

filter 1. An optical component through which light of only certain restricted wavelengths can pass, the other wavelengths being blocked. Two types are in general use: glass filters and *interference filters. Glass filters are made from disks of coloured glass. Interference filters are made from multiple layers or coatings on glass which cause interference between light waves, allowing only a limited range of wavelengths to pass. Glass filters are mostly used for *broad-band photometry, while interference filters are used for *intermediate-band and *narrow-band photometry. The *bandwidth* of the filters (i.e. the range of wavelengths which they pass) lies in the ranges 30–100nm for broad-band, 10–30nm for intermediate-band, and 3–10nm for narrow-band photometry.
2. An electronic device designed to transmit a certain range of frequencies and reject others. A *low-pass filter* transmits below a certain cut-off frequency; a *high-pass filter* transmits above a cutoff frequency; a *bandpass filter* transmits between two cutoff frequencies; a *notch filter* is designed to reject a narrow band of frequencies.

filtergram A photograph of the Sun taken through a filter with a particular wavelength range (bandpass), which is usually narrow and centred on part of some prominent spectral line. At visible wavelengths this is typically a Fraunhofer line such as *Hα, and the filter used is of the *birefringent (Lyot) or *etalon type that can be tuned to different parts of the spectral line. At extreme ultraviolet wavelengths multilayer mirror coatings can be used to give narrow bandpasses of 1–2nm that can be tuned to strong coronal emission lines.

find A meteorite that was not observed to fall, but was found and subsequently identified, sometimes many thousands of years after reaching the Earth. Over 95% of known meteorites are finds. Weathered iron and stony-iron meteorites are more readily identified than weathered stony meteorites, so these meteorites are over-represented in collections. Observed *falls more accurately reflect the true proportions of meteorite types.

finder A small telescope or alignment device fixed to a larger telescope to aid in locating a chosen object. It provides a comparatively wide field of view, often with cross-wires to mark the centre. Some finder devices simply display an illuminated marker against the night sky, allowing the telescope to be roughly aligned on an object.

fine structure Closely spaced spectral lines from a given element. Fine structure has two causes. One is the interaction of electrons in an atom with the magnetic field of the atom itself, and the second is relativistic effects on the electrons' kinetic energies due to their orbital motion. These combined effects create a fine splitting of the various principal energy levels and produce several closely spaced lines instead of one. For hydrogen the fine structure is so fine (0.006nm) that it is unobservable in stars. However, more complex atoms such as sodium can have more discernible fine structure because of the presence of many other electrons in the atom. The sodium *D lines are an easily observable example of fine structure, with a separation of 0.6nm. *Hyperfine structure is on a still smaller scale.

fireball A meteor whose apparent magnitude exceeds that of the planet Venus (magnitude −5 or brighter). Fireballs are comparatively rare, perhaps fewer than one per thousand meteors. During major annual meteor showers such as the *Perseids or *Geminids considerable numbers of fireballs may be seen. Fireballs are also associated with the atmospheric entry of large incoming bodies, which may result in meteorite falls. A fragmenting fireball with audible sound is termed a *bolide.

first contact At a solar eclipse, the moment at which the leading (eastern) limb of the Moon just touches the Sun's western limb, marking the beginning of the eclipse; or, at a lunar eclipse, the moment at which the Moon's eastern limb first enters the Earth's shadow.

first point of Aries (♈) The point on the celestial equator from which right ascension is measured. It is the same as the *vernal equinox. This point has zero right ascension and zero declination. It is defined as the point at which the Sun passes from south to north of the celestial equator, which happens on March 20 each year. Due to *precession it is not a fixed point on the star background but regresses around the ecliptic at an annual rate of approximately 50″. Consequently, despite the name, it is no longer in the constellation of Aries but currently lies in neighbouring Pisces and in the year 2597 will move into Aquarius.

first point of Libra (♎) The point on the celestial sphere diametrically opposite the *first point of Aries. It is the same as the autumnal equinox. It has right ascension 12h and declination zero. It is the point at which the Sun passes from north to south of the celestial equator, which happens on September 22 or 23 each year. Because of *precession, it no longer lies in Libra but in neighbouring Virgo.

first quarter The phase of the waxing Moon that occurs midway between new and full Moon, when half of the Moon is illuminated. At first quarter the Moon lies 90° east of the Sun.

Fish Mouth A dark nebula that separates the bright nebula M42 (the *Orion Nebula) from M43, which is part of the same cloud. The Fish Mouth appears as a dark indentation in the Orion Nebula just to the north of the stars of the *Trapezium.

fission A process in which a heavy atomic nucleus splits into smaller nuclei. The process is accompanied by the release of large amounts of energy. Fission may occur spontaneously or as the result of collision of neutrons with the nucleus.

fission track A trail of damage left in a sample of rock or mineral by charged atomic particles. The particles (atomic nuclei) that produce these tracks are either the spontaneous fission fragments of radioactive nuclei within the sample or cosmic rays. The age of a rock sample such as a meteorite may be calculated by counting the number of fission tracks; this is known as **fission-track dating**. *See also* EXPOSURE AGE.

Fitzgerald contraction *See* LORENTZ–FITZGERALD CONTRACTION.

Five-hundred-meter Aperture Spherical Telescope (FAST) A radio telescope with a diameter of 500 m being built in Guizhou province, southwest China, by the National Astronomical Observatories of China. It consists of a single dish suspended in natural hollow in the ground, similar to the *Arecibo radio telescope in Puerto Rico. FAST will have a reflecting surface consisting of around 4600 triangular segments which are adjusted to form a parabolic surface. A movable receiver allows the telescope to observe and track objects up to 40° from the zenith. The usable aperture at any one time will be 300 m. FAST is due for completion in 2014.

fixed stars An archaic expression for the background stars in general, as distinct from the planets which were known as *wandering stars*. As used today, the term refers to stars with no detectable *proper motion.

FK Abbr. for *Fundamentalkatalog*, one of a series of catalogues of positions and proper motions of fundamental stars published in Germany. The first appeared in 1879 and was intended to provide reference positions for the *AGK catalogues. The second FK (the NFK) appeared in 1907. FK3 followed in 1937–8 with FK4 in 1963 and FK5 in 1988, all three of which contained 1535 stars brighter than about visual magnitude 7.5. An extension to the FK5, containing a further 3117 stars down to magnitude 9.5, was published in 1991. The latest addition, the FK6, is a combination of Hipparcos data and ground-based data in four volumes, the first of which appeared in 1999. The FK4, FK5, and FK6 are published by the Astronomisches Rechen-Institut, Heidelberg.

(((●))) SEE WEB LINKS
• Detailed description and full catalogue downloadable from the CDS.

FK Comae Berenices star A giant of spectral type G–K which is also a rapidly rotating extrinsic variable with non-uniform surface brightness. The variations are caused by a region of cool spots (*starspots) localized on one hemisphere of the star; abbr. FKCOM. The rotation and variation periods are identical and are typically several days, with amplitudes up to several tenths of a magnitude. It is possible that these stars represent a later stage in the evolution of a *common envelope binary in which the cores have coalesced.

Flamsteed, John (1646–1719) English astronomer, the first Astronomer Royal, appointed in 1675 by King Charles II in response to the need to find a reliable means of measuring longitude at sea. Flamsteed had recommended obtaining more accurate measurements of the movements of the Moon and the positions of the stars. The Royal Observatory at Greenwich was built for him. Flamsteed's catalogue of 2935 stars, *Historia coelestis Britannica*, was published posthumously in 1725, and was the first major star catalogue compiled with the aid of a telescope. Four years later a set of star charts based on the catalogue, *Atlas coelestis*, appeared.

Flamsteed numbers Identification numbers applied to stars in a constellation in order of increasing right ascension. The stars so numbered are those included in the 1725 catalogue of J. *Flamsteed. Many stars are still known by their Flamsteed numbers (e.g. 61 Cygni).

flare, solar A sudden release of energy in the Sun's corona, lasting up to several hours or, exceptionally, more than a day. Flares usually occur within about 175000km of the Sun's surface (one-quarter of a solar radius). They emit radiation over the whole spectrum, from gamma rays to radio waves. They also throw out high-speed particles (electrons, protons, and atomic nuclei), at speeds up to about 70% of the speed of light, which reach the Earth in 15min or so, depending on their trajectory. Only the most energetic flares are visible in white light. Flares occur in *active regions with complex magnetic fields, the largest flares being in the most complex regions. Most of the energy may be released in the first few minutes, with an *impulsive stage* that may last only a few seconds. The total energy released can be up to 10^{27} joules; there is no well-defined lower energy limit.

Flares are classified in two ways: by their appearance in *Hα light, and by their soft X-ray emission. In Hα the term *subflare* is given to the smallest events, and the scale then runs from 1 to 4 with increasing area; a brightness code is added from faint (f), via normal (n), to bright (b). In soft X-rays (0.1–0.8nm) flares are classified as C, M, or X according to increasing strength, with subdivisions from 1 to 9.

In Hα, the flare may start with the disappearance of a *filament (i.e. a prominence seen from above), bright areas developing into *ribbons* either side of the *magnetic inversion line. There is generally a rapid expansion stage called the **flash** phase. There are often hard X-ray and microwave radio bursts forming an early *impulsive* stage, with soft X-rays rising more gradually to maximum a few minutes later, followed by a decline (the *decay* or *cooling* phase). Flares derive their energy from the energy stored in magnetic fields, although the exact mechanism is not known. According to one theory, flares occur when oppositely directed magnetic field lines reconnect. Particles are accelerated at the impulsive stage to give the hard X-ray emission, while the soft X-rays are emitted by a very hot (20 million K) plasma contained within coronal loops.

flare star A form of eruptive variable that exhibits abrupt, unpredictable flares with a rise time of seconds and a decay time of minutes. Flare stars have a Ke- or Me-type spectrum and a strong magnetic field. Although the term is generally taken to be synonymous with *UV Ceti stars, certain *BY Draconis stars also exhibit flare activity. These flares are, in principle, the same kind of phenomenon as solar flares, but with much greater energies involved, perhaps by a factor of 1000.

flash spectrum The emission-line spectrum of the Sun's chromosphere, which flashes out in the few seconds before totality in a solar eclipse. Slitless spectrographs show the flash spectrum as a series of thin crescents, each marking the position of a prominent emission line such as the hydrogen Balmer lines and the H and K lines of ionized calcium, which are emitted by the chromosphere.

flash star An obsolescent term for *flare stars found in T associations and young clusters. Such a star is now generally regarded as a subtype of *UV Ceti star (designated UVN), and is probably a form of *irregular variable (INB) that also exhibits flares.

flat A mirror with a plane surface, such as the secondary mirror of a Newtonian reflector.

flattening, polar Another term for *oblateness.

F layer The highest part of the Earth's *ionosphere. The F layer is split into two regions: a lower F1 layer around 170km altitude, and an upper F2 layer at 250km. The F2 layer disappears at night, and may occasionally be absent during daytime.

Fleming, Williamina Paton (1857–1911) Scottish astronomer who worked in the USA. At Harvard, working with E. C. *Pickering, she established herself from 1881 as the most prominent woman astronomer of her time. Her major achievement was the classification of 10351 stars according to their spectra in the *Draper Catalogue of Stellar Spectra* (1890), a forerunner of the *Henry Draper Catalogue. Her scheme had 17 categories, an enormous improvement on P. A. *Secchi's pioneering scheme (*see* SECCHI CLASSIFICATION); it was in turn refined by A. J. *Cannon.

flexus A meandering linear feature on a planetary surface; pl.*flexus*. The name, which means bending or winding, is not a geological term, but is used in the nomenclature of individual features, for example Delphi Flexus on Europa.

flickering Rapid, irregular fluctuations in a star's brightness over minutes or hours which, in some cases, may amount to several tenths of a magnitude. Flickering occurs in many *interacting binary systems, including *AM Herculis stars, and certain *symbiotic stars. It is thought to arise from instabilities in a *hot spot (1) or in the inner region of an *accretion disk, or, in AM Herculis stars, in the accretion column above a magnetic pole.

flint glass A basic type of optical glass, whose optical properties are modified by the addition of lead oxide and a smaller amount of potassium oxide. It usually has a *refractive index of about 1.6, higher than that of the other basic type of optical glass, *crown glass, and it is more easily scratched. Partly for this reason, flint glass is generally used as the inner component of a compound lens.

flocculi **1.** Another name for coarse *mottles. **2.** An obsolete name for *plages.

Flora group A complex grouping of asteroids near the inner edge of the main asteroid belt, at a mean distance of 2.2 AU from the Sun and with a mean orbital inclination of about 5°. The group is separated from the main belt by one of the *Kirkwood gaps, and the region is sometimes divided into several separate families. Of these, the members of the so-called Flora family may have originated from the break-up of a single object, the largest remaining fragment being (8) Flora itself, diameter 136km. Flora is an S-type asteroid, discovered in 1847 by the

English astronomer John Russell Hind (1823–95). Flora's orbit has a semimajor axis of 2.202 AU, period 3.27 years, perihelion 1.86 AU, aphelion 2.55 AU, and inclination 5°.9.

fluctus Flow terrain on a planetary surface; pl.*fluctus*. The name, which means flowing, is not a geological term, but is used in the nomenclature of individual features such as Mylitta Fluctus on Venus, or Tung Yo Fluctus on Io.

flumen A channel on the surface of Titan; pl.*flumina*. Flumina are distinguished from fissures in that they have a river-like morphology and are presumed to have carried fluids. Flumen is not a geological term, and no feature has yet been officially assigned this designation.

fluorescence The emission of light from a gas whose atoms have been excited by a source of higher-energy radiation, such as a nearby hot star. The atoms of the gas are first excited to higher energy levels when they absorb energetic photons, especially those at ultraviolet wavelengths. Subsequently, the electrons cascade down to lower energy levels, re-emitting the energy as a series of lower-energy photons, usually at visible or infrared wavelengths. Nebulae glow as a result of fluorescence. *See also* BOWEN FLUORESCENCE.

fluorite Calcium fluoride crystal, used as an optical material. Objective lenses made from fluorite are particularly free from chromatic aberration, since fluorite has approximately half the dispersion of crown glass. Its drawbacks are that it is very expensive and is attacked by water, so fluorite elements must be sealed between other elements.

flux (Φ) A measure of energy or number of particles passing through a given area of surface in unit time, usually per second. *Luminous flux* is the rate of flow of energy in the form of photons, measured in lumens; *magnetic flux* is a measure of the strength and extent of a magnetic field perpendicular to a surface, measured in webers; *particle flux* is the number of particles, for example in a stellar wind, passing through a unit area per second.

flux collector A reflecting telescope with a large collecting area, designed to bring the maximum amount of radiation to a focus for measurement rather than provide an image; also known as a *light bucket*. Generally speaking, flux collectors have a less precise surface figure than conventional telescopes, and are usually intended for use at infrared and submillimetre wavelengths.

flux density 1. (S) Spectral flux density, a quantity that indicates the observed strength of an astronomical source. It is a measure of the radiant power passing through unit area in unit interval of frequency or wavelength. The SI unit is W/m^2 per Hz, but the special unit *jansky is more common in radio or infrared astronomy. In infrared astronomy, the unit W/m^2 per μm is often used. The term is occasionally (and more correctly) used to mean radiant power per unit area (W/m^2).
  2. (B) Magnetic flux density, a quantity that indicates the strength and direction of a magnetic field. The SI unit is the tesla.

flux tube A cylindrical structure in a magnetic field whose sides are parallel to the local magnetic field lines along its whole length. In tenuous plasmas, such as the solar corona, plasma in a flux tube is confined within that tube, causing *coronal loops. Electrical currents flow easily along magnetic flux tubes, such as along the flux tubes that link Jupiter to its moon Io and also along the geomagnetic flux tubes that link the Earth's magnetosphere to the ionosphere, causing aurorae. These currents flowing along flux tubes generate radio waves with wavelengths of tens of meters at Jupiter and about a kilometre at Earth. A current flowing along a flux tube also twists the field lines to form a **flux rope**; this twist provides great stability to the structure. Examples of flux ropes are *coronal mass ejections.

flux unit An obsolete term for a *jansky.

f/number The number obtained by dividing the *focal length of a lens or mirror by its diameter; it is the same as the *focal ratio.

focal distance An alternative name for *focal length.

focal length (*F* or *f*). The distance between a lens or mirror and its focal point. Together with the aperture, the focal length is one of the main properties of a lens or mirror. The longer the focal length, the larger the image scale.

focal plane A flat surface at right angles to the optical axis of a lens or mirror which contains the image of a small, distant object and also the *focal point. If an extended image lies not in a plane but on a curved surface because of *curvature of field, it is said to be formed on the *focal surface* or *Petzval surface.*

focal point The point at which parallel light rays from a distant object are brought together by a lens or mirror; also known as the **focus**. It is the location of the sharpest image of a distant object. The distance of the focal point from the lens or mirror is the *focal length.

focal ratio The ratio of the *focal length of a lens or mirror to its aperture, usually expressed as a number, the *f/number. The smaller the focal ratio, the smaller the image scale and the brighter the image for a given aperture. Small focal ratios, below about $f/6$, are termed *fast*, while those greater than about $f/8$ are *slow*, a terminology adopted from photography.

focal reducer A converging lens, placed just before a telescope's *focal point, which has the effect of shortening the *effective focal length and giving a wider field of view.

focal surface *See* FOCAL PLANE.

focus, optical The point on the optical axis of an objective lens or mirror at which rays of light converge, and where an image is sharpest; also known as the **focal point**.

focus, orbital One of the two points within an ellipse whose distance apart determines the eccentricity of the ellipse; pl.*foci*. The two foci lie on the major axis of the ellipse, at equal distances either side of its centre. In an elliptical orbit, one focus is occupied by the body being orbited; the other is referred to as the *empty focus*. If the orbit is a parabola or hyperbola, there is only one focus, occupied by the body being orbited.

Fokker–Planck equation An equation from which the evolution of the orbits of stars in clusters or galaxies can be calculated. It includes descriptions of how stars orbit in the cluster or galaxy as a whole and how their orbits are affected by encounters with other stars. It is named after the Indonesian-born Dutch physicist Adriaan Daniel Fokker (1887–1972) and the German physicist Max Karl Ernst Ludwig Planck (1858–1947).

folded dipole A dipole antenna in which the two elements are folded back and joined to form a narrow rectangular loop.

following Referring to the side of an object, or member of a group of objects, that trails behind in motion across the sky or across the face of a rotating body. Examples are the following component of a double star or a group of sunspots; the following side of a planetary feature; the following limb of a planet; or the following side of a telescopic field of view. The leading side is said to be *preceding*.

Fomalhaut The star Alpha Piscis Austrini, magnitude 1.16. It is an A3 dwarf, 25 l.y. away.

footpoint *See* CORONAL LOOP.

forbidden line An emission line in a spectrum that is emitted only by a low-density gas, as in interstellar regions and nebulae. Such a line is said to be forbidden because it does not occur under normal conditions on Earth, where gases are denser. A forbidden line is produced when an electron jumps from an upper energy level, where it can remain for a long time, to a lower level; such a jump, or transition, is said to have a very low *transition probability*. In the Earth's atmosphere, the excited atom would collide with other atoms or free electrons and lose energy in the collision (without producing a photon) long before it could radiate the energy away. However, in the low densities of interstellar space and the regions around hot stars, collisions are extremely rare and there is time for the spontaneous decay to occur. Forbidden lines are denoted by square brackets, such as the [O III] lines of doubly ionized oxygen. Forbidden lines disappear above a certain critical density (typically about 10^8 atoms/cm^3), and so their existence is an indicator of density in interstellar gas. A *semi-forbidden line*, designated with a single square bracket, such as C III], occurs where the transition probability is about a thousand times higher than for a forbidden line.

Forbush effect A temporary decrease in the number of galactic cosmic rays reaching the Earth; also known as **Forbush decrease**. It occurs when an interplanetary disturbance, consisting of material ejected from the Sun (possibly a *coronal mass ejection) and preceded by a shock front, travels outwards. As a result, there are increases in the strength of the interplanetary magnetic field and the density of the solar wind, which scatters incoming cosmic rays away from the Earth. The effect is named after the American geophysicist Scott Ellsworth Forbush (1904–84).

fork mounting A type of *equatorial mounting consisting of an open-ended two-pronged fork, which forms the polar axis, with the telescope pivoted between the prongs of the fork on a declination axis. The design provides access to all parts of the sky and does not need a counterbalance, but is unsuitable for refractors or long-focus reflectors as the forks would need to be very long. A version of the fork mounting is often used on *Schmidt–Cassegrain telescopes.

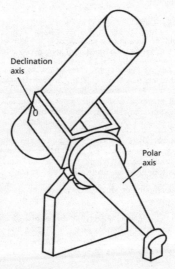

Declination axis

Polar axis

fork mounting

Fornax (For) (*gen.* Fornacis) A faint constellation of the southern sky, representing a chemist's furnace. Its brightest star, Alpha Fornacis, is a double of magnitudes 4.0 and 6.5. The constellation contains the radio source *Fornax A (a member of the *Fornax Cluster of galaxies) and the *Fornax Dwarf Galaxy.

Fornax A A powerful radio galaxy about 60 million l.y. away in the constellation Fornax, identified with the galaxy NGC 1316. Optically, it appears as a peculiar elliptical or lenticular galaxy, apparently the result of a merger between an elliptical and a spiral or two or more spirals. At magnitude 8.5 it is visually the brightest member of the *Fornax Cluster of galaxies, lying on the cluster's southwestern fringes, 3° from the cluster's centre.

Fornax Cluster A cluster of galaxies about 60 million light years away in the southern constellation Fornax. It is the largest nearby cluster of galaxies to us after the *Virgo Cluster. The Fornax Cluster is smaller and denser than the Virgo Cluster, with about one-tenth the mass, about one-fifth the number of galaxies and less than half the diameter. Prominent members include the supergiant elliptical NGC 1399 at the cluster's heart, the radio galaxy NGC 1316 (also known as *Fornax A) on the cluster's outskirts, and the barred spiral NGC 1365. A smaller group of galaxies, including NGC 1404 and NGC 1427A, appears to be colliding and merging with the main Fornax Cluster. 📷

Fornax Dwarf Galaxy A dwarf spheroidal galaxy (type dSph) in the Local Group of galaxies, about 450 000 l.y. away and nearly 6000 l.y. across. It has approximately 1.5×10^7 times the luminosity of the Sun and contains mostly old stars, but shows evidence of some younger stars about 3 billion years old. It was discovered in 1938 by H. *Shapley.

47 Tucanae The second-brightest globular cluster in the sky, appearing of 4th magnitude and the same diameter as the Moon, lying about 15 000 l.y. away in the constellation Tucana. It is also known as NGC 104. Its true diameter is about 150 l.y., and its mass about 10^6 solar masses. The density of stars at its centre (about 1000 stars per cubic light year) makes it one of the densest stellar environments known.

forward scattering Scattering in which photons emerge from the scattering medium travelling predominantly in the same direction as they entered. Forward scattering and *backscattering always occur together, but the proportion of forward scattering increases with particle size. The haloes around the Sun and Moon in wet weather are caused by forward scattering by water droplets in the Earth's atmosphere.

fossa A long, narrow, straight linear depression on a planetary surface; pl. *fossae*. The name, which means 'ditch' or 'trench', is not a geological term, but is used in the nomenclature of individual features, for example Erythraea Fossa on Mars, or Lakhmu Fossae on Ganymede.

Foucault test A method of determining the shape of a concave mirror; also known as the *knife-edge test*. A pinhole light source is placed at the mirror's *radius of curvature, and a straight edge (e.g. a knife-edge) adjacent to the pinhole. The mirror is illuminated by light from the pinhole. If the mirror is spheroidal it will produce a sharp image of the pinhole at the knife-edge, and the tester will see the mirror's illumination abruptly cut off as the knife-edge is moved across the image. Any small irregularities in the figure show up clearly as humps or hollows. A paraboloid appears like a doughnut, deeper at the centre than at the edges. The test was devised by the French physicist Jean Bernard Léon Foucault (1819–68).

four-colour photometry *See* STRÖMGREN PHOTOMETRY.

Fourier analysis A technique used to find the frequencies present in a complex signal; also known as **frequency analysis**. A series of measurements made at different times is broken down mathematically into the sum of simple oscillations of various frequencies. The lowest frequency present is called the **fundamental frequency**, and the higher frequencies are then *harmonics* (whole multiples) of the fundamental frequency. The brightness fluctuations of some variable stars, for example, can be broken down into two or three simple sinusoidal variations by the use of Fourier analysis. The technique was invented by the French mathematician (Jean-Baptiste) Joseph Fourier (1768–1830).

Fourier transform The mathematical breakdown of a non-repeating phenomenon into its component frequencies. It may be regarded as *Fourier analysis applied to a function with an infinite period. The technique is widely used in image processing to reduce the effects of periodic and random errors. It is also needed to convert the outputs from *aperture synthesis telescopes and *Fourier transform spectrometers into images and spectra respectively.

Fourier transform spectrometer (FTS) An instrument used for high-resolution spectroscopy, usually based on the principle of the *Michelson interferometer. A Michelson interferometer splits the incoming starlight and sends it to two mirrors, one stationary, the other movable. When the light recombines, an interference pattern is produced. Since the incoming light consists of a range of wavelengths (the star's spectrum), the interference pattern becomes very complicated as the mirror is moved. The input spectrum can, however, be recovered by taking the *Fourier transform of the output signal. The Fourier transform can be thought of as breaking down the output signal into a sum of sine waves of different amplitudes. This technique is particularly useful in the infrared, where conventional (dispersive) spectroscopy is difficult. A similar technique is also used in radio astronomy in the analysis of the 21-cm hydrogen line and other radio spectral lines. In this case the Fourier transformation splits the band of received frequencies into thousands of narrow channels, revealing the features of the spectral lines.

fourth contact   At a solar eclipse, the moment at which the trailing (western) limb of the Moon completely uncovers the Sun's eastern limb, marking the end of the eclipse; or, at a lunar eclipse, the moment at which the Moon's western limb exits the Earth's shadow; also known as *last contact*.

Fowler, William Alfred   (1911–95) American physicist and astrophysicist. In 1938, H. A. *Bethe had proposed the *proton–proton reaction as a source of energy in stars. Fowler demonstrated theoretically that the process was indeed possible. In the 1950s, he collaborated with G. R. and E. M. *Burbidge and F. *Hoyle on a theory of energy generation and nucleosynthesis in stars (known as the B^{2}FH theory). Further theoretical work with Hoyle in the late 1960s showed that a *hot Big Bang would produce the observed amount of helium in the Universe. He and S. *Chandrasekhar shared the 1983 Nobel Prize in Physics.

Fp star   A peculiar star (hence the suffix 'p') of spectral type F, a cooler extension of the *Ap stars to about spectral type F2.

fractional method   A visual method of estimating the brightness of a variable star, using two comparison stars. The interval between the comparisons is mentally divided into a small number of steps. Estimates such as A(2)V(3)B or C(4)V(1)D result, where the figures in parentheses represent the number of brightness steps between the variable, V, and pairs of comparison stars, A, B and C, D. The actual magnitude may be calculated from the known magnitude of the comparison stars.

frame dragging   The distortion of space and time by a rotating object as predicted by *general relativity. It is also known as the *Lense–Thirring effect*, or *Lense–Thirring precession*, after the Austrian physicists Josef Lense (1890–1985) and Hans Thirring (1888–1976), who first predicted it in 1918. Spacetime itself is dragged along with the rotating body, producing an effect on a spinning object near the rotating mass which causes the spin axis to precess. Evidence for frame dragging has been observed in a gas disk around a spinning black hole, and in precession of gyroscopes orbiting the Earth aboard *Gravity Probe B.

frame of reference   A set of axes, as used for example in systems of coordinates on the celestial sphere, that enables the position of a body or point in space to be defined at any instant.

Franklin-Adams charts   A pioneering photographic sky atlas by the English amateur astronomer John Franklin-Adams (1843–1912), published posthumously in 1913–14. It consists of 206 sections 15° square covering the entire sky, taken from Johannesburg, South Africa, and Godalming, England. It shows stars as faint as 17th magnitude on a scale of 15mm per degree.

Fraunhofer, Joseph von   (1787–1826) German physicist and optician. In 1814 he made the first spectrometer, to measure the dispersive power of lenses, using a yellow flame as the light source. He compared the flame spectrum with a spectrum of the Sun produced by a prism, noting and recording the positions of the Sun's *Fraunhofer lines. He later noted similar lines in the spectra of other stars. Fraunhofer also made achromatic lenses and the first diffraction grating, developed the equatorial mount, and constructed the 16-cm heliometer used by F. W. *Bessel to measure the parallax of 61 Cygni.

Fraunhofer lines   Absorption lines and bands in the spectrum of the Sun that were first labelled by J. von *Fraunhofer in 1814. In decreasing wavelength from the red, they include A (759nm), B (687nm), C (656nm), D (589nm), E (527nm), F (486nm), G (431nm), H (397nm), and K (393nm); Fraunhofer observed many other lines, labelling several weaker features a, b, etc. The A and B bands are due to absorption by oxygen molecules in the Earth's atmosphere, and the a band is due to absorption by terrestrial water vapour, but the remainder result from absorption in the Sun's photosphere. Most prominent are the *D lines of sodium, the *H and K lines of calcium, and the *G band caused by neutral iron and the CH molecule, all of which are features of stars of spectral types F, G, and K.

Fred Lawrence Whipple Observatory An observatory at an altitude of 2340 m on Mount Hopkins, 56 km south of Tucson, Arizona, owned and operated by the *Smithsonian Astrophysical Observatory. It was opened in 1968 as Mount Hopkins Observatory and was renamed in 1982 in honour of F. L. *Whipple, a former SAO director. Its main instruments are the 1.5-m Tillinghast reflector, opened in 1970; a 1.2-m reflector, opened in 1990; a 10-m optical gamma-ray reflector opened in 1968; and the 1.3-m Peters Automated Infrared Imaging Telescope (PAIRITEL), originally used for the *Two-Micron All-Sky Survey (2MASS). On the summit of Mount Hopkins is the *MMT Observatory, jointly operated by the SAO and the University of Arizona. An array of gamma-ray telescopes called the *Very Energetic Radiation Imaging Telescope Array System (VERITAS) began operation in 2007.

(⊕) SEE WEB LINKS
• Official observatory website.

free–bound transition The emission of radiation when a free electron (one unattached to an atom) is captured by an ion; also known as *radiative recombination*. The capture (recombination) may be to the *ground state, in which case a photon with an energy greater than the *ionization potential of the ion or atom is emitted, producing a band of continuous emission (a *continuous spectrum*). Alternatively, the recombination may be to an excited energy level, with the emission of a photon, after which the electron cascades down through the excited states to the ground state, producing emission lines characteristic of that ion or atom. *See also* RECOMBINATION LINE.

free–free transition The emission or absorption of radiation by an electron that is unbound to an atom before and after the event. As an electron passes an ionized atom it can be accelerated or decelerated, respectively absorbing or emitting a photon in the process. Since the photon can be of any wavelength, radiation emitted in this way has a continuous spectrum. The radiation emitted in this way is a type of thermal radiation, typically emitted by a hot plasma, and is also known as *thermal bremsstrahlung*. It is seen in many emission nebulae.

frequency (f, or ν) The number of waves (or some other regularly repeating event) per second. Frequency is expressed in hertz (Hz). For *electromagnetic radiation it is equal to the speed of the waves (c) divided by their wavelength, λ. *See also* ELECTROMAGNETIC SPECTRUM.

frequency analysis Another name for *Fourier analysis.

Friedmann universe A model describing an expanding universe containing matter and radiation, but without a *cosmological constant. Such a universe is both homogeneous and isotropic. There is, in fact, a family of such universes including those which expand for ever (*open universe), those that eventually collapse (*closed universe), and the particular example of the *Einstein–de Sitter universe which has a *critical density of matter. The geometry of *spacetime in these universes is described by the *Robertson–Walker metric and is, in the preceding examples, negatively curved, positively curved, and flat respectively (*see* CURVATURE OF SPACETIME). The Friedmann models, originated by the Russian mathematician Alexander Alexandrovich Friedmann (1888–1925), form the basis of the standard Big Bang theory.

fringe *See* INTERFERENCE PATTERN.

fringe pattern Another name for an *interference pattern, used particularly in radio astronomy.

f-spot The following member of a pair of sunspots.

F star A star of spectral type F, slightly hotter and more massive than the Sun, and white in colour. On the main sequence, F-type stars have temperatures of 6100–7200 K, while supergiants are a few hundred degrees cooler. Masses are 1.2–1.6 solar masses, and luminosities 2–6.5 times the Sun's on the main sequence, but F-type supergiants have masses up to 12 solar masses and luminosities up to 32000 times the Sun's. The hydrogen Balmer lines weaken dramatically from spectral type F0 to F9, while the calcium H and K lines increase in strength and many other metal lines start to become visible. Canopus, Polaris, and Procyon are F-type stars.

FTS Abbr. for *Fourier transform spectroscopy.

full Moon The phase of the Moon when it is fully illuminated as seen from the Earth and lies opposite the Sun in the sky.

full-wave dipole A dipole antenna of length approximately equal to the wavelength it is designed to receive. A full-wave dipole has a wider beamwidth than a *half-wave dipole.

full width at half maximum (FWHM) A measure of the width of a line in a spectrum, either emission or absorption. It is simply the width of the line at a point that is half the line's peak value. FWHM is usually quoted in nanometres, but can also be given as a velocity by applying the Doppler effect, since motions in the gas govern the width of the line. The absorption-line equivalent is **full width at half depth**. *See also* BEAMWIDTH.

fundamental catalogue A catalogue of positions and proper motions of stars, generally brighter than about magnitude 9, derived mostly from meridian observations made over many decades, which defines an accurate stellar reference frame. Apart from the German series (*see* FK), other fundamental catalogues, produced in the USA, are the *Preliminary General Catalogue* (PGC; 6188 stars; L. Boss, 1910), the *General Catalogue* (GC; 33342 stars; B. Boss, 1937), and the *N30* (5268 stars; H. R. Morgan, 1952). *See also* BOSS GENERAL CATALOGUE.

fundamental mode The lowest frequency at which an oscillation occurs, or the lowest component of a complex vibration. Whole-number multiples of the fundamental frequency are known as *overtones* or *harmonics*, and are present in many types of *pulsating variable. In stars such as RR Lyraes and Cepheids, for example, the fundamental mode of radial pulsation is the simple periodic expansion and contraction of the star's outer layers. More complex vibrational modes can exist, known in turn as the **first overtone** mode or harmonic, and so on. *See also* OVERTONE MODE; PULSATION MODE.

fundamental plane A correlation between the size of elliptical galaxies, their average surface brightness, and the velocity dispersion of the stars within them. When these three properties are plotted on a graph with three axes, they describe a narrow plane, which is the fundamental plane. Any theory of the origin of elliptical galaxies must explain why these three quantities are so closely related. Astronomers use the fundamental-plane relationship to estimate the distances of ellipticals from a combination of their size and surface brightness.

fundamental star A star whose position and proper motion are known precisely enough to be included in a fundamental catalogue.

fuor *See* FU ORIONIS STAR.

FU Orionis star A rare class of pre-main-sequence object which appears to be coupled with *cometary globules; abbr. FU. They are also known as **fuors** or **FUOrs**. Most probably they consist of a *T Tauri-like star surrounded by a very luminous accretion disk currently in a state of very high mass accretion (approx. 10^{-4} solar masses per year). FU Orionis stars exhibit a gradual rise of up to 6 magnitudes over several months, accompanied by mass ejection. The star subsequently remains constant or shows a slight decline over a period of years or decades, in contrast to the much shorter timescales in the related *EX Lupi stars.

FUSE Abbr. for *Far Ultraviolet Spectroscopic Explorer.

fusion The process in which two atomic nuclei join together to form a larger single nucleus, sometimes with the production of other particles as well. For nuclei with masses up to that of iron the process releases large amounts of energy (the *binding energy). Nuclear fusion takes place at very high temperatures and pressures, which are required to overcome the strong repulsive force between the positively charged nuclei. Such conditions exist within stars, where the energy released by fusion keeps the star shining.

fusion crust The melted, often glassy surface layer of a meteorite. It is caused by intense heating during the meteorite's passage through the atmosphere and is usually black or blue-black in colour. Stony meteorites usually have a thicker fusion crust than do irons.

future light cone *See* LIGHT CONE.

FWHM Abbr. for *full width at half maximum.

g Symbol for the *acceleration of free fall.

G Symbol for the *gravitational constant.

Gacrux *See* GAMMA CRUCIS.

Gaia An ESA astrometry satellite which will measure the distances and motions of over a billion stars down to 20th magnitude, building up a three-dimensional picture of our Galaxy. Gaia contains two identical telescopes with rectangular mirrors 1.45 by 0.5 m, pointing in two directions 106°.5 apart. They focus light to a common focal plane where arrays of CCDs will measure positions, proper motions, and parallaxes. Other detectors will perform multicolour photometry of the same stars and measure radial velocities down to 17th magnitude. Distances will be accurate to 10% as far away as the galactic centre. In addition, Gaia is expected to discover large numbers of asteroids, extrasolar planets, brown dwarfs, variable stars, and supernovae. Gaia is a successor to ESA's *Hipparcos mission and is planned for launch in 2013. It will be stationed at the L_2 *Lagrangian point of the Sun–Earth system, 1.5 million km from the Earth in the direction away from the Sun. The name originated as an acronym of Global Astrometric Interferometer for Astrophysics. Although Gaia is no longer an interferometer, the name has been retained.

(⊕) SEE WEB LINKS
• ESA mission website.

gain The factor by which a signal is amplified, often expressed in decibels (symbol dB). The gain of an antenna is a measure of how directional it is, and is related to *directivity. It is equal to the signal strength in the direction of maximum sensitivity divided by the signal strength that would be received, either by an antenna equally sensitive in all directions, or (in an alternative definition) by a half-wave dipole.

galactic anticentre The region of our Galaxy that lies in the opposite direction to the galactic centre, as viewed from the Earth. Its approximate coordinates are 5h 46m, +28° 56′, in southern Auriga.

galactic bulge The roughly spherical distribution of stars that forms the central hub of spiral and lenticular galaxies. The bulge is much less noticeable, relative to the disk, in the later spiral types Sc and Sd.

galactic centre The very central region or nucleus of our Galaxy, 26 000 l.y. away in the constellation Sagittarius; coordinates RA 17h 46m, dec. −28° 56′. There is a clumpy ring of gas (rich in molecules) and dust about 5–15 l.y. from the exact centre. Inside this is a dense star cluster around the bright radio source Sagittarius A*, which is believed to mark the very centre of the Galaxy. There is strong evidence for a supermassive black hole at the centre of our Galaxy. The orbits of stars in this region imply the existence of a central object of 3–4 million solar masses within a radius less than 0.01 parsec. No object or system other than a black hole could be stable at such a large density. Although highly obscured by dust from optical observation, the galactic centre has been extensively studied at infrared, radio, X-ray, and gamma-ray wavelengths. ◉

galactic cluster A largely obsolete name for an *open cluster.

galactic coordinates A system of coordinates that specifies the position of objects relative to the *galactic equator. Positions are given in terms of *galactic latitude and *galactic longitude.

galactic disk The major structural component of spiral, lenticular, and some irregular galaxies, containing stars and (for spirals and irregulars) gas and dust, orbiting the galaxy's centre. The thickness of the disk is small in relation to its diameter. In our own Galaxy the disk extends about 80 000 l.y. from the galactic centre, while its total thickness is about 1500 l.y. as measured by the distribution of older stars, and only 600 l.y. for young stars, gas, and dust.

galactic equator A great circle on the celestial sphere marking the plane of our Galaxy, running along the Milky Way. It is the reference plane for *galactic coordinates. The galactic equator is inclined at about 63° to the celestial equator, a result of the orientation of the Solar System within the Galaxy.

galactic halo Any material in a roughly spherical distribution around a galaxy, and extending out to beyond the visible regions. It can refer to a population of old (Population II) stars, including globular clusters, with little or no rotation about the galactic centre; to tenuous, highly ionized, high-temperature gas (10^6 K) around a galaxy; or to an extended distribution of invisible *dark matter.

galactic latitude (b) A coordinate that gives the angular position of an object north or south of the galactic equator. It is measured in degrees from 0° at the galactic equator to 90° at the galactic poles.

galactic longitude (l) A coordinate that gives the angular position of an object around the galactic equator. It is measured in degrees clockwise along the galactic equator starting from the point marking the direction of the *galactic centre, in Sagittarius.

galactic magnetic field The magnetic field in interstellar space in the disk of a galaxy. Magnetic field strengths between 10^{-9} and 10^{-10} tesla occur in the disks of spiral galaxies, with an overall pattern roughly following the spiral structure. Fields in elliptical galaxies are much more difficult to estimate, but in those that are strong radio emitters it appears that the fields may reach 10^{-7} tesla.

galactic nucleus The very central regions of a galaxy. A galaxy's nucleus often has a complex structure, with a dense cluster of stars within the inner few light years, and it can be the site of considerable activity. If the activity is high, with considerably more radiation emitted than can be accounted for by normal formation and evolution of stars, it is classed as an *active galactic nucleus.

galactic plane The mean plane of our Galaxy, marked by the band of the Milky Way. It is defined by specifying the positions of the galactic poles. *See also* GALACTIC EQUATOR.

galactic pole Either of the two points on the celestial sphere that lie 90° north and south of the galactic equator. The north galactic pole lies in Coma Berenices, and the south galactic pole in Sculptor.

galactic rotation The systematic rotation of stars and gas about the centre of a galaxy. Rotation is most obvious in spiral and lenticular galaxies, and is also seen in some (but not all) ellipticals and irregulars. The rotation maintains the structure of the galaxy against further gravitational collapse. Galaxies exhibit *differential rotation, the time taken for one revolution increasing with distance from the centre. The Sun orbits the centre of our Galaxy about once every 230 million years, at a speed of 220 km/s. *See also* ROTATION CURVE.

galactic window A region of the galactic plane in which there is less extinction of light by interstellar dust than elsewhere, thereby allowing a glimpse of more distant regions. Examples are *Baade's Window in Sagittarius, and windows in Circinus.

galactic year Another name for a *cosmic year.

Galatea The fourth-closest satellite of Neptune, distance 61 953 km, orbital period 0.429 days; also known as Neptune VI. Galatea orbits between the Arago and Adams rings, and probably acts as a *shepherd moon for the latter. There is in addition a faint ring of debris along Galatea's orbit. Its diameter is 158 km and it was discovered in 1989 from images taken with the Voyager 2 spacecraft.

Galaxy The spiral galaxy containing the Sun and all the stars visible to the naked eye at night; it is written with a capital 'G' to distinguish it from other galaxies. Its disk is visible to the naked eye as a faint band of light around the sky, the *Milky Way; hence the Galaxy itself is also often referred to as the Milky Way.

Our Galaxy has three principal components. One is a rotating disk of about 6×10^{10} solar masses consisting of relatively young stars (Population I), *open clusters, gas, and dust, with some concentration of the young stars and interstellar material into spiral arms. The disk is quite thin, about 1000 l.y., compared with its diameter of over 100 000 l.y. Active star formation continues in the disk, particularly in *giant molecular clouds. The second main component is a faint, roughly spherical halo with perhaps 15–30% the mass of the disk. The halo is composed of old stars (Population II), a few per cent of them in *globular clusters, plus small amounts of hot gas, and it merges into a more conspicuous central bulge of stars, also of Population II. The third main component is an unseen halo of *dark matter with a total mass of at least 4×10^{11} solar masses. In all, there are probably about 2×10^{11} stars in the Galaxy, most with masses less than the Sun's. The age of the Galaxy remains somewhat uncertain, but the disk is at least 10 billion years old, while the globular clusters and most of the halo stars are believed to be 12–14 billion years old.

The Sun lies about 26 000 l.y. from the centre, in the *Orion Arm, which lies between the *Sagittarius and *Perseus Arms. The *galactic centre itself lies in the constellation Sagittarius and is thought to harbour a supermassive black hole of 3–4 million solar masses. The Milky Way is a spiral, but observations of its structure and attempts to measure the dimensions of individual spiral arms are hampered by obscuring dust in the disk and by the difficulty of estimating distances. Infrared measurements have revealed a central bar some 28 000 l.y. long, angled at about 45° from the direct line of sight between the Sun and the galactic centre. The Galaxy has two main arms, the *Scutum-Centaurus Arm and the *Perseus Arm, which emerge from the ends of this bar. There are two minor arms, the *Norma Arm and *Sagittarius Arm, lying between the major arms. Hence our Galaxy is now thought to be a barred spiral, perhaps similar in appearance to M83 in Hydra. 📷

(⊕) SEE WEB LINKS
• New view of the structure of the Galaxy from the Spitzer Space Telescope.

galaxy A system of stars, often with interstellar gas and dust, bound together by gravity. Galaxies are the principal visible structures of the Universe. They range from dwarfs with less than one million stars to supergiants with over a million million stars, and in diameter from a few hundred to over 600 000 l.y. Galaxies may be isolated, or in small groups such as our *Local Group, or in large clusters such as the *Virgo Cluster.

Galaxies are usually classified according to appearance (*see* HUBBLE CLASSIFICATION). They come in two main shapes: spiral (with arms), and elliptical (without arms). *Elliptical galaxies (designated E) have a smooth, centrally concentrated distribution of stars, and very little interstellar gas or dust. *Spiral galaxies are subdivided into ordinary spirals (S) and *barred spirals (SB). Both types have disks containing interstellar material as well as stars. *Lenticular galaxies (S0) show a clear disk, but with no spiral arms visible.

*Irregular galaxies (Irr) have a rather amorphous, irregular structure, sometimes with evidence of a spiral arm or bar. A few galaxies look unlike any of the main types, and may be classified as peculiar (p or pec). Many of these are probably the results of the interaction or merger of pairs or small groups of galaxies. The most numerous type of galaxy may be the small, relatively faint *dwarf spheroidal galaxies (dSph) which are approximately elliptical in shape.

Galaxies are believed to have formed by gravitational accumulation of gas, some time after the *recombination epoch. Gas clouds would have begun to form stars, perhaps as a result of mutual collisions. The type of galaxy created may have depended on the rate at which the gas was turned into stars, ellipticals forming where the gas turned rapidly into stars, and spirals forming if star formation was slow enough to allow a significant disk of gas to grow. Galaxies evolve as they progressively convert their remaining gas into stars, but major evolution between different types in the Hubble classification probably does not occur. However, some elliptical galaxies may be created by the merger of spirals.

The relative numbers of galaxies of different types is closely related to their intrinsic brightness and the type of group or cluster to which they belong. In dense clusters, with hundreds to thousands of galaxies, a high proportion of the bright galaxies are ellipticals and lenticulars, with few spirals (5–10%). However, the proportion of spirals may have been higher in the past, spirals either having been stripped of gas so that they now resemble lenticulars, or undergone mergers with other spiral and irregular galaxies to become ellipticals. Outside clusters, most galaxies belong to groups containing from a few to several dozen members, and isolated galaxies are rare. Spirals account for 80% of bright galaxies in these low-density environments, with a correspondingly small proportion of ellipticals and lenticulars. At low luminosities there are no spirals: only irregulars, dwarf ellipticals, and dwarf spheroidals.

Some galaxies display unusual central activity, such as *Seyfert galaxies or *N galaxies. A *radio galaxy is an unusually strong emitter of radio energy.

galaxy cannibalism The ingestion of a small galaxy by a much larger one, in contrast to a *galaxy merger, in which the two galaxies are of comparable size. The extended envelopes of stars around giant and supergiant ellipticals are believed to result from cannibalism, which may occur quite frequently at the centres of galaxy clusters. Most average-sized galaxies may incorporate one or more satellite galaxies of up to about one-tenth their present size. The *Sagittarius Dwarf Galaxy is currently being cannibalized by our own Galaxy.

galaxy cluster *See* CLUSTER OF GALAXIES.

galaxy encounter The interaction of two galaxies when their paths become close enough for significant gravitational interaction, as is happening, for example, in the *Antennae. At suitable orientation and speed, the encounter may radically alter the structure of one or both of the galaxies. Physical collisions of stars in a galaxy encounter are rare, because of the great distances between stars. But the stars do interact gravitationally, and interstellar gas will collide.

galaxy evolution Any change in the properties of individual galaxies, or of populations of galaxies, with time. The properties of galaxies in the Universe today provide evidence that galaxy evolution does occur, and in more than one way. For example, *elliptical galaxies today contain mostly old, low-mass stars which do not emit much light. At the time these stars formed, about 10 billion years ago, the galaxies must also have contained many short-lived, massive stars of high luminosity, so these galaxies must then have been much brighter than they are today. This change in the intrinsic luminosity of a galaxy with time is known as *luminosity evolution*. There is also a second type of evolution, still at work today. Some galaxies with close neighbours have distorted shapes, implying that the galaxy is being disrupted by the gravitational pull of its neighbour. The strong gravitational attraction may eventually cause them to merge into a single galaxy. The number of galaxies is therefore gradually changing as a result of such galaxy mergers; this is known as *number-density evolution*.

In the most popular models of galaxy formation, which are often called 'bottom-up' or *hierarchical* models, the first galaxies which formed from the gas that filled the early Universe are thought to have been small. Under the influence of gravity, these small galaxies gradually merged to form larger ones. A consequence of such a model is that in the Universe today the largest galaxies should be younger than the smallest galaxies, and also that at early times there should have been fewer large galaxies than there are today.

It is possible to watch galaxy evolution at work, thanks to the finite speed of light. For example, when we observe a galaxy 2 billion light years away we are looking back in time to see what that galaxy looked like 2 billion years ago. Of course, galaxies at such large distances are quite faint, and this kind of cosmic archaeology became possible only in the mid-1990s with the construction of new large telescopes and the installation of sensitive new instruments. The first big step in this research came in 1992–4 with the Canada–France Redshift Survey (*see* REDSHIFT SURVEY), which discovered several hundred galaxies with *redshifts of up to 1. A galaxy with a redshift of 1 is about 8 billion light years away, and so appears as it did 8 billion years ago, over half the present age of the Universe. Follow-up images of these galaxies made with the Hubble Space Telescope (HST), however, showed that the Universe at that time was not dramatically different from the Universe today—the same types of elliptical and spiral galaxies were present, although there were more irregular galaxies than we see today.

Shortly after this survey, *Lyman-break galaxies, which have redshifts of approximately 3, were discovered. We see these galaxies as they appeared 12 billion years ago. This time, HST images did not reveal the familiar large spirals and ellipticals but instead rather smaller objects. These discoveries were at least roughly in agreement with expectations that large galaxies have gradually been assembled from smaller ones over the last 13 billion years or so. The most distant galaxies currently observable have redshifts of 6, which means that we are seeing them as they appeared less than a billion years after the Big Bang.

More recently, efforts have moved to wavebands other than the optical. Surveys at submillimetre wavelengths have revealed that billions of years ago there was a population of luminous galaxies that contained so much dust that they are almost invisible at optical wavelengths and can only be detected by the submillimetre radiation that the dust emits. It seems quite likely that these luminous dusty galaxies, which are inferred to contain large numbers of young stars, are the ancestors of present-day elliptical galaxies, although the details of how the one type of galaxy evolved into the other have yet to be worked out. Another interesting discovery is that there are far more large galaxies at early times than expected from theory. Indeed the observers have concluded that large galaxies are generally older than small ones—the opposite to what the models predict. Such inconsistencies remain to be resolved.

Galaxy Evolution Explorer (GALEX) A NASA satellite launched in 2003 April to study the star-formation history of galaxies. It carries a 0.5-m ultraviolet telescope to make a variety of imaging and spectroscopic surveys in the wavelength range 300–135 nm (near to far ultraviolet), producing a comprehensive database of UV images and spectra of nearby and distant galaxies.

(())) SEE WEB LINKS
• Official mission website.

galaxy formation The process or processes which gave rise to the galaxies. In the Big Bang theory, the early Universe was predominantly smooth, so something must have caused material to assemble into galaxies filled with stars. It is thought that small irregularities in density were present at the *recombination epoch. These density fluctuations were then amplified by the action of gravity, eventually leading to the formation of galaxy-sized objects. Calculations show that this process is assisted by the existence of *dark matter, particularly if it is *cold dark matter. There is still no complete theory of galaxy formation that includes an explanation of how stars form within them.

galaxy merger The combination of two or more galaxies of comparable size to form a single, larger galaxy. For example, the peculiar elliptical galaxy and strong radio source *Centaurus A is believed to result from the merger of an elliptical galaxy with a somewhat smaller, gas-rich spiral galaxy.

GALEX Abbr. for *Galaxy Evolution Explorer.

Galilean satellites The four largest satellites of Jupiter, namely Io, Europa, Ganymede, and Callisto, which were discovered by *Galileo in 1610. All four are easily visible in binoculars and small telescopes, and would be bright enough to be visible with the naked eye were it not for the glare of Jupiter. Some keen-sighted people have claimed to be able to see them without optical aid. ◙

Galilean telescope A refracting telescope with a *converging lens as the objective and a *diverging lens as the eyepiece. It is named after *Galileo, who used a telescope of this type. It has the advantage of an upright image, but the drawbacks of limited field of view, and poor *eye relief except at very low magnifications. Opera glasses and field glasses are usually of this design.

Galileo A NASA space mission to Jupiter, launched in 1989 October. It used a gravity-assist trajectory that involved one fly-by of Venus and two of the Earth. Galileo encountered two asteroids, *Gaspra in 1991 October and *Ida in 1993 August, before arriving at Jupiter in 1995 December. An atmospheric entry probe parachuted down through the clouds of Jupiter, transmitting data for 58 min, while the main craft went into orbit, surveying the planet's weather patterns and satellites for seven years, finally burning up in the planet's atmosphere in 2003 September.

(⊕) SEE WEB LINKS
• Official mission website.

Galileo Galilei (1564–1642) Italian astronomer, physicist, and mathematician. In 1609 he heard of the recent invention of the telescope, and began to make his own. He built several small refractors, up to 50 mm aperture, the highest magnification being about × 30. He published *Sidereus nuncius* ('Starry Messenger') in 1610, in which he outlined his early telescopic discoveries, including mountains on the Moon, the four satellites of Jupiter (known as the *Galilean satellites), and the innumerable stars of the Milky Way. Also in 1610 he observed the phases of Venus, and noted the unusual telescopic appearance of Saturn, although he did not recognize the rings' true nature. Galileo's discoveries added to his discontent with *Aristotle's world-view, at that time still widely believed, and he advocated the *Copernican system, notably in his book *Discourses and Mathematical Demonstrations on Two New Sciences* (1632). Conflict with the Church followed, and he was tried for heresy; he lived the rest of his life under house arrest.

Galileo National Telescope *See* TELESCOPIO NAZIONALE GALILEO.

Galileo Regio A large dark area about 5000 km across on Jupiter's satellite Ganymede, centred at +47° lat., 130° W long. It is crossed by an extensive parallel system of ancient ridged furrows 10–30 km wide with raised rims, and is prominent enough to be visible through large Earth-based telescopes. Galileo Regio has rounded hummocky terrain, suggestive of some erosive process.

Galle, Johann Gottfried (1812–1910) German astronomer. On 1846 September 23, observing at the Berlin Observatory with the German astronomer Heinrich Ludwig d'Arrest (1822–75), he was the first to locate the planet Neptune, having started his search from a position supplied by U. J. J. *Le Verrier. Galle observed Saturn's faint 'crêpe ring' (the C Ring) in 1838, over a decade before W. C. and G. P. *Bond. He proposed that the solar parallax could be established from the parallax of asteroids, to which end he observed (8) Flora in 1873; the method was later used successfully by, in particular, D. *Gill and H. Spencer *Jones. His computations of cometary orbits helped to establish the connection between comets and meteor showers.

Galle Ring The innermost ring of Neptune, named after J. G. *Galle. It is about 15 km wide, diffuse, dust-rich, and lies 41 900 km from the planet's centre.

Gamma Cassiopeiae star A type of rapidly rotating, irregular variable of spectral type Be with an outflow of material from its equatorial region; abbr. GCAS. The formation of equatorial rings or disks is often accompanied by brightness variations of up to 1.5 mag. The prototype is the subgiant *Be star Gamma Cassiopeiae.

Gamma Crucis An M3.5 giant of magnitude 1.6, also known as Gacrux, lying 89 l.y. away.

Gamma Doradus star A young main-sequence star of late A or early F spectral type, located on the Hertzsprung–Russell diagram between solar-like and Delta Scuti stars; abbr.

GDOR. Gamma Doradus stars have amplitudes up to 0.1 mag. in V and multiple periods ranging from 0.3 to 3 days due to *non-radial surface pulsations.

gamma-ray astronomy The study of electromagnetic radiation from space at the very shortest wavelengths and with the highest photon energies (*see* GAMMA RAYS). Gamma rays are produced in regions of extremely high temperature, density, and magnetic fields, sites of the most violent processes in the Universe.

Many hundreds of individual gamma-ray sources are known, as well as a general *gamma-ray background. Early experiments in the 1950s and 1960s used balloons to carry instruments to altitudes where the atmospheric absorption of gamma rays is low. Exploratory observations were also made with spacecraft, including Ranger and Apollo missions, during the 1960s. The first sky surveys were made by the satellites SAS-2 (*see* SMALL ASTRONOMY SATELLITE) and *COS-B, launched in 1972 and 1974. In the late 1970s two *High Energy Astrophysical Observatories (HEAO-1 and HEAO-3) carried gamma-ray experiments. The *Granat satellite was launched in 1990, the *Compton Gamma Ray Observatory in 1991, the International Gamma-Ray Astrophysics Laboratory (INTEGRAL) in 2002, *Swift in 2004, and the *Fermi Gamma-ray Space Telescope in 2008.

The large energy range involved in gamma-ray astronomy necessitates several observational techniques. Only the very highest energies (above 100 GeV) can penetrate the Earth's atmosphere, so most observations must be made from space. At the lowest energies (100 keV to 10 MeV) gamma-ray telescopes create images using the principle of the *Compton effect, *collimation, or the *coded mask. Between 20 MeV and 30 GeV gamma-ray detection relies on the production of electron pairs using *spark chambers and *NaI detectors. Above 100 GeV the low photon fluxes require larger instruments than can be carried on satellites. For these energies, the Earth's atmosphere is used as the detector, and optical telescopes record the *Cerenkov radiation from the secondary electrons produced by the primary gamma-ray photons.

gamma-ray background Diffuse celestial radiation extending up to energies of at least 200 MeV. On a coarse scale, the gamma-ray background appears to be uniformly distributed (isotropic), indicating that it originates outside our Galaxy. Its low intensity rules out the steady-state theory of the Universe. Three possible sources of at least part of the radiation are the annihilation of matter and antimatter, active galaxies, and the explosion of primordial black holes.

gamma-ray burst An intense burst of gamma-ray radiation, lasting from a few milliseconds to a few tens of minutes. Gamma-ray bursts were initially detected in the late 1960s by US Vela satellites designed to monitor nuclear explosions, and have since been studied in detail by specialized satellites. The bursts are distributed uniformly over the sky, indicating that they originate in distant galaxies and fall into two types, short and long, suggesting two distinct mechanisms. Short bursts last from a few milliseconds up to 2 seconds; these are believed to be caused by the merger of two compact objects such as black holes, neutron stars, or a white dwarf and a neutron star. Long bursts last from a few seconds to over a thousand seconds and are believed to be caused by massive stars exploding as supernovae, with the collapsing core of the star forming a rapidly rotating black hole. In both scenarios, an outflow of gas creates shells that collide internally at close to the speed of light, producing the burst of gamma rays. The combined shells expand and collide with the surrounding gas and dust of the interstellar medium, heating it so that it begins to emit *afterglow radiation at X-ray wavelengths. As the gas cools the afterglow becomes visible at optical, infrared, and radio wavelengths, and this emission may remain detectable for days to years. *See also* SOFT GAMMA-RAY REPEATER.

gamma rays (γ-rays) Electromagnetic radiation with wavelengths shorter than about 0.01 nm. Gamma rays are the highest-energy photons in the electromagnetic spectrum. Their energies range from 100 keV up to at least 10 GeV.

gamma-ray telescope An instrument for recording gamma rays from celestial objects. Direct detection of gamma rays at low energies (100 keV to 10 MeV) makes use of *coded masks or *collimators coupled with solid-state or gas-filled detectors, such as *NaI detectors or *proportional counters. Between 20 MeV and 30 GeV, *spark chambers and NaI detectors are used. At higher energies the gamma rays are detected indirectly, through the *Cerenkov radiation emitted by secondary electrons produced after the gamma rays strike the atmosphere.

Gamow, George (1904–68) Ukrainian-American scientist (originally Georgy Anthonovich Gamov). He applied his early training in nuclear physics to astrophysics and cosmology. He developed his version of the Big Bang theory in the late 1940s, with the American physicists Ralph Asher Alpher (1921–2007) and Robert Herman (1914–97). It proposed that the Universe underwent a rapid expansion from a hot concentration of atomic particles and radiation, which he called 'ylem', building the nuclei of the elements in the process. The theory predicted a Universe containing three-quarters hydrogen and one-quarter helium, roughly in accordance with present-day measurements, and that it would cool as it expanded. Alpher and Herman suggested that the present, cooled-down Universe should be pervaded by a *cosmic microwave background with a temperature of about 5 K; their suggestion was forgotten until the radiation was detected in 1964.

Ganymede The largest satellite of Jupiter, and the largest in the Solar System, at 5262 km in diameter; also known as Jupiter III. Ganymede orbits Jupiter in 7.155 days at a distance of 1 070 000 km. Its period of axial rotation is the same as its orbital period. Ganymede is the brightest of the Galilean satellites, reaching magnitude 4.6 at opposition. It has a surface rich in water ice and a mean density of $1.94 \, g/cm^3$, indicating that just over half its composition is rock. Irradiation of the surface ice by ultraviolet light and atomic particles is thought to generate the moon's tenuous oxygen atmosphere. Ganymede's surface is a mixture of high-albedo (young, smooth) and darker (older, more heavily cratered) terrains. The brighter regions are crossed by linear flow-like features called *sulci*, thought to result from partial resurfacing by ice followed by stretching and faulting. Some of the larger impact craters have become *palimpsests due to slow, glacial flow of the ice. Ganymede has a weak magnetic field with a strength of 750 nT at the surface, about 40 times weaker than that of the Earth.

Gaposchkin, Sergei Illarionovich *See* PAYNE-GAPOSCHKIN, CECILIA.

Garnet Star The name given by F. W. *Herschel to Mu Cephei on account of its prominent red colour. It is a red supergiant and a *semiregular variable of type SRC, ranging from magnitude 3.4 to 5.1 in a period of about 2 years.

gas giant Another name for a *giant planet.

Gaspra Asteroid 951, discovered in 1916 by the Russian astronomer Grigorii Nikolaevich Neujmin (1886–1946). It was the first asteroid to be seen in detail, when the Galileo probe flew within 1600 km of it in 1991. Gaspra, an S-class asteroid, is irregularly shaped with dimensions of 18.2 × 10.4 × 8.8km, and is probably a fragment of a much larger body. Its rotation period is 7.04 hours. As well as craters, Gaspra's surface has linear grooves similar to those on Mars's satellite Phobos. Gaspra's orbit has a semimajor axis of 2.210 AU, period 3.29 years, perihelion 1.83 AU, aphelion 2.59 AU, and inclination 4°.1. 📷

gas scintillation proportional counter An X-ray detector consisting of a chamber filled with a noble (i.e. unreactive) gas such as argon or xenon. Ionization of the gas by X-ray photons produces an avalanche of electrons. Atoms of gas excited by the avalanche emit an ultraviolet photon, producing a flash whose intensity is proportional to the energy of the X-ray. This technique gives an improved energy resolution compared with a conventional *proportional counter.

gas tail *See* TAIL, COMETARY.

Gauss, Carl Friedrich (1777–1855) German mathematician. He devised a method for calculating the orbit of a body from just three observations, enabling astronomers to recover the asteroid Ceres which had become lost behind the Sun after its discovery in 1801 by G. *Piazzi. Gauss made a detailed study of celestial mechanics, developing a theory of perturbations (later used by J. C. *Adams and U. J. J. *Le Verrier to calculate the position of Neptune) which he applied to the determination of cometary and planetary orbits. He invented the method of least squares for use in correcting observational errors.

Gaussian gravitational constant (k) A constant that appears in the precise form of Kepler's third law. Its value was calculated by C. F. *Gauss, who found it to be 0.017 202 098 95. Subsequent measurements have given a slightly different value for k, but Gauss's value is still used when calculating the *astronomical unit.

G band A broad *Fraunhofer line in the Sun's spectrum at about 431 nm, due to absorption by CH molecules and a closely spaced group of lines produced by neutral iron. The same feature is found in other stars, most notably of spectral types F to K.

G-class asteroid A subclass of the C-class asteroids. Its members have low albedos (0.05–0.09), and are distinguished from class C by having a very strong ultraviolet absorption feature at wavelengths shorter than 0.4 μm caused by *water of hydration in the surface material. The largest asteroid, *Ceres, is now classified as G class. Other members include (176) Iduna, diameter 121 km, and (640) Brambilla, diameter 81 km.

GCRS Abbr. for *Geocentric Celestial Reference System.

GCVS Abbr. for *General Catalogue of Variable Stars*.

gegenschein A faint, diffuse oval glow some 10° across, appearing in the midnight sky directly opposite the Sun; also known as the *counterglow* (its meaning in German). The gegenschein is produced by the reflection of sunlight from small *zodiacal dust particles, and has an extremely low surface brightness. It is visible only under the clearest and darkest sky conditions.

Geminga One of the brightest gamma-ray sources, lying in the constellation Gemini; the name is a contraction of 'Gemini gamma-ray source'. It was detected in 1972 by the satellite SAS-2, but not until 1988 was a faint optical counterpart of 25th magnitude identified. Geminga has since been found to be pulsating in X-rays and gamma rays every 0.237 s, suggesting that it is a spinning neutron star. In its gamma-ray, X-ray, and optical properties it is similar to the Vela Pulsar, although it emits no detectable radio waves. It is the nearest pulsar, 500 l.y. away, and is travelling at over 400 000 km/h perpendicular to our line of sight. Geminga is possibly the remnant of the supernova that formed the *local bubble in interstellar space.

Gemini (Gem) (*gen.* Geminorum) A constellation of the zodiac, representing the twins Castor and Pollux of Greek mythology. The Sun lies in Gemini for the last week of June and the first three weeks of July. The constellation's brightest stars are *Castor (Alpha Geminorum) and *Pollux (Beta Geminorum). Zeta Geminorum is a Cepheid that varies between magnitudes 3.6 and 4.2 in a period of 10.2 days. Eta Geminorum is a red-giant semiregular variable that ranges between magnitudes 3.1 and 3.9. M35 is a 5th-magnitude open cluster; NGC 2392 is the *Eskimo Nebula. The *Geminid meteors radiate from the constellation every December.

Geminid meteors One of the three most active annual meteor showers, reaching a maximum ZHR of about 100 on December 13 from a radiant at RA 7h 28m, dec. +32°, near Castor. Activity is seen from December 7 to 15 and is produced by debris in a common orbit with the asteroid *Phaethon. Geminid meteors are slow (geocentric velocity 35 km/s) and often bright. The Geminids' apparent asteroidal origin differs from the cometary source of most meteors, and may account for their longer luminous duration in flight, and apparent dearth of persistent trains.

Gemini Observatory A multi-national observatory that consists of two identical 8.1-m telescopes for optical and near-infrared astronomy on separate sites, one in the northern hemisphere and the other in the south, together providing all-sky coverage. Gemini North, known as the Frederick C. Gillett Gemini Telescope, opened in 1999, at an altitude of 4210 m on *Mauna Kea, Hawaii. Gemini South, opened in 2002, is at an altitude of 2715 m on *Cerro Pachón, a mountain in the Chilean Andes. The partners in the Gemini Observatory are the United States, United Kingdom, Canada, Chile, Australia, Brazil, and Argentina. The Gemini

Observatory is operated by the *Association of Universities for Research in Astronomy on behalf of the owner nations and has its headquarters in Hilo, Hawaii. 📷

SEE WEB LINKS
• Official observatory website.

Gemma Alternative name for the star *Alphecca (Alpha Coronae Borealis).

GEMS Abbr. for *Gravity and Extreme Magnetism Small Explorer.

General Catalogue of Variable Stars (GCVS) A listing of all known variable stars and their types, published by the Russian Academy of Sciences in Moscow. It first appeared in 1948. The fourth edition was published in three volumes in 1985–8. A fourth volume containing cross-identification tables appeared in 1990 and a fifth, dealing with extragalactic variable stars and supernovae, in 1995. Updated editions of the GCVS are now released electronically; the 2011 version contained a total of 43 519 variables. A companion listing of unconfirmed variables, the *New Catalogue of Suspected Variable Stars* (NSV), is also published by the same group, the most recent edition being in 1982 with 14 811 objects. A Supplement containing an additional 11 206 suspected variables appeared in 1998. These two catalogues of suspected variables have since been merged and are also now available electronically.

SEE WEB LINKS
• Official catalogue website with downloadable data, regularly updated.

general precession Another name for *precession of the equinoxes.

general theory of relativity A theory announced by A. *Einstein in 1915 that describes how space and time are affected by the gravitational fields of matter, and how space and time change as seen by an observer on an accelerating object. The theory predicts that gravitational fields change the geometry of space and time, causing it to become curved. This curvature is apparent in a number of ways. First, light is bent in a gravitational field, a prediction that was confirmed by photographic measurements of the positions of stars near the limb of the Sun made during a total solar eclipse in 1919. The same effect manifests itself in a delay in radio signals from distant space probes as the signals pass the limb of the Sun. The curvature of space near the Sun also causes the perihelion point of Mercury's orbit to move forward, by $43''$ per century more than predicted by I. *Newton's theory of gravity (*see* ADVANCE OF PERIHELION). In the orbits of pulsars in binary systems, the advance of periastron can amount to several degrees per year.
    Another effect predicted by general relativity is the redshift of light caused by gravity. This has been demonstrated in the redshift of lines in the spectra of the Sun and, more noticeably, white dwarfs. Other predictions of the general theory include the *expanding Universe; the *gravitational lens effect; *gravitational waves; *singularities; and the invariance of the universal *gravitational constant, G. General relativity was developed from the principle of equivalence between gravitational and inertial forces.

Genesis A NASA space probe launched in 2001 August to collect samples of the *solar wind and return them to Earth. Genesis was stationed at the L_1 *Lagrangian point, 1.5 million km on the sunward side of Earth, and began collecting particles on the surface of silicon wafers in 2001 December. It continued collection until 2004 April, then returned in 2004 September to Earth, where it dropped a sample-return capsule into the atmosphere for recovery.

SEE WEB LINKS
• Official mission website.

Geneva photometry An *intermediate-band photometric system developed at the Geneva Observatory, Switzerland. It uses seven glass filters centred on the following wavelengths: U, 344 nm; B_1, 402 nm; B, 425 nm; B_2, 448 nm; V_1, 541 nm; V, 551 nm; and G, 581 nm. The colours can be used to measure *interstellar absorption, stellar temperature, gravity, and chemical abundance. *See also* VILGEN PHOTOMETRY.

geo- Prefix referring to the Earth.

geocentric **1**. With the Earth at the centre, as in for example the geocentric system of cosmology (*see* PTOLEMAIC SYSTEM).
 2. As seen from the centre of the Earth, as in for example *geocentric coordinates.

Geocentric Celestial Reference System (GCRS) A set of spacetime coordinates with their origin at the centre of the Earth, and which take into account distortions in spacetime caused by the mass of the Earth. Its coordinates do not rotate with respect to those of the *Barycentric Celestial Reference System.

geocentric coordinates Any system of coordinates with their origin at the centre of the Earth. For bodies in the Solar System, geocentric coordinates differ slightly from the coordinates as actually measured by an observer on the surface of the Earth (*topocentric coordinates). Various forms of geocentric coordinates are defined: geocentric *equatorial coordinates, the most common, have the celestial equator as their reference plane, while geocentric *ecliptic coordinates are referred to the ecliptic. *Rectangular coordinates can also be geocentric in origin.

Geocentric Coordinate Time (TCG) The time-scale used in the *Geocentric Celestial Reference System, the unit of which is the SI second. TCG differs from *Terrestrial Time because a clock on the Earth's surface is moving with respect to the geocentric frame and feels the gravitational field of the Earth. Terrestrial Time is in effect a conveniently scaled TCG that differs from International Atomic Time (TAI) only by a fixed offset. The rate difference between TCG and Terrestrial Time is about 22 ms per year.

geocentric equatorial parallax The angle subtended by the equatorial radius of the Earth as seen from a planet at its actual distance from the Earth at any given time. The same angle when seen from a planet at its mean distance from Earth is known as the **geocentric mean equatorial parallax**.

geocentric parallax The difference between the direction of an object as seen from a point on the surface of the Earth (its topocentric direction) and the direction in which it would be seen from the Earth's centre (its geocentric direction); also known as *diurnal parallax*.

geocorona The extremely tenuous cloud of neutral hydrogen that surrounds the Earth to a distance of perhaps 1–2 Earth radii. It originates in the upper atmosphere, where water molecules are dissociated into hydrogen and oxygen atoms by the action of sunlight. Scattering of solar radiation by the geocorona interferes with ultraviolet observations at the Lyman-α wavelength. The geocorona can be regarded as an extension of the *exosphere.

geodesic The shortest distance in space between two points. On a plane surface a geodesic is a straight line; on a sphere it is part of a great circle. The term is usually used in the context of general relativity, where it represents the shortest route between two points in curved spacetime.

geodesy The study of the shape and dimensions of the Earth, or any other solid planetary body.

Geographos Asteroid 1620, a member of the *Apollo group, discovered by R. L. B. *Minkowski and the American astronomer Albert George Wilson (1918–) in 1951 but not seen again until 1969. It is of S class. Radar observations show that Geographos is one of the most highly elongated objects in the Solar System, 5.1 × 1.8 km, probably a splinter from a larger body. Its orbit has a semimajor axis of 1.245 AU, period 1.39 years, perihelion 0.83 AU, aphelion 1.66 AU, and inclination 13°.3.

geoid The figure the Earth would have if it were entirely covered by water at mean sea level. The geoid is approximately an ellipsoid, but there are departures from it caused by the gravitational attraction of mountains, as well as differences in density within the Earth. The concept of the geoid is now being applied to other planetary bodies as well. In such cases, an arbitrary datum is used in place of sea level, such as the planet's mean radius or, for gaseous worlds, the altitude at which a given atmospheric pressure is found.

geomagnetic storm A major (sometimes global) disturbance of the Earth's magnetic field that follows violent activity on the Sun, such as flares and *coronal mass ejections, by 36–48 hours. Rapid fluctuations in the magnetic field strength and direction follow the onset of the storm, with a gradual return to normal over perhaps 2–3 days. Geomagnetic storms are often accompanied by active displays of *aurora, extending to lower latitudes than normal. Electrical currents in the ground induced during geomagnetic storms can damage power grid systems. The associated auroral effects can wreak havoc with short-wave radio communication, while currents flowing in the ionosphere can damage artificial satellites.

geomagnetism A broad term used to describe phenomena of the Earth's magnetic field, which originates from dynamo effects associated with convection in the fluid outer parts of the planet's metallic core. The Earth's magnetic field is a comparatively strong $(3-6 \times 10^{-5}$ tesla) dipole field. For reasons as yet poorly understood, the geomagnetic poles wander over time, and have reversed polarity quite frequently over the course of geological history, as demonstrated by the *remanent magnetization* preserved in rocks of various ages. The north geomagnetic pole currently lies between Ellesmere Island and Greenland. Interactions between the geomagnetic field and the solar wind give rise to the *magnetosphere.

geometrical albedo A measure of the reflectivity of a surface, particularly that of a Solar System object such as a planet, a satellite, or an asteroid; also called *physical albedo*. The geometrical albedo is the ratio between the light or other radiation reflected from an object, as viewed from the direction of the Sun (i.e. at zero *phase angle), and that which would be reflected by a hypothetical white, perfectly diffusely reflecting sphere (i.e. with an albedo of 1.0); this hypothetical sphere is assumed to have the same apparent size and be at the same distance as the real object. The wavelength or range of wavelengths at which the geometrical albedo applies must be defined. The *bolometric geometrical albedo* refers to reflectivity over all wavelengths.

geometrical libration Another name for *optical libration.

georgiaite A type of tektite found in Georgia, USA. The *ablation age of georgiaites is 33–35 million years. This is very similar to the derived ages of the *bediasites and the Martha's Vineyard tektite; together, they make up the oldest group of tektites.

geostationary orbit A circular orbit 35 900 km above the Earth's equator, frequently used for communications satellites. A satellite in geostationary orbit moves at the same rate as the Earth spins, completing one revolution in 24 hours, so that it always remains at the same point above the Earth's equator. *See also* SYNCHRONOUS ORBIT.

geosynchronous orbit An orbit in which a satellite completes one revolution in 24 hours, the same rate at which the Earth spins; in other words, it is a *synchronous orbit around the Earth.

geotail The Earth's *magnetotail.

germanium detector A solid-state electronic detector for high-energy X-ray and gamma-ray spectroscopy. Germanium detectors are manufactured from either high-purity crystals of germanium, or from germanium crystals containing traces of lithium. They have much better spectral resolution than *NaI detectors, but are difficult to make with such large collecting areas.

German mounting A type of *equatorial mounting consisting of two axes forming a T-shape, the upright being the *polar axis and the bar the *declination axis. The telescope is attached to one end of the declination axis, with a counterweight at the other. It is a popular design for mounting small telescopes. It was invented by J. von *Fraunhofer.

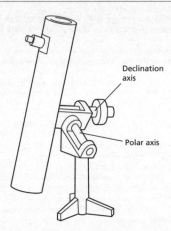

German mounting

GHA Abbr. for *Greenwich hour angle.

ghost crater A heavily eroded or submerged crater on a planetary surface. In some lunar examples, such as Lambert R, only the highest peaks of the crater protrude above a covering of lava that has all but submerged it. In others, such as Lamont on Mare Tranquillitatis, the original crater has been completely covered, but later subsidence has caused *wrinkle ridges to trace out the position of the old crater rim.

Ghost of Jupiter The 9th-magnitude planetary nebula NGC 3242 in Hydra, of similar apparent size to Jupiter. It lies about 2500 l.y. away.

Giacconi, Riccardo (1931–) Italian-American physicist and pioneer of X-ray astronomy. In the 1950s he began to build rocket-borne X-ray telescopes. By this means in 1962 he and B. *Rossi discovered Scorpius X-1, the first known cosmic X-ray source. Whether as designer, builder, or director, Giacconi has been associated with most of the major X-ray astronomy missions, including *Uhuru, the *Einstein Observatory, and the *Chandra X-ray Observatory. For this work he was awarded a share of the 2002 Nobel Prize in Physics.

Giacobinid meteors A periodic meteor shower associated with Comet *Giacobini–Zinner; also known as the *Draconids*. The radiant lies at RA 17h 23m, dec. +57°, near Beta Draconis. Activity is seen only in those years when the Earth crosses the descending node of the comet's orbit close to 21P/Giacobini–Zinner itself, usually around October 6–10. High Giacobinid rates (50–450 meteors/h) were noted in a 4.5-hour interval on 1933 October 9. Considerable activity was again recorded in 1946, 1985, and 1998. Giacobinid meteors have notably slow geocentric velocities (20 km/s), but many show persistent trains, possibly because they still contain volatile material, having been released comparatively recently from the comet's nucleus.

Giacobini–Zinner, Comet 21P/ A periodic comet discovered by the French astronomer Michel Giacobini (1873–1938) in 1900, and rediscovered two returns later, in 1913, by the German Ernst Zinner (1886–1970). Its orbital period is 6.6 years. Favourable and unfavourable apparitions alternate; at favourable returns, it may be as bright as 6th magnitude. Comet Giacobini–Zinner is the parent body of the *Giacobinid meteor shower. In 1985 it became the first comet to be visited by a spacecraft, the *International Cometary Explorer. Its orbit has a perihelion of 1.04 AU, eccentricity 0.71, and inclination 31°.8.

(()) SEE WEB LINKS

• Information page at Cometography website.

giant branch The region of the Hertzsprung–Russell diagram that extends diagonally to the upper right of the main sequence, containing *giant stars. The giant branch ranges from stars with a surface temperature similar to the Sun's, but a luminosity some 30 times greater, to cooler but even more luminous stars. Giants greater than about 1 solar mass are separated from the main sequence by the *Hertzsprung gap. Lower-mass giants are, however, linked to it by a *subgiant branch*.

Giant Magellan Telescope (GMT) A segmented-mirror telescope of extremely large diameter being built at *Las Campanas Observatory, Chile, by a consortium of ten American, Australian, and Korean universities and research institutes. The GMT's main mirror will consist of seven 8.4-m circular segments, six in a ring and the seventh at the centre, forming a collecting area equivalent to that of a single mirror 21.4 m in diameter and with the resolving power of a 24.5-m mirror. Each main-mirror segment reflects light onto one of seven secondary mirrors with *adaptive optics, and the multiple light paths are combined at the *Gregorian focus. The GMT consortium consists of the Carnegie Institution for Science, Harvard University, Smithsonian Astrophysical Observatory, Texas A&M University, Korea Astronomy and Space Science Institute, the University of Texas at Austin, the Australian National University, the University of Arizona, Astronomy Australia, and the University of Chicago. The GMT is planned for completion in 2019.

(((🌐))) **SEE WEB LINKS**
• Official telescope website.

Giant Metrewave Radio Telescope (GMRT) A radio interferometer at Khodad in Maharashtra state, about 80 km north of Pune, western India, owned and operated by India's National Centre for Radio Astrophysics. It consists of thirty fully steerable dishes, each 45 m in diameter. Fourteen of the dishes are arranged in a compact central array covering about 1 km^2 while the remainder are spread out in an approximate Y-shape giving baselines up to about 25 km. The array started operation in 1999.

(((🌐))) **SEE WEB LINKS**
• Official telescope website.

giant molecular cloud (GMC) A massive cloud of interstellar gas and dust composed mainly of molecules. The typical diameter is 100 l.y., and masses range from a few hundred thousand to 10 million solar masses. GMCs contain mostly hydrogen molecules (H_2, 73% by mass), helium atoms (He, 25%), dust particles (1%), neutral atomic hydrogen (H I, less than 1%), and a rich cocktail of interstellar molecules (less than 0.1%). Our Galaxy contains over 3000 GMCs, the most massive of which lies near the radio source Sagittarius B2 at the galactic centre. They comprise half the mass of all interstellar matter, although they occupy less than 1% of its volume. The average gas density is a few thousand molecules per cubic centimetre. GMCs are found mainly in the spiral arms of disk galaxies, and they are the birthplace of massive stars. GMCs exist for some 30 million years, during which time only a small fraction of their mass is converted to new stars. The nearest GMC lies in Orion, and is associated with the Orion Nebula. *See also* ASTROCHEMISTRY; INTERSTELLAR MOLECULE.

giant planet A planet of much larger mass and diameter than the Earth, particularly one consisting mostly of gas; also known as a **gas giant**. In the Solar System the giant planets are Jupiter, Saturn, Uranus, and Neptune, but there are now known to be similar objects around other stars.

giant star A star that has swollen in size towards the end of its life, having converted all the hydrogen in its core to helium. Such stars lie above the main sequence of the Hertzsprung–Russell diagram. A giant is brighter, larger, and cooler than a main-sequence star of the same mass. Giants have diameters 5–40 or more times that of the Sun, and luminosities from tens to thousands of times the Sun's. They are of *luminosity class II or III. *See also* BLUE GIANT; RED GIANT.

gibbous The phase of the Moon or a planet when it is between half and fully illuminated.

Gienah The star Epsilon Cygni, magnitude 2.5, a K0 giant 73 l.y. away.

Gill, David (1843–1914) Scottish astronomer. While Astronomer at the Cape (in modern South Africa), he derived a solar parallax of $8''.8$ from observations of three asteroids, a value that remained the standard until 1941. Photographs taken by Gill of the *Great September Comet of 1882 showed so many stars that he conceived the idea of producing a star catalogue based on photography. The result was the *Cape Photographic Durchmusterung*, compiled by J. C. *Kapteyn from Gill's photographic plates. A further consequence was the mammoth *Carte du Ciel* photographic project.

Ginga A Japanese X-ray astronomy satellite, known as Astro-C before its launch in 1987 February. It carried an exceptionally large array (0.4 m^2) of proportional counters provided by the UK, covering the energy range 1.5–37 keV (0.034–0.83 nm). An X-ray all-sky monitor and US gamma-ray burst detector were also carried. Ginga operated until late 1990 and re-entered in 1991 November.

(⊕) SEE WEB LINKS
• Information page at Goddard Space Flight Center.

Giotto The European Space Agency's first space probe, launched to Halley's Comet in 1985 July. It flew within 600 km of the sunward side of Halley's nucleus on 1986 March 14, taking close-up images and analysing the comet's dust and gas. Although battered by dust from the comet, Giotto survived and in 1990 July became the first spacecraft to use the Earth in a *gravity-assist manoeuvre, when it was retargeted to Comet *Grigg–Skjellerup; it passed 200 km from that comet's nucleus in 1992 July.

(⊕) SEE WEB LINKS
• ESA mission website.

Glauke Asteroid 288, discovered in 1890 by the German astronomer (Karl Theodor) Robert Luther (1822–1900). Glauke is an S-class asteroid of diameter 32 km. It is the slowest-spinning asteroid known, with a tumbling rotation period of 50 days. Its orbit has a semimajor axis of 2.756 AU, period 4.57 years, perihelion 2.18 AU, aphelion 3.33 AU, and inclination $4°.3$.

glitch An abrupt disturbance in the regular train of pulses from a pulsar, appearing as a sudden decrease in the pulsar's period (i.e. a speed-up in rotation), accompanied by an increase of the *spin-down rate which gradually returns to its former value over a few weeks or months. Glitches tend to occur in younger pulsars whose rate of spin is slowing rapidly, notably the Crab Pulsar and Vela Pulsar. They are believed to be caused by the sudden release of stress energy either in the crust (a *starquake) or between the crust and the superfluid interior.

Global Oscillation Network Group (GONG) A network of six world-wide observatories that monitors solar oscillations (*see* HELIOSEISMOLOGY). The oscillations are detected from precise measurements of the Doppler shifts in small areas of the Sun. The network started operation in 1995 at sites in the Canary Islands (Teide Observatory), Western Australia (Learmonth Solar Observatory), California (Big Bear Solar Observatory), Hawaii (Mauna Loa Solar Observatory), India (Udaipur Solar Observatory), and Chile (Cerro Tololo Inter-American Observatory). The network is managed by the US *National Solar Observatory and has its headquarters in Tucson, Arizona.

(⊕) SEE WEB LINKS
• Official GONG website.

globular cluster A roughly spherical group of old stars in the halo of a galaxy. Globular clusters contain from tens of thousands to millions of stars, and have diameters of 100–300 l.y. At the centre of the cluster, where most stars are concentrated, the density may be over 250 stars per cubic light year. About 150 globular clusters are known in our Galaxy, travelling on highly elongated orbits around the galactic centre. They are very old, about 10^{10} years, having formed early in the history of the Galaxy. Stars in globular clusters are members of Population II, with a low content of heavy elements (only a few per cent of the solar value), although a few show values approaching that of the Sun. On the Hertzsprung–Russell diagram of a globular cluster, the main sequence and the giant branch are smoothly joined via a subgiant branch; also conspicuous is an *asymptotic giant branch and a *horizontal branch, punctuated by an instability strip containing RR Lyrae variables. Globular clusters are found around all large galaxies, but are most abundant around giant elliptical galaxies. Some galaxies, particularly those that have undergone recent mergers, contain large numbers of young globular clusters.

SEE WEB LINKS
• Database of known globular clusters.

globule A small, dense cloud of dust and gas, usually rounded in shape, that appears dark through being seen in silhouette against a bright nebula or starfield. Globules are the smallest of the dark nebulae, and are usually less than 1 l.y. across; the smallest are only a few light days across and contain less than 1 solar mass. They are often called *Bok globules*, after B. J. *Bok, who drew attention to them and suggested that they may be *protostars. *See also* COMETARY GLOBULE.

GMAT Abbr. for *Greenwich Mean Astronomical Time.

GMC Abbr. for *giant molecular cloud.

GMST Abbr. for *Greenwich Mean Sidereal Time.

GMT Abbr. for *Greenwich Mean Time and *Giant Magellan Telescope.

gnomon The part of a sundial that casts a shadow. It may consist of a rod or a raised surface, as in designs where the edge of a decoration casts the shadow. If the gnomon is a triangular plate, the shadow-casting edge is known as the *style.

Goddard Space Flight Center (GSFC) A NASA establishment at Greenbelt, Maryland, founded in 1959, which performs Earth and space science research and satellite tracking. Goddard has designed and built numerous scientific satellites, including members of the Explorer series, and controls several astronomy satellites, including the Hubble Space Telescope. It houses the *National Space Science Data Center (NSSDC). Goddard also manages Wallops Flight Facility, at Wallops Island, Virginia, which coordinates NASA's sounding rocket and scientific balloon programmes.

SEE WEB LINKS
• Official website.

Gödel universe An unusual cosmological model which represents a rotating universe. This model possesses a number of strange mathematical features, including the fact that it allows time-travel to occur within it. It is due to the Austrian-American mathematician Kurt Gödel (1906–78).

Gold, Thomas (1920–2004) Austrian-American astronomer. In 1948, with H. *Bondi and F. *Hoyle, he proposed the *steady-state theory of the Universe in which matter is continuously created. In 1968 he and Hoyle explained the signals from the recently discovered pulsars as synchrotron radiation beamed out by rapidly rotating neutron stars.

Goldstone The location in the Mojave Desert, California, of one of the three deep-space communications complexes that make up NASA's world-wide *Deep Space Network.

GONG Abbr. for *Global Oscillation Network Group.

Goodricke, John (1764–86) English amateur astronomer, born in Holland. Although both deaf and dumb, at the age of 17 he started to observe bright variable stars, encouraged by his neighbour, the astronomer Edward Pigott (1753–1825). In 1782 Goodricke announced that Algol's variability was regular, and suggested that its periodic dimmings were caused by a dark body orbiting it. He went on to observe other such *eclipsing binaries. In 1784 he found that Beta Lyrae and Delta Cephei varied in brightness periodically, but did not realize that their variability had a different origin. In 1786, while observing Delta Cephei, he caught pneumonia and died, aged 21.

GoTo telescope A telescope under computer control that will automatically point to, and then track, a chosen object. The object is usually chosen from a built-in database of objects, although coordinates can be entered by hand. Many amateur telescopes now have this facility, which has long been standard on professional instruments.

Gould, Benjamin Apthorp (1824–96) American astronomer. In 1849 he founded the *Astronomical Journal*. In 1870 he moved to Argentina, founding the National Observatory at Córdoba, where he initiated the *Córdoba Durchmusterung*, a southern equivalent of the catalogue produced by F. W. A. *Argelander for the northern stars. *Gould's Belt is named after him.

Gould's Belt A band of hot, bright stars (types O and B) forming a circle around the sky. It represents a local structure of young stars and interstellar material tilted at about 16° to the galactic plane. Among the most prominent components of the belt are the bright stars in Orion, Canis Major, Puppis, Carina, Centaurus, and Scorpius, including the *Sco–Cen Association. The belt has a diameter of about 3000 l.y. (about one-tenth the radius of the Galaxy), and the Sun lies within it. Viewed from Earth, Gould's Belt projects below the plane of our Galaxy from the lower edge of the Orion Arm, and above the plane in the opposite direction. The belt is estimated to be about 50 million years old, but its origin is unknown. It is named after B. A. *Gould, who established its existence in 1879.

graben A strip of land, bounded by parallel faults, that has subsided between the faults; pl.*graben*. Many linear rilles on the Moon are graben, including the Rimae Sirsalis, about 3.5 km wide and 425 km long.

GRAIL Abbr. for *Gravity Recovery and Interior Laboratory.

grains, interplanetary Small particles of solid matter between the planets of the Solar System. They are produced mainly by collisions between asteroids and by the break-up of comets, but some enter the Solar System from interstellar space. There are trails of dust grains along the orbits of comets and asteroids, which are experienced as meteor streams if they cross the Earth's path. Interplanetary dust grains can also be detected when they scatter sunlight to form the *zodiacal light, and by the small amount of infrared radiation they produce.

grains, interstellar Small particles of solid matter between the stars. They are microscopic, the smallest being only 10 nm across. The dust grains cause extinction and reddening of starlight. Grains in molecular clouds have a thin coating of ice. Atoms colliding with the grains may stick to the icy surface, where they meet other atoms with which they can react to produce molecules.

Granat A Soviet satellite for X-ray and gamma-ray astronomy, launched in 1989 December. It carried seven X-ray and gamma-ray instruments. The ART-S and ART-P high-pressure proportional counters cover the energy range 3–150 keV. The French Sigma *coded-mask telescope operates in the energy range 30–1300 keV. Denmark provided the WATCH all-sky monitor. Granat ceased operation in 1998 November.

(⊕) SEE WEB LINKS
• Information page at Goddard Space Flight Center.

grand unified theory (GUT) An attempt to describe the weak and strong nuclear forces and electromagnetism in a single mathematical theory. Unification of the weak force with electromagnetism has been achieved in the *electroweak theory*. Before about 10^{-12} seconds after the Big Bang, by which time the Universe had cooled to about 10^{15} K, the electromagnetic and weak interactions acted as a single physical force; in the cooler temperatures since then, they have been distinct. Attempts to unify the electroweak force with the strong nuclear force have been only partially successful. It is thought that the temperature for their unification is of the order of 10^{27} K, which occurs only 10^{-36} s after the Big Bang. Particles surviving to the present day from this phase are possible candidates for non-baryonic *dark matter. Unification of the GUT interaction with gravity may take place at higher energies still, but there is no satisfactory theory which unifies all four physical forces. Such a theory would be called a *theory of everything* (TOE).

Gran Telescopio Canarias (GTC or GRANTECAN) A 10.4-m reflector with a mirror consisting of 36 hexagonal segments, opened in 2009 at an altitude of 2400 m at the *Roque de los Muchachos Observatory, Canary Islands. It is jointly owned by Spain, Mexico, and the University of Florida, and is operated by the Instituto de Astrofísica de Canarias. 📷

(((📶))) SEE WEB LINKS
• Official telescope website.

granulation A mottling of the Sun's photosphere caused by numerous small light areas called **granules**. Individual granules are as much as 1000 km across, and may have polygonal shapes. They are separated from each other by darker *intergranular lanes*, about 400 K cooler than granules. Granule lifetimes are around 20 min, their birth being from the fragments of previous granules and their demise from fragmentation, fading, or sometimes explosion into small fragments. Granules are elongated near sunspots, and also occur within sunspot umbrae (*umbral dots*). Spectroscopic observations suggest that granules are convection cells of rising hot gas (*see* CONVECTIVE OVERSHOOT), the intergranular lanes marking regions of descending cooler gas.

graticule Another name for a *reticle.

grating *See* DIFFRACTION GRATING.

grating spectrometer An instrument that uses a *diffraction grating to disperse light into a spectrum. Gratings may be placed at the focus of a telescope (a *focal-plane spectrometer*) or in front of a telescope (an *objective spectrometer*). Grating spectrometers are used for spectroscopy at wavelengths from X-rays to the far infrared.

gravitation The force of attraction that operates between all bodies. The size of the attraction depends on the masses of the bodies and the distance between them; gravitational force diminishes with the square of the distance apart according to the *inverse-square law. Gravitation is the weakest of the four fundamental forces in nature. I. *Newton formulated the laws of gravitational attraction and showed that gravitationally a body behaves as though all its mass were concentrated at its centre of gravity. Hence a gravitational force acts along a line joining the centres of gravity of two masses. In the *general theory of relativity, gravitation is interpreted as a distortion of space. Gravitational forces are significant between large masses such as stars, planets, and satellites, and it is this force which is responsible for holding together the major components of the Universe. However, on the atomic scale the gravitational force is about 10^{40} times weaker than the force of electromagnetic attraction.

gravitational acceleration *See* ACCELERATION OF FREE FALL.

gravitational collapse The collapse of a body that is unable to support itself against its own gravity. Gaseous bodies undergo such collapse if they are not hot enough for their gas pressure to balance gravity. This can happen in the early stages of star formation, or when nuclear burning ceases in a star's core. The time taken for such collapse decreases rapidly with increasing density, varying from about 100 000 years for the birth of a new star to less than a second for the formation of a neutron star. Star clusters may undergo a similar collapse if the random motions of their constituent stars are insufficient to offset gravitational effects, either during their formation (*see* VIOLENT RELAXATION) or at an advanced stage of their evolution (*see* CORE COLLAPSE 2).

gravitational constant (*G*) The constant that appears in Newton's law of gravitation. It is the attraction between two bodies of unit mass at unit distance apart. Its value is $6.673\,84 \times 10^{-11}$ N m^2/kg^2 when the distance is expressed in metres and the masses are in kilograms. Although it is described as a constant, in some models of the Universe *G* decreases with time as the Universe expands (*see* BRANS–DICKE THEORY), but there is no evidence for this.

gravitational deflection The bending of the path of a beam of light or any other electromagnetic radiation by the gravitational field of a body. The amount of bending depends on the mass of the body and how closely the beam of radiation passes it. The effect was first measured for stars close to the limb of the Sun at the total eclipse of 1919 May. The maximum deflection at the Sun's limb is $1''.75$ radially away from the Sun. Gravitational deflection also results in highly distorted images of distant objects when a closer, massive object lies along the line of sight (*see* GRAVITATIONAL LENS).

gravitational energy The energy released by a body falling in a gravitational field. It is a form of *potential energy*. When an interstellar gas cloud collapses to form a protostar, the gravitational energy released heats its interior to the point at which nuclear reactions can commence. Gravitational energy also supplies the energies for stars condensing into white dwarfs, and plays a major role in the processes operating inside supernovae. Large amounts of gravitational energy are released by material falling into the *accretion disks around black holes, and may provide the power source for quasars, Seyfert galaxies, and other *active galactic nuclei.

gravitational field The region of space around a body in which that body's gravitational force can be felt. Within this region, other bodies will experience a force of attraction that diminishes with distance from the body.

gravitational force The force of attraction, F, that exists between all bodies. For any two bodies, it is directly proportional to the product of the two masses and inversely proportional to the square of the distance between them. The constant of proportionality is G, the universal gravitational constant. Expressed in mathematical form, $F = Gm_1m_2/d^2$, where m_1 and m_2 are the two masses and d is their distance apart.

gravitational instability A phenomenon in which an object collapses because its internal gas pressure, magnetic pressure, or material strength is unable to support its own weight. For a gas, a region will become gravitationally unstable when its mass rises above a critical value, the *Jeans mass. In *molecular clouds, the Jeans mass can be low enough for stars and planets to form. In the early Universe, the instabilities may have been large enough to produce galaxies and clusters of galaxies. In Earth-like planets, gravitational instability leads to the precipitation of heavier elements downwards to form the core.

gravitational lens An effect in which light rays are bent by the gravitational field of a massive object, such as a galaxy or black hole. The Sun produces a slight gravitational lens effect (*see* GRAVITATIONAL DEFLECTION), but on cosmological scales the effect is seen as the formation of double or multiple images of a distant galaxy or quasar by a foreground object (as in, for example, the *Einstein Cross). More complicated lensing effects also occur, including the formation of *Einstein rings, *luminous arcs, and *microlensing. 📷

gravitational mass A measure of the quantity of matter in a body. It is measured in kilograms. Mass determines the strength of the gravitational force exerted by an object. *See also* INERTIAL MASS.

gravitational radiation A form of energy that is emitted by an accelerating mass in the form of waves that travel through spacetime at the speed of light. Pulses of gravitational radiation are expected from supernovae and objects falling into black holes, but have not yet been detected. *See also* GRAVITATIONAL WAVE.

gravitational redshift The redshift of light or other electromagnetic radiation caused by a strong gravitational field; also known as the *Einstein shift*. It arises because radiation loses energy as it passes out of the gravitational field of the emitting body. As a consequence, the frequency of the radiation decreases and its wavelength is shifted to the red end of the spectrum. The redshift at wavelength λ is given by $Gm\lambda/c^2r$, where m is the mass of the body, r is the distance of the emitting region from the centre of mass, c is the speed of light, and G is the universal gravitational constant. Small gravitational redshifts are observed in transmissions between receivers at different altitudes on Earth, and between satellites and receivers on the Earth. A gravitational redshift has been observed in the light from some white dwarfs, and would result in the rapid fading out of a black hole in the process of formation as seen from outside.

gravitational wave A wave-like motion in a gravitational field, produced when a mass is accelerated or otherwise disturbed. Gravitational waves travel through spacetime at the speed of light, and their amplitude is proportional to the rate of acceleration of the body producing them. The strongest sources are those with the strongest gravitational fields although the waves, like the force of gravity itself, would be very weak. Gravitational waves have not yet been observed directly. However, the decay in the orbital period of the *double pulsar J0737–3039AB is attributed to loss of energy through gravitational waves. Short-period *cataclysmic binaries (orbital periods less than about two hours) are also believed to evolve to shorter orbital periods as a result of the emission of gravitational waves.

graviton A hypothetical particle or quantum of gravitational energy, predicted by the general theory of relativity. Gravitons have not been observed but are predicted to travel at the speed of light and to have zero *rest mass and charge. A graviton is the gravitational equivalent of a photon.

Gravity and Extreme Magnetism Small Explorer (GEMS) A NASA mission planned for launch in 2014 to study the structure of space around black holes, and the magnetic fields around *magnetars and supernova remnants, by measuring the polarization of X-rays they emit. GEMS carries three telescopes with thin-foil mirrors to detect X-rays with energies between 2 and 10 keV.

(⊕) SEE WEB LINKS
• Mission page at Goddard Space Flight Center.

gravity assist The technique of using the gravitational field and orbital velocity of a planet to alter a spacecraft's trajectory and velocity; also known as a **gravitational slingshot**. As the spacecraft makes a close fly-by of a planet, its direction of travel is altered and it picks up additional speed from the planet's orbital velocity. The technique was first used by Mariner 10, which flew past Venus on its way to Mercury in 1974. The two Voyager probes made fly-bys of Jupiter, considerably shortening the time they took to reach Saturn. Voyager 2 subsequently used gravity assists from Saturn and Uranus to take it to Neptune. Other probes to use gravity assists were Giotto, Galileo, and Ulysses.

gravity gradient The direction of a gravitational field at a point within the field. In the vicinity of a massive body such as a star or planet, the gravity gradient points to the body's centre. An elongated object in orbit about the body will revolve with its long axis pointing towards the body's centre. For example, the Moon's longest axis lies along a line towards the Earth's centre. Artificial satellites can be oriented in orbit by making use of the gravity gradient.

Gravity Probe B A NASA satellite launched into polar orbit in 2004 April which carried four precisely aligned gyroscopes to test two predictions about gravity made by the general theory of relativity. Slight changes in the rotation axes of the gyroscopes will reveal how space and time are warped by the mass of the Earth, and how the Earth's rotation drags spacetime around with it (an effect known as *frame dragging). It ceased operation in 2005 September.

(⊕) SEE WEB LINKS
• Official mission website.

Gravity Recovery and Interior Laboratory (GRAIL) A NASA mission due for launch in 2011 to determine the internal structure of the Moon. It consists of two spacecraft, GRAIL A and GRAIL B, in low polar orbit. Tracking of the changing distance between the two craft will map the Moon's gravitational field in detail. The two spacecraft will each carry cameras to provide images of the lunar surface.

(⊕) SEE WEB LINKS
• Official mission website.

grazing-incidence telescope A telescope used at extreme ultraviolet and X-ray wavelengths, for which a conventional mirror is very inefficient because it absorbs photons. In a grazing-incidence telescope, incoming light is reflected off the mirror surface at very shallow angles. Several designs of grazing-incidence telescope have been used in satellites, including flat mirrors or combinations of parabolic and hyperbolic surfaces. To increase the collecting area a number of mirror elements are often nested inside one another.

grazing occultation A lunar *occultation in which the limb of the Moon just touches a star. Such events may produce numerous brief disappearances and reappearances, accurate timings of which can help in determining the Moon's limb profile. The most useful records of grazing occultations are obtained by teams of observers spread across the predicted ground track.

GRB Abbr. for *gamma-ray burst.

Great Attractor A purported concentration of matter in the direction of the constellations Hydra and Centaurus which may be pulling surrounding galaxies, including ours, towards it. The Great Attractor is calculated to have a mass about 5×10^{16} solar masses and to lie about 150 million l.y. from our Galaxy. Its existence was originally inferred from studies of the motions of galaxies with respect to the *Hubble flow. Although there clearly is a concentration of galaxies at the place where the Great Attractor supposedly lies, more recent studies indicate that the observed bulk flows of galaxies are probably due to the concerted pull of several distinct clusters.

Great Bear Popular name for the constellation *Ursa Major.

great circle A circle on a sphere whose plane passes through the sphere's centre. On the celestial sphere, the celestial equator, the ecliptic, and all lines of right ascension are great circles; lines of declination, other than the celestial equator, are *small circles*.

Great Dark Spot A large dark oval feature about $10\,000 \times 5000$ km in the clouds of Neptune, photographed by Voyager 2 in 1989. It lay at about 20° south latitude and appeared to consist of a huge rotating vortex, similar to Jupiter's Great Red Spot. However, the Great Dark Spot was not visible to the Hubble Space Telescope in 1994, having apparently dispersed. Since then, other similar features have been seen in the northern hemisphere of Neptune.

greatest brilliancy The greatest apparent magnitude of Venus at any particular apparition. It depends on both the planet's distance and its phase. Venus varies greatly in apparent size; at superior conjunction, its disk is fully illuminated but its diameter is only 10″, whereas at inferior conjunction it is over 60″ in diameter but it is a very thin crescent. Its brightness increases as it moves away from superior conjunction, because of the increasing apparent area of the illuminated disk, but eventually the rapidly thinning crescent counteracts the increase in apparent diameter, and the magnitude starts to fall again. Greatest brilliancy occurs about 36 days before and after inferior conjunction, when Venus is the brightest object in the sky after the Sun and Moon, reaching up to magnitude −4.7.

greatest elongation The occasion when either Mercury or Venus reaches its greatest angular separation from the Sun. **Greatest elongation east** occurs in the evening sky, and **greatest elongation west** in the morning sky. The greatest elongation of Mercury ranges from about 18° to 28°, depending on whether it lies near perihelion or aphelion, and for Venus from 45° to 47°.

Great Observatories A series of four NASA astronomy missions covering different regions of the spectrum. They are the *Hubble Space Telescope, the *Compton Gamma Ray Observatory, the *Chandra X-ray Observatory, and the *Spitzer Space Telescope.

Great Red Spot (GRS) A large oval feature in Jupiter's clouds at about 22° south latitude. Its dimensions are currently about $20\,000$ km east–west and $14\,000$ km north–south; it has been shrinking in length fairly steadily for the past century. It was first recorded by S. H. *Schwabe in 1831 but attracted little attention until 1878–82, when it became a striking dark red. Since then it has varied greatly in size, colour, and intensity, sometimes being so faint that it is detectable only by the hollow it makes in the South Temperate Belt (the *Red Spot Hollow*). A prominent, slightly smaller dark spot at a similar latitude to the Great Red Spot was seen in 1664 by the English scientist Robert Hooke (1635–1703), and lasted until 1713. This was probably an earlier manifestation of a similar phenomenon. Space probes have shown the Great Red Spot to be a vast vortex rotating anticlockwise (anticyclonically), equivalent to a storm or hurricane. At the time of the Voyager encounters in 1979 its rotation period was about 7 days, but 30 years later had speeded up to about 4.5 days. Its top lies a few kilometres above the surrounding cloud deck, and its red colour may be due to compounds such as phosphine (PH_3). A smaller version of the Great Red Spot arose at 34° south following the merger of three white oval storms in 1998 and 2000. Initially it remained white, like the storms from which it formed, but took on a ruddy colour in late 2005, presumably as complex chemicals were brought up from deeper within Jupiter's clouds. 📷

Great Rift A dark nebula that runs along the local spiral arm of our Galaxy, dividing the Milky Way. It starts in Cygnus, where it is also known as the Cygnus Rift or Northern Coalsack, and can be traced south through Aquila into Ophiuchus, where it broadens out. It consists of a large group of molecular clouds approximately 2200 l.y. away, containing almost 1 million solar masses of gas. The clouds of gas and dust extend for over 1000 l.y.

Great September Comet (C/1882 R1) A long-period comet, a member of the *Kreutz sungrazer group, discovered independently by many observers in the southern hemisphere in 1882 September; formerly designated 1882 II. Perihelion was on September 17 at 0.008 AU (1.2 million km), when the nucleus broke into at least four fragments. At its best the comet was visible in broad daylight (magnitude at least −10), and showed a tail exceeding 20°. Its orbit is very similar to that of *Ikeya–Seki (C/1965 S1), having a period of about 670 years, eccentricity 0.9999, and inclination 142°.0.

(⊕) SEE WEB LINKS
• Information page at Cometography website.

Great Wall A roughly two-dimensional concentration of galaxies, at least 200 million by 600 million l.y. in extent, but less than 20 million l.y. thick. It contains many thousands of galaxies and has a mass of at least 10^{16} solar masses. The Great Wall extends more than 120° across the sky (8–16 h RA) and is about 250 million l.y. away. This and other similar features, together with the many filamentary features in the distribution of galaxies, suggests that the *large-scale structure of the Universe may be cellular in nature, with other Great Walls forming the faces of the cells and filaments of galaxies forming where these faces intersect. *See also* REDSHIFT SURVEY.

Green Bank The location in West Virginia, USA, of one observing site of the *National Radio Astronomy Observatory.

Green Bank Telescope (GBT) The world's largest fully steerable radio telescope, with an elliptically shaped dish, 100 × 110 m, opened in 2000 at the *National Radio Astronomy Observatory site in Green Bank, West Virginia. It has an actively controlled surface to maintain its shape, and features an off-axis secondary held by an arm outside the dish so that the telescope's aperture is not obstructed. Its full title is the Robert C. Byrd Green Bank Telescope.

(⊕) SEE WEB LINKS
• Official telescope website.

green flash An optical effect, seen under favourable conditions at the moment of sunset or sunrise, in which the small uppermost fraction of the Sun's visible disk briefly appears intensely green as a result of the preferential refraction of light. The green flash is quite rare, requiring a flat clear horizon for visibility, perhaps coupled with special atmospheric conditions; a sea horizon is considered ideal.

greenhouse effect The elevation of a planet's surface temperature resulting from the absorption of long-wavelength (infrared) radiation by gases in the atmosphere. For instance, carbon dioxide, methane, chlorofluorocarbons, and water vapour in the Earth's atmosphere are all transparent to incoming short-wavelength solar radiation, but absorb outgoing infrared. The Earth is, on average, about 35 K warmer than it would be if its atmosphere were completely transparent to outgoing infrared. On Venus, the greenhouse effect raises the surface temperature by some 500 K to around 730 K.

Greenwich hour angle (GHA) The *hour angle of a celestial object as seen by an observer on the Greenwich meridian. GHA is a global standard. The hour angle for any other observer is simply GHA + λ, where λ is the observer's geographical longitude east. A star's Greenwich hour angle can be found by subtracting the star's right ascension from the Greenwich Sidereal Time.

Greenwich Mean Astronomical Time (GMAT) An obsolete form of Greenwich Mean Time, being the mean solar time on the Greenwich meridian but with the day beginning at noon rather than midnight. It is, therefore, the *Greenwich hour angle of the mean sun. It was used by astronomers before 1925 to avoid the need to change date during the night.

Greenwich Mean Sidereal Time (GMST) The mean sidereal time on the Greenwich meridian and, therefore, the Greenwich hour angle of the mean equinox. Greenwich mean sidereal time is derived from meridian observations of stars reduced to the Greenwich meridian. There is a strict relationship between GMST and *Universal Time (UT), so that UT is derived in practice from Greenwich Mean Sidereal Time.

Greenwich Mean Time (GMT) The mean solar time on the Greenwich meridian, now more generally known astronomically as *Universal Time (UT). It is the Greenwich hour angle of the mean sun + 12 hours, as the day starts at midnight.

Greenwich meridian The line of 0° longitude that passes through the Old Royal Observatory at Greenwich, London. It was adopted in 1884 as the *prime meridian, and still serves as the reference point for the world's timekeeping and navigation systems.

Greenwich Observatory *See* ROYAL GREENWICH OBSERVATORY.

Greenwich sidereal date The number of sidereal days that have elapsed at Greenwich since a given starting date. It is analogous to *Julian Date, and the initial epoch is chosen as the beginning of the sidereal day that was in progress at JD 0.0. The whole-number part of the Greenwich sidereal date is the number of sidereal days that have elapsed since that time (the **Greenwich Sidereal Day Number**), and the decimal part is the Greenwich Sidereal Time expressed as a fraction of a sidereal day.

Greenwich Sidereal Time (GST) The sidereal time on the Greenwich meridian; equivalently, the Greenwich hour angle of the equinox. The sidereal time will either be mean sidereal time or apparent sidereal time, depending on whether the *mean equinox or *true equinox is used.

Gregorian calendar The form of calendar that is now in almost world-wide civil use. It was devised with the help of the German Jesuit mathematics teacher Christopher Clavius (1537–1612), and was introduced by Pope Gregory XIII in 1582 to replace the *Julian calendar, which had got out of step with the seasons. In that year 10 days were omitted, Thursday October 4 being immediately followed by Friday October 15. This change was made in order to re-establish March 21 as the date of the vernal equinox. Britain and America did not adopt the Gregorian calendar until 1752, by which time 11 days had to be omitted. Subsequently the calendar has been kept in step with the date of the vernal equinox by introducing a leap year every four years, as the Julian calendar had done, but suppressing the leap year in century years, unless the year is divisible by 400. Thus 1700, 1800, and 1900 were not leap years, but 1600 and 2000 are. In the 400-year cycle there are therefore 97 leap years, giving an average length of the year of 365.2425 days, very close to the length of the *tropical year (365.2422 days).

Gregorian telescope A reflecting telescope with a concave paraboloidal primary mirror and a concave ellipsoidal secondary mirror. The secondary mirror reflects light to a focus behind the primary, which must therefore have a hole in it. The advantage of the system compared with the *Cassegrain is that the image is upright, but the drawback is that the Gregorian has a longer tube. Gregorians, proposed in 1663 by J. *Gregory, were popular in the 18th century but are now rarely seen.

Gregory, James (1638–75) Scottish mathematician. His *Optica promota* (1663) describes the first design of telescope (the *Gregorian reflector) to feature two mirrors as a means of circumventing chromatic aberration. In 1668 he described the *inverse-square law for the brightness of stars, and used it to estimate the distance of Sirius by assuming Sirius and the Sun to be of comparable luminosity (his result was far smaller than the true value). His nephew, the Scottish mathematician David Gregory (1661–1708), suggested that chromatic aberration could be overcome by using a lens made of two elements of different composition; the idea, published in 1695, may have come from I. *Newton.

Grigg–Skjellerup, Comet 26P/ A periodic comet discovered by the New Zealand amateur astronomer John Grigg (1838–1920) in 1902, and independently by the Australian amateur John Francis ('Frank') Skjellerup (1875–1952) in 1922. Its orbit is perturbed by Jupiter, which has led to an increase in its perihelion distance from 0.75 AU in 1902 to 1.1 now. Its orbital period is 5.3 years. Grigg–Skjellerup was visited by the *Giotto space probe in 1992. The comet was found to have a much lower dust production than expected, and a gas production only about 1% of that of Halley's Comet at its 1986 perihelion. Debris from Comet Grigg–Skjellerup produces the periodic Pi Puppid meteor shower, seen around April 23 but only when the comet is close to perihelion. The comet's orbit has an eccentricity of 0.63 and inclination 22°.4.

(((⊕))) SEE WEB LINKS

• Information page at Cometography website.

grism A combined *diffraction grating and *prism used in spectroscopy. Grisms consist of a grating ruled on a thin prism, and are commonly used to separate otherwise overlapping orders of diffraction spectra and increase the resolving power. Grisms are often used as *objective prisms.

grooved terrain Terrain on a planetary surface that consists of belts of near-parallel grooves. The term was first used to describe the grooves on Phobos. Since then it has been applied to grooved areas on Ganymede and on other satellites and asteroids. These terrains have different characteristics and origins, even on the same body. Some appear to be parallel lines of secondary impact craters, while others are tectonic features.

Grotrian diagram A diagram of the energy levels for a given atom or ion. It shows the transitions which are allowed between the various electron energy levels, and hence the spectral features that arise as a result. The diagram is an essential tool for understanding and identifying lines of elements such as iron that have a large number of energy levels and hence very complex spectra. It is named after the German physicist Walter Robert Wilhelm Grotrian (1890–1954).

ground state The lowest *energy levels which electrons can occupy in an atom, ion, or molecule. For hydrogen and helium, this is the energy level closest to the nucleus. However, this energy level can hold only 2 electrons, so for the next-heaviest element, lithium, the ground state has 2 electrons in the lowest energy level and 1 in the second level. The second energy level can contain a maximum of 8 electrons, the third level a maximum of 18 electrons, and so on ($2n^2$ electrons in the nth level). The ground states for the heavier elements may therefore have some of their electrons in quite high energy levels.

GRS Abbr. for *Great Red Spot.

Grubb, Howard (1844–1931) Irish optician. In 1865 he joined the optical company established by his father, Thomas Grubb (1800–78), where he assisted with construction of a 48-inch (1.2-m) reflector for Melbourne, Australia, completed in 1867. Under his control, the company developed a reputation for large telescopes of high quality. In 1914 Grubb moved his business to England where in 1925, as the result of a merger, it became Sir Howard Grubb, Parsons & Company.

Grus (Gru) (*gen.* Gruis) A constellation of the southern sky, representing a crane (bird). Its brightest star is *Alnair (Alpha Gruis). Delta Gruis is a wide double of unrelated stars, magnitudes 4.0 and 4.1, as is Mu Gruis, magnitudes 4.8 and 5.1.

GST Abbr. for *Greenwich Sidereal Time.

G star A star of spectral type G, which includes the Sun. G-type stars on the main sequence have temperatures of 5300–6000 K, and therefore appear yellow. G-type giants are about 100–500 K cooler than main-sequence stars. G-type supergiants are about 4500–5500 K. The spectrum of the Sun (type G2) is dominated by lines of singly ionized calcium (principally the *H and K lines) and neutral metals. In cooler G stars the molecular bands of CH and CN become visible. Main-sequence G stars, and giants such as Capella, have masses of 0.8–1.1 solar masses, while supergiants are of 10–12 solar masses. The luminosities of G-type giants are about 30–60 times the Sun's, and for supergiants 10 000–300 000 times the Sun's.

GTC Abbr. for *Gran Telescopio Canarias.

Guardians The stars Beta and Gamma Ursae Minoris (*Kochab and *Pherkad). They lie in the bowl of the Little Dipper, Ursa Minor.

guest star The name by which a temporary celestial object such as a comet, nova, or supernova was referred to in ancient Chinese records.

guider A device that allows a telescope to follow a guide star during photography or imaging. Guiding is necessary to compensate for small mechanical errors in the telescope and mounting, and changes in atmospheric refraction. There should be a means of monitoring a guide star and making small corrections to either the telescope's movement or the position of the camera. The monitoring may be carried out using a *guide telescope or an *off-axis guider, either visually or with an *autoguider.

guide star Any star used as a reference when keeping a telescope trained on a target object.

guide telescope A telescope designed for viewing a guide star when keeping a larger instrument correctly pointed as the Earth turns; also known as a **guidescope**. The actual guiding is done either by watching a guide star and keeping it precisely centred on cross-wires in the eyepiece, or using an *autoguider. An alternative to a guide telescope is an *off-axis guider.

Gum Nebula A large, almost circular emission nebula in the constellations Vela and Puppis, discovered by the Australian astronomer Colin Stanley Gum (1924–60). The Gum Nebula is 36° across and is estimated to be centred 1500 l.y. away, which gives it a true diameter of about 800 l.y. It is expanding at about 10 km/s, although the near side is expanding faster than the far side. The nebula is thought to have been formed by the supernova explosion of the former companion of the *runaway star Zeta Puppis about 1.5 million years ago. It contains the Vela OB2 association, the hot stars of which contribute to the photoionization along with Zeta Puppis. The *Vela Supernova Remnant and the *Vela Pulsar are embedded within the Gum Nebula, but are much younger.

GUT Abbr. for *grand unified theory.

GW Virginis star *See* ZZ CETI STAR.

gyrofrequency The number of revolutions per second described by a charged particle circling in a magnetic field at a velocity much less than the speed of light; also known as *cyclotron frequency*. It is independent of the particle's velocity, but depends on its mass and charge and on the magnetic flux density.

gyrosynchrotron radiation Electromagnetic radiation emitted by a charged particle moving in a magnetic field at an appreciable fraction of the speed of light. It is similar to *cyclotron radiation, except that relativistic effects cause most of the energy to be emitted at higher multiples of the gyrofrequency. The slowly varying radio emission received from active regions on the Sun may consist of gyrosynchrotron radiation. Gyrosynchrotron radiation is believed to be responsible for the intense radio emission received from solar flares occurring in the magnetic fields above groups of sunspots.

HA Abbr. for *hour angle.

Hadar The star Beta Centauri, the 11th-brightest star in the sky, also known as Agena. It is a B1 giant of magnitude 0.6, but is a variable of the Beta Cephei variety with an amplitude of only about 0.05 mag. It is a double, with a close companion of magnitude 3.9. Hadar lies 392 l.y. away.

Hadley circulation The movement of air from the equator towards the poles in a planet's atmosphere. Warm air near the equator rises and travels towards the poles, cooling as it does so until it becomes denser than the air below. It then sinks to the surface and returns to the equator, completing a cycle known as a **Hadley cell**. The tropical Hadley cells in the Earth's atmosphere extend to about 30° north and south of the equator, but there are two further sets of Hadley cells, one between 30° and 60° latitude, and another between 60° latitude and the poles, which are less prominent and less permanent than the tropical ones. On Venus, there is a single Hadley cell extending from the equator to the poles. The effect is named after the English meteorologist George Hadley (1685–1768).

Hadley Rille A sinuous rille at the foot of the Apennine Mountains on the Moon, also known as Rima Hadley, at +25° lat., 3° E long. It was formed early in the history of the Moon as a giant lava channel 80 km long, 1500 m wide, and 300 m deep. The plain to the east of the rille was the site of the Apollo 15 landing in 1971. 📷

hadron A heavy fundamental particle that participates in strong nuclear interactions. Hadrons are divided into *baryons, which include protons and neutrons, and *mesons, which include the pions.

hadron era The short period from about 10^{-6}s to 10^{-5}s after the Big Bang when heavy atomic particles such as protons, neutrons, pions, and kaons were formed. Before the start of the hadron era, *quarks behaved as free particles. The process by which hadrons formed from these quarks is called the *quark–hadron phase transition*. By the end of the hadron era, all the other hadron species had either decayed or annihilated, leaving only protons and neutrons. Immediately after this the Universe entered the *lepton era; both are subdivisions of the *radiation era.

HAEBE Abbr. for *Herbig Ae/Be star.

Haedi *See* KIDS.

Haig mount Another name for the *Scotch mount.

Hakucho The first Japanese X-ray astronomy satellite, launched in 1979 February, known before launch as Corsa-B (Corsa-A was a launch failure). A spinning satellite, it carried proportional counters which viewed along the spin axis and studied the luminosities of X-ray bursts. The mission ended in 1985 April.

(🌐) SEE WEB LINKS
• Information page at Goddard Space Flight Center.

halation The spreading of light through a photographic emulsion, which results in the recorded image of a star being larger than the actual image. The more light that falls on the emulsion, the larger the recorded image. Halation occurs because the light diffuses through the grains of the emulsion, and reflects off the surface of the glass or film on which the emulsion is coated. It has a practical use in that it allows a star's brightness to be found by measuring the size of its image.

HALCA A Japanese radio astronomy satellite, part of the *VLBI Space Observatory Programme, known as Muses-B before its launch in 1997 February. HALCA (Highly Advanced Laboratory for Communications and Astronomy) carried an 8-m radio dish. It is known in Japanese as Haruka, meaning 'far away'. It ceased observations in 2003 October.

Hale, George Ellery (1868–1938) American solar astronomer. In 1889 he invented the *spectroheliograph, which he used to study the Sun's prominences and surface features. In 1905 he found that sunspots were cooler than the surrounding photosphere, and in 1908 he showed from the *Zeeman effect in their spectra that they had strong magnetic fields. In 1925 he found that the Sun's magnetic field reversed polarity in each successive sunspot cycle. Hale planned and organized funding for three telescopes, each of which was in turn the world's largest. The 40-inch (1-metre) refractor at *Yerkes Observatory (still the largest refractor) was completed in 1897; the 100-inch (2.5-metre) Hooker Telescope at *Mount Wilson Observatory in 1917; and the 200-inch (5-metre) reflector (later named the Hale Telescope) at *Palomar Observatory in 1948.

Hale–Bopp, Comet (C/1995 O1) A long-period comet discovered independently on 1995 July 23 by the American amateur astronomers Alan Hale (1958–) and Thomas Joel Bopp (1949–). At that stage it was still over 7 AU from the Sun. It was noted for a series of jets and shells of dust thrown off by the 50-km-wide nucleus as it rotated in a period of 11.3 hours. The comet reached perigee, 1.32 AU, on 1997 March 22, and perihelion, 0.91 AU, on 1997 April 1. Its maximum magnitude was about −0.5. The gas tail extended to 20° and the dust tail to 25°, corresponding to actual lengths of over 1 AU. Its orbit has a period of about 2500 years, eccentricity 0.995, and inclination 89°.4.

(⊕) SEE WEB LINKS
• Information page at Cometography website.

Hale Observatories The name from 1970 to 1980 of the combined *Mount Wilson and*Palomar Observatories, the *Big Bear Solar Observatory, and *Las Campanas Observatory. During that time they were jointly owned and operated by the Carnegie Institution and the California Institute of Technology.

Hale's law See BIPOLAR GROUP.

Hale Telescope The 200-inch (5-m) reflector at *Palomar Observatory, California, opened in 1948. It is named after G. E. *Hale, who was responsible for its construction.

half-life In radioactive decay, the time taken for half the number of atoms initially present to disintegrate. For unstable elementary particles, it is the average time they take to transform spontaneously into other particles, as in for example the *beta decay of a free neutron into a proton and electron.

half-wave dipole A dipole antenna approximately half as long as the wavelength it is designed to receive. Half-wave dipoles are commonly used as a basic component of radio telescopes, particularly at longer wavelengths. A half-wave dipole has a narrower beamwidth than a full-wave dipole.

half-width See FULL WIDTH AT HALF MAXIMUM.

Hall, Asaph (1829–1907) American astronomer. Using the US Naval Observatory's 26-inch (0.66-m) refractor, he discovered the two small satellites of Mars in 1877 August, at a particularly close opposition. He named them Deimos and Phobos, respectively, and from their orbits he calculated the mass of Mars. Hall also studied double stars, proving that the two components of 61 Cygni comprise a true binary system.

Halley, Edmond (1656–1742) English scientist. From 1676 he spent two years observing the southern sky from St Helena, and in 1678 published a catalogue of 341 stars, the first southern star catalogue compiled from telescopic observations. He was the first to suggest that observations of transits of Venus could be used to measure the Sun's distance, which was eventually done by N. *Maskelyne long after his death. In 1683 Halley commenced a long series of lunar studies, discovering the Moon's *secular acceleration in 1693. In 1684, having deduced the inverse-square law, he visited I. *Newton and persuaded him to write the *Principia. In 1705 Halley published the *Synopsis of Cometary Astronomy* in which he concluded that the comet he had observed in 1682 was the same as those of 1531 and 1607. He predicted it would return in 1758, which it did and was named Halley's Comet. In 1718 he concluded that the brightest stars had changed position since the time of Ptolemy's *Almagest*, thus discovering *proper motion. As the second Astronomer Royal (from 1720) he initiated a series of lunar and solar observations spanning 18 years—a complete *saros. In 1721 he raised the problem of what has come to be called *Olbers' paradox.

Halley-family comet A member of a group of comets with short to intermediate periods (20–200 years) and aphelia between 7.4 and 40 AU; also known as a **Halley-type comet**. The prototype is Comet *Halley itself. Other well-known members are comets *Tempel–Tuttle and *Swift–Tuttle. Members of the Halley family are thought to derive from the inner *Oort Cloud by gravitational interaction with the giant planets. As of the end of 2010 over 50 members of the family were known.

Halley's Comet (1P/Halley) The best-known periodic comet, returning to perihelion at average intervals of 76 years on a retrograde orbit (the time between returns ranges from 74 to 79 years). Comet Halley has been observed at every return since 240 BC. It is named after E. *Halley, who in 1705 showed that comets seen in 1531, 1607, and 1682 were identical, and successfully predicted the return of 1758. In 1910 Halley's Comet passed 0.15 AU from Earth, reaching magnitude 0 and showing a 100° tail; the Earth passed through the tail on May 20. At its most recent return, in 1986 (perihelion February 9, closest to Earth 0.42 AU on April 11), it reached 3rd magnitude and had a 10° tail. On this occasion, five space probes were sent to investigate it: *Giotto, *Sakigake, *Suisei, and *Vega 1 and 2. It was found to have an irregular nucleus measuring 16×8 km with an albedo of 0.04. Halley's Comet is the parent of the *Eta Aquarid and *Orionid meteor showers. Its perihelion is 0.586 AU, aphelion 35.1 AU, eccentricity 0.967, inclination 162°.3. The comet will next return to perihelion in 2061 July.

((( ))) SEE WEB LINKS

• Information page at Cometography website.

halo, atmospheric A ring or arc of light appearing to surround the Sun or Moon, resulting from refraction and reflection of sunlight or moonlight by ice crystals in cirrus or cirrostratus clouds. The commonest solar and lunar halos have an angular diameter of 44°. The edge of the halo shows a prismatic effect, with blue light refracted to the outer edge and red to the inner. As a result of the preferential refraction of light to the halo's edge, the sky within a complete halo often appears darker than that outside. Lunar halos can be seen clearly only when the Moon is bright, typically within five days of full.

halo, galactic *See* GALACTIC HALO.

halo CME A class of *coronal mass ejection (CME) that appears as an expanding halo of gas around the Sun in *coronagraph images such as obtained by the *Solar and Heliospheric Observatory (SOHO). A halo CME occurs when the emitted gas is directed along the line of sight towards the observer. Because they are moving towards us, halo CMEs are more likely to produce effects at the Earth than other CMEs.

halo orbit An orbit in which a spacecraft will remain in the vicinity of a *Lagrangian point. A halo orbit actually takes the form of a circular or elliptical loop around the Lagrangian point. The first spacecraft to use a halo orbit was the *Solar and Heliospheric Observatory (SOHO), which was placed in an elliptical orbit with a semimajor axis of about 650 000 km around the L_1 Lagrangian point between the Earth and Sun.

halo population Those stars which belong to the spherical halo surrounding our Galaxy and others. Such stars have a low content of heavy elements, belong to Population II, and are believed to be older than most stars in the galactic disk, such as the Sun. Some are found within the disk but are simply in transit through it on their elongated orbits around the galactic centre, and can easily be distinguished by their high velocities with respect to the disk stars (*see* HIGH-VELOCITY STAR).

Hα The strongest line of hydrogen in the *Balmer series. Its wavelength is 656.3 nm, in the red part of the optical spectrum.

Hamal The star Alpha Arietis, magnitude 2.0. It is a K2 giant, 66 l.y. away.

H and K lines Two *Fraunhofer lines in the ultraviolet part of the spectrum, at 396.8 and 393.4 nm respectively, due to singly ionized calcium (Ca II). They are prominent in the spectrum of solar-type and cooler stars. Emission at the H and K lines is frequently found in stars where there is enhanced magnetic activity (e.g. flare stars and RS Canum Venaticorum stars).

Hanle effect The rotation of the plane of polarization of spectral lines due to a magnetic field. The Hanle effect has been used to measure the weak magnetic fields of solar prominences, which are found to be 10^{-3} tesla, or up to 10^{-2} tesla for active prominences. It is named after the German physicist Wilhelm Hanle (1901–93).

hard X-rays The higher-energy part of the X-ray spectrum, ranging from 3 to 100 keV (0.4–0.0124 nm).

Haro galaxy A type of galaxy with abnormally strong emission in the blue and violet region of the spectrum. Galaxies of this type were discovered in 1956 by the Mexican astronomer Guillermo Haro (1913–88). They are often elliptical or lenticular galaxies with strong emission lines. The emission may be from an active galactic nucleus, or from a burst of star formation. Haro galaxies are similar to *Markarian galaxies.

Hartebeesthoek Radio Astronomy Observatory (HartRAO) A radio observatory in Gauteng Province, South Africa, 60 km northwest of Johannesburg, owned and operated by the South African government's National Research Foundation. HartRAO was established in 1975, taking over the facilities of a former NASA ground station which was used for tracking probes to the Moon and planets from 1961 to 1974. The observatory has a 26-m dish, and participates in various networks using *very long baseline interferometry (VLBI).

(⊕) SEE WEB LINKS
• Official observatory website.

Hartmann test A method of finding the focal point or figure of a large lens or mirror. In the basic test, a mask with two holes, equally spaced on either side of the centre, is placed over the *objective. A star image is photographed inside and outside the focus position, resulting in two pairs of images. Measurement of these images then allows the focal point to be calculated. Repeating this for different zones around the lens or mirror provides information on its figure. It is named after the German astronomer Johannes Franz Hartmann (1865–1936), who devised it in 1900. An improved version, known as the Shack–Hartmann test, replaces the perforated mask with an array of tiny lenses (termed *lenslets*). These lenslets focus collimated light from the optical system under test onto a CCD detector, creating an array of spots; any aberrations in the optical system will cause the focused spots to be displaced from their expected reference positions. The Shack–Hartmann test is more precise than the Hartmann test and can be used for much fainter sources. It was devised in 1971 by the American optical engineer Roland Vincent Shack (1927–).

Haruka An alternative name for *HALCA.

Harvard classification A system of classifying stars according to the characteristics of their spectra, introduced at Harvard College Observatory in 1890 by E. C. *Pickering, which culminated in the *Henry Draper Catalogue*. At first, stars were ordered according to the strength of their hydrogen absorption lines (the *Balmer series), from A (strongest) to

P. Eventually several letters were merged or dropped, and the remaining spectral types were rearranged into the sequence O, B, A, F, G, K, M, in order of decreasing surface temperature. Early in the 20th century, spectral types R and N were applied to carbon-rich versions of types G, K, and M (now known as *carbon stars), and type S (M stars with heavy-metal lines) was subsequently added (*see* s star). The Harvard system has since been superseded by the *Morgan–Keenan classification system. *See also* SPECTRAL CLASSIFICATION.

Harvard College Observatory (HCO) The astronomical observatory of Harvard University at Cambridge, Massachusetts, founded in 1839. HCO is a partner in the *Magellan Telescopes project. HCO and the *Smithsonian Astrophysical Observatory jointly run the *Harvard–Smithsonian Center for Astrophysics.

((⊕)) SEE WEB LINKS
• Official observatory website.

Harvard Revised Photometry (HR Photometry) A catalogue of stars brighter than visual magnitude 6.5, published by E. C. *Pickering at Harvard College Observatory in 1908. In addition to 9096 stars, the catalogue contained nine novae or supernovae, four globular clusters, and the Andromeda Galaxy. This catalogue was the forerunner of the *Bright Star Catalogue.*

Harvard–Smithsonian Center for Astrophysics (CfA) A research organization at Cambridge, Massachusetts, founded in 1973, which combines the research activities of *Harvard College Observatory and the neighbouring *Smithsonian Astrophysical Observatory.

((⊕)) SEE WEB LINKS
• Official CFA website.

harvest Moon The full Moon closest to the autumnal equinox (September 23), when the Moon rises around the time of sunset for several nights in succession. This is because the ecliptic, and hence the Moon's orbit, is then at its minimum angle to the horizon at the time of moonrise in mid- or high northern latitudes. The harvest Moon is so named because its light in the evening was of help to workers at harvest time. *See also* RETARDATION.

Hat Creek Radio Observatory The radio observatory of the University of California at Berkeley, sited at an altitude of 1043 m at Cassel, California. The observatory was founded in 1960. It is the site of the *Allen Telescope Array (ATA), consisting of 350 antennae of 6.1-m diameter devoted to the Search for Extraterrestrial Intelligence (SETI). Hat Creek was previously the site of the Berkeley–Illinois–Maryland Association (BIMA) Array, moved in 2005 to become part of the *Combined Array for Research in Millimeter-wave Astronomy (CARMA).

((⊕)) SEE WEB LINKS
• Official observatory website.

Hathor Asteroid 2340, a member of the *Aten group, discovered in 1976 by the American astronomer Charles Thomas Kowal (1940–). It is of T class and its diameter is 300 m. Hathor can approach to within 0.006 AU (900 000 km) of the Earth's orbit. Its orbit has a semimajor axis of 0.844 AU, period 0.78 years, perihelion 0.46 AU, aphelion 1.22 AU, and inclination $5°.9$.

Haute-Provence Observatory (OHP) An observatory at an altitude of 660 m at St Michel, near Forcalquier in the Alpes de Haute Provence, southern France. It was founded in 1936 and is owned and operated by the Centre National de la Recherche Scientifique (CNRS). Its main telescopes are a 1.93-m reflector, opened in 1958; a 1.52-m reflector, opened in 1967; a 1.2-m reflector, opened in 1943; and a 0.8-m reflector, originally opened at Forcalquier by the Paris Observatory in 1932 and moved to OHP in 1945.

((⊕)) SEE WEB LINKS
• Official observatory website.

Hawking, Stephen William (1942–) British theoretical physicist. He has established a powerful reputation, despite being afflicted by a progressively disabling motor neurone disease for nearly all of his adult life. He has drawn on the *general theory of relativity and quantum mechanics to study *singularities in the *Big Bang and *black holes, showing that small black holes emit *Hawking radiation. He turned to popularization with *A Brief History of Time* (1988) which sold more copies than any other book on science. Much of Hawking's work is directed towards a single theory that will unite gravity with quantum mechanics.

Hawking radiation The emission of particles by a black hole. In the intense gravitational field around a black hole, pairs of *virtual particles may live long enough for one member of the pair to be pulled towards the black hole, while the other moves outwards until they are too far apart to destroy each other. This gives the appearance that the black hole is emitting radiation like a black body, with a temperature inversely proportional to its mass. When the Hawking radiation exceeds the amount of matter and energy entering the black hole, it will start to evaporate. For a black hole of the Sun's mass, the temperature is only about 10^{-7} K, which is less than the temperature of the *cosmic microwave background, so no evaporation would occur. However, mini black holes, with a mass of around 10^{12} kg and radius 10^{-15} m, would have a temperature of around 10^{11} K and would radiate strongly. The radiation was predicted to occur by S. W. *Hawking.

Hayabusa A Japanese asteroid sample-return probe, known as MUSES-C before its launch in 2003 May. It rendezvoused with asteroid (25143) Itokawa in 2005 September, surveying its surface from distances of 20 km and closer before landing on the asteroid twice in November. On the second occasion two small samples were taken from the surface. Hayabusa arrived back at Earth in 2010 June, ejecting its sample capsule containing dust particles from the surface of the asteroid.

( ) SEE WEB LINKS
• Official mission website.

Hayashi track The nearly vertical path on the Hertzsprung–Russell diagram that is followed by a fully convective star as it evolves (see diagram). Forming stars (*protostars*) descend such tracks on to the main sequence, becoming less luminous with time but maintaining a roughly constant surface temperature. The process occurs in reverse in old stars leaving the main sequence for the giant branch. It is named after the Japanese astrophysicist Chushiro Hayashi (1920–2010).

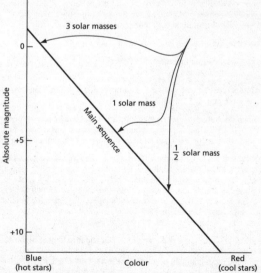

Hayashi and Henyey tracks: The Hayashi track (the near-vertical part of the evolutionary pathways) and Henyey track (the near-horizontal parts) for stars of different mass evolving on to the main sequence.

Haystack Observatory A radio observatory at Westford, Massachusetts, founded in 1960. Its main instrument is a 36.6-m dish, housed within a radome (protective dome), opened in 1964 and originally used for radar observations of planets but now mostly used for radio astronomy. It is operated by the Northeast Radio Observatory Corporation (NEROC), a consortium of nine educational institutions, on behalf of the Massachusetts Institute of Technology, which owns the facility. The existing antenna is being replaced with an improved version capable of working at wavelengths as short as 2 mm, expected to be in operation in 2012.

( ) SEE WEB LINKS
• Official observatory website.

HCG Abbr. for *Hickson compact group.

HD Catalogue Abbr. for *Henry Draper Catalogue.*

head, cometary The main part of a comet, consisting of the solid central *nucleus and the surrounding coma of gas and dust (*see* COMA, COMETARY). While the head may sometimes appear to contain a star-like point (the *pseudo-nucleus*), Earth-based observations do not reveal the true nucleus itself.

head–tail galaxy An elliptical galaxy in which strong radio emission in the nucleus (the head) is accompanied by an irregular tail of diffuse radio emission that may extend for hundreds of thousands of light years. The head and/or tail may be double. The radio emission is *synchrotron radiation from energetic electrons, as in double-lobed radio galaxies, but the motion of the galaxy through a tenuous intergalactic medium has apparently caused the plasma jets to be swept back in the wake of the galaxy.

HEAO Abbr. for *High Energy Astrophysical Observatory.

heat death of the Universe The long-term fate of the Universe, according to the laws of thermodynamics, in which all matter will eventually reach the same temperature. In this condition no energy is available for doing work and the *entropy of the Universe is at a maximum. This outcome was predicted by the German physicist Rudolf Julius Emmanuel Clausius (1822–88), who introduced the concept of entropy.

Heaviside layer Another name for the *E layer.

heavy element Any element heavier than hydrogen and helium. In astronomy, the heavy elements are often termed *metals*.

heavy-metal star A giant with unusual amounts of heavy elements in its spectrum, such as a *barium star or an *S star.

Hebe Asteroid 6, discovered in 1847 by the German amateur astronomer Karl Ludwig Hencke (1793–1866). Hebe is an S-class asteroid of diameter 185 km. Its orbit has a semimajor axis of 2.426 AU, period 3.78 years, perihelion 1.93 AU, aphelion 2.92 AU, and inclination 14°.8.

Hektor Asteroid 624, the largest known *Trojan asteroid, discovered in 1907 by the German astronomer August Kopff (1882–1960). It is a member of the group of Trojans at the L_4 *Lagrangian point 60° ahead of Jupiter. Hektor's brightness varies by a factor of three as it rotates with a period of 6.92 hours. The light curve indicates that Hektor is highly elongated, 150×300 km; alternatively, it might be dumbbell-shaped, two asteroids in contact, or a close binary. Hektor is a D-class asteroid. Its orbit has a semimajor axis of 5.238 AU, period 11.99 years, perihelion 5.12 AU, aphelion 5.36 AU, and inclination 18°.2.

Helene A satellite of Saturn, at a distance of 377 420 km; also known as Saturn XII. Helene's orbital period is 2.737 days. It shares the same orbit as the much larger satellite Dione, lying 60° ahead of it in the leading Lagrangian point. Helene is $43 \times 38 \times 26$ km in size. It was discovered in 1980 from ground-based observations.

heli-, helio- Prefixes referring to the Sun.

heliacal rising and setting The first visible rising of a celestial object in the morning sky after conjunction with the Sun; or the last visible setting of a celestial object in the evening sky before conjunction. The heliacal rising of Sirius was used by the Egyptians as a yearly marker for their calendar.

heliocentric 1. With the Sun at the centre, as in for example the heliocentric system of cosmology (*see* COPERNICAN SYSTEM).
2. As seen from the centre of the Sun, as in for example *heliocentric coordinates.

heliocentric coordinates Any system of coordinates with their origin at the centre of the Sun. They are often used for describing the positions of bodies in the Solar System. They can be either *spherical coordinates or *rectangular coordinates. Heliocentric spherical coordinates are given in terms of *heliocentric latitude and *heliocentric longitude. A variant system, *barycentric coordinates*, gives positions referred to the centre of mass of the Solar System (the *barycentre), which is slightly displaced from the centre of the Sun.

heliocentric latitude (*b*) A coordinate that gives the position of an object north or south of the ecliptic as it would be seen from the centre of the Sun. It is measured in degrees from 0° to at the ecliptic to 90° at the ecliptic poles.

heliocentric longitude (*l*) A coordinate that gives the position of an object around the ecliptic as it would be seen from the centre of the Sun. It is measured in degrees from 0° to 360° clockwise along the ecliptic, starting at the vernal equinox.

heliocentric parallax Another name for *annual parallax.

heliographic coordinates The latitude and longitude of a feature on the Sun's surface. Two systems of heliographic coordinates are in use: *Carrington heliographic coordinates and *Stonyhurst heliographic coordinates. They differ in terms of whether the lines of longitude are fixed relative to the Sun's rotation or fixed relative to the observer. In either case it is necessary to know the position angle of the solar rotation axis (P) and the latitude (B_0) and longitude (L_0) of the centre of the solar disk at the time of observation; these are given in almanacs.

heliometer An instrument formerly used for measuring the diameter of the Sun or the angular separation between close stars. A heliometer is a refracting telescope whose objective lens is split in half, so that the two halves can be moved along the dividing line. This allows two star images or opposite limbs of the Sun to be superimposed. The separation of the lenses and their position angle could be converted into angular measure. Heliometers were regularly used in the 19th century to measure the parallax of stars and angular diameters of planets, but have now been superseded.

heliopause The boundary of the *heliosphere, where the solar wind pressure balances the pressure of interstellar gas. It is thought to lie about 100 AU from the Sun.

helioseismology The study of the Sun's interior by observing the oscillations of its surface. Solar oscillations occur on both a global and a local scale. Local oscillations are apparent through small Doppler shifts in photospheric absorption lines at different places on the Sun's disk. The period of the waves averages 5 min, with maximum velocities of about 0.5 km/s. Typically, a pattern of oscillations appears over a few thousand kilometres and persists for 30 min. The oscillations are due to the superposition of many standing sound waves, or *p modes*, which travel around the Sun between the surface and relatively shallow layers of its interior. Global oscillations are observed by the Doppler shifts in photospheric absorption lines in the Sun's integrated light. Such large-scale oscillations are due to p modes which travel from the Sun's surface to the deepest parts of the interior; the oscillation period averages 5 min, with a range of about 4–8 min.

heliosheath The region of the heliosphere between the *termination shock and the *heliopause, where the solar wind flow is subsonic.

heliosphere The region of space around the Sun within which the solar wind flows. The heliosphere is thought to be about 100 AU in radius, and is bounded by the heliopause, beyond which interstellar gas exerts an equal pressure from outside. The shape of the heliosphere is unknown, but if there is a flow of interstellar material around it from a particular direction (an *interstellar wind*), the heliosphere may be like the Earth's magnetosphere: spherical on one side, but drawn out into a long tail on the other.

Helios probes Two German space probes, launched by NASA, to study the Sun and interplanetary space. Helios 1, launched in 1974 December, was put into an orbit that took it 45 million km from the Sun at perihelion, closer than any previous probe. Helios 2, launched in 1976 January, was about 43 million km from the Sun at closest approach.

heliostat A flat mirror on an *equatorial mounting, driven to follow the Sun across the sky and reflect its light into a stationary telescope. If the telescope is oriented parallel with the Earth's axis, only one mirror is needed. However, the Sun's orientation as seen through the telescope will change during the day.

helium burning The production of energy in stars by the fusion of helium to form carbon. It occurs by the *triple-alpha process.

helium flash The explosive event that occurs when helium burning begins in the core of a low-mass star. The core must be dense enough for *degeneracy effects to be important. A degenerate gas does not expand on heating, and is unable to cool through expansion in the way that a non-degenerate gas would. Thus, as helium begins to burn, the temperature rises rapidly, eventually becoming so high that the gas is no longer degenerate. At this point the core expands very rapidly, releasing its stored energy in less than 2 min.

helium shell flash A cycle of unstable nuclear burning in *asymptotic giant branch stars. It occurs when helium is being burnt in a thin shell around the star's core, and results from the strong dependence of the rate of energy generation on temperature. Any overheating causes considerable expansion followed by collapse, thus setting up large-scale pulsations.

helium star The core of a formerly massive star (originally more than 12 solar masses) which has evolved and lost its hydrogen-rich envelope. Hydrogen loss can happen either through a strong stellar wind, as in *Wolf–Rayet stars, or by mass transfer on to a companion, as in a close binary. A helium star is expected to evolve exactly as the core of a massive star would: by producing an iron core which collapses to cause a Type Ib or Ic *supernova explosion, depending on the mass of the star. 'Helium star' is also an obsolete term for a normal B-type star.

Helix Nebula A planetary nebula in Aquarius, also known as NGC 7293 or sometimes the Sunflower Nebula. It is the nearest planetary nebula, 700 l.y. away, and the largest in apparent size, nearly ¼° across; the true diameter is about 3 l.y. It is ionized by a very hot 13th-magnitude central white dwarf of spectral type DA and temperature 120 000 K. The nebula's name comes from the fact that on early photographs it appeared somewhat helical in shape, like two overlapping loops of a spiral. 📷

Hellas Planitia A circular lowland basin on Mars about 2200 km wide and 8 km deep, the site of an ancient impact, centred at −43° lat., 290°W long. Through Earth-based telescopes it frequently appears bright due to mists and cloud. The great dust storms that sometimes obscure the surface of Mars after perihelion often start from Hellas.

helmet streamer *see* STREAMER.

Helmholtz–Kelvin contraction The contraction of a star under gravity, the potential energy thus lost being converted into heat and radiated away. The **Helmholtz–Kelvin contraction time-scale** is defined as the time it would take for a star to collapse if nuclear burning were switched off and it continued to radiate at the same luminosity. For the Sun this time would be about 30 million years. It is named after the German scientist Hermann Ludwig Ferdinand von Helmholtz (1821–94) and Lord *Kelvin.

hemispherical albedo The fraction of light falling upon a non-luminous body, such as a planet, that is scattered by its surface, as a function of the angle of incidence. The body is assumed to be a sphere with a diffuse surface that reflects incoming parallel light in all directions.

Henry Draper Catalogue (HD) A catalogue listing the spectral types of 225 300 stars down to 8th magnitude, published in 1918–24. It was compiled at Harvard College Observatory by A. J. *Cannon, who assigned spectral types on the basis of the *Harvard classification system. The *Henry Draper Extension* (HDE), containing 47 000 spectra of fainter stars, was published in 1925–36, and a second supplement containing another 86 000 stars in 1949. The catalogue was named in honour of H. *Draper. Although the Harvard classification system has been superseded, stars in the *Henry Draper Catalogue* and *Extension* are still widely known by their HD or HDE numbers.

(⊕) SEE WEB LINKS
• Detailed description and full catalogue downloadable from the CDS.

Henyey track The roughly horizontal path on the Hertzsprung–Russell diagram followed by a pre-main-sequence star with a mass greater than about 0.4 solar masses. The track takes it on to the main sequence after it has descended the *Hayashi track. Stars on Henyey tracks are becoming hotter, and therefore bluer, as they contract. The contraction ends once the core is hot enough for hydrogen burning to begin and the star joins the main sequence. Whereas stars on Hayashi tracks are fully convective, those on Henyey tracks are radiative. It is named after the American astrophysicist Louis George Henyey (1910–70).

Hephaistos Asteroid 2212, a member of the *Apollo group, discovered in 1978 by the Russian astronomer Lyudmila Ivanovna Chernykh (1935–). Hephaistos is 5.7 km in diameter. Its orbit has a semimajor axis of 2.167 AU, period 3.19 years, perihelion 0.36 AU, aphelion 3.97 AU, and inclination 11°.7. These orbital elements are similar to those of Comet *Encke, and it has been speculated that Hephaistos and Comet Encke, as well as some other asteroids in similar orbits, may be remnants of a giant comet which split up within the past 20 000 years.

Heraclides of Pontus (388–315 BC) Greek philosopher and astronomer, born in modern Turkey. He made the earliest known suggestion that the apparent rotation of the sky is caused by the axial rotation of the Earth. Often attributed to him is the idea that Venus (and, by implication, Mercury) revolves around the Sun, although there is no firm evidence for this idea appearing in ancient Greece until over a century after his death.

Herbig Ae/Be star (HAEBE) A young star with mass between 1.5 and 10 times that of the Sun and intrinsically very luminous. HAEBE stars are the intermediate-mass counterparts of the more common *T Tauri stars. Being more massive than T Tauri stars, they are of earlier spectral type, B, A, or F. Like the classical T Tauri stars their spectra contain emission lines, especially the hydrogen *Balmer series. The existence of *P Cygni line profiles indicates mass outflow from the system. Excess infrared and millimetre emission shows that HAEBE stars are associated with abundant circumstellar dust. Significant variations in brightness indicate the existence of clumps in the dust, possibly protoplanets or planetesimals. More than 80% of HAEBE stars are thought to be in binary systems. They are named after the American astronomer George Howard Herbig (1920–) who first described them in 1960.

Herbig–Haro object (HH object) A small nebula with an emission-line spectrum, found in regions of star formation. HH objects have high velocities of several hundred kilometres per second relative to their surroundings. They are believed to be *bow shocks formed when fast-flowing jets of material from a young star encounter interstellar matter. The emission lines result from the recombination of ions and electrons in the cooling gas behind the bow shock. They are named after the American George Howard Herbig (1920–) and the Mexican Guillermo Haro (1913–88).

Hercules (Her) (*gen.* **Herculis**) The fifth-largest constellation, lying in the northern sky, representing the strong man of Greek mythology. Its brightest star is Beta Herculis, magnitude 2.8. Zeta Herculis is a close binary, magnitudes 2.9 and 5.5, with an orbital period of 34.5 years.

Alpha Herculis (*Rasalgethi) is both a variable star and a double. M13 is a 6th-magnitude globular cluster 25 300 l.y. away, rated as the finest globular in northern skies. M92 is a globular cluster of magnitude 6.5.

Hercules Cluster A loose, irregular cluster of galaxies some 500 million l.y. away in the constellation Hercules; also known as Abell 2151. The cluster is 1°.7 wide as seen from Earth, has a true diameter of 6 million l.y., is flattened in shape, and is notable for the high proportion of spirals among its bright members. There are about 75 bright galaxies, and many fainter members.

Hercules X-1 A low-mass X-ray pulsar about 15 000 l.y. away in the constellation Hercules. It is an eclipsing binary consisting of a visible star of about 13th magnitude accompanied by a neutron star. Material from the visible star is accreted by the neutron star, releasing its gravitational energy as X-rays. The optical counterpart, HZ Herculis, varies in spectral type from A to B during each orbit due to X-ray heating. The rapid spin of the neutron star produces X-ray pulses every 1.24 s, and the stars eclipse each other every 1.7 days.

Herculina Asteroid 532, discovered in 1904 by Max *Wolf. Observations of a stellar occultation by Herculina in 1978 indicate the presence of a companion about 50 km wide and 1000 km away. Herculina is elliptical, with a mean diameter of 222 km, and is of S class. Its orbit has a semimajor axis of 2.772 AU, period 4.61 years, perihelion 2.28 AU, aphelion 3.26 AU, and inclination 16°.3.

Hermes Asteroid 69230, a member of the *Apollo group, discovered by the German astronomer Karl Reinmuth (1892–1979) in 1937, when it approached within 0.006 AU (900 000 km) of the Earth. It was then lost until 2003. It is a binary consisting of two similar-sized objects each about 0.4 km across. Its orbit is perturbed by close approaches to Venus and the Earth but currently has a semimajor axis of 1.655 AU, period 2.13 years, perihelion 0.62 AU, aphelion 2.69 AU, and inclination 6°.1.

Herschel, Caroline Lucretia (1750–1848) German astronomer (in German, Karoline), sister of F. W. *Herschel, and the first woman astronomer to be widely recognized for her achievements. In 1772 she moved to England to become her brother's housekeeper and assistant, helping to grind mirrors, recording his observations, and preparing his catalogues. In 1782 she began her own observations, and went on to discover eight comets and several nebulae. After her brother's death in 1822, Caroline returned to Germany.

Herschel, John Frederick William (1792–1871) English scientist and astronomer, only son of F. W. *Herschel. He continued his father's observations of double stars and nebulae, and in 1834 began a survey of the southern sky from the Cape of Good Hope (in modern South Africa). The survey yielded over 2100 doubles plus 1700 nebulae and clusters which he combined with his father's discoveries in the *General Catalogue of Nebulae and Clusters of Stars* (which formed the basis for the *New General Catalogue*). His other contributions to astronomy included one of the first measurements of the *solar constant, and an early photometer. He was also a pioneer of photography. His second son, Alexander Stewart Herschel (1836–1907), was also an astronomer, best known for his work on meteor showers.

Herschel, (Frederick) William (1738–1822) German-born English astronomer and musician, originally Friedrich Wilhelm. He came to England in 1757, and became an organist. In 1773 William and his sister C. L. *Herschel, who had moved to England the previous year, took up astronomy, observing with the first of many telescopes that he made himself. His discovery of Uranus on 1781 March 13 made him famous, and he was appointed private astronomer to King George III. With the king's patronage he was able to build a telescope with an aperture of 1.2 m (48 inches), then the largest in the world. In 1787, Herschel discovered the two largest satellites of Uranus, and in 1789 the Saturnian satellites Mimas and Enceladus. During systematic surveys of the heavens he observed and catalogued many double stars and over 2000 nebulae and clusters; this work was continued by his son, J. F. W. *Herschel. His observations of double stars revealed that many are in orbital motion around each other. From counts of the stars in different parts of the sky, he reasoned that the Sun is part of a flattened system of stars which from our vantage point in its main plane we see as the Milky Way. From the proper motions of seven bright stars, he deduced that the Sun is moving towards a point in Hercules. In 1800, using a thermometer and prisms, he discovered infrared radiation.

Herschelian telescope A reflecting telescope in which the primary mirror is tilted so that it focuses light to the top edge of the tube. The eyepiece is located at the top edge, looking down at the mirror. This design avoids light loss caused by a second reflection, which was particularly important when mirrors were made of *speculum metal. But it results in *astigmatism unless the *focal ratio is large. The design was invented by F. W. *Herschel, who used it for his giant 48-inch (1.2-m), but it is now of historical interest only.

Herschel–Rigollet, Comet 35P/ A periodic comet discovered in 1788 by C. L. *Herschel. A comet discovered in 1939 by the French astronomer Roger Rigollet (1909–81) turned out to be a return of the same object. It was at that time the periodic comet with the longest period, currently 155 years, to have been seen more than once, although it has now been superseded by Comet *Ikeya–Zhang. Its perihelion is 0.748 AU, eccentricity 0.974, and inclination $64°.2$.

(⊕) SEE WEB LINKS
• Information page at Cometography website.

Herschel Space Observatory (HSO) An ESA spacecraft launched 2009 May carrying a 3.5-m telescope for imaging and spectroscopy at far-infrared and submillimetre wavelengths, from 55 to 672 μm, a part of the spectrum not previously covered. It is stationed at the L_2 *Lagrangian point, 1.5 million km from Earth in the direction away from the Sun. It carries three instruments: HIFI (Heterodyne Instrument for the Far Infrared), a very-high-resolution heterodyne spectrometer; PACS (Photodetector Array Camera and Spectrometer), an imaging photometer and medium-resolution grating spectrometer; and SPIRE (Spectral and Photometric Imaging Receiver), an imaging photometer and an imaging Fourier transform spectrometer. HSO is named after W. *Herschel, the discoverer of infrared light, and his sister and co-worker Caroline. HSO was launched on the same rocket as *Planck, another ESA scientific mission, but the two separated once in space and operate independently.

(⊕) SEE WEB LINKS
• ESA mission website.

Hertha family *See* NYSA FAMILY.

hertz (Hz) The unit of frequency. One hertz is the frequency of a cyclical event having a period of one second. It is named after the German physicist Heinrich Rudolf Hertz (1857–94).

Hertzsprung, Ejnar (1873–1967) Danish astronomer. In 1905 he showed how a star's luminosity was related to the width of lines in its spectrum, thus establishing *spectroscopic parallax as a means of distance-finding. He inferred the distinction between red giant and dwarf stars from the fact that, although they were of the same spectral type, their spectra had different linewidths. Hertzsprung also proposed the modern definition of absolute magnitude. He went on to plot the first *Hertzsprung–Russell diagram, for the stars of the Pleiades, in 1906. His work remained largely unknown, and in 1910 H. N. *Russell independently developed the diagram in a slightly different form. In 1911 Hertzsprung discovered that Polaris is a Cepheid variable, and in 1913 estimated the distance to the Small Magellanic Cloud from the brightness of Cepheids within it.

Hertzsprung gap An area of the Hertzsprung–Russell diagram containing no stars, between the top end of the main sequence (where stars are more massive than the Sun) and the *giant branch. Stars cross this region very quickly once they leave the main sequence because their outer layers are expanding rapidly. While the star's luminosity remains roughly constant, the surface temperature goes down so that the star follows an approximately horizontal path from left to right on the HR diagram.

Hertzsprung–Russell diagram (HR diagram) A graph on which a measure of the brightness of stars (usually their absolute magnitude) is plotted against a measure of their temperature (either spectral type or colour index). The diagram shows how the luminosities and surface temperatures of stars are linked. From a star's position on the diagram, astronomers can estimate its mass and the stage of its evolution.

Most stars lie on the *main sequence, a strip which runs from the upper left to the lower right of the diagram. A star on the main sequence is burning hydrogen in its core, and during this phase of its life will remain at a point on the diagram that is determined by its mass. Other areas of the HR diagram are populated by stars that are not burning hydrogen in their cores, but may be burning hydrogen in a thin shell. The most prominent of these areas is the *giant branch, consisting of stars which have exhausted the hydrogen fuel in their cores. Other features are the strips occupied by *supergiants, with luminosities 300 to 100 000 times that of the Sun, and *white dwarfs, dying stars with luminosities typically 10 000 times less than the Sun's. Theories of stellar evolution must explain the various features of the HR diagram. It is named after H. N. *Russell and E. *Hertzsprung, who independently devised it. *See also* COLOUR–LUMINOSITY RELATION; COLOUR–MAGNITUDE RELATION.

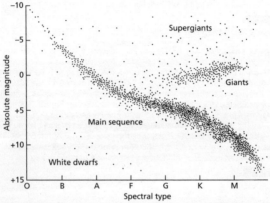

Hertzsprung-Russell diagram: Hertzsprung–Russell diagram for stars in the Sun's vicinity.

Hess, Victor Francis (1883–1964) Austrian-American physicist, originally Viktor Franz. In 1912 he began to investigate the recently discovered ionization of the atmosphere. In a series of balloon ascents, he used electroscopes to show that radiation was always present above 150 m, and increased steadily with altitude. Its intensity did not depend on the time of day or night, and so could not come from the Sun—it had to be of cosmic origin. Hess's discovery of these *cosmic rays was not recognized until after World War I; for it he shared the 1936 Nobel Prize in Physics.

HESSI *See* RAMATY HIGH-ENERGY SOLAR SPECTROSCOPIC IMAGER.

HETE-2 Abbr. for *High Energy Transient Explorer-2.

Hevelius, Johannes (1611–87) German astronomer, born in modern Poland (in German Johann Hewelcke, in Polish Jan Heweliusz). In the early 1640s he derived a reasonably accurate value for the Sun's rotation period, and described and named *faculae. His *Selenographia* (1647) contains the first fairly detailed map of the Moon with named features, although few of his names survive. He also discovered the Moon's libration in longitude. In the 1650s he began an observational programme that resulted in an atlas, *Firmamentum Sobiescianum sive Uranographia*, in which he introduced some new constellations, and an associated catalogue of 1564 stars, both published posthumously in 1690 by his second wife and assistant Elizabeth Hevelius, née Korpman (*c*.1646/7–*c*.1693). Hevelius was the last major astronomer to make positional measurements with naked-eye sighting instruments.

Hewish, Antony (1924–) English radio astronomer. In the 1950s at the *Mullard Radio Astronomy Observatory, Cambridge, he collaborated with M. *Ryle in surveys which led to the series of Cambridge catalogues of radio sources. In 1960 he and Ryle developed *aperture synthesis. Hewish began to study fluctuating radio sources, and in 1967 his student S. J. *Bell identified what proved to be the first signal from a pulsar. Hewish was awarded the 1974 Nobel Prize in Physics, shared with Ryle, for his work on pulsars.

hexahedrite A class of iron meteorite, containing not more than 4–6% nickel. Hexahedrites consist of large cubic crystals of kamacite (an iron–nickel alloy) over 50 mm across, and they can be cleaved in three perpendicular directions along the faces of a hexahedron, hence their name. They display a pattern of fine lines called *Neumann lines. The **hexa-octahedrite** meteorites are transitional between hexahedrites and *octahedrites. In these meteorites the thickness of the kamacite plates (the *band width*) varies from 3 to 50 mm, most of the kamacite grains are rounded, and little or no taenite remains.

HH object Abbr. for *Herbig–Haro object.

Hickson compact group (HCG) Any of 100 small, relatively isolated clusters of galaxies listed in a catalogue published in 1982 by the British-born Canadian astronomer Paul Hickson (1950–). A typical Hickson compact group contains four to five closely spaced members, up to half of which may be interacting or even merging. Well-known examples of Hickson compact groups are *Stephan's Quintet and *Seyfert's Sextet.

Hidalgo Asteroid 944, discovered in 1920 by W. *Baade. It is of D class, and its diameter is 38 km. Hidalgo has a highly elliptical 13.8-year orbit that carries it from perihelion at 1.95 AU, at the inner edge of the main asteroid belt, out to aphelion at 9.53 AU, the distance of Saturn; the semimajor axis is 5.742 AU. Until the discovery of *Chiron in 1977, it had the greatest aphelion distance and longest period of any known asteroid. Its orbit, which has the unusually high inclination of $42°.6$, is similar to those of periodic comets, and it could be an extinct cometary nucleus.

hierarchical cosmology A cosmological theory in which the Universe has structure on all scales, from individual galaxies, to clusters of galaxies, to superclusters, and so on without end. Each level of the hierarchy is supposed to look like an enlargement of the smaller scales. A hierarchical universe is not homogeneous on large scales and hence is not consistent with the *cosmological principle. Observations suggest that our Universe is approximately hierarchical on scales up to perhaps 100 million l.y., but thereafter becomes homogeneous.

High Energy Astrophysical Observatory (HEAO) A series of three NASA X-ray and gamma-ray astronomy satellites. HEAO-1, launched in 1977 August, surveyed the sky in the X-ray energy range 0.2–60 keV (0.02–6.2 nm) for discrete sources and studied the X-ray background. The satellite also studied low-energy gamma rays up to about 10 MeV. It ceased operation in 1979 January. HEAO-2, also known as the *Einstein Observatory, was launched in 1978 November and observed in the 0.1–4 keV (0.3–12.4 nm) range. It ceased operation in 1981 April. HEAO-3, launched in 1979 September, was devoted to gamma-ray astronomy. It ceased operation in 1981 May.

(⊕) SEE WEB LINKS
- Information page on HEAO 1 at Goddard Space Flight Center.
- Information page on HEAO 2 at Goddard Space Flight Center.
- Information page on HEAO 3 at Goddard Space Flight Center.

high-energy astrophysics The study of X-rays, gamma rays, and cosmic rays coming from the Universe. The energy of the radiation is usually expressed in terms of electronvolts (eV). X-rays and gamma rays range in energy from 100 eV to over 10^{10} eV. There is no sharp cutoff between the X-ray and gamma-ray regions of the spectrum. Cosmic rays are atomic and subatomic particles which move through space at velocities approaching that of light. Their energies range from about 10^8 to 10^{20} eV. *See also* COSMIC RAYS; GAMMA-RAY ASTRONOMY; X-RAY ASTRONOMY.

High-Energy Transient Explorer-2 (HETE-2) A joint US–French–Japanese satellite to detect gamma-ray bursts, launched in 2000 October. The satellite carries three instruments: a set of wide-field gamma-ray spectrometers (FREGATE); a wide-field X-ray monitor (WXM); and a set of soft X-ray cameras (SXC). These search for gamma-ray bursts, establish their location and alert observers on Earth. HETE-1 was a launch failure.

(((∰))) SEE WEB LINKS

• Mission page at Goddard Space Flight Center.

highlands, lunar Rough and heavily cratered bright terrain on the Moon. The highlands are generally 1–2 km higher than the dark lowland *mare areas. The highlands are older, dating from the last stages of planetary accretion before about 4 billion years ago. The old, impact-shocked rocks of the highlands are chemically different from the maria, being enriched in calcium and aluminium. Mineralogically they consist mostly of feldspar.

high-speed photometry A form of photometry in which the photons are counted at short time intervals, generally between 10 ms and 10s, to detect rapid changes in brightness of objects such as pulsars and variable stars. Exposures for normal photometry typically range anywhere between 10 and 100 seconds. *Photomultipliers or *area photometers are used as detectors. The same high-speed technique has also been applied to *polarimetry.

high-velocity cloud (HVC) A cloud of neutral hydrogen travelling at a velocity far greater than can be explained by the normal rotation of the Galaxy. Some HVCs are debris from a tidal interaction between our Galaxy and the neighbouring Magellanic Clouds. The most spectacular example is the *Magellanic Stream, which stretches half-way across the sky, but is visible only to radio telescopes. The distances and origin of other HVCs are still unknown.

high-velocity star A star moving faster than 65 km/s relative to the average motion of the stars in the Sun's neighbourhood (the *local standard of rest). High-velocity stars are members of the galactic halo, moving in highly elliptical orbits around the galactic centre. Their high relative velocities result from the fact that they are passing through the galactic disk and do not share the rotation of the Sun and its neighbouring stars around the galactic centre. Such stars may have formed early in the Galaxy's history, or they may be remnants of smaller galaxies that have merged with ours.

Hilda group A group of asteroids, often called the Hildas, marking the outer boundary of the main asteroid belt, concentrated close to the 3 : 2 resonance with Jupiter, at a mean distance of 4.0 AU from the Sun. In common with *Trojan asteroids, Hildas may have more elongated shapes than main-belt asteroids of the same sizes. The group is named after (153) Hilda, a P-class asteroid of diameter 171 km, discovered in 1875 by the Austrian astronomer Johann Palisa (1848–1925) in 1875. Hilda's orbit has a semimajor axis of 3.972 AU, period 7.92 years, perihelion 3.42 AU, aphelion 4.52 AU, and inclination 7°.8.

Hill, George William (1838–1914) American mathematician and astronomer. Inspired by the work of L. *Euler, he developed advanced mathematical methods for tackling problems in dynamical astronomy. His first major work was an application of the *three-body problem: the orbits of Jupiter and Saturn and the perturbations exerted by these planets on the Moon. In his study of the Moon's motion he introduced a differential equation now known as **Hill's equation**.

Himalia The eleventh-closest satellite of Jupiter, distance 11 443 000 km, orbital period 250.1 days; also known as Jupiter VI. Himalia is 170 km in diameter, and was discovered in 1904 by the American astronomer Charles Dillon Perrine (1867–1951). It is the largest member of the so-called Himalia prograde irregular group of Jovian satellites, the other members being Leda, Lysithea, and Elara.

Hind's Crimson Star The red-giant variable star R Leporis, named after the English astronomer John Russell Hind (1823–95), who remarked on its blood-red colour. It is a variable *Mira star, ranging from 6th to 12th magnitude in a period of about 14 months.

Hind's Variable Nebula A reflection nebula in Taurus, also known as NGC 1555, associated with the young, irregular variable star T Tauri. Both were discovered in 1852 by the English astronomer John Russell Hind (1823–95). Major variations in the brightness, extent, and form of the nebula occur over decades. More rapid changes may be caused by variations in T Tauri itself and shadowing by dense clouds of material near the star.

Hinode A joint Japanese–ESA–US satellite for solar observation, known as Solar-B before its launch into polar orbit around the Earth in 2006 September. Its prime purpose is to study the generation and development of solar magnetic fields and the dissipation of their energy in flares, coronal mass ejections, and coronal heating. It carries an 0.5-m Solar Optical Telescope (SOT) observing in the 388–668 nm range, an Extreme-ultraviolet Imaging Spectrometer (EIS) observing in the wavelength ranges 17–21 nm and 25–29 nm, and a high-resolution grazing-incidence X-Ray Telescope (XRT) observing between 0.2 and 20 nm. Hinode was a successor to the Japanese *Yohkoh solar observatory.

(⊕) SEE WEB LINKS
• Hinode science centre website.
• Mission website at Japan Aerospace Exploration Agency.

Hinotori A Japanese satellite for solar X-ray studies, launched in 1981 February; before launch it was known as Astro-A. It carried high-resolution soft X-ray spectrometers and hard X-ray instruments which obtained the first images of solar flares at energies above 25 keV (wavelengths shorter than 0.05 nm). Observations ended in 1982 October.

Hipparchus of Nicaea (*c.*190–*c.*120 BC) Greek astronomer, geographer, and mathematician, born in modern Turkey. He put Greek astronomy on a more scientific footing, introducing arithmetic and early trigonometric methods. His many accurate astronomical observations resulted in a catalogue of 850 stars, in which he gave their coordinates and divided them into six magnitude classes. Ptolemy incorporated the catalogue and other findings by Hipparchus in the *Almagest*. Hipparchus made surprisingly accurate measurements of the *precession of the equinoxes, the length of the year, and (from observations of eclipses) the Moon's distance. He may have been the inventor of the astrolabe.

Hipparcos An astrometry satellite launched by the European Space Agency in 1989 August. The name is an acronym for High Precision Parallax Collecting Satellite. It measured the position, brightness, proper motion, and trigonometric parallax of 118 218 stars down to 12th magnitude to an accuracy of better than $0''.002$. It did so by measuring angular separations between pairs of stars with a 0.29-m telescope as it scanned the sky continuously about a slowly varying rotation axis. Simultaneously, the satellite's star mappers measured the magnitude and colour of over a million stars down to about 10th magnitude, plus their positions to about one-tenth the accuracy of the main survey; these observations were used to construct the *Tycho Catalogue. Observations ended in 1993. The *Hipparcos Catalogue* was published in 1997.

(⊕) SEE WEB LINKS
• ESA mission website.

Hipparcos Catalogue A catalogue of 118 218 stars down to 12th magnitude observed by the *Hipparcos satellite between 1989 August and 1993 March. The catalogue gives positions, parallaxes, proper motions, magnitudes, and colour indexes of these stars for the epoch 1991.25, with supplementary volumes giving additional information on double and variable stars. In 2007 a new version of the catalogue was published, called **Hipparcos, the New Reduction**, which corrected for slight irregularities in the satellite's motion resulting in positional accuracies up to four times better than the original catalogue.

(⊕) SEE WEB LINKS
• Detailed description and full catalogue downloadable from the CDS.
• New reduction of the Hipparcos catalogue, downloadable from the CDS.

Hirayama family A group of asteroids whose members have very similar orbital characteristics (semimajor axis, eccentricity, and inclination), and are believed to share a common origin. Examples are the *Eos family, the *Koronis family, the *Themis family, the *Maria family, and members of the *Flora group. Hirayama families are thought to result from the break-up of larger asteroids several hundred kilometres in size following catastrophic collisions. They are named after the Japanese astronomer Kiyotsugu Hirayama (1874–1943), who demonstrated their existence in 1918.

Hiten A Japanese Moon probe, known as Muses-A before its launch in 1990 January. It went into a highly elliptical orbit around the Earth from which it released a smaller craft, Hagoromo, into lunar orbit in 1990 March, although no results were received because of a transmitter failure. Hiten itself was put into orbit around the Moon in 1993 February. It was the first Moon probe not launched by the USA or the former Soviet Union. Hiten was intentionally crashed into the Moon in 1993 April.

H line *See* H AND K LINES.

H magnitude The magnitude in the infrared H band of the *Johnson photometry system. The H-band filter has an effective wavelength of 1630 nm and a bandwidth of 280 nm. The importance of H (and the source of its name) is that it coincides with the minimum in the *opacity of the H⁻ ion, which is an important source of opacity in the atmospheres of cool stars.

Hoba West meteorite The world's largest known meteorite, which lies where it was found in 1920 at Hoba Farm, near Grootfontein, Namibia. It is an iron-nickel *ataxite with an estimated mass of 60 tonnes and a maximum width of 2.8 m. It produced no crater, but remains partly embedded in the ground.

Hobby–Eberly Telescope (HET) A segmented-mirror reflector at an altitude of 2025 m on Mount Fowlkes, Texas, opened in 1997. It is part of *McDonald Observatory, and is jointly owned by the University of Texas, Pennsylvania State University, Stanford University, the Ludwig Maximilian University in Munich, and the Georg-August University in Göttingen. Its mirror consists of 91 hexagonal segments each 1 m wide, forming an 11.1 × 9.8 m hexagon; however, its usable aperture is equivalent to a single mirror of 9.2 m. The telescope is designed primarily for spectroscopy rather than imaging. It is permanently angled at 55° altitude, but rotates freely in azimuth. The telescope remains stationary during observations; a movable secondary tracks objects as the Earth rotates. It is named after William P. Hobby and Robert E. Eberly, supporters of public education in Texas and Pennsylvania respectively. 📷

(⊕) SEE WEB LINKS
• Official telescope website.

Hohmann orbit A trajectory along which a spacecraft moves from one orbit to another with the minimum expenditure of energy, first calculated by the German engineer Walter Hohmann (1880–1945) in 1925. Such trajectories are used for changing the orbits of a satellite around the Earth, or for sending a probe from the Earth to another planet. A Hohmann orbit is elliptical, and just touches the orbits of origin and destination. It requires two firings of the spacecraft's rocket motor, one to break out of the original orbit and one to enter the destination orbit. The main disadvantage of the Hohmann orbit is the long flight times involved. From Earth to Mars the transfer time is 260 days; from Earth to Saturn would take about 6 years. For this reason *gravity assists are often used to boost the spacecraft's speed and hence reduce journey times.

Holmberg radius A measure of the size of a galaxy, based on its observed surface brightness. It is the radius at which the surface brightness is 26.5 magnitudes per square arc second in blue light, and represents about 1–2% of the brightness of the night sky. It is named after the Swedish astronomer Erik Bertil Holmberg (1908–2000).

homogeneity The property of being uniform throughout space. The assumption of homogeneity is part of the *cosmological principle. The Universe is clearly not homogeneous on the scale of planets, stars, or galaxies. However, astronomers assume that on a large enough scale the Universe is homogeneous: that is, a large enough box placed anywhere in the Universe would contain roughly the same numbers of galaxies and clusters of galaxies as any other.

homology A principle used in the design of parabolic radio telescopes to compensate for their unavoidable tendency to distort under their own weight. A **homologous** telescope is designed to sag in such a way that the reflecting surface is always in the shape of a paraboloid. Only the position of the focus varies, and the *feed position can be adjusted to compensate.

Homunculus Nebula A cloud of gas and dust that obscures the star *Eta Carinae, with a shape thought by early observers to resemble a small human. Its dimensions are about $17'' \times 12''$. The material was ejected by the star in 1843 and is now expanding at about 500 km/s. The nebula hides the star from view, and reradiates much of the absorbed energy at infrared wavelengths.

H I region A region of interstellar hydrogen that is in the form of neutral, unionized atoms. The notation H I refers to the fact that the hydrogen atoms (H) are not ionized (H II is ionized hydrogen). Each neutral hydrogen atom contributes just one particle to the gas. The density of H I regions is too low for hydrogen molecules to form, and starlight would dissociate any molecules that did, so the gas remains as atoms. Neutral hydrogen constitutes about half of all interstellar matter by mass and by volume, with an average density of 1 atom/cm^3. H I regions are cool (about 100 K) and emit no visible light, but they do give out the important radio spectral line at 21 cm wavelength.

Hooker Telescope The 100-inch (2.5-m) reflector at *Mount Wilson Observatory, opened in 1917. It is named after the American businessman John D. Hooker (1837–1910), who provided funds to buy the mirror. It was out of operation 1985–93, but was reopened when Mount Wilson Institute took over the running of the observatory.

horizon The great circle formed by the intersection of the plane perpendicular to the observer's zenith with the celestial sphere; also called the *astronomical horizon.*

horizontal branch A horizontal strip of stars in the Hertzsprung–Russell diagram of globular clusters, comprising stars somewhat bluer and fainter than those on the *giant branch. These stars are thought to be burning helium in their core and hydrogen in a surrounding shell. They are of 0.6–0.8 solar masses, of which more than half is in the helium-burning core, and have probably lost substantial mass through stellar winds during their giant phase. Stars typically spend 50–100 million years on the horizontal branch.

horizontal coordinates A system of coordinates that specifies the angular position of a celestial object relative to the observer's horizon at a given time. The coordinates used are *altitude, the object's angular distance above the horizon, and *azimuth, the bearing of the object clockwise from north measured parallel to the horizon. Sometimes the *zenith distance is used instead of altitude. The coordinates change with the observer's position and as the Earth rotates.

horizontal parallax The difference between the topocentric and geocentric positions of an object when it is on the astronomical horizon. The amount varies with latitude because the Earth is not exactly spherical, and is greatest at the Earth's equator (*equatorial horizontal parallax).

horizontal refraction The angular distance of an object below the horizon when it appears to lie on the horizon. The value 34' is used in calculating rising and setting times given in *The Astronomical Almanac.*

horn antenna An antenna in the shape of a flared horn. It is often fed by a *waveguide. A small horn is often used as the feed at the focus of a parabolic antenna. The *cosmic microwave background was first detected with a large horn antenna. Horn antennas have very low *side lobes and are thus less affected by radio interference than parabolic antennas.

Horologium (Hor) (*gen.* **Horologii**) An insignificant constellation of the southern sky, representing a pendulum clock. It contains little of interest. Its brightest star, Alpha Horologii, is of magnitude 3.9.

Horrocks or **Horrox, Jeremiah** (*c.* 1618–1641) English astronomer. He refined J. *Kepler's *Rudolphine Tables* of planetary positions, measured the apparent diameters of the planets, and obtained a value of the *solar parallax corresponding to an Earth–Sun distance of about 100 million km. In 1639 he observed a transit of Venus which he had predicted from his refinement of Kepler's tables, measuring the planet's diameter on a projected image. On the basis of Horrocks's lunar theory, J. *Flamsteed later constructed tables of the Moon's motion which remained the best available until the mid-18th century.

Horsehead Nebula A dark nebula in Orion shaped like the head of a horse, also known as Barnard 33. It is about 6′ long and projects into the bright nebula IC 434 south of the star Zeta Orionis, but is well seen only on long-exposure photographs. 📷

horseshoe mounting An *equatorial mounting in which the top end of a *fork mounting is supported by a large circular bearing resembling a horseshoe. The design has great stability and has been used for large telescopes, such as the 200-inch (5-m) Hale Telescope at Palomar Observatory.

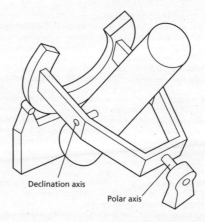

Declination axis

Polar axis

horseshoe mounting

Horseshoe Nebula An alternative name for the *Omega Nebula in Sagittarius.

horseshoe orbit The gravitationally perturbed path traced out by a body in a similar orbit to another body of comparable or greater mass. Whichever of the two bodies is moving on the slightly smaller and faster orbit will eventually catch up with the other. If the two bodies are of comparable mass, their mutual gravitational pull will raise the first object into a larger and slower orbit so that it falls behind again, while the second body will be pulled into a smaller and faster orbit so that it moves ahead. In effect, the two have swapped orbits. Next time it will be the turn of what was the second body to catch up the first, whereupon the two will exchange places again. Neither body actually passes the other during these manoeuvres; rather, they describe a horseshoe-shaped orbit relative to each other. Such behaviour is exhibited by the Saturnian satellites Janus and Epimetheus, whose orbits are only 50 km apart. If one object is much more massive than the other, such as a planet and an asteroid or a moon and a ring particle, only the orbit of the smaller body is significantly altered by the encounters, being alternately raised or lowered by the gravity of the more massive body.

horst A strip of land, bounded by parallel faults, that has been uplifted between them, the opposite movement to that in *graben. Examples of horsts occur in the Ceraunius Fossae on Mars.

hot Big Bang An alternative term for the standard Big Bang theory. The word 'hot' was initially used to distinguish it from a rival theory which had a cold initial phase. The existence of the *cosmic microwave background requires that the Universe must have been hot in the past if the Big Bang picture is correct.

hot dark matter A particular type of *non-baryonic particle that, according to some theories, is created in the early stages of the Big Bang and survives to the present time in sufficient numbers to contribute significantly to the present density of the Universe. The term 'hot' signifies that these particles are fast-moving (close to the speed of light), usually because they are of low mass. The favoured candidate for such a particle is a neutrino with a *rest mass of around 10 eV, which is 1/500 000 of the mass of the electron.

hot Jupiter A planet with a mass similar to Jupiter but lying close to its parent star, with an orbital period from a few days to a few weeks. Such planets are believed to have formed farther from their parent stars and subsequently moved inwards. The first *extrasolar planets to be discovered were hot Jupiters.

hot spot 1. A compact, highly luminous region (often called a *bright spot*) in a cataclysmic binary, located either where the stream of material hits the edge of the *accretion disk, or at the inner edge of the disk. There may be small-scale *flickering or larger variations in luminosity depending on fluctuations in the rate of *mass transfer. The presence of a hot spot is often indicated by a hump in the orbital light-curve.
2. A small, bright region in a lobe of a radio galaxy. It is believed to be the site where high-velocity material in a jet collides with the leading boundary between the lobe and the surrounding intergalactic medium.

hour angle (HA) The angle between the observer's meridian and the *hour circle of a celestial object, measured clockwise along the celestial equator. The observer's meridian (the great circle joining the celestial pole to the zenith) is therefore the meridian of zero hour angle. The hour angle of any object increases as the Earth rotates, and it can be expressed as a time by equating 360° with 24 hours. For example, an hour after an object crosses the meridian its hour angle is 1 h, and so on. It is sometimes used as a coordinate in place of *right ascension. The sum of the hour angle and right ascension is equal to the observer's *sidereal time. *See also* GREENWICH HOUR ANGLE.

hour circle A great circle on the celestial sphere passing through a celestial object and the celestial poles; also known as *meridian of hour angle*. It is perpendicular to the celestial equator.

Hourglass Nebula The brightest central part of the *Lagoon Nebula in Sagittarius, shaped like a figure-of-eight, near the 6th-magnitude star 9 Sagittarii.

howardite A class of calcium-rich achondrite meteorite; also known as *plagioclase-hypersthene achondrites*. They are named after the English chemist Edward Charles Howard (1774–1816). The howardites are *breccias whose texture resembles that of the breccias found in the lunar regolith, and were apparently formed on the surface of their parent body by impact processes. Most of the fragments found in howardites resemble eucrite and diogenite material. The howardites are apparently related to the stony-iron *mesosiderites. *See also* BASALTIC ACHONDRITE.

Hoyle, Fred (1915–2001) English astrophysicist and cosmologist. In 1948, with H. *Bondi and T. *Gold, he proposed the *steady-state theory of the Universe in which matter is continuously created. Subsequently abandoned by most astronomers in favour of the Big Bang (so named from a dismissive remark by Hoyle), the steady-state theory nevertheless stimulated much important astrophysical research. Particularly significant was the work by Hoyle, with W. A. *Fowler and G. R. and E. M. *Burbidge, on *nucleosynthesis in stars. As well as his noted work on stellar evolution, Hoyle propounded unorthodox ideas, suggesting for example that viruses and perhaps other life forms have been brought to Earth by comets. He was also a noted popularizer of astronomy.

HPBW Abbr. for half-power beamwidth. *See* BEAMWIDTH.

HR diagram Abbr. for *Hertzsprung–Russell diagram.

HR number The number of a star in the *Harvard Revised Photometry. The same numbers are used in the *Bright Star Catalogue*.

HSO Abbr. for *Herschel Space Observatory.

HST Abbr. for *Hubble Space Telescope.

H_2O maser A maser source in which the water (H_2O) molecule is excited to maser action. They are the most widely distributed of all the cosmic masers. There are many different H_2O maser lines. The first to be discovered, in 1969, was the powerful line at 22.2 GHz (13.5 mm) in the *Kleinmann–Low Nebula in Orion. Other H_2O lines at higher frequencies are difficult to observe with ground-based radio telescopes because of strong absorption by water vapour in the Earth's atmosphere. Water masers are found in star-forming regions, circumstellar envelopes, comets, and in the nuclei of some active galaxies in the form of *megamasers.

H II region A region of interstellar hydrogen that is ionized. The notation H II refers to the fact that the hydrogen atoms (H) are ionized (H I is neutral, un-ionized hydrogen). Each ionized hydrogen atom contributes two particles to the gas, namely a proton and an electron. H II regions are hot, with typical temperatures of 10 000 K, and densities of 10 to 100 000 atom/cm^3 (i.e. 10–100 000 times denser than H I regions). They are usually found around massive young O and B stars, the strong ultraviolet light from which ionizes the gas, causing it to glow. The Orion Nebula is a famous H II region. H II regions can be detected throughout the Galaxy by their strong radio and infrared emissions. The radio emission is *bremsstrahlung from the ionized gas, and the infrared is thermal emission from dust.

Hubble, Edwin Powell (1889–1953) American astronomer. He first studied nebulae, concluding in 1917 that the spiral-shaped ones (which we now know as galaxies) were different in nature from diffuse nebulae, which he found to be gas clouds illuminated by stars. From 1923, using the 100-inch (2.5-m) telescope at Mount Wilson Observatory, he resolved the outer regions of the spiral nebulae M31 and M33 into stars, identifying over 30 Cepheid variables in them. This proved that such 'nebulae' were truly independent star systems like our own—other galaxies. In 1925 he devised the so-called *tuning-fork diagram of galaxies, dividing them into ellipticals, spirals, and barred spirals, which he wrongly believed to indicate an evolutionary sequence. By 1929 Hubble had good distance measurements for over twenty galaxies, including members of the *Virgo Cluster. By comparing distances with their velocities, as revealed by the redshifts in their spectra, he concluded that galaxies were receding with speeds that increased with their distance, a relationship known as the *Hubble law. This was powerful evidence that the Universe is expanding.

Hubble classification A widely used system for classifying galaxies according to their visual appearance, illustrated on the *tuning-fork diagram. The sequence is based on three criteria: the relative sizes of the central bulge of stars and the flattened disk; the existence and character of spiral arms; and the resolution of the spiral arms and/or disk into stars and H II regions. The system was originated by E. P. *Hubble.

The sequence starts with round elliptical galaxies (E0) showing no disks. Increasing flattening of a galaxy is indicated by a number which is calculated from $10\,(a - b)/a$, where a and b are the major and minor axes as measured on the sky. No elliptical is known that is flatter than E7. Beyond this a clear disk is apparent in the lenticular or S0 galaxies. The classification then splits into two parallel sequences of disk galaxies showing spiral structure: ordinary spirals, S, and barred spirals, SB. The spiral types are subdivided into Sa, Sb, Sc, Sd (SBa, SBb, SBc, SBd for barred spirals). With each successive subdivision the arms become less tightly wound (but more easily resolvable into stars and H II regions), and the central bulge becomes less dominant. Two types of irregular galaxy are defined. Irr I galaxies show rather amorphous, irregular structure with perhaps a hint of a spiral arm or bar, and can be placed at the far end of the spiral sequence. Irr II galaxies are sufficiently unusual to defy assignment to any of the other types, although this category encompasses only about 2% of bright or moderately bright galaxies in the nearby Universe. The original, erroneous idea that the sequence might be an evolutionary one led to the ellipticals being referred to as *early-type galaxies*, and the spirals and Irr I irregulars as *late-type galaxies*.

Colour and amount of interstellar material vary systematically along the Hubble sequence: ellipticals are red and contain little interstellar gas or dust, whereas late spirals and Irr I galaxies are blue, with significant amounts of interstellar material. The relatively faint dwarf spheroidal galaxies (*see* DWARF GALAXY) were not recognized as a separate type in the Hubble classification. Some variants of the Hubble classification use plus and minus signs to subdivide classes, so that Sa^+ is later than Sa, but earlier than Sb^-.

Hubble constant (H_0) The figure that relates the speed of an object's recession in the expanding Universe to its distance in the *Hubble law. It represents the current rate of expansion of the Universe. This important cosmological parameter is usually measured in units of kilometres per second per megaparsec. In the Big Bang theory, H_0 varies with time and it is therefore more properly known as the **Hubble parameter**. In 2001 the Hubble Space Telescope Key Project team reported a final figure for H_0 of 72 ± 8 km/s/Mpc. In 2010 a figure of 71 ± 2.5 km/s/Mpc was announced from the findings of the *Wilkinson Microwave Anisotropy Probe.

Hubble diagram A graph that plots the recession velocity of a galaxy, measured by its *redshift, against its distance from us. The first such plot was made in 1929 by E. P. *Hubble. He found a linear relation between velocity and distance, implying that a galaxy's recession velocity is proportional to its distance from us; this is the *Hubble law. Hubble correctly interpreted this observation as meaning that the Universe is homogeneous and is expanding uniformly in all directions. He also realized that for very distant galaxies the observed relation would no longer be linear on account of the *curvature of spacetime and other effects caused by the expansion of the Universe. In principle, therefore, observations of distant galaxies can be used to constrain cosmological models.

However, light from very distant galaxies takes a significant fraction of the age of the Universe to reach us, so that high-redshift galaxies are seen when they were young while nearby galaxies are seen when much older. Hence allowance must be made for any evolution in the properties of galaxies with time (see GALAXY EVOLUTION). For most galaxy properties, such as luminosity, evolutionary uncertainties are larger than the differences between the models. In the late 1990s, astronomers made Hubble diagrams using Type Ia *supernovae in distant galaxies to infer that the Universe is currently undergoing an apparent acceleration in its rate of expansion.

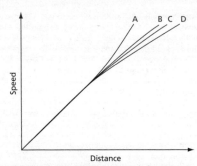

Hubble diagram: Distance of galaxies is plotted against their speed of recession. For nearby galaxies, there is a linear relationship between distance and speed (the *Hubble law). For great distances the departure of the relationship from a straight line can, in principle, be used to test cosmological theories. The four lines show the predictions of four different cosmological theories. If the galaxies follow curve A, then the Universe was once expanding much faster than today and will eventually collapse (i.e. the Universe is said to be closed). Curve C means that the Universe will continue to expand forever (i.e. the Universe is open). Curve B is the intermediate case, in which the expansion will stop infinitely far in the future (i.e. the Universe is spatially flat). Curve D is for a *steady-state universe. Unfortunately, galaxies far enough out in space for deviations from the straight line to be apparent are also being observed so far back in time that the effects of galaxy evolution are much greater than the predicted differences between the different theories.

Hubble flow The general outward motion of galaxies resulting from the uniform expansion of the Universe. All motions lie in a radial direction from the observer, and the velocities are proportional to the distance of the galaxies. The real pattern of galaxy motions is not exactly of this form, particularly close to us, because of the mutual gravitational interaction between galaxies; some nearby galaxies are even moving towards the Milky Way. At large distances, however, the discrepancies are small compared with the Hubble flow. *See also* HUBBLE LAW.

Hubble law The law that governs the expansion of the Universe. According to the law, the apparent recession velocity of galaxies is proportional to their distance from the observer. In mathematical terms, $v = H_0 r$, where v is the velocity, r the distance, and H_0 the *Hubble constant. The law was put forward in 1929 by E. P. *Hubble.

Hubble parameter *See* HUBBLE CONSTANT.

Hubble radius A distance defined as the ratio of the velocity of light, c, to the value of the Hubble constant, H_0. This gives the distance from the observer at which the recession velocity of a galaxy would equal the speed of light. Roughly speaking, the Hubble radius is the radius of the observable Universe. Current observations place the Hubble constant at around 71 km/s/Mpc, which gives a Hubble radius of 13.7 billion l.y.

Hubble–Sandage variable Any of a number of highly luminous stars identified in M31 and M33 by E. P. *Hubble and A. R. *Sandage in 1953. The name Hubble–Sandage variable has been used to refer to this type of star, but the name *S Doradus star is now preferred.

Hubble Space Telescope (HST) A 2.4-m reflector built by NASA and ESA, launched in 1990 April. It orbits at an altitude of about 600 km. An initial servicing mission by astronauts in 1993 December was necessary to install additional optics to correct the telescope's spherical aberration, caused by an error in manufacturing the main mirror. Subsequent servicing missions took place in 1997 February, 1999 December, and 2002 March, during which astronauts installed improved detectors and replaced the telescope's solar panels. After the fourth servicing mission, HST's scientific instruments consisted of the Advanced Camera for Surveys (ACS), the Wide Field and Planetary Camera 2 (WFPC2), the Near Infrared Camera and Multi–Object Spectrometer (NICMOS), and the Space Telescope Imaging Spectrograph (STIS). A fifth and final servicing mission took place in 2009 May, installing the Wide Field Camera 3

(WFC3) and the Cosmic Origins Spectrograph (COS), as well as repairing existing instruments. WFC3, which replaced WFPC2, can detect a wider range of wavelengths, from infrared to ultraviolet, than its predecessor. The servicing was intended to keep HST in operation at least until 2014. HST is controlled from NASA's *Goddard Space Flight Center. The *Space Telescope Science Institute plans the telescope's observing programme and collects its data. 📷

(⊕) SEE WEB LINKS
• Official mission website.

Hubble's Variable Nebula A reflection nebula in Monoceros, also known as NGC 2261, which is illuminated by the irregular variable R Monocerotis. The nebula's variability was detected by E. P. *Hubble in 1916 and probably arises from changes in the star's luminosity and shadowing by dense clouds within the nebula. R Monocerotis is a young *T Tauri star and a strong infrared source. It exhibits a vigorous outflow, and is apparently surrounded by a protoplanetary disk—a disk of material from which planets may form.

Hubble time The time required for the Universe to expand to its present size, assuming that the Hubble constant has remained unchanged since the Big Bang. It is defined as the reciprocal of the Hubble constant, $1/H_0$. For a Hubble constant of 71 km/s/Mpc, as given by current measurements, the Hubble time is around 14 billion years. In practice, the actual age of the Universe is slightly less than the Hubble time because the expansion rate was faster in the past, although it is now believed to be accelerating again.

Huggins, William (1824–1910) English amateur astronomer and spectroscopist. Inspired by G. R. *Kirchhoff's achievements, he carried out a series of pioneering spectroscopic observations, helped first by the chemist William Allen Miller (1817–70) and later by his wife, Margaret Lindsay Huggins, née Murray (1848–1915). By 1863 he had several stellar spectra, showing that the stars, like the Sun, consisted of incandescent gas and contained the same chemical elements as are found on Earth. In 1864 he detected green lines in the spectrum of the Orion Nebula which confirmed that it is gaseous, although he attributed the bright lines to an unknown substance which he termed 'nebulium'. These early observations were all made visually; later he turned to photography as photographic plates improved. Huggins investigated many other objects spectroscopically, including comets, meteors, and novae.

Hulst, Hendrik ('Henk') Christoffel van de (1918–2000) Dutch astronomer. His main field of study was interstellar matter. In 1944, while a student of J. H. *Oort, he calculated that *neutral hydrogen should emit radio waves with a wavelength of 21 cm. His prediction was borne out in 1951 when the American physicists Edward Mills Purcell (1912–97) and Harold Irving Ewen (1922–) detected 21-cm radio emission from interstellar hydrogen clouds. The same year, van de Hulst and Oort began to use Doppler shifts in hydrogen emission to map the Galaxy's spiral structure.

Humason, Milton Lasell (1891–1972) American astronomer. He worked at the Mount Wilson (where he was employed originally as janitor) and Palomar Observatories. In the 1920s he participated in long-term projects with W. S. *Adams, Alfred Harrison Joy (1882–1973), and others to measure the absolute magnitudes and radial velocities of thousands of stars. From 1930, first with the 100-inch (2.5-m) telescope and later the 200-inch (5-m), he measured the radial velocities of hundreds of faint galaxies, finding progressively larger redshifts.

hummocky terrain A type of terrain on a planetary surface consisting of irregular rounded hummocks in a chaotic pattern. The inner part of the ejecta blanket of a young impact crater is usually hummocky terrain, and it is also found in the furthest parts of large landslips.

Humphreys series A sequence of absorption or emission lines in the far-infrared part of the spectrum, due to hydrogen. They range from Humphreys-α at 12.37 μm, towards shorter wavelengths, the spacing between the lines diminishing as they converge on the Humphreys series limit at 3.28 μm. The Humphreys series is caused by electron jumps between the sixth energy level and higher levels. It is named after the American physicist Curtis Judson Humphreys (1898-1986). *See also* HYDROGEN SPECTRUM.

Hungaria group A group of asteroids at a mean distance of about 1.95 AU from the Sun, near the inner edge of the main asteroid belt but separated from it by the Kirkwood gap at the 4 : 1 resonance with Jupiter. Hungaria group members have low orbital eccentricities, but high inclinations of 22-24°. The group has been separated from the asteroid belt by the gravitational effects of the major planets, notably Jupiter, and contains a broad representation of asteroid classes. All Hungaria objects are small. The group is named after (434) Hungaria, an E-class asteroid of diameter 11.4 km, discovered in 1898 by Max *Wolf. Hungaria's orbit has a semimajor axis of 1.944 AU, period 2.71 years, perihelion 1.80 AU, aphelion 2.09 AU, and inclination 22°.5.

hunter's Moon The full Moon following the *harvest Moon, usually occurring in October. As at harvest Moon, the full Moon rises around the time of sunset for several nights in succession in mid- or high northern latitudes.

Huygenian eyepiece An eyepiece design invented by C. *Huygens *c.*1662. It consists of two *planoconvex lenses with their convex surfaces facing the objective lens. Because of its cheap and simple construction it is often used with small refractors, but will only perform well with large *focal ratios. The Huygenian eyepiece suffers from *aberrations if used on telescopes faster than about $f/10$. A variant known as the *Mittenzwey eyepiece uses convex meniscus lenses and is usable down to $f/8$.

Huygens, Christiaan (1629-95) Dutch mathematician, physicist, and astronomer. In 1655 he discovered Titan, the largest satellite of Saturn. Soon after, he announced that the planet was surrounded by a 'thin flat ring'; others had observed the rings, but misinterpreted what they saw. In 1659 he discovered dark markings on Mars, notably the wedge-shaped Syrtis Major. With his brother, Constantijn Huygens (1628-97), he constructed aerial telescopes (tubeless telescopes supported by cables) of very long focal length to overcome aberrations. He invented the *Huygenian eyepiece. To physics he contributed the principle of conservation of momentum (1656) and the wave theory of light (1678). He also made the first pendulum clock (1657).

Huygens Gap A clearing 300-400 km wide near the inner edge of the Cassini Division (which is not entirely empty but has ringlets and gaps within it). It was discovered in 1981 on images from Voyager 2 and named after C. *Huygens.

Huygens probe A European-built probe, part of the *Cassini–Huygens mission to Saturn. In 2005 January, Huygens landed on Saturn's largest moon, Titan. During its descent it studied the moon's clouds and atmosphere and photographed the surface below. It landed on a slushy area dotted with boulders composed of water ice, under orange clouds composed of complex hydrocarbons.

(())) SEE WEB LINKS
• Official mission website.
• ESA mission website.

Hyades A large, V-shaped open cluster in Taurus, spanning over 5° of sky. It contains about 200 stars, and is estimated to be 660 million years old. The cluster lies 150 l.y. away, and its true diameter is about 15 l.y. Comparison of the brightness of its stars with those in other clusters enables the distances to those other clusters to be calculated. The Hyades is thus an important yardstick in astronomical measurement.

Hyakutake, Comet (C/1996 B2) A long-period comet discovered on 1996 January 31 by the Japanese amateur astronomer Yuuji Hyakutake (1950-2002). It passed 0.10 AU from Earth on March 25, showing a bluish-green gas tail stretching up to 70°, while the head reached magnitude −1. Radar measurements revealed the comet's nucleus to be 1-3 km in diameter. Its rotation period was 6.23 h. Comet Hyakutake reached perihelion, 0.23 AU, on 1996 May 1. The comet's passage through the inner Solar System increased its orbital period from about 20 000 years to around 100 000 years. The comet's orbit has an eccentricity of 0.9999 and an inclination of 124°.9.

SEE WEB LINKS

• Information page at Cometography website.

Hydra (Hya) (gen. Hydrae) The largest constellation of all, representing a water-snake. Hydra meanders from the celestial equator into the southern celestial hemisphere, and is over 100° long. Its brightest star is *Alphard (Alpha Hydrae). R Hydrae is a *Mira star that varies between magnitudes 4 and 11 every 390 days, and U Hydrae is a semiregular variable that ranges between magnitudes 4.3 and 6.5. M48 is a 6th-magnitude open star cluster visible in binoculars, and M83 is an 8th-magnitude spiral galaxy. NGC 3242 is a planetary nebula known as the *Ghost of Jupiter.

hydrocarbon A compound which is composed entirely of carbon and hydrogen atoms. Examples of hydrocarbons include methane, ethane, and propane. They are found in meteorites, comets, and nebulae.

hydrogen (H) The most abundant chemical element in the Universe, comprising some 73% of its mass. Hydrogen is found in several forms: *molecular hydrogen*, H_2, the most familiar form on Earth, occurs in dense molecular clouds; *atomic hydrogen*, formed of individual hydrogen atoms, occurs in the more diffuse interstellar clouds (*H I regions); and *ionized hydrogen*, in which each atom is broken down into a proton and an electron, and which is the most common form of all, being found in stars and nebulae (*H II regions).

hydrogen burning A series of nuclear reactions in which hydrogen nuclei (protons) fuse to form a helium nucleus, with the associated release of nuclear energy. In stars with core temperatures below about 18 million K, hydrogen burning occurs primarily through the *proton–proton reaction. In hotter stars, the *carbon–nitrogen cycle is primarily responsible.

hydrogen emission region A region in interstellar space containing hot hydrogen gas which emits a characteristic red spectral line at 656.3 nm; such a region is the visible counterpart of an *H II region. The 656.3 nm red line is the brightest line in the optical spectrum of hydrogen, and is known as the Hα (hydrogen-alpha) line (*see* BALMER SERIES). It gives nebulae their characteristic red colour in photographs.

hydrogen ion A hydrogen atom that has gained or lost an electron, and thus has an electric charge. The positive hydrogen ion, H^+, has lost its only electron and therefore consists of a single proton. A nebula containing H^+ ions is known as an *H II region. The negative hydrogen ion, H^-, has gained a second electron. H^- ions are found in the outer layers of stars such as the Sun. Most of the light received from the Sun actually originates from the formation of H^- ions.

hydrogen line *See* 21-CENTIMETRE LINE.

hydrogen spectrum A distinctive pattern of spectral lines, either emission or absorption, produced by hydrogen, the simplest of all the atoms. The series of lines in the spectrum are labelled according to the lowest *energy level involved in the transitions that give rise to the lines. If an electron jumps to or from the ground state ($n = 1$), it produces a line in the *Lyman series; $n = 2$ corresponds to the *Balmer series; $n = 3$ to the *Paschen series; $n = 4$ to the *Brackett series; $n = 5$ to the *Pfund series; and $n = 6$ to the *Humphreys series. The Balmer series lines fall in the visible part of the spectrum, the Paschen, Brackett, Pfund, and Humphreys series are in the infrared, and the Lyman series is in the ultraviolet.

hydrostatic equilibrium The balance in a star between its gravitational force, which is directed inwards, and the outward forces of radiation pressure and gas pressure. Because these forces balance, the star neither collapses inwards upon itself nor flies apart.

hydroxyl (OH) A molecule composed of an oxygen atom and a hydrogen atom; also called the **hydroxyl radical**. Hydroxyl is one of the commonest interstellar molecules. It was the first interstellar molecule discovered by radio astronomers, in 1963. *See also* OH LINE; OH MASER.

Hydrus (Hyi) (*gen.* Hydri) An unremarkable constellation near the south celestial pole, representing a small water-snake. Beta Hydri is its brightest star, magnitude 2.8.

Hygiea Asteroid 10, discovered in 1849 by the Italian astronomer Annibale de Gasparis (1819–92). It is the fourth-largest main-belt asteroid, diameter 407 km, with a rotation period of 27.6 h, and is of C class. Its orbit has a semimajor axis of 3.140 AU, period 5.56 years, perihelion 2.77 AU, aphelion 3.51 AU, inclination 3°.8.

hyperbola A type of curve whose two 'arms' diverge and never rejoin, defined mathematically as a *conic section with an eccentricity greater than 1. The orbit of one body moving past another without being captured has the shape of a hyperbola.

hyperbolic comet A comet whose orbit around the Sun has an eccentricity greater than 1.0. Periodic comets may be injected into hyperbolic orbits following passage close to Jupiter, as with Comet *Lexell in 1779. Comets on such trajectories are lost to the Solar System. Some comets making their first visit to the inner Solar System from the *Oort Cloud may follow hyperbolic trajectories, never to return. The possibility that some hyperbolic comets come from outside the Solar System cannot be excluded.

hyperbolic orbit An orbit having the shape of a hyperbola. A body in a hyperbolic orbit about another makes only one approach to the other body, in theory approaching from infinity and then receding to infinity. A spacecraft making a fly-by of a planet follows a hyperbolic trajectory.

hyperbolic velocity The speed of an object that is moving faster than the *escape velocity of a planet or other massive body. Hyperbolic velocity is capable of taking the object to an infinite distance from the planet or body.

hyperboloid A surface or body obtained by rotating a hyperbola about its axis of symmetry. The secondary mirror in a Cassegrain telescope has a hyperboloidal surface.

hyperfine structure Very close splitting of spectral lines, due to interaction between the spin of the atomic nucleus and the spin of an orbiting electron. It is a much smaller effect than *fine structure and hence can be seen only at the very highest spectral resolutions.

hypergiant star A massive, evolved variable star with a luminosity around 10^5–10^6 times that of the Sun, close to the *Eddington limit. All hypergiants are unstable, exhibiting high mass loss, sometimes in violent outbursts. Two broad classes of hypergiant are identified: *S Doradus stars and cool hypergiants. The latter have spectral types ranging from late A to M, and may represent the evolutionary stage before becoming S Doradus variables.

Hyperion A satellite of Saturn, 1 481 100 km from the planet's centre, orbital period 21.277 days; also known as Saturn VII. It is irregular in shape, measuring $360 \times 266 \times 205$ km. Hyperion was discovered in 1848 by W. C. and G. P. *Bond, and independently two days later by W. *Lassell. Hyperion's surface is heavily covered with impact craters which show signs of collapse and infilling.

hypernova An extremely energetic explosion which, it is thought, may result after the collapse of a star that is too massive to be stable, possibly of the *Wolf–Rayet type. Initial shedding of material from the outer layers may be followed by a *gamma-ray burst from the core of the object as it collapses directly into a black hole.

hyperon A *baryon that is heavier than a proton or neutron. The term is becoming obsolete.

hypersensitization A method of treating photographic emulsions to reduce the effects of *reciprocity failure; also known as **hypering**. The most common modern technique involves soaking the film or plates in a chemically reducing gas—ideally, pure hydrogen. For amateur use, *forming gas* (8% hydrogen, 92% nitrogen) is preferred. Other methods include soaking the film or plates in water or ammonia; baking the emulsion for many hours at around 60°C; and briefly pre-exposing the emulsion to light (*pre-flashing*) to achieve a background level on top of which light from the sky will be recorded more efficiently.

hypersthene achondrite Another name for a *diogenite.

hypervelocity impact The impact of an object into a target at a speed that greatly exceeds the speed of sound in the struck body. Typically, in solids, this limit occurs at around 4 km/s. At these speeds solid rock behaves like a fluid. Hypervelocity impacts produce craters that are generally circular, and far larger than the impacting body.

Hz Symbol for *hertz.

HZ43 A hot white-dwarf star in the constellation Coma Berenices with a surface temperature of approximately 50 000 K. The SAS-3 satellite discovered that it was a strong emitter of soft X-rays, and it was also the first cosmic source of extreme ultraviolet (EUV) radiation to be discovered. Because of its simple, well-known spectrum, it is used as a standard for calibrating EUV astronomy missions.

h

Iapetus The third-largest satellite of Saturn, diameter 1469 km; also known as Saturn VIII. It orbits 3 561 850 km from the planet's centre every 79.33 days, keeping the same face turned towards Saturn. Its leading hemisphere is far darker than the trailing hemisphere and polar regions, the albedos being about 0.05 and 0.5 respectively, so that its mean opposition magnitude varies between 10.2 and 11.9. The cause of this one-sided darkening is not known. Impact craters mark its icy surface, the largest of which, called Turgis, is 580 km across. Iapetus was discovered in 1671 by G. D. *Cassini.

IAU Abbr. for *International Astronomical Union.

IBEX Abbr. for *Interstellar Boundary Explorer.

IC Abbr. for *Index Catalogue.

Icarus Asteroid 1566, a member of the *Apollo group, discovered by W. *Baade in 1949 during a close approach to Earth. In 1968, Icarus became the first minor planet to be observed by radar, when it came within 0.040 AU (6 million km) of Earth. Its diameter is about 1.4 km, and its rotation period 2.27 hours. Its orbit has a semimajor axis of 1.078 AU, period 1.12 years, perihelion 0.19 AU (well within the orbit of Mercury), aphelion 1.97 AU, and inclination 22°.8.

ICE Abbr. for *International Cometary Explorer.

ICRF Abbr. for *International Celestial Reference Frame.

ICRS Abbr. for *International Celestial Reference System.

Ida Asteroid 243, discovered in 1884 by the Austrian astronomer Johann Palisa (1848–1925). It has a satellite, *Dactyl. Ida, an S-class asteroid, measures 54 × 24 × 15 km and rotates in a period of 4.6 hours about its shortest axis. It is a member of the *Koronis family. If its parent body was broken into several pieces, one of these fragments may have become its satellite. Its orbit has a semimajor axis of 2.863 AU, period 4.84 years, perihelion 2.74 AU, aphelion 2.98 AU, and inclination 1°.1.

IGM Abbr. for *intergalactic medium.

igneous A description applied to a rock formed by the cooling and solidification of a magma (molten rock), either below or on the surface of a planetary body. Igneous rocks are said to be *extrusive* if they solidify on the surface of a planet, and *intrusive* if they solidify at depth beneath the crust.

IGY Abbr. for *International Geophysical Year.

Ikeya–Seki, Comet (C/1965 S1) A long-period comet, discovered independently on 1965 September 18 by the Japanese amateur astronomers Kaoru Ikeya (1943–) and Tsutomu Seki (1930–); formerly designated 1965 VIII. It is a member of the *Kreutz sungrazer group. Perihelion, 0.008 AU (1.2 million km), was on 1965 October 21, when its nucleus broke into three fragments. At perihelion the comet reached at least magnitude −10 and could be seen in broad daylight, while its tail reached about 60° in late October. Ikeya–Seki's orbit has a period of about 880 years, eccentricity 0.9999, and inclination 141°.9.

⊕ SEE WEB LINKS
• Information page at Cometography website.

Ikeya–Zhang, Comet (153P/) The comet with the longest period that has been seen to return to the inner Solar System. Discovered in 1661, when it was extensively observed by J. *Hevelius, it returned to perihelion in 2002 and was independently rediscovered by the amateur astronomers Kaoru Ikeya (1943–) of Japan and Zhang Daqing (1969–) of China. Currently it has a period of 366 years, perihelion 0.507 AU, aphelion 101.9 AU, eccentricity 0.990, and inclination 28°.1.

⊕ SEE WEB LINKS
• Information page at Cometography website.

illuminance (E_v) The energy in the form of visible radiation that falls on a surface in a given time. It is measured in *lux* (lumens per square metre). The energy incident at all wavelengths, not the visible, is termed *irradiance.

IMAGE Abbr. for *Imager for Magnetopause-to-Aurora Global Exploration.

Imager for Magnetopause-to-Aurora Global Exploration (IMAGE) A NASA satellite launched into high polar orbit in 2000 March to study how the Earth's magnetosphere reacts to changes in the solar wind. It carried instruments to image neutral atoms, ultraviolet emissions, and plasma motions. It ceased operation in 2005 December.

⊕ SEE WEB LINKS
• Official mission website.

imaging photometer Another name for an *area photometer.

I magnitude The magnitude of a star measured at infrared wavelengths through an I filter. The wavelength of the I filter can have three separate values, depending on the system used. The wavelength and bandwidth on the *Kron–Cousins RI photometry system are 798 and 154 nm; on the *Johnson photometry system, 878 and 218 nm; and on the *six-colour system, 1027 and 192 nm. The first is now the most widely used, and is sometimes denoted by the subscript KC or C.

Imbrium Basin An ancient impact basin on the Moon over 1100 km across, centred at about +33° lat., 16°W long. It is about 3.9 billion years old, and is the result of the largest impact on the Moon, which was nearly big enough to break the Moon apart. The Imbrium impact had far-reaching effects, sending ejected material over much of the Moon's surface and creating deep fractures within it. Following the impact, a long period of lava flooding occurred through these fractures, creating the Mare Imbrium and submerging much of the Imbrium Basin itself and a large area outside it.

immersion An alternative name for the disappearance of a star at an *occultation; or, the entry of an object into shadow at an eclipse.

impact crater A depression caused by the high-speed collision of a meteoroid, asteroid, or comet with a solid surface. Impact craters are found on the Moon, on all the terrestrial planets, and on planetary satellites except those where resurfacing processes are in operation, such as Io. They range in size from tiny pits a few micrometres across up to vast impact *basins thousands of kilometres wide. Young impact craters may be distinguished from volcanic craters by their circular shape, steep inner walls, and shallow outer slopes, as well as their *ejecta blankets, secondary craters, and bright *rays. Larger impact craters have *terracing on their inner walls, flat floors, and *central peaks. Old impact craters become progressively eroded and may lose some of these distinguishing features; others may be partially submerged or filled by lava, ice flows, or sediments (*see* GHOST CRATER).

impactite A substance composed of meteoritic material fused with melted rock, often in fragments smaller than a grain of rice. It results when a large meteorite strikes the Earth with sufficient energy to form a crater.

impersonal astrolabe An alternative name for a *Danjon astrolabe.

inclination (*i*) **1.** The angle between the orbital plane of a body and the reference plane centred on the object about which the body is revolving. For planets in the Solar System, the inclination of the orbit is usually given relative to the plane of the Earth's orbit, the ecliptic. For objects orbiting a planet, the inclination is usually given relative to the planet's equator. For double stars, the inclination is given relative to the plane of the sky. The inclination is one of the *elements of an orbit.
2. The angle at which the rotation axis of a planet or satellite is tilted with respect to the perpendicular to some reference plane. The reference plane is usually the plane of the body's orbit.

***Index Catalogue* (IC)** Either of two supplements to the *New General Catalogue* (NGC), compiled by J. L. E. *Dreyer. The first *Index Catalogue*, containing 1529 newly discovered non-stellar objects, was published in 1895, and the second (IC II), with 3857 objects, in 1908; both contained corrections to the NGC.

(⊕) SEE WEB LINKS

• Detailed description and full catalogue downloadable from the CDS.

• The original catalogue online.

indochinite A type of tektite found in Thailand and the area formerly known as Indo-China, comprising Laos, Cambodia, and Vietnam, one of many regions in Southeast Asia where tektites have been recovered.

Indus (Ind) (*gen.* Indi) A constellation of the southern sky, representing a native Indian of uncertain nationality. Its brightest star is Alpha Indi, magnitude 3.1. Epsilon Indi is a K5 dwarf star 11.8 l.y. away.

inequality A variation in the movement of a celestial object in its orbit about another which cannot be accounted for by their mutual gravitational attraction. Inequalities usually arise because of the perturbing forces of one or more other massive objects in the system. For example, the so-called **great inequality** in the orbital movements of the planets Jupiter and Saturn about the Sun is an oscillation in their heliocentric longitude with a period of some 900 years caused by their mutual perturbations and also by the nearly 2 : 5 *resonance in their mean motions. *See also* LUNAR INEQUALITY.

inert gas Another name for a *noble gas.

inertial coordinate system The system of coordinates that is used in an *inertial reference frame, normally in the special theory of relativity. The three space coordinates are usually *Cartesian coordinates (x, y, z), and the time coordinate is the time as measured by an observer at rest in the coordinate system. In *astrometry, an inertial coordinate system is a reference frame formed by assigning coordinates to specific observable objects, such as the positions and proper motions of stars in a *fundamental catalogue.

inertial mass A measure of a body's resistance to a change in its velocity or state of rest. Inertia is a direct property of the mass of a body: the greater the mass, the greater its inertia. Although mass is formally defined in terms of its inertia, it is usually measured by gravitation. *See also* GRAVITATIONAL MASS.

inertial reference frame A frame of reference in which any body initially at rest will remain at rest indefinitely, or in which a moving body moves in a straight line with constant speed indefinitely; in other words, it is free from any inertial forces. It can be defined as a frame of reference in which Newton's laws of motion apply exactly. The frame of reference fixed with respect to the Earth's surface is not inertial, because there are forces due to the Earth's rotation. The concept of an inertial frame of reference is particularly useful in the special theory of relativity, in which two different inertial frames have a fixed relative velocity but no relative acceleration. Inertial frames do not exist in general relativity, however, since, although a freely falling frame will appear inertial locally, differential gravitational effects prevent it from being globally inertial. In *astrometry, an inertial reference frame is a coordinate system defined by positions and proper motions of a set of stars or other distant objects. The ultimate inertial reference frame is that defined by the positions of extragalactic objects, which are so far away that they have negligible relative motions as seen from Earth.

infall velocity The motion of a galaxy or other cosmic object towards another, more massive object, resulting from its gravitational attraction. The *Local Group of galaxies, for example, is moving towards the centre of the *Virgo Cluster of galaxies with a velocity of around 200 km/s.

inferior conjunction The moment when an inferior planet (i.e. Mercury or Venus) lies on a straight line between the Earth and Sun. *See also* CONJUNCTION.

inferior planet A planet whose orbit is closer to the Sun (and hence smaller in radius) than that of the Earth (i.e. Mercury or Venus).

inflationary universe A variation on the *Big Bang theory that includes a period of accelerated expansion in its early stages. In this model, energy is released during a so-called **phase transition** about 10^{-35} s after the Big Bang, similar to the release of latent heat as a liquid freezes. The energy released acts in the same way as a *cosmological constant term and drives the Universe to expand much faster than in the standard Friedmann models. The ultra-rapid expansion stretches out any primordial 'wrinkles' in the *curvature of spacetime, rendering the Universe almost smooth and isotropic on the scales we can observe. Another feature of this theory is that it produces minute fluctuations in the density of the Universe which may be the seeds of *galaxy formation. Recent observations of the *cosmic background radiation imply that the *curvature of spacetime is close to zero, which provides some support for the theory.

Infrared Astronomical Satellite (IRAS) A joint US–Dutch–British satellite, launched in 1983 January, which made an infrared survey of the sky at wavelengths of 12, 25, 60, and 100 μm. It carried a 0.6-m telescope and an array of detectors cooled by liquid helium. In its 10-month lifetime it surveyed 95 % of the sky, detecting 250 000 infrared sources. Among its discoveries were bands of cometary dust in the Solar System; several new comets and asteroids; *infrared cirrus; possible forming planetary systems around certain stars; and a large number of galaxies emitting most of their energy in the infrared, including *starburst galaxies.

(⊕) SEE WEB LINKS
• Mission website at CalTech.

infrared astronomy The study of the Universe in the infrared part of the spectrum, at wavelengths of 1–300 μm. Infrared astronomy is hampered by the Earth's atmosphere, which is opaque and bright throughout much of the infrared band due mainly to water vapour and carbon dioxide. Another source of interference is warmth from a telescope's surroundings, including the telescope itself, which peaks around 10 μm. Ground-based infrared astronomy is restricted to the few *infrared windows in the Earth's atmosphere, especially in the near-infrared 1–5 μm region, and around 10 μm. Even then, infrared telescopes are placed on high, dry mountain tops. High-altitude balloons and aircraft have also been used, notably the *Kuiper Airborne Observatory (KAO). But unimpeded viewing of the infrared sky requires telescopes in space, such as the *Infrared Astronomical Satellite (IRAS), the *Infrared Space Observatory (ISO), the *Spitzer Space Telescope, and the *Herschel Space Observatory.

   Prominent infrared sources include red giants and supergiants with dust shells, H II regions, the galactic centre, star-forming regions, and active galaxies. Many active galaxies emit the bulk of their energy in the infrared, and the infrared luminosity of spiral galaxies has become a key element in the *Tully–Fisher relation method of measuring extragalactic distances. Infrared waves can readily penetrate interstellar dust, and infrared astronomy has played an important role in the study of obscured regions such as the galactic disk and dark nebulae. Spectroscopy at infrared wavelengths is an important source of information about interstellar molecules.

The type of detector used depends on the wavelength to be detected. At near-infrared wavelengths, photovoltaic detectors (such as indium antimonide) are common, while at far-infrared wavelengths *bolometers are used. Arrays of detectors are used for imaging. Infrared detectors are cooled by liquid helium (to 4 K) or liquid nitrogen (to 77 K) to reduce thermal noise.

infrared cirrus Wispy clouds of infrared emission first detected by the *Infrared Astronomical Satellite and prominent at 60 and 100 μm. Most infrared cirrus is believed to be emission from dust grains in hydrogen clouds, heated by ultraviolet radiation to 20–30 K. Some of the denser cirrus clouds have cooler cores. The clouds are brighter at 12 and 25 μm than expected, possibly due either to a component of very small dust grains or to line emission from complex molecules.

infrared excess Infrared emission from a star in excess of that expected from a black body with a temperature corresponding to the spectral type of the star. An infrared excess normally indicates that the star is surrounded by a shell or disk of warm dust heated by the star, for example a dusty shell around a red supergiant or a planetary system forming around a young star.

infrared photometry Photometry in the wavelength range 1.22–21 μm, that is, in the near- and mid-infrared parts of the spectrum. The detectors are cooled with liquid nitrogen or helium to reduce noise. It is usual to chop back and forth between the object and a neighbouring patch of sky to subtract the strong background signal from the telescope, its surroundings, and the sky itself (*see* CHOPPER). Further, this part of the spectrum contains many strong bands of water vapour and other molecules in the Earth's atmosphere, so filters must be used to select the windows between them. Observations free from this interference have been made from satellites such as the *Infrared Astronomical Satellite (IRAS), the *Infrared Space Observatory (ISO), the *Spitzer Space Telescope, and the *Herschel Space Observatory.

infrared radiation Electromagnetic radiation at longer wavelengths than visible light, but shorter than radio waves. The infrared region extends beyond the red end of the visible spectrum from approximately 700 nm to 1 mm. *See also* FAR INFRARED; NEAR INFRARED.

infrared source Any cosmic source of infrared radiation. All stars emit infrared radiation, and the cooler they are the more energy they emit in the infrared. Warm dust is also a major infrared source. Prominent infrared sources include red giants and supergiants with dust shells, H II regions, the galactic centre, star-forming regions, and active galaxies. Many active galaxies (especially *starburst galaxies) emit the bulk of their energy in the infrared.

Infrared Space Observatory (ISO) An ESA satellite for infrared astronomy, launched in 1995 November into a 24-hour elliptical orbit. ISO carried a 0.6-m helium-cooled *Ritchey–Chrétien telescope equipped with a camera (2.5–17 μm), an imaging photopolarimeter (2.5–240 μm), and two spectrometers (2.5–45 μm and 43–198 μm). It ceased operation in 1998 May.

(⊕)) SEE WEB LINKS
• ESA mission website.

infrared telescope A telescope designed specifically for observations in the infrared part of the spectrum. Infrared telescopes are similar to optical Cassegrain reflectors, but designed to minimize the amount of infrared radiation reaching the detector from the telescope itself. Since infrared wavelengths are longer than those of visible light, the optical quality of an infrared telescope is less critical. The main mirror can be made thinner and the supporting structure lighter than in an optical telescope. However, modern large optical telescopes are designed for good performance in both the visible and infrared parts of the spectrum. The two most important ground-based infrared telescopes are the *NASA Infrared Telescope Facility (IRTF) and the *United Kingdom Infrared Telescope (UKIRT), both at Mauna Kea Observatory, Hawaii.

infrared window A range of infrared wavelengths to which the Earth's atmosphere is relatively transparent, and at which observations can be made from the ground. Infrared windows are found near wavelengths of 1.25, 1.65, 2.2, 3.6, 5.0, 10, 20, and 30 µm, and beyond 300 µm.

initial mass function A mathematical description of the relative frequency with which stars of various masses are formed; also known as the *Salpeter function* after the American astrophysicist Edwin Ernest Salpeter (1924–2008). Massive stars are less numerous than lighter ones. The frequency of stars more massive than the Sun decreases slightly more steeply than the inverse square of the mass. The frequency of stars up to the Sun's mass falls off less steeply than the inverse square of their mass.

Innes, Robert Thorburn Ayton (1861–1933) Scottish astronomer, who worked mostly in South Africa. He introduced the orbital parameters for double stars now known as **Thiele–Innes constants** in conjunction with the Danish astronomer Thorvald Nicolai Thiele (1838–1910). Innes measured stellar proper motions, and introduced the blink comparator, which he used in 1915 to discover Proxima Centauri. His *Southern Double Star Catalogue* (1927) included over 1600 doubles he discovered.

Innisfree meteorite A meteorite fall observed in Canada on 1977 February 5. Photographs by the Canadian Meteorite Observatory and Recovery Project (MORP) allowed the fireball's trajectory to be determined, and 4.5 kg of the L5 *ordinary chondrite was recovered on the surface of snow near Innisfree, Alberta. Its orbit had an aphelion in the main asteroid belt.

insolation The amount of energy from the Sun that falls on a surface. The insolation at the top of the Earth's atmosphere is termed the *solar constant.

instability strip *See* CEPHEID INSTABILITY STRIP.

Institut de Radio Astronomie Millimétrique (IRAM) A Franco–German–Spanish organization for millimetre-wave astronomy, with its headquarters in Grenoble, France. It operates instruments in France and Spain. The IRAM 30-m Telescope was opened in 1985 at an altitude of 2870 m on Pico Veleta, in the Sierra Nevada mountains near Grenada in southern Spain. The IRAM Interferometer is situated on the Plateau de Bure, at an altitude of 2550 m in the French Alps 90 km south of Grenoble. It consists of six 15-m antennas movable on tracks arranged in a T-shape giving baselines up to 760 m. The interferometer was opened in 1990 with three dishes and expanded to six in 1996. IRAM is owned and operated by the French Centre National de la Recherche Scientifique, the German Max-Planck-Gesellschaft, and the Spanish Instituto Geográfico Nacional.

((⊕)) SEE WEB LINKS
• Official observatory website.

INT Abbr. for *Isaac Newton Telescope.

INTEGRAL Abbr. for *International Gamma-Ray Astrophysics Laboratory.

integrated magnitude The apparent magnitude an extended object would have if all of its light were concentrated into a point source. This is a difficult quantity to measure for galaxies, nebulae, or clusters which cover a large area of sky. Integrated magnitudes may be measured with standard filters such as B or V, or bolometrically.

integration time In data analysis, the interval over which data are averaged (*smoothed*) to reduce background noise and increase the signal-to-noise ratio.

intensity (I_ν or I_λ) A measure of the radiant power passing through unit area into unit solid angle in unit interval of frequency or wavelength. Intensity is also equal to the *flux density received from unit solid angle of an extended source (sometimes known as *surface brightness*). It is a characteristic of the source and does not depend on distance. For a given observing frequency, intensity is proportional to the *brightness temperature. The units are W/m^2 per Hz per steradian, or Jy/sr. In infrared astronomy, the unit W/m^2 per µm per steradian is often used.

intensity interferometer An optical interferometer for measuring the angular diameters of bright stars. An intensity interferometer operated from 1965 to 1972 at Narrabri in New South Wales, Australia, and consisted of two 6.5-m flux collectors on a circular railway track 188 m in diameter.

interacting binary A close binary in which there is mass transfer from the secondary to the primary component. In a *detached binary, the primary may accrete material from the secondary's stellar wind (*see* MIRA STAR; SYMBIOTIC STAR). In very close, *semidetached binary systems, a gas stream may impact on the primary (*see* ALGOL STAR), or, at greater separations, an *accretion disk may form (*see* CATACLYSMIC BINARY; DWARF NOVA; NOVA). When the primary has a strong magnetic field, it may gain material via an accretion column (*see* AM HERCULIS STAR).

interacting galaxy A galaxy whose structure is being altered by gravitational or other interaction with one or more nearby galaxies, for example the *Whirlpool Galaxy.

Interamnia Asteroid 704, discovered in 1910 by the Italian astronomer Vincenzo Cerulli (1859–1927). It is the fifth-largest main-belt asteroid, diameter 317 km. Interamnia is an F-class asteroid, with a rotation period of 8.73 hours. Its orbit has a semimajor axis of 3.061 AU, period 5.35 years, perihelion 2.60 AU, aphelion 3.52 AU, and inclination 17°.3.

Interface Region Imaging Spectrograph (IRIS) A NASA satellite due for launch in 2012 or later to obtain high-resolution images and spectra at ultraviolet wavelengths of the Sun's chromosphere, transition region, and lower corona. IRIS will be launched into a Sun-synchronous polar orbit from where it will make continuous observations through a 0.2-m telescope equipped with a multi-channel imaging spectrograph operating in the near and far ultraviolet.

((⊕)) SEE WEB LINKS
• Official mission website.

interference 1. A phenomenon in which two or more sets of electromagnetic waves may combine so as to increase their total amplitude at a given point (**constructive interference**), or decrease it (**destructive interference**), thereby forming an *interference pattern. For interference to occur, the waves must be coherent (i.e. the crests and troughs must be in step with each other). Constructive interference occurs where the crests coincide with each other, and destructive interference where crests coincide with troughs.
2. Unwanted radio signals received by a radio telescope that originate in human activity rather than natural phenomena.

interference filter A type of filter that uses the effect of interference of light to limit the wavelengths that pass through it. Ordinary filters are coloured, but interference filters consist of very thin multiple layers of alternating high and low refractive index (e.g. zirconium dioxide, magnesium fluoride) coated on glass. The incident light reflects backwards and forwards between the layers, thereby interfering with itself. By appropriate choice of thicknesses, the range of wavelengths that the filter transmits (the *passband*) can be adjusted. *Nebula filters and the anti-reflection coatings on camera and binocular lenses are simple interference filters.

interference pattern A pattern of alternating light and dark bands (*fringes*) caused by the interference of two or more beams of light; also known as a *fringe pattern*. By extension, the term is used in radio astronomy to describe the interference of radio signals received by two or more aerials of an interferometer.

interferometer An instrument designed to achieve high angular resolution by connecting two or more telescopes or antennas (*elements*) to operate as a single instrument. The signals from each element are combined to form an interference pattern. The maximum resolution of the array is then determined not by the size of the individual elements, but by their maximum separation. Interferometers are widely used in radio astronomy, and also in infrared and optical astronomy. At optical and infrared wavelengths, the incoming beam is split and then recombined with itself to form an interference pattern. This can give high resolution of spectral lines, as in a *Fabry–Perot interferometer. *See also* OPTICAL INTERFEROMETER; RADIO INTERFEROMETER.

intergalactic absorption The absorption of light in the space between galaxies. It is too weak to detect between our own Galaxy and its near neighbours. However, high-redshift (and hence distant) quasars frequently reveal many Lyman-α absorption lines at redshifts smaller than that of the parent quasar. This is the *Lyman-α forest, which arises from absorption by intervening clouds in intergalactic space.

intergalactic magnetic field The weak magnetic field of less than 10^{-12} tesla believed to be present in the space between galaxies. The intergalactic field may have arisen spontaneously in the very early Universe, or from the effect of atomic particles and magnetic fields from cosmic radio sources. It could be ten times stronger in clusters of galaxies.

intergalactic medium (IGM) The gas between the galaxies. In the early 1970s, the *Uhuru X-ray satellite showed that rich clusters of galaxies contain gas at a temperature of about 10^8 K. In a typical cluster the mass of this gas is actually larger than the total mass of all the galaxies in the cluster. The absorption lines in distant quasars—the *Lyman-α forest—show that there are also cool clouds of atomic hydrogen in intergalactic space. Our knowledge of the intergalactic medium, however, is seriously incomplete. For example, some astronomers argue that roughly half the normal matter in the Universe may be in the form of intergalactic gas with a temperature of about 10^6 K, but this gas would be virtually impossible to detect with current observational techniques.

intermediate-band photometry A general term for photometry with filters which have bandwidths of 10–30 nm. Examples are the *DDO, *Geneva, and *Strömgren photometry systems.

intermediate polar A type of *cataclysmic binary in which the magnetic field of the white-dwarf component is not strong enough to force the white dwarf to rotate synchronously (i.e. its rotation period is not the same as its orbital period); abbr. IP. Because the white dwarf does not rotate synchronously, the binary displays phenomena on a range of periods (including orbital, spin, and beat periods). Intermediate polars also differ from *polars* (*AM Herculis stars) by the presence of an accretion disk in some cases, depending on the strength of the magnetic field. However, in all IPs magnetic effects dominate the *mass transfer. The mass transfer predominantly occurs onto the polar caps of the white dwarf, via an *accretion curtain* from the inner edge of the accretion disk where such a disk exists, or via an *accretion column* if there is no disk. The *DQ Herculis* stars emit hard X-rays. The *DQ Herculis* stars (a term once synonymous with IPs) are now regarded as a sub-type of IPs, without hard X-ray emission and with a very rapidly rotating white dwarf. The type star, Nova Herculis 1934 (a famous *slow nova), has the shortest known white-dwarf rotation period (71 s), and a magnetic field some 30 times less than that of a typical polar.

intermediate-population star A star with properties between those of the old Population II stars in the galactic halo and the young Population I stars in the galactic disk. Their abundance of heavy metals is intermediate between that of the two populations, and they are distributed in a thick disk extending above and below the thin disk in which the stars of the disk population lie.

intermediate-type star A term sometimes used to describe stars of spectral types F, G, or K.

International Astronomical Union (IAU) The world governing body of astronomy, founded in 1919 to foster international cooperation in astronomical research. The IAU has its headquarters in Paris and holds a general assembly at different locations every three years. It appoints commissions on various aspects of astronomy, which consider matters concerning international agreements and standardization, including the naming of celestial bodies and

features. The IAU organizes the *Central Bureau for Astronomical Telegrams and the *Minor
Planet Center.

 SEE WEB LINKS
• Official IAU website.

International Atomic Time (TAI) The time-scale that has been used by international
agreement since 1972. It is ultimately based upon atomic oscillations, rather than
astronomical principles, and is the most accurate time-scale available today. Its fundamental
unit is the SI *second, defined in terms of properties of the caesium atom (*see* ATOMIC TIME).
TAI differs from *Coordinated Universal Time (UTC) by an exact number of seconds, because
*leap seconds are introduced in UTC to keep it in step with the changeable rotation rate of
the Earth. TAI was defined so as to be in agreement with UTI on 1958 January 1. *See also*
UNIVERSAL TIME.

International Celestial Reference Frame (ICRF) A highly accurate *frame of
reference for astrometric measurements adopted as the fundamental celestial reference frame
at the start of 1998, replacing the FK5 optical frame. It is based on the positions of 212 quasars
and radio galaxies measured by *VLBI (more sources were later added, bringing the total to
608). Distant radio sources are used because they do not exhibit the proper motions that
introduce long-term errors in the positions of nearby stars. At optical wavelengths, the positions
of stars in the *Hipparcos catalogue* are tied to the ICRF.

International Celestial Reference System (ICRS) An idealized coordinate system
that is independent of epoch, ecliptic, or equator. For ease of use it has been aligned close to the
mean equator and dynamical equinox of J2000.0. The origin of the coordinates is at the
centre of mass of the Solar System, i.e. they are *barycentric*. A reference system differs from
a reference frame in that it lays down the specifications for a coordinate system, whereas
a reference frame puts those specifications into practice; the *International Celestial
Reference Frame is the practical realization of the ICRS in the form of a list of coordinates
of specific celestial sources.

 SEE WEB LINKS
• Information page at the US Naval Observatory.

International Cometary Explorer (ICE) A NASA spacecraft, formerly called
*International Sun–Earth Explorer-3 (ISEE-3), which became the first spacecraft to be sent to
a comet. ISEE-3 was launched in 1978 August to study the solar wind. In 1983 December it
was retargeted to Comet *Giacobini–Zinner and renamed International Cometary Explorer.
In 1985 September it passed through the tail of Comet Giacobini–Zinner, 7800 km from the
nucleus. In 1986 March, ICE passed 28 million km from Halley's Comet.

International Date Line The imaginary line on the Earth's surface stretching from pole
to pole through the Pacific Ocean which, by international agreement in 1884, marks the
beginning or ending of a day. The line approximately follows the meridian of longitude 180°,
bending where necessary to avoid cutting through land. To the east of the line, the date is
one day behind the west of the line. On crossing this line in a westerly direction the calendar
must be put forward one day, while if it is crossed in an easterly direction the calendar is
put one day back.

International Gamma-Ray Astrophysics Laboratory (INTEGRAL) An ESA
gamma-ray astronomy satellite launched in 2002 October. INTEGRAL carries two main
instruments: an imager, IBIS, sensitive to the energy range 15 keV to 10 MeV (4×10^{-5} to
0.03 nm) with a resolution of 30 arc seconds; and a spectrometer, SPI, to measure gamma-ray
energies over the range 20 keV to 8 MeV. Two other instruments, an X-ray monitor (JEM-X)
and an Optical Monitoring Camera (OMC), help to identify the gamma-ray sources detected.
The imager, spectrometer, and X-ray monitor all use *coded-mask telescopes.

SEE WEB LINKS
• ESA mission website.

International Geophysical Year (IGY) An 18-month period, from 1957 July to the end
of 1958, in which world-wide collaborative observations were made to study the connection
between solar flares, geomagnetic disturbances, radio fade-outs, and particle emission from the
Sun. The IGY was designed to coincide with maximum sunspot activity, which turned out to be
the largest recorded since the invention of the telescope, and resulted in a greatly improved
picture of *solar–terrestrial relations. Sputnik 1, the first artificial satellite, was launched during
the IGY by the Soviet Union.

International Liquid Mirror Telescope A 4-m reflector with a mirror of liquid
mercury being built at Devasthal, India, by Belgium and Canada. The mirror consists of a dish
containing a thin layer of mercury spun at 7.5 rev/min to create a parabolic surface. The
telescope will scan a strip of sky half a degree wide at the zenith as the Earth rotates. The ILMT
is planned to begin operation in 2011. The partners in the project are the University of Liège,
Belgium; the University of British Columbia, Canada; and Laval University, Canada.

• Official telescope website.

International Sun–Earth Explorer (ISEE) A joint NASA–ESA series of probes to study
the Earth's magnetosphere and the Sun's effect on it. The ISEE-1 and ISEE-2 spacecraft were
both launched in 1977 October into orbits that carried them together into the magnetosphere.
The instruments on board included magnetometers and particle detectors. ISEE-3 was
launched in 1978 August into an orbit completely outside the magnetosphere to measure the
*solar wind flow independent of any influence from the Earth. ISEE-3 was later sent to intercept
Comet *Giacobini–Zinner, and was renamed the *International Cometary Explorer (ICE).

International Sunspot Number An index giving the number of spots on the Sun,
based on the *relative sunspot number system originated by R. *Wolf; symbol R_i. The Solar
Influences Data Analysis Center (SIDC, formerly the Sunspot Index Data Center), based at the
Royal Observatory of Belgium in Brussels, calculates the sunspot number from observations
made at over 25 sites around the world. The observatory at Locarno, Switzerland, is used as
a reference station to ensure consistency with the Zurich sunspot number used until 1981.

SEE WEB LINKS
• Solar Influences Data Analysis Center.

International Ultraviolet Explorer (IUE) A joint NASA–ESA–UK ultraviolet astronomy
satellite launched in 1978 January. It carried a 0.45-m telescope, with two spectrometers
covering the wavelength ranges 115–200 and 190–320 nm, both at low and high resolution.
From a geosynchronous orbit the satellite observed targets from Solar System objects to
bright stars and extragalactic objects as faint as 21st magnitude. IUE was switched off in 1996
September after 18 years, having become the longest-lived astronomical satellite.

SEE WEB LINKS
• ESA mission website.

International Virtual Observatory *See* VIRTUAL OBSERVATORY.

International Years of the Quiet Sun (IQSY) The period from 1964 January 1 to 1965
December 31, near *solar minimum, when solar and geophysical phenomena were studied
by observatories around the world and by spacecraft to improve our understanding of
*solar–terrestrial relations.

interplanetary matter Material found between the planets of our Solar System. It
includes the electrically charged particles of the solar wind, solid particles of interplanetary
dust, material from comets, and other gas and dust from interstellar space.

interplanetary scintillation The fluctuation in brightness ('twinkling') of a radio source
due to the scattering of radio waves by irregularities in the ionized gases of the *solar wind.

interpulse A small pulse that sometimes appears half-way between the main pulses from a pulsar. The interpulse is believed to represent emission from the pole of the neutron star opposite the one that produces the main pulse.

interstellar absorption The absorption of starlight by dust and gas in the space between the stars; also known as **interstellar extinction**. The dust and gas are strongly concentrated towards the plane of our Galaxy. Interstellar absorption increases towards the shorter-wavelength (blue) end of the spectrum, and hence makes stars appear redder, so that the terms **interstellar absorption** and **interstellar reddening** are used almost interchangeably. As well as continuous absorption there are *diffuse interstellar bands, such as that at 443 nm, due mostly to *polycyclic aromatic hydrocarbons and other large carbon-bearing molecules. There are also atomic absorption lines such as the calcium H and K lines at 393 and 397 nm, the sodium D lines at 589.0 and 589.6 nm, and other weaker lines.

Interstellar Boundary Explorer (IBEX) A NASA Small Explorer spacecraft launched in 2008 October to study the interactions between the solar wind and the interstellar medium. It does so by detecting so-called energetic neutral atoms (ENAs), created at the boundary of the heliosphere where the solar wind collides with interstellar gas. These atoms are accelerated into the inner Solar System where they are detected by IBEX in a highly elliptical orbit that takes it out to about 300 000 km from Earth, well outside the Earth's magnetosphere.

 SEE WEB LINKS
• Official mission website.

interstellar maser A maser in a star-forming region, in the dense clouds of gas and dust that surround young stars or protostars. They are the brightest maser sources in our Galaxy. Some interstellar masers are associated with violent ejections of matter from young stars, while others are located in disks of material around young stars from which planetary systems might be forming.

interstellar matter (ISM) The material between the stars, also called the **interstellar medium**. The ISM in our Galaxy consists of 99 % gas and 1 % fine dust particles by mass. There are also a few very energetic *cosmic-ray particles. The ISM is spread very thinly, with an average density of only 1 particle/cm^3. However, there are enormous variations about this average. The densest parts, which include *maser regions containing 10^{10} hydrogen molecules/cm^3, are also the coldest, with temperatures close to absolute zero. The hottest material has temperatures of 10^8 K, and densities of 1 particle/m^3 or less. The ISM is the reservoir of matter from which new stars form.

interstellar molecule Any molecule that occurs naturally in clouds of gas and dust in space. More than 150 such molecules have already been identified, chiefly by their emission or absorption of radio waves at particular wavelengths. The simplest molecule, H_2, is also the most abundant. Other known molecules include simple groupings such as carbon monoxide (CO), ammonia (NH_3), and water (H_2O); simple organic molecules such as ethyl alcohol (CH_3CH_2OH), formaldehyde (H_2CO), and acetic acid (CH_3COOH); and a variety of ions and radicals which are unstable on Earth, such as hydroxyl (OH), sulphur monoxide (SO), and the formyl ion (HCO^+). Many isotopic variants are also found, for example HDO, a form of water in which one hydrogen atom has been replaced by an atom of deuterium, a heavy form (isotope) of hydrogen. The largest molecule yet identified is cyanodecapentayne ($HC_{11}N$). More complicated molecules could be responsible for spectral lines that have been detected but not identified. Interstellar molecules regulate the temperature of the gas clouds by radiating away energy at radio wavelengths. This energy loss allows some of the densest regions of the cloud to collapse into stars. Molecules are also formed in the envelopes of gas and dust around old stars. These are usually termed *circumstellar molecules* because they have formed from material that came from the star, whereas interstellar molecules are formed from general interstellar matter.

interstellar reddening *See* INTERSTELLAR ABSORPTION.

interstellar scintillation The fluctuation in brightness ('twinkling') of a radio source due to the scattering of radio waves by irregularities in interstellar matter (ISM). Much of the fluctuation is caused by the relative motion of the Earth and the source rather than of the ISM, and occurs on a time-scale of minutes to hours. Measurements of scintillation rate can be used to determine the velocities of pulsars across the line of sight. Interstellar scintillation of pulsar signals is closely related to *pulse broadening.

interstellar wind The stream of interstellar gas and dust that appears to be moving past the Solar System because of the motion of the Sun around the Galaxy. It starts at the *heliopause, where the solar wind gives way to the interstellar wind. The Pioneer and Voyager probes now travelling out of the Solar System are expected to encounter the interstellar wind early in the 21st century.

intrinsic colour index The *colour index a star would have in the absence of *interstellar absorption. It is assumed that all stars of the same spectral type and luminosity class have the same intrinsic colour index.

intrinsic variable A variable star whose fluctuations in brightness are caused by actual changes in the luminosity of the star itself, not by external processes such as rotation or eclipses (as in an *extrinsic variable). The majority of variable stars fall into this broad category.

invariable plane A fixed plane passing through the centre of mass of the Solar System, and oriented perpendicular to the axis about which the angular momentum of the Solar System is measured. The invariable plane is inclined at $1°.58$ to the ecliptic, between the orbital planes of Jupiter and Saturn, but its precise location is not known because the masses of all the objects in the Solar System and their positions and velocities at a given time are not yet exactly known. Being fixed, it provides a permanent reference plane, whereas the ecliptic alters with time because of planetary perturbations.

inverse Compton effect The gain in energy of a photon when it is struck by a fast-moving electron; also known as **inverse Compton scattering** or **Compton upscattering**. The electron passes on a small proportion of its energy to the photon, and the photon's wavelength decreases. The electron has to suffer a large number of collisions before it loses an appreciable fraction of its energy. It is the opposite of the *Compton effect.

inverse Compton scattering *See* INVERSE COMPTON EFFECT.

inverse P Cygni profile *See* P CYGNI PROFILE.

inverse-square law A law that describes how the strength of a force (e.g. gravitation) or a flow of energy (e.g. light) weakens with distance from the source. According to the inverse-square law, the magnitude of a given quantity is inversely proportional to the square of the distance from the source. For example, if an object's distance were doubled, the strength of the light or gravity from it would fall by four times; if its distance were tripled, the strength of its light or gravity would be nine times less.

inversion layer *See* TRANSPARENCY, ATMOSPHERIC.

Io The third-largest satellite of Jupiter, diameter 3643 km, and the innermost of the four Galilean satellites; also known as Jupiter I. It orbits in 1.769 days at a distance of 422 000 km, keeping the same face turned towards Jupiter. Its geometric albedo is 0.6 and its opposition magnitude 5.0. The two Voyager spacecraft in 1979 revealed that Io has explosively erupting volcanoes which eject plumes of sulphur and sulphur dioxide over 300 km high, some of it landing up to 500 km from the eruption site. The sources of the eruptions are volcanic calderas or fissures, of which there are over 300. Huge lava flows radiate from many of the volcanoes, and the whole surface is yellowish in colour due to deposits of sulphur or sulphur oxides. There are extensive plains and mountainous regions on Io but no impact craters, indicating that its surface is geologically very young. Io's density, 3.53g/cm^3, suggests that it has an iron–sulphur core about 1500 km across and a silicate mantle. Io's volcanic activity is the result of heat released by tidal forces, which deform the satellite as it moves alternately closer to and farther away from Jupiter in its orbit. 📷

iodine cell A glass cell containing iodine gas placed in the light path through a spectrograph, superimposing many strong, sharp absorption lines on the spectra of astronomical sources. These lines provide a reference frame that allows extremely accurate velocity measurements, such as are required when searching for slight changes in radial velocities of stars due to the presence of orbiting planets.

ion An atom or molecule that has lost (or, more rarely, gained) one or more electrons. An ion is positively charged if it has lost one or more electrons, and negatively charged if it has gained one or more electrons. Ions are designated by the chemical symbol for the parent atom followed by a superscript indicating the charge; thus doubly ionized iron is represented as Fe^{2+}. Alternatively, the symbol is followed by a Roman numeral which is one more than the number of electrons lost (e.g. Fe III for doubly ionized iron, Fe I for neutral iron).

ionization The process by which an atom or molecule loses or gains electrons. Atoms that have lost or gained one electron are said to be **singly ionized**; if they have lost or gained two electrons they are **doubly ionized**; and so on. Ionization occurs at high temperatures (thermal ionization), as in a star, or by the impact of high-energy atomic particles (e.g. electrons, protons, alpha particles) or short-wavelength radiation (ultraviolet, X-rays, gamma rays). *See also* BOUND–FREE TRANSITION.

ionization equilibrium A balance achieved in a hot gas when the number of ionizations of a particular ion equals the number of recombinations into that ion. An example of a gas in ionization equilibrium is the Sun's corona. Thus, in ionization equilibrium, the number of ionizations of the ion Fe^{9+}, which emits the red *coronal line, to the Fe^{10+} ion equals the number of recombinations from Fe^{10+} to give Fe^{9+}.

ionization front A boundary separating electrically neutral gas from ionized gas. There are several astronomical situations in which light or ultraviolet radiation falls on neutral gas and ionizes it. For example, when a massive young star first begins to shine it ionizes the surrounding gas, first the nearest gas and then gas progressively farther away. The ionization front is the outward-moving boundary of the ionized region.

ionization potential (*I***)** The energy required to remove the least strongly bound electron from an atom or ion. It is usually measured in electronvolts (eV). The energy required increases as the level of ionization increases. Thus the ionization potential of neutron iron is 7.9 eV, that of singly ionized iron is 16.2 eV, doubly ionized iron, 30.7 eV, and so on.

ionization temperature The temperature of a gas or plasma obtained from the relative numbers of neutral atoms and ions. It is the temperature at which the *Saha ionization equation most nearly gives the observed levels of ionization in a star.

ionopause The upper boundary of the *ionosphere, lying at an altitude of around 250 km at the top of the *F layer.

ionosphere The collective term for the various layers of ionized particles and electrons found at altitudes of 80–250 km in the atmosphere. The principal regions are the *D layer, *E layer, and *F layer. Ionization is caused primarily by short-wavelength (X-ray and ultraviolet) solar radiation during the daytime.

ionospheric scintillation The fluctuation in brightness of a radio source due to the scattering of radio waves by irregularities in the Earth's ionosphere.

ion tail Another name for a comet's gas tail. *See* TAIL, COMETARY.

Iota Aquarid meteors A comparatively weak meteor shower, with ill-defined activity limits during July and August. The most prominent maximum falls on August 6, when the ZHR may reach 10. Iota Aquarids tend to be faint and fairly swift. The meteor stream, like many others that lie close to the ecliptic, is split into at least two components. At maximum, the Southern Iota Aquarids have a radiant at RA 22 h 10 m, dec. −15°, while the Northern Iota Aquarids emanate from RA 22 h 04 m, dec. −06°. Some authorities regard them not as a separate shower but as part of the general activity from the *anthelion radiant.

IQSY Abbr. for *International Years of the Quiet Sun.

IR Abbr. for infrared.

IRAM Abbr. for *Institut de Radio Astronomie Millimétrique.

IRAS Abbr. for *Infrared Astronomical Satellite.

IRAS–Araki–Alcock, Comet (C/1983 H1) A long-period comet detected by the *Infrared Astronomical Satellite (IRAS) on 1983 April 25, and found independently on May 3 by the Japanese amateur astronomer Genichi Araki (1954–) and the English amateur George Eric Deacon Alcock (1912–2000); formerly designated 1983 VII. It made an unusually close passage of 0.031 AU (4.6 million km) to Earth on 1983 May 11, when it was of 2nd magnitude and moving rapidly across the northern sky. Because of its proximity to Earth, the comet appeared diffuse, with a diameter of 2°. Perihelion, 0.99 AU from the Sun, was on 1983 May 21. Its orbit has a period of about 960 years, eccentricity 0.990, and inclination 73°.3.

Iris Asteroid 7, discovered in 1847 by the English astronomer John Russell Hind (1823–95). Iris is an S-class asteroid, of diameter 200 km. Its mean magnitude at opposition is 8.4; of the main-belt asteroids, only Vesta, Ceres, and Pallas can become brighter. Its orbit has a semimajor axis of 2.386 AU, period 3.68 years, perihelion 1.84 AU, aphelion 2.94 AU, and inclination 5°.5.

iron meteorite A meteorite composed primarily of nickel–iron; also known as a *siderite*. Some iron meteorites are probably material from the cores of asteroid-sized parent bodies, while others may have originated at shallower depths. Irons are the easiest meteorites to identify, because they are heavy, look metallic, are strongly magnetic, and are usually covered with pits, depressions, and grooves from their passage through the atmosphere. On impact they have a blue-black fusion crust, with traces of metal melted during the flight, but after exposure on Earth they rust, turning brown. Iron meteorites comprise only 4% of observed falls, but resist weathering better than stony meteorites and so are found more easily. Iron meteorites are divided into three main subgroups on the basis of the structure of their nickel–iron alloy: *hexahedrites, *octahedrites (six types), and the *ataxites. An iron meteorite which cannot be included in one of these groups is termed *anomalous*. The largest known meteorites are irons; examples are the *Cape York and *Hoba West meteorites. *Meteor Crater in Arizona was formed by the impact of an iron meteorite.

iron peak Elements whose nuclei contain a total of 56 protons and neutrons (i.e. they have a mass number of 56), for example ^{56}Fe, ^{56}Ni, or ^{56}Co. Their formation by nuclear fusion inside a star marks the end-point of the star's energy output. Further fusion to create elements with higher mass numbers would absorb energy rather than liberate it. Nuclei of ^{62}Ni have the maximum *binding energy, but those of ^{56}Fe have nearly the same binding energy and are more abundant as they are produced more readily in stellar *nucleosynthesis.

irradiance (E_e) The energy at all wavelengths that falls on a surface in a given time. It is measured in watts per square metre. The energy input at visual wavelengths alone is termed *illuminance.

irradiation 1. An optical contrast effect that makes a bright object appear larger than it really is when viewed against a darker background.
   2. Exposure to any form of electromagnetic radiation or atomic particles.

irregular galaxy A type of galaxy with ill-defined structure; Hubble type Irr or Ir. There are two main types in the *Hubble classification. Irr I galaxies are not as massive as large ellipticals or spirals, often have a high gas content, and are undergoing star formation. The categories Irr$^+$ and Irr$^-$ are sometimes used to denote the degree of resolution of the structure of the galaxy into star clusters, H II regions, and other features, with Irr$^+$ showing higher resolution than Irr$^-$. The high gas content of Irr I galaxies implies that they may not have undergone much evolution since their formation. A subdivision of Irr I, the so-called **Magellanic irregulars**, symbol Im, resemble the Magellanic Clouds in the Local Group. The Hubble type Irr II denotes galaxies of unusual appearance that are classified as irregular simply because they do not fit into any other class, and may in many cases represent interacting or merging systems.

irregular variable A star that exhibits an irregular light-curve. There are two types, with very different characteristics: *eruptive variables near the main sequence; and evolved *pulsating variables.

Irregular eruptive variables are divided into three broad groups. The poorly studied variables, designated I, are subdivided into IA for early spectral types (O–A) and IB for those with intermediate to late spectra (F–M). Stars associated with nebulosity (IN) may vary by several magnitudes with rapid changes (up to 1 magnitude in 1–10 days), and are divided by spectra into types INA and INB. Finally, there are IS stars with rapid variations (0.5–1 mag. in hours or days), subdivided into types ISA and ISB, although this group is poorly defined. Stars with specific emission features are classified as *T Tauri stars (INT), or in rare cases when nebulosity is not present, IT. When absorption indicates the infall of material, the star is a *YY Orionis star, type IN(YY).

Slow, irregular, pulsating giants or supergiants are designated type L. Many are poorly studied, and may subsequently be found to be *semiregular variables or other types. All have late spectral types (K–M, C, and S). Giants are generally classified as LB; supergiants with amplitudes of around 1 mag. are LC.

IRTF Abbr. for *NASA Infrared Telescope Facility.

Isaac Newton Group (ING) A group of telescopes at the *Roque de los Muchachos Observatory on La Palma in the Canary Islands, jointly owned and operated by the UK, Netherlands, and Spain. As well as the 4.2-m *William Herschel Telescope and the 2.5-m *Isaac Newton Telescope, the group includes the 1-m Jacobus Kapteyn Telescope (JKT), opened in 1984. The headquarters of the group is at Santa Cruz on La Palma island.

((⊕)) SEE WEB LINKS
• Official observatory website.

Isaac Newton Telescope (INT) A 2.5-m reflector at the *Roque de los Muchachos Observatory on La Palma in the Canary Islands, jointly owned and operated by the UK, Netherlands, and Spain. From 1967 to 1979 the INT was sited at Herstmonceux, Sussex. It was then given a new mirror and moved to La Palma in 1984.

ISEE Abbr. for *International Sun–Earth Explorer.

Ishtar Terra An upland area in the northern hemisphere of Venus, 5600 km long and 600 km wide. It includes the high volcanic plateau Lakshmi Planum, itself over 2000 km across, which rises to nearly 5 km above the planet's mean radius. This contains *Maxwell Montes, at 11 km the highest peak on Venus, and two large volcanic calderas, Colette Patera (149 km diameter) and Sacajawea Patera (233 km diameter). Lakshmi is surrounded by mountain ranges and tesserae.

ISM Abbr. for *interstellar matter or interstellar medium.

ISO Abbr. for *Infrared Space Observatory.

isochron A straight line on a graph from which the age of a rock can be derived. The graph plots the abundances of *parent isotopes against the abundance of their *daughter isotopes for several minerals from the same rock, or several rocks formed from the same source at the same time. The points lie on a straight line (the isochron) whose slope can be used to calculate the age of the rock. A steeper slope means an older rock, less of the parent isotope remaining. *See also* RADIOACTIVE AGE DATING.

isochrone In astronomy, a curve on the *Hertzsprung–Russell diagram that shows the predicted locations of stars of various masses at a specified time, assuming they were all born simultaneously. By fitting isochrones to observed HR diagrams of star clusters it is possible to estimate the age of the cluster.

isophotal wavelength The wavelength at which the *monochromatic magnitude of a star is equal to its magnitude measured through a broad-band filter. It varies with the colour of the star and with the amount of atmospheric extinction.

isophote A line joining points with the same surface brightness on a diagram or an image of a celestial object such as a galaxy or nebula. The surface brightness is usually measured in magnitudes per square arc second. The sum of all the light within a given isophote is termed the **isophotal magnitude**.

isotherm A line on a map or graph connecting points of equal temperature.

isothermal process A change or process that occurs at a constant temperature. During an isothermal process, energy enters or leaves the system to maintain a constant temperature. The collapse of a protostar down the *Hayashi track, and the collapse of a star at the end of its life to become a white dwarf, are examples of isothermal changes. *See also* ADIABATIC PROCESS.

isothermal region A region over which the temperature is constant. For example, the volume within a *Strömgren sphere (H II region) is close to isothermal because the rate of ionization of atoms balances the rate of recombination.

isotope One or more atomic variants of a chemical element which have the same number of protons in their nucleus but different numbers of neutrons. Most elements have several stable isotopes; in addition, a few elements have natural radioactive isotopes (**radioisotopes**) which are unstable. These radioactive nuclei (the **parent isotopes**) disintegrate spontaneously into different atoms, often of a different element (the **daughter isotopes**). The ratio of the parent and daughter isotopes is used in *radioactive age dating.

isotropy A characteristic of a substance or body in which physical properties are the same in all directions, such as the cosmic microwave background and the very large-scale distribution of matter throughout the Universe. An object such as a star that emits radiation equally in all directions is known as an **isotropic radiator**. An ideal antenna that is equally sensitive in all directions is known as an **isotropic antenna**. *See also* ANISOTROPY.

IUE Abbr. for *International Ultraviolet Explorer.

Izar The star Epsilon Boötis, a close double consisting of a K0 giant and an A0 dwarf of magnitudes 2.5 and 4.6, appearing orange and blue-green. Their colours give rise to the alternative name Pulcherrima, meaning 'most beautiful'. The pair lie 203 l.y. away.

James Clerk Maxwell Telescope (JCMT) A 15-m radio telescope opened in 1987 at an altitude of 4092 m at the *Mauna Kea Observatories, Hawaii, owned by the UK, Canada, and the Netherlands, and operated by the *Joint Astronomy Centre in Hawaii. It is designed to work at millimetre and submillimetre wavelengths (2 mm to 0.3 mm). It is named after J. C. *Maxwell.

(⊕) SEE WEB LINKS
• Official telescope website.

James Webb Space Telescope (JWST) A successor to the Hubble Space Telescope, planned for launch in 2014 or later. It is being built by NASA with contributions from ESA and the Canadian Space Agency. The telescope will have a lightweight mirror 6.5m in diameter consisting of 18 hexagonal segments which will be folded for launch and opened when in space. JWST will be positioned at the L_2 *Lagrangian point, 1.5 million km from Earth on the side away from the Sun. It is designed to operate at infrared wavelengths, to study heavily redshifted light from early in the history of the Universe. A sunshield will block the Sun's rays, keeping the four infrared instruments cool. The Near-Infrared Camera (NIR Cam) will take images in the wavelength range 0.6–5 μm; the Near-Infrared Spectrograph (NIR Spec) will collect spectra of more than 100 objects simultaneously in the range 1–5 μm; and the Mid-Infrared Instrument (MIRI) will perform imaging and spectroscopy at wavelengths of 5–27 μm, and the Tunable Filter Imager (TFI), a narrow-band camera that provides imagery over a wavelength range of 1.5 to 5 μm. The JWST is named after James Edwin Webb (1906–92), NASA's second administrator who served from 1961 to 1968. It will be operated by the *Space Telescope Science Institute. 📷

(⊕) SEE WEB LINKS
• Official mission website.

jansky (Jy) A unit of flux density used in radio and infrared astronomy. It was formerly known as a *flux unit*. One jansky equals 10^{-26} W/m^2 per Hz. It is named after K. G. *Jansky.

Jansky, Karl Guthe (1905–50) American radio research engineer. In 1931 he began to study atmospheric 'static' that interfered with telecommunications. By the end of 1932 he had accounted for all but one persistent signal, which he concluded came from outside the Solar System in the direction of the constellation Sagittarius, towards the centre of our Galaxy. Thus was born radio astronomy, although Jansky himself never followed up his discovery. In radio astronomy, the unit of flux density is now named the *jansky in his honour.

Janssen, (Pierre) Jules César (1824–1907) French spectroscopist. In 1862 he discovered and named the *telluric lines in the Sun's spectrum which originate in the Earth's atmosphere, and realized that similar lines in the spectra of planets would reveal the composition of their atmospheres. In 1868, independently of J. N. *Lockyer, he observed the spectra of solar prominences at a total eclipse, and went on to invent the *spectrohelioscope. He found a new line in the solar spectrum, which Lockyer attributed to a new element later named helium. In a celebrated adventure in 1870, during the Franco-Prussian War, Janssen escaped by balloon from a besieged Paris and flew to the Atlantic coast to observe a solar eclipse.

Janus A satellite of Saturn, in virtually the same orbit as *Epimetheus between the F and G Rings (the tenuous rings outside the A Ring), at a mean distance of 151 460 km;

also known as Saturn X. Its mean orbital period is 0.695 days. Janus was discovered in 1966 by the French astronomer Audouin Charles Dollfus (1924–2010). It is an irregular body, 203 × 185 × 153 km in size.

JCMT Abbr. for *James Clerk Maxwell Telescope.

JD Abbr. for *Julian Date.

Jeans, James Hopwood (1877–1946) English mathematician, physicist, and astronomer. By considering the stability of a rotating mass of fluid, and building on earlier work by J. H. *Poincaré and others, he developed theories of the formation and evolution of stars, double-star systems, and spiral galaxies. In 1917 he proposed a theory, subsequently expanded on by the English mathematician Harold Jeffreys (1891–1989), in which the planets condensed from filaments pulled out of the Sun by a passing star. In 1928 Jeans advanced a cosmological theory in which matter was continually being created—a cornerstone of the later *steady-state theory. He then turned to popular science writing and broadcasting.

Jeans length The minimum size of a cloud of gas, of given temperature and density, which is capable of collapsing under its own gravity. Gas clouds smaller than this do not collapse because their gas pressure more than balances their gravitational force. The Jeans length is proportional to the square root of the temperature, and inversely proportional to the square root of the density. For the dense gas characteristic of star-forming clouds it is a few tenths of a light year. It is named after J. H. *Jeans.

Jeans mass The mass contained in a sphere whose radius is the *Jeans length, and thus the minimum mass that can collapse under its own gravity. For the dense gas characteristic of star-forming clouds, it is about a solar mass.

jet A narrow, bright feature associated with several types of astronomical object. Jets are seen mainly at radio but occasionally at other wavelengths. They emerge from the core of certain active galaxies and may extend for many hundreds of kiloparsecs. In some sources, notably *radio galaxies, jets are commonly seen on both sides of the nucleus (**double-sided jet**) while in others, such as quasars, only one jet is normally seen (**single-sided jet**). Some jets exhibit the phenomenon of *superluminal velocity. Optical stellar jets are seen emerging from young stars such as *T Tauri stars and *FU Orionis stars, and are associated with *Herbig–Haro objects. The X-ray binary system *SS433 has two jets.

Jet Propulsion Laboratory (JPL) A NASA establishment in Pasadena, California, founded in 1944 for missile development but which became part of NASA in 1958. It is operated for NASA by the California Institute of Technology (Caltech). JPL tracks and controls space probes, as well as developing new planetary missions. JPL manages the world-wide *Deep Space Network, including its own antennas at Goldstone in the Mojave Desert. JPL also operates an observatory at Table Mountain, California, altitude 2290 m, equipped with a 1.2-m reflector and smaller instruments.

(⊕) SEE WEB LINKS
• Official website.

Jewel Box A 4th-magnitude open cluster in Crux, near the Coalsack dark nebula, also known as NGC 4755 or the Kappa Crucis Cluster. Most of the bright stars in the cluster are blue-white in colour, including Kappa Crucis, a magnitude 5.9 supergiant, but an 8th-magnitude red supergiant lies near the centre. The cluster is just under 5000 l.y. away. It was named by J. F. W. *Herschel who compared it to a collection of jewels. The name Jewel Box is sometimes also applied to M6 in Scorpius, better known as the *Butterfly Cluster.

Jilin meteorite The greatest recorded fall of stony meteorites, which occurred near Jilin (formerly known as Kirin) in Manchuria, northeastern China, on 1976 March 8. About 4 tonnes of fragments of the H5 chondrite meteorite were collected. One fragment, weighing 1.77 tonnes, is the largest single piece of stony meteorite ever recovered.

J magnitude The magnitude of a star as measured through the infrared J filter in the *Johnson photometry system. The J filter has an effective wavelength of 1250 nm and a bandwidth of 300 nm.

Job's Coffin A quadrilateral of stars in Delphinus, formed by Alpha, Beta, Gamma, and Delta Delphini.

Jodrell Bank Observatory The radio astronomy observatory of the University of Manchester at Jodrell Bank in Cheshire, England, founded by A. C. B. *Lovell in 1945. Its major instruments include the 76-m *Lovell Telescope, the 25 × 38-m elliptical Mark II telescope, built in 1964, and the *Multi-Element Radio-Linked Interferometer Network.

(⊕) SEE WEB LINKS
• Official observatory website.

Johnson photometry Photometry based on the work of the American astronomer Harold Lester Johnson (1921–80). It falls into three natural parts: UBV, RI, and *infrared photometry. UBV photometry was based on a photomultiplier sensitive to the wavelength range 300–600 nm. The U and B magnitudes were formed by splitting the old *photographic magnitude into blue (B) and ultraviolet (U) parts. The V magnitude is similar to but more accurate than the old visual system. Johnson later extended this work into the red and infrared with two further filters designated R and I, using a photomultiplier sensitive to wavelengths 0.3–1.1 μm. The third innovation was the extension into *infrared photometry with J, H, K, L, M, and N filters covering the wavelength range 1.25–10.4 μm.

Johnson Space Center (JSC) A NASA establishment near Houston, Texas, founded in 1961 to design and develop manned spacecraft, to select and train astronauts, and to conduct their missions. JSC houses the Mission Control for Shuttle flights and also a separate control centre for the International Space Station. Lunar samples collected by the Apollo missions are stored at JSC.

(⊕) SEE WEB LINKS
• Official website.

Joint Astronomy Centre (JAC) An establishment in Hilo, Hawaii, owned by the UK, which operates the *United Kingdom Infrared Telescope and the *James Clerk Maxwell Telescope, the latter jointly with Canada and the Netherlands. The JAC was originally founded in 1978 as the headquarters for the UK Infrared Telescope, but its role was enlarged and it was renamed in 1988.

(⊕) SEE WEB LINKS
• Official JAC website.

Jones, Harold Spencer (1890–1960) English astronomer. From over a thousand observations of the asteroid *Eros at its opposition in 1930/31, he calculated a value of 8″.790 for the *solar parallax. The task, which took him ten years, was one of the most impressive calculational feats of the pre-computer era. Jones was appointed tenth Astronomer Royal in 1933. In this role he introduced improvements to the timekeeping system at Greenwich (and demonstrated that the rotational speed of the Earth is not constant) and instigated the Royal Observatory's move to Herstmonceux, Sussex.

Joule (J) The unit of work and energy. It is defined as the work done when a force of one newton is moved through a distance of one metre in the direction of the force. It is named after the English physicist James Prescott Joule (1818–89).

Jovian planet Any of the four giant planets in the Solar System—Jupiter, Saturn, Uranus, and Neptune; also known as a *gas giant*. The Jovian planets have thick atmospheres, low densities, large diameters, and are primarily composed of hydrogen and helium.

JPL Abbr. for *Jet Propulsion Laboratory.

Julian calendar The form of calendar first introduced in 46 BC by the Roman emperor Julius Caesar, after whom it is named. It was prepared in consultation with the Greek astronomer Sosigenes (1st century BC). Each month was assigned the number of days it has today, and a normal year had 365 days. It was intended to be a solar calendar in which the date remained in step with the seasons. To preserve this link, every fourth year was a leap year with February having an additional day. The average length of a year is then 365.25 days, close to but not exactly equal to the *tropical year. The Julian calendar was superseded by the more precise *Gregorian calendar in 1582.

Julian Date (JD) A calendar and timekeeping system introduced to give unambiguous dates and times of celestial events, unaffected by changes in the civil calendar. The system was begun in 1582 by the French scholar Joseph Justus Scaliger (1540–1609), who named it to honour his father, the Italian-born French scholar Julius Caesar Scaliger (1484–1558). The starting date is chosen as January 1 of the year 4713 BC, sufficiently far in the past that it predates all known recorded astronomical observations. Time is measured from *mean noon* (12 h UT) on that date by the number of days and fractions of a day elapsed. For example, an observation made on 1962 June 24 at 18 h UT was made at JD 2 437 840.25. The integral part of the Julian date is called the Julian day number, and the fractional part is the universal time expressed as a fraction of a day. *See also* MODIFIED JULIAN DATE.

Julian day number (JD) The whole-number part of the *Julian Date, and therefore the number of days that has elapsed since 12 h UT on January 1, 4713 BC.

Julian epoch A method of specifying the date as a year with a decimal part, as for example in 1995.5. The unit is the Julian year of 365.25 days and the time-scale used is *Barycentric Dynamical Time (TDB). It is often convenient to use the standard epoch, J2000.0, for example, in star catalogues. The prefix 'J' is added to distinguish this epoch from the more complicated *Besselian epoch used prior to 1984. The standard epoch is defined as

J2000.0 = 2000 January 1.5 = 2000 January 1 12 h TDB

The date for other epochs can be calculated from this. For example, J1999.0 is exactly 365.25 days earlier, that is, 1999 January 1.25.

Juliet The sixth-closest satellite of Uranus, distance 64 360 km, orbital period 0.493 days; also known as Uranus XI. Its diameter is 84 km, and it was discovered in 1986 on images from the Voyager 2 spacecraft.

Juno Asteroid 3, the third asteroid to be discovered, by the German astronomer Karl Ludwig Harding (1765–1834) in 1804. Its diameter is 234 km. Juno is of S class, with a rotation period of 7.21 hours. Its orbit has a semimajor axis of 2.670 AU, period 4.36 years, perihelion 1.99 AU, aphelion 3.35 AU, and inclination 13°.0.

Juno mission A NASA Jupiter orbiter, due for launch in 2011 August. After undergoing an Earth gravity assist in 2013 October, Juno will arrive at Jupiter in 2016 July and go into a highly elliptical polar orbit, bringing it as close as 4500 km from Jupiter's cloud tops. Juno will map Jupiter's gravitational field and magnetosphere and analyse its atmospheric composition and structure.

(⊕) SEE WEB LINKS
• Official mission website.

Jupiter (♃) The fifth planet from the Sun, and the largest planet in the Solar System, with a mass two and a half times greater than all the other planets combined. Its maximum opposition magnitude is −2.9, making it normally the second-brightest planet after Venus. Jupiter is distinctly ellipsoidal in shape (equatorial diameter 142 984 km, polar diameter 133 708 km). The rotation period of the visible surface is about 9 h 50 m in equatorial regions and about 9 h 56 m for the rest of the planet (*see* SYSTEMS I AND II). The rotation of the solid surface, derived from radio observations, is thought to be 9 h 55 m 29s, known as System III.

Jupiter has a thick atmosphere composed of about 90 % hydrogen and 10 % helium (molecular percentages), plus traces of methane, ammonia, water, ethane, ethyne, phosphine, carbon monoxide, and germanium tetrahydride. Near the top of the atmosphere the temperature is around −143°C. Internally, Jupiter is thought to have an iron-rich rocky and perhaps icy core about 10 times the mass of the Earth. The pressure at the centre is thought to be about 10^8 bar. Surrounding this is a layer of dense hydrogen and liquid helium. Nearly 20 000 km below the surface the pressure reaches 3 million bar. Under these pressures the hydrogen begins to behave like a liquid metal, with convection currents that are probably responsible for Jupiter's strong magnetic field. The metallic hydrogen layer, about 50 000 km thick, is surrounded by a layer of normal molecular liquid hydrogen, which gradually merges into gaseous hydrogen near the surface. Giant aurorae and electrical storms occur in the planet's turbulent atmosphere. Jupiter's centre is estimated to be very hot, about 20 000 K. This internal heat is left over from the kinetic energy of impacts during accretion, and from the conversion of gravitational potential energy into heat when the core formed.

Jupiter

Physical data

Diameter (equatorial)	Oblateness	Inclination of equator to orbit	Axial rotation period (sidereal)	
142 984 km	0.065	3°.13	9.842 hours	
Mean density	Mass (Earth = 1)	Volume (Earth = 1)	Mean albedo (geometric)	Escape velocity
1.33 g/cm^3	317.8	1321	0.52	60.2 km/s

Orbital data

Mean distance from Sun

10^6 km	AU	Eccentricity of orbit	Inclination of orbit to ecliptic	Orbital period (sidereal)
778.412	5.203	0.048	1°.3	11.863 years

Jupiter's visible surface consists of dark belts or bands where the atmospheric gases are descending, with bright zones of rising gas between. Dark and bright spots are frequently seen near the belts, together with wisps and festoons suggestive of turbulence in the atmosphere. Most spots appear and disappear in a few days. Some last for months, and a few for much longer. The best-known feature is the *Great Red Spot, first recorded in 1831. Another long-lived feature was the South Tropical Disturbance, a dark bridge across the bright south tropical zone, which was first seen in 1901 and finally disappeared 39 years later. Three white ovals south of the south temperate belt were formed in 1940–42; between 1998 and 2000 these merged into one. This new oval initially remained white like the spots that formed it, but in late 2005 it began to darken and in 2006 became the same colour as the Great Red Spot, although only about half the size. The Great Red Spot itself may have originated from the merger of smaller ovals in this way.

An extraordinary event known as an *SEB revival* occurs from time to time in Jupiter's southern hemisphere. The southern component of the south equatorial belt (SEB) fades over a period of months until it is virtually invisible, leaving the Great Red Spot isolated. Then a sudden outbreak of dark spots begins from one point on the northern component of the SEB, and spreads at high velocity around the planet, creating an almost explosive turmoil in this zone. Sometimes the disturbance affects much of the planet, as in 1975. At the end of the disturbance, the southern component of the SEB returns to its former prominence.

Jupiter has over 60 known moons, the four brightest of which, the *Galilean satellites, can be seen through binoculars. In 1979, the Voyager probes discovered a very faint ring of particles around Jupiter with an albedo of about 0.05. There are now known to be three parts to the ring: the Halo ring, 100 000–122 800 km from the planet's centre; the Main ring (radius 122 800–129 200 km) and the Gossamer ring (129 200–214 200 km).

Jupiter-family comet A member of the group of periodic comets whose orbits have been modified by close passages to Jupiter. Jupiter-family comets have orbital periods less than 20 years and direct orbits with inclinations below 40°. An example is Comet 16P/Brooks 2, whose orbit was shortened from an initial period of 29 years to only 7 years after passing within 0.001 AU of Jupiter in 1886. The comet's perihelion distance was decreased from 5.48 to 1.95 AU. Tidal disruption by Jupiter's gravity split the nucleus of Comet Brooks 2 into several fragments. Other celebrated Jupiter-family comets are *Encke, *Giacobini–Zinner, *Grigg–Skjellerup, *Tuttle–Giacobini–Kresák, 67P/Churyumov–Gerasimenko (the target of the *Rosetta probe), and 81P/Wild 2 (visited in 2004 by the *Stardust mission). Comets in the Jupiter family probably originated from the Kuiper Belt. As of the end of 2010 over 400 members of the family were known.

JWST Abbr. for *James Webb Space Telescope.

K Symbol for kelvin, the unit of temperature on the *thermodynamic temperature scale. The kelvin unit is equal in size to the Celsius degree. It is named after Lord *Kelvin. *See* KELVIN SCALE.

Kaguya A Japanese Moon probe, known as Selene before its launch in 2007 September. In 2007 October Kaguya was placed into polar orbit around the Moon where it released two subsatellites. One was a relay satellite called Okina; the other was a VLBI Radio (VRAD) satellite called Ouna. Both the main orbiter and Ouna use the relay satellite to communicate with Earth when they are on the far side of the Moon. Kaguya itself carries 13 instruments including imagers, a radar sounder, laser altimeter, X-ray fluorescence spectrometer, and gamma-ray spectrometer to study the origin, evolution, and tectonics of the Moon from orbit.

(⊕) SEE WEB LINKS
• Official mission website.
• Mission overview at Japan Aerospace Exploration Agency.

Kant, Immanuel (1724–1804) German philosopher. He proposed a cosmogony, published in 1755, in which the Solar System formed from a disk which had condensed out of primordial material. The Solar System was part of a larger system (what we would now call a galaxy), and many of the nebulae seen by astronomers were in fact other galaxies, which he termed *island universes*. Kant was influenced by I. *Newton's theories and also by the English philosopher Thomas Wright of Durham (1711–86).

Kappa Crucis Cluster Another name for the *Jewel Box cluster.

Kappa Cygnid meteors A weak meteor shower, active August 3–25. Maximum (maximum ZHR 5) occurs around August 18 from a radiant at RA 19h 04m, dec. +59°. The Kappa Cygnids are reputedly rich in *fireballs.

kappa mechanism The process that drives the pulsations of many types of variable star, including Cepheid variables and RR Lyrae stars. It is named from the Greek letter kappa (κ), the symbol for the coefficient of absorption (or *opacity) of stellar material. In an ionization zone within a star, any small, chance rise in density produces increased opacity and thus increased absorption of energy from the stellar interior. This causes heating and expansion of the layer, which overshoots its rest position, leading to a drop in pressure, density, and temperature (and opacity). The effect then reverses, setting up an oscillation and causing the star's outer layers to pulsate.

Kapteyn, Jacobus Cornelius (1851–1922) Dutch astronomer. He developed statistical methods for reducing and handling large amounts of observational data. Kapteyn compiled the *Cape Photographic Durchmusterung*, a catalogue of southern stars, from photographs taken in South Africa by D. *Gill. A programme of measuring proper motions (which turned up *Kapteyn's Star in 1897) revealed two preferred directions of motion (*star streams*), a result of the rotation of the Galaxy. In 1906 he began his plan of measuring the positions and brightnesses of stars in 206 areas distributed regularly over the sky (the *Kapteyn Selected Areas). With his compatriot Pieter Johannes van Rhijn (1886–1960), Kapteyn attempted to estimate the size and shape of the Galaxy. However, he wrongly concluded that the Sun lay near the centre of the Galaxy, being unable to gauge correctly the effect of interstellar absorption which obscures the Galaxy's true extent.

Kapteyn Selected Areas Areas of sky approximately $1° \times 1°$ square which were chosen by J. C. *Kapteyn in 1906 for his study of the structure of the Galaxy. The 206 areas were uniformly spaced at about 15° intervals over the whole sky. In addition, 46 areas were chosen for their special significance in the Galaxy, such as the galactic pole regions. In each area the magnitudes, colours, spectral types, and proper motions of the stars were measured to as faint an apparent magnitude as possible.

Kapteyn's Star A 9th-magnitude red dwarf, 12.8 l.y. away in Pictor; also known as VZ Pictoris. It is a variable of *BY Draconis type, fluctuating by a few tenths of a magnitude. It has the second-largest proper motion of any star, $8''.67$ per year. It is named after J. C. *Kapteyn, who discovered its fast motion in 1897.

Karoo Array Telescope (MeerKAT) A radio telescope array being built in the Karoo region of Northern Cape province, South Africa. The MeerKAT Precursor Array, also known as KAT-7, consisting of seven 12-m dishes, began operation in 2011. The main MeerKAT array, consisting of 64 dishes of 13.5 m aperture, is due to be completed in 2016. The full array will eventually work at wavelengths from 0.2 to 52 cm. MeerKAT is a prototype instrument for the *Square Kilometre Array.

(((⊕))) SEE WEB LINKS
• Official telescope website.

Kaus Australis The star Epsilon Sagittarii, magnitude 1.8. It is a bright giant of spectral type A0 lying 143 l.y. away.

K corona The bright inner part of the Sun's corona, caused by sunlight scattered from electrons. It is the true corona, as opposed to the *F corona which is due to light scattered by dust particles. Owing to the extremely large speeds of the free electrons (averaging about 10 000 km/s for a coronal temperature of 2 million K), the Fraunhofer lines of the photospheric spectrum are smeared out so that the spectrum of the K corona is almost a pure continuum (the K stands for the German **Kontinuum**). The K corona is brighter than the F corona out to 1.5 solar radii from the Sun's surface.

***K*-correction** An adjustment made to the photometric magnitudes and colours of distant galaxies; also known as *K-term*. It takes account of the effect of the redshift on a galaxy's spectrum.

Keck Telescopes Two identical 10-m reflectors with mirrors composed of hexagonal segments at the *W. M. Keck Observatory, Hawaii.

Keeler Gap A division in Saturn's rings, very close to the outer edge of the A Ring. It lies 136 530 km from Saturn's centre and is between 32 and 47 km wide. Its boundaries are not circular, and show radial wave-like variations as a result of resonances with the moon Prometheus which orbits just outside the A Ring. Its existence was confirmed by the Voyager probes; it is named after the American astronomer James Edward Keeler (1857–1900), who saw it in 1888.

Kellner eyepiece An eyepiece design which is in effect an *achromatic version of the *Ramsden eyepiece, and is therefore sometimes known as the *achromatic Ramsden*. It has a *planoconvex field lens and an achromatic eye lens with a flat surface facing the eye. Its field of view is fairly large, 45–50°, and is free from aberrations. For this reason the Kellner eyepiece is often used in binoculars. It is a good all-purpose eyepiece, giving acceptable results even with low *f*-numbers, although it is prone to ghost images caused by internal reflections. It was invented in 1849 by the German optician Carl Kellner (1826–55).

Kelvin, Lord (William Thomson) (1824–1907) Scottish physicist, born in Ireland. He originated the *thermodynamic temperature scale, and considered the consequences of energy dissipation in the Universe. Kelvin made one of the first scientific attempts at estimating the Earth's age, based on known cooling rates of materials, although his result (20–400 million years) was far too low. He also calculated the *solar constant.

Kelvin–Helmholtz contraction *See* HELMHOLTZ–KELVIN CONTRACTION.

kelvin scale A temperature scale in which the zero point is defined to be equal to −273.15° Celsius. This zero point is also known as *absolute zero*. The *thermodynamic temperature is expressed in kelvin, symbol K.

Kennedy Space Center (KSC) A NASA establishment on Merritt Island at Cape Canaveral, Florida, responsible for launching manned and unmanned spacecraft. Launch facilities built for the Apollo Moon missions were modified for use by the Space Shuttle. KSC also has a landing runway for Space Shuttles. KSC is adjacent to Cape Canaveral Air Force Station, where the launch pads for unmanned rockets are located. Cape Canaveral began operation as a missile test centre in 1950. KSC opened on its present site in 1965.

(())) SEE WEB LINKS
• Official website.

Kepler A NASA satellite launched in 2009 March to search for extrasolar planets. It monitors a field of 100 000 main-sequence stars in Cygnus for four years or more with a 0.95-m Schmidt telescope, looking for dips in their light caused by the transit of planets. Kepler will be able to detect planets as small as Mercury. The spacecraft was launched into an Earth-trailing heliocentric orbit with a period of 372.5 days, in which the spacecraft slowly drifts away from Earth.

(())) SEE WEB LINKS
• Official mission website.

Kepler, Johannes (or Johann) (1571–1630) German mathematician and astronomer. In 1600 he became Tycho *Brahe's assistant in Prague where he undertook to complete the tables of planetary motion Tycho had begun. Kepler first calculated the orbit of Mars. He spent much time trying to reconcile Tycho's accurate observations of the planet with a circular orbit, but concluded (in *Astronomia nova*, published in 1609) that Mars moved instead in an elliptical orbit. Thus he established the first of his laws of planetary motion (*see* KEPLER'S LAWS). *Astronomia nova* contained the rudiments of the second law; the third was first stated in *Harmonice mundi* (1619). All three laws stemmed from his idea that the Sun controlled the planets by magnetic force. Although erroneous, this is significant as an attempt to look for a physical cause for planetary motion. Kepler's *Rudolphine Tables* (named after Tycho's patron, the Holy Roman Emperor Rudolph II) of planetary motion appeared in 1627 and were still in use in the 18th century. Kepler also wrote *De stella nova*, on the supernova of 1604 (*Kepler's Star), and *Dioptrice* (1611) on optics and the theory of the telescope.

Keplerian telescope A basic refracting telescope which has simple convex lenses for both objective lens and eyepiece, devised by J. *Kepler in 1615. It suffers from *chromatic aberration, which can be reduced by increasing the *focal ratio. For this reason, early telescopes were made extremely long. It was the main form of refractor before the *achromatic lens was invented.

Kepler's equation An equation that relates the *eccentric anomaly of a body in an elliptical orbit to its *mean anomaly. The equation is

$$E - e \sin E = M,$$

where E is the eccentric anomaly, M the mean anomaly, and e the eccentricity of the orbit. It is important as one of the mathematical relations enabling the position of a planet about the Sun, or a satellite about its planet, to be calculated from the orbital elements for any time.

Kepler's laws Three laws governing the orbital motions of the planets, discovered by J. *Kepler. The first law states that the orbit of a planet is an ellipse with the Sun at one *focus of the ellipse. The second law states that the *radius vector joining planet to Sun sweeps out equal areas in equal times. The third law states that the square of the orbital period of each planet in years is proportional to the cube of the semimajor axis of the planet's orbit. The first law gives the shape of the planet's orbit; the second describes how the planet must continuously vary its speed as it follows its orbit, moving fastest at perihelion and slowest at aphelion. The third law gives the relationship between the planets' average distances from the Sun and their periods of revolution.

From his law of gravitation and three laws of motion, I. *Newton generalized Kepler's first law, verified the second law, and showed that the third law should be amended to the form

$$4\pi^2 a^3 / T^2 = G(m + m_p),$$

where T and a are the period of revolution and semimajor axis of the orbit of a planet of mass m_p about the Sun of mass m, and G is the *gravitational constant.

Kepler's Star A supernova in Ophiuchus, first observed on 1604 October 9, and described by J. *Kepler in his book *De stella nova*. Observations from Europe and Korea show that it reached a maximum apparent magnitude of -3 in late October. The star remained visible for almost a year. The light-curve is that of a Type Ia *supernova. The visible supernova remnant consists of a few faint filaments and brighter knots at a distance of approximately 30 000 l.y. It is the radio source 3C 358.

Kerr black hole A rotating black hole, as distinct from a non-rotating *Schwarzschild black hole, named after the New Zealand mathematician Roy Patrick Kerr (1934–), who first described their properties in 1963. Black holes are expected to rotate rapidly, since the stars that formed them would have been rotating; hence they will be Kerr black holes. Several consequences arise from the addition of rotation to a black hole. First, the *event horizon becomes elliptical, and its surface area becomes less than that for a static black hole of the same mass. If the black hole were rotating sufficiently quickly, the area of the event horizon would reduce to zero, leaving the central *singularity visible from outside (a *naked singularity). Second, there is a region around a rotating black hole, the *ergosphere, in which objects are forced to spin around the black hole. The outer edge of the ergosphere is the *static limit. Third, a new, inner event horizon forms, and it becomes possible to travel through the black hole, and emerge into a new universe or perhaps another part of our own Universe, through this second event horizon. Rotating black holes with electric charge are called **Kerr–Newman black holes**, named after Kerr and the American mathematical physicist Ezra Ted Newman (1929–), but in practice black holes are unlikely to have any significant electric charge.

Keyhole Nebula A dark nebula in the brightest part of the *Eta Carinae Nebula, named by J. F. W. *Herschel from its keyhole shape. It lies next to the star Eta Carinae itself. The name Keyhole Nebula is sometimes applied to the whole Eta Carinae Nebula.

Keystone A quadrilateral-shaped asterism in the constellation Hercules formed by the stars Epsilon, Zeta, Eta, and Pi Herculis.

Kids, the Popular name for the stars Zeta and Eta Aurigae (the Haedi in Latin). They represent two young goats, offspring of the she-goat that is personified by the bright star Capella. All three goats are visualized as being carried by Auriga, the charioteer.

kinematic parallax An estimate of an object's distance derived by comparing its observed proper motion with a known or assumed linear velocity. *See also* MOVING CLUSTER.

kinetic energy The energy possessed by a body by virtue of its motion in space. It is equivalent to the work that would be done if the moving body were brought to rest. When the speed of a body is much less than the speed of light, kinetic energy is equal to $\frac{1}{2}mv^2$, where m is the mass of the body and v is its velocity. A rotating body has kinetic energy $\frac{1}{2}I\omega^2$, where I is its moment of inertia and ω is its angular velocity.

kinetic temperature The temperature of a gas defined by the average velocity of its particles; the faster the particles move, the higher the temperature. In bodies in *thermal equilibrium, the kinetic temperature will be the same as other measures of temperature, such as *effective temperature. In other circumstances the various temperatures may have very different values. For example, in the solar corona the kinetic temperature of the electrons is 1 to 2 million K, while the effective temperature is about 100 K.

King model A mathematical model that describes the distribution of stars within a *globular cluster as a function of their distance from the cluster's centre; also known as a **King profile**. The model uses two scales: a core radius, r_c, at which the density of stars as seen projected on the sky (and hence the cluster's surface brightness) drops to half its central value; and a tidal radius, r_t, at which stellar density becomes zero. The King model is named after the American astronomer Ivan Robert King (1927–), who first published it in 1966.

Kirchhoff, Gustav Robert (1824–87) German physicist. With the chemist Robert Wilhelm Bunsen (1811–99) he established the principles of spectral analysis. In 1859 he reasoned that the *Fraunhofer lines in the solar spectrum indicated that light from the photosphere was being absorbed at those wavelengths by the Sun's atmosphere. Furthermore, he realized that the Fraunhofer D lines were produced by sodium in the Sun's atmosphere, and other Fraunhofer lines would therefore reveal which other elements were present in the Sun. From then on, astronomical spectroscopy developed rapidly in the hands of others such as P. A. *Secchi in Italy and W. *Huggins in England.

Kirchhoff's laws Three laws concerning spectra, stated in 1859 by the German physicist G. R. *Kirchhoff:
   1. A solid, liquid, or gas under high pressure, when heated to incandescence, produces a continuous spectrum.
   2. A gas under low pressure, but at a sufficiently high temperature, produces a spectrum of bright emission lines.
   3. A gas at low pressure (and low temperature), lying between a hot continuum source and the observer, produces an absorption line spectrum, i.e. a number of dark lines superimposed on a continuous spectrum.

Kirin meteorite *See* JILIN METEORITE.

Kirkwood, Daniel (1814–95) American astronomer. He is best known for his discovery (published in 1866) of the *Kirkwood gaps. He also explained how the Cassini and Encke Divisions in Saturn's rings resulted from orbital resonances with the planet's larger satellites. He pointed out that certain groups of asteroids shared very similar orbital elements, such as (153) Hilda and other asteroids with periods two-thirds of Jupiter's (*see* HIRAYAMA FAMILY). In 1880 Kirkwood was the first to suggest the existence of a group of sungrazing comets.

Kirkwood gaps Any of several narrow regions within the main asteroid belt where few bodies are found, as a result of gravitational perturbations by Jupiter. These gaps occur at mean distances from the Sun which correspond to zones of *commensurability. In these regions of the belt, an asteroid orbits the Sun in a simple fraction of Jupiter's orbital period. These commensurabilities are generally written in the form of a ratio, for example 3:2 for objects which orbit in exactly two-thirds of Jupiter's orbital period. A few zones of commensurability contain isolated concentrations of asteroids in stable orbits, but most are virtually devoid of asteroids. Examples are the gaps at the 2:1, 5:2, and 3:1 commensurabilities (mean distances of 3.28, 2.82, and 2.50 AU from the Sun). Such gaps were first noted by D. *Kirkwood in 1857.

Kitt Peak National Observatory (KPNO) An observatory at an altitude of 2120 m on Kitt Peak in the Quinlan Mountains 90 km southwest of Tucson, Arizona. It was founded in 1958 and is now part of the *National Optical Astronomy Observatory (NOAO). Its largest telescope is the 4-m *Mayall Telescope, opened in 1973. The 3.5-m *WIYN Telescope was opened in 1994. Other KNPO instruments are a 2.1-m reflector for optical and infrared observations, opened in 1964, and a 0.9-m reflector opened in 1960 but operated by the WIYN consortium since 2001. Also on Kitt Peak are two facilities of the *National Solar Observatory, originally the solar division of KPNO but which became a separate organization in 1984: the *McMath–Pierce Solar Telescope and SOLIS (Synoptic Optical Long-term Investigations of the Sun), a trio of solar instruments which began operation in 2004. Telescopes owned by other institutions are also sited at KPNO. These include the 1.3-m Robotically Controlled Telescope (RCT), a KPNO telescope from 1965 to 1995 but subsequently transferred to the RCT

Consortium of five institutions and reopened in 2003; a 1.2-m reflector of the privately owned Calypso Observatory, opened in 1999; a 0.9-m reflector which was a KPNO telescope from 1966 to 1990 but was then transferred to the Southeastern Association for Research in Astronomy (SARA) consortium and reopened in 1995; the 0.6-m Burrell Schmidt of Case Western Reserve University, Ohio (originally opened in 1946 and moved to Kitt Peak in 1979); and a 25-m dish of the *National Radio Astronomy Observatory's VLBA network. The University of Arizona's *Steward Observatory and the *MDM Observatory are also sited on Kitt Peak, but they are not part of KPNO.

((⊕)) SEE WEB LINKS

• Official observatory website.

Kleinmann–Low Nebula (KL Nebula) A powerful extended source of infrared radiation behind the Orion Nebula, discovered in 1967 by the American astronomers Douglas Erwin Kleinmann (1942–) and Frank James Low (1933–2009). The total luminosity is 100 000 times that of the Sun. The KL region consists of a cluster of infrared sources, presumably young stars, of which the most powerful is designated IRc2. The source of high-velocity outflows from the region lies close to IRc2.

K line *See* H AND K LINES.

K magnitude The magnitude of a star as measured through the infrared K filter of the *Johnson photometry system. The K filter has an effective wavelength of 2200 nm and a bandwidth of 540 nm.

knife-edge test Another name for the *Foucault test.

Kochab The star Beta Ursae Minoris, magnitude 2.1. It is a K4 giant 131 l.y. away. With *Pherkad, it forms the so-called Guardians of the Pole.

Kohoutek, Comet (C/1973 E1) A long-period comet, discovered by the Czech astronomer Luboš Kohoutek (1935–) on 1973 March 18; formerly designated 1973 XII. When discovered, Comet Kohoutek lay beyond Jupiter's orbit and was unusually bright for an object so distant. This led to some over-estimates of the comet's likely magnitude close to perihelion, 0.14 AU, on 1973 December 28. Astronauts aboard the Skylab space station observed the comet with a coronagraph during its perihelion passage at magnitude −3. Comet Kohoutek emerged into the evening sky early in 1974 January. Although then of 4th magnitude with a tail up to 25° long and an *antitail, it was much less spectacular than originally anticipated. Its orbit is calculated to be hyperbolic (eccentricity over 1) with an inclination of 14°.3.

((⊕)) SEE WEB LINKS

• Information page at Cometography website.

König eyepiece A modification of the *Erfle eyepiece.

Koronis family A well-defined *Hirayama family of asteroids in the outer region of the main asteroid belt at a mean distance of 2.87 AU from the Sun. The members are mainly S-class asteroids with very similar colours, spectra, and albedos. They are believed to have originated in the break-up of a highly homogeneous parent body, presumably of S class. The largest family members are (208) Lacrimosa, diameter 41 km, and (167) Urda, diameter 40 km. *Ida is also a member. The family is named after (158) Koronis, an S-class asteroid of diameter 35 km, discovered in 1876 by the German astronomer Victor Carl Knorre (1840–1919). The orbit of Koronis has a semimajor axis of 2.872 AU, period 4.87 years, perihelion 2.71 AU, aphelion 3.03 AU, and inclination 1°.0.

KPNO Abbr. for *Kitt Peak National Observatory.

Kracht group A family of *sunskirting comets seen by the *SOHO spacecraft and first identified in 2002 by the German amateur astronomer Rainer Karl Otto Kracht (1948–). Members of this group have a mean perihelion distance of 0.046 AU and inclination around 13°.5. Like the *Marsden group, they are thought to be an offshoot of Comet 96P/Machholz.

Kramers opacity A formula for the *opacity in a star's interior. It is valid at temperatures around 10^4–10^6K when the opacity is mostly due to absorption of photons by free electrons in the interior, and it declines as the temperature increases. For stars up to 1 solar mass the formula applies through much of the interior, but for more massive stars it is valid only in the surface layers. Kramers opacity is named after the Dutch physicist Hendrik Anthony Kramers (1894–1952).

KREEP Crystalline rock from the highlands of the Moon, enriched in potassium (chemical symbol K), rare earth elements (abbr. REE), and phosphorus (chemical symbol P). The presence of KREEP rocks suggests that chemical separation has occurred within the Moon. Such rocks usually have high concentrations of uranium and thorium, which are radioactive and therefore produce heat, so KREEP rocks were important to the Moon's internal development.

Kreutz sungrazer A group of apparently related long-period comets with very small perihelion distances (less than 0.01 AU). They are named after the German astronomer Heinrich Carl Friedrich Kreutz (1854–1907), who studied them in 1888. Sungrazer orbital elements fall into two subgroups, travelling in retrograde orbits with periods of about 500–1000 years. It thus appears that the sungrazers are derived from the break-up of a single large progenitor (perhaps similar to *Chiron), and the further disintegration at subsequent returns of two or more of its fragments. Notable sungrazers include the *Great September Comet of 1882 and Comet *Ikeya–Seki in 1965. Numerous sungrazers have been discovered by coronagraphs aboard spacecraft, over one thousand of them by the *Solar and Heliospheric Observatory (SOHO). *See also* SUNSKIRTER. 📷

Kron–Cousins RI photometry The most commonly used red (i.e. 600–1100nm) photometric system with two filters, called R and I, of effective wavelengths and bandwidths 641 and 158nm, and 798 and 154nm, respectively. R − I is more sensitive to temperature than B − V for cool stars because of the number of lines and molecular bands in the B and V passbands. It was originated by the American astronomer Gerald Edward Kron (1913–) and modified by the South African Alan William James Cousins (1903–2001).

Kruskal diagram A diagram used to plot trajectories in spacetime near a black hole. The vertical and horizontal coordinates, called **Kruskal coordinates**, are two complicated functions of time and distance from the black hole. Lines of constant time radiate from the origin of the diagram, with steeper slopes corresponding to later times. Photons always travel along diagonal lines at $\pm 45°$ to the vertical. The trajectory of an object falling into the black hole is shown as a curving line moving upwards on the diagram at less than 45° to the vertical. The diagram is named after the American physicist Martin David Kruskal (1925–2006).

K star A star of spectral type K, somewhat cooler than the Sun and appearing orange in colour. K-type stars on the main sequence have surface temperatures in the range 3900–5200K, while giants are about 100–400K cooler, and supergiants a few hundred degrees cooler still. Masses of main-sequence K stars are 0.5–0.8 solar masses, and their luminosities 0.1–0.4 times the Sun's. Giants are of 1.1–1.2 solar masses and have luminosities 60–300 times the Sun's, while K supergiants can be up to 13 solar masses and have luminosities as much as 40000 times the Sun's. The dominant spectral features of K-type stars are the neutral metal lines of iron and titanium, with calcium (both Ca I and Ca II) being particularly strong. The molecular bands due to cyanogen (CN) and titanium oxide (TiO) strengthen considerably from K0 to K9. The best-known stars of this type are Arcturus, a K1 giant, and Aldebaran, a K5 giant.

Kueyen The second 8.2-m Unit Telescope (UT2) of the European Southern Observatory's *Very Large Telescope in Chile, opened in 1999. Its name means 'the Moon' in the local Mapudungun language.

Kuiper, Gerard Peter (1905–73) Dutch-American astronomer. In a search for new planetary satellites he discovered Miranda (in 1948, orbiting Uranus) and Nereid (in 1949, orbiting Neptune). His spectroscopic studies revealed methane in the atmospheres of Uranus and Neptune. He also found methane bands in Titan's spectrum, demonstrating that the satellite has an atmosphere. He suggested the existence of the *Kuiper Belt as the source of short-period comets. Kuiper was adviser on many American lunar and planetary missions, and proposed the idea of flying infrared telescopes on board high-altitude aircraft, which led to the *Kuiper Airborne Observatory.

Kuiper Airborne Observatory (KAO) A NASA Lockheed C-141 transport aircraft modified to carry a 0.91-m Cassegrain reflector for infrared astronomy. The KAO operated from 1975 to 1995 and flew at altitudes of up to 13.7 km. Its replacement is the *Stratospheric Observatory for Infrared Astronomy (SOFIA). The KAO was named after G. P. *Kuiper.

Kuiper Belt A region of the outer Solar System containing an estimated $10^7 - 10^9$ icy planetesimals, or comet nuclei. The Kuiper Belt is an inner, flattened extension of the *Oort Cloud. It lies more or less in the same plane as the planets and extends outwards from around 30 AU (the orbit of Neptune) to perhaps 1000 AU. Members of the Kuiper Belt are also known as *trans-Neptunian objects. Such a vast reservoir of comets beyond Neptune was proposed in 1951 by G. P. *Kuiper. The Irish engineer and astronomer Kenneth Essex Edgeworth (1880–1972) wrote papers about objects beyond Pluto in 1943 and 1949, so it is also known as the **Edgeworth–Kuiper belt**. In 1992, the British-born American astronomer David Clifford Jewitt (1958–) and the Vietnamese-born American astronomer Jane Luu (1963–) discovered the first Kuiper Belt object, 1992 QB$_1$, now numbered (15760). This has a diameter of about 200 km, semimajor axis 44.2 AU, orbital period about 294 years, perihelion 40.9 AU, aphelion 47.5 AU, and inclination $2°.2$. Since then, over a thousand have been found. The Kuiper Belt is thought to be the source of most *periodic comets. Members of the Kuiper Belt that pass close to Neptune can be diverted inwards by gravitational perturbations to become *Centaurs, or outwards to become *scattered-disk objects. *See also* CUBEWANO; PLUTINO. 📷

k

labes A landslide on a planetary surface; pl.*labes*. The name is not a geological term, but is used in the nomenclature of individual features, for example Ophir Labes on Mars.

labyrinthus A complex, intersecting network of linear depressions on a planetary surface; pl.*labyrinthi*. The name is not a geological term, but is used in the nomenclature of individual features, for example Noctis Labyrinthus on Mars.

Lacaille, Nicolas Louis de (1713–62) French astronomer. From the Cape of Good Hope (in modern South Africa) he surveyed the southern skies in 1751–3. He discovered 24 new nebulae and clusters, and charted the positions of nearly 10 000 stars. While there he measured the position of the Moon simultaneously with J. J. de *Lalande in Berlin to obtain an accurate value of its distance. Lacaille published a star chart which introduced 14 new southern constellations; his full southern star catalogue, *Coelum australe stelliferum*, was published posthumously in 1763.

Lacerta (Lac) (gen. Lacertae) A faint constellation of the northern sky, representing a lizard. Its brightest star is Alpha Lacertae, magnitude 3.8. The constellation contains *BL Lacertae, the prototype of a class of peculiar galaxies.

Lacertid An alternative name for a *BL Lacertae object.

lacus A small, irregular dark patch on a planetary surface; pl.*lacus*. The name is not a geological term, but is used in the nomenclature of individual features, for example Lacus Somniorum on the Moon and Solis Lacus on Mars. The word means 'lake', and was first applied to lunar features in the 17th century, when it was thought the dark patches might be bodies of water.

LADEE Abbr. for *Lunar Atmosphere and Dust Environment Explorer.

Lagoon Nebula The diffuse nebula M8 in Sagittarius, also known as NGC 6523. It is about 1½° long and ½° wide. The Lagoon Nebula gets its name from a dark lane that divides it in two. The brightest part of the nebula is called the Hourglass. The Lagoon Nebula contains the 5th-magnitude star cluster NGC 6530. The main illumination of the nebula comes not from this cluster but from the 6th-magnitude blue supergiant 9 Sagittarii. The Lagoon Nebula lies 5000 l.y. away. 📷

Lagrange(-Tournier), Joseph Louis de (1736–1813) French mathematician, born in Italy. In celestial mechanics he studied perturbations and stability in the Solar System. He examined the *three-body problem for the Earth, Moon, and Sun (1764) and the motion of Jupiter's satellites (1766). In 1772 he found the particular solutions to the problem that give rise to the equilibrium positions now called *Lagrangian points. Lagrange also studied the Moon's libration.

Lagrangian point One of five points at which small bodies can remain in the orbital plane of two massive bodies; also known as **libration points**. Three of the points lie on the line joining the two massive bodies: L_1 lies between them, while L_2 and L_3 have the two bodies between them. These three points are unstable, slight displacements of a body from them resulting in its rapid departure. The fourth and fifth points (L_4 and L_5) each form an equilateral triangle with the two massive bodies, 60° ahead of and behind the smaller body in its orbit around the larger one. A well-known example of bodies lying at the L_4 and L_5 Lagrangian points are the *Trojan asteroids in Jupiter's orbit. Among Saturn's satellites, Telesto and Calypso lie at the L_4 and L_5 Lagrangian points in the orbit of the much larger Tethys. In similar fashion, tiny Helene precedes Saturn's satellite Dione, keeping 60° ahead of Dione. The Lagrangian points are named after the French mathematician J. L. de *Lagrange, who first calculated their existence. *See also* EQUIPOTENTIAL SURFACE.

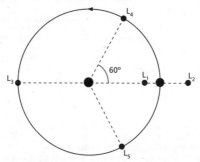

Lagrangian point: The Lagrangian points are five equilibrium points in the orbit of one body around another, such as a planet around the Sun

l'Aigle meteorite shower A shower of 2000–3000 stony meteorites weighing 37 kg that fell near the village of l'Aigle, northern France, on 1803 April 26. A commission of the French Academy of Sciences, headed by J.-B. *Biot, concluded that the meteorites were fragments of a single body that had entered the Earth's atmosphere from space. This was the first conclusive evidence that stones of extraterrestrial origin could fall from the sky. The meteorites were *ordinary chondrites of L6 type.

Lalande, Joseph Jérême (Le Français) de (1732–1807) French astronomer and mathematician. In 1751, in conjunction with N. L. *Lacaille, he measured the Moon's distance. He helped organize expeditions to observe the transits of Venus in 1761 and 1769, and used the results to calculate the solar parallax. In 1801 he published *Histoire céleste française*, which contained a catalogue of over 47 000 stars. One of these 'stars' was Neptune, which he recorded in 1795, 51 years before it was found by J. G. *Galle.

Lalande 21185 The fourth-closest star to the Sun, 8.3 l.y. away in the constellation Ursa Major. It is an M2 dwarf of apparent magnitude 7.5.

Lambda Boötis star An A-type star with abnormally weak metallic lines and slow rotation. The cause of the metallic deficiencies is unknown. The stars have low space velocities and thus form a subgroup of Population I.

Lambda Eridani star A *Be star that exhibits photometric or spectroscopic variations (or both) with a period of 0.4–2 days. The periodicity is very pronounced, and in some cases the light-curve exhibits two minima per cycle. The cause of the variation is unknown, but may arise from non-uniform surface brightness or from a form of pulsation.

La Palma *See* ROQUE DE LOS MUCHACHOS OBSERVATORY.

Laplace, Pierre Simon de (1749–1827) French mathematician and astronomer. He established celestial mechanics as a mathematical discipline, applying Newton's theory of gravitation to the study of orbits, perturbations, and stability in the Solar System. He explained the small, long-term variations in the orbital speeds of Jupiter and Saturn (1786) and the Moon (1787). His nebular hypothesis of the Solar System's origin (1796), in which planets condense from different zones in a nebula, is similar to I. *Kant's theory, of which he was apparently unaware. Laplace's *Traité de mécanique céleste* (published 1799–1825) is a summary of 18th-century celestial mechanics.

Laplacian plane A plane about a planet upon which a satellite's orbital plane precesses because of perturbations. In the usual case where the satellite's orbit is perturbed both by the oblateness of the planet and by the Sun, the Laplacian plane lies between the planet's equatorial plane and the planet's orbital plane about the Sun. It is named after the French mathematician P. S. de *Laplace.

Large Binocular Telescope (LBT) A joint US–Italian–German telescope consisting of twin 8.4-m mirrors on a common mounting at *Mount Graham International Observatory. The first mirror began operation in 2005, with the second following in 2008. The telescope has the light-gathering area of a single 11.8-m reflector. The mirror centres are 14.4 m apart. The LBT has its headquarters at Tucson, Arizona. ◙

(((⊕))) SEE WEB LINKS
• Official telescope website.

Large Magellanic Cloud (LMC) The larger of the two irregular galaxies which accompany our own Galaxy; also known as the Nubecula Major. The LMC is visible to the naked eye as a hazy, elongated patch of light spanning about 8° of sky in the constellations Dorado and Mensa. The LMC is about 20 000 l.y. in diameter and 160 000 l.y. from Earth. Its visible mass, in stars and gas, is about one-tenth that of our Galaxy, and it has a relatively higher gas content. There is some weak spiral structure, and a noticeable bar-like distribution of stars. The LMC contains both old and young clusters, including several young *blue populous clusters*, unknown in our Galaxy, which resemble younger and smaller versions of globular clusters. Clusters of intermediate age appear to be lacking, suggesting that the LMC has undergone early and late bursts of star formation. A major feature is the *Tarantula Nebula, a large complex of gas and young stars. A blue supergiant star in the LMC exploded as *Supernova 1987A. *See also* MAGELLANIC CLOUDS. ◙

Large Millimeter Telescope (LMT) A 50-m radio telescope for millimetre-wave observations sited at an altitude of 4600 m on Sierra Negra in the state of Puebla, Mexico. It is jointly owned by the Instituto Nacional de Astrofísica, Óptica, y Electrónica (INAOE) of Mexico and the University of Massachusetts at Amherst. The LMT observes at wavelengths of 0.85–4 mm. It began operation in 2008, initially using only the inner 32 m of the dish.

(((⊕))) SEE WEB LINKS
• Official telescope website.

large-scale structure The distribution of galaxies on scales larger than around 30 million l.y. Galaxies are not scattered randomly in space but are grouped in clusters and superclusters of various sizes. They also sometimes lie in extended chains called filaments, or in flattened structures such as the *Great Wall. Contrasting with these areas are vast regions where galaxies are scarce, known as *voids, which appear to be roughly spherical. The largest structures known are around 300 million l.y. across. If the *cosmological principle is correct, astronomers should eventually reach a scale at which the Universe becomes roughly homogeneous.

Large Synoptic Survey Telescope (LSST) An 8.4-m wide-field telescope at an altitude of 2680 m on Cerro Pachón, Chile, that will repeatedly survey the entire visible sky every three nights. Its images, each covering a field of view 3.5 degrees across, will be recorded by a digital camera with 3.2 billion pixels, each exposure lasting 15 seconds. They will be valuable for discovering and following objects such as variable stars, novae and supernovae, comets, near-Earth objects, and Kuiper Belt objects, as well as for mapping the distribution of remote galaxies. The data will be made publicly available via the Internet. The telescope is owned and operated by the LSST Corporation, a consortium of American universities, observatories, and research institutions with its headquarters in Tucson, Arizona. LSST is scheduled to begin operation in 2020.

(((⊕))) SEE WEB LINKS
• .Official telescope website

Larissa The fifth-closest satellite of Neptune, distance 73 548 km, orbital period 0.555 days; also known as Neptune VII. Its size is 208 × 178 km. Larissa was discovered in 1989 on images taken by Voyager 2.

Las Campanas Observatory An observatory at an altitude of 2280 m on Cerro Las Campanas, 110 km northeast of La Serena, Chile. It was founded in 1971, and is owned and operated by the Carnegie Institution of Washington. It contains the 2.5-m Irénée du Pont Telescope, opened in 1976; the 1-m Henrietta Swope Telescope, opened in 1971; and the 1.3-m Warsaw University Observatory Telescope, opened in 1996. It is also the site of the twin 6.5-m *Magellan Telescopes and the forthcoming *Giant Magellan Telescope.

SEE WEB LINKS
• Official observatory website.

laser A device for producing a beam of radiation that is *monochromatic* (all of one wavelength) and *coherent* (all the waves are in step), usually in the infrared, visual, or ultraviolet part of the spectrum. The name is an acronym for light amplification by stimulated emission of radiation. The radiation is produced when excited electrons in an atom or molecule are stimulated into emitting radiation by the nearby passage of a photon. The new photon is emitted in step with (*coherent* with) the passing photon, with the same wavelength, and in the same direction. When many such photons are produced they produce a highly intense, parallel beam of radiation. Stimulated emission is the inverse of *absorption. The microwave equivalent of a laser is a *maser.

Laser Interferometer Space Antenna (LISA) A planned European Space Agency mission to detect gravitational waves, a follow-on from *LISA Path finder. LISA's initial design envisaged three separate spacecraft flying in triangular formation with lasers measuring relative movement caused by passing gravitational waves. Originally intended as a joint mission between ESA and NASA, it is being redesigned as a simplified ESA-led mission. Launch is expected in the 2020s.

SEE WEB LINKS
• Official mission website.
• ESA mission website.

La Silla Observatory An observing site of the *European Southern Observatory, founded in 1964, at an altitude of 2350 m on La Silla mountain in the Atacama desert of Chile about 90 km northeast of La Serena. ESO's main optical instruments on La Silla are a 3.6-m reflector, opened in 1976; the 3.5-m New Technology Telescope (NTT), opened in 1989; a 2.2-m reflector jointly owned by the Max-Planck-Institut für Astronomie, opened 1984. A 1.52-m reflector, opened in 1968, was closed in 2002 and the ESO 1-m Schmidt, opened in 1971, was closed in 1998 but reopened in 2009 and is now operated by Yale University. Instruments at La Silla owned by individual nations include a Danish 1.54-m reflector opened in 1979 and the 1.2-m Leonhard Euler Telescope owned by Geneva Observatory, opened 2000. La Silla was also the site of the 15-m Swedish–ESO Submillimetre Telescope (SEST), opened in 1987 but closed in 2003.

SEE WEB LINKS
• Official observatory website.
• Facilities at La Silla Observatory.

Lassell, William (1799–1880) English brewer and amateur astronomer. In 1845 he completed a 24-inch (0.6-m) telescope, the first sizeable reflector to be mounted equatorially, with which he discovered several planetary satellites. He found Neptune's largest, Triton, in 1846, just 17 days after the planet's discovery. In 1848 he discovered the Saturnian satellite Hyperion independently of (but two days after) W. C. and G. P. *Bond, and in 1851 he discovered Ariel and Umbriel orbiting Uranus. From 1862 to 1864, observing from Malta with his 48-inch (1.2-m) reflector, he and the German-born astronomer Albert Marth (1828–97) found 600 new nebulae.

Lassell ring A ring of Neptune, third in order from the planet, named after W. *Lassell and formerly known as the Plateau ring. It is about 4000 km wide, spanning the gap between the Le

Verrier ring 53 200 km from the planet's centre and the Arago ring, 57 600 km from the planet's centre.

last contact Another name for *fourth contact at an eclipse.

last quarter The phase of the waning Moon that occurs midway between full and new Moon, when half of the Moon is illuminated; also known as *third quarter*. At last quarter the Moon lies 90° west of the Sun.

La Superba The star Y Canum Venaticorum. It is a *semiregular variable red supergiant of type SRB ranging between 5th and 7th magnitudes with a period of just over 5 months. It was named by P. A. *Secchi for its remarkable spectrum. It lies about 1000 l.y. away

late heavy bombardment A period about 4 billion years ago during which planetary bodies suffered intense bombardment by fragments left over from the formation of the planets. As the fragments were swept up by collision with the planets their numbers diminished, and after about 3.9 billion years ago the late heavy bombardment came to an end. Impacts have continued to occur on planets and satellites, but at a much lower rate.

late-type galaxy A spiral or irregular galaxy. The name comes from the conventional position of these galaxies in the *tuning-fork diagram of galaxy types. For similar reasons, an Sc or Sd spiral galaxy may be referred to as a late-type spiral, as opposed to an early-type Sa or Sb spiral. *See also* HUBBLE CLASSIFICATION.

late-type star A star with a surface temperature cooler than the Sun, with a spectral type K, M, C, or S; G stars are often included as well. Late-type stars can be either of low mass, if they are on the main sequence, or more massive than the Sun if they are giants or supergiants. The designation 'late' derives from the time when it was wrongly thought that stars with K or M spectra were old and evolved. *See also* EARLY-TYPE STAR.

latitude The angle north or south of some reference plane. In astronomy, the equivalent of terrestrial latitude is termed *declination. *See also* CELESTIAL LATITUDE; GALACTIC LATITUDE; HELIOCENTRIC LATITUDE.

latitude variation The slight change in the geographical latitude of an observing site which results from motion of the Earth's poles due to the *Chandler wobble of the Earth's rotational axis. This may be detected by observations of stars at the zenith, since the declination of the zenith point is equivalent to the latitude of the observing site.

LBV Abbr. for *luminous blue variable.

L dwarf A spectral classification applied to cool, faint stars at the bottom end of the *main sequence. Objects that fall into this category are the lowest-mass red dwarfs and the sub-stars known as *brown dwarfs. L dwarfs have surface temperatures in the range about 2200–1300 K. Their spectra show absorption caused by hydrides, i.e. hydrogen attached to atoms of metals such as iron and chromium, as well as features due to neutral alkali metals such as potassium, caesium, and rubidium. The presence of strong lithium absorption lines in the spectrum is proof that the object is a brown dwarf. Masses are estimated in the range 0.06–0.09 Suns (i.e. 60–90 Jupiters).

leap second The second that is occasionally added to *Coordinated Universal Time (UTC) to keep it within 0.9 s of *Universal Time (UT). This is necessary usually once, but sometimes twice, in the year. The leap second is added by extending the final minute of the day by one second either on December 31 or June 30, although leap seconds can also be added at the end of March and September if necessary.

leap year A year with one extra day (February 29) and therefore 366 days long. The day is added to keep the vernal equinox on or around March 21. In the *Julian calendar, every fourth year was a leap year. In the *Gregorian calendar, century years are not leap years unless divisible by 400; hence three leap years are omitted in a 400-year period.

least circle of confusion *See* ASTIGMATISM; SPHERICAL ABERRATION.

Leavitt, Henrietta Swan (1868–1921) American astronomer. She worked with E. C. *Pickering at Harvard College Observatory, measuring the brightness of stars on photographic plates. From her studies of Cepheid variables in the Small Magellanic Cloud, she established the *period–luminosity law in 1912. In all, Leavitt discovered 2400 variable stars, more than half the total known in her lifetime. She also established the magnitudes of the stars in the original *North Polar Sequence.

Leda The tenth-closest satellite of Jupiter, distance 11 150 000 km, orbital period 240.5 days; also known as Jupiter XIII. Its diameter is about 10 km. It was discovered in 1974 by the American astronomer Charles Thomas Kowal (1940–). Leda is a member of the *Himalia group of Jovian satellites.

Lemaître, Georges Édouard (1894–1966) Belgian priest and cosmologist. His model of an expanding Universe (1927) was superior to that of W. *de Sitter in that it took into account mass, gravitation, and the curvature of space. Similar models had been proposed in the early 1920s by the Russian mathematician Alexander Alexandrovich Friedmann (1888–1925). Lemaître argued further (1931) that the quantum theory supported an origin in the explosion of a 'primeval atom' or 'cosmic egg' into which was originally concentrated all mass and energy. As modified by A. S. *Eddington, Lemaître's model provided the springboard for G. *Gamow's Big Bang theory.

Lemaître universe A model of the universe containing a *cosmological constant term, named after G. É. *Lemaître. In this model, space has a positive curvature but expands for ever. The Lemaître universe is both homogeneous and isotropic. The most interesting aspect of such a universe is that it undergoes a so-called *coasting phase* in which the *cosmic scale factor is roughly constant with time.

lens A transparent optical element, either glass, crystalline, or plastic, that refracts light to form an image. A lens has either concave or convex surfaces, so that parallel light which strikes it is refracted either towards the *focal point, as in a *converging lens, or away from it, as in a *diverging lens. A lens which is thin compared with its diameter will have a more distant focal point (i.e. a longer *focal length) than a thicker one, will be easier to manufacture, and will suffer less from *chromatic aberration and *spherical aberration. In practice, to reduce these and other *distortions, combinations of lenses, known as **compound lenses**, are used. *See also* ACHROMATIC LENS.

lens, gravitational *See* GRAVITATIONAL LENS.

Lense-Thirring effect An alternative name for *frame dragging.

lenticula A dark and somewhat reddish kilometre-scale patch on Europa; pl.*lenticulae*. Lenticulae are complex and can take the form of shallow domes, broad flat spots, or regions of chaotic pitting. They are thought to form from underlying ice convecting or melting through to Europa's surface.

lenticular galaxy A type of galaxy with a definite disk of stars and a central bulge, but showing no sign of spiral arms and little or no interstellar material. Lenticulars are classified as S0 galaxies in the *Hubble classification. Their name comes from the lens-like appearance when viewed edge-on. There is very little evidence for current star formation in them.

Leo (Leo) (*gen.* Leonis) A constellation of the zodiac, representing a lion. The Sun passes through Leo from the second week of August to the third week of September. Leo's brightest star is *Regulus (Alpha Leonis); Beta Leonis is *Denebola; Gamma Leonis (*Algieba) is a handsome double star. R Leonis is a deep-red *Mira Star, ranging from 6th to 10th magnitude with a period of about 10 months. Leo contains two pairs of spiral galaxies visible in small telescopes: M65 and M66, and M95 and M96. Every November the *Leonid meteors radiate from near Gamma Leonis. Leo also contains the third-closest star to the Sun, 7.8 l.y. away, an M6 dwarf called Wolf 359, magnitude 13.5.

Leo Minor (LMi) (*gen.* **Leonis Minoris**) An insignificant constellation of the northern sky, representing a little lion. It contains little of interest. Its brightest star, 46 Leonis Minoris, is of magnitude 3.8.

Leonid meteors A meteor shower showing weak activity (maximum ZHR 15) in most years, although meteor storms sometimes occur when the parent comet, *Tempel–Tuttle, returns to perihelion, at approximately 33-year intervals. Leonid storms were seen in 1799, 1833, 1866, 1966, 1999, 2001, and 2002. During the storm of 1966 November 17, the Leonid ZHR may have reached 100 000 for 40 min. The most recent storms, in 1999, 2001, and 2002, had peak ZHR up to about 4000. Activity is usually substantial for several years either side of the storms. Leonids are seen from November 10 to 23, with maximum on November 17, when the radiant lies at RA 10 h 12 m, dec. +22°, in the Sickle of Leo. Leonids have the highest geocentric velocities of any meteor stream, 71 km/s, and a large proportion leave persistent trains.

lepton One of a class of elementary particles including the electron, muon, and neutrino. Leptons do not participate in the strong nuclear interactions.

lepton era The interval, commencing about 10^{-5} s after the Big Bang, in which the various kinds of *lepton were the main contributors to the density of the Universe. Pairs of leptons and antileptons were created in large numbers in the early Universe but, as the Universe cooled, most lepton species were annihilated. The lepton era immediately followed the *hadron era, and is a subdivision of the *radiation era. The end of the lepton era is usually taken to have occurred when most electron–positron pairs were annihilated, at a temperature of 5×10^9 K, about 1 second after the Big Bang.

Lepus (Lep) (*gen.* **Leporis**) A constellation of the southern sky, representing a hare. Its brightest star is Alpha Leporis (*Arneb). R Leporis is *Hind's Crimson Star, a deep-red variable. M79 is a globular cluster; NGC 2017 is a small open cluster, actually a complex multiple star with five components visible in moderate-sized telescopes.

Leto family A small, well-defined family of asteroids in the main asteroid belt about 2.8 AU from the Sun. The three largest members of the family, (68) Leto, (236) Honoria, and (858) El Djezair, are of S class. Leto itself, diameter 123 km, was discovered in 1861 by the German astronomer (Karl Theodor) Robert Luther (1822–1900). Leto's orbit has a semimajor axis of 2.781 AU, period 4.64 years, perihelion 2.26 AU, aphelion 3.30 AU, and inclination 8°.0.

Le Verrier (or Leverrier), Urbain Jean Joseph (1811–77) French celestial mechanician. From 1838 he studied the long-term stability of the Solar System, in particular the causes of perturbations, leading to a revision of the masses of the planets, the dimensions of the Solar System, and the velocity of light. In 1845 D. J. F. *Arago alerted him to irregularities in observed positions of Uranus. The next year he ascribed these to the presence of an undiscovered planet, and supplied a predicted position to J. G. *Galle who located Neptune in 1846 November. In 1859 Le Verrier suggested that a small planet or belt of asteroids within Mercury's orbit would account for irregularities in Mercury's observed positions (*see* VULCAN).

Le Verrier Ring One of the rings of Neptune, second in order from the planet, 53 200 km from the centre and 15 km wide. It is named after U. J. J. *Le Verrier.

Lexell, Comet (D/1770 L1) A lost periodic comet, discovered by C. J. *Messier on 1770 June 14. On July 1 it passed 0.015 AU (2.2 million km) from Earth, the closest approach of a comet on record, appearing of 2nd magnitude with a coma over 2° wide. Calculations in 1779 by the Swedish astronomer Anders Johan Lexell (1740–84), after whom the comet is named, indicated that it was in an elliptical orbit of period 5.6 years. The comet's orbit has been considerably modified by close approaches to Jupiter. A close passage to Jupiter in 1779 increased its perihelion distance and may even have placed it into a hyperbolic orbit, effectively ejecting it from the Solar System. It has not been seen since.

 SEE WEB LINKS
• Information page at Cometography website.

LHA Abbr. for *local hour angle.

Libra (Lib) (*gen*. **Librae**) A constellation of the zodiac, representing a pair of scales. The Sun lies in Libra for the first three weeks of November. Alpha Librae (*Zubenelgenubi) is a wide double. Beta Librae (Zubeneschamali) is the constellation's brightest star, magnitude 2.6. The two star names come from the Arabic meaning 'southern claw' and 'northern claw', and arise because Libra once represented the claws of the scorpion, Scorpius, which it adjoins. Delta Librae is an *Algol star, an eclipsing binary that varies between magnitudes 4.9 and 5.9 in a period of 2 days 8 hours.

Libra, first point of *See* FIRST POINT OF LIBRA.

libration A periodic wobble of a celestial body (see diagram). The most familiar librations are those of the Moon as seen from Earth. In **libration in longitude**, the Moon appears to swing slightly from side to side (east–west) by up to 8° 08′ in each direction. This occurs because the Moon's speed along its elliptical orbit varies with its distance from Earth, while its axial rotation remains constant. In **libration in latitude**, the Moon appears to nod from north to south by up to 6° 53′ in each direction. This is because the Moon's axis of rotation is not perpendicular to its orbital plane. As a result of these two librations, we can see up to 59% of the Moon's surface. A third libration, *diurnal libration, occurs because we view the Moon from different sides of the Earth at moonrise and moonset, so we can again see slightly round the eastern and western limbs, although by only about 1°. Libration occurs when two motions are locked in resonance (a *synchronous orbit). For this reason, the planet Mercury also exhibits a libration, as do some planetary satellites.

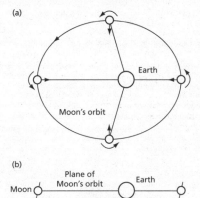

libration: (a) Libration in longitude is due to the elliptical shape of the Moon's orbit. The Moon rotates on its axis at constant rate, but its orbital velocity about the Earth varies, so we can see slightly round its east and west edges during the course of an orbit. (b) Libration in latitude is due to the Moon's equator being inclined to the plane of its orbit, so we can see slightly over its north and south poles

libration point Another name for a *Lagrangian point.

Lick Observatory The observatory of the University of California, opened in 1888 at an altitude of 1290 m on Mount Hamilton, 21 km east of San Jose, California. Its headquarters are in Santa Cruz. Its most celebrated telescope is the 36-inch (0.91-m) Lick refractor, opened in 1888 and named, like the observatory, after the landowner James Lick (1796–1876), who donated money for its construction. Another historic instrument at the observatory is the 36-inch Crossley reflector, opened in 1896. In 1959 the 3-m Shane reflector was opened, named after C(harles) Donald Shane (1895–1983), a former director of the observatory. Other instruments on Mount Hamilton are the 1-m Anna L. Nickel reflector, opened 1979, and the 0.76-m Katzman Automatic Imaging Telescope (KAIT), opened 1996.

(((⊕))) SEE WEB LINKS
• Official observatory website.

light Electromagnetic radiation that can be seen by the human eye. It lies between the ultraviolet and infrared regions of the electromagnetic spectrum. Different wavelengths of light appear as different colours. Visible radiation ranges from wavelengths of about 750 nm at the red (long-wavelength) end to around 380 nm at the violet (short-wavelength) end.

light, speed of (c) The speed of light in a vacuum is 299 792.5 km/s. The speed is less in other media. All electromagnetic radiation travels at this speed in a vacuum.

light bucket Another name for a *flux collector.

light cone A means of illustrating the past and future of an event in spacetime. Spacetime diagrams in three dimensions are conventionally depicted with time as the vertical axis, and the other two axes representing dimensions in space. The axes are scaled such that one unit along the time axis is one second, and one unit along either of the spatial axes is 300 000 km (because the speed of light is approximately 300 000 km/s). On such a diagram, the paths of rays of light from an event slant upwards at 45° along the sides of a cone whose vertex is at the event and whose axis is vertical. This is the **future light cone** of the event. Any and all events occurring after the first event which are to be affected in any way by that first event must lie within the cone. Events outside the cone would require faster-than-light travel if they were to be affected by the initial event. Conversely, any earlier events which are to affect the event in question must be contained within a 45° cone going down from the event. This is the **past light cone**.

light-curve A graph of the variations in brightness of an object, particularly a variable star or a rotating asteroid, plotted against time.

light pollution Artificial lighting that shines where it is neither needed nor wanted. Such inefficient lighting has many detrimental effects, including on human health, wildlife, the natural environment, and energy consumption. Light pollution is a significant problem for astronomers in or near urban areas. Light shining into the sky is reflected by particles and water vapour in the atmosphere (*skyglow*), overpowering the light from celestial objects and degrading the performance of telescopes. From many urban locations, only the brightest stars are now visible. The skyglow from a town or city can be visible over distances of 100 km or more. Increasing awareness of the energy wastage involved, coupled with improved designs of lighting, has led to some limitation in the spread of light pollution. 📷

light pressure *See* RADIATION PRESSURE.

light time The time taken for light to travel from an object to the observer. Light time must be taken into account when calculating the real times of occurrence of events such as eclipses, transits, and occultations of planetary satellites. The light time corresponding to one astronomical unit is 499 s. *See also* EQUATION OF LIGHT.

light year (l.y.) The distance travelled by light, or other electromagnetic radiation, in one *tropical year through space. One light year equals 9.4605×10^{12} km, or 63240 astronomical units, or 0.3066 parsecs.

limb The rim of the visible disk of a celestial body as seen from Earth. A body's leading edge as it moves across the sky is the **preceding limb**, while the trailing edge is the **following limb**. The edges of the Moon's visible face are known as the **limb regions**.

limb brightening A phenomenon in which an object appears to become brighter towards the edges. Limb brightening can be observed at certain wavelengths in objects that are surrounded by hot but tenuous gas, such as the Sun's corona. It is particularly noticeable at ultraviolet wavelengths, and to a small extent at centimetre and millimetre radio wavelengths.

limb darkening The decrease in brightness of an object towards its edges. Limb darkening is most noticeable at optical wavelengths in the Sun, where we see into deeper, hotter (and therefore brighter) layers at the centre of the Sun's disk than at the limb. Characteristic features in the light curves of certain eclipsing binaries demonstrate the existence of limb darkening in these stars. Limb darkening is also noticeable in the gaseous giant planets.

limiting magnitude **1**. The faintest apparent magnitude that can be detected with a given instrument under prevailing conditions.
  2. The faintest magnitude of object included in an astronomical catalogue.

Lincoln Near-Earth Asteroid Research (LINEAR) A search programme for *near-Earth asteroids and comets operated by the Lincoln Laboratory of the Massachusetts Institute of Technology (MIT). The search uses two 1-m wide-field telescopes at Lincoln Laboratory's Experimental Test Site on the White Sands Missile Range in Socorro, New Mexico. The first telescope started operation in 1996 and was joined by the second in 1999.

• Official project website.

Lindblad, Bertil (1895–1965) Swedish astronomer. His early work on luminosity criteria made possible a photometric survey which revealed the distribution of stars in the Sun's part of the Galaxy. He went on to study stellar statistics and dynamics on an increasingly larger scale, and showed how J. C. *Kapteyn's discovery of two star streams provided evidence that the entire Galaxy was rotating, with *differential rotation (soon confirmed by J. H. *Oort). Lindblad also showed that the Galaxy is a spiral, and went on to study the dynamics of other galaxies, discovering what is now called *Lindblad resonance. In the 1940s he proposed the *density wave theory to explain spiral structure. In his galactic work he was aided by his son Per Olof Lindblad (1927–).

Lindblad resonance A resonance that can affect stars in a galaxy's disk at certain distances from the galaxy's centre. It occurs when the natural frequency of the radial (i.e. in and out in the direction of the centre) component of motion of a star in its orbit is the same as the frequency of passage of the star through the maxima of the gravitational potential associated with the spiral pattern. If the star is moving around the centre faster than, and overtakes, the spiral pattern, then an **inner Lindblad resonance** occurs; if the pattern is moving faster than the star around the centre, then an **outer Lindblad resonance** occurs. At an inner resonance, energy is fed into the orbits of the stars from the spiral pattern, and vice versa at an outer resonance. The effect is named after B. *Lindblad.

linea A linear feature or elongated marking on a planetary surface; pl.*lineae*. The name, which means 'thread' or 'plumb-line', is not a geological term, but is used in the nomenclature of individual features, for example Minos Linea on Europa and Palatine Linea on Dione.

LINEAR Abbr. for *Lincoln Near-Earth Asteroid Research.

line blanketing The dimming of a star's spectrum by the presence of many hundreds or thousands of weak absorption lines too faint and close together to be individually resolved. In cool stars there can be so many different absorbing atoms or molecules with lines in the visible or ultraviolet part of the spectrum that the spectrum appears to be 'blanketed' with them. The energy absorbed must be re-radiated, but this is usually at lower energies (i.e. longer wavelengths); hence the red or infrared part of the spectrum will appear enhanced relative to a star with a non-blanketed spectrum.

line broadening The broadening of absorption or emission lines in a spectrum by one of four mechanisms: *Doppler broadening, due to the motion of the emitting gas; *pressure broadening or *Stark broadening, due to collisions with other atoms and molecules in the star's atmosphere; or *Zeeman splitting, as a result of the presence of a strong magnetic field.

line of apsides Another name for the major axis of an orbit. *See also* APSIDES.

line of inversion *See* MAGNETIC INVERSION LINE.

line of nodes The straight line joining the ascending and descending *nodes of an orbit.

line profile The shape of an absorption or emission line in a spectrum across the narrow wavelength range it occupies, usually requiring high spectral resolution to be studied. The shape is different for the different *line-broadening mechanisms, so the line profile can be used to infer the physical properties of the region where the line is formed.

liner The nucleus of a galaxy that shows a characteristic emission-line spectrum; it is an acronym for low-ionization nuclear emission region. A liner's spectrum is dominated by emission lines from low ionization states (e.g. O II, N II, S II), with only weak emission lines from higher ionization states (e.g. O III, Ne III, He II). The spectrum is evidence of unusual activity in the nucleus, probably not associated with normal stars, and may result from the heating of interstellar matter, either by radiation from a central source or by shock waves generated by supernova explosions. The linewidths are similar to those seen in *Seyfert galaxies and, like Seyferts, the activity seems to occur more often in S0, Sa, and Sb galaxies than in other types.

line ratio The ratio of the strengths (*equivalent widths) of two particular absorption or emission lines. Because these lines correspond to different energy levels or arise from entirely different elements, they can be sensitive to the density and/or temperature of the gas. For example, in a hot gas, the ratio of the $H\alpha$ to the $H\beta$ line strength is sensitive to the density of the gas.

line receiver A radio receiver designed to detect radio waves in a narrow band around a known spectral line, such as those from interstellar molecules and the 21-cm hydrogen line.

line spectrum A spectrum consisting of bright emission lines, or dark absorption lines superimposed on a continuous spectrum. Such a line spectrum will convey information about the physical nature of the emitting or absorbing gas.

linewidth The width of an absorption or emission line in a spectrum. It is usually quoted as *full width at half maximum (FWHM), and gives an indication of the temperature or velocity of the emitting or absorbing region.

line wing The region of an absorption or emission line that is far removed from the line's peak. Extended line wings indicate the presence of high-velocity material (relative to the main region), or that the line formed in a high-density gas.

lingula A rounded finger-like extension of a plateau on Titan; pl.*lingulae.*

LISA Abbr. for *Laser Interferometer Space Antenna.

LISA Pathfinder A joint ESA/NASA mission to test technology for detecting gravitational waves which will be used in the subsequent *Laser Interferometer Space Antenna (LISA) mission. The launch of LISA Pathfinder is planned for 2013. It will be placed at the L_1*Lagrangian point of the Earth's orbit.

(∰) SEE WEB LINKS
• ESA mission website.

Lissajous orbit A path around the L_1 or L_2 *Lagrangian points of a two-body system. Lissajous orbits are utilized by certain spacecraft that are required to be in a stable position relative to the Earth and Sun while making long-term observations. In a Lissajous orbit the spacecraft follows a natural (but complex) motion that requires the minimum amount of energy for station-keeping, unlike a *halo orbit, in which the craft follows a simple circular or elliptical path. Spacecraft that use a Lissajous orbit around the Sun-Earth L2 point are the *Herschel Space Observatory and *Planck. The *James Webb Space Telescope will also be placed in a Lissajous orbit at the same location. Such orbits are named after the French mathematician

Jules Antoine Lissajous (1822–80), who studied the types of curves followed by an object in such a position.

lithium star A giant star of spectral type G, K, or M that shows unusually strong lines of lithium in its spectrum. Nuclear reactions in or near the core of the evolving star produce beryllium, which is transported by convection to higher layers where it captures an electron to become lithium. The term is also sometimes used to refer to a *T Tauri star (which is very young and still forming); in these cases the lithium is likely to have been present in the gas from which the star formed, and will soon be destroyed once the star reaches the main sequence.

lithosiderite An alternative, now largely obsolete, name for a *stony-iron meteorite.

lithosphere The outer rigid shell of a planetary body, including the crust and part of the upper mantle. On the Earth it is distinguished from the *asthenosphere beneath, which is a weaker, deformable part of a planetary body. The Earth's lithosphere, which includes the crust and the upper mantle, is about 150 km deep beneath the continents, and 80 km beneath the oceans.

Little Dipper Popular name for the shape formed by the seven main stars of the constellation Ursa Minor, which resembles a smaller version of the Big Dipper, or Plough, formed by seven stars in Ursa Major.

Little Dumbbell A planetary nebula in Perseus, also known as M76 or NGC 650-1. It is a smaller and fainter version of the *Dumbbell Nebula in Vulpecula and, at 12th magnitude, is the faintest *Messier object. It lies about 3500 l.y. away.

Liverpool Telescope A robotically controlled 2-m reflector owned and operated by Liverpool John Moores University, England, sited at the *Roque de los Muchachos Observatory, La Palma. It was opened in 2003. The telescope is used for educational purposes by schools as well as for research.

(((●))) SEE WEB LINKS
• Official telescope website.

L magnitude The magnitude of a star as measured through the infrared L filter of the *Johnson photometry system. The L filter has an effective wavelength of 3450 nm and a bandwidth of 472 nm. More recent observers have redefined the L filter to avoid the worst of the absorption bands in the Earth's atmosphere. The new filter has a longer central wavelength of 3830 nm, bandwidth 510 nm, and is called L′. There is also an L magnitude in *Walraven photometry, but the context should make clear whichever is intended.

LMC Abbr. for *Large Magellanic Cloud.

LMT Abbr. for *Large Millimeter Telescope.

lobate ridge A ridge on a planetary surface with projecting lobes on one or both sides. Many lunar *wrinkle ridges are lobate; this has been interpreted as evidence of lava extrusion from the ridges, but it now seems more likely that the lobes are caused by buckling and folding of the surface during tectonic movement. Lobate ridges can also be caused by non-volcanic means, and are found at the boundaries of some Martian ejecta blankets.

lobate scarp A scarp on a planetary surface that takes the form of a series of lobes. Lobate scarps are a notable feature on Mercury, where they are interpreted as low-angle thrust faults, and may be evidence of global contraction.

lobe 1. A bright, diffuse area of radio emission seen on one or both sides of the nucleus of a *radio galaxy or other active galaxy and situated far outside the visible confines of the galaxy. The emission is mainly synchrotron radiation. The lobe is believed to consist of material ejected from the nucleus of the galaxy and transported into intergalactic space along a *jet.
  2. A region of high sensitivity appearing as a bulge in the antenna pattern of a radio telescope.

local arm An alternative name for the *Orion Arm of our Galaxy, the arm in which the Sun lies.

local bubble A region of low density (approximately 0.07 atom/cm^3) in the interstellar matter surrounding the Solar System. The bubble is approximately 200 pc across and encompasses the stars of the immediate solar neighbourhood. The Solar System seems to lie about 10–20 pc from the bubble's edge. The local bubble is thought to be caused by shock waves from several supernova explosions in the region within the past 10 million years.

Local Group The group of galaxies about 3 million l.y. in diameter which contains our Galaxy. There are around 50 known members of the Local Group, most of them *dwarf spheroidal companions of our own Galaxy or the Andromeda Galaxy. The nearest other prominent galaxies (Sculptor and M81 groups) are considerably farther away, about 10 million l.y. The total mass of the Local Group is estimated at about 2×10^{12} solar masses. The brightest members are the three spirals: the Andromeda Galaxy, our Galaxy, and M33. Other faint dwarf galaxies doubtless remain to be discovered.

local hour angle (LHA) The *hour angle of a celestial object as measured with reference to the observer's local meridian.

local mean time The *mean solar time on the meridian of an observing site. It is the *local hour angle of the *mean sun plus 12 hours.

local sidereal time (LST) The sidereal time determined with reference to the observer's own local meridian. It is the *local hour angle of the vernal equinox. It differs from the *Greenwich Sidereal Time by the observer's longitude, being 4 min greater than GST for every degree east of Greenwich, and 4 min less for every degree west.

local standard of rest (LSR) The average of the motions of stars in the Sun's neighbourhood. In the local standard of rest, the *space velocities of stars average out to zero. The LSR is moving around the centre of the Galaxy at a speed of about 250 km/s, and hence orbits the Galaxy once in about 200 million years.

local supercluster The supercluster of galaxies that contains the *Local Group, which lies near its edge. It is also known as the **Virgo Supercluster** since it is centred on the Virgo Cluster of galaxies. The local supercluster has a flattened shape and is probably around 100 million l.y. across. It probably contains at least 10 000 galaxies.

local thermodynamic equilibrium (LTE) A balance achieved in a hot gas, in which the radiation emitted is determined by local values of the gas temperature and density. LTE is assumed to apply in stellar atmospheres where strict thermodynamic equilibrium does not hold because of radiation loss to space.

local time The time at an observing site determined with reference to the observer's own local meridian. The local time could be the *local mean time, the *local sidereal time, or even the local *apparent solar time. All involve *hour angles measured with reference to the observer's meridian.

Lockman hole An area of sky with minimal absorption by neutral hydrogen gas. The Lockman hole is approximately 15 square degrees in size, centred on RA 10h 45m, dec. +58°, in Ursa Major. Because of the lack of absorption it is the ideal location for sensitive studies of extragalactic objects. Hence this region has been studied in detail over a wide range of wavelengths, from radio to gamma rays. It is named after the American astronomer Felix James Lockman (1947–), who drew attention to it in 1986.

Lockyer (Joseph), Norman (1836–1920) English scientist and solar physicist. He was the first to study sunspots spectroscopically, finding Doppler shifts caused by convection currents in the Sun's gases. In 1868, independently of P. J. C. *Janssen, he observed the spectra of solar prominences and developed a *spectrohelioscope. Janssen found a new line in the solar spectrum, which Lockyer attributed to a hitherto unknown element. He named it helium (*hēlios* is Greek for 'Sun'); it was discovered in the Earth's atmosphere in 1895. Lockyer also studied *archaeoastronomical sites such as Stonehenge, and founded and edited the science journal *Nature*.

lodranite A very rare class of stony-iron meteorite, named after the town of Lodran in Pakistan, where this type was first identified. Lodranite meteorites consist primarily of olivine and bronzite (orthopyroxene) mixed with metallic nickel–iron. Lodranites show some similarity to the *ureilites.

LOFAR Abbr. for *Low Frequency Array.

LONEOS Abbr. for *Lowell Observatory Near-Earth Object Search.

long-baseline interferometry (LBI) A technique in which radio telescopes hundreds of kilometres apart can be joined electronically to form an array. Signals received by the various antennas are brought to a central *correlator by means of radio or fibre-optic links. A well-known example is the *MERLIN array. *See also* VERY LONG BASELINE INTERFEROMETRY.

longitude The angle around some reference plane from an adopted starting-point. In astronomy, the equivalent of longitude on Earth is *right ascension. *See also* CELESTIAL LONGITUDE; GALACTIC LONGITUDE; HELIOCENTRIC LONGITUDE.

longitude at the epoch (*L*) The longitude that a planet has at a particular date and time. It is the angle measured along the ecliptic from the vernal equinox eastwards to the ascending node (i.e. to the longitude of the ascending node), and then eastwards along the orbital plane to the planet's position at that instant. In the diagram, the longitude at the epoch is the angle γSN plus the angle NSP. *See also* ELEMENTS, ORBITAL.

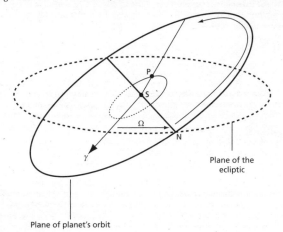

longitude at the epoch:

N   ascending node
P   planet
S   Sun
γ   vernal equinox
Ω   longitude of the ascending node

longitude of perigee For the Moon or an artificial Earth satellite, the angle measured from the direction of the vernal equinox eastwards along the celestial equator to the ascending node of the satellite's orbit, and then continued eastwards along the orbital plane to the satellite's perigee.

longitude of perihelion (ϖ) The angle measured from the vernal equinox eastwards along the ecliptic to the ascending node of a planet's orbit, and then continued eastwards along the orbital plane to the perihelion. In other words, it is the *longitude of the ascending node added to the *argument of perihelion. *See also* ELEMENTS, ORBITAL.

longitude of the ascending node (Ω) The angle measured along the ecliptic from the vernal equinox eastwards to the ascending node of a planet's orbit. It defines the point at which an object's orbit crosses the plane of the Earth's orbit. *See also* ELEMENTS, ORBITAL.

long-period comet A comet with an orbital period greater than 200 years. Examples include Comets *Donati, *Tebbutt, and *West. The *Kreutz sungrazers are also long-period comets. The shortest computed period for a comet in this category is 226 years (C/1905 F1 Giacobini). *See also* PERIODIC COMET.

long-period variable (LPV) A red-giant pulsating variable, with a well-defined period in excess of 100 days, otherwise known as a *Mira star. The term was previously applied to any late-type variable with a long period, including the types currently known as *RV Tauri stars and *semiregular variables, but it is now generally restricted to the Mira type.

Long Wavelength Array (LWA) A radio telescope array observing at wavelengths from 3.4 to 30 m being constructed in New Mexico by a consortium of US universities and research institutions. When finished, it will consist of 53 stations each containing 256 individual pyramid-shaped elements. The first station, LWA-1, came into operation in 2011 near the centre of the *Very Large Array at Socorro, New Mexico. Additional stations are being added as funding permits, eventually providing baselines up to 400 km. The partners in the project are the University of New Mexico, the Naval Research Laboratory, Jet Propulsion Laboratory, Los Alamos National Laboratory, Virginia Tech, National Radio Astronomy Observatory, Air Force Office of Scientific Research, and the University of Iowa. Its headquarters are at the University of New Mexico in Albuquerque.

((⊕)) SEE WEB LINKS
• Official telescope website.

lookback time The time taken for light from a distant object to reach the Earth; sometimes also known as light travel time. Light from nearby galaxies takes several million years to reach us, but the lookback time to very distant galaxies and quasars may be of the same order as the *Hubble time. Galaxies which are so far away that their lookback time would be greater than the age of the Universe can never be observed.

((⊕)) SEE WEB LINKS
• Online calculator for converting redshift to lookback time and other parameters.

loop prominence A very bright *active prominence in the form of a loop above the site of a very large flare, seen on the Sun's limb in *Hα light. A series of loop prominences usually occurs, progressively larger loops forming above one another. The smaller, earlier ones eventually fade from sight after an hour or so. The maximum height attained is 100 000 km, and a complete loop prominence system may last more than a day. X-ray loops, observed simultaneously on occasion, are located on the outer edge of Hα loops, giving the impression that loop prominences somehow condense out of the hot coronal material.

Lorentz–Fitzgerald contraction The contraction in length of a moving body in its direction of motion. The effect was proposed independently by the Dutch physicist Hendrik Antoon Lorentz (1853–1928) in 1895 and the Irish physicist George Francis Fitzgerald (1851–1901) in 1893 to account for the inability of the *Michelson–Morley experiment to detect the Earth's motion through the ether. It was later incorporated into the special theory of relativity. The contraction becomes noticeable only at velocities near that of light; at the speed of light itself, the object's length would in theory contract to zero.

Lorentz transformation A set of equations relating the time and position of an event as seen by one observer to the time and position of that same event seen by a second observer who is moving at constant velocity relative to the first. The equations are named after the Dutch theoretical physicist Hendrik Antoon Lorentz (1853–1928), who initially derived them. They were later incorporated in *special relativity.

Lost City meteorite The second meteorite fall to be photographed by a camera network, thus enabling its trajectory and orbit to be determined and the meteorite to be recovered. A magnitude −15 fireball was photographed by the Smithsonian Prairie Network over Oklahoma on 1970 January 3. Four fragments, totalling 17 kg, of the H5 *ordinary chondrite meteorite were subsequently found near the predicted impact site, about 4.5 km east of Lost City, Oklahoma. Its orbit had an aphelion in the asteroid belt.

Lovell, (Alfred Charles) Bernard (1913–) English radio astronomer. His daytime radar observations of the Giacobinid meteor shower in 1946 proved the worth of the technique. From 1950 he studied radio emission from space, showing that its fluctuations were atmospheric effects akin to the scintillation of starlight. He supervised the building of the Mark I radio telescope at Jodrell Bank (now known as the *Lovell Telescope), and attracted publicity by using it to monitor the first Soviet space probes. Lovell also studied radio emission from flare stars.

Lovell Telescope A 76-m fully steerable radio telescope at the *Jodrell Bank Observatory in Cheshire, England. The telescope was completed in 1957, and was known as the Mark I until 1971 when it was substantially rebuilt with a new bowl and supporting structure, and became the Mark IA. In 1987 it was named the Lovell Telescope in honour of A. C. B. *Lovell. It was resurfaced again in 2001–02.

Lowell, Percival (1855–1916) American astronomer. He established the *Lowell Observatory in 1894 to search for signs of intelligent life on Mars, and drew maps depicting a network of Martian *canals as originally reported by G. V. *Schiaparelli. He calculated the position of an unknown 'Planet X' that he believed was the cause of irregularities in the orbits of Uranus and Neptune, and began to search for it, unsuccessfully. Pluto was found in 1930 by C. W. *Tombaugh at Lowell Observatory, but turned out not to be the large planet that Lowell had predicted.

Lowell Observatory An observatory at an altitude of 2210 m on Mars Hill, Flagstaff, Arizona, founded in 1894 by P. *Lowell. It contains a 24-inch (0.61-m) refractor installed by Lowell in 1896, and the 0.33-m astrograph with which Pluto was discovered. In 1961 a dark-sky outstation was opened on Anderson Mesa, 19 km southeast of Flagstaff, altitude 2200 m. This contains the 1.83-m Perkins reflector operated jointly with Boston University, moved here from Ohio in 1961; the 1.1-m John S. Hall reflector opened in 1968; a 0.8-m reflector; and a 0.6-m Schmidt used for the *Lowell Observatory Near-Earth Object Search. The Navy Prototype Optical Interferometer (NPOI), a joint project of the Naval Research Laboratory, the US Naval Observatory, and Lowell Observatory, began operation on Anderson Mesa in 1996, performing high-resolution astrometry and imaging. Lowell Observatory will also operate the 4.2-m *Discovery Channel Telescope, due for completion in 2011.

(( )) SEE WEB LINKS
• Official observatory website.

Lowell Observatory Near-Earth Object Search (LONEOS) A search for nearth-Earth asteroids and comets undertaken by *Lowell Observatory, using a 0.6-m Schmidt camera on the Observatory's site at Anderson Mesa near Flagstaff, Arizona. LONEOS ran from 1998 to 2008, during which time it discovered 289 near-Earth asteroids and 42 comets.

(( )) SEE WEB LINKS
• Official project website.

lower culmination The point on an observer's meridian at which a celestial body reaches its lowest altitude. *See* CULMINATION.

Low Frequency Array (LOFAR) An aperture synthesis radio telescope consisting of numerous *phased arrays observing at 10–240 MHz (wavelengths of 30–1.25 m) being built by a consortium of European countries led by the Netherlands. When complete, LOFAR will consist of 44 stations each containing up to 192 antennas. The central core of the LOFAR array, 3 × 2 km across and containing 18 stations, is at Exloo in the northern Netherlands. An additional 18 stations will be sited in the Netherlands at distances up to 50 km from the core. Four other countries are building eight more stations: Germany (5 stations), the UK (1), France (1), and Sweden (1), giving baselines up to 1500 km. The simple omnidirectional antennas are of two types: low-band antennas (LBA) sensitive to the range 10–80 MHz and high-band antennas (HBA) sensitive to 120–240 MHz. The Dutch stations will have 48 HBAs and 96 LBAs, while the international stations will have 96 of each. Signals from the stations are combined in a central processor at Groningen, Netherlands. Seismic geophones, bio-sensors, and weather instruments installed at various stations will make LOFAR not only a radio telescope but also a sensor platform for geophysicists and agricultural scientists.

(⊕) SEE WEB LINKS
• Official telescope website.

low-luminosity star A loose term that can encompass *red dwarfs, *subdwarfs, *white dwarfs, and *brown dwarfs. The difficulty of detecting stars of low luminosity means that their total numbers are uncertain. However, they may constitute a significant fraction of the total mass of the Galaxy.

low-mass star A loose term, sometimes including stars of mass a little greater than that of the Sun, and sometimes used only of stars of less than a few tenths of a solar mass, but still with sufficient mass to burn hydrogen in their cores (i.e. at least 0.08 solar masses). The first definition separates stars with radiative cores from higher-mass stars with convective cores; the latter restricts the term to red dwarfs.

low surface brightness galaxy (LSB galaxy) A type of galaxy whose density of stars is so low that it is difficult to detect against the sky background. The proportion of low surface brightness galaxies relative to normal galaxies is unknown, and they may represent a significant component of the Universe. Many of these dim galaxies are dwarfs, found particularly in clusters of galaxies; some are as massive as large spirals, for example *Malin-1.

low-velocity star A star whose speed relative to stars in the solar neighbourhood is small, and which is therefore in a similar orbit around the galactic centre.

LRO Abbr. for *Lunar Reconnaissance Orbiter.

LSR Abbr. for *local standard of rest.

LSST Abbr. for *Large Synoptic Survey Telescope.

LST Abbr. for *local sidereal time.

LTP Abbr. for *lunar transient phenomenon.

lucida The brightest star in a constellation or in a cluster of stars.

luminosity (L) The amount of radiation a star emits, corrected for *interstellar absorption. It is expressed in watts, or in terms of the luminosity of the Sun ($L_\odot$), approximately 3.9×10^{26} watts. Luminosity is related to the absolute bolometric magnitude (*see* ABSOLUTE MAGNITUDE) of the star, M_{bol}, by the equation

$$M_{bol} - 4.72 = -2.5 \log (L/L_\odot).$$

The luminosities of stars range from over $10^5 L_\odot$ for the brightest supergiants to less than $10^{-5} L_\odot$ for feeble red dwarfs. For X-ray sources it is usual to define luminosity not bolometrically but in particular passbands.

luminosity class A classification of stars according to their luminosity, which can vary widely for a given *spectral type. Luminosity class indicates, for example, whether a star is a supergiant, a giant, or a dwarf. The luminosity class is assigned by examination of the star's spectrum, looking at luminosity-sensitive lines or line ratios. The luminosity classes generally used are:

Ia–0   Extreme supergiants (sometimes called hypergiants)
Ia   Luminous supergiants
Iab   Normal supergiants
Ib   Underluminous supergiants
II   Bright giants
III   Giants
IV   Subgiants
V   Dwarfs (main-sequence stars)

The accompanying diagram shows the luminosity classes on a *Hertzsprung–Russell diagram. *See also* SPECTRAL CLASSIFICATION.

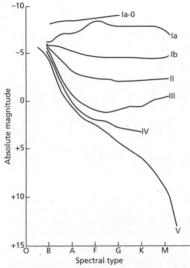

luminosity class: A plot of the absolute magnitude of stars against their spectral type shows different luminosity classes.

luminosity density The average integrated light per unit volume of the Universe, which provides a useful constraint on the overall star-formation history since the Big Bang. The present-day luminosity density in optical light of the local Universe (and hence, according to the *cosmological principle, the entire Universe) is around 50 million solar luminosities per cubic megaparsec. In other words, an average cube one million parsecs on a side contains the light of 50 million Suns. For comparison, a typical galaxy is around 10 000 parsecs in radius and contains a billion or more stars; hence most of space is devoid of stars.

luminosity evolution *See* GALAXY EVOLUTION.

luminosity function The number of stars, galaxies, or other objects of different luminosity within a given volume of space (see the accompanying diagram). The luminosity function of a star cluster may be found by counting the number of stars of different *apparent magnitudes, since all the stars in the cluster are at roughly the same distance.

luminosity–volume test A method of testing whether or not galaxies evolve with time. It involves calculating the volume of a sphere with the Earth at its centre and a given galaxy or quasar on its perimeter. The calculated volume is then divided by the volume of a sphere for which the object, if it were on the sphere's perimeter, would be just too faint to detect. The calculation is repeated for each member of a sample of galaxies or quasars. The average value of this ratio for the complete sample should be 0.5 if there is no evolution in the properties of the sources. Quasar samples give a value of about 0.7, which indicates that there are many more quasars at high redshift than would be expected without evolution.

luminous arc A phenomenon produced by a special kind of *gravitational lens. Light from a distant galaxy or quasar is distorted by a foreground lensing object (usually a *cluster of galaxies) into an arc shape. From the geometry of such arcs, the mass of the cluster can be estimated.

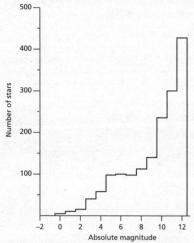

luminosity function: The luminosity function for stars within 20 parsecs of the Sun from Hipparcos data. The number of stars is plotted against their absolute magnitude. The flattening of the slope around 6th–7th magnitude is known as the *Wielen dip.

luminous blue variable (LBV) An extremely hot, massive star at the very top of the Hertzsprung–Russell diagram, with a luminosity up to 10 million times or more that of the Sun; also known as an *S Doradus star or a *Hubble–Sandage variable. Such stars display variability in brightness and colour on various scales, ranging from rapid microvariations of 0.1–0.2 mag. to rare outbreaks of catastrophic mass loss on a timescale of centuries when the brightness can increase by 3 magnitudes or more. One famous example of the latter type is *Eta Carinae. LBVs represent a very short-lived (perhaps as little as 40 000 years) phase in the evolution of young, massive stars, during which they undergo deep erosion of the outer layers before they enter the *Wolf–Rayet phase. The spectra typically exhibit prominent emission lines of H, He, and iron, often with *P-Cygni profiles. During quiescence the spectrum may be B-type, while at maximum brightness the star may appear as an A- or F-type supergiant. The luminosity and temperature are also highly variable, due to surrounding shells of ejected matter which absorb the star's visible and bluer radiation and re-radiate it at longer wavelengths.

Luna A series of Moon probes launched by the former Soviet Union, the first few of which were originally known as Lunik. Luna 1 was the first space probe, although it missed its target; Luna 2 was the first man-made object to hit another body in space; Luna 3 took the first pictures of the Moon's farside. Luna 9 made the first soft landing on the Moon and sent back the first pictures from its surface, while Luna 10 was the first probe to go into orbit around the Moon. Lunas 16, 20, and 24 brought lunar samples back to Earth; Lunas 17 and 21 delivered the *Lunokhod rovers to the Moon's surface (see accompanying table).

Successful Luna Probes[a]

Probe	Launch date	Results
Luna 1	1959 January 2	Missed Moon by 6000 km
Luna 2	1959 September 12	Hit Moon September 13
Luna 3	1959 October 4	Photographed Moon's farside
Luna 9	1966 January 31	Landed on Oceanus Procellarum February 3
Luna 10	1966 March 31	Entered lunar orbit April 3
Luna 11	1966 August 24	Entered lunar orbit August 27
Luna 12	1966 October 22	Entered lunar orbit October 25
Luna 13	1966 December 21	Landed in Oceanus Procellarum December 24
Luna 14	1968 April 7	Entered lunar orbit April 10
Luna 16	1970 September 12	Landed in Mare Fecunditatis September 20; returned to Earth September 24 with 100 g of lunar soil
Luna 17	1970 November 10	Landed in Mare Imbrium November 17; carried Lunokhod 1 rover
Luna 19	1971 September 28	Entered lunar orbit October 3
Luna 20	1972 February 14	Landed in Mare Fecunditatis February 21; returned to Earth February 25 with 30 g soil sample
Luna 21	1973 January 8	Landed in Mare Serenitatis January 15; carried Lunokhod 2 rover
Luna 22	1974 May 29	Entered lunar orbit June 2
Luna 23	1974 October 28	Landed in Mare Crisium November 6; damaged drill prevented sample return
Luna 24	1976 August 9	Landed in Mare Crisium August 18; returned to Earth August 22 with 170 g of soil

[a] Lunas 4–8, 15, and 18 were failed landers.
All dates are UT.

Lunar Atmosphere and Dust Environment Explorer (LADEE) A NASA lunar orbiting mission that will study the extremely tenuous gas and dust environment of the Moon. The orbiter will carry a neutral mass spectrometer, an ultraviolet/visible spectrometer, and a dust detector. It is planned for launch in 2013.

 SEE WEB LINKS
• NASA information page.

lunar eclipse The passage of the Moon through the Earth's shadow, resulting in a darkening which may range from barely detectable in a *penumbral eclipse to very dark when the Moon enters the Earth's *umbra. Lunar eclipses can occur only at full Moon when the Moon is close to the node of its orbit around the Earth. When the Moon is slightly to the north or south of the node, it does not pass centrally through the umbra, and may undergo only a *partial eclipse. During umbral eclipses the Moon appears darkened to varying degrees (*see* DANJON SCALE), determined principally by the clarity of the Earth's atmosphere. Refraction of sunlight through the atmosphere on to the eclipsed Moon produces a coppery or reddish coloration in some events, while darker eclipses may appear steel-grey. Coloured fringes may be seen around the edges of the umbra during the partial phases. A lunar eclipse is visible over the entire hemisphere of the Earth from which the Moon is above the horizon. The maximum duration of totality for a lunar eclipse, when the Moon passes centrally through the umbra, is 1 h 47 m.

(⊕) SEE WEB LINKS
• Lunar eclipse predictions, reports, and information.

lunar inequality A departure in the Moon's orbital motion from its behaviour according to Kepler's laws. It results not only from the perturbing presence of the Sun and the planets, but also from tidal forces and the non-spherical shapes of the Earth and Moon. **Lunar theory** takes all these factors into account to produce a mathematical description of the Moon's past, present, and future geocentric positions and velocities over thousands of years. Such a description contains not only terms describing the elliptical orbit of the Moon, but also hundreds of periodic terms—*inequalities—due to those other factors. Among those factors are *evection, *variation (1), and the *annual equation.

Lunar Orbiter A series of five NASA space probes that photographed the Moon from orbit. Their main targets were potential landing sites for the Apollo missions. The first three of the series orbited around the Moon's equator; the last two were put into polar orbit from which they photographed the entire lunar surface. Tracking the Lunar Orbiters revealed the existence of *mascons, areas of denser rock within the Moon.

Lunar Orbiter Probes

Probe	Launch date	Results
Lunar Orbiter 1	1966 August 10	Entered lunar orbit August 14
Lunar Orbiter 2	1966 November 6	Entered lunar orbit November 10
Lunar Orbiter 3	1967 February 5	Entered lunar orbit February 8
Lunar Orbiter 4	1967 May 4	Entered lunar orbit May 8
Lunar Orbiter 5	1967 August 1	Entered lunar orbit August 5

lunar parallax The angle subtended at the Moon by the Earth's equatorial radius. The parallax corresponding to the Moon's mean distance from the Earth is $57' \, 02''.608$.

Lunar Prospector A NASA space probe launched in 1998 January that went into a 100-km-high polar orbit around the Moon, later reduced to as low as 15 km, from which it mapped the Moon's surface composition and measured its gravitational field. Data from the probe's neutron spectrometer instrument suggested the existence of large subsurface ice deposits at the lunar poles. The mission ended in 1999 July when Lunar Prospector was deliberately crashed into an area near the Moon's south pole.

(⊕) SEE WEB LINKS
• NASA information page.

Lunar Reconnaissance Orbiter (LRO) A NASA Moon probe, launched in 2009 June, to improve knowledge in preparation for future human landings. It studies the Moon from low polar orbit (30–50 km) with the following instruments: the Lunar Reconnaissance Orbiter Camera (LROC), which takes wide-angle and narrow-angle photographs of possible landing sites with resolutions down to 1 m and looks for areas of permanent sunlight or shade at the lunar poles; the Lunar Orbiter Laser Altimeter (LOLA), to measure landing site slopes and search for possible locations for surface ice at the poles; the Lunar Exploration Neutron Detector (LEND), to search for ice near the surface and study radiation around the Moon; Diviner Lunar Radiometer Experiment (DLRE), an infrared radiometer to measure lunar surface temperatures and identify possible ice deposits; Lyman-Alpha Mapping Project (LAMP), an imaging ultraviolet spectrometer to observe the surface at ultraviolet wavelengths, including permanently shadowed regions; Cosmic Ray Telescope for the Effects of Radiation

(CRaTER), to investigate radiation around the Moon and its possible biological effects; and Mini-RF, a miniature radar. In addition, a subsatellite called Lunar Crater Observation and Sensing Satellite (LCROSS) was launched on the same rocket as LRO. On 2009 October 9 the upper stage of the launch rocket was crashed into the Moon in the permanently shadowed floor of the crater Cabeus. LCROSS observed the impact for signs of water vapour before itself crashing into the Moon. 📷

(⊕) SEE WEB LINKS
• Official mission website

lunar theory *See* LUNAR INEQUALITY.

lunar transient phenomenon (LTP) A temporary event observed on the Moon's surface, taking the form of a local abnormal brightening, obscuration, or colour change; also known as a transient lunar phenomenon (TLP). Such events have been reported by many visual observers from the time of F. W. *Herschel, and have been taken as evidence of volcanic activity or outgassing at the lunar surface. However, there has been no unequivocal photographic or photometric record of such an event, and some authorities doubt their reality.

lunation The period of time taken for the Moon to go through a complete cycle of phases. It is the same as a *synodic month.

Lundmark, Knut Emil (1889–1958) Swedish astronomer. In the 1920s he argued that the lack of significant rotational motion in some 'spiral nebulae' (galaxies) indicated their extreme distance. Using redshift data for over 40 galaxies, he derived an early form of distance–velocity relation (now known as the *Hubble law). He was one of the first to point out that many of the brighter galaxies were distributed around a great circle on the celestial sphere, later shown by G. H. *de Vaucouleurs to reflect groupings in clusters and superclusters. In 1925 Lundmark introduced an 'upper class' of novae (later named supernovae) for the brightest novae observed in other galaxies.

lunisolar precession The effect of the attractions of the Sun and Moon on the Earth's equatorial bulge, which causes the Earth's pole to sweep out a circle on the sky around the pole of the ecliptic. Simultaneously, the position of the equinoxes moves westwards along the ecliptic. *See also* NUTATION; PRECESSION.

lunitidal interval The time between the transit of the Moon across the local meridian and the time of the next high tide.

Lunokhod Two unmanned lunar rovers, launched by the former Soviet Union. The eight-wheeled vehicles carried TV cameras and instruments to analyse the Moon's soil, and were driven over the lunar surface by remote control from Earth. Lunokhod 1 was delivered to Mare Imbrium aboard Luna 17 in 1970 November. It had covered a total of 10½ km by the end of its mission 11 months later. Lunokhod 2, an improved version with twice the top speed, landed aboard Luna 21 in 1973 January in the partially ruined crater Le Monnier on the edge of Mare Serenitatis, and covered a total of 37 km in its five-month lifetime.

Lupus (Lup) (*gen.* **Lupi)** A constellation of the southern sky, representing a wolf. Its brightest star is Alpha Lupi, of magnitude 2.3. The constellation contains a variety of double stars, notably Epsilon, Eta, Kappa, Mu, Xi, and Pi Lupi; NGC 5822 is a 6th-magnitude open cluster, and NGC 5000 a 7th-magnitude globular cluster.

Lutz–Kelker bias A statistical effect caused by random errors in the observed parallaxes of stars which leads to an underestimation of the stars' distances and hence a corresponding error in their derived luminosities. It occurs because the number of stars increases with distance from Earth, and hence more stars in any given sample are likely to have their distances underestimated due to observational error than overestimated. The effect was first calibrated in 1973 by the American astronomer Thomas Edward Lutz (1940–95) and the American-Canadian statistician Douglas Herson Kelker (1940–).

Luyten, Willem ('William') Jacob (1899–1994) Dutch-American astronomer, born in Java, in modern Indonesia. In the early 1920s he began a search for stars with large proper motion in the hope of detecting nearby white dwarfs, only three of which were known in 1920. The search eventually yielded the proper motions of 120 000 stars, and added considerably to the number of known white dwarfs, which by 1960 had risen to 400, most of them found by Luyten. In 1947 he discovered a star in Cetus (UV Ceti) with high proper motion; this became the first *flare star to be identified when photographs taken for measuring its parallax showed it to brighten and fade suddenly.

LWA Abbr. for *Long Wavelength Array.

l.y. Abbr. for *light year.

Lydia family A small family of asteroids at a mean distance of about 2.75 AU from the Sun. The family is unusual in that its members are of varying composition. The largest member is the T-class (308) Polyxo, diameter 141 km. The family is named after (110) Lydia, an M-type asteroid, diameter 86 km, discovered in 1870 by the French astronomer Louis Alphonse Nicolas Borrelly (1842–1926). Lydia's orbit has a semimajor axis of 2.734 AU, period 4.52 years, perihelion 2.52 AU, aphelion 2.95 AU, and inclination 6°.0.

Lyman-α forest A close series of narrow absorption lines seen at shorter wavelengths than the strong Lyman-α emission line in the spectra of quasars. They are believed to be caused by absorption in cool hydrogen clouds, perhaps in the haloes of galaxies which are too faint to be seen directly, between us and the quasar. All these absorption lines are Lyman-α lines of hydrogen, but redshifted by different amounts depending on the distance of the absorbing cloud. Since the clouds are between us and the quasar, their redshifts are less than that of the Lyman-α emission from the quasar itself. *See also* LYMAN SERIES.

Lyman-break galaxy A galaxy with a very high *redshift discovered from its red colour. Hydrogen is very effective at absorbing radiation with wavelengths shorter than 91.2 nm (the *Lyman limit), and all galaxies contain large amounts of hydrogen; hence galaxies are virtually dark at wavelengths shorter than 91.2 nm. This dividing point in a galaxy's spectrum is termed the **Lyman break**. For a galaxy at a redshift of about 3, the Lyman break falls between the U and B photometric bands. The galaxy should therefore be seen in B but be virtually invisible in U, an effect called the *U-band dropout*. By looking for galaxies with just these properties, astronomers have found many at very high redshifts. About 1000 Lyman-break galaxies are now known. They have been important in investigations of *galaxy evolution.

Lyman limit The short-wavelength end of the hydrogen *Lyman series, at 91.2 nm. It corresponds to the energy required for an electron in the hydrogen ground state to jump completely out of the atom, leaving the atom ionized.

Lyman series A sequence of absorption or emission lines in the ultraviolet part of the spectrum, due to hydrogen. They range from Lyman-α at 121.6 nm towards shorter wavelengths, the spacing between the lines diminishing as they converge on the **Lyman limit** at 91.2 nm. The Lyman series is caused by electron jumps between the ground state and higher levels of the hydrogen atom. The term is also used to describe certain lines in the spectrum of singly ionized helium. The He II Lyman lines have almost exactly one-quarter the wavelength of their hydrogen equivalents: for example, He II Lyman-α is at 30.4 nm, and the corresponding Lyman limit is at 22.7 nm. It is named after the American physicist Theodore Lyman (1874–1954). *See also* HYDROGEN SPECTRUM.

Lynx (Lyn) (*gen.* **Lyncis**) A constellation of the northern sky, representing a lynx. Its brightest star, Alpha Lyncis, is of magnitude 3.1. The other stars are faint, but include several interesting doubles and triples.

Lyot, Bernard Ferdinand (1897–1952) French solar astronomer and scientific instrument designer. In the 1920s he designed and built a highly sensitive polariscope for studying the Moon and planets in polarized light. Among his findings were that Venus has a high-altitude haze layer, and that Mars has a thin atmosphere and dusty surface. In 1930 he invented the *coronagraph for observing the Sun's corona, and various filters for solar observation. In 1935 he made the first cine films of the corona.

Lyot filter Another name for a *birefringent filter.

Lyra (Lyr) (*gen.* **Lyrae**) A prominent constellation of the northern sky, representing a lyre. Its brightest star is *Vega. Beta Lyrae (Sheliak) is an eclipsing binary, the prototype of the *Beta Lyrae stars; it varies between magnitudes 3.3 and 4.4 in a period of 12.9 days. It is also a double star, with a fainter component of magnitude 7.2. Delta Lyrae is a wide double, magnitudes 4.2 and 5.6, while Epsilon Lyrae is a famous quadruple star, the *Double Double. Zeta Lyrae is another double star, magnitudes 4.4 and 5.7. RR Lyrae is the prototype of the *RR Lyrae stars, an important class of pulsating variables. M57 is a well-known planetary nebula, the *Ring Nebula. The *Lyrid meteors radiate from the constellation every April.

Lyrid meteors A meteor shower active from April 16 to 25, with maximum on April 22 from a radiant at RA 18h 04m, dec. +34°, southwest of Vega. In most years the shower is not particularly prolific, with a peak ZHR of 10–15. Short-lived outbursts were seen in 1922 and 1982. Lyrids are fast (geocentric velocity 49 km/s), with a fair proportion of bright events and meteors leaving persistent trains. The Lyrids are associated with Comet Thatcher (C/1861 G1). They are the oldest recorded meteor shower, seen by the Chinese in 687 BC.

Lysithea A satellite of Jupiter, twelfth from the planet; also known as Jupiter X. Its distance is 11 700 000 km, close to that of *Elara, and its orbital period is 258.5 days. Its diameter is 24 km. Lysithea was discovered in 1938 by the American astronomer Seth Barnes Nicholson (1891–1963). It is a member of the *Himalia group of Jovian satellites.

m The symbol for *apparent magnitude.

M Prefix applied to objects listed in the *Messier Catalogue.

M The symbol for *absolute magnitude.

McDonald Observatory The observatory of the University of Texas, on the adjacent peaks of Mount Locke (altitude 2070 m) and Mount Fowlkes (altitude 1980 m), near Fort Davis in western Texas. It was founded in 1932 and is named after William Johnson McDonald (1844–1926), a banker who gave money for its construction. Its largest instruments are the 9.2 m *Hobby-Eberly Telescope on Mount Fowlkes; the 2.72 m Harlan J. Smith reflector opened in 1968 on Mount Locke; the 2.1 m Otto Struve reflector, opened in 1939 on Mount Locke; and a 0.76 m reflector opened in 1970 on Mount Locke. It is also the site of the 1.2 m MONET/North robotic telescope owned by the University of Göttingen, Germany, opened in 2006 on Mount Locke.

((∰)) SEE WEB LINKS
• Official observatory website.

MACHO Abbr. for massive compact halo object, a hypothetical object presumed to exist in the halo of galaxies, including our own. MACHOs are invisible but they make a significant contribution to the overall mass of a galaxy; they are therefore a form of *dark matter. MACHOs could be low-luminosity stars, Jupiter-like planets, or black holes. Observed *microlensing of stars in the Large Magellanic Cloud could be due to MACHOs around our Galaxy. *See also* WIMP.

Mach's principle The proposal that the inertial mass of a particle is determined by the gravitational effect of all the other matter in the Universe. Hence the concept of mass would be meaningless in an empty universe. The principle is named after the Austrian physicist Ernst Mach (1838–1916). A. *Einstein was influenced by Mach's principle, but his theory of general relativity failed to incorporate it. Many theorists have sought to remedy this failing, but with only limited success.

McIntosh scheme See SUNSPOT.

McMath–Pierce Solar Telescope The world's largest solar telescope, part of the *National Solar Observatory, opened in 1962 at an altitude of 2100 m on Kitt Peak, Arizona. The main telescope is housed in a diagonal shaft, 152 m long, aligned parallel with the Earth's axis. A heliostat on a tower 30.5 m high reflects sunlight down this shaft to a 1.6-m mirror, some 50 m below ground level. Rays are bounced from this mirror part-way up the shaft again to a flat mirror, which diverts the light vertically downwards to form a solar image 0.8 m wide in an observation room below ground. It is named after the American solar physicists Robert Raynolds McMath (1891–1962) and A(ustin) Keith Pierce (1918–2005).

McNaught, Comet (C/2006 P1) A long-period comet discovered on 2006 August 7 by the Scottish-born Australian astronomer Robert Houston McNaught (1956–). It reached perihelion, 0.17 AU, on 2007 January 12 and was at its brightest a day or so later when it was estimated to be magnitude −5 or −6. The comet was prominently visible from the southern hemisphere and was noted for its broad, streaked dust tail, similar to that of Comet *West in

1976, which stretched for up to 35 degrees. Its orbit has an eccentricity slightly greater than 1.0, making it hyperbolic; hence it may never return.

((∰)) **SEE WEB LINKS**
• Information page at Cometography website.

macrospicules Large spicule-like features that radiate from the limb at the Sun's polar regions. Macrospicules extend to much greater heights than ordinary spicules (40 000 km instead of 9000 km). They are visible in spectroheliograms of the ionized helium ultraviolet line at 30.4 nm.

macroturbulence The broadening of a star's spectral lines when viewed from a distance, or at low resolution, so that the star's individual surface features are no longer distinguishable. The broadening is caused by Doppler shifts from motions of different parts of the star's surface.

macula A dark spot on a planetary surface; pl.*maculae*. The name is not a geological term, but is used in the nomenclature of individual features, for example Tyre Macula on Europa.

Maffei Galaxies Two galaxies just beyond the Local Group, discovered by the Italian astronomer Paolo Maffei (1926–2009) in 1968. Both lie about 10 million l.y. away in Cassiopeia, near the border with Perseus, and are heavily obscured by dust in the plane of our Galaxy. Maffei 1 is an elliptical galaxy and Maffei 2 a barred spiral. They belong to a group of at least a dozen members which includes the spiral galaxy IC 342, Dwingeloo 1 (*see* DWINGELOO GALAXY), and two probable companions to Maffei 1: MB 1, a late-type spiral, and MB 2, a dwarf irregular galaxy.

Magdalena Ridge Observatory (MRO) An observatory at an altitude of 3244 m in the Magdalena mountains about 50 km west of Socorro, New Mexico, owned and operated by the New Mexico Institute of Mining and Technology (NMT) in Socorro. Its facilities are a 2.4-m reflector capable of tracking rapidly moving objects, which began operation in 2008; and an interferometer consisting of ten 1.4-m telescopes in a Y-shape with baselines up to 340 m, observing at visible and near-infrared wavelengths between 0.6 and 2.4 μm, due to begin operation in 2012.

((∰)) **SEE WEB LINKS**
• Official observatory website.

Magellan A NASA space probe to Venus, launched from the Space Shuttle *Atlantis* in 1989 May. It went into a highly elliptical near-polar orbit around Venus in 1990 August and mapped the entire planet by radar with an average resolution better than 300 m. In 1993 its orbit was deliberately lowered and circularized by atmospheric drag, a process termed *aerobraking, the first time such a manoeuvre had been carried out at another planet. In its new orbit, Magellan completed a map of the gravity field of Venus. The probe burned up in the planet's atmosphere in 1994 October.

((∰)) **SEE WEB LINKS**
• NASA information page.
• Mission website at JPL

Magellanic Clouds The two irregular galaxies that are satellites of our own Galaxy, easily seen with the naked eye in the southern hemisphere like detached portions of the Milky Way. They are named after the Portuguese explorer Ferdinand Magellan (1480–1521), who described them during his voyage around the world. Both Clouds are believed to orbit our Galaxy in a plane nearly perpendicular to its disk, and may eventually spiral into the Galaxy. *See also* LARGE MAGELLANIC CLOUD; MAGELLANIC STREAM; SMALL MAGELLANIC CLOUD.

Magellanic Stream A thin trail of neutral hydrogen which stretches in a great circle around the sky for at least 110° from the Magellanic Clouds, passing near the south galactic pole. It may be debris pulled out in a close encounter between the Magellanic Clouds and our Galaxy about 200 million years ago.

Magellan Telescopes A pair of 6.5-m reflectors sited 60 m apart at the *Las Campanas Observatory, Chile, jointly owned and operated by the Carnegie Institution of Washington, the Universities of Arizona, Harvard, Michigan, and the Massachusetts Institute of Technology. The first of the pair, named the Walter Baade Telescope, was opened in 2001. The second, the Landon Clay Telescope, opened in 2002.

(((()))) SEE WEB LINKS

• Official telescope website.

• Information page at Las Campanas Observatory.

MAGIC Abbr. for *Major Atmospheric Gamma-ray Imaging Cherenkov telescope.

magma Molten rock within a planetary body. When magma erupts onto the surface it separates into lava and gas. Non-silicate fluids, such as flows of water and ice, may be referred to as *cryomagma* (*see* CRYOVOLCANISM).

magnetar A type of neutron star with an extremely powerful magnetic field, around 10^{11} tesla. This magnetic field can decay, radiating vast amounts of energy away in the form of X-rays or gamma rays. Magnetars are believed to be the cause of *soft gamma-ray repeaters.

magnetic braking A process by which a single rotating star, or a binary star, can lose angular momentum via a magnetically controlled stellar wind. A post-common-envelope binary (i.e. a compact binary that has emerged from a period as a *common envelope binary) may evolve into a cataclysmic binary by losing angular momentum via magnetic braking. *See also* PRE-CATACLYSMIC VARIABLE.

magnetic inversion line The line separating areas of opposite magnetic polarity on the Sun (sometimes referred to as the *neutral line*). Magnetic inversion lines are often marked by *filaments, either quiescent filaments in quiet-Sun regions or active-region filaments.

magnetic monopole A hypothetical point-like defect in the fabric of *spacetime, produced in some models of the early Universe. They are so named because they would behave like the isolated north or south pole of a magnet. A magnetic monopole is the one-dimensional analogue of the *cosmic string, *domain wall, and *cosmic texture. Monopoles, as well as the other types of defect, are a consequence of some forms of *grand unified theory (GUT). No monopoles have yet been detected.

magnetic reconnection A physical process in which a highly stretched or twisted magnetic field can rapidly relax to a less-stressed configuration, thereby releasing stored energy. The process is somewhat analogous to cutting a stretched or twisted rubber band. The concept was introduced to explain solar flares, but is now understood as a more general process that couples magnetic fields and plasmas. Magnetic reconnection can also change how magnetic fields are connected; for example, magnetic reconnection between the Earth's and the interplanetary magnetic fields creates magnetic field lines that link the Earth to interplanetary space and ultimately the Sun.

magnetic star A star with a strong magnetic field, as demonstrated by the *Zeeman splitting of lines in its spectrum. Prominent examples are the peculiar A stars (*Ap stars), in which the strength of the field is also variable. More recently, the term 'magnetic star' has been applied to the *AM Herculis stars, a class of cataclysmic variables which contains white dwarfs with extremely powerful magnetic fields (about 100 tesla). *See also* MAGNETIC VARIABLE.

magnetic storm *See* GEOMAGNETIC STORM.

magnetic variable An alternative term for an *Alpha^2 Canum Venaticorum star or *SX Arietis star, types of variable that exhibit low-amplitude optical variations and a variable magnetic field. The term is also sometimes incorrectly applied to any variable in which magnetic processes are important, such as an *AM Herculis star (a *polar*), an *RS Canum Venaticorum star, and certain types of *RR Lyrae star.

magnetobremsstrahlung A class of radiation mechanisms in which a charged particle (usually an electron) follows a spiral path in a magnetic field and thereby radiates energy in the form of electromagnetic waves. The name means 'magnetic braking radiation'. If the velocity of the particle is much less than the speed of light, the emission is known as *cyclotron radiation. If the particle's speed is an appreciable fraction of the speed of light the emission is known as *gyrosynchrotron radiation, and for speeds very close to that of light it is termed *synchrotron radiation. In some usages, magnetobremsstrahlung is regarded as synonymous with synchrotron radiation.

magnetogram A map of the magnetic field across the Sun's photosphere, measured by the *Zeeman splitting of photospheric absorption lines. Daily magnetograms are produced by a number of observatories.

magnetohydrodynamics (MHD) A theory that treats the motion of a plasma (ionized gas) in the presence of a magnetic field as a fluid with a high electrical conductivity. Although approximate, the theory works well for tenuous plasmas on large scales. Under these conditions magnetic forces effectively tie the charged particles (usually protons and electrons) that compose the plasma to the magnetic field; the plasma and magnetic field are then said to be 'frozen' together and the magnetic field moves with the plasma. Depending on the relative energies of the magnetic field and plasma, the magnetic field may control the motion of the plasma, as in a solar coronal streamer or solar prominence, or the plasma may drag the magnetic field with it, as in the solar photosphere or solar wind. Stellar and planetary *dynamos are understood using MHD. The theory of MHD was developed by H. O. G. *Alfvén in 1942. MHD fluid waves are known as *Alfvén waves.

magnetopause The boundary between a planet's *magnetosphere and the external magnetic field of the solar wind. The upwind region of the magnetopause usually lies below a bow shock, whose distance from the planet depends on the solar wind's intensity and field direction. The Earth's magnetopause normally lies about 64 000 km (10 Earth radii) upwind, but may be driven closer by disturbed solar wind conditions following *flares or *coronal mass ejections.

magnetosheath A region of a planet's *magnetosphere close to the boundary marked by the *magnetopause in which significant plasma circulation occurs.

magnetosphere The volume of space surrounding a planet, within which its magnetic field dominates external fields. Significant magnetospheres exist around Mercury, Earth, Jupiter, Saturn, Uranus, and Neptune. The *solar wind compresses a planet's magnetosphere in the sunward direction and draws it out downwind into a long *magnetotail. Planetary magnetospheres can be regarded as teardrop-shaped cavities in the solar wind.

Magnetospheric Multiscale Mission (MMS) A NASA mission to study the Earth's magnetosphere, consisting of four identical spacecraft with variable spacing ranging from 1 km to several Earth radii. Instruments will study the processes of *magnetic reconnection and the acceleration of charged particles in the magnetosphere. MMS is scheduled for launch in 2014.

(()) SEE WEB LINKS
• Official mission website.
• Mission website at Goddard Space Flight Center.

magnetotail The part of a planet's *magnetosphere that is elongated in the direction of the solar wind (i.e. away from the Sun). The Earth's magnetotail (the *geotail*) extends beyond the Moon's orbit, and that of Jupiter extends beyond Saturn's orbit. Reconnection between magnetic field lines in the Earth's magnetotail under disturbed geomagnetic conditions results in the acceleration of electrons into the high atmosphere, producing active and sometimes extensive displays of the *aurora.

magnification The apparent increase in angular size of an object when viewed through a telescope or binoculars, compared with the direct view of the same object. The magnification

of a telescope is found by dividing the *focal length of the objective lens or mirror by that of the eyepiece. The higher the magnification on a given telescope, the dimmer the image. There is a practical limit to the magnification of a telescope, which is approximately twice its aperture in millimetres. For example, a 100-mm objective has a practical magnification limit of 200, set by the effects of *diffraction on the image. There is a lower practical limit, set by the size of the *exit pupil. When this becomes larger than the pupil of the eye, light is wasted and the image does not appear any brighter as the magnification is reduced.
A magnification of, say, 100 is often referred to as a power of 100, and written as ×100.

magnitude A measure of the brightness of a star. Ancient Greek astronomers defined the brightest stars as being of the first magnitude because they were the first to appear after sunset. The magnitude scale continued in steps of decreasing brightness down to sixth magnitude, for those stars which were visible only in total darkness. From its crude beginnings, the magnitude scale has been extended and is now on a strictly defined scale (*see* POGSON SCALE) so that a difference of one magnitude corresponds to a difference in brightness of a factor of 2.512, and 5 magnitudes equals a brightness difference of exactly a hundredfold. Ancient magnitude estimates depended solely on the human eye, corresponding roughly to the modern *V magnitude. The *apparent magnitude of a star is its brightness as seen from Earth, whereas the *absolute magnitude is a measure of its actual (i.e. intrinsic) brightness; the two differ because the intensity of light falls off with distance, and because of *interstellar absorption. When the brightness is measured over all wavelengths, rather than just visible wavelengths, it is known as the *bolometric magnitude.

magnitude of an eclipse A measure of the extent of an eclipse. In a solar eclipse, the magnitude is the fraction of the Sun's diameter that is covered by the Moon. If the eclipse is total, magnitude is replaced by the ratio of the apparent diameters of the Moon and Sun, which is always 1.00 or more during totality. The magnitude of a lunar eclipse is the fraction of the Moon's diameter covered by the Earth's umbra; at a total eclipse it can be much larger than 1.00 (or 100%) because the Earth's shadow is much larger than the Moon. Eclipse magnitudes are expressed either as a decimal or as a percentage; the magnitude of a partial eclipse may be denoted as, for example, 0.59 or 59 %.

main beam The direction in which a radio telescope is most sensitive; also called **main lobe**. Technically, it is the central lobe of the telescope's *antenna pattern.

main-belt asteroid An asteroid that orbits within the main *asteroid belt.

main-belt comet A body that orbits within the main belt of asteroids yet shows activity similar to that of a cometary nucleus. The first such body to be recognized was 133P/Elst–Pizarro. When originally discovered in 1979 it appeared to be an ordinary main-belt asteroid and was numbered 7968, but in 1996 it was seen to be exhibiting cometary activity; as a result, it was reclassified as a periodic comet. This comet and 176P/LINEAR have orbits that show them to be members of the *Themis family of asteroids. The cometary activity exhibited by main-belt comets is thought to be due to pockets of subsurface ice exposed by impacts. Currently, only a handful of main-belt comets are known.

main lobe Another name for *main beam.

main sequence A diagonal strip on the *Hertzsprung–Russell diagram, from top left to lower right. A star is said to be on the main sequence at the stage of its life when it shines by converting hydrogen to helium at its centre. A star's position on the main sequence depends on its mass, with the most massive stars at the upper left and the least massive at the lower right. The Sun, an average star, lies about half-way along the main sequence. Whatever its mass, a star on the main sequence is termed a *dwarf. Stars spend most of their lives on the main sequence, remaining at roughly constant temperature and luminosity, but the time they spend there depends on their mass. For very massive stars, the main-sequence stage lasts only about a million years, but for the least massive stars it is potentially longer than the age of the Universe. *See also* ZERO-AGE MAIN SEQUENCE.

Major Atmospheric Gamma-ray Imaging Cherenkov telescopes (MAGIC)

A pair of 17-m gamma-ray telescopes at the *Roque de los Muchachos Observatory, Canary Islands. They detect the *Cerenkov radiation produced when gamma rays strike the Earth's atmosphere. The MAGIC telescopes are owned and operated by a consortium of institutions from Germany, Italy, Spain, Switzerland, Bulgaria, Croatia, Finland, Poland, and Romania. MAGIC-I began operation in 2004, and MAGIC-II in 2009.

(⊕) SEE WEB LINKS
• Official telescopes website.

major axis The longest diameter of an ellipse, passing through the two foci (*see* FOCUS, ORBITAL); also known as the *line of apsides*.

Maksutov telescope A *catadioptric telescope with a spheroidal primary mirror of short focal length and a *meniscus lens as a corrector plate. The secondary mirror is also spheroidal, and is often produced simply by aluminizing a spot on the convex centre of the corrector plate. This design is compact and gives superlative performance, but the steep curves on the meniscus corrector are difficult to make for large apertures and it is used mostly for amateur instruments. The field of view free from coma (*see* COMA, OPTICAL) is comparatively small. A modern Maksutov has an aspheric primary mirror, which produces a flatter field of view. The design was published in 1944 by the Russian optician Dmitrii Dmitrievich Maksutov (1896–1964). *See also* BOUWERS TELESCOPE.

Malin-1 A huge, *low surface brightness galaxy, about 1 billion l.y. away in the constellation Coma Berenices. It has a diameter of nearly 600 000 l.y. and a total mass of over 2×10^{12} solar masses, but the gas-rich disk is so faint that it was discovered only in 1986 on specially processed photographic plates by the British astronomer David Frederick Malin (1941–).

Malmquist bias A statistical selection effect that arises in astronomical surveys that are complete to some apparent magnitude limit. At large distances from the observer, only objects that are intrinsically luminous can be seen. Nearer the observer, objects with average or below-average luminosity can also be seen. The statistical properties of the sample therefore depend on distance from the observer in a complicated way. This form of bias, first described in 1924 by the Swedish astronomer Karl Gunnar Malmquist (1893–1982), can be avoided by forming a more restricted *volume-limited sample*.

Nowadays, the term Malmquist bias is often used to describe the systematic bias on a measured quantity due to random observational errors. For example, random errors in magnitude measurements will lead to an overestimate of the number of galaxies to a given magnitude limit, because there are more galaxies fainter than the limit which are scattered into the sample by measurement errors than there are galaxies brighter than the limit which are scattered out. A random measurement error thus leads to a systematic bias.

manganese–mercury star *See* MERCURY–MANGANESE STAR.

manganese star A chemically peculiar star with an unusually high ratio of manganese to iron, and a temperature corresponding to a late B spectral type. They are main-sequence stars, similar to *Ap stars, but without evidence for a strong magnetic field.

mantle A thick layer within a planetary body, underneath the crust but overlying the core, and differing in composition from both. The Earth's mantle is about 2900 km thick. Its top lies about 7 km beneath the oceans, and about 30 km beneath the continents.

many-body problem Another name for the *n-body problem.

mare A large dark or low area on a planetary surface; pl.*maria*. The name, which means 'sea', is not a geological term; it was originally used in the 17th century, when the dark plains of the Moon were thought to be water, and was subsequently applied to the dark patches on Mars in the 19th century. The lunar maria are dark, smooth lowland plains of lava that erupted between the end of the *late heavy bombardment 3.9 billion years ago and about 2 billion years ago. Samples of the maria have been obtained by the Apollo missions and unmanned Luna landers. The maria consist of solidified basaltic lavas, chemically and mineralogically different from the highlands, being enriched in iron and titanium, and with abundant pyroxene. The lunar mare lavas differ from those on Earth in that they were very fluid, with a viscosity similar to that of motor oil, so that they flowed for great distances. The Martian maria do not consistently correspond to any kind of topographical feature or geological province, and in most cases appear to consist of dark surface dust.

Maria family A *Hirayama family of asteroids at a mean distance of 2.55 AU from the Sun, differing from other main-belt asteroids in their high orbital inclination, about 15°. Its members are mainly S class. The largest family members are (170) Maria, (472) Roma, (660) Crescentia, (695) Bella, and (714) Ulula, all with diameters of 40–50 km. Maria itself, an S-class asteroid of diameter 44 km, was discovered in 1877 by the Frenchman (Henri) Joseph (Anastase) Perrotin (1845–1904). Its orbit has a semimajor axis of 2.555 AU, period 4.08 years, perihelion 2.39 AU, aphelion 2.71 AU, and inclination 14°.4.

Mariner A series of US planetary probes. Mariner 2, which flew past Venus, was the first space probe to reach another planet. Mariner 4 provided our first close-up views of Mars, and Mariner 9 was the first probe to go into orbit around another planet. Mariner 10, the last of the series, was the first probe to visit two planets, Venus and Mercury.

Successful Mariner Probes[a]

Probe	Launch date	Results
Mariner 2	1962 August 27	Flew past Venus 1962 December 14
Mariner 4	1964 November 28	Flew past Mars 1965 July 15
Mariner 5	1967 June 14	Flew past Venus 1967 October 19
Mariner 6	1969 February 25	Flew past Mars 1969 July 31
Mariner 7	1969 March 27	Flew past Mars 1969 August 5
Mariner 9	1971 May 30	Went into orbit around Mars 1971 November 14
Mariner 10	1973 November 3	Flew past Venus 1974 February 5. Flew past Mercury 1974 March 29. Re-encountered Mercury 1974 September 21 and 1975 March 16

[a] Mariners 1, 3, and 8 were launch failures.
All dates UT.

Mariner Valley *See* VALLES MARINERIS.

Markab The star Alpha Pegasi, magnitude 2.5. It is a giant of type A0, lying 133 l.y. away.

Markarian galaxy A type of galaxy with unusually strong emission at near-ultraviolet wavelengths; most also show strong emission lines in the optical region. Galaxies in which the source of the ultraviolet radiation lies in the nucleus, which is often the case, can be subdivided into *Seyfert galaxies, *N galaxies, and *quasars. Alternatively, the ultraviolet radiation may originate from a more extended region, as in dwarf galaxies undergoing intense bursts of star formation. The galaxies are named after the Armenian astronomer Benjamin Eghishe Markarian (1913–85), who first detected them in the 1960s.

Mars (♂) The fourth planet from the Sun. It appears distinctly reddish to the naked eye. Its mean opposition magnitude is –2.0, although at perihelic oppositions it can reach –2.9, brighter than all planets except Venus and Jupiter. It is slightly ellipsoidal in shape (equatorial diameter 6792 km, polar diameter 6752 km).

Mars has a thin atmosphere composed (by volume) of about 95% carbon dioxide, 2.7% nitrogen, 1.6% argon, 0.1% oxygen, 0.1% carbon monoxide, and small variable traces of water vapour. The average atmospheric pressure at the surface is about 6 mbar, with a seasonal variation of over 1 mbar. Surface temperatures range from 20° C to –140°C, averaging about –60°C. White clouds of condensed water vapour or carbon dioxide are relatively common, particularly near the terminator and in polar latitudes. There are two permanent water-ice caps at the poles which never melt. In winter these are overlain by caps of frozen carbon dioxide several metres thick, extending to latitude 60°. Dust storms occur from time to time, particularly just after perihelion when they may spread to cover the entire planet in a yellow haze, hiding the familiar surface markings.

Mars

Physical data

Diameter (equatorial)	Oblateness	Inclination of equator to orbit	Axial rotation period (sidereal)	
6792km	0.0059	25°.19	24.623 hours	
Mean density	Mass (Earth = 1)	Volume (Earth = 1)	Mean albedo (geometric)	Escape velocity
3.94 g/cm^3	0.11	0.15	0.15	5.03 km/s

Orbital data

Mean distance from Sun				
10^6 km	AU	Eccentricity of orbit	Inclination of orbit to ecliptic	Orbital period (sidereal)
227.937	1.524	0.093	1°.9	686.980 days

Mars's surface is a volcanic basalt with a high iron content. Oxidation of this iron gives Mars its distinctive rust-red colour. Dark and bright markings can be seen through telescopes, but these do not always correspond to topographic features or terrain types; dark patches appear to be due to areas of dark surface dust. These may slowly change over the years as the dust is transported by winds. The most prominent dark marking, *Syrtis Major, is an unremarkable east-facing slope with a gradient of less than 1°. There are many areas of sand dunes; the largest surround the polar caps, and constitute the largest dune field of the Solar System.

Extensive volcanic activity has occurred on Mars. *Tharsis Montes is the largest volcanic region, with *Olympus Mons to the northwest and the vast collapsed structure of Alba Patera to the north. Together, these volcanic areas make up nearly 10% of the planet's surface. No volcanoes are presently active on Mars, but in the past they produced plains of lava stretching for hundreds of kilometres.

Impact craters are widespread on Mars, but there is an almost continuously cratered upland area, similar to the lunar highlands, which makes up about half the planet's surface, mainly in the southern hemisphere. Many of the fresher impact craters, known as *rampart craters, have steep slopes at the edges of their ejecta blankets, suggesting that the surface was damp or muddy when the impacting bodies struck. The best-preserved large impact basins are Argyre and Hellas (*see* ARGYRE PLANITIA; HELLAS PLANITIA). Although there is now no liquid water on the surface, there are signs that rivers and lakes once existed when the atmosphere was presumably denser, warmer, and wetter than at present. Dried-up water channels include Ma'adim Vallis, over 800 km long and several kilometres wide. Direct evidence that liquid water existed at the surface in the distant past was found by the *Mars Exploration Rovers in the form of deposits of salts associated with the weathering of rocks by water, plus the identification of jarosite, a mineral that forms on Earth by prolonged exposure to water. The likely presence of liquid water in the past opens the possibility that complex organic chemistry, perhaps leading to simple forms of life, once occurred on Mars.

Internally, Mars probably has a lithosphere hundreds of kilometres thick, a rocky asthenosphere, and a metallic core about half the planet's diameter. There is no significant magnetic field on Mars, but its surface displays an irregular patchwork of regions showing faint north and south magnetic poles. Mars has two small satellites, *Phobos and *Deimos. 📷

Mars Atmosphere and Volatile Evolution (MAVEN) A NASA Mars orbiter to study the upper atmosphere of the red planet and its interactions with solar radiation and the solar wind. MAVEN will measure the current loss of Mars's upper atmosphere to space in an attempt to estimate how much of the planet's atmosphere has been lost over time, which in turn has caused long-term changes to the planet's climate. MAVEN is due for launch in late 2013 and will arrive at Mars 10 months later.

(⊕) SEE WEB LINKS
- Official mission website.
- Mission page at University of Colorado.

Marsden group A family of *sunskirting comets seen by the SOHO spacecraft and first identified in 2002 by the English astronomer Brian Geoffrey Marsden (1937–2010). They have a mean perihelion distance of 0.048 AU and inclination around 26°. Like the *Kracht group, they are thought to be an offshoot of comet 96P/Machholz.

Mars Exploration Rover (MER) Two identical NASA Mars rovers, named Spirit and Opportunity, launched in 2003 June and July. They landed on opposite sides of the planet, in Gusev crater and on Meridiani Planum, in 2004 January to explore the surrounding surface. Each carried a panoramic camera (Pancam), instruments to analyse selected rocks, and a Rock Abrasion Tool (RAT) for grinding away the weathered surfaces of rocks to expose their interiors.

(⊕) SEE WEB LINKS
- Official mission website.

Mars Express An ESA space probe launched in 2003 June which went into polar orbit around Mars in 2003 December to study the planet's atmosphere and geology, including searching for signs of subsurface water. Mars Express carries a high-resolution stereoscopic camera (HRSC) to image the entire planet in full colour and in 3D at a resolution of 10 metres, and 2 metres in selected areas; a visible and infrared mineralogical spectrometer (OMEGA) to map the surface composition; two spectrometers, SPICAM and PFS, to analyse the composition and structure of the atmosphere; a sensor for neutral and charged particles, ASPERA, to study the interaction of the outer atmosphere with the solar wind; and a radar (MARSIS) to probe the upper 2–3 km of crust. Mars Express released the Beagle 2 lander but nothing was heard from it.

(⊕) SEE WEB LINKS
- ESA mission website.

Mars Global Surveyor (MGS) A NASA space probe launched in 1996 November to study the surface and atmosphere of Mars. MGS went into polar orbit around Mars in 1997 September, using *aerobraking to circularize its orbit. In 1999 March the Mars Orbiter Camera (MOC) began a photographic survey, while the Mars Orbiter Laser Altimeter (MOLA) created a topographic map and the Thermal Emission Spectrometer (TES) studied the composition of the Martian surface, clouds, and atmospheric dust. MGS ceased operation in 2006 November.

(⊕) SEE WEB LINKS
- Official mission website.

Marshall Space Flight Center (MSFC) A NASA establishment at Huntsville, Alabama, founded in 1960 by the transfer to NASA of part of the missile work of the US Army's Redstone Arsenal. MSFC developed the Saturn series of launch rockets and the *Skylab space station under the leadership of Wernher Magnus Maximilian von Braun (1912–77), and subsequently played a leading role in the development of the Space Shuttle and its rocket engines. It is now involved in developing future space launchers. MSFC houses the Payload Operations Control Center for scientific experiments aboard the International

Space Station. It also specializes in producing optics for orbiting observatories including the *James Webb Space Telescope.

(⊕) SEE WEB LINKS
• Official website.

Mars Odyssey (2001 Mars Odyssey) A NASA Mars orbiter, launched in 2001 April. It went into polar orbit around Mars in 2001 October, using *aerobraking to circularize the orbit. The craft carried three main instruments: a Thermal Emission Imaging System (THEMIS) to map the surface mineralogy; a Gamma Ray Spectrometer (GRS) to analyse the surface chemistry; and the Mars Radiation Environment Experiment (MARIE) to assess the radiation risk to future human explorers.

(⊕) SEE WEB LINKS
• Official mission website.

Mars Pathfinder (MPF) A NASA space probe, launched in 1996 December, that landed in a lowland area on Mars, Ares Vallis, in 1997 July. It deployed a micro-rover called Sojourner that roamed the surface, analysing the composition of Martian rocks. The lander and rover operated for nearly three months until communication was lost in 1997 September.

(⊕) SEE WEB LINKS
• Official mission website.

Mars probes A series of probes to the planet Mars, launched by the former Soviet Union. Technical problems meant that most of them were complete or partial failures, although Mars 5 did send some photographs from orbit around the planet.

Mars Probes		
Probe	Launch date	Results
Mars 1	1962 November 1	Radio contact lost *en route*
Mars 2	1971 May 19	Entered Mars orbit November 27; ejected lander which crashed
Mars 3	1971 May 28	Entered Mars orbit December 2; ejected lander, but transmissions failed
Mars 4	1973 July 21	Intended Mars orbiter; passed Mars 1974 February 10 after braking rocket failure
Mars 5	1973 July 25	Entered Mars orbit 1974 February 12
Mars 6	1973 August 5	Flew past Mars 1974 March 12 and ejected lander, which crashed
Mars 7	1973 August 9	Flew past Mars 1974 March 9 and ejected lander, which missed the planet

Mars Reconnaissance Orbiter A NASA space probe to Mars, launched in 2005 August. It entered polar orbit around Mars in 2006 March and used *aerobraking to circularize its orbit. The mission's main objectives are to look for evidence of past or present water, to study Martian weather and climate, and to identify landing sites for future missions. The Orbiter carries a wide-angle camera (CTX), a stereo imaging camera (HiRISE) capable of resolving details as small as 0.2 m wide, a camera to monitor clouds and dust storms (MARCI), and a visible/near-infrared spectrometer (CRISM) to examine the surface composition. Also on board is an infrared radiometer (MCS) to study the atmosphere, an accelerometer, and a shallow subsurface sounding radar (SHARAD) provided by the Italian Space Agency to search for underground water. The orbiter will also serve as a telecommunications link for future missions.

• Official mission website.

Mars Science Laboratory A NASA mission due for launch in late 2011 that will land a rover called *Curiosity on Mars in 2012 August inside the crater Gale.

• Official mission website.

mascon A region of a planetary body that is either made of, or underlain by, denser-than-average material, as evidenced by an increased gravitational pull above it. The word is a contraction of **mass concentration**. Mascons first revealed their presence by the effects they had on the motion of spacecraft orbiting the Moon; they lie beneath many of the major lunar mare basins. Mars also has mascons associated with its impact basins.

maser Acronym for microwave amplification by stimulated emission of radiation, the microwave equivalent of a *laser. In a maser, radiation at a certain frequency causes excited atoms, ions, or molecules of a gas to emit further radiation in the same direction and at the same wavelength, resulting in amplification. Artificial masers are used as amplifiers in some sensitive radio-astronomy receivers. Radio astronomers also study naturally occurring cosmic maser sources.

maser amplifier An amplifier used in radio astronomy which employs the principle of *maser amplification. Maser amplifiers have extremely low noise and high sensitivity.

maser source A radio source in which the spectral lines of an atom, ion, or molecule are greatly amplified by maser action to produce an intense source of radio emission. The first maser to be identified in space was the *OH maser at 1665 MHz, in 1965. Maser emission is now known from many other molecules, including water (H_2O), silicon monoxide (SiO), methanol (CH_3OH), ammonia (NH_3), formaldehyde (H_2CO), hydrogen cyanide (HCN), and silicon sulphide (SiS), and from hydrogen ions. Sometimes many spectral lines from the same molecule can be observed as masers. For example, over 30 different maser lines of SiO have been observed in the circumstellar envelope of the star VY Canis Majoris. Masers occur in star-forming regions (*interstellar masers), in the circumstellar envelopes of red giant stars (*circumstellar masers), in comets, in some planetary atmospheres, and in some active galactic nuclei, where they may be so bright that they are classed as *megamasers. *See also* CYCLOTRON MASER; H_2O MASER; METHANOL MASER; SIO MASER.

Maskelyne, Nevil (1732–1811) English astronomer. On a voyage to St Helena in 1761 he successfully tested tables of the Moon's position compiled by the German astronomer Johan Tobias Mayer (1723–62) for finding longitude at sea. When in 1765 he became the fifth Astronomer Royal, Maskelyne included tables of lunar positions for navigation in *The *Nautical Almanac*, which he founded in 1766. He observed the 1769 transit of Venus and from it calculated the Sun's distance to an accuracy of 1%. In 1774 he measured the deflection of a plumbline near the Scottish mountain Schiehallion, from which he calculated the Earth's average density to be $4.7 g/cm^3$, the first good approximation to the true value ($5.52 g/cm^3$).

mass A measure of the amount of matter in a body. Mass gives rise to the inertia of an object, i.e. its resistance to change in its motion or state of rest; this is known as the *inertial mass*. It also produces a gravitational force (*see* GRAVITATION). Although mass is formally defined in terms of its inertia, it is usually measured by the effects of its gravitation. The *equivalence principle of general relativity asserts that inertial mass and gravitational mass are equal for all bodies, and experiments have verified this equality to one part in 10^{12}. The *weight* of a body is the force by which it is gravitationally attracted to the Earth. Although the terms 'mass' and 'weight' are often used interchangeably on Earth, in space a body may be weightless but still retain its inertia, i.e. it still requires a force to change its motion. According to the special theory of relativity, the mass of a body increases as its velocity approaches that of light. At the speed of light itself the mass would be infinite.

mass defect The difference between the mass of an atomic nucleus and the masses of its individual protons and neutrons when in an unbound state. The mass defect arises because energy is given up when protons and neutrons bind together to form a nucleus. The energy equivalent of the mass defect is known as the *binding energy.

mass discrepancy The difference between the amount of directly observed luminous matter in an object and the mass inferred from the object's speed of rotation or its motion. Many objects, from galaxies up to superclusters, have a significant mass discrepancy, and this is usually taken to be evidence for *dark matter.

mass function A mathematical relationship between the masses of the components in a spectroscopic binary and the orbital inclination. It is the only information about the masses that can be derived when the spectrum of just one star is visible. If that star is of mass m_1 (in solar units), then the mass function is given by $(m_2 \sin i)^3/(m_1+m_2)^3$, where m_2 is the mass of the unseen star and i is the inclination of the orbit.

mass loss The loss of material from a star, which may occur at different evolutionary stages and by various processes. A protostar may be the source of a *bipolar flow. A T Tauri, giant, or supergiant star may lose mass through a vigorous *stellar wind. Ejected material creates a disk around a *Be star, or a shell around a *Gamma Cassiopeiae star (a *shell star); at a late stage of stellar evolution it may give rise to a *planetary nebula. Violent ejection occurs in a nova or supernova explosion. Mass loss may also occur through the outer *Lagrangian point in a *contact binary or *overcontact binary system.

mass–luminosity relation The relationship between the luminosity of a star on the main sequence and its mass. Factors involved include whether radiation or convection is the dominant energy transport process, and, if the former, the type of *opacity involved. For stars of less than 0.4 solar masses, which are fully convective, the luminosity varies as the square of the mass. For stars of mass similar to or slightly less than the Sun, the luminosity is approximately proportional to the mass to the fifth power (e.g. a star with twice the mass will have 32 times the luminosity). For more massive stars, the luminosity varies approximately as the mass cubed (e.g. a star with twice the mass will have about 8 times the luminosity).

m

mass–radius relation The relationship between the radius of a main-sequence star and its mass; it shows that radius is proportional to mass to the power 0.7. Thus a star of spectral type B, of 16 solar masses, has a radius of about seven times the Sun's, while a low-mass red dwarf of 0.1 solar masses has a radius about one-fifth of the Sun's.

mass ratio An expression of the relative masses of the components of a *binary star when both spectra are visible. It is derived from the radial *velocity curves for both components and may be written as $M_2/M_1 = a_1 \sin i/a_2 \sin i$, where M_1 and M_2 are the masses of the components, and a_1 and a_2 are the amplitudes of the velocity curves. If the orbital inclination, i, can be determined (as in an eclipsing binary) the actual masses of the components, and the dimensions of the stars and their orbits, can be obtained.

mass-to-light ratio The ratio of the mass of an object to its total luminosity. The mass and luminosity are usually measured in terms of solar mass and solar luminosity; hence the mass-to-light ratio of the Sun is 1. Most extragalactic objects have mass-to-light ratios greater than 1, indicating that not all their mass is in the form of visible stars. The high values inferred for galaxies (up to 30) and clusters of galaxies (up to 300) indicates the existence of considerable amounts of *dark matter.

mass transfer The transfer of gas between stars in a binary system. The component that is losing material may have evolved to fill its *Roche lobe (in which case the system is a *semidetached binary), or it may be the source of a strong *stellar wind. The component accreting material does so either directly from the stellar wind or a stream of gas, or indirectly through an *accretion disk. The transferred gas is generally hydrogen-rich because it originates in the outer layers of the donor star.

Mather, John Cromwell (1946–) American astrophysicist and cosmologist. He was the driving force behind the *Cosmic Background Explorer (COBE), coordinating both the proposal and the thousand-strong mission team, and had specific responsibility for the Far-Infrared Absolute Spectrophotometer on board COBE, the instrument which showed that the cosmic microwave background radiation has a black-body spectrum. For this work he shared the 2006 Nobel Prize in Physics with G. *Smoot.

matter era In the *Big Bang theory, the era that started when the gravitational effect of matter began to dominate the effect of radiation pressure. Although radiation is massless, it has a gravitational effect which increases with the intensity of the radiation. Moreover, at high energies, matter itself behaves like electromagnetic radiation because it is moving at a speed close to that of light. In the very early Universe, the expansion rate was dominated by the gravitational effect of radiation pressure but, as the Universe cooled, this effect became less important than the gravitational effect of matter. Matter is thought to have become predominant at a temperature of around 10^4K, roughly 30 000 years after the Big Bang. This marked the start of the matter era. *See also* RADIATION ERA.

Mauna Kea Observatories A complex of observatories and telescopes at an altitude of approximately 4200 m on the summit of Mauna Kea island, Hawaii. The site, founded in 1964, is owned and managed by the University of Hawaii's Institute for Astronomy. It is the world's highest observatory site and is generally regarded by astronomers as the best location for ground-based optical, infrared, and millimetre/submillimetre observations. More major telescopes are located on Mauna Kea than at any other site. The University of Hawaii has its own 2.2-m and 0.9-m reflectors on Mauna Kea, opened in 1970 and 2010 respectively (the 0.9-m replaced a 0.6-m opened in 1968). Optical and infrared telescopes from other organizations are the *NASA Infrared Telescope Facility; the *Canada–France–Hawaii Telescope; the *United Kingdom Infrared Telescope; the Gemini North telescope (*see* GEMINI OBSERVATORY); the Japanese *Subaru Telescope; and the *W. M. Keck Observatory. Radio and millimetre-wave instruments on Mauna Kea include the *Caltech Submillimeter Observatory; the *James Clerk Maxwell Telescope; the *Smithsonian Astrophysical Observatory's Submillimeter Array; and a 25-m antenna of the *Very Long Baseline Array.

(⊕) SEE WEB LINKS

• Official observatory website.

Maunder, (Edward) Walter (1851–1928) English solar astronomer. From the Royal Observatory at Greenwich he observed the Sun and recorded sunspots on every possible day for forty years, assisted by his second wife, Annie Scott Dill Maunder, née Russell (1858–1947). He established the Sun's differential rotation from the motions of sunspots at different latitudes, and identified the relationship between solar activity and disturbances in the Earth's magnetic field. The Maunders also studied the history of astronomy; his research, published in 1894 and 1922, revealed what is now called the *Maunder minimum.

Maunder minimum The period from 1645 to 1715 when scarcely any sunspots or aurorae were seen. E. W. *Maunder (and, before him, G. F. W. *Spörer) concluded that there had been a real decline in solar activity then. Further evidence for the Maunder minimum is provided by an increased carbon-14 content in tree rings during that period, since the cosmic rays that produce carbon-14 reach the Earth in greater numbers when the Sun's activity is low. There was also a lengthy cold spell on the Earth from 1550 to 1700, known as the *Little Ice Age*, which, roughly corresponds to a period including the Maunder minimum and the earlier *Spörer minimum. This cold period could be explained by a decrease in solar output of about 1%. The existence of the minimum was confirmed in 1976 by the American solar physicist John Allen ('Jack') Eddy (1931–2009).

MAVEN Abbr. for *Mars Atmosphere and Volatile Evolution.

maximum-entropy method A mathematical technique used in image processing to extract reliable information from noisy and incomplete data. It is designed to ensure that features appearing in the final image (such as a radio map) are likely to be real rather than spurious. It is especially useful for images of extended objects which contain no point sources.

Maxwell, James Clerk (1831–79) Scottish physicist. In the late 1850s he explained that Saturn's rings must consist of many small objects following independent concentric orbits, and that a solid or fluid ring would be unstable. Of his many contributions to theoretical physics, the most important is perhaps the concept of *electromagnetic radiation. In 1865 he published what are now called *Maxwell's equations*, which unify the phenomena of electricity and magnetism into a single theory.

Maxwell Gap A division in Saturn's rings, in the outer part of the C Ring (the crêpe ring). It is about 87 500 km from Saturn's centre, and is 270 km wide. The gap was discovered in 1980 on images taken by Voyager 1 and was named after J. C. *Maxwell.

Maxwell Montes A range of mountains on Venus at around +65° lat., 3°E long. Maxwell Montes are nearly 800 km wide and lie near the centre of *Ishtar Terra. They contain the highest peak on Venus, 11 km above the mean surface level. On the eastern slopes of Maxwell Montes lies the volcanic caldera Cleopatra, 100 km in diameter. Maxwell Montes showed up on the first Earth-based radar maps of Venus, and were named after J. C. *Maxwell.

Maxwell Telescope *See* JAMES CLERK MAXWELL TELESCOPE.

Mayall Telescope The largest telescope at *Kitt Peak National Observatory, Arizona, aperture 4 m, opened in 1973. It is named after a former director of Kitt Peak, Nicholas Ulrich Mayall (1906–93).

M-class asteroid A class of asteroid, fairly common in the main asteroid belt, having featureless reflectance spectra, flat to slightly reddish over the wavelength range 0.3–1.1 µm. M-class asteroids are distinguished from the spectrally identical classes E and P by their moderate albedos (0.10–0.18). The 'M' is for metallic, since they are believed to have metallic (nickel–iron) compositions. Members of this class include (16) Psyche, diameter 253 km, and (21) Lutetia, diameter 96 km.

MDM Observatory An observatory on Kitt Peak, Arizona, jointly owned by the University of Michigan, Dartmouth College, the Ohio State University, Columbia University, and Ohio University; the initials MDM arise from the names of the three initial owners, Michigan, Dartmouth, and MIT. The MDM Observatory operates two telescopes: the 2.4-m Hiltner Telescope, altitude 1938 m, opened in 1986, and the 1.3-m McGraw–Hill Telescope, altitude 1925 m, opened at Dexter, Michigan, in 1969 and moved to Kitt Peak in 1975.

(⊕) SEE WEB LINKS
• Official observatory website.

mean anomaly (M) The angle between the periapsis of an orbit and the position of an imaginary body that orbits in the same period as the real one but at a constant angular speed. The angular speed assigned to the imaginary body is the average angular velocity (the *mean motion) of the real orbiting body. The angle of mean anomaly is measured from the body being orbited, in the direction of orbital motion. *See also* KEPLER'S EQUATION.

mean daily motion (n) The average movement of a body along its orbit in one day, usually expressed in degrees. The mean daily motion of a planet about the Sun can be accurately calculated from theory, but it can also be found by direct observation. For the Moon, ancient eclipse records extending back 2000 years or more have given us a very accurate value of its mean daily motion.

mean density of matter The density of material that would be obtained if all the matter contained in galaxies were smoothed out across the Universe. Although stars and planets have densities greater than the density of water (about 1 g/cm^3), the cosmological mean density is extremely low (less than 10^{-29} g /cm^3, or 10^{-5} atoms/cm^3) because the Universe consists mostly of virtually empty space between galaxies. The mean density of matter determines whether the Universe will continue to expand. *See also* CRITICAL DENSITY; DENSITY PARAMETER.

mean equator The great circle perpendicular to the direction of the *mean pole.

mean equinox The direction to the equinox at a particular epoch, with the effect of nutation subtracted. The mean equinox therefore moves smoothly across the sky due to precession alone, without short-term oscillations due to nutation. *See also* TRUE EQUINOX.

mean motion (*n*) The average *angular velocity of a body in an elliptical orbit, often measured in degrees per day. If the orbit is perturbed, allowance must be made for changes in the period of revolution. For an unperturbed orbit, the mean motion is the same as the angular velocity that the body's radius vector would have if the orbit were circular and of radius equal to the semimajor axis of the ellipse.

mean parallax *See* STATISTICAL PARALLAX.

mean place The coordinates of a celestial object referred to the mean equator and equinox at some standard epoch, such as the beginning of a year. In calculating a star's mean place, the effects of atmospheric refraction, parallax, aberration, and the gravitational deflection of light must be removed.

mean pole The direction on the sky towards which the Earth's axis points at a particular epoch, with the oscillations due to nutation removed. It moves smoothly across the sky, affected only by *lunisolar precession. *See also* TRUE POLE.

mean position *See* MEAN PLACE.

mean sidereal time The time since the *mean equinox crossed the meridian, i.e. the hour angle of the mean equinox. The equinox is not fixed with respect to the stars, but changes its position due to precession and nutation. It is convenient to define a mean equinox with the effect of nutation removed. The mean sidereal time defined by this equinox will, unlike *apparent sidereal time, proceed uniformly.

mean solar day The average length of the apparent solar day; also, the interval between successive transits of the *mean sun. The apparent solar day is not constant in length due to the changing value of the *equation of time. The number of mean solar days in the year is exactly one fewer than the number of sidereal days.

mean solar time The time shown on a clock; technically, it is the time since the *mean sun crossed the meridian with 12 hours added to make the day begin at midnight rather than noon. As the variations of *apparent solar time due to the *equation of time have been averaged out, mean solar time proceeds uniformly except for small irregularities in the gradual slowing of the Earth's rotation. Mean solar time is closely related to *mean sidereal time, and one can be calculated if the other is known. The mean solar time on the Greenwich meridian is *Universal Time.

mean sun A fictitious body whose position on the celestial sphere defines the *mean solar time. It is defined as moving around the equator in one year at a regular rate. The hour angle of the mean sun plus 12 hours gives mean solar time. The mean sun is necessary for uniform timekeeping since the motion of the true Sun along the ecliptic varies due to the eccentricity of the Earth's orbit around the Sun. Moreover, the ecliptic is inclined to the celestial equator, so that even a uniform motion on the ecliptic would not produce a uniform time-scale. For these two reasons the true Sun is not a suitable reference body for timekeeping.

mean time A smoothly progressing form of time, with the short-term irregularities of apparent time averaged out. In *mean sidereal time the effect of nutation is removed; in *mean solar time a correction is made for the irregular movement of the Sun due to the Earth's elliptical orbit and the inclination of the ecliptic to the celestial equator. *Greenwich mean time*, which is the mean solar time at Greenwich, is the same thing as *Universal Time.

medium-band photometry A little-used synonym for *intermediate-band photometry.

MeerKAT Popular name for the *Karoo Array Telescope.

megamaser An extremely powerful maser source in the nucleus of an active galaxy. Megamasers can be more than a million times as powerful as maser sources within our own

Galaxy. The most powerful are the OH and H_2O megamasers, which can generate more than 10^{30} watts of radio power (enough to power 10 000 Suns).

megaregolith A layer of fractured or fragmented material that underlies the surface regolith of a rocky body, such as a planetary satellite or asteroid. A megaregolith may be formed by a major collision that causes extensive fracturing. On the Moon, the megaregolith is the thicker layer of ejecta from basin-forming impacts that underlies the surface regolith. It varies in thickness with distance from the lunar basins from a few hundred metres to a kilometre or more.

Megrez The star Delta Ursae Majoris, magnitude 3.3. It is an A2 dwarf, 81 l.y. away.

Melipal The third 8.2-m Unit Telescope (UT3) of the European Southern Observatory's *Very Large Telescope in Chile, opened in 2000. Its name means 'the Southern Cross' in the local Mapudungun language.

meniscus lens A lens with one concave and one convex surface, so that the glass forms part of a shell. A *positive meniscus lens* is thicker at the centre than the edge and is a *converging lens. A *negative meniscus lens* is thinner in the centre than at the edge and is a *diverging lens. A **meniscus mirror** is used in lightweight telescopes, and is usually supported only by its centre as it has a comparatively thin edge.

meniscus Schmidt telescope A reflecting telescope with a spheroidal primary mirror, a *meniscus lens whose surfaces are concentric with the primary mirror, and an additional *corrector plate in front of the meniscus lens. This provides a wide field of view and small focal ratio.

Menkalinan The star Beta Aurigae. It is an eclipsing binary consisting of a pair of A2 subgiants that varies between magnitudes 1.9 and 2.0 with a period of 3.96 days. It lies 81 l.y. away.

Menkar The star Alpha Ceti, magnitude 2.5. It is an M1.5 giant, 249 l.y. away.

Mensa (Men) (*gen.* **Mensae**) A small, faint constellation near the south celestial pole, representing Table Mountain at the Cape of Good Hope, South Africa. Its brightest star, Alpha Mensae, is of magnitude 5.1. It contains part of the Large Magellanic Cloud.

mensa A small plateau or tableland on a planetary surface; pl.*mensae*. The name, which means 'table', is not a geological term, but is used in the nomenclature of individual features, for example Deuteronilus Mensae on Mars.

Menzel, Donald Howard (1901–76) American astrophysicist and solar astronomer. In the 1920s, with H. N. *Russell, he established the field of theoretical astrophysics. In the 1930s he used quantum physics to interpret the Sun's *flash spectrum, ascertaining the composition of the chromosphere and demonstrating the preponderance of hydrogen. Also in the 1930s he used similar techniques to study nebulae, discovering much about planetary nebulae and the later stages of stellar evolution. After World War II, as an astronomy administrator, he was responsible for founding many institutions and observatories, including the *Harvard–Smithsonian Center for Astrophysics and the *Kitt Peak National Observatory.

Merak The star Beta Ursae Majoris, magnitude 2.3. It is an A0 subgiant or dwarf and lies 80 l.y. away. With *Dubhe, it forms the Pointers that indicate the direction of the Pole Star, Polaris.

Mercury (☿) The closest planet to the Sun. It has the most elliptical orbit (eccentricity 0.206) of all the major planets, so that at perihelion it is only 46 000 000 km from the Sun's centre, but 69 820 000 km at aphelion. Its mean geometric albedo, 0.11, is similar to the Moon's, and its overall colour is grey. Mercury's mean magnitude at greatest elongation is 0.0, but it keeps close to the Sun in the sky and so is seldom visible to the naked eye. Its period of axial rotation is exactly two-thirds of its orbital period, an example of *spin–orbit coupling. As a result, two lines of longitude, spaced by 180°, experience the Sun overhead at perihelion, making these two regions the hottest on Mercury.

Mercury

Physical data

Diameter	Oblateness	Inclination of equator to orbit	Axial rotation period (sidereal)	
4879 km	0.0	0°.01	58.646 days	
Mean density	Mass (Earth = 1)	Volume (Earth = 1)	Mean albedo (geometric)	Escape velocity
5.43 g/cm³	0.06	0.06	0.11	4.25 km/s

Orbital data

Mean distance from the Sun

10⁶ km	AU	Eccentricity of orbit	Inclination of orbit to ecliptic	Orbital period (sidereal)
57.909	0.387	0.206	7°.0	87.969 days

Mercury has no permanent atmosphere, although some hydrogen and helium from the solar wind is temporarily captured. The surface temperatures average about 170°C, but Mercury has the most extreme temperature range of any planet in the Solar System, becoming extremely hot during the day, over 450°C at the subsolar point at perihelion, and rapidly dropping below –183°C during the long night. Dark and bright surface markings can be glimpsed through a telescope, but they have much lower contrast than the markings on Mars or the Moon. The Mariner 10 probe photographed half the planet in detail, revealing a lunar-like landscape heavily scarred with impact craters, many with bright rays. The largest known crater is Beethoven, 640 km wide. There are some slight differences in cratering style: on Mercury, secondary craters fall closer to the main crater than they do on the Moon because of the higher gravity, and inner rings are seen in smaller craters than on the Moon. The largest impact structure on Mercury, the *Caloris Basin, is 1300 km across, similar in size to the Moon's Imbrium Basin. The very small axial tilt of Mercury means that, like the Moon, it is likely to have craters in its polar regions that are never sunlit. Such craters are cold enough that they can preserve water ice, perhaps from impacting comets. Evidence for this has come from terrestrial radar studies that indicate the presence of water ice in some craters at both the south and north poles.

There are no obvious volcanoes on the planet, nor any sinuous rilles to indicate lava eruption, nor are there any dark maria. However, the widespread smooth plains material, which has obscured part of the rim of the Caloris Basin, for example, and infilled many impact craters to make them flat-floored, is probably lava, although it could be ejecta or impact melt from the large basins. *Lobate scarps* up to 500 km long are found in many areas on Mercury, and appear to be thrust faults resulting from sideways compression. These scarps and the general lack of tensional features such as the graben found on the Moon are evidence that the planet has contracted, probably as a result of cooling.

Mercury's high density suggests that it is composed of about 70% iron, probably concentrated in a central core, and 30% rock. The iron-rich core probably has a diameter 75% of that of the planet, proportionally the largest of any planetary body known. The planet has a weak magnetic field with a strength of 3×10^{-7} tesla, around 1% that of the Earth. It has no natural satellites. 📷

mercury–manganese star A form of *manganese star that has a spectral line at wavelength 398.4 nm, identified with ionized mercury; also known as a **manganese–mercury star**.

Mercury Surface, Space Environment, Geochemistry, and Ranging (MESSENGER) A NASA space probe to Mercury, launched in 2004 August. It used two Venus flybys, in 2006 October and 2007 June, to adjust its trajectory. It flew past Mercury three times, in 2008 January and October and 2009 September, before going into near-polar orbit around the planet in 2011 March. The probe studies Mercury's surface composition, geologic history, interior, magnetic field, and tenuous atmosphere, as well as searching for water ice at the poles. Its instruments are the Mercury Dual Imaging System (MDIS), Gamma-Ray and Neutron Spectrometer (GRNS), X-ray Spectrometer (XRS), Mercury Laser Altimeter (MLA), Atmospheric and Surface Composition Spectrometer (ASCS), Energetic Particle and Plasma Spectrometer (EPPS), and a magnetometer.

(⊕) SEE WEB LINKS
• Official mission website.
• Mission page at NASA.

meridian, celestial A great circle on the celestial sphere that joins the north and south points on an observer's horizon, passing through the zenith and the celestial poles. When a celestial object crosses an observer's meridian it is said to be at *culmination or in *transit.

meridian, terrestrial A plane through the Earth's poles, defining a particular longitude on Earth. *See also* PRIME MERIDIAN.

meridian angle An alternative name, only occasionally used, for *hour angle.

meridian astronomy The measurement of positions of celestial objects based on observation of the times of their transit across the meridian and of their zenith distance at those times. The intention is to obtain star positions which are self-consistent over large areas of sky, but accuracy is limited by the thermal and mechanical instability of the instrument, and uncertainty in the calculation of the effect of atmospheric refraction. *See also* MERIDIAN CIRCLE; TRANSIT CIRCLE; TRANSIT INSTRUMENT.

meridian circle A telescope with an accurately calibrated circle which is constrained to rotate in the plane of the meridian about an east-west horizontal axis. The altitude at which the telescope is aimed at the time of an observation is obtained by measuring the circle divisions with a series of fixed microscopes, usually four or six. *See also* TRANSIT CIRCLE.

meridian passage Another name for a meridian *transit.

meridian telescope *See* MERIDIAN CIRCLE; TRANSIT CIRCLE; TRANSIT INSTRUMENT.

meridian transit *See* TRANSIT, MERIDIAN.

MERLIN Abbr. for *Multi-Element Radio-Linked Interferometer Network.

meson A class of elementary particle. Mesons are a subdivision of the *hadrons, and are composed of two *quarks. Mesons are involved in the processes that hold atomic nuclei together, and are also a major component of secondary cosmic-ray showers. They are unstable, with half-lives in the range 10^{-8} to 10^{-16} s, and decay into stable particles. They can be positively or negatively charged, or electrically neutral. Mesons have masses intermediate between those of the electron and proton, and include the kaon, pion, and psi particles. The mu-meson was an old name for a *muon; it is not a meson at all, but a *lepton.

mesopause The upper boundary of the *mesosphere, 85 km above Earth's surface. Temperatures close to the mesopause are low, typically around −110°C. Temperatures at this level are at their lowest in the summer, and values as low as −162°C have been recorded by rocket-borne instruments flown into *noctilucent clouds, whose condensation close to this atmospheric level during the summer months is favoured by such conditions.

mesosiderite A class of stony-iron meteorite, apparently related to the *howardites. Mesosiderites resemble lunar surface *breccias, and were apparently formed on the surface of their parent body by impact processes. They consist of fragments of *eucrite and *diogenite material, mixed with a large proportion of metallic nickel–iron alloy (typically 40–60% by mass). Their textures indicate that they were subjected to shock melting and metamorphism at temperatures up to 1000°C. *See also* BASALTIC ACHONDRITE.

mesosphere The middle layer of Earth's atmosphere, lying above the *stratosphere and below the *thermosphere, at altitudes of 50–85 km. Temperatures in the mesosphere fall with increasing altitude, reaching a minimum at the *mesopause. Measurements from sounding rockets indicate the mesosphere to be extremely dry, but sufficient traces of water vapour are present during the summer months for *noctilucent clouds to form near the mesopause.

MESSENGER Abbr. for *Mercury Surface, Space Environment, Geochemistry, and Ranging.

Messier, Charles Joseph (1730–1817) French astronomer. From Paris, where he had started out as assistant to Joseph Nicolas Delisle (1688–1768), Messier located Halley's Comet on its 1758/9 return. He went on to discover some fifteen comets, plus six co-discoveries. During his searches he noted various objects (nebulae and star clusters) that could be mistaken for comets, starting with the Crab Nebula in 1758. His list of these objects became the *Messier Catalogue.

Messier Catalogue A list of celestial objects that might be mistaken for comets, compiled by the French comet-hunter C. J. *Messier. His list first appeared in 1771, containing 45 objects; a revised list in 1780 added another 23 objects; and his final list, with 103 objects, appeared in 1781. Many of the objects were actually discovered by others, notably his compatriot Pierre François André Méchain (1744–1804). The objects catalogued by Messier include star clusters, nebulae, and galaxies. Astronomers still refer to these so-called **Messier objects** by their Messier, or M, numbers. Later observers have extended the catalogue beyond the 103 objects listed by Messier. (*See* Table 6, Appendix.)

Me star A star of spectral type M which shows emission lines of hydrogen (hence the suffix 'e'). Me stars are often *Mira stars (giants), but such emission lines are frequently found in M dwarfs (*dMe stars).

metagalaxy An archaic term for the entire Universe, deriving from the time before it was realized that the spiral nebulae were in fact separate galaxies of stars beyond the Milky Way.

metal In astronomical terminology, any element heavier than hydrogen or helium. The abundance of such heavy elements in celestial objects is termed their **metallicity**.

metallic hydrogen A form of hydrogen in which the atoms are highly compressed by intense pressure, as in the interiors of the massive gaseous planets Jupiter and Saturn. Under such conditions, hydrogen behaves like a liquid metal and hence can conduct electricity and generate a magnetic field.

metal-poor star A star with a far smaller proportion of heavy elements than the Sun, perhaps as little as 1% or less. They are *Population II stars, and are usually found in galactic haloes or in globular clusters; they are very old stars, formed before the galaxies were chemically enriched by the first generations of supernova explosions.

metal-rich star A star with a high proportion of heavy elements such as calcium, iron, and titanium. They are members of *Population I, and are found in the disk and spiral arms of galaxies. *Exoplanets are much more likely to be found around metal-rich stars than metal-poor ones.

metastable state An excited energy level in an atom, ion, or molecule, from which transitions to lower energy levels are said to be *forbidden*. A forbidden transition is one of such low probability that it is not observed under laboratory conditions. However, in the low-density conditions of interstellar space and nebulae, electrons cannot escape from the metastable state by upward excitation to a higher level by collisions. The metastable state can then become highly overpopulated, and emission lines from the forbidden transitions can be the strongest lines in the spectrum of a nebula. *See* FORBIDDEN LINE.

meteor A brief streak of light in the Earth's upper atmosphere between altitudes of 85 and 115 km, produced by the high-speed entry of a small fragment of interplanetary debris (a *meteoroid). An estimated 100 million meteors are visible to the naked eye over the whole Earth in an average 24-hour period. Meteoroids enter the atmosphere at velocities of 11–72 km/s. A typical naked-eye meteor of magnitude +2 is produced by a meteoroid about 8 mm in diameter. Over 0.1–0.2 s, its kinetic energy is converted principally to heat and ionization; only a small proportion is converted to visible light. The surface of a meteoroid is rapidly vaporized by the process of *ablation. Material eroded from the meteoroid's surface goes on to collide further with atmospheric particles, producing excitation and ionization along a column perhaps 20–30 km long. The excess energy imparted to the atmospheric particles is re-emitted in a fraction of a second as visible light.

Meteors may be produced by particles sharing an orbit around the Sun (a *meteor stream), or by solitary, random particles (*sporadic meteors). During a *meteor shower, more meteors will be seen. Most meteors are faint. The naked eye can detect events down to about magnitude +5, while binoculars show meteors as faint as magnitude +8. Smaller material gives rise to *radio meteors or *radar meteors, perhaps equivalent to visual magnitudes around magnitude +12. Occasionally, the arrival of a more substantial piece of debris produces an extremely bright *fireball.

Meteor Crater One of the best-preserved and the most famous terrestrial meteorite impact crater, 55 km east of Flagstaff, Arizona; also known as *Barringer Crater*. The crater is 1.2 km in diameter, 175 m deep, and its rim rises an average of 45 m above the surrounding plains. It was formed about 50 000 years ago when a nickel–iron meteorite, about 50 m across and weighing several hundred thousand tonnes, hit the desert at about 16 km/s. The energy of the impact, equivalent to about 20 megatonnes of TNT, excavated over 175 million tonnes of limestone and sandstone to produce the crater. The meteorite vaporized or melted almost completely on impact, any remaining material being scattered around the crater. Over 10 000 nickel–iron *octahedrite meteorite fragments have been recovered, up to 7 km from the crater. The crater's impact origin was recognized in the 1930s, mainly through the efforts of the American mining engineer Daniel Moreau Barringer (1860–1929).

(⊕) SEE WEB LINKS

• Meteor Crater Visitor Center and information website.

meteorite A natural object from space that hits the surface of the Earth or other planetary body. Impacts by large meteorites are believed to have created most of the craters on the planets and their satellites. An estimated 50–100 tonnes of cosmic debris enters the Earth's atmosphere every day, but only about a tonne reaches the ground. The fate of a body entering the Earth's atmosphere depends mainly on its mass and velocity. The smallest objects (*micrometeorites) are decelerated and drift slowly down to the surface. Objects with masses between about 10^{-6} g and 1 kg burn up to produce *meteors. Bodies with masses between about 1 kg and 1000 tonnes are substantially slowed by atmospheric drag, but penetrate the atmosphere. The atmosphere has no significant slowing effect on bodies of over 1000 tonnes. The average entry velocity of incoming bodies which fall as meteorites is about 20 km/s. Incoming bodies with high velocities are more likely to disintegrate in the atmosphere than those with low velocities; those with an entry velocity in excess of 30 km/s suffer more than 99% ablation. However, the composition of an incoming body also affects whether it survives to reach the ground.

Meteorites were observed and collected for thousands of years, but their extraterrestrial origin was not accepted until J.-B. *Biot investigated the *l'Aigle meteorite shower in 1803. A meteorite fall is preceded by a brilliant fireball, often accompanied by hissing noises and detonations like thunder. As the meteorite is slowed down it may break up, showering fragments in a *scatter ellipse. Meteorites range in mass from a few grams to 60 tonnes or more. A meteorite seen to hit the ground is known as a *fall*, whereas one discovered by chance at some later date is known as a *find*. In both cases, the meteorite is named after the place where it is picked up. The great majority (nearly 80%) of known meteorites come from Antarctica where they have been preserved in deep-freeze since they fell.

There are three main classes of meteorite, divided according to their composition: *iron meteorites, *stony meteorites, and *stony-iron meteorites. From observed falls, stony meteorites appear to be about twenty times more abundant in our part of the Solar System than are iron and stony-iron meteorites combined. However, the true ratio is probably even higher because stones tend to be more friable (crumbly) than irons, and so disintegrate more readily in the atmosphere. However, stony meteorites are under-represented in collections because once on the ground they are more susceptible to weathering and disintegration. Stones also tend to resemble terrestrial rocks and so are more difficult to recognize than are irons and stony-irons.

About 1200 meteorites have been observed to fall, but this is only a small fraction of the total number of incoming objects, most of which fall unseen in the oceans or unpopulated areas. By contrast, over 30 000 meteorites have been found, many of them since 1969 when it was discovered that meteorites are preserved on the surface of the Antarctic ice sheet. The large numbers and new types found among Antarctic meteorites have greatly stimulated research.

Meteorites are the most ancient rocks known, about 4.5 billion years old, about the same as the age of the Solar System. Hence, they may carry clues to the formation of the Solar System and the bodies within it. Although most meteorites are believed to be fragments of asteroids or asteroid-sized bodies, recent studies have identified a small number which appear to have come from the Moon. Another group, the *SNC meteorites, probably originated on Mars. These meteorites were presumably ejected from their parent bodies by massive impacts.

(⊕) SEE WEB LINKS
• Meteorite information database.

meteoroid A small particle from a comet or asteroid in orbit around the Sun. While the distinction between a large meteoroid and a small asteroid may be somewhat blurred, the term meteoroid most commonly refers to a particle which gives rise to a *meteor on entering Earth's atmosphere. Typical meteoroids in a meteor stream, such as that which produces the *Perseid meteor shower, are a few millimetres in diameter, and have low densities (0.2–0.3 g/cm^3). Meteoroids in the *Geminid stream, whose parent body is the asteroid *Phaethon, have a higher density, 2 g/cm^3.

meteor shower An increase in meteor activity produced when the Earth passes through a trail of debris (a *meteor stream) in orbit about the Sun. Meteors from a given shower appear to emanate from a common area of sky, the *radiant. Meteor showers recur annually, and range from weak displays barely detectable above the background of *sporadic meteors to major activity such as that of the *Perseids or *Geminids. During such strong showers, up to one meteor per minute can be seen for a day or so. Shower activity may be seen for only a few days in the case of a young meteor stream, or may persist for a number of weeks in the case of an older, more spread-out stream. The main showers are listed in the table.

Main Meteor Showers

Shower	Date of maximum	Radiant		ZHR (approx.)
		RA	dec.	
Quadrantids[a]	January 3/4	15.3h	+49°	100
Lyrids	April 22	18.1h	+34°	10
Eta Aquarids	May 5	22.3h	−01°	35
Delta Aquarids (south)	July 29	22.6h	−17°	25
(north)	August 6	23.1h	+02°	10
Perseids	August 12	03.1h	+58°	100
Orionids	October 20–22	06.3h	+16°	25
Taurids (south)	November 5	03.5h	+15°	10
(north)	November 12	03.9h	+22°	10
Leonids[b]	November 17	10.2h	+22°	10
Geminids	December 13	07.5h	+32°	100
Ursids	December 23	14.5h	+76°	10

[a] Unusually sharp maximum.
[b] Major storms every 33 years.

meteor storm An extremely rare and typically short-lived event, usually associated with a regular meteor shower, during which rates exceed 1000 meteors/h. Such events can occur when the Earth runs through a *meteor swarm, or a denser filament within a meteor stream, normally located relatively close to the parent body. For instance, the *Leonid meteor storm of 2001 closely followed the passage of the parent comet, *Tempel–Tuttle, through the descending node of the stream orbit, and resulted from recently ejected debris not yet dispersed into the main body of the stream.

meteor stream A trail of debris laid down by a larger parent body, comprising small particles (*meteoroids) which pursue a common orbit around the Sun. Most meteor streams originate from periodic comets, whose nuclei release dust when near perihelion. Over time, the meteoroids spread out along the comet's orbit. When a meteor stream's orbit approaches within about 0.1 AU of the Earth, a meteor shower may occur. Filaments of debris released by the parent body over several successive returns give rise to variable activity from year to year, particularly in younger showers such as the *Perseids; older streams such as the *Taurids show a smoother, wider distribution of material, and less variable activity.

meteor swarm A denser concentration of meteoroid particles within a meteor stream. Particles comprising a meteor swarm will usually have been released from the parent body comparatively recently, and have not yet become dispersed by gravitational and other perturbations. Examples of streams in which meteor swarms are found include the *Giacobinids and *Leonids. Passage of the Earth through these more highly populated regions in the streams gives rise to periodic *meteor storms.

methanol maser A maser source in which the methanol molecule (CH_3OH) is excited to maser action. Methanol masers are found only in star-forming regions, where they are second only to H_2O (water) masers in brightness. There are two classes of methanol masers: class I, excited by collisions, and class II, excited by infrared radiation. The most important class I masers are at a frequency of 44.1 GHz, while the most important class II masers are at a frequency of 6.7 GHz.

Metis The innermost known satellite of Jupiter, distance 127 980 km, orbital period 0.295 days; also known as Jupiter XVI. It lies less than one Jovian radius above the planet's cloud tops, and within Jupiter's main ring. Metis is noticeably elongated, 60 × 40 × 34 km. As in the case of *Adrastea, particles knocked off the surface of Metis by micrometeoroid impacts probably contribute to the ring. Metis was discovered in 1979 on images taken by the two Voyager spacecraft.

Metonic cycle A period of 19 years after which the Moon's phases repeat on approximately the same calendar dates. The cycle was discovered by the ancient Greek astronomer Meton in the 5th century BC and is used in constructing lunisolar calendars. It occurs because there are almost exactly 235 *synodic months (lunations) in 19 tropical years.

Meudon Observatory A branch of the Paris Observatory, 4 km south of Paris, founded in 1876 by P. J. C. *Janssen. It is the site of the 0.83-m Grande Lunette refractor, opened in 1895.

Meyer group A family of *sunskirting comets seen by the SOHO spacecraft and first identified in 2002 by the German amateur astronomer Maik Meyer (1970–). They have a mean perihelion distance of 0.036 AU and inclination around 73°.

MHD Abbr. for *magnetohydrodynamics.

Miaplacidus The star Beta Carinae, magnitude 1.7, an A1 giant 113 l.y. away.

Mice, the A pair of interacting 14th-magnitude spiral galaxies some 300 million l.y. away in Coma Berenices; also known as NGC 4676A and B, or IC 819 and 820. The gravitational interaction around 160 million years ago has drawn off long tails of stars and gas from each galaxy, hence the name.

Michelson, Albert Abraham (1852–1931) German-American physicist, born in modern Poland. In 1878 he constructed a rotating-mirror apparatus for measuring the velocity of light, on which he collaborated with S. *Newcomb. The interferometer he built for the *Michelson–Morley experiment of the 1880s led him to construct highly accurate spectroscopes and diffraction gratings for astronomy, and also the *Michelson interferometer. In 1920, Francis Gladheim Pease (1881–1938) measured the angular diameter of Betelgeuse with the *stellar interferometer that Michelson built and installed on the 100-inch (2.5-m) telescope at Mount Wilson Observatory. For the instruments he made and the measurements made with them, Michelson received the 1907 Nobel Prize in Physics.

Michelson interferometer An apparatus used to measure very precise lengths, such as the wavelength of light, and to analyse details of spectral lines. It consists of a half-silvered flat mirror which divides a beam of light into two. These beams are returned along the same paths by additional mirrors, then recombined. The interference of light creates dark and light bands or fringes whose positions depend on the lengths of the two light paths. It is named after its inventor, A. A. *Michelson.

Michelson–Morley experiment An experiment performed in 1887 by the American physicists A. A. *Michelson and Edward Williams Morley (1838–1923) to establish the presence or absence of an ether, the medium through which light was supposed to travel. The experiment attempted to detect the motion of the Earth relative to the ether by measuring the speed of light in two directions at right angles to each other. No such motion was found. An explanation was provided by the *Lorentz–Fitzgerald contraction, and subsequently the *special theory of relativity showed that the concept of an ether was unnecessary.

Michelson stellar interferometer *See* STELLAR INTERFEROMETER.

microchannel plate detector A device for producing images of high-energy photons. It consists of one or more arrays of tiny hollow glass tubes, typically 12.5–25 μm in diameter and 1–2 mm long, each operating as an individual photomultiplier. Around 10 million tubes are stacked side by side to form a thin plate structure. These microchannel plates can be used singly or stacked in pairs or triplets to detect ultraviolet, extreme ultraviolet, and X-ray photons directly, when coated with a suitable material. They are also used as optical image intensifiers in combination with a phosphor screen.

microdensitometer An instrument for measuring the photographic density (degree of blackening) at different points on a photographic emulsion or paper. For light passing through the emulsion, the *transmitted density* is measured; for prints the *reflected density* is measured. Modern computer-controlled measuring machines such as APM and SuperCOSMOS are sophisticated versions of the microdensitometer, designed to measure accurately the centres of star images and to distinguish stars from galaxies.

microlensing A small-scale *gravitational lens effect. In microlensing, the gravitational field of the lensing object is not strong enough to form distinct images of the background source; instead, it causes an apparent brightening of the source. Stars are expected to vary in brightness in a characteristic manner if low-mass stars or planets pass in front of them, and this effect has been detected for stars in the Large Magellanic Cloud and in the central bulge of our Galaxy.

micrometeorite A cosmic dust particle that has survived its passage through the Earth's atmosphere without burning up as a *meteor. Particles smaller than 0.1 mm in diameter, with masses under 10^{-6} g, are slowed by the upper atmosphere without melting, and drift down to the Earth's surface. About one micrometeorite of diameter 0.1 mm falls on each square metre of the Earth's surface per year, while for diameters of 0.01 mm the rate rises to about one per square metre per day. Micrometeorites can be collected in the stratosphere, providing samples of material from comets and asteroids for analysis. On the Moon and asteroids, the lack of an atmosphere means that micrometeorites impact the surfaces of these bodies at high speed.

micrometer An instrument used in astronomy for measuring angular distances and relative positions on the sky. Micrometers are used particularly for measuring the orbital motions of double stars, but can also be used to measure the locations of features on planets. There are many types of micrometer, but all introduce standard reference points into the field of view of an eyepiece. These references may be illuminated *reticles or wires, although the wires may be spider thread and are referred to as *webs*. The webs are located in the same plane as the field stop, so that they are always in focus. The reticle may be fixed, as in the *cross-wire micrometer, in which case the daily movement of the sky is used for reference; or there may be a movable portion, as in the *filar micrometer, with scales which should be calibrated against the sky's daily motion. Other types of micrometer provide movable artificial stars or double images of the stars in the field of view.

micrometre (μm) A measurement of length equal to 10^{-6} m.

micron An alternative term for *micrometre; it is becoming obsolete.

micropore optics A type of X-ray focusing optic, made up of a large number of square holes 0.05 mm wide (*pores*) in a glass or silicon wafer. One wall of each pore acts as the X-ray reflecting surface, while the other walls provide stiffness to the structure.

Microscopium (Mic) (*gen.* **Microscopii**) An insignificant constellation of the southern sky, representing a microscope. Its brightest stars are Gamma and Epsilon Microscopii, both of magnitude 4.7.

microturbulence Turbulence of an emitting gas in which the turbulent eddies are very small and unresolvable, giving rise to the broadening of a spectral line in a star's spectrum by Doppler shifts. If a Fraunhofer line is broader than expected from the motions of emitting atoms and other known causes, the excess broadening is sometimes attributed to unseen microturbulence, regardless of whether it actually exists.

Microvariability and Oscillations of Stars (MOST) A small Canadian satellite launched in 2003 June that carries a 0.15-m telescope to monitor stars down to 6th magnitude for surface oscillations. Its other goals are to detect light reflected from giant planets orbiting Sun-like stars and to study mass loss from Wolf–Rayet stars.

(())) SEE WEB LINKS
- Mission page at Canadian Space Agency.
- Official mission website.

microvariable A variable star in which the amplitude of the fluctuations amounts to a few thousandths of a magnitude. The variations may arise from a number of causes (e.g. rotation in conjunction with non-uniform surface brightness). The Sun may be considered a microvariable.

microwave background radiation *See* COSMIC MICROWAVE BACKGROUND.

microwaves Electromagnetic waves at the short end of the radio range, with wavelengths from about 1 mm to 30 cm.

Midcourse Space Experiment (MSX) A satellite of the US Ballistic Missile Defense Organization launched in 1996 April to test technology for identifying and tracking ballistic missiles while in flight. As part of its work it studied celestial background sources at wavelengths from mid-infrared to far-ultraviolet with the Spatial Infrared Imaging Telescope (SPIRIT), a 0.33-m infrared telescope; the SBV (Space-Based Visible) Instrument, a 150-mm visible-band telescope; and a set of Ultraviolet and Visible Imagers and Spectrographic Imagers (UVISI). It performed astronomical observations until 1997 October, and was then used to monitor other orbiting spacecraft until 2008 June.

(())) SEE WEB LINKS
- Official mission website.

Mie scattering Scattering of light by particles whose size is similar to the wavelength of light. It occurs in interstellar space and in the Earth's atmosphere. For wavelengths much shorter than a given particle size, Mie scattering is a complex function of wavelength. Scattering drops to a minimum at a wavelength half the particle size, rises to a maximum when it is the same as the particle size, and then decays to zero towards longer wavelengths. Mie scattering can thus make objects appear either redder or bluer according to the size of the scattering particle. In the Earth's atmosphere there is usually a wide range of particle sizes, so that these colour effects are blurred into a nondescript grey. Very occasionally, atmospheric effects will combine so that the dust is mostly 900 nm in size and the minimum in scattering is at 450 nm. This is the wavelength of blue light that causes the phenomenon of the *blue Moon (1). Mie scattering is named after the German physicist Gustav Adolf Feodor Wilhelm Ludwig Mie (1868–1957).

Milky Way A band of faint light across the sky, visible to the naked eye on a moonless night, consisting of stars and glowing gas in the disk of our own Galaxy. From Cassiopeia to Cygnus the Milky Way varies in width, and between Cygnus and Sagittarius it appears to split into two, separated by the *Great Rift. It is faint from Cassiopeia to Canis Major, particularly near Taurus. In the southern hemisphere from Sagittarius to Carina the Milky Way is spectacular. Dark areas stand out in greater contrast in this region, notably the *Coalsack nebula in Crux. The *galactic centre lies in Sagittarius, where star clouds are particularly bright and dense. The plane of our Galaxy runs along the Milky Way and is inclined at about 63° to the celestial equator, a result of the orientation of the Solar System within the Galaxy. 🖼

millimetre-wave astronomy The observation of electromagnetic waves from the Universe at wavelengths of about 1–10 mm, a part of the spectrum rich in lines emitted by interstellar molecules. Millimetre-wave astronomy is one of the newest fields in astronomy and is used to study the chemistry of interstellar matter, and the complex processes in star-forming regions and dark clouds. The *cosmic microwave background is brightest in this part of the spectrum. Major millimetre-wave facilities include those of the *Institut de Radio Astronomie Millimétrique (IRAM) and *Nobeyama Radio Observatory. *See also* SUBMILLIMETRE ASTRONOMY.

millimetre waves Electromagnetic radiation with wavelengths in the approximate range 1–10 mm. Wavelengths shorter than 1 mm are known as *submillimetre waves*.

millisecond pulsar A pulsar that flashes every few thousandths of a second, i.e. with a period of 1–10 ms. The first to be discovered, PSR 1937+21, has a period of 1.56 ms. Even faster is a pulsar called PSR J1748-2446ad, in the globular cluster Terzan 5, which has a period of 1.40 ms, very close to the theoretical minimum for a spinning neutron star. Around 150 millisecond pulsars have been discovered, many of them in globular clusters. Millisecond pulsars are extremely stable rotators and keep time better than atomic clocks. *See also* RECYCLED PULSAR.

Mills cross A design of radio interferometer consisting of numerous antennas arranged in two arms in the form of a cross oriented north–south and east–west. It can be configured to produce either a pencil beam or a *fan beam. The original Mills Cross, in Sydney, had 500 half-wave dipoles placed along two arms 457 m long. The Mills cross is named after its designer, the Australian radio astronomer Bernard Yarnton Mills (1920–).

Milne, Edward Arthur (1896–1950) English mathematical astrophysicist and cosmologist. His work on radiative equilibrium and stellar atmospheres in the 1920s yielded a temperature scale for the sequence of spectral types. With the English mathematician Ralph Howard Fowler (1889–1944) he developed equations which allowed C. H. *Payne-Gaposchkin to show that stars are composed largely of hydrogen. In later work on stellar structure he was one of the first to associate nova outbursts with stellar collapse and the emission of a gas shell. In 1932 he turned to relativity and cosmology, showing that there are valid, non-Einsteinian ways of modelling the expanding universe and spacetime.

Milne–Eddington approximation A simple model of the outer layers of a star, used in studying the formation of absorption lines in stellar spectra. It assumes that absorption occurs throughout the outer region of the star, rather than in just the cool outermost layers, and that the ratio of line absorption to continuum absorption is constant. It successfully predicts the strength of lines of ionized metals such as calcium and iron in solar-type and similar stars. The model is named after E. A. *Milne and A. S. *Eddington.

Mimas The seventh-closest satellite of Saturn, distance 185 540 km, orbital period 0.942 days; also known as Saturn I. Its axial rotation period is the same as its orbital period. It is 416 × 393 × 381 km in diameter, and was discovered in 1789 by F. W. *Herschel. Most of its surface is densely covered with impact craters. One of these, named Herschel, is 139 km across, nearly one-third of the satellite's diameter, and has a large, high central peak. The impact that caused this crater must have nearly split the satellite apart.

Mimosa Name sometimes used for the star *Beta Crucis.

Minkowski, Hermann (1864–1909) German mathematician, born in modern Lithuania. He was the first to see that the principle of relativity set out by the Dutch physicist Hendrik Antoon Lorentz (1853–1928) and A. *Einstein implied that space and time were not separate entities, but components of a four-dimensional *spacetime. His mathematical model of spacetime provided the foundation for all subsequent developments of relativity, and allowed Einstein to move from special to general relativity.

Minkowski, Rudolph Leo Bernhard (1895–1976) German-American astrophysicist, born in Strasbourg (now in France). In 1941 he classified supernovae into Types I and II according to their spectra. In 1954 he and W. *Baade reported the first optical counterpart to an extragalactic radio source, *Cygnus A. Minkowski went on to specialize in the optical identification of radio sources with high redshift. He also carried out a search for planetary nebulae, doubling the number known.

minor axis The shortest diameter of an ellipse, at right angles to the *major axis.

minor planet Another name for an *asteroid.

Minor Planet Center The organization responsible for collecting and disseminating positional measurements and orbital data for asteroids (minor planets) and comets. It was established by the International Astronomical Union at Cincinnati Observatory in 1947, and moved to the *Smithsonian Astrophysical Observatory in 1978. The Minor Planet Center allocates preliminary designations to new discoveries and gives permanent numbers when satisfactory orbits have been determined. The information is published through the *Minor Planet Circulars*. The work of the Minor Planet Center complements that undertaken at the Institute for Theoretical Astronomy in St Petersburg, Russia, which publishes annual volumes of ephemerides of numbered asteroids.

(⊕) SEE WEB LINKS
• Official website, with links to many information pages on minor planets.

Mintaka The star Delta Orionis, one of the stars of Orion's belt. It is an eclipsing binary that varies between magnitudes 2.1 and 2.3 in a period of 5.7 days. The main star is an O9.5 bright giant. Mintaka is estimated to be about 700 l.y. away. It forms a wide double with an unrelated background star of magnitude 6.9.

minute (min) A non-SI unit of time, equivalent to 60 seconds. In astronomy it is often abbreviated to 'm'.

minute of arc (arcmin) (′) A small unit of angular measure, equivalent to one-sixtieth of a degree.

Mira The star Omicron Ceti, the prototype *Mira star or *long-period variable. It was first observed by D. *Fabricius in 1596, and was originally believed to be a nova. It was found to be periodic by the Dutch astronomer Jan Fokkens (Latinized as Johannes Phocylides) Holwarda (1618–51) in 1638, and was named Mira (Latin for 'Wonderful') by J. *Hevelius. Mira has a period of 331.96 days, a range of 2.0–10.1 mag., and is a red giant of spectral type M5e–M9e. Its diameter is 400–500 times that of the Sun. It forms a binary with VZ Ceti, a white dwarf that is accreting material from the giant's stellar wind. Mira is 300 l.y. away.

Mirach The star Beta Andromedae, magnitude 2.1. It is an M0 giant 197 l.y. away.

Miranda The fifth-largest satellite of Uranus, 481 × 468 × 466 km in diameter; also known as Uranus V. Its distance is 129 870 km and its orbital period 1.413 days, the same as its axial rotation period. Miranda was discovered in 1948 by G. P. *Kuiper. As revealed by Voyager 2 in 1986, its icy surface has three very young areas of oblong parallel ridges and furrows unlike any other terrain in the Solar System, in some places heavily faulted at the edges, with cliffs some 10 km high. The rest of the visible surface is much older and cratered. The contrasts between the different types of terrain suggest that Miranda might have been broken up by a giant impact and then re-formed, with subsequent deformation producing faulting and ice flows.

Mira star A type of red giant or supergiant pulsating variable; abbr. M. Mira stars have highly characteristic late-type spectra (Me, Ce, or Se) with molecular bands, and there is extensive *mass loss. Their light-curves may be approximately sinusoidal, but most show a steep rise to maximum and slower decline; amplitudes range from 2.5 to as much as 11 magnitudes, with periods from 80 to 1000 days. Because of their relatively high luminosities, large amplitudes, and well-defined features, Mira stars are readily detectable. More of them are known than any other type of variable star.

Mirfak The star Alpha Persei, also spelt Mirphak; an alternative name is Algenib. It is an F5 supergiant of magnitude 1.8, 506 l.y. away, and is surrounded by a scattering of stars known as the Alpha Persei Cluster or Melotte 20.

mirror An optical element that reflects light. A mirror may be flat (a *plane mirror*), which simply laterally inverts an image on reflection, or a concave or convex mirror, which forms an image. Mirrors in an optical path are always coated by *aluminizing their front surface, unlike domestic mirrors which are coated on their backs. Astronomical images usually have to be very sharp, so the surface shape of the mirror must be precise. In general, the shape of an astronomical mirror should alter the reflected wavefront of light by less than a quarter of a wavelength of light across its entire surface (the *Rayleigh criterion). *See also* SEGMENTED MIRROR.

mirror blank A disk of glass or other suitable material which is to be turned into a mirror. It may be cast as a disk or cut from a sheet. The ratio of diameter to thickness is usually at least 6 : 1 so that the mirror will not deform. Low-expansion glass such as Pyrex is advisable, to prevent distortion of the finished mirror's precise curve as a result of temperature changes.

Mirzam The star Beta Canis Majoris. It is a B1 giant of magnitude 2.0, and a variable of Beta Cephei type, changing by less than 0.1 mag. in a period of 6 hours. It lies 493 l.y. away.

missing mass The additional, unseen matter whose gravitational effect is needed to explain the rotation speeds of galaxies, and also to keep clusters of galaxies together. Such mass is thought to reside in massive dark haloes around galaxies. In addition, if gravity is to prevent the Universe from expanding indefinitely, there must be considerably more matter in intergalactic space than is contained in visible stars and galaxies. The term *dark matter is now generally used instead of missing mass.

Mitchell, Maria (1818–89) American astronomer. The first American woman astronomer of note, she achieved world fame with her discovery of a comet in 1847 (now designated C/1847 T1). She was elected the first woman member of the American Academy of Arts and Sciences in 1848. From 1849 to 1868 she was employed by the US Government Almanac Office to compute ephemerides for Venus. From 1865 until her death she worked at Vassar College at Poughkeepsie, NY, as both professor of astronomy and director of the college observatory.

Mittenzwey eyepiece A modified form of *Huygenian eyepiece consisting of convex *meniscus lenses. It is named after its inventor, the German chemist and amateur astronomer Moritz Mittenzwey (1836–89).

mixing ratio A relationship which indicates how effectively energy is transported by convection in stars. It is defined as the length of eddies in the convective current divided by the distance over which the gas pressure varies significantly. Convection in stars is not fully understood, so calculations of the sizes of convective stars are uncertain by about 15%.

Mizar The star Zeta Ursae Majoris, an A1 dwarf of magnitude 2.2, distance 86 l.y. It forms a naked-eye double with *Alcor, but the two are not a binary pair. However, a closer companion of magnitude 4.0 is connected. In 1889 E. C. *Pickering found that Mizar is a spectroscopic binary. Mizar was therefore the first telescopic binary and the first spectroscopic binary to be discovered. The 4th-magnitude companion is also a spectroscopic binary.

MJD Abbr. for *Modified Julian Date.

MK classification Abbr. for *Morgan–Keenan classification.

MKK classification *See* MORGAN–KEENAN CLASSIFICATION.

M magnitude The magnitude of a star as measured through the infrared M filter of the *Johnson photometry system. It has an effective wavelength of 5030 nm and a bandwidth of 1060 nm.

MMT Observatory An observatory jointly owned by the Smithsonian Institution and the University of Arizona, located at an altitude of 2606 m on Mount Hopkins, Arizona, in the grounds of the *Fred Lawrence Whipple Observatory. It operates the MMT, a telescope with a 6.5-m mirror, opened in 2000. This replaced the original Multiple Mirror Telescope at the same observatory, which operated from 1979 to 1998. The original MMT consisted of six 1.8-m mirrors on a single mounting, giving a collecting area equivalent to one 4.5-m mirror. Although the new telescope uses a single mirror, the MMT name has been retained.

(⊕) SEE WEB LINKS
• Official observatory website.

mock Moon Popular name for a *parselene.

mock Sun Popular name for a *parhelion.

Modified Julian Date (MJD) A more convenient form of the *Julian date, in which the zero point is 1858 November 17.0. Hence Modified Julian Date = Julian Date − 2 400 000.5 days. In the modified Julian date, the day starts at midnight. Orbital data for artificial Earth satellites are often expressed in Modified Julian Date Numbers.

moldavite A type of tektite found in the Czech Republic. The name derives from the Moldau (Vltava) River in the province of Bohemia, near where the first of them were found. It has been suggested that the moldavites originated in the impact that produced the 24-km diameter Ries crater, around 15 million years old, some 500 km away near Nordlingen, Germany.

molecular cloud A cloud of interstellar gas and dust composed mainly of molecules. The smallest known molecular clouds contain less than 1 solar mass of gas, whereas *giant molecular clouds have masses of up to 10^7 solar masses. Most of the molecular gas is very cold, with typical temperatures of around 20 K. Molecular clouds can be detected optically as dark nebulae. Infrared telescopes detect the small amount of heat emitted by the dust. Radio telescopes detect numerous spectral lines from the interstellar molecules in such clouds.

molecular hydrogen (H_2) The normal form of hydrogen gas found on Earth, with two hydrogen atoms bound together by two shared electrons. Hydrogen molecules cannot survive in most regions of space, because they are easily dissociated (split) by ultraviolet light from stars, but they do occur in cold, dense molecular clouds where they are shielded by dust particles. Molecular hydrogen is the main constituent of molecular clouds, but it is difficult to observe directly as it is a symmetrical molecule with no strong radio or millimetre-wave spectral lines.

molecular line A narrow spectral emission or absorption feature produced by an *interstellar molecule. Most molecules of astrophysical interest emit and absorb in the infrared, millimetre, and radio parts of the spectrum. *See also* MASER SOURCE; OH LINE; 21-CENTIMETRE LINE.

molecule The smallest part of a compound, which is a substance formed by the chemical combination of one or more types of atom. For example, water consists of two atoms of hydrogen and one atom of oxygen, hence it has the formula H_2O.

Molonglo Radio Observatory A radio observatory at Hoskinstown, New South Wales, 30 km east of Canberra, Australia. It was founded in 1962 and is owned and operated by the University of Sydney. It contains the Molonglo Observatory Synthesis Telescope (MOST), opened in 1981. This is an *aperture synthesis telescope that consists of two parabolic reflectors, each 778 m long and 11.6 m wide, oriented east–west, originally one arm of a *Mills cross that operated 1965–78.

(((⊕))) **SEE WEB LINKS**
• Official observatory website.

moment of inertia (I) For a system of several bodies, the mass of each body multiplied by the square of its distance from the centre of the system, the products all being added together. For a single solid body, the moment of inertia is the sum of the mass of each particle in that body multiplied by the square of its distance from the body's axis of rotation. The shape of the Moon, for example, is an ellipsoid with three different axes, the longest of them pointing towards the Earth. The moments of inertia of the Moon about these three axes are slightly different, which can produce measurable effects in the orbit of an artificial lunar satellite. Moment of inertia is a measure of a body's ability to resist a change in its state of rest or angular velocity.

MONET Abbr. for *Monitoring Network of Telescopes.

Monitoring Network of Telescopes (MONET) Two 1.2-m robotic telescopes owned by the University of Göttingen, Germany. MONET/North, opened in 2006, is located at *McDonald Observatory, Texas, and MONET/South, opened in 2011, is at the *South African Astronomical Observatory.

⊕ SEE WEB LINKS
• Official MONET website.

monocentric eyepiece An eyepiece design consisting of three thick glass elements cemented together, with no air gaps, the outer two being of *flint glass and the inner one of *crown glass. The design, like other solid eyepieces such as the Tolles, is free from ghost reflections of bright objects in the field of view and gives bright images with good contrast. However, the field of view is rather small, about 30°, which makes such eyepieces suitable only for planetary and double-star observations. An adaptation is known as the *Hastings triplet*.

Monoceros (Mon) (*gen.* **Monocerotis)** A constellation on the celestial equator, representing a unicorn. It contains the massive binary *Plaskett's Star. Its brightest star is Alpha Monocerotis, magnitude 3.9. Beta Monocerotis is a superb triple star, magnitudes 4.6, 5.4, and 5.6. M50 and NGC 2232 are open clusters. NGC 2244 is a star cluster within the *Rosette Nebula. NGC 2261 is *Hubble's Variable Nebula. NGC 2264 is a star cluster associated with the *Cone Nebula, including the highly luminous star S Monocerotis, magnitude 4.7 but slightly variable.

monochromatic light Light of one colour only, having a single wavelength. A common source of monochromatic light is an emission lamp, such as a mercury or low-pressure sodium streetlight, filtered to remove unwanted emission lines.

monochromatic magnitude The brightness of a star at a single wavelength. Such magnitudes are important in assessing the variations in magnitude to be expected when a given star is observed through different filters. The monochromatic magnitude is a mathematical ideal that cannot be measured in practice, but can be approximated by *narrow-band photometry, provided there are no strong spectral features in the band.

monochromator A device for selecting a very narrow range of wavelengths of light. It can be a spectrograph in which incoming light is dispersed by a diffraction grating, and a small part of the resulting spectrum selected by passing it through a narrow exit slit. Alternatively, it can be an *interference filter that allows only a narrow band of wavelengths to pass.

Monogem Ring A region of soft X-ray emission about 25° wide extending from Gemini into Monoceros; also known as the Gemini–Monoceros X-ray enhancement. It is thought to be a supernova remnant about 1000 l.y. away with a true diameter of about 150 l.y. Pulsar PSR B0656+14 lies near its centre and is thought to be the core of the star that exploded about 100 000 years ago.

monopole *See* MAGNETIC MONOPOLE.

mons A mountain on a planetary surface; pl.*montes*. The name, which was first used for features on the Moon in the 17th century, is not a geological term, but is used in the nomenclature of individual mountains irrespective of their origin. For example, the Montes Apenninus on the Moon are part of the walls of a large impact basin, while *Olympus Mons on Mars is a giant volcano.

month A period of time based on the orbit of the Moon around the Earth. Several types of month are defined. The *synodic month is equivalent to one lunation, the time taken for the Moon to go through one cycle of phases; its mean length is 29.530 59 days. The *anomalistic month is the time between successive passages of the Moon through perigee; it lasts 27.554 64 days. The *sidereal month is the time taken for the Moon to return to the same position against the star background; it lasts 27.321 66 days. The *tropical month is the time between successive passages of the Moon through the vernal equinox; it lasts 27.321 58 days. The *draconic month (or *nodical month*) is the time between successive passages of the Moon through its ascending node; it lasts 27.2122 days. The *calendar month* is an artificial unit consisting of a whole number of days; *see* CALENDAR.

Monthly Notices of the Royal Astronomical Society A publication of the *Royal Astronomical Society, founded in 1827 and now issued three times a month. It contains papers reporting original research in various branches of astronomy and astrophysics.

(((⊕))) SEE WEB LINKS

• Official website with contents and abstracts of recent issues and free online access to older issues.

Mont Mégantic Observatory *See* OBSERVATOIRE DU MONT-MÉGANTIC.

Moon The only natural satellite of the Earth, orbiting at an average distance of 384 400 km. The magnitude of the full Moon is −12.7, but its surface is actually dark, with a mean geometric albedo of only 0.12, lower than for all the planets except Mercury. It is the fifth-largest satellite in the Solar System (diameter 3475 km), over a quarter the diameter of the Earth and about 1/81 the Earth's mass. Being so similar in size, the Earth and Moon are often considered a double planet. The Moon's sidereal period of axial rotation, 27.322 days, is the same as its orbital period, so that it keeps the same face towards the Earth. Its equator is inclined by 1°.53 to the plane of the ecliptic. Surface temperatures vary from extremes of 123°C during the day down to −233°C at night; typical values are 107°C (day) and −175°C (night). Polar regions of the Moon contain craters with permanently shadowed floors, where signs of ice have been detected by spacecraft.

The Moon shows two distinctly different types of terrain with very different densities of impact craters: the brighter highlands and the darker lowland mare areas. The lunar highlands have an albedo of 0.11–0.18, and are saturated with large craters of 50 km diameter and greater; the maria have an albedo of 0.07–0.10, and consist of younger plains of basaltic lava with few large craters. The mare basalts are enriched in iron and titanium, and have abundant pyroxene. The highland rocks are chemically different from the maria, being enriched in calcium and aluminium and consisting mainly of feldspar. The highlands date from before 4 billion years ago, whereas the maria were mostly erupted between 2 and 3.9 billion years ago. The farside of the Moon has few dark mare areas but does contain the largest and oldest impact structure on the Moon, the *South Pole–Aitken Basin. The prevalence of maria on the nearside may be due to the impact that formed the *Imbrium Basin, which created deep fractures within the Moon through which erupting lava later poured.

Moon

Physical data

Diameter	Oblateness	Inclination of equator to orbital plane	Axial rotation period (sidereal)	
3475 km	0.0	6°.69	27.322 days	
Mean density	**Mass**	**Volume (Earth = 1)**	**Mean albedo (geometric)**	**Escape velocity**
3.35 g/cm³	7.348 × 10²² kg	0.02	0.12	2.37 km/s

Orbital data

Mean distance from Earth	Eccentricity	Inclination of orbit to ecliptic	Orbital period (sidereal)	
384 400 km	0.055	5°.15	27.322 days	

The Moon has an exceedingly tenuous atmosphere consisting of outgassed elements such as radon arising from radioactive decay in the lunar interior, plus temporarily trapped solar wind particles. Because of the lack of any effective atmosphere, the main erosive process is impact cratering. Lunar craters vary in size from tiny pits less than 1 mm across to major impact basins over 1000 km in diameter. Young impact craters, such as Tycho, are very bright, with prominent central peaks, terraced walls, and bright rays radiating far across the surface. Older craters are gradually worn down and smoothed over by tiny impacts, or obscured by bigger ones or lava flooding. The constant churning of the surface by small impacts has created a soil layer, or *regolith, 5–15 m deep over the entire Moon.

Lunar volcanic craters are rare and comparatively small, only a few kilometres in diameter at most. Lunar domes with shallow slopes and summit pits appear to be the equivalent of shield volcanoes on Earth. There are a few tiny cinder cones, plus some bigger collapse pits and calderas. Many of the calderas are the sources of the *sinuous rilles, the channels that supplied the extensive fluid lavas of the mare plains. *Wrinkle ridges and *rilles bear witness to forces of compression and tension on the Moon. They are frequently found in concentric patterns within or around the impact basins.

The interior of the Moon consists of a thick lithosphere down to about 800 km. Below it is an asthenosphere, with perhaps a small core less than 700 km in diameter. Moonquakes are minor events compared with earthquakes, and tremors occur regularly in the same places each month as a result of tidal forces. There is no significant magnetic field.

It is now thought that the Moon formed when the Earth was struck a glancing blow by a passing body similar in size to Mars, sending ejecta from the Earth and the impactor into orbit, where it accreted to form the Moon. 📷

Mopra Observatory The site near Coonabarabran, New South Wales, of a 22-m dish of the *Australia Telescope National Facility. The Mopra dish is used primarily for millimetre-wave observations and very long baseline interferometry.

(((⊕))) SEE WEB LINKS
• Official observatory website.

Moreton wave A shock wave in the Sun's chromosphere moving outwards from a large solar flare at a velocity of about 1000 km/s. A Moreton wave is visible in *Hα light as an arc, sometimes moving for several hundred thousand kilometres, and may cause the *winking* of dark filaments in its path. Metre-wave radio bursts invariably accompany Moreton waves. They are named after the American solar astronomer Gail Ernest Moreton (1930–82).

Morgan, William Wilson (1906–94) American astronomer. With Philip Childs Keenan (1908–2000) he developed the *Morgan–Keenan classification of stellar spectra, set out in *An Atlas of Stellar Spectra* (1943). Morgan and Keenan devised a series of standard spectra against which a star's spectrum may be identified by visual comparison to yield its mass and luminosity (and hence distance). In 1951, Morgan inferred the spiral structure of the Galaxy from the distribution of O- and B-type stars near the galactic equator. In 1953 he helped to originate the system of UBV photometry by which a star's colour index may be determined (*see* JOHNSON PHOTOMETRY). He also developed classification systems for galaxies and clusters of galaxies (*see* MORGAN'S CLASSIFICATION; BAUTZ–MORGAN CLASS).

Morgan–Keenan classification A system of classifying stellar spectra developed at Yerkes Observatory by W. W. *Morgan, Philip Childs Keenan (1908–2000), and Edith Marie Kellman (1911–2007), and published in 1943; also known as the *MK* (or *MKK*) *classification* or the *Yerkes system*. The Morgan–Keenan system retained the sequence of stellar spectral types O, B, A, F, G, K, M introduced in the *Harvard classification, but with a more precise observational definition of each type. To these spectral types were added a range of *luminosity classes which indicate whether the star is a supergiant, giant, dwarf, or some intermediate class. Unusual stars which do not fit this system, such as the *carbon stars (types R and N in the Harvard system), the *S stars, white dwarfs (*see* D STAR), and *Wolf–Rayet stars, have their own individual classification schemes. Recently the scheme has been extended at the cool end with the introduction of types L and T for brown dwarfs. *See also* SPECTRAL CLASSIFICATION.

Morgan's classification A classification scheme for galaxies; also known as the *Yerkes system*. It is based primarily on the spectrum of the galaxy's central regions, which correlates with the degree of concentration of light towards the centre. A notation is also given for the simple structural form of the galaxy. Lower-case letters af, f, fg, g, gk, and k indicate the spectral type of star that most closely matches the spectrum of the galaxy's nuclear region. Structural forms are E (elliptical), S (spiral), B (barred spiral), I (irregular), D (dustless elliptical with extended envelope of stars), L (low surface brightness), and N (small brilliant nucleus on faint background). An index from 1 (face-on) to 7 (edge-on) shows the inclination of the plane of the galaxy to the line of sight. For example, M33 is classified as fS3. The scheme was devised by W. W. *Morgan.

morning star Popular name for the planet Venus when it is visible in the morning sky.

MOST Abbr. for *Microvariability and Oscillations of Stars.

mottles Features that form the *chromospheric network on the Sun, visible in *Hα and calcium K-line spectrograms. They cover the entire Sun, quiet and active regions alike. Coarse mottles are bright, up to about 20 000 km across, and often elongated. They merge to form *plages. Fine mottles are narrow (700 km wide, up to 7000 km long), dark or bright, and are probably *spicules seen against the disk. Fine mottles form clusters with their bases rooted in the coarse mottles. All coincide with regions of strong magnetic field on the photosphere.

Mount Graham International Observatory (MGIO) An observatory at an altitude of 3260 m on Mount Graham in the Pinaleño mountains near Safford, Arizona, 125 km northeast of Tucson, founded in 1988. It is owned and operated by the University of Arizona's *Steward Observatory, and hosts telescopes from various other institutions. Its largest instrument is the *Large Binocular Telescope. It is also the site of the *Vatican Advanced Technology Telescope and the 10-m Submillimeter Telescope of the *Arizona Radio Observatory.

(⊕) SEE WEB LINKS
• Official observatory website.

mounting The structure that supports a telescope and allows it to be pointed at a chosen part of the sky. A mounting should hold the telescope firmly and allow it to be moved so that it can follow objects across the sky as the Earth rotates. This can be done either with a motor drive or with controls known as slow motions. Telescope mountings fall into two main categories: *altazimuth and *equatorial. The part that allows the telescope to move is often referred to as the *head*, which may be mounted on either a pillar or a tripod.

Mount Palomar Observatory *See* PALOMAR OBSERVATORY.

Mount Stromlo and Siding Spring Observatories (MSSSO) Two observatories owned and operated by the Australian National University, Canberra. Mount Stromlo Observatory, founded in 1924 as a solar observatory, lies 11 km west of Canberra at an altitude of 770 m. In 2003 January, fire damaged much of the observatory and its historic telescopes. As a replacement, a new robotically controlled 1.35-m wide-field survey telescope called SkyMapper was opened in 2010. SkyMapper is located at *Siding Spring Observatory but controlled from Mt Stromlo. Siding Spring Observatory was founded in 1962 as a field station of Mount Stromlo.

(⊕) SEE WEB LINKS
• Official observatory website.

Mount Wilson classification *See* SUNSPOT.

Mount Wilson Observatory An observatory at an altitude of 1740 m on Mount Wilson in the San Gabriel Mountains about 30 km northwest of Los Angeles, founded by G. E. *Hale in 1904 for solar observation. Mount Wilson Observatory is owned by the Carnegie Institution of Washington and has been operated since 1991 by the Mount Wilson Institute. Its main instrument is the 100-inch (2.5-m) Hooker Telescope, opened in 1917. There is also a 60-inch (1.5-m) reflector (1908). Two tower telescopes for solar observation, 18 m and 46 m high, were opened in 1907 and 1909; they have been operated since 1986 by the University of Southern California and the University of California at Los Angeles, respectively. Mount Wilson is also the site of the Infrared Spatial Interferometer (ISI) consisting of three 1.65-m mirrors, operated by the University of California at Berkeley, and of the *CHARA Array.

(⊕) SEE WEB LINKS

• Official observatory website.

moustache *See* ELLERMAN BOMB.

moving cluster A physically related group of stars which share a common motion in space, such as the Hyades. Because of perspective, the proper motions appear to be directed towards a single point, called the *convergent point*. If the linear velocity of the cluster is known from observations of radial velocity, then the distance of each star can be estimated from the total proper motion; this technique is known as **moving cluster parallax**.

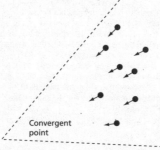

moving cluster: As a result of perspective, a group of stars with a common motion away from us appear to be heading towards a convergent point.

MRAO Abbr. for *Mullard Radio Astronomy Observatory.

M region The name originally given to areas on the Sun that were assumed to cause geomagnetic disturbances not associated with observed active regions; 'M' stands for magnetically effective. Data from the Skylab space station in the early 1970s revealed that M regions are actually *coronal holes at low solar latitudes.

MRO Abbr. for either *Magdalena Ridge Observatory or *Murchison Radio-astronomy Observatory.

MSSSO Abbr. for *Mount Stromlo and Siding Spring Observatories.

MS star 1. Abbr. for main-sequence star.
   2. *See* S STAR.

M star A star of spectral type M, with a very cool surface (temperature below 3900 K), appearing reddish in colour and emitting most of its radiation in the infrared. M-type dwarfs, popularly known as red dwarfs, lie near the bottom end of the main sequence. They have masses under 0.5 solar masses, and luminosities up to 0.08 times the Sun's, too faint for even the closest, *Proxima Centauri and *Barnard's Star, to be seen with the naked eye. They have potential lifetimes longer than the present age of the Universe. Many are *flare stars. M-type giants, however, are of 1.2–1.3 solar masses and luminosities of 300 or more times the Sun's. M-type supergiants, such as Betelgeuse and Antares, are of 13–25 solar masses and luminosities 40 000–500 000 times the Sun's. The low temperatures of M-type supergiants are a consequence of their inflated physical size, as large as the orbit of Jupiter; they also tend to vary in size and brightness. M-type spectra are completely dominated by broad molecular absorption bands, particularly titanium oxide (TiO), but neutral metal lines are also present. *See also* DME STAR; ME STAR.

MSX Abbr. for *Midcourse Space Experiment.

Mu Cephei star An obsolete term originally applied to a variable star with a late-type spectrum and irregular variations. Such a star is now classified as a *semiregular late-type supergiant (SRC). The type star, Mu Cephei, is also known as the *Garnet Star.

Mullard Radio Astronomy Observatory (MRAO) The radio astronomy observatory of the University of Cambridge, founded in 1957 at Lord's Bridge near Cambridge, England. The technique of *aperture synthesis was developed at MRAO. Many radio sources are referred to by their designations in the third, fourth, fifth, etc. Cambridge sky surveys (3C, 4C, 5C, etc.). The One-Mile Telescope, which operated 1964–99, was a pioneering aperture-synthesis instrument, consisting of three 18-m dishes in an east–west line. It was succeeded in 1972 by the *Ryle Telescope (originally the Five-Kilometre Telescope), which operated until 2006. The Cambridge Low Frequency Synthesis Telescope (CLFST), operated 1980–2000, was a survey instrument consisting of 60 *Yagi antennas arranged along the line of the Ryle Telescope. The Interplanetary Scintillation Array of 4096 dipole antennas covering 3.6 hectares studied the scintillation of radio sources 1978–1992; an earlier version of this instrument, half the area, detected the first pulsars in 1967. The Arcminute MicroKelvin Imager (AMI) is a twin array for studying features in the cosmic microwave background. The AMI Small Array, consisting of ten 3.7-m antennae with baselines of 4–20 m, came into operation in 2005. The AMI Large Array, which started operation in 2008, consists of the eight 13-m dishes of the former Ryle Telescope rearranged to give baselines of 18–120 m. MRAO also hosts a 32-m dish of the *Multi-Element Radio-Linked Interferometer Network (MERLIN), and the *Cambridge Optical Aperture Synthesis Telescope (COAST).

(())) SEE WEB LINKS
• Official observatory website.

Müller, Johann *See* REGIOMONTANUS.

multi-beam receiver An array of feed horns and receivers mounted at the focus of a radio telescope which form a number of independent beams on the sky. Multi-beam receivers considerably reduce the time taken to complete a radio source survey, the actual time saving depending on the numbers of beams. A 13-beam receiver was installed in 1997 at *Parkes Observatory, Australia, and receivers with 100 beams for use at very high frequencies are being designed.

Multi-Element Radio-Linked Interferometer Network (MERLIN) A long-baseline interferometer operated by *Jodrell Bank Observatory consisting of seven radio telescopes distributed across England. It first came into operation in 1980. The telescopes are at Jodrell Bank in Cheshire (76-m Lovell Telescope and 25 × 38-m Mark II); Tabley (25-m dish) and Darnhall (25-m dish) both in Cheshire; Knockin (25-m dish) in Shropshire; Defford (25-m dish) in Worcestershire; and Cambridge (32-m dish), added in 1990. The longest baseline is 217 km. An upgraded version of the array called e-MERLIN, with over ten times the sensitivity of the original system, came into operation in 2010.

(())) SEE WEB LINKS
• Official MERLIN website.

multiple star A system of three or more stars bound by their mutual gravitational attraction. It is estimated that about one-third of all known binaries are actually triple. The proportions decrease with greater multiplicity. Systems with six components are known, but they are rare (less than 1 % of all multiples). In triple systems there is often one relatively close pair, with the third component in a wider orbit. Quadruple systems, however, occur approximately equally in 2 : 2 (2 pairs) and 1 : 1 : 2 hierarchies.

multiplet A group of closely spaced spectral lines formed by the same element or ion, caused when electrons with different directions of spin jump from a common lower energy level. *See also* FINE STRUCTURE.

multi-ringed basin *See* BASIN, IMPACT.

multiverse A collection of universes which some speculative theories suggest could exist. If true, our own Universe would be only one member of the larger **multiverse**. If these different universes had values of the fundamental constants that are different from those in our own Universe, this might explain why our Universe appears so well-suited for the existence of life (*see* ANTHROPIC PRINCIPLE).

muon An elementary particle having the same charge and spin as an electron, but a mass 207 times greater. It is a *lepton which decays into electrons and neutrinos with a half-life of two microseconds. Muons are present in cosmic-ray showers which are detected on Earth. Muons are produced by the decay of *pions. Muons were once known as mu-mesons, but they are not mesons.

mural circle An instrument used before the invention of the telescope to measure the angle above the horizon of astronomical objects as they crossed the meridian. It consisted of a large arc engraved with an angular scale, mounted vertically on a north–south wall. If the arc covered 90° (a quarter-circle) it was known as a **mural quadrant**. A sighting device allowed the altitude above the horizon of a chosen star to be measured.

Murchison Radio-astronomy Observatory (MRO) A radio observatory in the outback of Western Australia, some 315 km northeast of Geraldton. It is the site of the *Australian Square Kilometre Array Pathfinder (ASKAP) and the *Murchison Widefield Array.

(⊕) SEE WEB LINKS
• Official observatory website.

Murchison Widefield Array (MWA) A radio telescope consisting of 512 antenna 'tiles', each of which is a *phased array containing 16 crossed *dipole antennas. The array observes at wavelengths from 1 to 3.7 m. A core region some 1.5 km wide contains 496 tiles, with the remaining 16 tiles more widely spaced, giving baselines up to 3 km. The MWA, due for completion in 2011, is an international project headed by the Massachusetts Institute of Technology in conjunction with institutes from the United States, Australia, and India. It is located at the Murchison Radio-astronomy Observatory in Western Australia.

(⊕) SEE WEB LINKS
• Official telescope website.

Musca (Mus) (*gen.* **Muscae**) A small constellation of the southern sky, representing a fly. Its brightest star is Alpha Muscae, magnitude 2.7. Theta Muscae is a double star, magnitudes 5.6 and 7.6, the companion being the second-brightest *Wolf–Rayet star.

MWA Abbr. for *Murchison Widefield Array.

nadir The point on the celestial sphere directly beneath an observer. It is the direction in which a plumb-line points. The nadir is 180° from the *zenith.

Nagler eyepiece An ultra-wide-field eyepiece designed in 1980 by the American optician Albert Hirsch Nagler (1935–). Characteristically it is a complex seven-element design (the Nagler-2 has eight elements) that produces an 82° apparent field of view while correcting for spherical and chromatic aberration, coma, astigmatism, and field curvature. Its *eye relief is very generous, and the performance at the edge of field is excellent. The design is optimized to work with focal ratios as fast as $f/4$.

Naiad The innermost satellite of Neptune, orbiting between the planet's Galle and Le Verrier rings every 0.294 days at a distance of 48230 km from the planet's centre; also known as Neptune III. Its diameter is about 58 km. Naiad was discovered in 1989 on images from the Voyager 2 spacecraft.

NaI detector An instrument for detecting high-energy X-rays and gamma rays. Crystalline sodium iodide (NaI) converts an X-ray or gamma-ray photon into a pulse of visible light, the brightness of which is proportional to the energy of the incoming photon. The optical pulse is detected by a photomultiplier tube. To obtain images, a two-dimensional array of photomultipliers is used.

naked eye The eye used without any optical aid other than normal spectacles or contact lenses. The **naked-eye limiting magnitude** is that of the faintest star visible to the eye alone.

naked singularity A singularity that is not hidden by an *event horizon. A singularity is a point of infinite density that theories predict should exist at the centre of a black hole. If the black hole is rotating (a *Kerr black hole), then the area of the surrounding event horizon can be reduced to zero, leaving the singularity exposed to view. Some theorists have invoked the concept of *cosmic censorship*, which requires that all singularities be hidden behind event horizons at all times. If this is true, some unknown process must be operating in very rapidly rotating black holes to retain the event horizon.

nakhlite A rare type of achondrite meteorite, named after the meteorite that fell at Nakhla, Egypt, in 1911, the first known fall of this type, and which is reputed to have killed a dog. The nakhlites (also known as *augite–olivine achondrites*) consist of approximately 80% by weight of the mineral augite, and about 14% by weight of iron-rich olivine. Their textures suggest they formed within cooling magmas. Nakhlites belong to the class of *SNC meteorites, which are thought to come from Mars. In common with the Chassigny meteorite, the nakhlites have a formation age of 1.3 billion years and an *exposure age of about 12 million years.

Nançay Radio Astronomy Observatory The radio astronomy station of Paris Observatory, founded in 1953, 200 km south of Paris. Its main instrument is the Nançay Radio Telescope, opened in 1965, which consists of a fixed antenna 300 m long and 35 m high, shaped like a portion of a sphere, into which radio waves are reflected by a tiltable flat antenna 200 m long and 40 m wide. A radioheliograph, opened in 1982, consists of a T-shaped array with arms

3.2 and 1.25 km long. A Decametric Array was constructed in 1975–8 for detecting radio bursts from Jupiter and the Sun.

 SEE WEB LINKS
• Official observatory website.

nanometre (nm) A unit of length equal to 10^{-9} m. One nanometre is equal to 10 *angstroms.

Naos The star Zeta Puppis. It is an O5 supergiant, one of the hottest stars known (surface temperature 40000 K), of magnitude 2.2. Its distance is about 1100 l.y.

narrow-band photometry Photometry using filters with a bandwidth of 3–10 nm. For wavelength ranges narrower than 3 nm a *spectrophotometer must be used, since it is difficult to make filters with such a narrow bandwidth. Narrow-band systems are mostly used when it is necessary to isolate an individual spectroscopic line or molecular band.

NASA Abbr. for *National Aeronautics and Space Administration.

NASA Infrared Telescope Facility (IRTF) A 3-m reflector for infrared astronomy, particularly of Solar System objects, opened in 1979 at *Mauna Kea Observatory, Hawaii. It is operated for NASA by the University of Hawaii.

 SEE WEB LINKS
• Official telescope website.

Nasir (or Nasser) Eddin See ṬŪSĪ.

Nasmyth focus A focal position for a reflecting telescope, to which light is reflected along the hollow *declination axis or altitude axis of its mounting. At the Nasmyth focus, the image position remains stationary and heavy equipment can be used without affecting the balance of the telescope. The configuration was devised by the Scottish engineer James Nasmyth (1808–90).

National Aeronautics and Space Administration (NASA) A US government agency founded in 1958 for civil aeronautical research and space exploration, superseding the National Advisory Committee for Aeronautics (NACA). NASA headquarters are at Washington, DC. NASA operates several Field Centers: *Ames Research Center; Dryden Flight Research Facility at Edwards, California, used for flight testing and as a landing site for the Space Shuttle; Glenn Research Center at Cleveland, Ohio, concerned with aircraft and rocket propulsion; *Goddard Space Flight Center; *Jet Propulsion Laboratory; *Johnson Space Center; *Kennedy Space Center; Langley Research Center at Hampton, Virginia, which carries out research in aeronautics and space technology; *Marshall Space Flight Center; the *Space Telescope Science Institute; Stennis Space Center, near Bay St Louis, Mississippi, for testing rocket engines; and Wallops Flight Facility on Wallops Island, Virginia, which manages NASA's sounding rocket and scientific balloon programmes.

 SEE WEB LINKS
• Official website.

National Astronomy and Ionosphere Center (NAIC) See ARECIBO OBSERVATORY.

National Optical Astronomy Observatory (NOAO) A group of optical observatories for night-time astronomy, operated by the *Association of Universities for Research in Astronomy on behalf of the US National Science Foundation. NOAO was founded in 1982 and currently has three divisions: *Cerro Tololo Inter-American Observatory, Chile; *Kitt Peak National Observatory, Arizona; and the NOAO System Support Center, which is the US partner in the *Gemini Observatory. NOAO's headquarters are in Tucson, Arizona.

 SEE WEB LINKS
• Official NOAO website.

National Radio Astronomy Observatory (NRAO) The collective title for the government-owned radio astronomy facilities at various sites in the USA. NRAO was founded in 1956 and is administered by Associated Universities, Inc., a consortium of universities, on behalf of the US National Science Foundation. NRAO's oldest site is at Green Bank, West Virginia. Its first major aerial there was the 26-m Tatel Telescope, which began operation in 1959. In 1964 it was joined by a second 26-m dish to form the Green Bank Interferometer, with a third 26-m dish being added in 1967. The Green Bank Interferometer ceased operations in 2000. Its main instruments currently in operation there are the 100 × 110-m *Green Bank Telescope, opened in 2000; a 43-m equatorially mounted dish opened in 1965; a 26-m dish, originally the third element of the Green Bank Interferometer, now used for pulsar monitoring; and a 13.7-m dish for space VLBI and detecting solar radio bursts. NRAO also operates the *Very Large Array and the *Very Long Baseline Array, and is participating in the *Atacama Large Millimeter Array project. NRAO's headquarters are in Charlottesville, Virginia.

(⊕) SEE WEB LINKS
• Official NRAO website.

National Solar Observatory (NSO) An organization for solar astronomy, founded in 1984 and operated by the *Association of Universities for Research in Astronomy. NSO conducts research at two main sites: Kitt Peak, Arizona, and *Sacramento Peak Observatory, New Mexico. It also manages the *Global Oscillation Network Group (GONG). The NSO instruments on Kitt Peak are the *McMath–Pierce Solar Telescope; and the Synoptic Optical Long-term Investigations of the Sun (SOLIS) facility, opened in 2004. SOLIS consists of three telescopes on one mounting: a 0.5-m Vector-Spectromagnetograph (VSM) to measure magnetic field strength; a 0.14-m Full-Disk Patrol (FDP) to image the Sun in various spectral lines; and an 8-mm Integrated Sunlight Spectrometer (ISS) to record high-precision spectra in visible light. SOLIS is installed on the tower of the former Kitt Peak Vacuum Telescope (KPVT), which operated from 1973 to 2003; this is now known as the Kitt Peak SOLIS Tower (KPST).

(⊕) SEE WEB LINKS
• Official observatory website.

National Space Science Data Center (NSSDC) A department of NASA's Goddard Space Flight Center, founded in 1966, that is the repository for data from US space science missions—astronomy and astrophysics, solar physics, space plasma physics, and lunar and planetary science.

(⊕) SEE WEB LINKS
• Official website, with links to many information pages on spacecraft and space exploration.

Nautical Almanac, The A yearly publication containing tables of the Sun, Moon, planets, and stars, plus other data for marine navigation, prepared and issued jointly by the US Naval Observatory and HM Nautical Almanac Office. Founded by N. *Maskelyne, it was first published in Britain in 1766 (for the year 1767) under the title *The Nautical Almanac and Astronomical Ephemeris*. In 1914 it was divided into two versions, one for astronomers and an abridged version for navigators; the astronomical version is now known as *The *Astronomical Almanac*. In 1960 *The Nautical Almanac* merged with *The American Nautical Almanac*, published separately in the USA since 1914 (for the year 1916). Related publications are *The Air Almanac* for air navigators and *The Star Almanac* for land surveyors.

(⊕) SEE WEB LINKS
• Official website.

nautical twilight The period before sunrise and after sunset when the centre of the Sun's disk is between 6° and 12° below the horizon. It begins in the evening when the brightest stars become visible. During nautical twilight the sea horizon is visible so that altitudes of celestial objects can be measured. *See also* TWILIGHT.

n-body problem The mathematical problem of finding the velocities and positions of any number (*n*) of objects moving under their mutual gravitational attraction for any time in the past or future; also known as the *many-body problem*. It applies, for example, to the members of the Solar System and the members of a star cluster. As with the *three-body problem, no general mathematical solution holding for all time has ever been found, but certain general results are known, true for any number of bodies. The centre of mass of the system of particles travels with constant velocity; the total energy of the system is constant; and the total angular momentum of the system does not change. With modern electronic computers, the bodies' velocities and positions can be calculated to any desired accuracy for finite lengths of time into the past or future but, for reasons such as round-off error (the error accumulating from slight initial inaccuracies when a large number of calculations are performed) and the chaotic nature of the problem, the accuracy deteriorates as the time increases.

NEA Abbr. for *near-Earth asteroid.

neap tide The tide raised in the Earth's oceans when the Moon is at either first or third quarter phase, and is pulling on the Earth at right angles to the Sun. The amplitude of neap tides (i.e. the difference in water level between high and low tide) is the smallest in the monthly tidal cycle.

near-Earth asteroid (NEA) Any asteroid belonging to the *Apollo, *Amor, or *Aten groups. Such asteroids have perihelion distances of less than 1.3 AU. The term *near-Earth objects* (NEOs) is also used, allowing for the fact that some may be extinct short-period comets. Members of the Amor group cross the orbit of Mars but do not quite reach the Earth's orbit, while the Apollo and Aten groups do cross the Earth's orbit. The three groups are not completely separate, since planetary perturbations may cause an Amor asteroid to become an Apollo, and vice versa. Atens are the least common NEAs, while Apollos and Amors are found in roughly equal numbers. Only three NEAs are larger than 10 km in diameter: (1036) Ganymed, (433) Eros, and (4954) Eric, all members of the Amor group. Near-Earth asteroids have finite lifetimes (typically 10 million years); most will be destroyed by collision with one of the inner planets, and the remainder ejected from the Solar System. At least 100 000 NEAs larger than 100 m in diameter are estimated to exist, and perhaps 1000 larger than 1 km. As of 2010 over 7000 were known, with new discoveries running at around 800 a year. *See also* POTENTIALLY HAZARDOUS ASTEROID.

Near-Earth Asteroid Tracking (NEAT) A search for asteroids and comets operated by the *Jet Propulsion Laboratory in conjunction with the US Air Force. NEAT began in 1995 using a 1-m telescope at the USAF's Ground-based Electro-Optical Deep Space Surveillance (GEODSS) site on Mt Haleakala, Maui, Hawaii. In 2000 the search was transferred to the 1.2-m telescope of the Maui Space Surveillance Site (MSSS). In 2001 the 1.2-m Schmidt at Palomar Observatory also joined in the project. NEAT ended in 2007, having discovered 442 near-Earth asteroids.

near infrared That part of the electromagnetic spectrum covering the shorter infrared wavelengths, but with no well-defined limit at the longer-wavelength end. Broadly, the near infrared extends from the red end of the visible spectrum (approximately 0.7 μm, or 700 nm) to about 35 μm. In practice, the term is often used to describe the *infrared windows in the atmosphere between 1 and 5 μm. The 0.7–1.0 μm region is sometimes known as the *photographic infrared*.

NEAR Shoemaker A NASA space probe launched in 1996 February to the asteroid *Eros. Originally known as Near-Earth Asteroid Rendezvous (NEAR), it was renamed NEAR Shoemaker after entering orbit around Eros in honour of E. *Shoemaker. On the way to Eros it passed and photographed (253) Mathilde, a *C-class (carbonaceous) asteroid, at a distance of under 1200 km in 1997 June, and underwent an Earth gravity-assist in 1998 January. The spacecraft flew past Eros at a distance of 3827 km in 1998 December. After making one orbit of the Sun, the spacecraft re-encountered Eros and went into orbit around it in 2000 February, photographing its surface in detail and studying it with an X-ray/gamma-ray spectrometer, a near-infrared imaging spectrograph, a laser rangefinder, and a magnetometer. The mission ended when NEAR Shoemaker made a low-speed descent to the surface in 2001 February, continuing to transmit data for over two weeks. It was the first of NASA's Discovery class of low-cost missions.

(((⊕))) SEE WEB LINKS
• Official mission website.

near ultraviolet The part of the ultraviolet spectrum adjacent to the visible band. It covers the wavelength range from approximately 200 to 350 nm.

NEAT Abbr. for *Near-Earth Asteroid Tracking.

nebula A cloud of gas and dust in space. The term was originally applied to any object with a fuzzy telescopic appearance, but with the advent of larger instruments many 'nebulae' were found to consist of faint stars. In 1864 W. *Huggins discovered that true nebulae could be distinguished from those composed of stars on the basis of their spectra. Nowadays the term 'nebula' means a gaseous nebula. The term *extragalactic nebula*, originally used for galaxies, is now obsolete. There are three broad types of gaseous nebula: *emission nebulae, which shine by their own light; *reflection nebulae, which reflect light from nearby bright sources such as stars; and *dark nebulae (or absorption nebulae) which appear dark against a brighter background. This broad classification scheme has been carried over to other wavelengths, giving rise to terms such as *infrared reflection nebula* and *infrared dark cloud*. Emission nebulae include the diffuse nebulae or H II regions around young stars, planetary nebulae around old stars, and supernova remnants. 📷

nebula filter A filter that transmits only selected visual spectral lines from emission nebulae. A typical filter transmits a band 25 nm wide, including the O III (495.9 and 500.7 nm) and Hβ (485.6 nm) lines. Narrow-band filters transmit bands just 10 nm wide, isolating either O III or Hβ.

nebular line An emission line that is found in the spectra of diffuse nebulae, notably the green lines due to doubly ionized oxygen, [O III], at 495.9 and 500.7 nm. Nebular lines are also *forbidden lines, since they do not occur under conditions on Earth.

nebular variable Any of the various types of *eruptive variable intimately associated with bright or dark diffuse nebulae, or observed in the region of those nebulae; abbr. IN. They are also known as *Orion variables. Most are young stars of the *FU Orionis or *T Tauri type, together with the UVN subtype of *UV Ceti stars.

nebulosity A general term for any fuzzy area in the sky, usually a cloud of gas and dust. The use of the term may imply that the gas cloud is part of a larger nebula, is associated with a more readily resolvable star cluster, or that its form does not allow a more exact classification as a particular type of nebula.

negative eyepiece An eyepiece whose focal point lies within it and which cannot, therefore, be used as a magnifying glass. The *Huygenian eyepiece is an example.

negative lens Another name for a *diverging lens.

Neptune (♆) The eighth planet from the Sun. It appears distinctly blue, due to absorption of red light by methane in its atmosphere. Its mean opposition magnitude is +7.8, too faint to be seen with the naked eye. It was discovered in 1846 by J. G. *Galle, after its position had been predicted mathematically by U. J. J. *Le Verrier. Its shape is distinctly ellipsoidal (equatorial diameter 49528 km, polar diameter 48682 km). The rotation period of the visible surface varies between about 12 hours near the poles and 20 hours near the equator, but radio bursts indicate that the core rotates in 16h 07m.

Neptune has a thick atmosphere composed of 80% hydrogen, 19% helium, and 1.5% methane (molecular percentages). The temperature near the top of the atmosphere is around −220°C. Internally there is thought to be a small rocky core at a high temperature, probably surrounded by a layer of icy materials, and above that a layer of hydrogen and helium. The internal heat is thought to have been released by differentiation, when denser material separated out and sank to the core. Neptune has a magnetic field slightly weaker than the

Earth's, with an average strength at the equator of around 10^{-5} tesla. As with Uranus, the magnetic axis does not run through the planet's core but is offset half-way to the surface, and is tilted at 47° to the planet's axis of rotation.

Neptune

Physical data

Diameter (equatorial)	Oblateness	Inclination of equator to orbit	Axial rotation period (sidereal)	
49528km	0.017	28°.32	16.11 hours	
Mean density	Mass (Earth = 1)	Volume (Earth = 1)	Mean albedo (geometric)	Escape velocity
1.64g/cm³	17.15	58	0.41	23.6km/s

Orbital data

Mean distance from Sun				
10⁶km	AU	Eccentricity of orbit	Inclination of orbit to ecliptic	Orbital period (sidereal)
4498.253	30.07	0.009	1°.8	164.79 years

Neptune's atmosphere has dark belts or bands with bright zones between, similar to those on Jupiter and Saturn but fewer in number and less prominent. Dark and bright spots also occur, the most prominent being the *Great Dark Spot discovered in 1989 by Voyager 2, similar to Jupiter's Great Red Spot but not as long-lived. There are also bright wispy clouds of methane ice resembling cirrus clouds on Earth, 50–100km higher than Neptune's main cloud tops. One of them in the southern hemisphere, photographed by Voyager 2, had a much higher velocity than other clouds and was dubbed the Scooter. Wind speeds reach 2000km/h, faster than on any other planet.

Neptune has six rings. The Galle, Le Verrier, Lassell, Arago, and Adams rings are named after those involved in Neptune's prediction and discovery. A sixth ring, as yet unnamed, lies along the orbit of the moon Galatea. The outer (Adams) ring has most of its matter in four bright concentrations, known as *ring arcs. Neptune currently has thirteen known natural satellites.

Nereid A satellite of Neptune, eighth from the planet at a distance of 5513400km, orbital period 360.13 days; also known as Neptune II. Its orbit is one of the most elliptical of any satellite in the Solar System, with an eccentricity of 0.75, and is inclined at 6°.7 to Neptune's equator. Nereid's diameter is 340km. It was discovered in 1949 by G. P. *Kuiper.

Neumann lines A rectangular pattern of fine striations appearing when a slice of *hexahedrite iron meteorite is cut, polished, and etched with dilute nitric acid; also called **Neumann bands**. The pattern is aligned along the faces of a cube. Neumann lines are produced by compressional shock waves. Although most obvious in hexahedrites, Neumann lines are present in other iron meteorites as well. They are named after their discoverer, the German mineralogist Franz Ernst Neumann (1798–1895).

neutral hydrogen Hydrogen gas made up of electrically neutral (not ionized) atoms. *See* H I REGION.

neutral hydrogen line *See* 21-CENTIMETRE LINE.

neutral point A point at which a particle would experience no net gravitational force. A neutral point exists between two static massive bodies. Its position depends on the bodies' separation and their masses, being closer to the more massive body. If the bodies are orbiting each other, as in reality, the problem is more complicated, and there exist five points (the *Lagrangian points) at which the total acceleration due to gravitational and centrifugal forces is zero.

neutrino An elementary particle with zero charge and a very small *rest mass. Three types of neutrino are known: the electron neutrino, the muon neutrino, and the tau neutrino. Neutrinos have only a weak interaction with matter and consequently neutrinos produced in nuclear reactions at the centres of stars can escape without colliding with the overlying material. Neutrinos may account for some of the *dark matter in the Universe. Neutrinos are *leptons.

neutrino astronomy The observation of neutrinos emitted by celestial objects. Neutrinos pass through large quantities of matter without significant absorption. For example, neutrinos produced by the nuclear processes in the Sun's core have only one chance in 10^{10} of being absorbed in escaping from the Sun and so reach the Earth in huge numbers (*see* SOLAR NEUTRINO UNIT). Bursts of neutrinos are predicted to be produced in supernova explosions, and such a burst was detected from Supernova 1987A. Neutrinos can be detected in several ways. One uses the neutrino's interaction with the chlorine isotope ^{37}Cl, which produces radioactive argon, ^{37}Ar. Detectors have also been built which utilize the conversion of gallium to germanium by a neutrino (^{71}Ga to ^{71}Ge). **Neutrino telescopes** detect the direction from which the neutrino comes, as well as its existence. These rely on the neutrino colliding with an electron inside a large tank of water. The electron is then detected via its *Cerenkov radiation.

neutron An elementary particle that is present in the nuclei of all atoms except the lightest isotope of hydrogen. It has a slightly greater mass than the proton and zero charge. Neutrons are stable in the atomic nucleus, but outside it they undergo beta decay to produce a proton, an electron, and an antineutrino. Neutrons are *hadrons.

neutron degeneracy The state of *degeneracy attained when the density of matter is so high that neutrons cannot be packed any more closely together. The effect is analogous with that of *electron degeneracy, but the density required for neutron degeneracy (about 10^{17} kg/m^3) is much higher, and is encountered only in neutron stars.

neutron star An extremely small, superdense object believed to be formed when a massive star undergoes a Type II supernova explosion. During the explosion, the core of the massive star collapses under its own gravity until, at a density of about 10^{17} kg/m^3, electrons and protons are so closely packed that they combine to form neutrons. The resultant object, consisting only of neutrons, is supported against further gravitational collapse by the *degeneracy pressure of the neutrons, provided its mass is not greater than about 2 solar masses (the *Oppenheimer–Volkoff limit). If the object were more massive it would collapse further into a black hole. A typical neutron star, with a mass a little greater than the Sun's, would have a diameter of only about 30 km, and a density such that the mass of the entire human race would occupy the volume of a sugar cube. The greater the mass of a neutron star, the smaller its diameter. Neutron stars are believed to have an interior of superfluid neutrons (i.e. neutrons behaving like a fluid of zero viscosity), surrounded by a solid crust about 1 km thick composed of elements such as iron. *Pulsars are spinning, magnetized neutron stars. Massive X-ray binaries are also thought to contain neutron stars.

Newcomb, Simon (1835–1909) Canadian-born American mathematical astronomer. At the US Naval Observatory's Nautical Almanac office, of which he became head in 1877, he initiated an extensive project to refine the orbits of the Moon and planets. (G. W. *Hill was assigned the motions of Jupiter and Saturn.) The project involved the use of historical data, from which he discovered the non-Newtonian component of the *advance of the perihelion of Mercury's orbit that was later accounted for by the *general theory of relativity. Newcomb improved the value of the solar parallax and other astronomical constants, and worked with A. A. *Michelson on a method for measuring the speed of light.

New Frontiers Program A NASA series of medium-sized spacecraft to explore the Solar System. They are larger, more expensive, and less frequent than members of the *Discovery Program. The first of the series was *New Horizons, launched in 2006, and the second was the *Juno mission to Jupiter in 2011. Missions in the New Frontiers Program are intended to be launched every three years, on average.

(⊕) SEE WEB LINKS
• Official New Frontiers program website.

New General Catalogue (NGC) Shortened title of the *New General Catalogue of Nebulae and Clusters of Stars* compiled by J. L. E. *Dreyer and published in 1888. It contained 7840 galaxies, nebulae, and star clusters, the majority of them discovered by F. W. and J. F. W. *Herschel. In fact, Dreyer's work was a revised and enlarged successor to J. F. W. Herschel's *General Catalogue of Nebulae and Clusters of Stars* (1864). Additional objects were listed in two *Index Catalogues*. Objects were listed in order of right ascension with a brief description of their appearance. The objects in Dreyer's catalogues are still known by their NGC or IC numbers.

(⊕) SEE WEB LINKS
• Detailed description and full catalogue downloadable from the CDS.
• The original catalogue online.

New Horizons A NASA space probe to Pluto and the Kuiper Belt, launched in 2006 January. It swung past Jupiter for a gravity assist in 2007 February, and is due to reach Pluto and its moon, Charon, in 2015 July. The spacecraft will then head into the Kuiper Belt to encounter one or more of its members. New Horizons will photograph these bodies, analyse their surface compositions, and study any atmospheres they may have with the following instruments: the Long Range Reconnaissance Imager (LORRI), a high-resolution camera; Ralph, an instrument which combines the Multispectral Visible Imaging Camera (MVIC) and the Linear Etalon Imaging Spectral Array (LEISA, a near-infrared imaging spectrometer), to study surface features and composition; Alice, an ultraviolet imaging spectrometer to examine Pluto's atmosphere; REX, a radio-science experiment to probe Pluto's atmosphere and measure its nightside temperature; the Solar Wind Analyzer around Pluto (SWAP), to measure charged particles from the solar wind near Pluto; the Pluto Energetic Particle Spectrometer Science Investigation (PEPSSI) to search for neutral atoms from Pluto's atmosphere; and a student-built dust counter, SDC, which will count and measure the sizes of dust particles along the craft's entire trajectory.

(⊕) SEE WEB LINKS
• Official mission website.

New Millennium Program (NMP) A NASA series of small, low-cost space probes designed to test new technologies for future missions, such as solar-electric propulsion ('ion drive') and artificial intelligence. The first of the series was *Deep Space 1, launched in 1998 October.

(⊕) SEE WEB LINKS
• Official New Millennium Program website.

new Moon The phase of the Moon when none of its illuminated side is visible from the Earth. At new Moon, the Moon has the same celestial longitude as the Sun.

New Style date (NS) The date of a historical or astronomical event on the *Gregorian calendar, as distinct from dates on the *Julian calendar which are known as Old Style (OS). Since the Gregorian calendar was introduced at different times in different countries, confusion can arise if the form of calendar being used is not specified.

New Technology Telescope (NTT) A 3.5-m reflector at the *European Southern Observatory, Chile, opened in 1989. It uses *active optics to produce particularly sharp images, the first large telescope to do so, hence its name.

(⊕) SEE WEB LINKS
• Official telescope website.

newton (N). A unit of force, defined as the force required to accelerate a mass of one kilogram by one metre per second per second. It is named after I. *Newton.

Newton, Isaac (1642–1727) English physicist and mathematician. He developed his principal theories of gravitation, optics, and mathematics in 1665 and 1666. In 1668 he made the first working reflecting telescope. Most of his findings remained unpublished for long periods, partly because of criticisms by C. *Huygens and the English scientist Robert Hooke (1635–1703) of his early work on the corpuscular theory of light. However, in 1684 E. *Halley persuaded him to organize his work on the celestial mechanics of the Solar System, which was published as the *Principia*. Newton's other major book, *Opticks*, was not published until 1704. It contains his corpuscular theory of light and the theory of the telescope. His greatest mathematical achievement was his invention of calculus, independently of the German mathematician Gottfried Wilhelm Leibniz (1646–1716). His profound influence on physics and astronomy is reflected in the phrase 'Newtonian revolution'.

Newtonian–Cassegrain telescope A *Cassegrain telescope with an additional flat diagonal mirror which intercepts the converging beam from the secondary and reflects it to the side of the tube. Unlike a standard Cassegrain, the primary mirror need not have a central hole, and the focal position can be made to coincide with the declination or altitude axis of the mounting, thus giving a stationary observing position.

Newtonian focus A focal position for a reflecting telescope at the side of its tube, at right angles to the incoming light. Light reflected from the main mirror is diverted to the Newtonian focus by a flat diagonal mirror, or a prism, higher up inside the tube. In a standard *Newtonian telescope, the focal ratio and focal length are the same for the Newtonian focus as for the prime focus.

Newtonian telescope A reflecting telescope in which light is collected by a paraboloidal primary mirror and then reflected by a secondary mirror or prism within the tube to a focal point at the side of the tube, at right angles to the direction of view of the telescope. The image is inverted but not laterally reversed. The focal ratio should be larger than *f*/3, otherwise coma (*see* COMA, OPTICAL) becomes excessive. The design was invented in 1668 by I. *Newton.

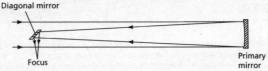

Newtonian telescope

Newton's law of gravitation The law put forward by I. *Newton in 1687 which states that any two bodies attract each other with a force that depends on the product of their masses divided by the square of the distance between them. Put mathematically, the gravitational force of attraction, *F*, between two masses m_1 and m_2, separated by distance *r*, is given by $F = Gm_1 m_2/r^2$, where *G* is a constant known as the *gravitational constant. *See also* GRAVITATION.

Newton's laws of motion Three laws published in 1687 by I. *Newton concerning the motion of bodies:
1. A body continues in a state of uniform rest or motion unless acted upon by an external force.

2. The acceleration produced when a force acts is directly proportional to the force and takes place in the direction in which the force acts.
3. To every action there is an equal and opposite reaction.

N galaxy A type of galaxy with a bright, star-like nucleus that has a strong emission-line spectrum, surrounded by a faint nebulous envelope. They have *active galactic nuclei, and may represent lower-luminosity versions of quasars. The designation N galaxy comes from the *Morgan classification.

NGC Abbr. for *New General Catalogue.

nickel–iron meteorite See IRON METEORITE.

Nicol prism A device for producing plane-polarized light. It consists of a rhombus-shaped prism of calcite crystal, cut diagonally and recemented with Canada balsam, a resin with similar optical properties to glass. Calcite naturally splits light into two separate beams, polarized at right angles to each other. The layer of Canada balsam within the prism cuts out one beam through total internal reflection. The advantage of a Nicol prism over Polaroid material as a polarizer is that all wavelengths of light are polarized and transmitted equally. It was devised by the Scottish geologist and physicist William Nicol (1768–1851).

nightglow Another name for *airglow.

night sky brightness The very low background level of sky illumination present even in the absence of artificial or other *light pollution. It results from atmospheric *airglow emissions and, to a lesser extent, from starlight and the zodiacal light.

night vision An alternative name for *dark adaptation.

Nix Olympica The name by which *Olympus Mons was known before 1973.

NLC Abbr. for *noctilucent clouds.

nm Symbol for *nanometre.

N magnitude The magnitude of a star as measured through the infrared N filter of the *Johnson photometry system. It has an effective wavelength of 10.4 µm and a bandwidth of 4.19 µm.

NOAO Abbr. for *National Optical Astronomy Observatory.

Nobeyama Radio Observatory The radio observatory of the National Astronomical Observatory of Japan, 120 km west of Tokyo at an altitude of 1350 m, founded in 1978. Its main instruments are a 45-m millimetre-wave dish opened in 1982, and the Nobeyama Millimeter Array (NMA), an aperture-synthesis array of six 10-m antennas movable along two intersecting 500-m baselines. The millimetre array was opened in 1986 with five dishes; the sixth dish was added in 1993. The NMA and the 45-m dish can work together to form an interferometer called RAINBOW. Also at the same site is the Nobeyama Radioheliograph, which consists of 84 dishes of 0.8-m aperture in a T-shaped array, opened in 1992. The Observatory also operates the *Atacama Submillimeter Telescope Experiment in Chile.

(⊕) SEE WEB LINKS
• Official observatory website.

noble gas One of the inert gases helium, neon, argon, krypton, xenon, and radon. Each has a stable configuration of electrons in its atoms that renders it unreactive. Apart from argon, only traces of the noble gases are present in the Earth's atmosphere.

noctilucent clouds (NLC) Literally, 'night-shining' blue–silver clouds seen in late or early twilight during the summer from high temperate latitudes. Noctilucent clouds result from water vapour condensing on nuclei (possibly meteoric or volcanic debris) at altitudes of around 82–85 km, close to the *mesopause. The clouds are extremely tenuous, and are seen only when the Sun lies between 6° and 16° below the horizon, appearing bright by contrast with the twilight sky. Noctilucent clouds lie five times higher than the loftiest clouds in the *troposphere, and are distinct from cirrus clouds to which they bear a superficial resemblance. In particular, noctilucent clouds may be identified by their delicate, feathery, interwoven herringbone structure. They are more common in years of low solar activity, and there is evidence that they are becoming more frequent.

nocturnal A historical hand-held device for telling the time at night using the orientation of northern circumpolar constellations. It consisted of a dial marked with the hours, with indicators to be aligned with specific stars, notably the Pointers of Ursa Major. When oriented with the stars and with Polaris in the centre, the time was shown against the appropriate date.

nodal line The straight line joining the *nodes of an orbit.

node Either of the two points at which an orbit intersects a reference plane such as the plane of the ecliptic or of the celestial equator. The line joining these points is known as the *line of nodes* or *nodal line*. At the *ascending node* an orbiting body moves from south to north of the reference plane; at the *descending node* it moves from north to south. *See also* REGRESSION OF NODES.

nodical month An alternative name for the *draconic month.

non-baryonic matter A hypothetical form of matter not containing *baryons—that is, without protons or neutrons. An example would be the positron–electron 'atoms' that may constitute much of the Universe in the very distant future if protons decay. Non-baryonic matter has been suggested as a possible component of the *dark matter in the Universe. In this case it could take the form of neutrinos, if they have non-zero *rest mass, or of hypothetical particles called *WIMPs (weakly interacting massive particles).

non-gravitational force An effect that accelerates or decelerates a comet's motion, changing its orbital period. Such forces are caused by jets of gas emerging from active regions on the surface of the nucleus, giving a rocket-like effect. Non-gravitational forces are most marked when the nucleus is highly active, close to perihelion, leading to uncertainties in subsequent returns for some *periodic comets. They are responsible for systematic changes in the perihelion time of Comet *Encke, and the apparently delayed 1992 return of Comet *Swift–Tuttle.

non-radial pulsation A form of pulsation in which waves run in all directions across the star's surface, as opposed to spherically symmetrical expansion and contraction (*radial pulsation). There are normally multiple periods, resulting in a complex pattern of nodes and antinodes over the surface. Non-radial pulsations are particularly prominent in white-dwarf *ZZ Ceti stars.

non-thermal radiation Radiation due to a cause other than the temperature of the emitting body. It has a different spectrum from that of *black-body radiation. Examples of non-thermal radiation are *synchrotron radiation, *maser radiation, and artificially generated radio and TV signals.

Nordic Optical Telescope (NOT) A 2.56-m reflector opened in 1989 at the *Roque de los Muchachos Observatory on La Palma in the Canary Islands, which uses a thin mirror figured to high accuracy. It is owned and operated by a consortium of research councils of Denmark, Finland, Iceland, Norway, and Sweden.

 SEE WEB LINKS
• Official telescope website.

Norma (Nor) (*gen.* **Normae**) A constellation of the southern sky, representing a set square. Its brightest star, Gamma² Normae, is of magnitude 4.0 and forms an optical double with the unrelated Gamma¹ Normae, magnitude 5.0. NGC 6087 is a 5th-magnitude open star cluster.

Norma Arm One of the two minor spiral arms of our Galaxy, according to recent studies of galactic structure. It lies between the main *Scutum–Centaurus Arm and the galactic centre, about 15 000 l.y. from the local spiral arm (the *Orion Arm) in which we live.

normal astrograph A photographic refractor with aperture 0.33 m and focal length 3.438 m, giving a *plate scale of 60″ to 1 mm. The usable field is about $2° \times 2°$. This type of telescope was designed in Paris by the brothers Paul Pierre Henry (1848–1905) and Prosper Matthieu Henry (1849–1903) in about 1880. Many such telescopes were constructed and were used for the *Astrographic Catalogue and *Carte du Ciel.

North America Nebula A diffuse nebula in Cygnus, also known as NGC 7000, shaped like the continent of North America. Its dimensions are about $2° \times 1½°$ and it lies 1500 l.y. away. Its illuminating star is thought to be a 6th-magnitude hot, blue star within the nebula, HR 8023. Next to the North America Nebula lies the *Pelican Nebula, actually part of the same enormous cloud which stretches for 100 l.y.

Northern Coalsack Another name for part of the dark nebula known as the *Great Rift.

Northern Cross A popular name for the constellation *Cygnus.

northern lights Popular name for the aurora borealis (*see* AURORA).

North Galactic Spur An arc of radio emission about 80° long, projecting from the Milky Way towards the north galactic pole; also known as the **North Polar Spur**. It is believed to be part of the expanding shell of an ancient supernova remnant a few hundred light years from the Sun.

north polar distance (NPD) The angular distance of an object from the north celestial pole, measured along the *hour circle passing through the object. It is 90° minus the object's declination.

North Polar Sequence A series of stars within 2° of the north celestial pole that were used to provide standard magnitudes and an arbitrary zero point for the magnitude scale. The original sequence (published 1914–22) included 96 stars with photographic magnitudes down to 20.1; it was extended by a supplementary list of 56 other stars. A **North Polar Sequence of Photovisual Magnitudes** down to 17th magnitude was subsequently introduced. The sequences have been superseded by the use of UBV photometry to establish the zero point, used in conjunction with lists of accurately measured standard stars.

North Star A popular name for *Polaris.

nova A type of cataclysmic variable that exhibits a sudden, unpredictable outburst, with a typical amplitude of 11–12 mag.; abbr. N. Novae are interacting binaries, and usually consist of a main-sequence or slightly evolved secondary and a white-dwarf primary. *Mass transfer occurs from the secondary to an *accretion disk and thence to the white dwarf. The accumulation of hydrogen-rich gas on the white dwarf's surface eventually leads to a thermonuclear runaway (a sudden initiation of nuclear reactions), producing the outburst and ejecting much of the outer envelope from the system.

Novae are divided into three subtypes, based on their speed of decline from maximum (*see also* FAST NOVA): fast (NA), declining by 3 magnitudes in less than 100 days; slow (NB), 3 magnitudes in 150 days or more; very slow (NC), persisting at maximum for years. Very slow novae probably have a giant or supergiant secondary, and are sometimes referred to as *symbiotic novae* or *RR Telescopii stars.

No *pre-nova has been studied in detail, but there may be a pre-outburst rise of some magnitudes, perhaps with fluctuations and activity like that of a *dwarf nova. The initial rise is rapid (less than 1 day in most cases, 2–3 days in slow novae) to a *pre-maximum halt*, about 2 mag. below maximum. After a few hours to days this is followed by the final rise in 1–2 days or (in slow novae) a few weeks. About 3–4 mag. below maximum, a *transition region* lasting 2–3 months may occur; some novae (e.g. Nova Herculis 1934 = DQ Herculis) undergo a deep fade of 7–10 mag. and a subsequent recovery, while others exhibit quasi-periodic oscillations. The final decline into the *post-nova phase is generally steady.

The spectrum of a nova at maximum shows a continuum, overlain by a sequence of absorption lines arising in the expanding envelope which indicates velocities of hundreds to thousands of kilometres per second. In the final stages the continuum fades, leaving a nebular spectrum with forbidden lines. There is an extended period of mass loss in which at least 10^{-4} solar masses is ejected. The mass transfer rate through the accretion disk is estimated at about 10^{-8} solar masses a year, so the system should undergo a further outburst after about 10000 years. The ejecta may become visible after a period of time.

Estimates suggest that 25 to 50 novae occur each year in our Galaxy, most of which go undetected because of interstellar extinction and other factors. They occur throughout the Galaxy (including the outermost halo) and among systems of all ages from old Population II to extremely young Population I stars.

nova-like variable A heterogeneous type of poorly studied cataclysmic binaries; abbr. NL. There are either small-amplitude outbursts or minor fluctuations, which together with the spectral details resemble those of a *post-nova at minimum. Closer study usually reveals features that enable a nova-like object to be more firmly classified as a post-nova, a *dwarf nova, a *UX Ursae Majoris star, or some other related group.

NPD Abbr. for *north polar distance.

NRAO Abbr. for *National Radio Astronomy Observatory.

NS Abbr. for *New Style date.

NSO Abbr. for *National Solar Observatory.

NSSDC Abbr. for *National Space Science Data Center.

N star See CARBON STAR.

NTT Abbr. for *New Technology Telescope.

Nubecula Major, Nubecula Minor Other names for the *Magellanic Clouds.

nuclear fusion The fusing together of two atomic nuclei of low atomic number to form a nucleus of higher atomic number, with the release of energy (e.g. hydrogen to helium). Nuclei are positively charged, so high kinetic energies are required to overcome the mutual repulsive forces; this implies temperatures of the order of 10^8 K. Nuclear fusion takes place inside stars and is the process by which energy is released that makes the star shine. See also CARBON–NITROGEN CYCLE; PROTON–PROTON REACTION; TRIPLE-ALPHA PROCESS.

nuclear reaction A reaction involving one or more nuclei of atoms in which there is a change to the nuclei. Nuclear reactions can release enormous amounts of energy; examples are *nucleosynthesis, *fission, *fusion, and radioactive decay. However, other nuclear reactions require an input of energy to build up elements heavier than iron, such as the *r-process and *s-process.

nuclear time-scale The time taken for a star to exhaust its supply of a particular nuclear fuel. The nuclear time-scale may be very short, as with the oxygen-burning phase of some massive stars, which can be less than a year, or very long, as with the hydrogen-burning phase of low-mass stars, which is potentially longer than the age of the Universe. The nuclear time-scale for hydrogen burning in the Sun is about 7 billion years, but becomes shorter with increasing stellar mass because the larger amount of fuel in massive stars is more than offset by the faster rate of burning.

nucleon A proton or a neutron present in the nucleus of an atom.

nucleosynthesis The process of creating elements by nuclear reactions. Helium was produced by nucleosynthesis in the few minutes following the Big Bang. Helium and heavier elements are built up by nucleosynthesis inside stars. First, hydrogen is converted to helium by the *proton–proton reaction or the *carbon–nitrogen cycle. When the hydrogen-to-helium phase ends, the *triple-alpha process takes over. Successively heavier elements up to iron are then synthesized, each in turn using the product of the previous reaction. If a star becomes a supernova, the heaviest possible nuclei are formed and are ejected into interstellar space. New generations of stars forming from the enriched medium have a higher heavy element content than old stars. *See also* R-PROCESS; S-PROCESS.

nucleus, atomic The central core of an atom that contains most of its mass. The nuclei of atoms are positively charged and consist of one or more protons and neutrons. The number of protons in a nucleus defines the *atomic number* of the element. The simplest nucleus is that of the lightest isotope of hydrogen, consisting of one proton only. All other nuclei also contain one or more neutrons, which contribute to the mass but not the charge of the nucleus. The most massive nucleus that occurs naturally on Earth is ^{238}U, containing 92 protons and 146 neutrons.

nucleus, cometary The small solid body, composed of frozen water and gas plus embedded dusty material, at the centre of a comet's head. It is the source of activity in a comet, and contains essentially all its mass. The nucleus of *Halley's Comet, the first to be directly observed (by the *Giotto space probe in 1986), is irregularly shaped, 16 × 8 km, with a dark crust; most nuclei are probably smaller. When very distant from the Sun, a comet's nucleus is inert. Within 3–4 AU, however, solar heating leads to gas sublimation, and production first of a coma, then a tail (*see* COMA, COMETARY; TAIL, COMETARY). Near perihelion, surface temperatures may reach 25°C. Cometary nuclei have low densities (typically 0.2g/cm^3), and are prone to fragmentation. Nuclei which have made several perihelion passages are thought to be covered by a dark crust, as Halley's Comet. Gas emerges through fissures in the crust, carrying off dust which pursues its own orbit as a meteor stream. Perhaps only 10% of the nucleus consists of such active regions at any one time. Emerging jets are the cause of *non-gravitational forces which affect the orbits of comets. Pristine nuclei from the *Oort Cloud may lack a dark crust and experience less heating than those of well-evolved comets, as happened with Comet *Kohoutek. Some *Apollo asteroids may be old, outgassed nuclei. ◙

nucleus, galactic *See* GALACTIC NUCLEUS.

null geodesic A path or *geodesic joining two events in spacetime along which the separation is zero. The separation in spacetime can be zero even if the two events actually occur at different points in space. This apparent contradiction can arise because in the formula for determining the separation of events in spacetime, the time interval between the events has the opposite sign to that for their distance apart in space. Light in a vacuum travels along a null geodesic.

number evolution *See* GALAXY EVOLUTION.

Nunki The star Sigma Sagittarii, magnitude 2.1. It is a B3 subgiant, 228 l.y. away.

NuSTAR The Nuclear Spectroscopic Telescope Array, a NASA X-ray observatory due for launch in 2012. It carries twin *grazing-incidence telescopes to image the sky in the high-energy range 5–80 keV. The images from the two telescopes are combined on the ground to provide greater sensitivity than would be possible with one telescope alone.

((⊕)) SEE WEB LINKS
• Official mission website.

nutation A periodic oscillation of the Earth's pole about its mean position on the celestial sphere, due to the attractions of the Sun and Moon on the Earth's equatorial bulge. It causes a small periodic variation in the positions of stars, and is superimposed on the much larger effect of precession. The main oscillation has an amplitude of about ±9″ and a period of 18.6 years, equal to the revolution period of the Moon's nodes around the ecliptic. In addition there are many smaller oscillations, some with periods as small as a few days. Nutation is conventionally resolved into **nutation in longitude** and **nutation in obliquity**, which are respectively parallel and perpendicular to the ecliptic.

nutation in right ascension An alternative name for *equation of the equinoxes.

Nysa family One of two subfamilies of asteroids at a mean distance of 2.42 AU from the Sun, the other being the *Hertha family*. They are close neighbours with a small but distinct difference in orbital inclination. Both consist of a single large asteroid (namely (44) Nysa, diameter 41 km, orbital inclination 3°.7, and (135) Hertha, diameter 79 km, inclination 2°.3) plus many small objects 20 km or less in diameter. Nysa is of E class and Hertha M class, while the minor members of the Nysa subfamily are of the rare F class. Some family members are very close to the Kirkwood gap at the 3 : 1 resonance with Jupiter at 2.50 AU, and could be thrown into the gap to be subsequently ejected into Earth-crossing orbits. This family may therefore be an important source of meteorites. Nysa itself was discovered in 1857 by the German astronomer Hermann Mayer Salomon Goldschmidt (1802–66). Its orbit has a semimajor axis of 2.423 AU, period 3.77 years, perihelion 2.07 AU, aphelion 2.78 AU, and inclination 3°.7. Nysa is notable for its high albedo, over 0.5, the highest of any known asteroid.

OAO Abbr. for *Orbiting Astronomical Observatory.

OB association A group of dozens of young, massive, and hot stars of spectral types O and B found in the spiral arms of galaxies. OB associations can be tens or hundreds of light years across. The stars are moving apart at several kilometres per second, so associations disperse among the background stars within about 30 million years. Famous examples are the *Orion Association and the *Sco–Cen Association.

Oberon The second-largest satellite of Uranus, diameter 1523 km; also known as Uranus IV. It orbits at a distance of 583 520 km every 13.46 days, keeping one face turned permanently towards the planet. Oberon was discovered in 1787 by F. W. *Herschel, soon after the discovery of Uranus itself. The Voyager 2 flyby of 1986 showed Oberon to have a surface densely covered with impact craters, many of which have dark material in their floors, suggesting that underlying material has been uncovered during or after the craters' formation. The largest craters, Hamlet and Macbeth, are over 200 km wide.

object glass (o.g.) The main lens of a refracting telescope, which brings light to a focus. *See also* LENS.

objective The main lens or mirror of a telescope, which brings light to a focus. The term is usually taken to mean the main lens of a refractor.

objective grating A coarse grating or series of parallel rods placed over the front lens (the objective) of a telescope to create fringes alongside star images. These fringes are short spectra, but if the spacing of the grating is very coarse (about 10 mm) they will effectively be so short that they are subsidiary star images at a known distance from a star. The grating can be rotated, enabling the position angle and separation of double stars to be measured. A fine grating may be used to create spectra, in the same manner as an *objective prism.

objective prism A wedge of glass placed over the aperture of a telescope, which disperses the light from each star in the field of view into a short spectrum. Low-resolution spectra of all the stars in the field of view can therefore be obtained in one exposure. In many cases, an objective prism is really a *grism.

oblateness The flattening of a planet or star at the poles, caused by its rotation; also known as *ellipticity* or *polar flattening*. The oblateness of a body is found by subtracting its polar diameter from its equatorial diameter, then dividing by the equatorial diameter.

oblate spheroid *See* SPHEROID.

oblique rotator A star whose magnetic axis does not coincide with its rotational axis, but is inclined at an angle to it. The magnetic field strength therefore appears to vary as the star rotates. These fluctuations may be accompanied by small brightness variations, as in *Alpha^2 Canum Venaticorum stars. The *Blazhko effect in certain RR Lyrae stars is thought to arise because they are oblique rotators.

obliquity of the ecliptic (ε) The angle between the Earth's equator and the ecliptic. It is the same as the Earth's axial tilt. The mean obliquity, corresponding to the mean equator, is currently just over 23° 26′, but is decreasing slowly at a rate of 47.5″ per century because of planetary perturbations of the Earth's orbit. The obliquity oscillates between limits of about 22° and 24°.6 with a mean period of some 41 000 years. In the current cycle it was at a maximum of 24°.2 some 9500 years ago and will reach a minimum of 22°.6 in another 10200 years before starting to increase again. The true obliquity at any epoch is the sum of the mean obliquity and the *nutation in obliquity* (*see* NUTATION).

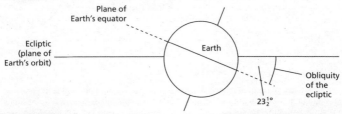

obliquity of the ecliptic

Observatoire du mont Mégantic An observatory at an altitude of 1111 m on Mont Mégantic, about 250 km east of Montreal, Canada. It contains a 1.6-m telescope opened in 1978 and is operated jointly by the Université de Montréal and Université Laval.

( ) SEE WEB LINKS
• Official observatory website.

Observatorio del Teide (OT) *See* TEIDE OBSERVATORY.

observatory A place from which astronomical observations are made; sometimes the name is also used for an administration building completely separate from the observing site. An observatory contains instruments for detecting electromagnetic radiation from celestial objects at various wavelengths. Before the invention of the telescope, observatories such as those of *Ulugh Beg and Tycho *Brahe had only simple equipment for naked-eye sightings of positions. The development of optical telescopes in the 17th century led to the need for a permanent protective building, usually with a dome that could be opened to view the sky. To ensure the best possible observing conditions, astronomers began in the late 19th century to site their telescopes on high mountains, above the densest parts of the atmosphere. This trend has continued by placing telescopes in orbit, where they can also observe wavelengths such as X-rays, ultraviolet, and infrared that do not penetrate the atmosphere. Radio telescopes, developed after World War II, are not usually housed in observatory buildings, because they are too large and also because they do not need as much protection from the elements as optical telescopes.

occultation The passage of one astronomical body in front of another, usually the obscuration of a star by the Moon; strictly speaking, a solar eclipse, when the Moon passes in front of the Sun, is a special form of occultation. Precise timings of lunar occultations of stars help to refine our knowledge of the Moon's orbit. Stars can be occulted by asteroids or planetary satellites, which can provide improved diameters for the occulting bodies. Perhaps the best-observed such event was the occultation of the star 28 Sagittarii by Saturn's satellite Titan in 1989. The Galilean satellites of Jupiter are regularly occulted by the planet. Jupiter's satellites may occult one another when Earth lies close to their orbital plane. *See also* GRAZING OCCULTATION.

occulting bar An obstruction placed in the field of view of an eyepiece to hide or occult bright objects so that fainter objects may be seen more easily.

occulting disk A small disk placed centrally in the field of view of an eyepiece, or at the focal point of a telescope, to hide or occult a bright object so that fainter objects can be seen more readily. An occulting disk may be used, for example, to reveal faint satellites of a planet, or to hide the Sun's disk in a coronagraph.

oceanus A very large lowland plain on the surface of a Solar System. The only named example is Oceanus Procellarum on the Moon.

OC star A supergiant of spectral type O in which the absorption lines of carbon are unusually strong relative to the nitrogen lines, hence the suffix 'C'. They are probably stars in which nitrogen produced by the *carbon–nitrogen cycle of nuclear burning in the star's core has not reached the surface in any great quantities, as is the case with most type O supergiants. Two factors, age and speed of rotation, are thought to be behind the difference in composition between OC and O stars. The OC stars are younger than normal O stars, so there has been less time for nitrogen to reach the surface; and they are slower rotators, so that material from their interiors is not brought to the surface as quickly as in faster rotators.

octahedrite The most common type of iron meteorite, containing 6–12% nickel. The meteorite that formed *Meteor Crater was an octahedrite. In addition to having a greater nickel content than *hexahedrites, octahedrites have an additional mineral, taenite, which is nickel-rich. Octahedrite meteorites have a structure which consists of plates of the mineral kamacite orientated parallel to the sides of an octahedron (hence their name), with the intervening spaces filled with taenite. This arrangement of the two minerals forms a *Widmanstätten pattern. Octahedrites are classified according to the thickness of the kamacite plates (the *band width*). *Coarse octahedrites* have band widths of 1.5–3 mm; *medium octahedrites* have band widths of 0.5–1.5 mm; and *fine octahedrites* have band widths of 0.2–0.5 mm. Below a band width of 0.2 mm, three different types of octahedral structure are apparent. The *finest octahedrites* have band widths of less than 0.2 mm. The *ataxites have spindles of kamacite rimmed with taenite which may intersect, but do not overlap. The *plessitic octahedrites* are transitional between finest octahedrites and ataxites.

Octans (Oct) (*gen.* Octantis) The constellation containing the south celestial pole, representing an octant (a sighting instrument similar to a sextant). The closest naked-eye star to the pole is *Sigma Octantis. The brightest star in the constellation is Nu Octantis, of magnitude 3.7.

ocular Another term for an *eyepiece.

Oe star A main-sequence star of spectral type O but with emission lines of hydrogen (hence the suffix 'e'). They rotate rapidly, and are a hotter version of *Be stars.

off-axis guider A device for following a guide star which is just outside the normal field of view of a telescope. The telescope is positioned so that the guide star is over a small mirror or prism, in the light path close to the *focal plane. The mirror or prism reflects the guide star's light to an eyepiece with an illuminated *reticle. The advantages of an off-axis guider over a guide telescope are that it eliminates the risk of flexure between the guidescope and the main instrument, it is lighter, and fainter guide stars can be used because of the usually larger aperture of the main instrument. However, the object being used for guiding cannot be photographed at the same time.

Of star A star of spectral type O which shows emission lines due to doubly ionized nitrogen (N III) at 464 nm and singly ionized helium (He II) at 468.6 nm, in addition to the normal absorption-line spectrum. Bright examples include 2nd-mag. Zeta Puppis and 5th-mag. Lambda Cephei.

o.g. Abbr. for *object glass.

OH–IR source A highly evolved star obscured by a thick cocoon of gas and dust which it has ejected. Most of the energy from the star is absorbed by the dust and reradiated at infrared (IR) wavelengths. The gas is rich in the hydroxyl (OH) radical, which is excited by infrared radiation to produce a powerful *maser. Other circumstellar masers of water (H_2O) and silicon monoxide (SiO) are also excited.

OH line An emission or absorption line caused by the *hydroxyl (OH) molecule. The most important are a group of four lines near 1.7 GHz (at a wavelength of about 18 cm).

OH maser A maser source in which the hydroxyl (OH) molecule is excited to maser action.
There are four OH spectral lines at frequencies of 1612, 1665, 1667, and 1720 MHz (a wavelength
of about 18 cm). Other OH spectral lines at higher frequencies are also sometimes excited as
masers. OH masers are found in star-forming regions, circumstellar envelopes, supernova
remnants, and comets, and in active galactic nuclei in the form of *megamasers.

OHP Abbr. (in French) for *Haute-Provence Observatory.

Olbers, Heinrich Wilhelm Matthäus (1758–1840) German physician and amateur
astronomer. The method he developed to calculate the orbit of a comet he discovered in 1796
(now designated C/1796 F1) became standard in the 19th century. He later suggested that
comets' tails are somehow expelled from their heads by the Sun, thereby anticipating the
discovery of *radiation pressure. In 1802, a year to the day after its discovery, Olbers
recovered the asteroid *Ceres in the position predicted by C. F. *Gauss. Olbers subsequently
discovered *Pallas (1802) and *Vesta (1807). In 1823 he first pointed out what has come to
be called *Olbers' paradox.

Olbers' paradox The apparent contradiction between the simple observation that the
night sky is dark and the theoretical expectation that an infinite, static Universe, filled more or
less uniformly with stars and galaxies, should be as bright as the surface of a star. The first
correct discussion of this paradox was published in 1744 by the Swiss astronomer (Jean)
Philippe Loys de Chéseaux (1718–51); H. W. M. *Olbers published his discussion of it in 1826.
The paradox can be resolved by identifying its incorrect assumptions. Most important of
these is that, as shown by the Big Bang theory, the Universe is not infinite, having come into
being around 15 billion years ago. An additional, less important, effect is that the expansion
of the Universe weakens the light from distant galaxies, but this on its own cannot fully
explain the paradox.

Old Style date (OS) The date of a historical or astronomical event on the *Julian
calendar, as opposed to the modern Gregorian calendar. Thus the Russian revolution of 1917,
whose anniversary is November 7 on the Gregorian calendar, was referred to in Russia as
the October Revolution, since Russia still used the Julian calendar at the time and its
Old Style date was 1917 October 25.

Oljato Asteroid 2201, a member of the *Apollo group, discovered by the American
astronomer Henry Lee Giclas (1910–2007) in 1947, then lost until 1979. Its diameter is 1.8 km.
Oljato's highly elliptical orbit ($e = 0.7$) is chaotic, its semimajor axis varying violently due to
its many close approaches to Earth and Venus. Its current semimajor axis is 2.176 AU, period
3.21 years, perihelion 0.62 AU, aphelion 3.73 AU, and inclination 2°.5. Oljato has a unique
reflectance spectrum that does not resemble that of any other known asteroid, meteorite, or
comet; it could be a dormant or extinct cometary nucleus. Oljato is associated with several
meteoroid streams.

Olympus Mons The largest volcano in the Solar System, lying on Mars at +19° lat., 134° W
long. From Earth, it is visible as a white spot previously known as Nix Olympica. The summit of
Olympus Mons is 21 km above the mean surface level of Mars, and contains a caldera 80 km
across. From the caldera, several lava flows have issued down to a scarp about 4 km high which
surrounds the volcano. The diameter of Olympus Mons at the scarp is about 650 km. Around
the scarp are many ancient landslips, themselves the largest known in the Solar System, some of
which have produced an apron of debris extending out another 600 km or more. The whole
complex of volcano and scarps is in places 1600 km across. Olympus Mons is thought to be
comparatively young, and was probably active within the last billion years.

O magnitude The magnitude of a star measured with a filter with wavelength 11.5 μm
and a bandwidth of 2 μm. This infrared magnitude is now little used because it has a
wavelength close to the N filter of the *Johnson photometry system, which admits more light.

Omega Centauri The brightest globular cluster in the sky, of apparent magnitude 3.7
and width 0°.6, in Centaurus. It is also known as NGC 5139. Omega Centauri is markedly
elliptical in shape, with a short axis only 80% the length of its long axis. Its actual diameter

is about 180 l.y., and it is the most luminous globular known, with an absolute magnitude of −10.3. At a distance of 17 000 l.y., it is one of the closest globulars to us. ◙

Omega Nebula The diffuse nebula M17 in Sagittarius, also known as NGC 6618; alternative names are the Horseshoe Nebula and the Swan Nebula. The Omega Nebula is ¾° across at its widest, and is variously described as being shaped like a Greek letter omega, a horseshoe, a swan, or the figure 2. It lies about 5000 l.y. away.

Onsala Space Observatory A radio observatory at Onsala in Sweden, 45 km south of Gothenburg, founded in 1955. It is operated by Chalmers University of Technology, Onsala. Its main instruments are a 25.6-m antenna, opened in 1964, and a 20-m millimetre-wave telescope in a radome (protective dome), opened in 1976 and upgraded in 1992. Onsala Space Observatory is a partner in the *Atacama Pathfinder Experiment (APEX).

((())) SEE WEB LINKS
• Official observatory website.

ON star A star of spectral type O in which the absorption lines of nitrogen are unusually strong relative to those of carbon, hence the suffix 'N'. This enhanced abundance is probably due to material that has undergone nuclear processing by the *carbon–nitrogen cycle in the star's core; the nitrogen-rich material is thought to have been brought to the surface by circulation currents induced by rotation in the star's interior.

Oort, Jan Hendrik (1900–1992) Dutch astronomer. He studied with J. C. *Kapteyn, whose work on star streams he continued, finding an overall net motion of the Sun with respect to other stars. In 1927 he showed that *high-velocity stars appeared to be rotating about the galactic centre, and went on to estimate the Sun's distance from the centre and also the Galaxy's diameter and mass. He also suggested that the Galaxy had *missing mass. In the 1950s he and others (including B. J. *Bok and H. van de *Hulst) used the 21-cm line of interstellar hydrogen to map the Galaxy, revealing its spiral structure. In 1950 he suggested that comets originate in a region now called the *Oort Cloud.

Oort Cloud A roughly spherical halo of comet nuclei surrounding the Sun out to perhaps 100 000 AU (over one-third of the distance to the nearest star). Its existence was proposed in 1950 by J. H. *Oort to account for the fact that new comets approach the Sun on highly elliptical orbits at all inclinations. The Oort Cloud remains a theoretical concept, since we cannot currently detect inert comets at such great distances. The cloud is estimated to contain some 10^{12} comets remaining from the formation of the Solar System. The most distant members are fairly loosely bound by the Sun's gravity. There may be a greater concentration of comets relatively close to the ecliptic, at 10 000–20 000 AU from the Sun, extending inwards to join the *Kuiper Belt. Oort Cloud comets are affected by the gravitational influence of passing stars, occasionally being perturbed into orbits which take them through the inner Solar System. ◙

Oort's constants Two parameters defined by J. H. *Oort to describe the major features of our Galaxy's *differential rotation in the Sun's neighbourhood. They are usually expressed in units of kilometres per second per kiloparsec. The two parameters are given the symbols A and B. Values derived from Hipparcos proper motions are A = 14.82 km/s/kpc and B = −12.37 km/s/kpc. Subtracting B from A gives the angular velocity of the *local standard of rest around the centre of the Galaxy, which corresponds to a period of about 200 million years.

Oosterhoff group A category in a classification of globular clusters, in which they are divided into Group I or Group II according to their abundances of heavy metals. Group I clusters have higher metal abundances, somewhat brighter *horizontal branch stars, and slightly shorter periods for their RR Lyrae variables (0.55 days compared with 0.65 days) than Group II clusters. Omega Centauri is an Oosterhoff Group II cluster, the iron abundance of its stars being only a few per cent of the Sun's value. They are named after the Dutch astronomer Pieter Theodorus Oosterhoff (1904–78).

Ootacamund Radio Astronomy Centre A radio astronomy observatory in Tamil Nadu state, southern India, founded in 1968, owned and operated by India's National Centre for Radio Astrophysics; its name is popularly abbreviated to Ooty. The main instrument is the Ooty Radio Telescope, a parabolic cylinder 530 m long and 30 m wide, opened in 1970. Uniquely, its long axis is aligned north–south on a hill with a slope equal to the latitude of the station (about 11° north). Radio sources are tracked by rotating the cylinder along its long axis.

(((●))) SEE WEB LINKS
Official telescope website.

opacity The property of a medium which determines how it absorbs or scatters electromagnetic radiation. It depends on the composition of the medium, its temperature, its density, and—unless an averaged value is being taken, as in the *Rosseland mean opacity—on the wavelength of the radiation. In astronomical contexts, in a gas at a temperature below a few thousand degrees, molecules, grains, and ices suspended in the gas may all be significant causes of opacity. Above about 8000 K, the opacity rises very steeply with temperature as the gas is ionized, then above 10^4-10^5 K it declines slowly (*see* KRAMERS OPACITY), reaching a plateau at around 10^6 K as *electron scattering opacity becomes dominant. Calculations of opacity are important in determining stellar structure.

open cluster A group of stars formed together in the spiral arms of a galaxy; sometimes called a *galactic cluster*. Open clusters are usually irregular in shape and contain anything from a few dozen to several hundred relatively young stars in a volume up to 50 l.y. across. The Hyades and the Pleiades are well-known examples. Open clusters are divided into various types according to the *Trumpler classification. They are more loosely packed than globular clusters, but may still have a central star density of up to 10 000 times that in the Sun's neighbourhood. Over 2000 are known, all in the galactic disk. Their ages range from a few million to several billion years, and the youngest are still surrounded by traces of the nebula from which they formed. The less dense of them are gradually disrupted by gravitational interaction with the rest of the Galaxy.

(((●))) SEE WEB LINKS
• Database of open clusters.

open universe A universe which expands for ever and has an infinite lifetime. A *Friedmann universe with a density less than the *critical density is an example. Recent observations show both that the density of our Universe is well below the critical density and that there is a *cosmological constant. It therefore now seems very likely that we live in this kind of universe. *See also* COSMOLOGY.

Ophelia The second-closest satellite of Uranus, distance 53 760 km, orbital period 0.376 days; also known as Uranus VII. Its mean diameter is 30 km, and it was discovered in 1986 with the Voyager 2 spacecraft. Ophelia and *Cordelia are *shepherd moons for the Epsilon ring of Uranus.

Ophiuchid meteors A weak meteor shower with ill-defined activity limits during May and June. The maximum ZHR is only around 5, with possible peaks around June 9 from a radiant at RA 17h 56m, dec. −23 °, and June 19 from RA 17h 20m, dec. −20°. The shower is not recognized by some authorities, and the meteors may simply be due to the general activity from the *anthelion radiant.

Ophiuchus (Oph) (*gen.* **Ophiuchi**) A large constellation on the celestial equator, representing a man holding a serpent (the constellation *Serpens). It contains the second-closest star to the Sun, *Barnard's Star. Although Ophiuchus is not officially a constellation of the zodiac, the Sun nevertheless passes through it in the first half of December. Its brightest star is *Rasalhague (Alpha Ophiuchi). Rho Ophiuchi is a quadruple star of magnitudes 5.0, 5.7, 6.7, and 7.3, surrounded by the faint *Rho Ophiuchi Nebula. 70 Ophiuchi is a double star with an orbital period of 88 years, magnitudes 4.2 and 6.0, colours yellow and orange, 16 l.y. away. RS Ophiuchi is a recurrent nova that has flared up to naked-eye brightness six times, in 1898, 1933, 1958,

1967, 1985, and 2006. NGC 6633 and IC 4665 are large open clusters. Ophiuchus contains numerous globular clusters, including M9, M10, M12, M14, M19, M62, and M107.

Öpik, Ernst Julius (1893–1985) Estonian astronomer. He established the process by which meteoroids ablate on entry into the Earth's atmosphere. From a study of the orbits and perturbations of comets, he predicted in 1932 that the Solar System's comets reside in a cloud extending out to a radius of 60 000 AU, an idea later revived by J. H. *Oort. Öpik pioneered the measurement of meteoroid size distributions, and suggested that *Apollo asteroids are 'burnt-out' comets.

Oppenheimer–Volkoff limit The maximum mass that a neutron star can have without it being overwhelmed by its own gravity. Calculations put this between 1.6 and 2 solar masses, although the exact figure is uncertain. A neutron star with a mass greater than this is expected to collapse further into a black hole. The limit is named after the American physicist (Julius) Robert Oppenheimer (1904–67) and the Russian-born Canadian George Michael Volkoff (1914–2000).

Opportunity One of two identical NASA *Mars Exploration Rovers. Opportunity was launched in 2003 July and landed in 2004 January on Meridiani Planum near the Martian equator. Its instruments confirmed that liquid water had been present at that location in the past. 📷

opposition The occasion when a body in the Solar System lies opposite the Sun in the sky, and hence has a *celestial longitude of 180°. At opposition, a body is highest in the sky at midnight and is visible all night. It is the best time for observing the outer planets, since they are then at their closest to Earth. The inner planets, Mercury and Venus, cannot come to opposition.

opposition effect An increase in the brightness of a rough surface when viewed with the source of illumination almost directly behind the observer. It arises partly because sunlight is reflected directly back out of the tiny pores in the surface. A second process, known as *coherent backscattering*, can contribute to the effect, as a result of multiple scattering from an irregular surface. The brilliance of the full Moon or of Mars at opposition is due to this effect, which is sometimes also known as the *dry heiligenschein*.

optical axis The line passing through an optical system, around which all image-forming properties are symmetrical. Usually, the optical axis is at right angles to principal optical components (lenses or mirrors) and passes through their centres.

optical depth (τ) A measure of how far light will travel through a partially transparent medium, such as the atmosphere of a star or a planet, before it is absorbed or scattered. A completely transparent medium has an optical depth of zero. A medium with a low optical depth is described as **optically thin**, whereas one with a high optical depth is **optically thick**. Absorption and scattering vary with wavelength (λ), so optical depth is usually specified for a particular wavelength and written as τ_λ.

optical double Two stars that happen to lie close to the same line of sight, and thus appear near each other on the sky. The components of an optical double are at different distances and are not related. An optical double is therefore distinct from a physical double, whose components are bound by gravity. Optical doubles are rarer than true binaries.

optical flat A piece of glass with one surface that is flat to a high accuracy, used for testing the flatness of other plane mirrors.

optical glass Glass manufactured so as to have a uniform *refractive index throughout, and *annealed slowly so as to be suitable for lenses.

optical interferometer An instrument that combines two separate light beams from the same object to form an interference pattern. Examples include the *stellar interferometer, for measuring the diameters of stars. *See also* FABRY–PEROT INTERFEROMETER; SPECKLE INTERFEROMETRY.

optical libration An apparent wobble in the movement of a celestial body caused by the changing geometry between it and the body from which it is observed; also known as *geometric libration*. The Moon's librations are optical librations. *See* LIBRATION.

optical pathlength The length of the optical path through a refracting medium, such as the filter in a photometer. This differs from the geometrical distance travelled because the light is refracted on entering and leaving the filter, so if a filter is placed in front of the focal plane of a telescope then the focus must be lengthened.

optical pulsar A pulsar that exhibits pulsations in the visible part of the spectrum, as well as at radio and other wavelengths. The first to have its optical pulsations discovered, in 1969, was the *Crab Pulsar, followed in 1977 by the *Vela Pulsar.

optical thickness *See* OPTICAL DEPTH.

optical wedge A strip of glass or other transparent material whose light transmission decreases evenly from one end to the other. The change in light transmission is accomplished by applying a coating whose thickness increases along the wedge, or by an increase in photographic density. The wedge is used at the eyepiece to reduce the intensity of a star in the field of view until it appears the same brightness as another, fainter, comparison star. The magnitude difference between the stars can then be measured.

optics Any optical components, such as the objective lens or mirror, secondary mirrors, and eyepieces of a telescope.

orbit The path of a body around another in space. Planets moving around the Sun, and large satellites moving around planets, follow orbits that are approximately ellipses, governed by *Kepler's laws. Other possible orbital shapes are a parabola and a hyperbola. The size and shape of an orbit is defined by its elements (*see* ELEMENTS, ORBITAL). The changes in those elements with time due to perturbing forces such as the gravitational influence of other bodies can be predicted by *celestial mechanics.

orbital elements *See* ELEMENTS, ORBITAL.

orbital resonance See RESONANCE, ORBITAL.

orbital velocity The velocity an object has at a given point in its orbit. Unless the orbit is circular, the object's velocity continually changes in accordance with Kepler's second law (*see* KEPLER'S LAWS). The velocity of an object defines not only speed but direction. The *circular velocity of an object in a circular orbit is always perpendicular to the radius vector. An object moving faster than circular velocity will enter an elliptical orbit. *Escape velocity puts the object into a parabolic orbit so that it departs from the body it was orbiting, never to return (*see also* PARABOLIC VELOCITY). *Hyperbolic velocity is a velocity greater than escape velocity.

Orbiting Astronomical Observatory (OAO) A series of NASA astronomy satellites operating mainly in the ultraviolet. The first successful mission, OAO-2, was launched in 1968 December. It carried Celescope, a group of four 0.32-m telescopes, plus several other telescopes for ultraviolet observations. The last of the series, OAO-3, was launched in 1972 and was later named the *Copernicus satellite.

( ) SEE WEB LINKS
- NASA information page for OAO 2.
- NASA information page for OAO 3.

Orbiting Solar Observatory (OSO) A series of NASA satellites, which studied ultraviolet, X-ray, and gamma-ray emission from the Sun during an entire solar cycle. Eight were launched between 1962 and 1975. Among their discoveries were the first observations of gamma rays from solar flares, and bursts of hard X-rays and ultraviolet emission at the impulsive stage of flares.

ordinary chondrite The most common of the three main classes of chondrite meteorites, and the most numerous of the observed falls. Ordinary chondrites consist of 10–15% (by volume) of fine-grained matrix material, 65–75% of chondrules (embedded in the matrix), and less than 1% of inclusions. The presence of chondrules, with diameters of 0.3–0.9 mm, in the ordinary chondrites shows that they have not been melted since they formed. The ordinary chondrites are divided into three subgroups on the basis of their content of iron and related elements, such as nickel. The H or high-iron group contains 25–30% iron; the L or low-iron group contains 20–25%; and the LL or low-iron, low-metal group contains only 18–20%. A substantial fraction of all three groups are impact *breccias, resembling lunar breccias. The distinct chemistry of the three groups indicates that they originated from separate parent bodies. Each of the three subgroups may be further divided into a number of types on the basis of texture and mineralogy (e.g. H3, L6, and LL5).

organic molecule A molecule containing carbon atoms. Organic molecules are present in living materials, or have the potential to form the fundamental building blocks of living materials. Many such molecules have been detected in interstellar matter, in the atmospheres of the giant planets, in meteorites, in comets, on the surfaces of some asteroids and satellites, and now in brown dwarfs.

Orgueil meteorite A meteorite of carbonaceous chondrite type, which fell at Orgueil, near Montauban in southern France, on 1864 May 14. Twenty fragments totalling 11 kg were collected, and it was identified as of CI1 type. CI carbonaceous chondrites are the most chemically primitive of meteorites and are rare. The Orgueil meteorite was found to have a substantial content of organic compounds.

Orientale Basin A major impact basin on the Moon, 930 km across, centred at −19° lat., 93° W long.; the central part is known as Mare Orientale. It is the best-preserved and probably the youngest of the large lunar basins, formed 3.8–3.9 billion years ago. At least three concentric rings of mountains can be seen within it. Long chains of secondary craters, caused by ejecta from the impact, extend to distances of more than 1000 km. Volcanic lava and small shield volcanoes cover part of the basin's interior. 🞄

Orion (Ori) (*gen.* Orionis) A magnificent constellation on the celestial equator, representing a great hunter of Greek mythology. Orion is outlined by the prominent stars *Betelgeuse (Alpha Orionis), *Rigel (Beta Orionis, the constellation's brightest star), *Bellatrix (Gamma Orionis), and Saiph (Kappa Orionis). A line of three stars, *Alnilam (Epsilon Orionis), *Alnitak (Zeta Orionis), and *Mintaka (Delta Orionis), make up the Belt of Orion. The famous *Orion Nebula, M42, contains the multiple star known as the *Trapezium. On the southern edge of the Orion Nebula is the double star Iota Orionis, magnitudes 2.8 and 6.9. North of the Orion Nebula is another bright nebula, NGC 1977, and further north again is the 5th-magnitude open cluster NGC 1981. This complex of nebulosity and clusters forms the Sword of Orion, hanging from Orion's belt. Sigma Orionis is an impressive multiple star of magnitudes 3.8, 6.6, 6.7, and 8.8. Eta Orionis is a close double, magnitudes 3.8 and 4.8. The dark *Horsehead Nebula intrudes into a faint strip of nebulosity, IC 434, which runs south of Alnitak. The *Orionid meteors radiate from the constellation every October.

Orion Arm The local spiral arm of our Galaxy, also called the **Orion Spur** as it is a partial arm located between the larger *Sagittarius Arm and *Perseus Arm. The Sun lies near its inner edge, that is, towards the galactic centre. All the main naked-eye stars are part of the Orion arm, including those of Orion itself. The arm stretches towards Cygnus in one direction, and towards Puppis and Vela in the opposite direction. Features in the Orion Arm include the *Gum Nebula, the *Orion Nebula, the *North America Nebula, the *Cygnus Loop, and the *Great Rift.

Orion Association A large association of very young stars of spectral types O and B, centred on the *Orion Nebula, nearly 1500 l.y. away. It is about 400 l.y. across and contains all the main stars of Orion, except Betelgeuse. The area is still partly shrouded in gas from which stars continue to form.

Orionid meteors A meteor shower showing moderate activity between October 2 and November 7. The activity pattern is complex, with several peaks and troughs, presumably reflecting the filamentary nature of the meteor stream laid down over several perihelion returns by *Halley's Comet. A broad maximum, with ZHR reaching 30, is seen for a couple of days around October 21, from a radiant at RA 6h 20m, dec. +16° (between Betelgeuse and Gamma Geminorum). Orionid meteors are very swift (geocentric velocity 66km/s), and a high proportion leave persistent trains, but most are faint.

Orion Molecular Clouds A group of large clouds of gas and dust around 1500 l.y. away in the constellation Orion, the nearest examples of *giant molecular clouds. Each cloud is over 100 l.y. across and contains more than 100 000 solar masses of gas, mostly in the form of hydrogen molecules. The young stars of Orion formed from these clouds within the last 10 million years. New stars are still forming in dense *molecular cores* within the clouds, hidden from sight by the dust, but detectable as sources of infrared and molecular radiation. The best-known molecular cores are designated OMC-1 and OMC-2.

Orion Nebula A large and bright nebula, 1350 l.y. away in the constellation Orion; also known as M42 or NGC 1976. The nebula covers more than 1° × 1° of sky and is faintly visible to the naked eye as a fuzzy patch around the multiple star Theta^1 Orionis, also known as the *Trapezium. The ultraviolet light from the Trapezium stars, particularly Theta^1 C (spectral type O6), ionizes the nebula. The Orion Nebula is over 20 l.y. in diameter, and contains several hundred solar masses of ionized gas. It forms a hot blister on the near side of one of the massive *Orion Molecular Clouds. 📷

Orion Spur Alternative name for the *Orion Arm of our Galaxy, given when it was thought to be a branch of the *Perseus Arm.

Orion variable Any of the different types of eruptive variable associated with nebulosity, also known as a *nebular variable. The majority are *irregular variables of the various IN subtypes (including *T Tauri stars), but the *FU Orionis stars and the UVN subtype of *flare star are also included.

ORM Abbr. (in Spanish) for *Roque de los Muchachos Observatory.

orrery A working model of the Solar System which represents mechanically the movements of the planets. The name is from the fourth Earl of Cork and Orrery, Charles Boyle (1676–1731), for whom one of the first was made. Popular in the 18th century, they ranged from 'grand orreries' a metre in diameter which re-created the rotational and orbital motions of planets and satellites, down to small portable devices featuring just the Earth, Moon, and Sun. An orrery of the latter type is also known as a *tellurium* or *tellurian*.

orthoscopic eyepiece A design of eyepiece that usually consists of a cemented triplet *field lens and a planoconvex eye lens. This combination provides good correction for *chromatic aberration, *spherical aberration, coma (*see* COMA, OPTICAL), and *distortion, at reasonable cost. Orthoscopic eyepieces have a field of view of 35–50° and good eye relief. The design was introduced in 1880 by the German physicist Ernst Karl Abbe (1840–1905).

OS Abbr. for *Old Style date.

oscillating universe A modification of the Big Bang theory, in which an expanding universe eventually reverses and collapses back to a Big Crunch before starting to expand again, forming one cycle of an endless series. In such a model, each cycle has a finite duration but the universe itself can last for an infinite time. It is not known whether our Universe is of this type, or whether it will expand for ever. *See also* CLOSED UNIVERSE; OPEN UNIVERSE.

osculating elements The orbital elements at a given time of a body in an orbit that is perturbed by the gravitational pulls of other bodies, such as the planets. Osculating elements describe the orbit that the body would follow if those perturbations were to cease. However, the real orbit gradually changes because of perturbations, and so the osculating elements alter. Predictions based on osculating elements are accurate only for times close to the epoch of osculation. The way in which osculating elements change with time is a useful way of demonstrating the effects of perturbations on a body's orbit.

osculating orbit The orbit that one celestial body would pursue about another if no other celestial object or forces existed. At any given time the osculating orbit touches the real orbit at a point where the body would have the same position and velocity in both orbits. An osculating orbit would be a perfect ellipse; in practice, real orbits are not because of perturbations from the planets, so the real and osculating orbits gradually depart from each other with time.

OSO Abbr. for *Orbiting Solar Observatory.

O star A star of spectral type O, the brightest, hottest, and most massive of all normal, hydrogen-burning stars on the main sequence, appearing blue in colour and emitting most of its energy in the ultraviolet. Their temperatures range from 30 000 K to above 50 000 K, and they have luminosities from 100 000 to more than a million times the Sun's. With masses of 20–100 solar masses or more, O stars burn their nuclear fuel at a prodigious rate and have lifetimes of only 3–6 million years. Because of their high temperatures, hydrogen lines in their spectra are weak, the dominant lines being those of singly ionized helium (the *Pickering series). O stars form infrequently and have short lifetimes, so they are very rare; only a handful of O2, O3, and O4 stars are known (no stars are classified earlier than O2). The brightest naked-eye O stars are Delta and Zeta Orionis, both O9.5 supergiants. O stars are often found with B stars in *OB associations. *See also* OE STAR; OF STAR.

OT Abbr. (in Spanish) for *Teide Observatory.

outgassing **1**. The loss of gas from within a planet through processes such as volcanism. Outgassing is thought to have been responsible for much of the atmospheres of Earth and the other terrestrial planets.
   2. In a more general sense, the loss of gas from any solid, such as the outgassing and recondensation of water vapour and other light molecules from spacecraft structures. This form of outgassing can have serious effects on spacecraft mechanisms and telescope optics.

overcontact binary A close binary system in which each star has filled and expanded well beyond its *Roche lobe. The system therefore consists of two stellar cores surrounded by a common, dumbbell-shaped convective envelope. Such a system is often an eclipsing binary of the *W Ursae Majoris type. *See also* COMMON ENVELOPE BINARY.

oversampling The circumstance in which a detector has more resolution elements (e.g. *pixels*) than are required to sample the finest details of the object or spectrum under study. For example, if an image is taken under seeing conditions of 1″, but the CCD detector has a scale of 0″.1 per pixel, then the image is said to be **oversampled**.

overshooting The phenomenon in which a region of a star in which convection currents would not normally occur is penetrated by convection currents 'overshooting' from underlying or overlying regions. Such overshooting may be important in transporting heavy elements, synthesized within the star, into regions they would not otherwise reach. Evidence of convective overshoot in the Sun is found in the granulated appearance of its surface.

overtone mode A more complex vibration in pulsating stars than the *fundamental mode. For example, in the *first overtone* mode of radial pulsation, a shell of material within the star periodically swells and contracts. The *second overtone* arises when two concentric shells are expanding and contracting. Most Cepheid variables pulsate in the fundamental mode but some also pulsate in the first overtone as well; these are termed *beat Cepheids. Some rare cases, such as *Polaris, pulsate only in the first overtone. Cepheids which are thought to pulsate only in the second overtone have been found in the Small Magellanic Cloud. *See also* PULSATION MODE.

OVV quasar Abbr. for optically violently variable quasar. *See* BLAZAR.

Owens Valley Radio Observatory (OVRO) The radio astronomy observatory of the California Institute of Technology (Caltech), located at an altitude of 1222 m at Big Pine, California, and founded in 1956. It has two 27.4-m dishes opened in 1959, later joined by five 2-m dishes to form the Owens Valley Solar Array, operated since 1997 by the

New Jersey Institute of Technology, and a 39.6-m dish opened in 1968. It also hosts a 25-m dish of the National Radio Astronomy Observatory's *Very Long Baseline Array. The six 10.4-m dishes of the former OVRO Millimeter Array were moved in 2004 to a new site to form part of the *Combined Array for Research in Millimeter-wave Astronomy (CARMA).

((∰)) SEE WEB LINKS

• Official observatory website.

Owl Nebula The 11th-magnitude planetary nebula M97 in Ursa Major, also known as NGC 3587. It gets its name because in large apertures it shows two large dark patches like the eyes of an owl. It lies 1300 l.y. away.

oxygen burning A set of nuclear reactions inside stars through which oxygen is converted to heavier elements such as silicon. It occurs in stellar cores (and sometimes later, when oxygen in the core is exhausted, in a surrounding shell) at temperatures over 2×10^9 K.

ozonosphere A level of the *stratosphere at an altitude of 10–50 km where substantial amounts of ozone (O_3) are found. The highest concentrations occur between 20 and 25 km in the **ozone layer**. Stratospheric ozone is generated by the action of sunlight, which splits oxygen molecules into single atoms. Atomic oxygen (O) can combine with molecular oxygen (O_2) to produce the highly reactive triatomic form (O_3). The ozone layer is an important barrier to ultraviolet light from the Sun at wavelengths of 230–320 nm. The term is not in regular use by meteorologists.

PA Abbr. for *position angle.

PAH Abbr. for *polycyclic aromatic hydrocarbon.

pair annihilation The mutual destruction of a particle and its antiparticle. The energy reappears as photons or other particle–antiparticle pairs. For example, an electron and positron combine to produce two 511 keV gamma-ray photons. Annihilation of a proton and antiproton forms pions.

pair production The production of a particle and its antiparticle. This can happen in several ways. For example, an electron–positron pair can be produced from a gamma-ray photon of energy greater than 1.022 MeV in the electric field surrounding an atomic nucleus, or the pair may be produced by the collision of two photons whose total energy exceeds 1.022 MeV. The energies of photons in the early Big Bang were high enough to produce many varieties of pairs of particles. Protons and antiprotons could be produced during the first 2×10^{-6} s after the Big Bang when the temperature exceeded 10^{13} K, and positive and negative muon pairs until about 2×10^{-4} s. Electrons and positrons could be produced until the temperature fell below 10^{10}K, about 2s after the Big Bang. Pair production is also important in secondary cosmic-ray showers.

palimpsest An impact crater whose topography has been smoothed by glacier-like flow of the surface. The term was coined for features on the icy surfaces of Ganymede and Callisto. Palimpsests there consist of bright circular features, sometimes at the centre of concentric rings. 📷

Pallas Asteroid 2, the second asteroid to be discovered, by H. W. M. *Olbers in 1802. Its shape is a triaxial ellipsoid, $570 \times 525 \times 482$ km, giving a mean diameter similar to that of Vesta; its rotation period is 7.81 h. Its mass is 3.2×10^{20}kg, also similar to that of Vesta, and its density is 4.2. Pallas is of B class, with a spectrum similar to that of the carbonaceous chondrite meteorites. Its mean magnitude at opposition is 8.0; only Vesta and Ceres can become brighter. Its orbit has a semimajor axis of 2.772 AU, period 4.62 years, perihelion 2.13 AU, aphelion 3.41 AU, and an unusually high inclination of 34°.8.

pallasite A class of stony-iron meteorite named after the German naturalist Peter Simon Pallas (1741–1811), who found a 680-kg specimen near Krasnojarsk, Russia, in 1772. In pallasites, cavities in the nickel–iron metal are filled with crystals of the hard, glassy silicate mineral olivine; the proportions of metal and silicate are about the same.

Palomar Observatory An observatory at an altitude of 1706m, 80km northeast of San Diego, California, owned and operated by the California Institute of Technology (Caltech), Pasadena. It was founded in 1934 but completion was delayed by World War II. Its main instrument is the 200-inch (5-m) Hale Telescope opened in 1948, named after the observatory's founder, G. E. *Hale. Also in operation since 1948 has been the 1.2-m Oschin Schmidt Telescope, named after Samuel Oschin, a benefactor of the observatory, used for the *Palomar Observatory Sky Survey (POSS); a new corrector plate was installed in 1985. Also on the mountain are a 1.5-m reflector jointly operated with the Carnegie Institution of Washington, opened in 1970, and a 0.46-m Schmidt, opened in 1936.

(⊕) SEE WEB LINKS
• Official observatory website.

Palomar Observatory Sky Survey (POSS) A photographic survey of the sky from the north celestial pole to declination −33°, made using the 1.2-m Oschin Schmidt Telescope at *Palomar Observatory, California. The original survey, undertaken 1949–56, consists of 936 pairs of plates 0.355 m square, each pair consisting of one plate sensitive to blue light and one to red light. The field of view of each plate is 6°.5. It was financed by the National Geographic Society. A second survey, POSS II, was made in 1985–2000 from the north celestial pole to the equator, using modern photographic plates. It consists of 897 fields in blue, red, and near infrared light, with greater overlap between fields than the original survey. A digitized version, DPOSS, is now available.

(⊕) SEE WEB LINKS
• Information and data retrieval page.

palus A mottled area on a planetary surface; pl.*paludes*. The name, which means 'marsh' or 'swamp', was first used for features on the Moon in the 17th century. It is not a geological term, but is used in the nomenclature of individual features, for example Palus Putredinis on the Moon, or Oxia Palus on Mars.

Pan The innermost known satellite of Saturn, orbiting within the *Encke Gap in the A Ring at a distance of 133 580 km; also known as Saturn XVIII. Its orbital period is 0.575 days and its mean diameter about 28 km. Pan was discovered in 1990 on photographs taken nine years earlier by the Voyager 2 spacecraft.

Pandora The fourth-closest satellite of Saturn, distance 141 720 km, orbital period 0.629 days; also known as Saturn XVII. It was discovered in 1980 on images from Voyager 1, between Saturn's F and G Rings. Pandora is irregularly shaped, 104 × 81 × 64 km in size. Its lightly cratered surface appears to have a thin cover of dust, softening the outlines of craters and ridges. Pandora is the outer *shepherd moon of the F Ring.

Pan-STARRS A planned wide-field survey telescope with four 1.8-m mirrors to detect and track *potentially hazardous asteroids; the name is short for Panoramic Survey Telescope & Rapid Response System. A prototype telescope, PS1, which has a single 1.8-m mirror, became operational on Haleakala, Hawaii, in 2010. The telescope has a 3° field of view and takes exposures of 30 to 60 seconds in four of the wavelength bands used in *SDSS photometry plus a near-infrared band. The final four-mirror telescope, PS4, which may be sited at *Mauna Kea Observatories, will scan all the sky visible from Hawaii three times per month. In addition to near-Earth objects, Pan-STARRS will locate many other asteroids, comets, and Kuiper Belt objects, as well as building up a massive database of stars and galaxies. The Pan-STARRS telescopes are owned by the University of Hawaii and operated by them in conjunction with a consortium of institutions from the US, Germany, the UK, and Taiwan.

(⊕) SEE WEB LINKS
• Official telescope website.

parabola A type of curve whose 'arms' become parallel as they approach infinity, so that the curve never quite closes in on itself, defined mathematically as a *conic section with an eccentricity of 1. It may be regarded as an ellipse in which the two foci are infinitely far apart, and is the limiting case between an ellipse and a hyperbola. Some comets have elliptical orbits that are so extended they are indistinguishable from parabolas.

parabolic antenna A bowl-shaped aerial with a parabolic cross-section that reflects incoming radio waves to a focal point above the bowl; also known as a *dish antenna*. At the focal point is a collector known as a *feed*. High-performance parabolic antennas are often of the Cassegrain design, in which a secondary reflector directs radio waves to a focal point low down inside the bowl. The principle of a parabolic antenna is identical to that of the primary mirror in an optical telescope.

parabolic comet A comet whose orbit around the Sun has an eccentricity of exactly 1.0. Many cometary orbits are initially calculated on the basis that they are parabolic, since the observed arc close to perihelion is only a small part of the true ellipse, and a parabola represents the best first approximation.

parabolic orbit An orbit having the shape of a parabola. A body pursuing a parabolic orbit about another makes only one approach to the other body, in theory coming from infinity and then receding to infinity.

parabolic velocity The speed of an object that is moving at exactly the *escape velocity of another body. The relation between parabolic velocity and circular velocity at any point in the orbit is that parabolic velocity equals $\sqrt{2} \times$ circular velocity.

paraboloid A surface swept out by a parabola rotated 360° about its axis. The main mirrors in large telescopes are usually paraboloidal in cross-section, since such mirrors are free from *spherical aberration.

parallactic angle The angle at a point on the celestial sphere between the great circles passing through the point and the celestial pole, and the point and the zenith.

parallactic ellipse The path on the sky of the apparent position of a star as seen from the Earth, due to the Earth's annual motion around the Sun. The shape of the parallactic ellipse ranges from a circle for a star at the ecliptic pole to a line for a star on the ecliptic. Its size depends on the distance of the star, becoming smaller with increasing distance.

parallactic inequality A variation in the Moon's orbital motion caused by the Sun's gravitational attraction. As a result of the Sun's attraction, the Moon is about 2′ ahead of its expected orbital position at first quarter, and a similar amount behind at last quarter.

parallactic motion The proper motion of a star due to the effect of the Sun's motion relative to the *local standard of rest.

parallax (π) The angular difference between an object's direction as seen from two points of observation, such as opposite sides of the Earth's orbit; also commonly known as *trigonometric parallax. It is thus a form of triangulation. Parallax can also be defined as the angular distance between two points as seen from a third point in space, such as the radius of the Earth's orbit as seen from a star. By extension, the term parallax can also be used to mean a star's distance, even if measured by some indirect method such as photometry (*photometric parallax) or spectroscopy (*spectroscopic parallax). *See also* ANNUAL PARALLAX; DYNAMICAL PARALLAX; GEOCENTRIC PARALLAX; HORIZONTAL PARALLAX; KINEMATIC PARALLAX; LUNAR PARALLAX; PARALLACTIC ELLIPSE; SECULAR PARALLAX; SOLAR PARALLAX; STATISTICAL PARALLAX.

parametric amplifier An amplifier used in radio astronomy for the frequency range of about 1–100 GHz. They are cooled to 15–20 K to reduce noise. Parametric amplifiers operate by transferring energy to the signal from a highly stabilized laboratory oscillator known as the *pump*.

Paranal Observatory An observing site of the *European Southern Observatory, founded in 1987, at an altitude of 2635 m on Cerro Paranal mountain in the Atacama desert of northern Chile, about 120 km south of Antofagasta. It is the site of the *Very Large Telescope (VLT), plus the associated 2.6-m VLT Survey Telescope (VST), which came into operation in 2011, and *Visible and Infrared Survey Telescope for Astronomy (VISTA).

((⊕)) SEE WEB LINKS
• Official observatory website.

paraxial Describing light rays which are both close to and parallel to the optical axis of an optical system.

parent isotope An isotope that undergoes radioactive decay, its nuclei disintegrating spontaneously to form a *daughter isotope (often of a different element). For example, rubidium-87 is the parent isotope of strontium-87, into which it decays with a half-life of 4.88×10^{10} years. The amount of the parent isotope in a sample decreases with time, while that of any daughter isotope (and its own decay products) increases; the relative proportions of the two can be used to derive the age of the sample (*see* RADIOACTIVE AGE DATING).

parfocal Having the same focus position. A set of parfocal eyepieces, for example, are manufactured so that they can be interchanged without refocusing the telescope.

parhelion An optical phenomenon resulting from the refraction and reflection of sunlight within ice crystals in cirrus or cirrostratus cloud; also known as a *mock Sun* or *sundog*. Parhelia appear 22° either side of the Sun, and at the same altitude above the horizon, but disappear when the Sun's altitude exceeds 60°. Red light is refracted to the edge of the parhelion closer to the Sun, blue to the farther edge. Parhelia are often associated with an atmospheric *halo.

Paris Observatory The national observatory of France, founded in 1667, and the first national observatory of any country. G. D. *Cassini was its first head. In 1926 it took over Meudon Observatory, 9 km to the south of Paris, founded in 1876 for solar studies. Paris Observatory also operates the *Nançay Radio Astronomy Observatory.

• Official observatory website.

Parkes Observatory A radio observatory near the town of Parkes in New South Wales, Australia, owned and operated by the Commonwealth Scientific and Industrial Research Organisation (CSIRO). It consists of a 64-m diameter radio dish opened in 1961. As well as radio astronomy, it is used for tracking space probes. Since 1988 it has been part of the *Australia Telescope National Facility.

• Official observatory website.

parsec (pc) A basic unit of stellar distance, corresponding to a *trigonometric parallax of one second of arc (1″). In other words, it is the distance at which one astronomical unit subtends an angle of one second of arc. One parsec equals 3.2616 light years, 206 265 astronomical units, or 30.857×10^{12} km. For distances on galactic and intergalactic scales, kiloparsec (kpc) and megaparsec (Mpc) are used. *See also* LIGHT YEAR.

parselene An optical phenomenon resulting from the refraction and reflection of moonlight within ice crystals in cirrus or cirrostratus cloud; also known as a *mock Moon*. It is the lunar equivalent of the *parhelion. Parselenae appear 22° from the Moon and at the same altitude as the Moon above the horizon. They occur only when the Moon is bright, typically within five days of full. Being fainter than parhelia, they are usually colourless to the naked eye.

Parsons, William *See* ROSSE, THIRD EARL OF.

partial eclipse A solar eclipse in which the Moon does not completely obscure the Sun's disk, which then appears as a crescent; or, a lunar eclipse in which the Moon is not entirely immersed in the Earth's umbra, so that some of the Moon remains sunlit.

particle *See* ELEMENTARY PARTICLE.

pascal (Pa) A unit of pressure equal to a pressure of one newton per square metre. One pascal equals 10^{-5} bar. Hence, the Earth's atmospheric pressure at sea level is approximately 10^5 pascal. It is named after the French mathematician and theologian Blaise Pascal (1623–62).

Paschen–Back effect The splitting of lines in a spectrum when the source is located in a strong magnetic field and the magnetic splitting exceeds the normal multiplet splitting. It is the strong-field form of the *Zeeman effect. The effect is named after the German physicists

(Louis Carl Heinrich) Friedrich Paschen (1865–1947) and Ernst Emil Alexander Back (1881–1959).

Paschen series A sequence of absorption or emission lines in the near-infrared part of the spectrum, due to hydrogen. They are caused by electron jumps between the third energy level and higher levels. The Paschen-α line has a wavelength of 1875 nm, and the spacing between the lines diminishes as they converge on the series limit at 820 nm. The series is named after the German physicist (Louis Carl Heinrich) Friedrich Paschen (1865–1947). *See also* HYDROGEN SPECTRUM.

Pasiphae A retrograde outer satellite of Jupiter, mean distance 23 658 000 km; also known as Jupiter VIII. It orbits the planet in 744.2 days at an inclination of 151°.4, and has an orbital eccentricity of 0.41, one of the greatest eccentricities of any Jovian moon. Pasiphae is 36 km in diameter and was discovered in 1908 by the English astronomer Philibert Jacques Melotte (1880–1961). It is the largest of the so-called Pasiphae retrograde irregular group of Jovian moons, the other members of which are Helike, Eurydome, Autonoe, Sponde, Megaclite, Sinope, Hegemone, Aoede, Callirrhoe, Cyllene, and Kore.

passband The range of wavelengths transmitted by a filter. It is often quoted as the wavelength limits within which the transmission exceeds 50% of the peak value.

past light cone *See* LIGHT CONE.

patera A crater on a planetary surface with irregular, fluted, or scalloped edges; pl.*paterae*. The name, which means 'shallow dish', is not a geological term, but is used in the nomenclature of individual features, for example Ulysses Patera on Mars, or Loki Patera on Io.

pathlength *See* OPTICAL PATHLENGTH.

Paul Wild Observatory A radio observatory at Culgoora, near the town of Narrabri in New South Wales, Australia. It is the site of the Compact Array of the *Australia Telescope National Facility, and consists of a line of six 22-m antennas that can operate as one instrument by *aperture synthesis. Five of these antennas can be moved along a 3-km stretch of rail track running east–west, with a 200-m north–south spur added in 1998. A sixth antenna lies 3 km to the west of the main group, on its own short track. The array began operation in 1988. The observatory is named after the English-born Australian radio astronomer (John) Paul Wild (1923–2008). Also at the observatory is the Sydney University Stellar Interferometer (SUSI), an optical interferometer with baselines up to 640 m; the Culgoora Solar Observatory; and a station of the *Birmingham Solar Oscillations Network (BiSON).

(⊕) SEE WEB LINKS
• Official observatory website.

Pavo (Pav) (*gen.* **Pavonis**) A constellation of the southern sky, representing a peacock. Its brightest star is called *Peacock (Alpha Pavonis). Kappa Pavonis is a bright Cepheid variable that ranges between magnitudes 3.9 and 4.8 with a period of 9.1 days. NGC 6752 is a 6th-magnitude globular cluster.

Payne-Gaposchkin, Cecilia Helena (1900–79) English astronomer who worked in America. She developed a temperature scale for the various spectral types of stars based on the strengths of their spectral lines. In 1925, in a study of the relative abundances of elements in stars of various ages, she established that hydrogen is the major constituent of stars. With the Russian-American astronomer Sergei Illarionovich Gaposchkin (1898–1984), whom she married in 1934, she began a huge programme of measuring the magnitudes of variable stars from photographic plates, which resulted in a catalogue of variables published in 1938. A similar joint study of variables in the Magellanic Clouds appeared in 1971.

pc Abbr. for *parsec.

P-class asteroid A class of asteroid, fairly common in the outer main asteroid belt, with a distribution that peaks at a mean distance from the Sun of 4.0 AU. Members of this class have a featureless reflectance spectrum, flat to slightly reddish over the wavelength range 0.3–1.1 μm. Their spectra are intermediate between classes C and D. P-class asteroids are distinguished from the spectrally identical E and M classes by their lower albedos (0.02–0.06). Members of this class include (87) Sylvia, diameter 261 km, and (153) Hilda, diameter 171 km.

P Cygni line profile A feature in a spectrum that indicates a strong outflow of matter from a star. It is named after P Cygni, a Be star whose spectrum shows strong hydrogen and helium emission lines with absorption lines on their blueward side. These dual characteristics are caused by an expanding shell or wind of material that is being blown off the star either by radiation pressure or rapid rotation. Where the expanding shell is directly between us and the star, it produces absorption at a velocity which is blueshifted relative to the star. The other parts of the shell produce the emission lines. Conversely, if material is falling on to the star, the absorption component can be redshifted relative to the emission lines, causing an *inverse P Cygni line profile*.

P Cygni star *See* S DORADUS STAR.

Peacock The star Alpha Pavonis. It is a B2 subgiant, magnitude 1.9, distance 179 l.y.

peculiar galaxy A galaxy that shows some exceptional features in addition to the type assigned to it in schemes such as the *Hubble classification (and is given the suffix p or pec), or a galaxy whose structure is so unusual that it defies classification. M32, the small elliptical companion to the Andromeda Galaxy, is classified as peculiar because it appears to have had its outer regions removed in gravitational encounters with its massive neighbour. M82 in Ursa Major appears to be an edge-on spiral encountering a massive cloud of gas and dust and undergoing a consequent burst of star formation.

peculiar motion The proper motion of a star due to its velocity relative to the *local standard of rest. The peculiar motion is that part of a star's total proper motion which is not accounted for by *parallactic motion.

peculiar star A star whose spectrum has unusual features compared with the majority of stars of its spectral type; the letter p is appended to the spectral type to distinguish this fact. The term is applied most usually to stars between spectral types B and F, where the spectra show signs of chemical anomalies, notably enhanced abundance of elements such as manganese, silicon, europium, chromium, and strontium.

peculiar velocity The space velocity of a star relative to the *local standard of rest.

pedestal crater An impact crater formed in weak surface materials which have subsequently been eroded away, except where protected by the surrounding ejecta blanket. The term was coined for certain craters on Mars, which in some areas of the planet form prominent circular plateaux above the surrounding terrain.

Peekskill meteorite A meteorite that fell on 1992 October 9, damaging a parked car at Peekskill, New York. The fireball preceding the fall of the 12.6-kg H6 *ordinary chondrite was observed over Virginia and New Jersey. Video and other records of its trajectory allowed the meteorite's orbit to be determined. It was found to have had an aphelion in the main asteroid belt.

Pegasus (Peg) (*gen.* **Pegasi**) The seventh-largest constellation, lying in the northern sky, and representing the winged horse of Greek mythology. The stars *Markab (Alpha Pegasi), *Scheat (Beta Pegasi), and *Algenib (Gamma Pegasi) form three corners of the Square of Pegasus, which is completed by the star *Alpheratz (Alpha Andromedae) from neighbouring Andromeda. *Enif (Epsilon Pegasi) is a double star. M15 is a 6th-magnitude globular cluster.

Pele An active volcano on Jupiter's satellite Io, at -19° lat., 255° W long. It lies in a volcanic complex about 500 km across, which includes lava flows and tablelands. It was active in 1979 March during the Voyager 1 encounter, but had ceased by the time Voyager 2 passed four months later. In 1996 it was seen by the Hubble Space Telescope to generate a plume over 400 km high in the course of a few tens of hours. Vaporized sulphur might be responsible for eruptions of this size, the heat being provided by an intrusion of molten silicate magma.

Pelican Nebula A diffuse nebula in Cygnus, so named because of its resemblance to a pelican on long-exposure photographs; also known as IC 5067 and 5070. The Pelican Nebula is about 1° wide. It lies next to the larger and brighter *North America Nebula, and is part of the same cloud.

Penrose process A process for extracting energy from a rotating black hole (i.e. a *Kerr black hole). The process requires sending a mass into a trajectory in the *ergosphere around the black hole, against the direction of rotation of the black hole. While inside the ergosphere the mass splits into two parts, one of which enters the black hole while the other escapes. Given a suitable trajectory, the emerging fragment may possess a total energy (i.e. *rest mass plus kinetic energy) greater than the total energy of the mass that went in. The extra energy has been extracted from the rotational energy of the black hole, which must therefore slow down slightly. The process is named after the English mathematician Roger Penrose (1931–), who discovered it in 1969.

penumbra **1**. The outer region of the shadow cast in space by a planet or satellite; an observer within the penumbra would witness a partial eclipse. The penumbra surrounds the dark, narrower *umbra (1), within which the Sun is completely obscured. The penumbra is darkest immediately adjacent to the umbra. During a lunar eclipse the Moon passes through the penumbra of the Earth's shadow before and after its umbral passage.
   2. The lighter, outer region of a sunspot, with a temperature of around 5500 K. Small spots often do not have a penumbra, but in mature sunspots the penumbra is well developed and occupies about 70% of the total spot area. The penumbra consists of filaments radiating from the central umbra. These filaments are brighter than the umbra, but less bright than the surrounding photosphere. They consist of light grains lasting about an hour, which drift in towards the umbra where they become *umbral dots*, apparently smaller versions of photospheric granules. The horizontal outward flow of gas called the *Evershed effect occurs in the penumbra and a little beyond it. The magnetic field intensity is about 0.1 tesla, less than in the umbra, and the field is aligned more nearly horizontally than in the umbra. Outside the penumbra of mature spots is a radial pattern of *fibrilles—the *superpenumbra*—visible in *Hα light.

penumbral eclipse A lunar eclipse during which the Moon passes slightly to the north or south of the umbra, and only through the less intense outer region (the penumbra) of the Earth's shadow, usually showing only negligible dimming. When the Earth's atmosphere is heavily laden with dust (following volcanic eruptions, for example), the penumbra may be darker than normal.

Penzias, Arno Allan (1933–) German-American physicist. In the early 1960s he and R. W. *Wilson were using a low-noise horn antenna in connection with satellite communications, and picked up a mysterious, persistent signal. They consulted R. H. *Dicke and colleagues, and concluded that they had detected the *cosmic microwave background. This, the single most important piece of observational evidence for the Big Bang, gained Penzias and Wilson the 1978 Nobel Prize in Physics.

perfect cosmological principle An extension of the *cosmological principle which proposes that the Universe is not only the same in all places and in all directions, but also at all times. The principle is the cornerstone of the *steady-state theory, but is incompatible with observations that show that the Universe is evolving with time.

peri- Prefix referring to the closest point in the orbit of one object around another, as in *perihelion and *perigee.

periapsis The point in an elliptical orbit that lies closest to the centre of the object being orbited.

periastron The point in an elliptical orbit around a star that is closest to the centre of the star.

periastron effect An increase in the brightness of a binary star that has a highly eccentric orbit, when the separation between the components is at a minimum. It is, in fact, an enhancement of the *reflection effect at the time of periastron, and arises from the same cause: irradiation of one star by the other.

pericentre The point in an elliptical orbit that is nearest to the centre of mass of the orbiting system, such as a binary star or a planet and satellite.

perigee The point in an elliptical orbit around the Earth that is nearest the Earth's centre. 📷

perihelion The point in an elliptical orbit around the Sun that is nearest the Sun's centre.

perihelion passage, time of (T or τ) The date and time at which an object orbiting the Sun is at its closest to the Sun. *See also* ELEMENTS, ORBITAL.

period, orbital (P) The time that a celestial body takes to make one orbit of another body. The period from one perihelion to the next is called the *anomalistic period*. The time the orbiting body takes to revolve through 360° is called the *sidereal period*. These two periods are different in duration if the orbit is perturbed by other bodies or forces.

period–age relation A relationship between the inferred evolutionary age of a Cepheid variable and its period. High-mass stars evolve more rapidly and are therefore younger when they enter the *Cepheid instability strip. At the same time, their high mass is reflected in a longer period of pulsation.

period–colour relation A statistical relationship between the mean colour of a *pulsating variable and its period. It reflects the fact that the period of pulsation depends on various factors, including a star's mass, radius, density, and temperature, with the temperature directly affecting the colour.

period–density relation A relationship between the period of a *pulsating variable and its mean density. The lower the density, the longer the period.

periodic comet A comet that has been observed on more than one return to perihelion, allowing its orbital period to be established reliably. Originally the term referred to comets with periods up to 200 years, but this arbitrary cut-off is becoming obsolete as comets with ever-longer periods are recovered. Two sub-classes of periodic comet are now defined: *short-period comets*, with periods less than 30 years, and *intermediate-period* (or *Halley-type) comets*, with periods between 30 and 200 years; however, there is no real physical difference between the comets in each category. Most periodic comets have direct orbits with inclinations of less than 30°. The typical example has a period of seven years, perihelion 1.5 AU, and inclination 13°. The names of periodic comets are prefixed by P/ (or D/ if they have disintegrated or disappeared), preceded by a number indicating the order in which their orbit was established (e.g. 1P/Halley, 2P/Encke, 3D/Biela, 109P/Swift–Tuttle). The periodic comet with the longest established period, 366 years, is *Ikeya–Zhang. Most periodic comets are thought to come from the *Kuiper Belt.

periodic orbit An orbit that is closed so that the orbiting body follows the same trajectory repeatedly. An elliptical orbit is a periodic orbit. The orbit of planets about the Sun and the major satellites about their planets are almost periodic, although the mutual perturbations of the planets prevent the orbits from being exactly periodic.

periodic perturbation A disturbance in a body's orbit that is oscillatory in nature, having a certain period and amplitude. Such perturbations can occur in all six elements of the orbit (*see* ELEMENTS, ORBITAL) and are usually termed long-term or short-term depending on whether the period is much larger than the body's orbital period, or of the same order as it.

period–luminosity–colour relation (PLC relation) A relationship between the period and colour of a *pulsating variable and its luminosity. It is an extension of the *period–luminosity relation that takes differences in colour (i.e. temperature) into account.

period–luminosity relation (PL relation) A statistical relationship between the period of a pulsating variable and its luminosity. For classical Cepheids and W Virginis stars, the relationship is approximately linear: the longer the period, the greater the luminosity. Classical Cepheids show a greater colour dependence, so it is preferable to use a *period–luminosity–colour relation. Mira stars have a non-linear PL relation, but it is well-established and useful for determining distances within the Galaxy. The Cepheid PL relation is the basis of the traditional method of determining distances to nearby galaxies.

period–mass relation A relationship between the mass of a *pulsating variable and its period. In approximate terms, the greater the mass of a star, the larger its radius, the lower its mean density, and thus the longer its period of pulsation (*see* PERIOD–DENSITY RELATION).

period–radius relation A mathematical relationship between the period and mean radius of a pulsating variable. The exact equation depends on the type of variable.

period–spectrum relation An approximate relationship between the mean spectral type of a *pulsating variable and its period. Generally, the later the spectral type, the longer the period. The larger a star, the cooler its surface temperature and the longer it will take to pulsate. Period–spectrum relations are found in Cepheid variables and Mira stars. In Mira stars a **period–spectrum–luminosity relation** can be established.

Perseid meteors A meteor shower, one of the three most active of the year, detectable between July 17 and August 24. Peak activity normally falls on August 12, when the radiant lies at RA 3h 04m, dec. +58°, near the Double Cluster. Activity at maximum is high, with ZHRs ranging from 80 to 140. Perseid meteors are fast (geocentric velocity 60 km/s); many are bright, and a high proportion leave persistent trains. The Perseids are produced by debris from Comet *Swift–Tuttle.

Perseus (Per) (*gen.*** Persei)** A prominent constellation of the northern sky, representing the mythical Greek hero who rescued Andromeda. Its brightest star, *Mirfak (Alpha Persei), is surrounded by a loose open star cluster, Melotte 20. Beta Persei is *Algol, the prototype eclipsing binary. Rho Persei is a semiregular variable red giant that ranges between magnitudes 3.3 and 4.0 with a period of about 7 weeks. NGC 869 and 884 are the famous *Double Cluster, also known as h and χ (Chi) Persei. M34 is a 5th-magnitude open cluster. M76 is a planetary nebula known as the *Little Dumbbell. NGC 1499 is the *California Nebula, and NGC 1275 is a peculiar galaxy known as the radio source *Perseus A, part of the *Perseus Cluster of galaxies. The year's brightest meteor shower, the *Perseids, radiates from the constellation every August.

Perseus A A radio source in the constellation Perseus, identified with the 12th-magnitude supergiant elliptical galaxy NGC 1275. It is the brightest member of the Perseus Cluster of galaxies, about 250 million l.y. away.

Perseus Arm The spiral arm of our Galaxy that is about 5000 l.y. farther from the centre than the local *Orion Arm in which the Sun lies, and roughly parallel to it. The Perseus Arm can be traced from the constellation of Cassiopeia via Perseus to Auriga, Gemini, and Monoceros, behind stars in our local arm. It contains the *Double Cluster, the *Crab Nebula, and the *Rosette Nebula.

Perseus Cluster A cluster of galaxies about 250 million l.y. away, covering 4° of sky in the constellation Perseus; also known as Abell 426. At its centre lies the supergiant elliptical galaxy NGC 1275, also known as the radio source Perseus A. The Perseus Cluster is dominated by elliptical galaxies.

personal equation The systematic bias of an observer when making measurements. For example, when timing a lunar occultation, there will always be a certain delay between seeing the event and making the timing. In variable-star observing, the observer may consistently underestimate bright stars by a fixed amount. A personal equation can be calculated for individual observers and applied as a correction factor to their observations.

perturbation A change in the orbit of a body, usually due to the gravitational attraction of another body. Thus the planets in their elliptical orbits about the Sun mutually perturb one another, producing changes in their orbital elements. The orbits of satellites are perturbed by their mutual gravitational attractions and also by the Sun. Perturbations also arise in the orbits of satellites that lie close to their planets because the planets are not perfectly spherical in shape. The orbits of some artificial Earth satellites are also perturbed by atmospheric drag. *See also* PERIODIC PERTURBATION; SECULAR PERTURBATION.

Petzval surface The curved focal surface produced by an uncorrected optical system, named after the Hungarian physicist and optician Józeph Miksa Petzval (1807–91). *See also* FOCAL PLANE.

Pfund series A sequence of absorption or emission lines in the far-infrared part of the spectrum, due to hydrogen. They are caused by electron jumps between the fifth energy level and higher levels. Pfund-α is at a wavelength of 7.46 μm, and the spacing between the lines diminishes as they converge on the series limit at 2.28 μm. The series is named after the American physicist (August) Herman Pfund (1879–1949). *See also* HYDROGEN SPECTRUM.

PHA Abbr. for *Potentially Hazardous Asteroid.

Phad An alternative name for the star *Phecda.

Phaethon Asteroid 3200, a member of the *Apollo group, discovered in 1983 by the Infrared Astronomical Satellite. Phaethon's orbit is very similar to that of the *Geminid meteor shower, and it may be their parent body. Before the discovery of Phaethon, meteor showers were generally believed to be associated only with comets, but infrared observations suggest that Phaethon has a rocky surface, not the dusty crust expected for an extinct cometary nucleus. Phaethon is an F-class asteroid with a diameter of 5.1 km. Its orbit has a semimajor axis of 1.271 AU, period 1.43 years, perihelion 0.14 AU, aphelion 2.40 AU, and inclination 22°.2. It can approach to within 0.02 AU (3 million km) of the Earth's orbit.

phase 1. The proportion of the visible disk of the Moon or a planet that is illuminated as seen from Earth.
   2. A measure of position along a cyclically varying quantity, such as a wave or a periodic vibration. Phase is measured as an angle, where one complete cycle is equivalent to a phase of 360° (or 2π radians), or occasionally as a number between 0.0 and 1.0. Two or more waves of the same frequency are said to be *in phase* when their maxima and minima occur at the same instants. Otherwise, they are said to be *out of phase*, or to have a *phase difference. If their phases differ by exactly 180° they are said to be in *antiphase*.

phase angle The angle between the Sun and an observer as seen from the centre of a body in the Solar System. At a phase angle of 0° the Sun and observer lie in exactly the same direction, and so the observer sees the object fully illuminated. At a phase angle of 180° the Sun and observer lie on opposite sides of the body, and the whole of its unilluminated side is facing the observer.

phased array An array of closely spaced radio antennas in which the beam can be formed and steered by adjusting the time delay between each antenna and the receiver. As the antennas may be simple *dipole antennas with no moving parts, large effective areas can be achieved relatively cheaply. Phased arrays are widely used in radar due to the rapid speed with which the beam may be steered. They also have the ability to form simultaneous multiple beams and are used in the central element of the *Square Kilometre Array.

phase defect The amount by which the illuminated disk of the Moon appears to depart from a complete circle at full Moon, because of the inclination of the Moon's orbit to the ecliptic. It is measured in degrees.

phase diagram A graph showing the range of temperatures and pressures at which a substance can exist as a solid, liquid, or gas. These different physical states are known as **phases**, and the temperatures and pressures under which each phase exists are represented by an area on the diagram. A line between two areas defines the conditions under which two phases can exist in equilibrium. The *triple point* is the only point on the diagram where conditions are such that the solid, liquid, and gas phases can coexist in equilibrium.

phase difference (ϕ) The amount by which two cyclical motions of the same frequency, such as electromagnetic waves, are out of step with each other. It can be measured in degrees, radians, or seconds of time.

phase rotator A device used in radio astronomy to adjust the phase of an incoming signal. It is often used in interferometry to compensate for the slow turning of the baseline caused by the rotation of the Earth.

phase-switching interferometer A radio interferometer with two aerials in which the signal from one antenna is periodically reversed in phase before being multiplied by the signal from the other antenna. The output is a square wave whose amplitude is proportional to the product of the two signals. As the source moves across the sky, interference fringes are produced in a similar way to those from a *correlation receiver, except that any steady background emission is removed.

phase transition *See* INFLATIONARY UNIVERSE.

Phecda The star Gamma Ursae Majoris, magnitude 2.4. It is an A0 dwarf, 83 l.y. away. An alternative name is Phad.

Pherkad The star Gamma Ursae Minoris, magnitude 3.0. It is an A3 giant, lying 487 l.y. away. It is one of the two so-called Guardians of the Pole, the other being *Kochab.

Phillips bands Features in the spectrum of the carbon molecule C_2, starting in the red at 772 nm and extending into the near-infrared where the strongest bands occur, at 1.21 and 1.55 µm. Phillips bands are prominent in the atmospheres of carbon stars. They are named after the American astrophysicist John Gardner Phillips (1917–2001), who published the first measurements of them in 1948.

Phobos The inner satellite of Mars, 9380 km from the planet's centre and only 6000 km above its surface. Phobos orbits in 0.319 days, faster than Mars rotates on its axis, keeping one face turned permanently towards the planet. Phobos is irregularly shaped, $26 \times 23 \times 18$ km in size, and is covered with impact craters, the largest, Stickney, being 9 km in diameter. There are also several families of parallel grooves or chains of craters, possibly secondary craters from large impacts on Mars. Phobos was discovered in 1877 by A. *Hall, and may be a captured asteroid. 📷

Phobos-Grunt A Russian Space Agency probe to land on the larger of the two moons of Mars, Phobos, and return samples of it to Earth. The word 'Grunt' is Russian for ground. The mission will also carry a Chinese Mars orbiter, Yinghuo-1. Phobos-Grunt is planned for launch in 2011 or 2012. Its upper stage will return to Earth in 2014, leaving instruments on the surface of Phobos.

Phobos probes Two space probes to the planet Mars, launched by the former Soviet Union in 1988 July. Contact was lost with Phobos 1 *en route*, but Phobos 2 went into orbit around Mars in 1989 January, studying the planet and its larger satellite, Phobos, although it failed before it could drop its instrumented landers on to the satellite's surface.

(⊕) SEE WEB LINKS
- NASA information page for Phobos 1.
- NASA information page for Phobos 2.

Phocaea group A group of asteroids at a mean distance of about 2.36 AU from the Sun, with orbital inclinations of 23–25°. The group has been separated from the main asteroid belt by the gravitational effects of the major planets, notably Jupiter, and contains a broad selection of asteroid classes. The largest member of the group is the C-class (105) Artemis, diameter 119 km. Certain members, among them (323) Brucia, (852) Wladilena, (1568) Aisleen, and (1575) Winifred, may form a true family within the Phocaea group. The group is named after (25)

Phocaea, an S-class asteroid, diameter 75km, discovered in 1853 by the French astronomer Jean Chacornac (1823–73). Phocaea's orbit has a semimajor axis of 2.400 AU, period 3.72 years, perihelion 1.79 AU, aphelion 3.01 AU, and inclination 21°.6.

Phoebe A retrograde satellite of Saturn, distance 12 893 240 km; also known as Saturn IX. It orbits Saturn in 548.2 days and has a mean diameter of 213 km. Phoebe's surface is heavily cratered (the largest crater, Jason, is 101 km wide) and significantly darker than all but the smallest Saturnian moons, with an albedo of 0.08. Phoebe was discovered in 1898 by W. H. *Pickering.

Phoenicid meteors A weak southern-hemisphere meteor shower, generally showing low rates (maximum ZHR 5) between November 28 and December 9. At maximum on December 6, the radiant lies at RA 1h 12m, dec. −53° (northwest of Achernar). Unusually high activity (ZHR 100) occurred for a limited period around maximum in 1956.

Phoenix (Phe) (*gen.* **Phoenicis**) A constellation of the southern sky, representing the phoenix. Its brightest star is Alpha Phoenicis, magnitude 2.4. Beta Phoenicis is a close double, magnitudes 4.0 and 4.2. Zeta Phoenicis is an eclipsing binary that varies between magnitudes 3.9 and 4.4 every 1.67 days.

Phoenix (Mars mission) A NASA Mars lander, launched 2007 August, to study the icy north polar region of Mars. It landed 2008 May in a lowland plain called Vastitas Borealis at 68°.2 N, 234°.3 E. It carried a robotic arm to scoop up samples of soil and ice for analysis by instruments on board the lander. The Surface Stereo Imager (SSI) camera photographed the surrounding landscape, while a set of meteorological instruments recorded temperature, pressure and other atmospheric conditions. Phoenix operated until 2008 November.

SEE WEB LINKS
• Official mission website.

Pholus Asteroid 5145, an unusual object in the outer Solar System, discovered in 1992 by the American astronomer David Lincoln Rabinowitz (1960–). Pholus lies well outside the main asteroid belt, moving from near the orbit of Saturn out to beyond the orbit of Neptune. It was the second of the *Centaur group to be discovered. Pholus is estimated to be 190km in diameter. Its extremely red spectrum is suggestive of a comet but, unlike *Chiron, Pholus seems inactive near perihelion. Its orbit has a semimajor axis of 20.30 AU, period 91.48 years, perihelion 8.69 AU, aphelion 31.91 AU, and inclination 24°.7.

photino A hypothetical fundamental particle that, according to certain theories, is related to the photon. In so-called *supersymmetry* theories, all *bosons have a partner which is a fermion; the partner of the photon (a boson) would be the photino (a fermion). Photinos may have been produced in sufficient numbers in the Big Bang to account for part of the *dark matter in the Universe.

photocathode The light-sensitive surface of a *photomultiplier. It is a coating of a material such as caesium, rubidium, or potassium applied to the inside of a clear window of an evacuated tube.

photocell *See* PHOTOELECTRIC CELL.

photocentre 1. The marked peak in intensity at the centre of an *Airy disk.
2. The point of maximum intensity of the combined light from an unresolved binary star. The position of the photocentre does not correspond to that of either component, nor to that of the *barycentre. Motion of the photocentre is evidence that the system in question is an *astrometric binary.

photoconductive cell An electronic component that varies in electrical conductivity with the amount of light falling on it. A voltage applied across the cell will therefore vary according to the light intensity. Devices using semiconducting materials such as cadmium sulphide or germanium are known as *photodiodes* and have enough sensitivity to be used for measuring the brightness of variable stars.

photodissociation The dissociation (breaking up) of molecules by exposure to light. This process occurs when light from stars shines into a cloud of cool molecular gas. If the cloud is diffuse, all the molecules may be dissociated by the light. More usually the cloud is so dense that the light penetrates only a small distance before being absorbed. *Photodissociation regions are thereby formed at the edges and boundaries of the cloud.

photodissociation region A region at the boundary of a molecular cloud where ultraviolet light dominates, dissociating (breaking up) the molecules into fragments such as radicals and ions. These fragments are very reactive and generate a rich and complex chemistry. The ultraviolet light may come from an external source such as a star or nebula, or from an embedded source such as a young star within the cloud.

photoelectric cell A device that gives an electrical output in response to light; also known as a **photocell**. It may be **photovoltaic**, in which case the light generates an electric current, a *photoconductive cell, or a *photomultiplier. Only the last two are sensitive enough to be used in astronomy.

photoelectric magnitude The brightness of a star or other object measured with a *photoelectric photometer. This measurement requires correction for *atmospheric extinction and comparison with at least one photometric *standard star. The magnitude is expressed in terms of a system such as *Johnson photometry or *Strömgren photometry.

photoelectric photometer An instrument for measuring the brightness of a star by means of the electric current produced when its light falls on a light-sensitive surface (the **photoelectric effect**). The light gathered by a telescope passes through a filter on to a light-sensitive surface known as the *cathode*. The cathode emits electrons which are multiplied within a *photomultiplier so that the signal is easily measurable. Each photon detected by the cathode generates a pulse, and the number of pulses per second is directly proportional to the star's brightness. This is known as *pulse-counting photometry*. In *high-speed photometry* the pulses are counted at intervals as short as 10ms. The best photometers can yield magnitudes to accuracies approaching 0.001 mag.

photographic amplification The technique of making faint detail on a photographic emulsion more easily visible by copying it on to a high-contrast emulsion. The copy is made by contact printing the original using diffuse illumination which preferentially copies the outermost, most sensitive layer of the original emulsion. In this way, otherwise invisible features can be seen on the copy.

photographic emulsion The light-sensitive coating given to photographic materials. It consists of crystals (usually called *grains*) of silver halides, suspended in gelatin which is coated on to a base of glass or film. The emulsion's sensitivity to light depends on the size of the grains, the most sensitive (fastest) emulsions having the largest grains. The basic emulsion is sensitive only to ultraviolet and blue light, but by adding organic dyes the grains can be made sensitive to other colours. On exposure to light, small specks of metallic silver are created on the surfaces of the grains. These then trigger the conversion of the whole grain to silver when the emulsion is placed in developer.

photographic magnitude (m_{pg}) The brightness of a star measured with a blue-sensitive (*orthochromatic*) photographic plate. This magnitude is poorly defined because different amounts of ultraviolet light are included depending on whether a refractor or an aluminized reflector is used. Although this magnitude is now obsolete, it can be roughly related to the B magnitude by the approximation

$$B \approx m_{pg} + 0.11$$

photographic zenith tube (PZT) A telescope with its objective lens fixed pointing vertically upwards. Stars are photographed as they pass the zenith, both directly through the telescope and after their light has been reflected from a mercury mirror beneath the objective lens. The PZT is designed to measure the direction of the instantaneous vertical relative to the stars, and hence accurately determine the longitude and latitude of the observatory.

photoheliograph A telescope in which an image of the Sun in white light is recorded on a photographic plate. The first photoheliograph was designed by the English scientist Warren De la Rue (1815–89) in 1857. Daily solar photography with modern photoheliographs is carried out at observatories around the world.

photoionization The loss of an electron by an atom, ion, or molecule by the absorption of a photon. For an electron in the *ground state, a photon can remove an electron if the photon's energy equals the *ionization potential of the atom or molecule. Lower-energy photons may ionize the atom or molecule if it has electrons in an excited energy level. Photoionization is of importance primarily in stellar interiors and atmospheres, where the number of photons is high.

photometer An instrument for measuring the brightness of stars or other objects. In its widest sense the term can include the human eye or the photographic plate. Conventionally, the term is confined to instruments such as the *photoelectric photometer, or *area photometers based on CCDs or infrared arrays, whose output is directly proportional to the incident radiation.

photometric binary A binary star whose duplexity is detectable because it is variable in brightness, with a light-curve that has certain specific characteristics. *See* ECLIPSING BINARY; ELLIPSOIDAL VARIABLE; REFLECTION VARIABLE.

photometric parallax The distance of a star as inferred from its position on the lower main sequence, where colour and absolute magnitude are tightly correlated by the *colour–magnitude relation. If the colour is known, the absolute magnitude may be inferred. The difference between the apparent magnitude, m, and the absolute magnitude, M, is related to the parallax, π, in seconds of arc by the formula

$$\log \pi = 0.2(M - m - 5).$$

photometric standard *See* STANDARD STAR.

photometry The science of the measurement of light. The earliest estimates of star brightnesses were made with the naked eye, which formed a basis for the *magnitude scale used in astronomy. Now, professional astronomers use photometers to compare objects under study with *standard stars. Broad-band filters are used for the faintest objects to achieve sufficient signal, while brighter stars can be measured with intermediate-band or narrow-band filters. The filters are carefully chosen to reveal characteristics of the spectrum such as absorption lines in stars and galaxies, and emission lines and bands in nebulae. Magnitudes from different filters are usually compared to obtain a *colour index. Satellites have extended astronomical photometry into the ultraviolet and infrared.

photomultiplier An evacuated electronic tube which converts light into a measurable electric current. Light falling on a *photocathode releases electrons, which are accelerated by an electric field and attracted to the first *dynode* (positive electrode), where they liberate more electrons which are attracted to the second dynode, and so on. A common type of photomultiplier in astronomical use has ten dynodes, each at an increasingly positive electric potential. The flow of electrons arriving at the final anode is proportional to the amount of light falling on the photocathode. Photomultipliers are widely used for photometric measurements in astronomy, such as of variable stars.

photon A particle of electromagnetic radiation. A photon has zero *rest mass, zero charge, and travels at the speed of light. The energy, E, of a photon is related to its frequency, f, by the formula $E = hf$, where h is the *Planck constant. Hence a photon at a radio frequency is of much lower energy than a gamma-ray photon.

photon sphere A sphere surrounding a black hole, at the surface of which light follows a circular path around the hole. Inside the photon sphere light spirals in towards the black hole. Outside the sphere the path of light is bent, but the photons can still escape back into the Universe.

photopolarimeter *See* POLARIMETER.

photosphere The visible surface of a star, from which most of its energy is emitted in the form of visible and infrared radiation. The name means 'light sphere'. The Sun's photosphere is a thin layer about 500 km deep. Its temperature decreases steadily from about 6400 K at its base to 4400 K at the *temperature minimum, where it merges with the chromosphere above. This drop in temperature with height causes *limb darkening. The photosphere has a rice-grain texture called *granulation, caused by rising convection cells of hot gas. Other photospheric features include *sunspots, *faculae, and *filigree structures, all associated with strong magnetic fields. Almost all the features of the Sun's visible-light spectrum originate in the photosphere, including the dark Fraunhofer lines.

photovisual magnitude (m_{pv}) The brightness of a star measured on a *panchromatic* (yellow-sensitive) photographic plate exposed through a yellow filter to mimic the wavelength response of the eye. Such magnitudes are similar to photoelectric *V magnitudes, although not so accurate. The term is now obsolete; many astronomers are content to use the technique and call the resulting magnitude V.

photovoltaic detector A detector in which the voltage changes in response to incident radiation. In a simple p–n semiconductor junction, incident radiation leads to a flow of current over the junction, and the device acts as a **photoconductive** detector. But if the diode is in series with a very high resistance, the voltage across that resistance changes with the intensity of the incoming radiation, so that the diode then acts as a **photovoltaic** detector. Various semiconductor materials such as silicon, indium antimonide, and gallium arsenide can be used for the detectors. They can operate from visual wavelengths to 10 μm or longer.

physical albedo Another term for *geometrical albedo.

physical double A double star in which the components are physically linked by their mutual gravitational attraction and therefore form a true *binary star, as opposed to an optical double.

physical libration A real periodic variation in the rotation rate of a celestial object, as distinct from its *optical libration. In addition to its familiar optical librations, the Moon also possesses a very small physical libration in its axial rotation, amounting to less than 2′. This small physical libration allows the *moments of inertia of the Moon to be found.

Piazzi, Giuseppe (1746–1826) Italian astronomer and monk. In 1789 he acquired a high-quality transit instrument made by the English optician Jesse Ramsden (1735–1800). From accurate positional measurements made with it, he compiled a star catalogue at the Observatory of Palermo, and in the process discovered the first asteroid, *Ceres, on 1801 January 1. Piazzi's proposed term 'planetoid' lost out to F. W. *Herschel's suggestion of 'asteroid'.

PICARD A French satellite launched in 2010 June to measure the *total solar irradiance, perform helioseismology observations, and measure the Sun's diameter and shape to within a few milliarcseconds. The satellite is named after the French astronomer Jean-Felix Picard (1620–1682), who first accurately measured the Sun's diameter.

(⊕) SEE WEB LINKS
• Official mission website.

Pic du Midi Observatory An observatory at an altitude of 2860m in the Pyrenees mountains of southwestern France, owned by the French Centre National de la Recherche Scientifique (CNRS) and operated by Paul Sabatier University, Toulouse; its headquarters are in Bagnères-de-Bigorre. Pic du Midi Observatory was founded in 1878 primarily for meteorology, with solar observations starting in 1892. Its main instruments are the 2-m Bernard Lyot reflector, opened in 1979, a 1-m reflector opened in 1963, and several solar telescopes.

(((⊕))) SEE WEB LINKS

• Official observatory website.

Pickering, Edward Charles (1846–1919) American physicist and astronomer, elder brother of W. H. *Pickering. As director of Harvard College Observatory, he produced a catalogue of brightnesses for 4260 stars, the Harvard Photometry, in 1884, and extended the work in 1908 with the *Harvard Revised Photometry. He was an early exponent of astrophotography, and built up an extensive library of photographic plates at Harvard. In 1903 he published the first photographic map of the whole sky. In the 1880s, Pickering began a programme to classify stellar spectra, discovering in the process the first *spectroscopic binary, Mizar A. He invented a method for recording several spectra on one plate by placing a large, low-dispersion prism in front of the telescope. The routine work was largely carried out by a team of women assistants, including A. J. *Cannon, W. P. *Fleming, and Antonia Caetana de Paiva Pereira Maury (1866–1952). The programme yielded the *Harvard classification and the *Henry Draper Catalogue* of stellar spectra.

Pickering, William Henry (1858–1938) American astronomer, younger brother of E. C. *Pickering. He assisted P. *Lowell in establishing the Flagstaff Observatory, but later disagreed with Lowell's exotic theories about life on Mars. From Harvard College Observatory's southern station in Peru, Pickering discovered Saturn's satellite Phoebe in 1898. His photographic lunar atlas was published in 1903. Pickering made many predictions of trans-Neptunian planets; a photographic search based on one of them recorded Pluto in 1919, but it was not identified at the time.

Pickering series A sequence of emission and absorption lines in the visible part of the spectrum, caused by the singly ionized helium atom, He II. Such an ion is very similar to a hydrogen atom in that both contain only a single electron. However, the helium nucleus is four times heavier and has twice the nuclear electric charge. For these reasons, alternating lines in the Pickering series almost align with lines in the hydrogen *Balmer series: Pickering-β, for example, is at 656nm, close to the Hα line. The series limit is at 364.4nm (almost coincident with the *Balmer limit). Four times as much energy is needed to ionize a helium atom as a hydrogen atom, so He II lines are seen only in very hot stars (spectral type O and Wolf–Rayet stars) and accretion disks. The Pickering series is named after E. C. *Pickering.

Pico Veleta The site in the Sierra Nevada mountains near Granada, Spain, of the 30-m millimetre-wave telescope of the *Institut de Radio Astronomie Millimétrique (IRAM).

Pictor (Pic) (*gen.* Pictoris) A constellation of the southern sky, representing a painter's easel. Its brightest star, Alpha Pictoris, is of magnitude 3.2. *Beta Pictoris is surrounded by what may be a planetary system in the process of formation. Pictor contains the fast-moving *Kapteyn's Star.

pincushion distortion An optical defect in which the magnification of a lens decreases with distance from the optical axis, giving the image of a square object with concave sides, like a pincushion.

Pinwheel Galaxy A name sometimes applied to either of the Sc spiral galaxies M33 (NGC 598) in Triangulum or M101 (NGC 5457) in Ursa Major.

pion An unstable elementary particle which exists in three forms: neutral, positively charged, and negatively charged. The charge is equal to that of the electron. Charged pions decay into muons and neutrinos. The neutral pion decays into two gamma-ray photons. The pion is a *meson, and is also known as a *pi-meson.*

Pioneer A series of US space probes. The first Pioneers, launched in 1958–9, were intended as Moon probes, but none succeeded. Pioneers 5–9, in 1960–8, were put into orbit around the Sun, monitoring solar activity and conditions in interplanetary space. Pioneers 10 and 11, the last and best-known of the series, were the first probes to reach Jupiter, photographing the planet and studying its environment. Pioneer 10, launched in 1972 March, flew past Jupiter in 1973 December. Pioneer 11, launched 1973 April passed Jupiter in 1974 December and then became the first probe to reach Saturn, in 1979 September. Both Pioneers are currently on their way out of the Solar System. Pioneer 10 crossed the orbit of Pluto in 1983 June, the first spacecraft to do so. Contact with it was finally lost in 2003 January at a distance of 82 AU. Contact with Pioneer 11 was lost in 1995 November.

(⊕) SEE WEB LINKS
- NASA information page for Pioneer 10.
- NASA information page for Pioneer 11.

Pioneer Venus Two NASA space probes to Venus. Pioneer Venus 1 (also known as Pioneer Venus orbiter) went into orbit around Venus, photographing the planet's clouds and mapping its surface by radar. Pioneer Venus 2 (also known as the multiprobe) ejected four sub-probes that analysed the planet's clouds and atmosphere during their descent. One of the sub-probes survived its impact with Venus, continuing to transmit from the surface for over an hour.

Pioneer Venus Probes

Probe	Launch date	Results
Pioneer Venus 1	1978 May 20	Entered orbit around Venus 1978 December 4
Pioneer Venus 2	1978 August 8	Ejected four smaller probes which entered the atmosphere of Venus 1978 December 9

Pipe Nebula A dark nebula in Ophiuchus, with the curving shape of a pipe. It is one of the largest dark nebulae in the Milky Way, extending for several degrees. Sections of the pipe's stem have the designations Barnard 59 and 65–7, while the bowl is Barnard 78.

Pisces (Psc) (*gen.* **Piscium)** A constellation of the zodiac, representing a pair of fishes. The Sun passes through Pisces between mid-March and the third week of April, and so is in the constellation at the *vernal equinox. The constellation's brightest star is Eta Piscium, magnitude 3.6. *Alrescha (Alpha Piscium) is a close double star. TX Piscium (also known as 19 Piscium) is a red giant that varies irregularly between magnitudes 4.8 and 5.2. M74 is a 9th-magnitude spiral galaxy seen face-on.

Piscid meteors A meteor shower producing low rates (maximum ZHR 10) during September and October from a multiple radiant near the ecliptic. Maxima occur, possibly, around September 8 (radiant at RA 0h 36m, dec. +07°), September 21 (RA 0h 24m, dec. 00°), and October 13 (RA 1h 44m, dec. +14°). Piscids are typically slow and sometimes comparatively long-lasting. The Piscids are not recognized as a separate shower by some authorities, and the activity may simply be due to general background from the *anthelion radiant.

Piscis Austrinid meteors A meteor shower active during July and August; also known as the *Piscis Australids*. At maximum (ZHR 5), on July 28, the radiant lies at RA 22h 44m, dec. −30°, west of Fomalhaut. The shower is best observed from more southerly latitudes, and has received relatively little attention.

Piscis Austrinus (PsA) (*gen.* **Piscis Austrini)** A constellation of the southern sky, popularly known as the Southern Fish. It contains the bright star *Fomalhaut (Alpha Piscis Austrini) but little else of note.

pitch angle A measure of how tightly the spiral arms of a galaxy are wound, defined as the angle between the tangent to a spiral arm and a circle about the galactic centre at the same radius.

pixel An individual element in an array which makes up an electronic image, as produced, for example, by a *charge-coupled device. The size of the pixels in a detector governs its resolution. The word is a contraction of 'picture element'.

plage A brighter, hotter patch in the Sun's chromosphere, visible in *Hα light and the calcium K line. Plages are the chromospheric equivalent of faculae on the photosphere, as can be seen when an active region is near the limb. They are regions of particularly strong magnetic field.

Planck An ESA spacecraft launched in 2009 May to study fluctuations in the *cosmic microwave background, caused by slight density differences in the early Universe from which galaxies formed. Planck, named after the German physicist Max Planck (1858–1947), carries a 1.5-m mirror to focus microwaves onto two detectors which register variations in the temperature of the background radiation of a few millionths of a degree, with an angular resolution better than 10 arcminutes. Planck was launched on the same rocket as the *Herschel Space Observatory, another ESA scientific mission, but the two separated once in space and operate independently at the L_2 *Lagrangian point, 1.5 million km from Earth in the direction away from the Sun.

(((()))) SEE WEB LINKS
• ESA mission website.

Planck constant (*h*) A constant that relates the energy of a *photon to its frequency. It has the value 6.626 069 57 × 10^{-34} Js. It is named after the German physicist Max Karl Ernst Ludwig Planck (1858–1947).

Planck era In the *Big Bang theory, the fleeting period between the Big Bang itself and the so-called **Planck time** when the Universe was 10^{-43} s old and the temperature was 10^{34} K. In this period, quantum gravitational effects are thought to have dominated. Theoretical understanding of this phase is virtually non-existent. It is named after Max Planck (1858–1947).

Planck's law A mathematical description of the energy radiated at different wavelengths by a black body: $E = hf$, where E is the energy of a *photon and f its frequency. It was formulated in 1900 by Max Planck (1858–1947), who realized that energy is radiated in discrete packets, which he called *quanta*, and it formed the basis of *quantum theory. The quantum of light is a photon, the energy of which depends on its wavelength.

planemo An object of planetary mass that does not orbit a star but floats freely in space. The word is an abbreviation of *planetary mass object*.

planet A non-luminous body in orbit around the Sun, or another star, which has sufficient mass to have become rounded by its own gravity and which has significantly cleared its orbital neighbourhood of smaller objects. Planets can consist of rock and metal, as do the inner planets of the Solar System, or predominantly of liquid and gas, as do the giant outer planets. The term does not include comets or other small objects such as meteoroids. Asteroids, however, are sometimes referred to as *minor planets*. A planet can have a mass up to about 10 times that of Jupiter, above which it would become a *brown dwarf. In 2006 the International Astronomical Union introduced the term *dwarf planet* to describe objects that are rounded in shape but which have not cleared the neighbourhood around their orbit; this category includes the largest member of the asteroid belt and the largest *trans-Neptunian objects.

planetarium A device for projecting a simulated view of the heavens on to the inside of a dome; or the building that houses such a projector.

planetary aberration The angle between the geometric direction to an object in the Solar System at the instant of observation and its apparent direction as seen by a moving observer. It is the combined effect of aberration due to the observer's motion, and the movement of the planet during the time that its light takes to reach the observer.

planetary aspect *See* ASPECT.

planetary migration The movement of a planet towards or away from its parent star as a result of aerodynamic drag from gas, random gravitational interactions with planetesimals, or resonant interactions with other coorbiting bodies. Depending on the mechanism at work and the starting location of the planet, its orbit may decrease or increase in size. The large number of *hot Jupiters seen around other stars is a consequence of planetary migration at work.

planetary nebula A bright cloud of glowing gas and dust surrounding a highly evolved star. A planetary nebula forms when a red giant ejects its outer layers at speeds of about 10 km/s. The ejected gas is then ionized by ultraviolet light from the hot core of the star. As mass is lost this core is progressively exposed, and it ultimately turns into a white dwarf. Planetary nebulae are typically 0.5 l.y. across, and the amount of ejected material is 0.1 solar mass or more. Because the core is so hot the gas in the nebula is highly ionized. The planetary nebula exists for up to 100 000 years, during which time a sizable fraction of the star's mass is returned to interstellar space. Planetary nebulae were so named because they appeared to early observers to resemble a planetary disk. In fact the detailed shapes of planetary nebulae revealed by modern telescopes cover many different types, including ring-shaped (as in the Ring Nebula), dumbbell-shaped, or irregular. The various apparent shapes are now thought to be due to the angle at which we are viewing two lobes or cylinders of gas ejected in opposite directions from the central star in a *bipolar outflow. 📷

planetary precession The perturbation of the Earth's orbital plane by the attractions of the planets on the Earth's centre of mass. It causes the equinox to move eastwards along the celestial equator (i.e. in the opposite direction to *lunisolar precession) at a rate of about 0″.12 per year. Its effect is much smaller than that of lunisolar precession.

planetesimal A 0.1–100 km body of rock and/or ice that is presumed to have formed in the early history of the Solar System. The planets are thought to have grown from the accumulation of planetesimals. Most planetesimals left over from planetary accretion were ejected by perturbations of the planets into the *Kuiper Belt and *Oort Cloud beyond Neptune.

planetoid An obsolete name for an *asteroid, originally proposed by G. *Piazzi.

planetology The study of the planets, including their surfaces, interiors, and atmospheres; also known as planetary science.

planisphere A circular map of the heavens, with an overlying mask that can be rotated to reveal the stars visible from a given latitude at any chosen date and time.

planitia A large, low plain on a planetary surface; pl.*planitiae*. The name is not a geological term, but is used in the nomenclature of individual features, for example Guinevere Planitia on Venus or Sarandib Planitia on Enceladus.

planoconcave lens A lens having one flat face and one concave face; it is a *diverging lens.

planoconvex lens A lens with one flat face and one convex face; it is a *converging lens.

planum A large plateau or high plain on a planetary surface; pl.*plana*. The name is not a geological term, but is used in the nomenclature of individual features, for example Lakshmi Planum on Venus, or Planum Australe on Mars.

Plaskett, John Stanley (1865–1941) Canadian engineer and astronomer. He designed a new spectrograph for measuring stellar radial velocities, which was used from 1918 with the *Dominion Astrophysical Observatory's 72-inch (1.85-m) reflector, a telescope largely of his design and now named after him. The numerous radial velocity measurements revealed many spectroscopic binaries, including *Plaskett's Star, as well as the rotation of the Galaxy and the location of its centre. Plaskett also showed that lines of calcium in stellar spectra came from interstellar matter. His son, Harry Hemley Plaskett (1893–1980), was an accomplished solar spectroscopist.

Plaskett's Star A 6th-magnitude spectroscopic binary star in Monoceros named after J. S. *Plaskett, who discovered in 1922 that it was the most massive binary known; it is also known as V640 Mon. According to current measurements, the individual stars are both blue supergiants of 51 and 43 solar masses. Plaskett's Star lies about 5000 l.y. away.

plasma A state of matter consisting of ions and electrons moving freely. Stars consist of plasma, and plasmas exist in interstellar space; the *solar wind is a plasma. Because a plasma is highly ionized, its behaviour differs from that of a normal gas. External magnetic and electric fields can affect a plasma, and the charged particles themselves can interact magnetically and electrically.

plasmapause The outer boundary of the *plasmasphere, marked by a sharp decrease in plasma density.

plasmasphere A region within the Earth's *magnetosphere containing relatively cool, low-energy plasma at a temperature of 2000 K extending to a distance of about 4 Earth radii (25000 km), bounded by the inner *Van Allen Belt. The particles in the plasmasphere are believed to originate in the ionosphere.

plasma tail *See* TAIL, COMETARY.

Plateau de Bure The location south of Grenoble, France, of the millimetre-wave interferometer of the *Institut de Radio Astronomie Millimétrique (IRAM).

plate centre The celestial coordinates of the centre of the field of an astronomical photographic plate.

plate constants The coefficients for converting measurements on a photographic plate to celestial coordinates. In their simplest form these include the zero point, plate scale, and orientation.

plate-measuring machine A device for the accurate measurement of positions and images on a photographic plate. In the simplest form of machine, the plate is driven in one coordinate direction only by a single micrometer screw, and the measurements are recorded manually. Large automatic machines are now used for fast measurement and digital recording of coordinates.

plate scale The scale factor for converting linear measure on a photographic plate to angular measure on the sky. It is equal to the inverse of the focal length of the telescope.

plate tectonics The process of continental drift, involving the movement of crustal **plates** across the surface of the Earth, or other planet, in response to convective movements below. On the Earth, the lighter continental crust floats upon the denser oceanic crust, which is in constant motion. Convection causes the hotter zones within the asthenosphere to rise until they reach the surface. The hot rock cools and forms oceanic crust at a *spreading axis*, usually in the middle of an ocean. The oceanic crust then slowly moves away from the spreading axis at up to 100 mm/year until it reaches a *subduction zone*, usually at the edge of a continent, where it descends into the asthenosphere again. Plate tectonics does not appear to operate on the other terrestrial planets, although the *Tharsis Montes on Mars might be a 'failed' spreading axis. Some of the larger icy satellites such as Ganymede and Europa show evidence of a similar surface process operating in water ice.

Platonic year The time taken for the Earth's poles to describe one complete circle on the celestial sphere as the result of *precession: 25800 years.

PLC relation Abbr. for *period–luminosity–colour relation.

Pleiades A prominent open cluster in Taurus, popularly termed the Seven Sisters, and also known as M45. The cluster spans over 1½° of sky and contains about 100 stars, the brightest of which is 3rd-magnitude Alcyone. Several other members are visible to the naked eye. Its true diameter is about 13 l.y., although there may also be scattered outliers. The cluster lies nearly 400 l.y. away and is about 80 million years old. Its stars are embedded in a reflection nebula, which is now thought to be the result of a chance encounter rather than the nebula being the remains of the gas cloud from which the stars formed. 📷

plerion A rare form of supernova remnant in which radiation is being emitted from the central region as well as from the expanding shell; also known as a *filled-centre supernova remnant*. Plerions tend to be relatively young remnants containing a pulsar, the energy from which makes central regions glow by synchrotron radiation. The best-known example is the *Crab Nebula.

Plössl eyepiece A design of eyepiece, similar to a *Kellner or achromatic Ramsden but with a two-element *field lens. Aberrations, particularly astigmatism, are less than with the Kellner, and it gives good *eye relief. Plössls will give good performance even with telescopes having a *focal ratio as short as *f*/4. The design was invented in 1860 by the Austrian optician (Georg) Simon Plössl (1794–1868).

Plough Popular name for the shape formed by the stars Alpha, Beta, Gamma, Delta, Epsilon, Zeta, and Eta Ursae Majoris, which resembles the outline of an old horse-drawn plough. An alternative name for the same shape is the Big Dipper.

PL relation Abbr. for *period–luminosity relation.

plume 1. A cloud of gas, liquid, aerosols, or particles originating from within a planet. The term is most commonly used for the gas and aerosols emitted by a volcano that drift downwind. The plumes above the erupting volcanoes of Io are the largest and most spectacular such examples, some rising to 280 km altitude and depositing material 500 km from the vent. The plumes seen rising from the surface of Triton by Voyager 2 appear to be from geyser-like eruptions.

2. A ray-like feature seen in the Sun's atmosphere. *See* CORONAL PLUME.

Plutino A member of the *Kuiper Belt with an average distance from the Sun of around 39.5 AU, the same as that of Pluto, hence the name Plutino ('little Pluto'). An object at such a distance orbits the Sun twice in the time that Neptune takes to complete three orbits, and so is in a 3:2 resonant orbit with Neptune. About 25% of the known Kuiper Belt objects are Plutinos, although this percentage will probably fall as more distant members of the Kuiper Belt are discovered. The first object confirmed to be in such an orbit was 1993 SC, subsequently numbered (15789), which has a semimajor axis of 39.76 AU, perihelion 32.26 AU, aphelion 47.26 AU, inclination 5°.1, and period 250.7 years.

Pluto (♇) The smallest of the nine 'traditional' planets in the Solar System, and the farthest from the Sun. In 2006 the International Astronomical Union introduced a new definition which categorizes Pluto as a *dwarf planet* rather than one of the major planets; Pluto is now assigned the minor planet number (134340). This reclassification recognizes the fact that Pluto has many characteristics which distinguish it from the eight planets from Mercury to Neptune. It is far smaller than any of those, with a diameter of only 2390 km, less than that of our Moon. Its orbit has a greater inclination than any of the major planets, 17°.1 to the ecliptic, and its orbit is also the most elliptical (eccentricity 0.25). At aphelion Pluto lies 7375 million km from the Sun, but only 4425 million km at perihelion, inside the orbit of Neptune; it last reached perihelion in 1989. Its mean opposition magnitude is +15. Pluto was discovered in 1930 by C. W. *Tombaugh. Its rotation axis is tilted at 122°.5 to its orbital plane, so that its rotation is retrograde, and it presents its poles and its equator alternately towards the Sun and the Earth as it moves around its orbit. Its axial rotation period, 6.387 days, is the same as the orbital period of its largest satellite, *Charon, so that Pluto always keeps the same face towards Charon. Two much smaller and more distant moons, Hydra and Nix, were discovered by the Hubble Space Telescope in 2005, and a fourth in 2011.

Pluto

Physical data

Diameter	Oblateness	Inclination of equator to orbit	Axial rotation period (sidereal)	
2390 km	0	122°.5	6.387 days	
Mean density	Mass (Earth = 1)	Volume (Earth = 1)	Mean albedo (geometric)	Escape velocity
1.8 g/cm³	0.0022	0.007	0.3	1.3 km/s

Orbital data

Mean distance from Sun

10⁶ km	AU	Eccentricity of orbit	Inclination of orbit to ecliptic	Orbital period (sidereal)
5906.4	39.48	0.25	17°.1	247.9 years

Pluto has an extremely thin atmosphere with a surface pressure of about 10 μbar, composed of methane, possibly with some nitrogen and carbon monoxide. This atmosphere may be seasonal, forming when the planet heats up and releases surface volatiles around the time of perihelion. Methane may escape from the atmosphere near perihelion, so that Pluto behaves somewhat like a comet. Its mean surface temperature is estimated at −220°C. Pluto is thought to have a large rocky core, probably surrounded by a layer of frozen water and other icy materials, and a surface layer of methane. Pluto is now regarded as simply the largest of the sub-group of trans-Neptunian objects known as *Plutinos.

plutoid A *dwarf planet orbiting the Sun beyond Neptune. The name was introduced by the International Astronomical Union in 2008. In their official definition, a plutoid is a celestial body in orbit around the Sun at a semimajor axis greater than that of Neptune that has sufficient mass for its self-gravity to overcome rigid body forces so that it assumes a hydrostatic equilibrium (near-spherical) shape, and that has not cleared the neighbourhood around its orbit. Satellites of plutoids are not themselves plutoids, even if they are massive enough to be rounded in shape. The first four named plutoids are Pluto, Eris, Haumea, and Makemake.

P magnitude The magnitude of a star measured with a filter with wavelength 12.2 μm and a bandwidth of 1.0 μm. This infrared magnitude is now little used.

Pockels cell A device used to change the nature of polarized light for analysis. It consists of a crystal of potassium, hydrogen, and phosphorus with a fine grid of gold electrodes on its surface. Depending on the voltage applied to the electrodes, light polarized in one direction can be arranged to emerge from the device 90° or 180° out of phase with the light polarized at right angles to it. The Pockels cell then acts as a *quarter-wave plate* or a *half-wave plate* respectively (*see* WAVE PLATE). It is named after the German physicist Friedrich Karl Alwin Pockels (1865–1913). Pockels cells are used in some *polarimeters and spectropolarimeters.

Pogson, Norman Robert (1829–91) English astronomer. He discovered eight asteroids, the first, (42) Isis, in 1856. It was in connection with the light curves of variable stars, his other main interest, that he introduced a mathematically rigorous definition of stellar magnitudes based on a logarithmic scale (*see* POGSON RATIO; POGSON SCALE). Others, notably the German scientist Carl August von Steinheil (1801–80), had proposed similar scales, but Pogson's advocacy led to its adoption as a universal standard.

Pogson ratio The brightness ratio between two objects that differ by one magnitude. Because of the way in which the *Pogson scale is defined, this ratio is 2.512.

Pogson scale The standard scale of magnitude, which was formulated mathematically by N. R. *Pogson in 1856. He proposed that a difference of 5 magnitudes should be defined as corresponding to a difference of exactly 100 in the intensities of the stars concerned. A difference of 1 magnitude therefore corresponds to the fifth root of 100, which is 2.512.

Pogson step method A visual method of estimating the magnitude of variable stars, based on training the eye to recognize the difference between the variable and a comparison star in steps of 0.1 magnitude. Consistent estimates by this method are generally possible only after considerable experience. The comparison star should not be more than 0.5 magnitude different in brightness from the variable. *See also* ARGELANDER STEP METHOD; FRACTIONAL METHOD.

Poincaré, (Jules) Henri (1854–1912) French mathematician. From 1889 he worked on the *three-body problem in both its general and restricted forms, and showed that there are no exact solutions to the *n-body problem. He expanded the field of celestial mechanics, for example demonstrating the possibility of *chaotic orbits. Poincaré's mathematical studies of special relativity were the first of importance.

Pointers The stars Alpha and Beta Ursae Majoris (*Dubhe and *Merak). A line drawn from Merak through Dubhe points towards the north pole star, Polaris.

point source A source with an angular size less than the resolution of the instrument used to observe it, and therefore unresolved. Stars appear as point sources to the human eye. An object that appears as a point source at some wavelengths (e.g. long radio wavelengths) may be resolvable at shorter wavelengths, or with other instruments. *See also* EXTENDED SOURCE.

point-spread function A mathematical description of the image of a point source formed by a telescope and associated instrumentation either before or after any image processing. Examples of point-spread functions include the diffraction rings of an optical telescope and the beam pattern of a radio telescope.

polar *See* AM HERCULIS STAR; INTERMEDIATE POLAR.

polar axis One of the axes of an *equatorial mounting. The polar axis is aligned parallel to the Earth's axis, and thus points towards the celestial pole.

polar cap A bright surface layer of ice (either frozen water or some other volatile compound) covering the poles of a planetary body, usually varying in size during the year. On Earth, the polar caps are made of water ice, whereas on Mars they consist mostly of carbon dioxide with a layer of water ice underneath that persists from year to year. The transient Martian polar caps of carbon dioxide can extend down to latitude 60° in winter, but retreat back to 85–87° latitude at the height of summer, revealing the permanent cap of water ice. Of the other bodies in the Solar System only Triton has a definite polar cap, which appears to consist of solid nitrogen and may vary seasonally.

polar diagram A graph showing the variation in sensitivity of a radio antenna with direction, which is also generally the same as the pattern of energy radiated by the same antenna when used as a transmitter (*see* ANTENNA PATTERN). For many antennas, the pattern will vary in three dimensions, so that two polar diagrams at right angles are needed to show the complete variation. Polar diagrams for arrays of antennas usually show a main lobe, along the centre of which the sensitivity is highest, but also side lobes which show that the antenna will detect some sources at large angles from its central axis.

polar distance The angular distance of an object from the celestial pole, measured along a line at right angles to the celestial equator (the *hour circle). It is 90° minus the object's declination, and is sometimes used as a coordinate in place of *declination.

polar flattening Another term for *oblateness.

polarimeter An instrument for measuring the *polarization of light. In a typical polarimeter the light passes through a polarizing device which is rotated between readings. The intensity of the emergent light in various planes depends on the orientation of the polarizer with respect to the direction of polarization of the source. The intensity is measured with a *photoelectric photometer with filters. For stars and other point sources the results are stated as the percentage of the light that is polarized. For extended sources, a false-colour image may be used, or the direction and strength of the polarization may be shown by arrows superimposed upon a normal image. A photometer combined with a polarimeter is termed a **photopolarimeter**. *See also* SPECTROPOLARIMETRY.

polarimetry The study of the *polarization of light.

Polaris The north pole star, Alpha Ursae Minoris, magnitude 2.0. It is an F-type supergiant, 433 l.y. away. Polaris is currently less than 1° from the north celestial pole, and the distance is gradually decreasing due to precession; it will be closest to the pole, just under ½°, around the year 2100. Polaris is a Cepheid variable with a period of 4 days and a small range, originally about 0.1 mag. but which decreased during the 20th century to only a few hundredths of a magnitude now. Polaris is also a spectroscopic binary with a period of 30 years. From observations with the Hubble Space Telescope this companion appears to be a dwarf F star. There is a more distant companion of magnitude 8.2 which can be seen with a small telescope. 📷

polarization The phenomenon in which electromagnetic waves, such as light waves, vibrate in a preferred plane or planes; or the process of confining the vibrations to certain planes. In unpolarized light the vibrations are equally distributed in all directions perpendicular to the direction of propagation of the wave. If all the vibrations are confined to one plane, the light is said to be **plane-polarized** (or *linearly polarized*). If the light in one plane is out of phase with the light in the plane at right angles to it (i.e. if the peaks and troughs of the waves are not in step), then the light is said to be *circularly polarized*. If all these phenomena occur together, the light is said to be *elliptically polarized*. Plane polarization is usually caused by scattering, and circular polarization by strong magnetic fields. Circularly and elliptically polarized light can also be produced by a *wave plate. *See also* STOKES PARAMETERS.

polar motion The displacement within the Earth of the axis of rotation. The principal components are an annual term, and a 14-month term known as the *Chandler wobble. The maximum displacement is about 0″.1, which corresponds to 10 metres on the surface of the Earth.

polar orbit An orbit with an inclination close to 90°, passing over, or almost over, the poles of a planet. A satellite in polar orbit will in due course pass over every point on the planet below as the planet rotates on its axis.

polar plume *See* CORONAL PLUME.

polar ring galaxy A rare type of galaxy, almost always a *lenticular galaxy, that has a luminous ring of stars, gas, and dust orbiting over the poles of its disk. Hence, the rotation axes of the ring and disk are almost at right angles. Such a system may be the result of a collision, tidal capture, or merger of a gas-rich galaxy with the lenticular galaxy.

polar sequence *See* NORTH POLAR SEQUENCE.

polar wandering The irregular movement of the Earth's geographical poles due to the *Chandler wobble.

pole The direction perpendicular to a given plane, such as the plane of the Earth's equator or the plane of the ecliptic. For a rotating body, the poles lie at the ends of the body's axis of rotation. *See also* CELESTIAL POLE; ECLIPTIC POLE; GALACTIC POLE.

pole star The naked-eye star nearest the north or south celestial pole. The current north pole star is *Polaris, and the south pole star is *Sigma Octantis. However, the position of the celestial pole (and hence the pole star) changes with time due to the effect of *precession.

Pollux The star Beta Geminorum, magnitude 1.16. It is a K0 giant 34 l.y. away.

polycyclic aromatic hydrocarbon (PAH) A large molecule containing carbon atoms arranged in a honeycomb lattice of hexagonal groups, with hydrogen atoms attached. Emission from PAHs is responsible for the widespread mid-infrared emission known as *infrared cirrus. These large molecules behave differently from small dust particles. In particular, they can reach very high temperatures immediately following the absorption of light waves, which leads to emission at shorter infrared wavelengths.

Pond, John (1767–1836) English astronomer. He was appointed the sixth Astronomer Royal in 1811. Pond installed new instruments at Greenwich Observatory, having previously revealed errors in star positions caused by warping of the observatory's ageing equipment, and in 1833 produced a star catalogue of unprecedented accuracy. That same year he introduced the first public time signal by dropping a time ball each day at 1 p.m.

Pons, Jean Louis (1761–1831) French astronomer. His knowledge of the night sky, coupled with acute vision and great patience, made him highly successful at comet-hunting. His first comet, in 1801 (now designated C/1801 N1), was the first of 37 (including co-discoveries)—still a record for visual discoveries. Most of these discoveries were non-periodic comets, but he did co-discover the periodic comets 12P/Pons–Brooks (1812) and 7P/Pons–Winnecke (1819). Two comets he found in 1805 were later identified as Comets *Biela and *Encke.

population, stellar A classification of stars on the basis of certain physical characteristics, such as their location in the Galaxy, the types of orbits they have around it, and their content of heavy elements. Each of these properties is believed to depend on the age of the Galaxy when the star formed, so that *Population I, *Population II, and *Population III contain stars formed at progressively earlier epochs.

Population I Those stars that, like the Sun, lie in the disk of our Galaxy and follow roughly circular orbits around its centre. They have a high content of heavy elements and have probably been formed continuously during the lifetime of the disk, from gas enriched by the debris from supernovae in Population II stars. *See also* DISK POPULATION; EXTREME POPULATION I STAR.

Population II Those stars that are found in the halo of our Galaxy and in its central bulge. The stars in the galactic halo, including globular clusters, describe highly elliptical orbits around the Galaxy's centre. Members of Population II have a significantly lower content of heavy elements than stars of Population I. It is believed that Population II stars formed in the first billion years or so of the Galaxy's life, before the formation of the galactic disk. *See also* HALO POPULATION.

Population III A hypothetical generation of stars, no longer observable, assumed to have been formed before those of Population II. They are presumed to have existed because their supernovae would have been needed to supply the heavy elements observed in Population II stars. The neutron stars or black holes produced by such supernovae might be candidates for the *dark matter in the galactic halo.

pore A small, dark area on the Sun's photosphere, from which a *sunspot may develop. Pores last less than an hour, and are up to 2000 km across. In an alternative usage, a pore is a sunspot without a penumbra.

Porrima The star Gamma Virginis. It is a binary, consisting of a pair of F-type dwarfs each of magnitude 3.5, which orbit each other in a period of 169 years. Together they appear to the naked eye as a star of magnitude 2.7. Porrima lies 38 l.y. away.

Porro prism A prism with one 90° and two 45° apexes, used in binoculars. Light entering via the long face is totally internally reflected and emerges through the same face. A pair of such prisms are used in prismatic binoculars to both invert the image and fold the light path, making the binoculars more compact. This system was devised in 1851 by the Italian geodesist and optician Ignazio Porro (1801–75).

Portia The seventh-closest satellite of Uranus, distance 66100 km, orbital period 0.513 days; also known as Uranus XII. Its diameter is 108 km, and it was discovered in 1986 on images taken by the Voyager 2 spacecraft.

positional astronomy *See* ASTROMETRY; MERIDIAN ASTRONOMY.

position angle (PA) The direction in which one object lies relative to another on the celestial sphere, such as the two components of a double star or the axis of rotation of the Sun or a planet. Position angle is measured in degrees from north via east, although for the axial inclination of the Sun it is measured positive (+) to the east and negative (−) to the west (see diagram).

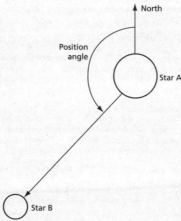

position angle: The direction of one object, star B, relative to another, star A, on the celestial sphere (on which east is anticlockwise from north)

position-angle effect An error affecting visual estimates of variable stars. It causes a difference in the perceived brightness of a star, relative to another, depending on its position in the field of view. Generally, the star that is either lower in the field or closer to the observer's nose appears relatively brighter. The effect is caused by the variation of sensitivity across the retina, and is independent of variations caused by colour differences such as the *Purkinje effect.

position circle A circle on the Earth's surface on which the observer is situated at some point. The centre of the circle is directly under a celestial body, and its radius is equal to the zenith distance of the star. The position of the circle on the Earth is calculated by measuring the altitude of the celestial body at a particular time. In practice it is possible to select a small part of the circle, called the **position line**, on which the observer is situated. If a second position line is found from another observation (either of the same object at a

different time, or of a different object at the same time) the intersection of these lines
identifies the observer's location.

position micrometer *See* FILAR MICROMETER.

positive lens Another term for a *converging lens.

POSS Abbr. for *Palomar Observatory Sky Survey.

post-AGB star *See* ASYMPTOTIC GIANT BRANCH STAR.

post-nova The stage following a nova outburst, when the star has returned to a quiescent
state. Some decades after the eruption the spectrum remains characteristic, with H I and
He I emission, accompanied by high-excitation lines such as He II, C III, and N III. Such
a spectrum is sometimes encountered in a star for which no outburst has been observed
(a *quiescent nova*).

potassium–argon method A technique for dating rocks which utilizes the
weak radioactive decay of potassium-40, the only naturally occurring radioactive
isotope of potassium. The isotope potassium-40 decays to argon-40 with a half-life of
1.3 billion years. The method is important in determining the age of specimens such
as meteorites.

potentially hazardous asteroid (PHA) An asteroid with an orbit that is predicted to
bring it within about 0.05 AU of the Earth (i.e. about 20 times the distance of the Moon),
and a size large enough to cause significant damage should it hit the Earth, usually taken as
a diameter of at least 150 m. By the end of 2010, nearly 1200 PHAs were known.

(()) SEE WEB LINKS
• NASA information page.

power spectrum A plot of the way in which the intensity of the frequency components
of a signal (or other function) vary with frequency. Usually a power spectrum is produced
from a Fourier transform of the signal.

Poynting–Robertson effect An effect in which dust particles in space are slowed down
in their orbits and spiral into the Sun. It occurs because the particles absorb solar radiation
and then re-radiate it in all directions, which slows them down. The effect is particularly marked
for small (micrometre-sized) particles. However, for the smallest particles (under about a
micrometre), *radiation pressure is greater. The effect is named after the English physicist
John Henry Poynting (1852–1914) and the American mathematician and cosmologist Howard
Percy Robertson (1903–61).

PPM Star Catalogue A catalogue listing the positions and proper motions (hence PPM)
of 378910 stars across the whole sky for epoch 2000.0, published by the Astronomisches
Rechen-Institut, Heidelberg. It was based on the same selection of stars as the *AGK3, and
superseded both the AGK3 and the *Smithsonian Astrophysical Observatory Star Catalog. PPM
was the first large star catalogue of positions and proper motions to be based on the FK5
reference system (the AGK3 and SAO catalogues used the FK4 system). PPM North, containing
181731 stars north of declination −2.5°, was published in 1991. PPM South, a southern
hemisphere extension containing 197179 stars, was published in 1993. A supplement
containing an additional 89676 southern stars was published electronically in 1994.

(()) SEE WEB LINKS
• Detailed description and full catalogue of PPM North downloadable from the CDS.
• Detailed description and full catalogue of PPM South downloadable from the CDS.
• Supplement to the PPM Catalogue downloadable from the CDS.

p-process A type of nuclear reaction that has been proposed to explain the formation of
rare proton-rich isotopes of certain elements, such as tin. This involves the capture of

protons by nuclei already produced by the *r-process and *s-process. It can occur only in the envelopes of supernovae where the temperature exceeds 10^9K.

Praesepe The 3rd-magnitude open cluster M44 in Cancer, also known as NGC 2632, the Beehive Cluster, or the Manger. It is 1½° across and contains about 50 stars of 6th magnitude and fainter. It is centred about 590 l.y. away.

pre-cataclysmic variable A *detached binary consisting of a red dwarf and a white dwarf that has a short enough period that it may evolve into a *cataclysmic binary through loss of angular momentum on a timescale of 10^8 to 10^9 years. Also known as a *post-common-envelope binary*. *See also* MAGNETIC BRAKING; GRAVITATIONAL WAVE; COMMON ENVELOPE BINARY.

preceding Referring to the side of an object, or member of a group of objects, that leads in motion across the sky or across the face of a rotating body. Examples are the preceding component of a double star or a group of sunspots; the preceding side of a planetary feature; the preceding limb of a planet; or the preceding side of a telescopic field of view. The trailing side is said to be *following*.

precession The wobbling motion of a spinning top or gyroscope in which the axis of rotation gradually sweeps out a conical shape. The spinning Earth undergoes a slow precession, due to the combined gravitational attractions of the Sun, Moon, and planets. The Earth's pole takes about 25800 years to describe one complete circle on the celestial sphere; this circle has a radius of approximately 23°.5, i.e. the inclination of the Earth's axis. The equinoxes make one circuit of the ecliptic in the same time. As a result of precession, the right ascension and declination of stars change with time, so the date or *epoch for which these coordinates apply must always be stated. *See also* LUNISOLAR PRECESSION; PLANETARY PRECESSION; PRECESSION OF THE EQUINOXES.

precession constant The annual rate of precession of the equinoxes. It has the value 50″.29 at epoch 2000.0, but is not quite constant because it depends on the eccentricity of the Earth's orbit, which changes slowly.

precession of the equinoxes The motion of the equinoxes along the ecliptic, arising from the combined motion of the equator (*lunisolar precession) and the ecliptic (*planetary precession); also known as general precession. It was first detected by the Greek astronomer *Hipparchus in about 130 BC from the apparent increase in the observed celestial longitudes of stars. It amounts to about 50″.3 per year. Hence the equinoxes move westwards on the celestial sphere by 1° in about 72 years, and take 25800 years to complete one circuit.

precursor pulse A small pulse that appears shortly in advance of the main pulse from a pulsar.

pre-nova The precursor to a nova, which is expected to be a form of *cataclysmic binary. To date, no object that has been studied in detail has subsequently become a nova, but some novae have been identified on earlier photographs. Of the few pre-outburst light-curves that can be constructed, some appear to show a rise in brightness of 1–4 mag. (perhaps with oscillations), over a period of 1–5 years before the explosion.

pressure broadening The broadening of spectral lines from a star due to the high pressure of gas increasing the number of collisions between atoms in a star's atmosphere. In stellar atmospheres, high pressures are associated with strong surface gravity; thus main-sequence stars show intrinsically broader lines than do supergiants, while white dwarfs have extremely broad hydrogen lines because of this effect. *See also* STARK BROADENING.

Příbram meteorite The first meteorite fall to be photographed by a camera network, enabling its trajectory and orbit to be determined and the meteorite to be recovered. Cameras operated by the Ondrejov Observatory in the Czech Republic photographed a brilliant fireball on 1959 April 7. Nineteen fragments, totalling 9.5 kg, of an H5 *ordinary chondrite meteorite were subsequently found at the impact site, close to the town of Příbram, near Prague. The largest fragment weighed 4.3 kg. Its calculated aphelion was in the outer main asteroid belt.

primary cosmic ray An atomic particle moving close to the speed of light, with extremely high energy (10^8–10^{20} eV). On collision with atoms in the Earth's atmosphere, a primary cosmic ray produces showers of secondary cosmic rays. These secondary particles decay into electrons, positrons, and neutrinos. Deceleration of the electrons and positrons by the atmosphere produces a flash of light which can be observed from the ground with a special telescope, and gives information about the primary cosmic ray.

primary minimum The deepest minimum in the light-curve of an eclipsing binary. It occurs when the **primary** (the star with the greater surface brightness) is eclipsed by the fainter *secondary* star.

primary mirror The main light-gathering mirror of a reflecting telescope, which collects and focuses incoming light.

prime focus The focal point of the primary mirror or objective lens of a telescope. In a reflector or Schmidt camera, the prime focus lies inside the top end of the tube.

prime meridian A fixed circle of longitude on the surface of a body, used as a reference from which other longitudes can be measured west or east. The prime meridian on Earth passes through the *transit circle established by G. B. *Airy at Greenwich, and defines the zero of terrestrial longitude. On other bodies the prime meridian is defined with respect to a known surface feature, such as a small crater.

prime vertical The great circle that passes through the observer's zenith, perpendicular to the meridian. It intersects the horizon at the west and east points. The parts of the circle going from zenith through east and west are known as the **prime vertical east** and the **prime vertical west**.

primordial fireball An alternative term for the *radiation era in the Big Bang theory, during which the Universe was hot and dense and was dominated by the effects of radiation. The existence of such a phase is strongly suggested by the observed properties of the *cosmic microwave background.

primordial galaxy A galaxy that formed very early in the history of the Universe and which today would be observed at very high redshift. Galaxies have now been found out to redshifts of about 6, which means we can see what they looked like back to within a billion years after the Big Bang. *See also* GALAXY EVOLUTION.

Principia The short form of the title of I. *Newton's treatise *Philosophiae naturalis principia mathematica* ('Mathematical Principles of Natural Philosophy', first edition 1687, second edition 1713). One of the most important scientific books ever written, it deals with the celestial mechanics of the Solar System and for the first time weaves together mathematical analysis and physical observation. It sets out *Newton's law of gravitation and *Newton's laws of motion, presents derivations of *Kepler's laws, and shows how the inverse-square law of gravitation produces motion in an ellipse. There are mathematical treatments of many other phenomena, including the tides and cometary orbits.

principle of equivalence *See* EQUIVALENCE PRINCIPLE.

prism A solid block of glass or other transparent substance, with at least two flat faces inclined to each other at an angle called the *refracting angle*. The intersection of these two faces is called the *edge* of the prism, and the cross-section at right angles to the edge is known as the prism's *principal section*. The main use of prisms in astronomy is to disperse light into a spectrum for spectroscopy; they are also used in binoculars to fold the light path. *See also* PORRO PRISM; ROOF PRISM.

prismatic astrolabe A horizontal telescope for measuring the instant at which a star reaches a specific altitude on the celestial sphere. A prism is mounted with one face vertical immediately in front of the telescope's objective lens. Below the prism is a horizontal mirror formed by a dish of mercury. The star's image, and its reflection in the mercury mirror, are reflected internally in the prism. The two images coincide in the focal plane of the telescope when the star's zenith distance is half the angle at the leading edge of the prism (30° for a 60° prism).

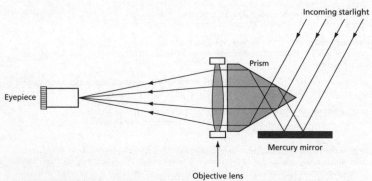

prismatic astrolabe: Light from a star is directed into a telescope by a prism. Some starlight falls directly on to the prism, and some is reflected into the prism via a mercury mirror. The two images converge when the star reaches a precise altitude that is fixed by the angle of the prism, usually 30° from the zenith. In practice, the light path is folded inside the astrolabe for compactness

probable error A measure of the precision with which a value has been obtained from a series of measurements. For example, the parallax, π, of a star may be given as $\pi = 0''.096 \pm 0''.003$. Here, $0''.096$ is the value and $0''.003$ is the probable error. It means that the probability that the true value of π lies between $0''.093$ and $0''.099$ is equal to the probability that it lies outside the range $0''.093$ to $0''.099$.

Procyon The star Alpha Canis Minoris, magnitude 0.40, the eighth-brightest star in the sky. It is an F5 subgiant or dwarf and lies 11.5 l.y. away, among the nearest stars to us. Procyon has a white dwarf companion of spectral type DA, visual magnitude 10.9, and absolute magnitude 13.2, which orbits it in a period of 41 years. The masses of the two stars are 1.5 and 0.6 that of the Sun.

prograde motion Another name for *direct motion.

prolate spheroid *See* SPHEROID.

Prometheus The third-closest satellite of Saturn, distance 139380 km; also known as Saturn XVI. It orbits Saturn in 0.613 days just outside Saturn's A Ring and acts as a shepherd moon to the F Ring beyond it. Prometheus is irregularly shaped and $136 \times 79 \times 59$ km in size. It was discovered in 1980 on images taken by Voyager 1.

prominence A cloud-like feature, visible especially in *Hα light, located in the Sun's corona but cooler and denser than the corona. Prominences have temperatures of around 10 000 K, typical of the solar chromosphere, and densities 100 times higher than the corona. They are often seen around the Sun's limb at total solar eclipses. In *Hα light they can be seen silhouetted against the Sun's disk, when they are termed *filaments. They are categorized as *quiescent prominences or *active prominences, according to their behaviour. Active-region prominences have rapid motions and last for only a few days, whereas quiescent prominences last for at least a month (one solar rotation). Between the relatively cool prominence material and the hot corona is a *transition region* or *sheath* in which temperatures range from 15000 to 60000K. Prominences closely follow the *magnetic inversion line, and are thought to be supported by magnetic fields. They are most frequent during the rising part of the solar cycle.

prominence spectroscope A spectroscope by which solar prominences can be observed. This is usually done by observing the *Hα line and opening the spectroscope slit to reveal any prominences present on the Sun's limb.

promontorium A headland or cape extending from a larger upland on the Moon; pl.*promontoria*. The name is not a geological term, but is used in the nomenclature of individual features, for example Promontorium Heraclides.

proper motion (μ) The progressive change in position of a star due to its motion relative to the Sun. *Absolute proper motion* is the change in position relative to an inertial reference frame such as that defined by extragalactic objects. *Relative proper motion* is obtained by measurements relative to selected reference stars at different epochs. The average absolute proper motion of these reference stars is known as the *reduction from relative to absolute* proper motion. Relative proper motions can be used to separate members of a star cluster from *field stars in the same line of sight. Proper motions are usually listed in star catalogues as changes in right ascension and declination per year or century. The star with the largest known proper motion is *Barnard's Star. *See also* COMMON PROPER MOTION; REDUCED PROPER MOTION.

proper time The time as measured by any observer from a clock at their location that is stationary with respect to them. In the theory of relativity, time is said to be proper to each observer. This relativity of time is an essential feature of the special theory of relativity, and makes a moving clock appear to run slow (*see* TIME DILATION). Each observer will record the clock as ticking at a different rate, because each is comparing it with their own proper time. The time recorded on the clock face, which is something all observers can agree upon, is the clock's own proper time. In the general theory of relativity a gravitational field also makes a clock appear to run slow. In both general and special relativity the proper time interval between two neighbouring events can be defined as the time measured by an observer present at both of them. *Terrestrial Time, the time-scale used in *The Astronomical Almanac* for geocentric predictions, is the proper time for an observer on the Earth's surface.

proplyd A disk of gas and dust orbiting a young star from which a planetary system may form; the word is a contraction of *protoplanetary disk*. The first proplyds were photographed in the Orion Nebula in 1992 by the Hubble Space Telescope.

proportional counter A detector for X-rays and low-energy gamma rays, consisting of a chamber filled with a noble (i.e. unreactive) gas such as argon or xenon. Ionization of the gas by an X-ray or gamma-ray photon in an electrical field produces an avalanche of electrons. Since the size of the electrical pulse is proportional to the energy of the photon, the detector is able to give spectral information. The instrument can also be used as an imaging device when a wire grid arrangement is incorporated into the construction.

Prospero A retrograde satellite of Uranus outermost but one at a distance of 16 089 000 km; also known as Uranus XVIII. It orbits every 1948.1 days at an inclination of 146° to Uranus's equator. Prospero has a diameter of about 50 km and was discovered in 1999 from ground-based observations.

Proteus The second-largest satellite of Neptune, $436 \times 416 \times 402$ km in size; also known as Neptune VIII. Proteus was discovered in 1989 on images from the Voyager 2 spacecraft, and orbits the planet in 1.122 days at a distance of 117647 km. It has a rough surface covered with impact craters, including one over 200 km in diameter, named Pharos.

protogalaxy The precursor of a galaxy. In the Big Bang model, galaxies are thought to form from large gas clouds which collapse under their own self-gravity. A protogalaxy is such a cloud which is just beginning to collapse. No protogalaxy has yet been unambiguously identified from astronomical observations.

proton An elementary particle that has a positive charge, equal and opposite to that of the electron. It is a *hadron, with a *rest mass of 1.673×10^{-27} kg, which is 1836.15 times that of the electron. It is present in the nuclei of all atoms; the hydrogen nucleus is a single proton.

proton–proton reaction A chain of nuclear reactions inside stars that converts hydrogen into helium, with the associated release of energy. In the reaction, four hydrogen nuclei (protons) fuse to form one nucleus of helium, with the production of a number of intermediate nuclei such as deuterium and isotopes of lithium, beryllium, and boron. At temperatures below 18 million K the proton–proton reaction is more important than the *carbon–nitrogen cycle, and thus operates chiefly in stars of less than 2 solar masses.

protoplanet A body presumed to have formed in the early history of the Solar System, from which the major planets grew. Protoplanets are believed to have formed from the aggregation of *planetesimals. They eventually became the major planets by sweeping up other bodies that crossed their orbits, in a process of *accretion*.

protoplanetary nebula 1. An early stage in the formation of a *planetary nebula. In this phase the central star has shed its outer layers, exposing the hot stellar core. Ultraviolet light from the core starts to ionize the surrounding cloud of gas and dust, and for a brief phase the circumstellar envelope contains both hot ionized material near the star and cool molecular material far from the star.
   2. A cloud from which planets are formed around a newborn star, as in the *solar nebula.

protostar A star in the earliest phase of its life, condensing out of a cloud of gas and dust, before the onset of nuclear burning. Its mass grows over about 100000 years as material falls on to it from the surrounding cloud. A protostar is not visible at optical wavelengths because the infalling material obscures it, but it is bright at infrared wavelengths.

protosun The Sun in the initial stages of its formation, nearly 5 billion years ago. The protosun had less mass than the Sun has today, and was larger, with a radius comparable to that of the orbits of the inner planets. Over about a million years the protosun's mass increased through the addition of gas from a surrounding disk, the *solar nebula*, while the weight of this infalling material made it shrink to only a few times its present size.

Proxima Centauri The closest star to the Sun, 4.23 l.y. away, the third member of the *Alpha Centauri system. It is an M5 dwarf of magnitude 11.0, but is also a *flare star and undergoes sudden brightness increases of up to 1 mag. lasting several minutes. Intrinsically it is 20000 times fainter than the Sun. It is thought to be gravitationally bound to the other two members of Alpha Centauri, which lie 0.09 l.y. farther away. Its orbital period around them must be a million years or more.

p-spot The preceding member of a pair of sunspots.

Ptolemaic system The ancient Greek geocentric model of the Solar System, as described by *Ptolemy. It may be traced back through the work of, for example, *Hipparchus, *Apollonius, *Callippus, and *Eudoxus. The Earth is placed at the centre of the Universe, and around it revolve the Moon, Mercury, Venus, the Sun, Mars, Jupiter, and Saturn; beyond Saturn is the sphere of the fixed stars. In the basic model each body moves along the circumference of a small circle, the *epicycle, whose centre in turn follows the circumference of a larger circle, the *deferent, centred on the Earth. In later refinements, Ptolemy introduced two points equally spaced on either side of the Earth: the *eccentric* and the *equant*. The centre of the epicycle revolved around the eccentric, not the Earth, and the orbiting body moved uniformly with respect to the equant. As a computational device the Ptolemaic system predicted planetary movements, including their retrograde motion, tolerably well, and survived with minor amendments until displaced by the *Copernican system in the 16th century.

Ptolemy (Claudius Ptolemaeus) (2nd century AD) Egyptian astronomer and geographer. He produced the *Almagest*, a compendium of contemporary astronomical knowledge, drawing on writers, such as Plato and *Hipparchus, whose works were kept in the great library at Alexandria. His *Ptolemaic system was a geocentric model of the Universe. Highly contrived as it now appears, it accounted for the observed apparent motions of the planets reasonably well, and remained largely unquestioned until the 16th century, when it was challenged by N. *Copernicus. Ptolemy's *Geography* enjoyed a similar period of dominance (it convinced Columbus that he could sail westwards to India); his *Tetrabiblos* was an astrological treatise.

Publications of the Astronomical Society of the Pacific (PASP) A journal of research papers and review articles, founded in 1889 by the *Astronomical Society of the Pacific and now published monthly.

(🌐) SEE WEB LINKS
• Official website with contents and abstracts of recent issues and free online access to older issues.

Puck A satellite of Uranus, tenth from the planet at 86000 km, and with an orbital period of 0.762 days; also known as Uranus XV. It is 154 km in diameter, and was discovered in 1985 on images taken by the Voyager 2 spacecraft.

Pulcherrima Alternative name for the star *Izar (Epsilon Boötis).

pulsar A radio source from which is received a highly regular train of pulses. As of the end of 2010 nearly 2000 pulsars had been catalogued since the first was discovered in 1967. Pulsars are rapidly spinning *neutron stars, 20–30 km in diameter. The stars are highly magnetized (about 10^8 tesla), with the magnetic axis inclined to the spin axis. The radio emission is believed to arise from the acceleration of charged particles above the magnetic poles. As the star rotates, a beam of radio waves sweeps across the Earth and a pulse is seen, much like the beam from a lighthouse. Pulse periods are typically 1s, but range from 1.4 ms (*millisecond pulsars) to over 10s. The pulse periods are lengthening gradually as the neutron stars lose rotational energy, but a few young pulsars are prone to abrupt disturbances known as *glitches. Precise timing of pulses has revealed the existence of *binary pulsars, and two pulsars, PSR 1257+12 and PSR B1620-26, have been shown to be accompanied by objects of planetary mass. Optical flashes have been detected from a few pulsars, notably the Crab and Vela Pulsars.

Most pulsars are believed to have been created in supernova explosions by the collapse of the core of a supergiant star, but there is now considerable evidence that at least some of them originate from white dwarfs that have collapsed into neutron stars following accretion of mass from a companion star (*see* RECYCLED PULSAR). The great majority of known pulsars are members of the Milky Way and are concentrated in the galactic plane. There are estimated to be about 100000 pulsars in the Galaxy. Observations of interstellar *dispersion (2) and the *Faraday effect in pulsars provides information about the distribution of free electrons and magnetic fields in the Milky Way.

Pulsars are denoted by the prefix PSR followed by the approximate position in right ascension (4 digits) and declination (2 or 3 digits), usually for equinox 1950.0. The figures may be preceded by B if the coordinates are for epoch 1950.0, or J for epoch 2000.0.

(🌐) SEE WEB LINKS
• Searchable database of all known pulsars, regularly updated.

pulsating variable Any intrinsic variable in which the flow of energy from the interior varies in a more or less rhythmic manner, causing large-scale motion in the outer layers and resulting in brightness changes. The pulsation may be essentially symmetrical (*radial pulsation), or take the form of waves travelling across the star's surface (*non-radial pulsation), or combine both forms. As stars evolve they all pass through one or more pulsating stages.

pulsation mode The way in which pulsations occur in a star. It may be described in terms of either the frequency of pulsation, or the physical motions of the star's surface. Pulsation at the lowest (forcing) frequency is known as the *fundamental mode; at twice the frequency, the *first overtone* mode; and so on. A star may pulsate either with approximately spherical symmetry (*radial pulsation), or as a series of waves running across the surface (*non-radial pulsation). Pulsation may occur in a single mode or in multiple modes, depending on the type of star.

pulse broadening A phenomenon in which irregularities in interstellar matter broaden the pulses from a pulsar by scattering the radio waves. Pulse broadening is more pronounced at lower radio frequencies, and takes the form of a characteristic 'tail' to the pulse. It is closely related to *interstellar scintillation.

pulse profile The variation with time of the signal strength from a pulsar, usually plotted against **pulse phase** (also known as **pulse longitude**), where one complete rotation is equal to 360° of longitude. Pulse profiles vary erratically from pulse to pulse, but the *integrated pulse profile*, the average of a large number of pulse profiles, is characteristic of each individual pulsar. The shape of the profile is believed to reflect the structure of the pulsar's emitting region.

pulse width A measure of the duration of a pulse from a pulsar, measured either in milliseconds or degrees. Pulse widths vary widely between pulsars, but are typically a few per cent of the pulse period.

Pup, the Popular name for Sirius B, the white-dwarf companion of *Sirius (the Dog Star).

pupil The aperture in the iris, the coloured area of the eye. The iris is a diaphragm which opens and closes to vary the amount of light admitted to the eye. Pupil size varies between 2.5 mm and nearly 9 mm, depending on the light conditions and the individual. The maximum, dark-adapted pupil size decreases with a person's age from an average of 7 mm to about 5 mm. Observers should avoid using low-power eyepieces which give an *exit pupil larger than their eye can accept, thus wasting light.

Puppid–Velid meteors A complex of poorly studied southern-hemisphere radiants, producing moderate activity between December 1 and December 15. The peak ZHR may be as high as 15 with maximum rates around December 7, when the radiant lies at RA 8h 12m, dec. −45°.

Puppis (Pup) (*gen.* **Puppis)** A significant constellation of the southern sky, representing the stern of the ship *Argo Navis. Its brightest star is *Naos (Zeta Puppis). Xi Puppis is an optical double, magnitudes 3.3 and 5.3. L Puppis is another optical double, one star of magnitude 4.9 and an unrelated semiregular red-giant variable that ranges between magnitudes 2.6 and 6.2 in a period of about 140 days. V Puppis is an eclipsing binary, ranging from magnitude 4.4 to 4.9 in a period of 1.45 days. M46, M47, NGC 2451, and NGC 2477 are all large, bright open clusters.

Purkinje effect An error affecting visual estimates of variable stars that arises when comparing stars of different colours, especially when one is red. It is caused by the differing sensitivity of the rods and cones in the retina to light of various wavelengths. It varies between observers, and also depends on the aperture of instrument used (i.e. on the amount of light gathered by the system), because the cones of the eye—unlike the rods—do not function at low light levels. It is named after the Czech physiologist Johannes Evangelista Purkinje (1787–1869).

PV Telescopii star A type of *pulsating variable that consists of a helium supergiant, with a Bp spectrum; abbr. PVTEL. About a dozen such stars are currently known. They pulsate with periods of approximately 0.1–1.0 day, but some may vary in brightness by about 0.1 mag. in a period of a year or so.

Pyxis (Pyx) (*gen.* **Pyxidis)** An unremarkable constellation of the southern sky, representing a mariner's compass. Its brightest star is Alpha Pyxidis, magnitude 3.7. It contains the recurrent nova T Pyxidis that has flared up in 1890, 1902, 1920, 1944, and 1966.

PZT Abbr. for *photographic zenith tube.

Q-class asteroid A rare class of asteroid whose members are distinguished by moderately high albedos and reflectance spectra that display a strong absorption feature at wavelengths shorter than 0.7µm and a modest absorption feature near 1µm. The spectra are interpreted as being similar to ordinary chondrite meteorites. At present, only (1862) Apollo and a few other *near-Earth asteroids have been identified as belonging to this class.

Q magnitude An infrared magnitude with effective wavelength 21.0µm and bandwidth of 5.5µm. There is also an index termed Q, formed by combining the optical magnitudes U, B, V:

$$Q = (U - B) - 0.72(B - V),$$

which is independent of interstellar reddening (*see* INTERSTELLAR ABSORPTION) and is strongly correlated with spectral type on the upper main sequence.

QSO Abbr. for *quasi-stellar object.

QSS Abbr. for *quasi-stellar radio source.

quadrant An obsolete navigational instrument used for measuring angles. It consisted of an engraved arc of a quarter of a circle, with a plumb line suspended from the centre of the circle. Stars were sighted along one arm and their elevation was read off the scale against the plumb line.

Quadrantid meteors A meteor shower active between January 1 and 6, but with marked activity confined mainly to a 12-hour interval around January 3/4. At maximum, the radiant lies at RA 15h 20m, dec. +49°, a few degrees northwest of Tau, Phi, and Nu Herculis in what used to be the constellation of Quadrans Muralis. The ZHR at maximum can be as high as 120, placing the Quadrantids among the three most active regular annual showers. Quadrantid meteors are moderately fast, with a geocentric velocity of 41 km/s. The stream orbit oscillates up and down relative to the ecliptic over the centuries. As a result, the stream will fail to encounter the Earth after about AD 2200. Attempts to identify a parent body are also confounded by this rapid motion of the orbit, although a strong connection with minor planet 2003 EH$_1$ has been suggested. This object may be the remaining fragment of a comet that broke up a few hundred years before 1600, possibly Comet 1490 Y1. The Quadrantids may be an end member of a complex that includes the *Delta Aquarid meteors, the *Kracht and *Marsden group comets, the daytime Arietid meteors, and Comet 96P/Machholz.

quadrature The occasion when a body in the Solar System has an angular separation, or *elongation*, of 90° east or west of the Sun.

quadrupole A fourfold system. An electric quadrupole, for example, would be two dipoles with one reversed with respect to the other. Quadrupole interactions between electrons in atoms lead to some of the *forbidden lines in the spectra of nebulae. Mass distributions such as an oblate spheroid or a binary star system will also have a quadrupole moment. General relativity predicts that a rotating mass system having a quadrupole moment will radiate gravitational waves which are quadrupole in nature. Thus if a gravity wave were to pass by a system with four masses at the corners of a square, the two along one diagonal would move towards each other, while the other two would move apart. The *cosmic microwave background, although very close to being isotropic, shows a very small quadrupole variation.

quantum The minimum amount by which quantities such as the energy or angular momentum of a system can change through the emission or absorption of radiation. The energy of a quantum represents the energy of one particle of that radiation. Hence, changes to a particular quantity do not occur continuously but in multiples of that quantum. The quantum of electromagnetic radiation is the photon, and that of gravitation is the hypothetical graviton.

quantum cosmology The study of the very earliest stages of the Universe after the Big Bang, including the properties of the initial *singularity and of the *Planck era in general. This field is highly speculative since it requires an understanding of quantum gravitational effects, and there is as yet no satisfactory theory of quantum gravity.

quantum efficiency The efficiency of a detector in recording radiation. It is the ratio of the number of photons usefully detected to the number (of a given frequency) that strike the detector. An ideal detector would have a quantum efficiency of 1. Photographic emulsion records only a small proportion of incoming photons, and thus has a low quantum efficiency of 0.001–0.01. Electronic detectors, such as CCDs and photomultipliers, have quantum efficiencies of 0.6–0.8 and hence can record faint objects during short exposures. The quantum efficiency of the human eye is about 0.01–0.05. Quantum efficiency is frequently expressed as a percentage.

quantum gravitation A theory in which gravitational interactions between bodies are described by the exchange of hypothetical elementary particles called *gravitons*. The graviton is the quantum of the gravitational field. Gravitons have not been observed, but are presumed to exist by analogy with photons of light.

quantum theory A theory of physics in which energy exists only in discrete quantities, called **quanta**. It was originated in 1900 by the German physicist Max Karl Ernst Ludwig Planck (1858–1947), who suggested that electromagnetic radiation is **quantized**, i.e. it can be emitted or absorbed only in tiny packets, not continuously. Each quantum of radiation, called a photon, has an energy equal to hf, where h is the *Planck constant and f the frequency of the radiation. This enabled Planck to explain the wavelength distribution of energy from hot bodies, known as *black-body radiation. Quantum theory led to the modern theory of the interaction between matter and radiation known as **quantum mechanics**.

Quaoar One of the largest known members of the Kuiper Belt, approximate diameter 1250 km, i.e. half that of Pluto; also known as minor planet (50000). It was discovered in 2002 by the American astronomers Chadwick Aaron Trujillo (1973–) and Michael Edwards Brown (1965–). Its orbit has a semimajor axis of 43.41 AU, perihelion 41.70 AU, aphelion 45.12 AU, inclination 8°.0, period 285.97 years.

quark A fundamental particle, of which *hadrons are believed to be composed. Quarks have charges that are either $+\frac{2}{3}$ or $-\frac{1}{3}$ of the electron's charge, and combine to make elementary particles. Baryons consist of three quarks, and mesons of two. For example, a proton consists of quarks with charges $+\frac{2}{3}+\frac{2}{3}-\frac{1}{3} = 1$ and a neutron consists of three with charges $+\frac{2}{3}-\frac{2}{3}-\frac{1}{3} = 0$. Eighteen different quarks with various properties are thought to exist, with a corresponding number of antiquarks. An isolated quark has never been observed; quarks appear to exist only in combination as baryons or mesons.

quark star A hypothetical star intermediate in density between a neutron star and a black hole. Such a star would consist of free quarks. The forces between the quarks balance the gravitational forces. Quark stars are unlikely to exist in reality, but some models for the cores of neutron stars suggest that the neutrons (and any protons) break up into a broth of quarks.

quasar An object with a high *redshift which looks like a star, but is actually the very luminous active nucleus of a distant galaxy. The name is a contraction of **quasi-stellar**, from their star-like appearance. The first quasars discovered were strong radio sources (**quasi-stellar radio sources**, or QSSs), but many more are now known which are relatively radio-quiet (**quasi-stellar objects**, or QSOs). At the distance implied by the redshift, the nucleus must be up to 100 times brighter than the whole of a normal galaxy. Yet some quasars vary in brightness on a time-scale of weeks, indicating that this huge amount of radiation originates in a volume only a few light weeks across. The source may therefore be an accretion disk around a black hole with 10^7 or 10^8 solar masses. Some quasars show little change in their light output, while others are much more variable: for example, 3C 279 has varied by a factor of nearly 500 in four months.

The first quasar to be identified as such in 1963 was the radio source 3C 273 at a redshift of 0.158, and it remains the optically brightest quasar as observed from Earth, at 13th magnitude. Thousands of quasars have since been found, many with high redshifts which imply that we see them as they were when the Universe was only about a tenth of its present age. The redshifted spectra of quasars show strong, often very broad, emission lines as well as continuum radiation. A rich absorption-line spectrum is also seen in the ultraviolet region (shifted into the optical region in high-redshift quasars), caused by clouds of intergalactic gas or interstellar material in galaxies between the quasar and Earth. The numerous absorption lines due to hydrogen in the clouds are together known as the *Lyman-α forest.

The large distances to quasars and the dominance of light from the central regions makes it difficult to observe the surrounding galaxy with Earth-based telescopes, but the Hubble Space Telescope has clearly resolved the host galaxies, both spiral and elliptical. In many cases the host galaxies are seen to be interacting or merging with one or more neighbours. Such interactions supply the massive black-hole nucleus with gas or stars, which ultimately fuels the quasar outburst. Some apparently normal galaxies may contain the remnants of quasar activity in their nuclei, and some *Seyfert galaxies and *Markarian galaxies have nuclei that are intrinsically as bright as some quasars. Quasars which vary greatly in their light output are termed *optically violently variable* (OVV) and are classified as *blazars along with *BL Lacertae objects. *See also* ACTIVE GALACTIC NUCLEUS. 📷

quasi-stellar object (QSO) A *quasar which resembles a *quasi-stellar radio source in many optical properties, but is relatively radio-quiet. QSOs appear to be about 100 times more numerous than the strong radio quasars.

quasi-stellar radio source (QSS) A *quasar which is a strong radio source. This was the first type of quasar to be discovered. It has been found that the intrinsically brightest quasars are the most likely to be strong radio sources, while only a few per cent of lower-luminosity quasars show strong radio activity.

quiescent prominence A long-lived solar prominence that lasts up to several months, changing little in appearance. Quiescent prominences are arch-shaped, several hundred thousand kilometres long, a few thousand kilometres thick, and up to 50000 km high. When seen against the solar disk they appear as darker features called *filaments. Quiescent prominences are found to the poleward side of *active regions. Over the course of the 11-year sunspot cycle they gradually migrate to higher latitudes, eventually forming a *polar crown* at sunspot minimum. Quiescent prominences are most frequent on the rising portion of the sunspot cycle.

quiet sun The Sun when it is at or near minimum activity in its 11-year cycle, and the numbers of sunspots and active regions are at their lowest. At such times, activity is still present in the form of small *X-ray bright points, prominences, and some coronal features.

quintessence A proposed explanation for the force commonly known as *dark energy that causes the apparent acceleration in the expansion of the Universe. Quintessence literally translates as 'fifth essence', a fifth fundamental force of nature that dominates over gravity on very large scales. Quintessence models typically predict a slow decrease in the density of dark energy as the Universe evolves, in contrast to the *cosmological constant explanation for dark energy which maintains a constant density. This results in a different expansion history, which may be observable with future supernova surveys.

RA Abbr. for *right ascension.

radar astronomy The study of bodies in the Solar System by reflecting radio pulses off them. Radar work requires very large (and hence sensitive) telescopes, such as the 305-m radio dish at *Arecibo Observatory. Radar astronomy can be used to determine the accurate distances of the planets (by measuring the time delay of the reflected signal), rotation rates (by Doppler broadening of the signal), and to map surface features (by detailed analysis of the echoes). Notable achievements in radar astronomy include the accurate measurement of the astronomical unit; determination of the rotation periods of Mercury and Venus; and the mapping of the cloud-covered surface of Venus. Earth-based radar has also been used to study the surface of Mars, the larger moons of Jupiter and Saturn, the rings of Saturn, asteroids, comets, and meteor trails. Several spacecraft sent to Venus have carried radar mapping equipment, notably Pioneer Venus Orbiter, Veneras 15 and 16, and Magellan.

radar meteor A meteor detected by a pulsed radio signal reflected (backscattered) from its ionized trail to a receiver at the same location as the transmitter. Radar observations allow accurate determination of meteor velocities. Meteors are preferentially detected by radar around 90° from the radiant, where they present a broader reflective surface. Radar can detect meteoroids far smaller than those seen visually. *See also* RADIO METEOR.

radial pulsation A form of pulsation in which a star expands and contracts symmetrically over its whole surface. The changes in radius are accompanied by variations in brightness, surface temperature, and spectrum. The majority of giant and supergiant variables (e.g. Cepheid variables, Mira stars, RR Lyrae stars, RV Tauri stars, and semiregular variables) pulsate in this manner. *See also* NON-RADIAL PULSATION.

radial velocity A measure of the velocity of an astronomical object along the observer's line of sight. The radial velocity is determined by comparing the wavelengths of lines in the object's spectrum with their laboratory values, the differences being due to the *Doppler shift. Corrections are usually applied for the Earth's rotation on its axis and its orbital motion around the Sun. By convention, velocities away from us (*redshifts*) are positive; those towards us (*blueshifts*), negative. *See also* REDSHIFT.

radial-velocity spectrometer A device for determining the radial velocities of stars of known spectral type. One design consists of a spectrometer which focuses a star's spectrum on to a mask which has slots at the laboratory positions of the stellar absorption lines. The total light transmitted through the mask is measured with a photometer, and the mask is moved until the amount of light collected reaches a minimum, which indicates that the alignment with the target spectrum is exact. The displacement of the mask then gives the radial velocity. Alternatively, a mask can be superimposed digitally (i.e. using a computer) on a normally recorded spectrum, and aligned with the observed absorption lines to determine their offset. In another method, light is passed through an *iodine cell before being dispersed; the iodine introduces a few sharp absorption lines in the red part of the spectrum which act as highly accurate markers for measuring radial velocities.

radian (rad) A unit of angle used in geometry. It is defined as the angle subtended at the centre of a circle by an arc of equal length to the radius of the circle. The whole circumference of the circle, of length 2π times the radius, subtends an angle of 2π radians; thus, $360° = 2\pi$ radians, and 1 radian = $57°.2958$. A useful approximation for astronomers is that 1 radian is about 200000 seconds of arc (actually 206265). As a consequence of the definitions of the units, the number of seconds of arc in a radian equals the number of astronomical units in a parsec.

radiant The region of sky from which a meteor shower appears to emanate, as a result of perspective. In reality, meteoroids in a meteor stream pursue parallel orbits around the Sun, and enter the atmosphere along parallel trajectories. As an observational convenience, a radiant diameter of $8°$ is adopted. A meteor shower is named for the constellation in which its radiant lies (e.g. the Perseids emanate from Perseus). Because of the Earth's orbital motion, a radiant appears to drift eastwards by about $1°$ each day. Photographic and telescopic studies can reveal true radiant diameters and a degree of structure; splitting of a meteor stream by planetary perturbations, for instance, can result in showers with multiple sub-radiants.

radiation 1. Energy propagating in the form of electromagnetic waves or photons (*see* ELECTROMAGNETIC RADIATION).
   2. A stream of particles, most commonly protons and electrons.

radiation belt A toroidal region within a planet's *magnetosphere where charged particles become trapped for long periods, spiralling backwards and forwards between 'mirror points' at opposite ends of magnetic field lines. The *Van Allen Belts surrounding the Earth are examples. Extensive radiation belts also surround Jupiter. Energetic particles in planetary radiation belts can present a hazard to spacecraft, causing damage to electronic components.

radiation era The period from about 10^{-43} s (the *Planck era) to 30000 years after the Big Bang. During this time, the expansion of the Universe was dominated by the effects of radiation or high-speed particles (at high energies, all particles behave like radiation). The *lepton era and the *hadron era are both subdivisions of the radiation era. The radiation era was followed by the *matter era, during which slow-moving particles dominated the expansion of the Universe.

radiation laws *See* PLANCK'S LAW; STEFAN–BOLTZMANN LAW; WIEN'S DISPLACEMENT LAW.

radiation pressure A small force exerted by photons of electromagnetic radiation on an illuminated surface. Radiation pressure is inconsequential for large bodies, but it can have a significant effect on small particles in orbit around the Sun. Zodiacal cloud dust particles with diameter less than 1 μm are pushed outwards from the Sun by radiation pressure, countering the *Poynting–Robertson effect which leads to orbital decay of larger particles. Radiation pressure also affects small dust particles in comet tails.

radiation temperature The calculated surface temperature of a celestial body, assuming that it behaves as a *black body. The radiation temperature is the same as the *effective temperature but is usually measured over a narrow portion of the electromagnetic spectrum, such as the visible range; this gives the *optical temperature*.

radiative equilibrium The state attained by a gas in which the rate of energy generation (such as by nuclear reactions in stars) is exactly balanced by the rate at which energy is transported outwards by radiation. For this to occur, convection must be insignificant. Radiative equilibrium is achieved in the outer regions of massive stars and the cores of low-mass stars. *See also* CONVECTIVE EQUILIBRIUM.

radiative recombination *See* FREE–BOUND TRANSITION.

radiative transfer The transport of energy by electromagnetic waves through a gas. It depends on factors such as the gas temperature, the abundances of the elements present, and their degree of ionization.

radiative zone The region of a star in which radiation, and not convection, is the main process by which energy is transferred outwards. Such zones occur in the cores of low-mass stars and the envelopes of high-mass stars.

radioactive age dating A technique for measuring the ages of rocks and minerals from the decay of certain radioactive elements within them; also known as **radiometric dating**. The technique involves comparing the amount of a long-lived radioactive parent isotope in a sample with the amount of the daughter isotope into which it decays. Isotopes used in radioactive age dating include uranium-238, which decays to thorium-230 with a half-life of 4.5×10^9 years; uranium-235, which decays to protactinium-231 with a half-life of 7.07×10^8 years; potassium-40, which decays to argon-40 with a half-life of 1.27×10^9 years; and rubidium-87, which decays to strontium-87 with a half-life of 4.88×10^{10} years.

radio astrometry The use of radio astronomy techniques to measure accurate positions of astronomical objects on the sky. The most precise positions, measured by *very long baseline interferometry, are superior to those made by optical methods.

RadioAstron A Russian satellite, with international collaboration, for *very long baseline interferometry, launched in 2011 July. It carries a 10-m dish, consisting of a solid 3-m diameter centre with 27 surrounding petals which unfurl in orbit. RadioAstron works in conjunction with radio telescopes on the ground to provide baselines up to 350000 km, the satellite's apogee.

(⊕) SEE WEB LINKS
• Official mission website.

radio astronomy The study of the Universe in the radio part of the electromagnetic spectrum. There is a wide radio window covering wavelengths from about 1 mm to 30 m, almost all of it accessible from ground-based observatories, both day and night. Emission mechanisms are either *thermal* (*black-body radiation, *free–free transition) or *non-thermal* (mainly *synchrotron radiation). *Maser sources are also common. Study of emission and absorption lines from molecules provides information about conditions in interstellar space. The 21-cm hydrogen line has proved to be a particularly valuable probe of the structure of our Galaxy and others.

Radio astronomy began in the 1930s with the pioneering work of K. G. *Jansky and G. *Reber, but it was not until after World War II that major research groups were set up. The first sources to be identified were the Sun, the Milky Way, and supernova remnants such as the Crab Nebula. As radio telescopes improved in sensitivity and resolution, *radio galaxies and *quasars were identified. New understanding of the late stages of stellar evolution came with the discovery of *pulsars. The counting of radio sources of different brightnesses (and hence presumably different distances) has helped astronomers understand how the Universe has evolved, and the discovery of the *cosmic microwave background has provided direct support for the *Big Bang theory of cosmology.

Radio telescopes employ several different techniques to collect the very weak radio waves coming from the sky. The most familiar is the circular parabolic dish that gathers radio waves and brings them to a focus, but the simplest radio telescopes may consist of little more than a metal rod (a *dipole antenna). Several antennas or telescopes may be used together as an *interferometer to achieve high resolution, notably in the sophisticated techniques of *aperture synthesis and *very long baseline interferometry.

radio galaxy A galaxy that is an unusually powerful emitter of radio waves. The output of a radio galaxy can be up to 10^{38} watts, a million times greater than a normal galaxy such as our own. Radio galaxies have a compact radio nucleus coincident with the core of the visible parent galaxy, a pair of opposed *jets emerging from the nucleus, and a pair of *lobes (1) far outside the visible confines of the galaxy. The galaxy is almost always a giant elliptical, which may be the result of the collision and merger of two or more smaller galaxies. The source of the radio galaxy's energy is believed to be a massive black hole in the galactic nucleus from which the jets emerge, delivering energy to the lobes. Notable radio galaxies include Centaurus A, Cygnus A, and Virgo A.

radiogenic A term that describes something produced by radioactive disintegration. For example, radiogenic heating is the heating produced by energy released from the radioactivity of substances such as potassium, thorium, and uranium, usually in the interiors of planets. A radiogenic substance is one produced by the fission of a heavier substance.

radioheliograph A radio telescope designed to observe and map the Sun.

radio interferometer An array of antennas operating as a single instrument of high angular resolution. Radio interferometers have many different designs, ranging from two simple dipoles connected together to sophisticated multi-element arrays such as the *Very Large Array, the *Multi-Element Radio-Linked Interferometer Network, and the *Mills cross. *See also* APERTURE SYNTHESIS; LONG-BASELINE INTERFEROMETRY; VERY LONG BASELINE INTERFEROMETRY.

radio jet *See* JET.

radio meteor A meteor detected by *forward scatter*, in which its ionization trail momentarily reflects the signal from a transmitter beyond the observer's horizon. Raw counts of reflections can be processed to give an index of meteor shower activity, even in cloudy or daylit conditions. As with *radar meteors, most events detected by forward scatter are produced by much smaller meteoroids than those which cause visual meteors. Small particles produce short, sharp signals, while larger meteoroids give longer-lasting reflections.

radiometer A device for measuring the intensity of electromagnetic radiation. Radiometers operate across the whole spectrum, but in the visual region they are usually termed *photometers. Many Earth-resources satellites carry radiometers operating in the infrared and visual parts of the spectrum to produce false-colour maps of the planet showing, for example, crops, geological forms, and weather systems. Space probes carry radiometers to survey other Solar System bodies in similar fashion. At radio wavelengths, radiometers tend to be comparison devices, repeatedly comparing the signal from the object with that from an artificial source to improve their stability. *See* DICKE RADIOMETER.

radiometric dating Another name for *radioactive age dating.

radionuclide A radioactive atomic nucleus. The presence in stars of radioactive elements with short half-lives, such as technetium, is evidence for the existence of *nucleosynthesis in stars.

radio source Any cosmic source of radio waves. Major celestial radio sources include the Sun, Jupiter, flare stars, novae, pulsars, supernova remnants, H II regions, the Milky Way, the galactic centre, radio galaxies, quasars and other active galaxies, and the *cosmic microwave background. Radio sources are designated by a number of methods. The first sources were named after the constellation in which they lay (e.g. Sagittarius A, Virgo A), but later they were labelled with a serial number and a prefix denoting the survey in which they were catalogued (e.g. 3C for the Third Cambridge Survey). Modern practice is to identify sources by their coordinates, either right ascension and declination, or galactic coordinates. Radio *source counts can yield important information about the early Universe.

radio star An obsolete term for a cosmic source of radio waves. It is now known that the vast majority of so-called 'radio stars' are galaxies and other non-stellar objects, and for them the term *radio source is now used instead. However, some stars, such as novae and flare stars, do emit radio waves.

radio telescope An instrument for collecting and measuring radio waves from astronomical sources. Single-dish radio telescopes can be either used alone or joined into interferometers and arrays. Single-dish telescopes usually have a parabolic reflector, which works in a similar way to the main mirror of an optical telescope. Because radio waves are much longer than light waves, radio reflectors need have only relatively moderate surface accuracy. But, for the same reason, even the largest single-dish radio telescopes cannot match the angular resolution of optical telescopes. *Radio interferometers, made of two or more spaced antennas (elements) connected together, are used to achieve high angular resolution. Resolutions better than 0″.001 can be achieved in this way, far exceeding the performance of even the Hubble Space Telescope. *Aperture synthesis telescopes are interferometers in which the whole or part of a large, imaginary aperture is built up by the rotation of the Earth and, in some cases, by the movement of the dishes. Parabolic dishes are normally steerable, the larger ones being on altazimuth mountings, but many interferometers consist of arrays of static antennas which either have a fixed beam or can be steered electronically.

radio waves Electromagnetic radiation of wavelength longer than about 1 mm (30 GHz). The longest radio waves observable by radio astronomers have a wavelength of about 30 m (10 MHz) (*see* RADIO WINDOW). The shortest radio wavelengths, from about 1 mm to 30 cm, are known as *microwaves*.

radio window A region of the electromagnetic spectrum in which the Earth's atmosphere is relatively transparent to radio waves. It extends over a wavelength range of approximately 1 cm to 30 m. At the short-wave end of the window, radio waves are blocked by water vapour, and at the long-wave end by the Earth's ionosphere. Most of the radio spectrum is occupied by broadcasting and telecommunications services and only a few narrow bands are kept free for radio astronomy, notably those coinciding with the position of important spectral lines, such as the 21-cm hydrogen line at 1420 MHz.

radius–luminosity relation A relationship between the radius of a pulsating variable and its luminosity. The radius is derived from the radial-velocity curve, which generally has a different phase from the luminosity curve; the amount of phase lag depends on the specific type of variable. In a Beta Cephei star, for example, the luminosity lags behind the radial-velocity curve by almost exactly one-quarter of the period, so that maximum luminosity corresponds to minimum radius. In Cepheid variables and RR Lyrae stars, by contrast, maximum light corresponds to an intermediate radius at the maximum velocity of expansion.

radius of curvature The radius of the spheroid that forms the basic curve of a concave mirror. The *focal length of a mirror is half its radius of curvature.

radius vector An imaginary line joining the centres of any two bodies in orbit about each other, for example between a planet and the Sun.

Raman effect The scattering of a photon by an atom or molecule that is accompanied by an excitation (or de-excitation) of the atom or molecule, and hence a change in the wavelength of the photon. It is named after the Indian physicist Chandrasekhara Venkata Raman (1888–1970), who discovered it in 1928. The effect can be important in the spectra of molecules.

Ramaty High-Energy Solar Spectroscopic Imager (RHESSI) A NASA satellite which takes images of solar flares in X-rays and gamma rays (energy range 3 keV–17 MeV), launched in 2002 February. It is named in honour of the Romanian-born American astrophysicist Reuven Ramaty (1937–2001), a specialist in solar flare research.

r

(⊕) SEE WEB LINKS
• Official mission website.

rampart crater A crater surrounded by a ring of low hills. Such craters are thought to be formed by the impact of a meteorite on a planet with an atmosphere or a wet or muddy surface. The impact produces a splash pattern and surface flow within the ejecta blanket, forming a rampart at its edge. An example is the 18-km Martian crater Yuty.

Ramsden disk The disk of light seen through the eyepiece of a telescope pointed at bright sky, as observed from some distance behind the eyepiece. It is actually the image of the objective, reduced in size. The magnification given by an eyepiece can be estimated by dividing the diameter of the objective lens by that of the Ramsden disk when the telescope is focused on infinity. It is named after the English instrument-maker Jesse Ramsden (1735–1800).

Ramsden eyepiece A type of eyepiece consisting of two *planoconvex lenses of the same *focal length with their flat sides facing outwards. Its main advantage is its simplicity and low cost. It has poor *eye relief, and suffers from *chromatic aberration and *spherical aberration, although not coma (*see* COMA, OPTICAL). It was invented in 1782 by the English instrument-maker Jesse Ramsden (1735–1800). *See also* KELLNER EYEPIECE.

Ranger A series of NASA Moon probes, originally intended to photograph the Moon as they approached and then to eject an instrument package on to the surface. After a sequence of failures the Rangers were changed into photographic-only missions. Three probes were successful, sending back increasingly detailed pictures before they crashed into the Moon. The Ranger pictures of the lunar surface assisted the design of the later *Surveyor soft landers.

(⊕) SEE WEB LINKS
• NASA information page.

Successful Ranger Probes		
Probe	Launch date	Results
Ranger 7	1964 July 28	Hit Moon July 31
Ranger 8	1965 February 17	Hit Moon February 20
Ranger 9	1965 March 21	Hit Moon March 24

RAS Abbr. for *Royal Astronomical Society.

Rasalgethi The star Alpha Herculis, an M5 supergiant or bright giant that is one of the largest stars known, with an estimated diameter some 400 times the Sun's. It is a semiregular variable that ranges between magnitudes 2.7 and 4.0 with a period of 100 days or so, although there are also longer-period variations lasting about 6 years. It is a binary, with a companion of magnitude 5.4 that appears greenish by contrast with the orange primary; this star is itself a spectroscopic binary. The orbital period of the visual pair is estimated at around 3600 years. Rasalgethi lies 360 l.y. away.

Rasalhague The star Alpha Ophiuchi, magnitude 2.1. It is an A5 dwarf, 49 l.y. away.

Ra-Shalom Asteroid 2100, the second-largest of the *Aten group, diameter 2.3 km. It was discovered in 1978 by the American astronomer Eleanor Kay Helin, née Francis (1932–2009). Its name comes from the Egyptian Sun-god Ra, and Shalom, the Hebrew greeting meaning 'peace'. It is of C class. Ra-Shalom's orbit has a semimajor axis of 0.832 AU, period 0.76 years, perihelion 0.47 AU, aphelion 1.20 AU, and inclination $15°.8$.

R association A stellar association containing a number of reflection nebulae (hence the R). In such associations the stars are less than a million years old and are still embedded in the nebula from which they formed. Because of this they may be obscured, and their light is visible only after reflection from nearby dust.

raster A pattern used for scanning or making an image, consisting of repeated lines. A black-and-white television screen, for example, shows a raster pattern of lines.

raster scan A technique for surveying areas of the sky in which the telescope is driven back and forth along adjacent strips of sky. The scans can be aligned whichever way is required, such as along lines of galactic latitude in order to survey the galactic plane. *See also* DRIFT SCAN.

RATAN-600 A radio telescope at the *Special Astrophysical Observatory near Zelenchukskaya in southern Russia, which began operation in 1975. Its name is an abbreviation of Radio Astronomy Telescope of Academy Nauk (Sciences). It consists of 895 tiltable panels arranged in a ring 576 m wide, which can be used as a unit or as four individual sections.

(⊕) SEE WEB LINKS
• Official telescope website.

Rayleigh criterion The limiting resolution of a telescope set by diffraction; also known as the **Rayleigh limit**. The image of a star in a telescope consists of the *Airy disk surrounded by diffraction rings. The English physicist Lord Rayleigh (1842–1919) defined the limit of resolution as the separation of two stars when the centre of one image lies on the dark interval between the Airy disk of the other and its first diffraction ring. This separation is 1.22 λ/a radians, where λ is the observing wavelength and a is the telescope aperture. For visual light this corresponds to $13.8/a$ seconds of arc, where a is the telescope aperture in centimetres. Thus a 30-cm telescope should resolve double stars with separations $0''.45$ or greater, given good optics, steady air, and adequate magnification. (For comparison, the *Dawes limit above which two 6th-magnitude stars should be resolved with this aperture is $0''.39$.) Rayleigh also investigated how accurately the objective should be figured to avoid serious degradation of the size of the Airy disk; he found the limit to be a quarter of the wavelength of light.

Rayleigh–Jeans formula An approximation to a black-body spectrum for low frequencies. The Rayleigh–Jeans approximation is widely used in radio astronomy since most astrophysical black-body sources peak at frequencies much higher than those found in the radio spectrum.

Rayleigh limit Another name for the *Rayleigh criterion.

Rayleigh number (*Ra*) A parameter that determines when convection may occur in a gas or liquid. Convection can occur when more energy is gained by moving material than is lost to friction. The Rayleigh number is dimensionless and depends on the local gravitational acceleration and the properties of the material. Convection occurs when the Rayleigh number exceeds a critical limit. The Rayleigh number is used in modelling stellar and planetary interiors and planetary atmospheres. It is named after Lord Rayleigh.

Rayleigh scattering The scattering of light by particles much smaller than the wavelength of the light. The amount of scattering increases as the wavelength gets shorter, so blue light is scattered more than red light. The incoming beam is scattered equally forwards and backwards. Rayleigh scattering of sunlight by molecules in air is the cause of the blue colour of the daylight sky. It is named after the English physicist Lord Rayleigh (1842–1919).

rays, crater Bright streaks radiating from young impact craters. They are prominent on the Moon. Rays appear to consist of crushed surface rock thrown out by the impact. Clusters of secondary impact craters are often found inside rays. Some ray systems, such as those from the large lunar craters Tycho and Copernicus, stretch for hundreds of kilometres. Similar rays are found on Mercury and on the satellites Ganymede, Callisto, and Oberon.

R Canis Majoris star A binary star with components of widely differing masses, and somewhat ambiguous spectral features. The type star is a *semidetached *Algol star, with a late-A-type primary (1.5 solar masses) and a G-type subgiant of very low mass (about 0.1 solar mass) and evidence of circumstellar material. The stars in this group have periods of the order of 1 day that are variable, and thus imply *mass transfer or *mass loss.

R-class asteroid A rare class of asteroid, distinguished by a moderately high albedo and a reflectance spectrum with a strong absorption feature at wavelengths shorter than $0.7\,\mu m$ and a fairly strong absorption feature near $1\,\mu m$. At present the only confirmed member of this class is (349) Dembowska, diameter 140 km.

R Coronae Borealis star A type of eruptive variable that undergoes occasional fades of variable amplitude, with a sudden onset and generally slower recovery, often with major fluctuations; abbr. RCB. The fades have depths of 1–9 mag., and last from months to years. The onset, depth, and duration of the fades are unpredictable. They occur when clouds of carbon particles in the outermost atmospheric layers are ejected in the direction of the observer, and absorb the light from the underlying photosphere. When the photosphere is obscured the star enters a *chromospheric phase* in which the residual light originates in the deep overlying chromosphere. The stars are high-luminosity supergiants of spectral types B–M and R, hydrogen-deficient but rich in helium and carbon. The majority also appear to be pulsating variables with amplitudes of up to a few tenths of a magnitude, and periods of 30–100 days. Although few in number (approximately two dozen are known), R Coronae Borealis stars are significant for studies of stellar evolution. It is now thought that they arise from the

merger of two white dwarfs. The prototype, R Coronae Borealis itself, is of spectral type
F or G and varies between mags. 5.7 and 14.8.

Reber, Grote (1911–2002) American radio astronomer. In 1932 he learnt of K. G. *Jansky's
detection of radio emission from beyond the Solar System, and by 1938 had built a paraboloidal
dish 9.4m in diameter, movable in declination like a *transit instrument. From then until after
World War II he was the world's sole radio astronomer. He mapped the sky at radio
wavelengths, and detected many sources ('radio stars'), including *Cassiopeia A and *Cygnus A,
which did not correspond to visible stars. He also found that the Sun and the Andromeda
Galaxy emit radiation at radio wavelengths.

receiver 1. An electronic device that detects and amplifies radio signals captured by an
antenna. Receivers used in radio astronomy are generally similar in principle to those used
for other purposes, such as a domestic radio, although they are usually cooled to near absolute
zero to reduce thermal noise and hence increase their sensitivity. On radio telescopes the
incoming signal, after amplification, is mixed with a local oscillator to produce a lower
intermediate frequency that suffers lower loss when passing along cables from the telescope
to the observing laboratories. This is sometimes termed a *superheterodyne receiver. *See also*
DICKE RADIOMETER.
   2. The equipment used to process signals received by a radio telescope. There are three main
types: *line receivers* are used to observe spectral lines, where high stability is required together
with the ability to scan in frequency or to observe at several closely spaced frequencies;
continuum receivers are used for carrying out radio source surveys at a wide band of
frequencies; and *timing receivers* are used to measure the arrival times of the radio pulses
emitted by *pulsars. *See also* CORRELATION RECEIVER.

reciprocity failure The decreasing efficiency of photographic emulsions with longer
exposure times. In daylight there is a reciprocity between exposure time and image brightness:
a halving of the image brightness can be compensated for by doubling the exposure time.
At low image brightnesses, however, this reciprocity fails and very much longer exposure times
are needed. This is thought to be because the effects of individual photons on the grains of
an emulsion are easily overcome by thermal motions. Cooling the emulsion in a *cooled
camera helps prevent this. Alternatively, the film may be *hypersensitized.

recombination The process in which an electron combines with a positive ion. The neutral
atom so formed is usually in an excited state, and subsequently emits energy in the form of a
photon, producing a *recombination line. Recombination is the opposite of *ionization. The
two processes are in equilibrium in an *H II region. *See also* DIELECTRONIC RECOMBINATION.

recombination epoch The era reached about 300000 years after the Big Bang when the
Universe cooled to a temperature of a few thousand degrees, allowing electrons and protons
to combine to form hydrogen atoms. The occurrence of recombination is closely associated
with the *decoupling of matter and radiation, so observations of the *cosmic microwave
background provide a means of studying the Universe when it was roughly this age.

recombination line An emission line in a spectrum arising from the *recombination of a
free electron with an ion. As the electron drops down through the energy levels of the atom,
it emits recombination lines at wavelengths that depend on the difference in energy between
the levels. For hydrogen, these range from radio lines (caused by jumps between the outer,
lowest-energy levels), via infrared, to optical lines produced by jumps down to level 2.
Recombination to level 1 (the ground state) produces an ultraviolet photon which itself
causes further ionization. From recombination lines, the gas temperature and density
in ionized nebulae can be estimated.

rectangular coordinates A system for specifying the position of an object relative to two
or three mutually perpendicular axes from some specified origin. The axes are identified as *x*, *y*,
and (where a third axis is used) *z*. *Cartesian coordinates are rectangular coordinates. In
astronomy, rectangular coordinates are sometimes used for Solar System objects, with the
length of each axis given in astronomical units. They are usually heliocentric or geocentric in
origin, but other points of origin can be used.

recurrent nova A *cataclysmic binary that has undergone repeated outbursts similar to those found in novae. Five or six stars are assigned to this class but only two, T Coronae Borealis (the *Blaze Star) and RS Ophiuchi have much in common, giant secondaries and long orbital periods (227.6 and 230 days respectively). These two stars appear to be similar to an *RR Telescopii star or *symbiotic nova*.

recycled pulsar A pulsar of unusually low magnetic field (1–100 tesla) and low spin-down rate, frequently of very short pulse period, and often found in a binary system. Recycled pulsars are believed to be ordinary pulsars which have lost energy and faded out, and then been spun up again by accretion of gas from a companion star in a binary system. A high proportion of recycled pulsars are found in the cores of globular clusters, where the high density of stars makes the capture of an old neutron star into a binary system more likely. The first recycled pulsars to be discovered had very short pulse periods and are known as *millisecond pulsars, but later examples had much longer periods.

reddening The change in the colour of light when it traverses interstellar space or the Earth's atmosphere. It occurs because absorption and scattering are nearly always stronger at shorter (blue) wavelengths. *See also* INTERSTELLAR ABSORPTION.

red dwarf A cool, faint, low-mass star lying at the lower end of the main sequence. Red dwarfs have masses and diameters less than half those of the Sun. They are red because of their low surface temperatures, less than 4000 K, and are of spectral type K or M. Red dwarfs are the most common type of star, and also the longest-lived, with potential lifetimes greater than the current age of the Universe. Because of their low luminosity, no more than 10% of the Sun's, they are inconspicuous. Barnard's Star and Proxima Centauri are nearby examples. Many red dwarfs are *flare stars.

red giant A cool, large, and highly luminous star. Red giants are stars that have left the main sequence, having exhausted the hydrogen fuel in their cores, and are powered by nuclear reactions between elements heavier than hydrogen. They owe their colour to their cool surface temperatures (below 4000 K, spectral type K or M). Because of their large diameters, 25 times that of the Sun or more, they are highly luminous, many hundreds of times that of the sun. Massive main-sequence stars may evolve into red giants after passing through a *blue-giant phase, while lower-mass stars, such as the Sun, will evolve directly into a red giant. Many red giants are pulsating variable stars.

Red Planet Popular name for the planet Mars, arising from its distinctive ruddy colour.

redshift (z) The amount by which the wavelength of light from a receding object is lengthened (i.e. moved to the red) by the *Doppler shift. Redshift is calculated from the formula

$$z = \Delta\lambda/\lambda,$$

where λ is the original wavelength (as measured in the laboratory) and $\Delta\lambda$ the observed change in wavelength. A redshift of 0.1, for example, means that the light has been redshifted by 10% in wavelength, whereas a redshift of 1 means a change of 100% (i.e. a doubling in wavelength). All galaxies at large distances from our own have redshifts resulting from the expansion of the Universe. At redshifts less than about 0.1, z is related to the velocity of the object, v, by the simple expression $z = v/c$, where c is the velocity of light. At larger redshifts, this relationship is no longer true, and it is better to think of the redshift as being caused by the expansion of space rather than a Doppler shift. While light is on its way to us from a distant object the Universe is expanding, and this expansion 'stretches' the wavelength of the light. For example, when the light left a galaxy with a redshift of 1 the Universe was only half its present size; during the time taken for the light to travel from the galaxy to our telescopes the Universe has doubled in size, causing the wavelength of the light to increase by the same factor.

A galaxy's redshift can be measured quite easily from its spectrum, unlike its distance, which is very hard to measure directly. At small redshifts, astronomers can calculate the velocity of the galaxy from its redshift and then use *Hubble's law, which relates velocity and distance, to estimate the distance of the galaxy. At larger redshifts, the expansion of the Universe means that distance becomes an ambiguous quantity, but astronomers can use a galaxy's redshift to estimate the age of the Universe at the time the light was emitted, which is the most important method they have for investigating the history of the Universe (*see* GALAXY EVOLUTION). The largest redshifts currently known are over 6, meaning that the light has been shifted in wavelength by 600% or more so that ultraviolet lines appear in the red part of the spectrum. *See also* GRAVITATIONAL REDSHIFT; REDSHIFT SURVEY.

(⊕) SEE WEB LINKS

• Online calculator for converting redshift to lookback time and other parameters.

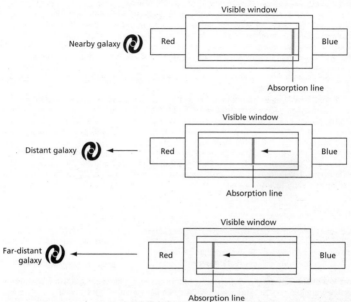

redshift: Light from receding galaxies is lengthened in wavelength (i.e. shifted to the red), as indicated by the position of absorption lines in their spectra. Redshift is shown here in relation to the visible window of wavelengths to which the eye is sensitive. The more distant the galaxy, the faster its motion away from us, and hence the greater its redshift.

redshift–distance relation A formula relating the observed redshift of an extragalactic object to its distance in the expanding Universe. For redshifts of about 0.1 or less, this relationship is linear and appears as a straight line on the *Hubble diagram. The relationship does not hold for objects so close together that their mutual gravitational attraction overrides the expansion of the Universe (e.g. galaxies in a cluster). For redshifts much larger than 0.1, the relation becomes non-linear, and depends in a complicated way on cosmological parameters such as the *curvature of spacetime.

redshift–magnitude relation The relationship between the observed redshift and apparent magnitude of an object with known absolute magnitude in the expanding Universe. The predicted relationship is different for different cosmological theories (*see* HUBBLE DIAGRAM).

redshift survey A compilation of the *redshifts of a large number of galaxies, or other extragalactic objects, selected according to certain criteria. Such surveys may be used to map out the *large-scale structure of the Universe or to investigate *galaxy evolution. Redshift surveys are important because it is very difficult to measure true distances of all but the closest galaxies. It is relatively easy, however, to measure a galaxy's redshift and hence its velocity of recession. For nearby galaxies (redshift 0.1 and less), there is a simple relation between the velocity of a galaxy and its distance — the *Hubble law — and so astronomers can estimate the distance of a galaxy from its redshift. The distance is only an estimate since galaxies possess *peculiar velocities relative to the *Hubble flow – any component of this velocity along the radial direction to the galaxy will affect the measured redshift. These peculiar velocities are limited to a few hundred km/s, significantly smaller than the Hubble flow velocities for all but the nearest galaxies. For more distant galaxies, with redshifts much larger than 0.1, the Hubble law breaks down and the *redshift–distance relation is used instead.

The two largest redshift surveys to date are the Two-Degree Field Galaxy Redshift Survey (2dFGRS, 1997–2002) which measured nearly 250 000 galaxy redshifts using the *Anglo-Australian Telescope, and the *Sloan Digital Sky Survey (SDSS, 2000–08) which measured redshifts (in addition to five-band imaging) for almost 1 million galaxies. The results of these and previous redshift surveys show that galaxies do not lie in spherical clusters and groups, but rather lie in filaments and sheets (including the so-called *Great Wall) surrounding *voids in which there are very few galaxies. This remarkable structure is often called the *cosmic web*.

Although very large in volume, the 2dFGRS and SDSS surveys are limited to moderate redshifts, with few galaxies beyond a redshift of 0.3. Other surveys have covered much smaller areas of sky but with very long integration times to obtain redshifts of 2 and beyond. Recent examples include the Deep Extragalactic Evolutionary Probe (DEEP) project using the *Keck Telescopes and the VIMOS VLT Deep Survey (VVDS) on the European Southern Observatory's Very Large Telescope. Such surveys are powerful for constraining the evolution of galaxies because the *lookback time for high redshift galaxies is a significant fraction of the age of the Universe. For example, it has taken roughly two-thirds of the age of the Universe, around 9 billion years, for light from a galaxy at redshift 2 to reach us. In that time, the Universe has trebled in size.

Red Spot *See* GREAT RED SPOT.

reduced proper motion Observed proper motion multiplied by a scaling factor to reduce it to a standard distance. For example, the relative distances of a sample of stars such as Cepheid variables can be derived from photometry, and their observed motions can be scaled to some adopted standard distance. This gives a scale model of linear velocities.

red variable Any giant or supergiant variable star with a late-type spectrum. The term is normally applied only to *Mira stars and *semiregular variables, but it can also include slow *irregular variables of types LB and LC. Flare stars are not included.

reference frame *See* FRAME OF REFERENCE.

reference star A star whose position and/or proper motion are known, so that it can be used to define a local *frame of reference for relative positions or proper motions of other stars in the same area of sky.

reflectance spectrum A spectrum which shows the proportion of sunlight reflected from the surface of a planet, planetary satellite, or asteroid over a range of wavelengths. Various minerals on the surface reflect sunlight in characteristic ways, so the reflectance spectrum of a body provides information about its mineralogical composition.

reflecting telescope A telescope that uses a concave mirror to focus light. The main advantage of reflection (by a mirror) over refraction (by a lens) is that all colours of light are reflected equally, so the images do not suffer from *chromatic aberration. Additionally, there is no practical upper limit to the size of a mirror and only one surface needs to be optically worked, thus making large mirrors cheaper than large lenses. It is also comparatively easy to make reflecting telescopes with focal ratios smaller than $f/6$. All telescopes with apertures greater than about 1 m are reflectors.

The principal drawback of reflectors is that some obstruction of the light path is inevitable, resulting in a loss of light and image contrast. Furthermore, a mirror's coating loses its reflectivity over time and must periodically be renewed; and normally the tube of a reflector is open at the top end, so that air currents (*tube currents*) can circulate within it and disturb the quality of the image. The largest reflectors are now made with *segmented mirrors. The first design for a reflecting telescope was published by J. *Gregory in 1663, but the first person to build a reflector was I. *Newton in 1668. The main designs of reflector are the *Cassegrain, *Newtonian, and *Ritchey–Chrétien telescopes.

reflection effect The reflection (more strictly, the absorption, scattering, and re-emission) of the light from one star by a close companion. The effect is seen in many close binaries in which the cooler, fainter secondary is irradiated by the primary.

reflection grating A *diffraction grating in which the grooves are ruled in a reflective coating. The grating may be either plane or concave, the latter being able to focus light. The advantage of a reflection grating over a *transmission grating is that it produces a spectrum extending from the ultraviolet to the infrared, since the light does not pass through the grating material.

reflection nebula A cloud of interstellar gas and dust that appears bright because it reflects or scatters starlight. The light from a reflection nebula has the same spectral lines as the starlight it reflects, but it is usually bluer and may be polarized. Reflection nebulae often occur together with emission nebulae in regions of recent star formation. The *Pleiades cluster is surrounded by a reflection nebula.

reflection variable A type of close binary system in which the variations arise largely from non-uniform surface brightness caused by the *reflection effect; abbr. R. The amplitude may be 0.5–1.0 mag. even when (as in many cases) eclipses are absent.

reflector *See* REFLECTING TELESCOPE.

refracting telescope A telescope that produces images by refraction using a lens. The advantage of this over reflection by a mirror is that the aperture remains completely unobstructed, thus yielding images of the highest contrast and brightness. The light transmission of a coated lens is better than 90%, and remains better than that of a set of mirrors for apertures up to about 0.4 m. Refractors are considered ideal for observing fine, low-contrast details, as on planets. Their drawbacks are that in order to overcome *chromatic aberration each lens must have several optical *elements, and that the weight of a large lens is difficult to support. In addition, it is difficult to make lenses of focal ratios smaller than about $f/8$, because of the steepness of the curves required, so tubes become very long in large-aperture instruments.

Refractors were the first type of telescope to be invented, in the early 17th century; the invention is usually attributed to the Dutch optician Hans Lippershey ($c.$1570–$c.$ 1619), although *Galileo was the first to use the telescope for serious astronomy. The practical upper size limit of refractors is represented by the 40-inch (1-m) $f/19$ refractor at *Yerkes Observatory. Virtually all large refractors were made before the end of the 19th century. However, small refractors of aperture 50–100 mm continue to be popular for amateur use.

refraction, atmospheric *See* ATMOSPHERIC REFRACTION.

refractive index (n) A measure of the refractive power of a transparent medium such as glass. Mathematically, the refractive index is the ratio of the sines of the angles of incidence and refraction as light passes into the medium. These angles are measured from the normal (a line at right angles to the edge of the medium). The *angle of incidence*, at which the light enters the medium, is greater than the *angle of refraction*, at which it travels through the medium. A typical refractive index for glass is about 1.5.

refractor *See* REFRACTING TELESCOPE.

refractory A term describing an element or compound that melts or boils (vaporizes) at a relatively high temperature, or (equivalently) condenses from a gas at a high temperature. Aluminium, calcium, and uranium are examples. A refractory substance is the opposite of a *volatile one.

regio An area, sometimes of distinctive colour or topography, on a planetary surface; pl.*regiones*. The name is not a geological term; it is used in the nomenclature of major areas such as Eistla Regio on Venus and Galileo Regio on Ganymede.

Regiomontanus Name by which Johann Müller (1436–76), German astronomer and mathematician, was known. His tutor, the Austrian mathematician and astronomer Georg von Peurbach (1423–61), had begun a Latin translation of Ptolemy's *Almagest* which Regiomontanus completed in 1463. He added some more recent observations and criticisms which were later to influence Copernicus in rejecting the *Ptolemaic system. He compiled, printed, and published scientific books, including his own ephemerides and trigonometric tables.

✓ **regmaglypt** An elliptical or polygonal depression, resembling a thumbprint in soft clay, found in a semiregular pattern on the surfaces of meteorites larger than about 100mm. Regmaglypts are probably caused by the flow of air and the loss of molten fragments by ablation during the meteorite's passage through the atmosphere.

regolith A layer of loose and broken rock and dust on a planetary surface. It is analogous to a soil, but has no biological content. The lunar regolith, 5–15m deep, is formed by meteoritic and micrometeoritic bombardment. Mercury is thought to have a regolith produced by the same process, whereas on Mars the regolith may include deposits from the erosional effects of wind and ice, and volcanic ash. *See also* MEGAREGOLITH.

regression of nodes The westward movement of the nodes of an orbit, caused by the gravitational pull of other bodies, particularly the Sun. For example, the nodes of the Moon's mean orbit regress once around the ecliptic in 18.6 years.

Regulus The star Alpha Leonis, magnitude 1.36, the faintest of the first-magnitude stars. It is a B7 dwarf 79 l.y. away, and has a wide companion of magnitude 7.7.

reionization epoch The era between about 150 million and 1 billion years after the Big Bang, when the Universe went from consisting of neutral atoms to once again being an ionized *plasma as it had been immediately after the Big Bang. This epoch is seen as the next major step in the development of the Universe following the *recombination epoch. The recombination epoch occurred about 300000 years after the Big Bang, when the Universe had cooled sufficiently for the positive ions and negative electrons to combine into neutral atoms. However, once the first stars were formed their ultraviolet radiation began to ionize the atoms again, hence initiating the reionization epoch. Several hundred million years would have been needed for the *intergalactic medium to become fully ionized, and it remains so today.

relative orbit The orbit of a celestial object with respect to the body it is orbiting. The Moon and Earth perform elliptical orbits about their centre of mass, a point about 1600km below the Earth's surface. The Moon's relative orbit about the Earth is of the same shape, but with the Earth's centre at a focus of the ellipse. Similarly, the planets have relative orbits about the Sun and each component of a binary star has a relative orbit with respect to the other.

relative sunspot number (R) A measure of the number of spots on the Sun. It was introduced in 1848 by R. *Wolf, and is calculated by taking into account both the total number of individual spots (f) and the number of sunspot groups (g), plus a factor k that represents the efficiency of the observer and telescope: $R = k(10g + f)$. The *Wolf sunspot number* refers to Wolf's initial observations at Bern Observatory, Switzerland, while the *Zürich relative sunspot number*, R_z, refers to the index produced at the Swiss Federal Observatory in Zürich, where Wolf moved in 1855. The Zürich sunspot number included observations from several locations, thus leading to the introduction of the k factor for observer efficiency. In 1981 responsibility for generating the relative sunspot number was transferred to the Solar Influences Data Analysis Center (SIDC, known before 2000 as the Sunspot Index Data Center) at the Royal Observatory of Belgium in Brussels. The index was renamed the *International Sunspot Number and is designated R_i. Other sunspot indices include the Boulder sunspot number, published by the Space Environment Center of the US National Oceanic and

Atmospheric Administration (NOAA), and the American relative sunspot number, published by the solar physics division of the American Association of Variable Star Observers.

relativistic beaming A phenomenon in which a charged particle moving close to the velocity of light will emit electromagnetic radiation in a narrow beam in the direction of motion. The faster the particle, the narrower the beam.

relativistic velocity A velocity approaching that of light. For example, high-energy cosmic-ray particles move at relativistic velocities. At such velocities, the mass of an object becomes significantly greater than its *rest mass.

relativity A collective term for two theories, *special relativity* and *general relativity*, developed by A. *Einstein. The *special theory of relativity, published in 1905, is concerned with the laws of physics as viewed by observers moving relative to each other at constant speed (i.e. not subject to acceleration). It describes how the motion of one observer relative to another affects measurements made by these observers. At low speeds special relativity reduces to situations describable by classical physics, embodied in *Newton's laws of motion. Differences between Newton's and Einstein's physics become apparent only at velocities towards that of light. The *general theory of relativity (1915) describes how the relationship between space and time is affected by the gravitational effects of matter (*see* GRAVITATION) and how space and time change as seen by an observer on an accelerating object. The theory concludes that gravitational fields created by the presence of matter cause spacetime to become curved. This curvature controls the motion of bodies in space.

relaxation time The time expected for the orbit of a star in a cluster or galaxy to be significantly changed through gravitational perturbation by other stars. For a cluster of N stars this is about $0.1N$ times the time a typical star takes to cross the cluster (the *crossing time). Globular clusters are far older than their relaxation times, and hence are said to be relaxed. For the Galaxy, by contrast, the relaxation time is longer than the age of the Universe.

remote sensing Any technique for obtaining information about an object without coming into direct contact with it. In this broad sense it includes all the techniques used in ground-based and space astronomy. In terrestrial studies it refers particularly to satellite-borne instrumentation designed to observe features on or above the Earth's surface. Sophisticated remote-sensing techniques, such as synthetic-aperture radar, have been responsible for enormous improvements in our knowledge of the surface of Venus, for example. Spectroscopic methods such as infrared radiometry and colorimetry have been used to study the surface compositions of asteroids.

réseau An array of small crosses or dots exposed on to a photographic film or plate before processing, or overlain on the photocathode of an electronic detector, to provide a standard basis for measurement of positions of astronomical objects.

residuals The difference between observed and calculated values, such as of the position of a planet or comet in its orbit. They can result from observational errors and perturbations not included in the calculations; in the case of comets, such perturbations include *non-gravitational forces. From a study of residuals in the orbit of Uranus the existence of the then unknown planet Neptune was calculated.

resolution *See* RESOLVING POWER.

resolving power The finest detail an optical instrument will show; also known as **resolution**. For a telescope it is the smallest separation between two objects which allows them just to be distinguished as separate. However, the criterion for this depends on the method used to observe the objects and on their nature. In practice, the test object is taken to be a double star whose components are equally bright. One criterion is when the centre of the *Airy disk of one star coincides with the adjacent dark space before the first *diffraction ring. This is the resolution given by *diffraction-limited optics. In practice, observers can separate double stars closer than this, leading to the criterion known as the *Dawes limit. For a spectroscope, the resolving power is its ability to discriminate between two adjacent bright spectral lines.

resonance, orbital An effect in the orbits of celestial objects due to a *commensurability in their orbital periods. If the ratio of their orbital periods is close to a small fraction, such as ⅓, ⅔, or ¾, the bodies are said to be in resonant orbits. Because of this, the bodies' mutual perturbations repeat at nearly the same points in their orbits, thereby building up to produce large oscillations, or even driving one of the bodies into another orbit. The *Kirkwood gaps in the asteroid belt are places where asteroids would be in resonance with Jupiter. In Jupiter's satellite system, the mean motions of Io, Europa, and Ganymede are in resonance with each other, such that the three satellites can never line up on the same side of Jupiter. Other pairs of satellites such as Mimas and Tethys, Dione and Enceladus, and Titan and Hyperion are in resonance, the effects being to stabilize their orbits and produce in them oscillations of large amplitude and long period.

resonance line A spectral line caused by an electron jumping between the ground state and the first energy level in an atom or ion. It is the longest-wavelength line produced by a jump to or from the ground state. Because the majority of electrons are in the ground state in many astrophysical environments, and because the energy required to reach the first level is the least needed for any transition, resonance lines are the strongest lines in the spectrum for any given atom or ion.

resonant scattering In optical and ultraviolet spectroscopy, the absorption and prompt re-emission of photons of a particular wavelength by an atom. In this process, a photon of exactly the right wavelength (i.e. energy) excites an electron in the atom from one energy level to another. There is said to be a **resonance** between the energy of the photon and the energy difference in the atom. The electron then drops back down to its original energy level more or less immediately, emitting a photon of almost identical energy to the one that was absorbed in the first place, but in some random direction. Resonant scattering applies only to line radiation, unlike other forms of scattering which are of continuous radiation.

rest mass The mass of an atomic particle that is at rest relative to an observer. A particle moving relative to an observer becomes more massive, particularly if its velocity approaches that of light (in which case its mass is termed **relativistic mass**).

restricted three-body problem A version of the *three-body problem in which one of the bodies is taken to be essentially massless and so does not affect the *relative orbit of the other two massive bodies. The problem is thus simplified to finding the behaviour (position and velocity) of the massless body at any time in the combined gravitational field of the two other bodies. In the *circular restricted three-body problem* the two massive bodies follow circular orbits about their common centre of mass; in the *elliptical restricted three-body problem* they follow elliptical orbits about the centre of mass. Other variations include the *coplanar restricted three-body problem*, in which the massless body moves entirely in the plane of the two massive bodies' orbits; and the *three-dimensional restricted three-body problem*, in which the massless body is free to move in all three dimensions. Practical applications include three-body systems such as the Sun, a planet, and a small planetary satellite; the Sun, Jupiter, and an asteroid; the Sun, a planet, and a comet; the Earth, the Moon, and an artificial satellite; and a binary star with a planet.

retardation The difference between the times of moonrise on successive nights, due to the Moon's motion in its orbit. It has an average value of 50.4 min, but the actual figure can vary widely depending on the Moon's declination. On about September 23, the autumnal equinox, the full Moon rises only 18 min later on successive evenings. This is known as the *harvest Moon*. A similar effect in October is called the *hunter's Moon*.

retardation plate Another name for a *wave plate.

reticle Fine lines or webs in the focal plane of an eyepiece, used for centring objects when guiding or for making angular measurements; also known as a *graticule*. The word implies a net or grid, but in practice a reticle may be a simple cross-wire or a series of concentric circles. The lines or the field of view are often illuminated.

Reticulum (Ret) (*gen.* **Reticuli**) The seventh-smallest constellation, lying in the southern sky. Reticulum represents the reticle used in an eyepiece for measuring star positions. Its brightest star is Alpha Reticuli, magnitude 3.3. Zeta Reticuli is a wide double of stars similar to the Sun, magnitudes 5.2 and 5.5.

reticulum A net-like (reticular) pattern on a planetary surface; pl.*reticula*. It is not a geological term; it was introduced for the nomenclature of features on Venus, but has not yet been assigned to any feature.

retrograde motion The movement of a body such as a planet from east to west on the celestial sphere; or the movement of a body in its orbit in a clockwise sense as seen from above the Sun's north pole; or the rotation of a body on its axis in a clockwise sense as seen from above the Sun's north pole. Motion in the opposite direction is termed *direct or prograde motion. Objects with retrograde motion have an orbital or axial inclination greater than $90°$. The outer planets appear to move in a retrograde direction temporarily when the Earth catches them up and overtakes them; they are said to describe a **retrograde loop**.

reversing layer A supposed region in the atmosphere of a star, particularly the Sun, where the dark Fraunhofer lines of the star's spectrum were once presumed to originate. At the end of the 19th century the solar chromosphere was thought to be the reversing layer (its spectrum being the reverse of the Fraunhofer spectrum), but it is now known that the Fraunhofer lines are produced in the photosphere, along with the continuum radiation, so that a specific reversing layer does not exist.

revolution The movement of one body in orbit around another, or around a common centre of mass, such as the monthly revolution of the Moon around the Earth or the yearly revolution of the Earth around the Sun. For artificial satellites, revolutions are counted relative to a given point on the planet's surface. Because the Earth is rotating from west to east, a satellite in a west-to-east orbit takes slightly longer to complete one revolution of the Earth than to complete one orbit relative to a fixed point in space. Hence a spacecraft such as the International Space Station may make 16 orbits of the Earth in a day but only 15 revolutions.

RFT Abbr. for *richest-field telescope.

RGO Abbr. for *Royal Greenwich Observatory.

RGU photometry Photographic photometry with filters of the following wavelengths and bandwidths: U, 359 and 49nm; G, 466 and 52nm; and R, 641 and 40nm. The filters and photographic emulsions were selected so that the *colour excesses $E(U − G)$ and $E(G − R)$ were approximately equal. There are no standard stars for RGU photometry; instead, UBV standards are converted into U, G, and R respectively. The system has been used extensively to deduce the structure of the Galaxy from star counts.

Rhea The second-largest satellite of Saturn, 1527km in diameter; also known as Saturn V. Rhea orbits Saturn in 4.518 days at a distance of 527370km. Its axial rotation period is the same as its orbital period. It was discovered in 1672 by G. D. *Cassini. Rhea has an icy surface saturated with impact craters, and shows broad, elongated wispy features of unknown origin on its trailing hemisphere. 📷

RHESSI Abbr. for *Ramaty High-Energy Solar Spectroscopic Imager.

Rho Cassiopeiae star One of a small group of yellow, cool, *hypergiant, *pulsating variables. The type star is itself a semiregular variable (subtype SRD) that has undergone shell episodes, with low-velocity ejection and corresponding fades. Stars in the group have periods of 60–400 days, and they may all exhibit *non-radial pulsation. The stars have masses slightly below 40 solar masses.

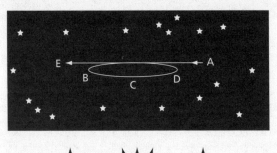

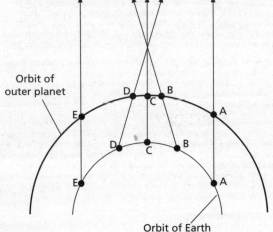

Orbit of Earth

retrograde motion: Superior planets appear to move in a retrograde direction as the Earth catches up and then overtakes them.

A   planet in direct motion
B   stationary point
C   planet in middle of retrograde loop
D   stationary point
E   planet resumes direct motion

Rho Ophiuchi Nebula A group of nebulae associated with recent and current star formation, about 400 l.y. away in the constellation Ophiuchus. They take their collective name from one of the young stars within the nebulosity, Rho Ophiuchi. Emission nebulae, reflection nebulae, and dark nebulae all appear close together. It is one of the closest regions of star formation to us, about one-third the distance of the Orion Nebula.

Richardson–Lucy algorithm A technique for estimating corrections to data in statistical astronomy. It is used for removing the effects of diffraction in radio observations and seeing in ground-based optical images, and was used for sharpening the aberrated images from the Hubble Space Telescope before its new optics were installed. The technique is named after its inventors, William Hadley Richardson, an American optician, and the British astronomer Leon Brian Lucy (1938–).

richest-field telescope (RFT) A telescope that combines the widest possible field of view with the maximum usable *exit pupil. Its magnification is just low enough for the exit pupil to be the same as the observer's maximum dark-adapted pupil size. If the maximum pupil size is taken to be 7.5 mm, the magnification will be approximately 0.13 times the aperture in millimetres. This is usually provided by a telescope of small focal ratio and wide-field eyepiece, and is particularly suited to observing the Milky Way, extended deep-sky objects, and comets. A typical RFT has an aperture of 100–180 mm, a focal ratio of f/4, and a field of view of 2–5°.

Rigel The star Beta Orionis, magnitude 0.18, the seventh-brightest star in the sky. It is a B8 supergiant 863 l.y. away and has a luminosity 40 000 times the Sun's. It has a companion of magnitude 6.8, which is itself a spectroscopic binary.

right ascension (RA) (α) A coordinate on the celestial sphere, the equivalent of longitude on Earth. Right ascension is measured anticlockwise around the celestial equator, usually in hours, minutes, and seconds of sidereal time (although sometimes in degrees), starting from 0h at the *vernal equinox. The 0h line of right ascension is the celestial equivalent of the Greenwich meridian on Earth. An hour of right ascension is equivalent to 15 degrees of arc, so that 24 h of RA is equivalent to 360°. Right ascension is an *equatorial coordinate.

Rigil Kentaurus Alternative name for the star *Alpha Centauri.

rille A well-defined, long, narrow valley on the Moon with steep walls and roughly parallel sides. There are three main types: *straight* (or *linear*) *rilles* which appear to be *graben, where the surface has dropped between two parallel faults; *arcuate rilles*, which are similar but curved in plan view, and often concentric with a circular mare basin; and *sinuous rilles, which are different in origin. *See also* RIMA.

rima A long, well-defined narrow furrow on a planetary surface; pl.*rimae*. The name, which means crack or fissure, was first used for features on the Moon in the 17th century. It is applied to features of various origins. For example, Rima Hadley (Hadley Rille) on the Moon is a *sinuous rille, formed as a lava channel in a vast lava flow, whereas the Rimae Hippalus are *graben surrounding a large impact basin.

ring, planetary A disk of matter encircling a planet, consisting of numerous particles in orbit ranging in size from dust grains up to objects tens of metres across. In the Solar System, Saturn has the brightest and most extensive rings, composed mainly of ice. In 1979 the Voyager probes detected a ring around Jupiter. Rings were found around both Uranus and Neptune by stellar occultations, and confirmed when Voyager 2 visited them. The various planetary rings have different proportions of ice and dust, and different distributions of particle sizes. Planetary rings may consist of debris from impacts on the planets' satellites.

(🌐) SEE WEB LINKS
• Names and data of planetary rings.

ring arcs Planetary rings that are considerably fainter or discontinuous along part of their length. The F Ring of Saturn, a very thin ring outside the A Ring discovered by the *Voyager probes, has brighter areas within it. Ring arcs were also suspected to exist around Neptune as a result of occultations in which the star dimmed in some places but not in others. Voyager 2 found that most of the matter in Neptune's outer ring (the *Adams Ring) was concentrated in four bright arcs.

ring galaxy An unusual galaxy with a well-defined luminous ring around a bright core. The ring can appear smooth and regular, or knotted or warped, and may include gas and dust in addition to stars. Such an arrangement may result from a collision of a disk galaxy with another smaller galaxy, which passes through the disk. An example is the *Cartwheel Galaxy. 📷

Ring Nebula A 9th-magnitude planetary nebula about 2000 l.y. away in Lyra, also known as M57 or NGC 6720. The hot 15th-magnitude central star is surrounded by a shell of luminous gas and dust some three-quarters of a light year in diameter. In small telescopes the shell looks like an elliptical smoke ring $70'' \times 150''$ across. The ring is expanding at 19 km/s. Observations by the Hubble Space Telescope have demonstrated that the 'ring' is in fact a cylinder of gas aligned end-on to us. 📷

rise time The time taken for a variable star to rise from minimum to maximum. In an eclipsing binary it is the time either from mid-eclipse or (for total eclipses) from third contact to maximum brightness. In novae and supernovae, it is usually taken to be the time to any pre-maximum halt, rather than to the extreme maximum.

rising The moment at which a celestial body appears above an observer's horizon. For bodies with an observable disk, particularly the Sun and Moon, rising is taken as the moment when their upper limb lies exactly on the observer's horizon. Refraction in the Earth's atmosphere means that an object appears to rise before it actually does, and this effect must be taken into account when calculating observed rising and setting times.

Ritchey, George Willis (1864–1945) American astronomer and optician. At Mount Wilson Observatory he made the mirror for the 60-inch (1.5-m) telescope and supervised the work on the mirror for the 100-inch (2.5-m). With these instruments he discovered dark lanes in the Andromeda Galaxy, and in 1917 he took the first photographs of novae and supernovae in the Andromeda Galaxy, providing the first means of estimating its distance. He also took the first photographs to show globular clusters around other galaxies. He later collaborated on the design of the *Ritchey–Chrétien telescope, the first of which was completed in 1930.

Ritchey–Chrétien telescope A *Cassegrain telescope in which both the primary and secondary mirrors have hyperboloidal surfaces. This has the advantage over the standard Cassegrain system of giving images free from coma and spherical aberration over wide fields. Typically the primary mirror has a focal ratio of about 2.5, giving a final focal ratio of about 8. The short tube length and good optical performance make the Ritchey–Chrétien very popular for modern large instruments; the *Keck Telescopes and the *Very Large Telescope are of this design. It is named after its co-inventors, G. W. *Ritchey and the French optician Henri Chrétien (1879–1956).

R magnitude The magnitude of a star measured with an R filter. The filter used has several different definitions. The central wavelength and bandwidth of the R filter in *Kron–Cousins RI photometry are 641 and 158nm; in *Johnson photometry they are 710 and 210nm; in the *six-colour system, 714 and 165nm; and in *RGU photometry, 641 and 40nm. The R most frequently used nowadays is that of the Kron–Cousins system. R is usually measured photoelectrically, except in the RGU system, which is photographic.

Robertson–Walker metric A mathematical function describing the geometry of *spacetime in a model which incorporates the *cosmological principle. In general, a metric relates physical distances or intervals between events separated in space and/or time to the coordinates used to describe their position. General relativity deals with a four-dimensional spacetime in which the separation between space and time coordinates is not obvious. In a homogeneous and isotropic cosmology, however, it is possible to define a unique time coordinate, called *cosmic time*, and three spatial coordinates. The Robertson–Walker metric is the most general possible four-dimensional metric function compatible with homogeneity and isotropy. In general, it describes a curved space which is either expanding or contracting with cosmic time. It is named after the American mathematician and cosmologist Howard Percy Robertson (1903–61) and the English mathematician Arthur Geoffrey Walker (1909–2001).

Roche limit The distance from the centre of a planet within which any large satellite would be torn apart by tidal forces. The Roche limit lies at approximately 2.46 times the planet's radius, if the planet and satellite have similar densities. Within the limiting distance no major satellite could form, although small, strong objects such as artificial satellites can survive well within this distance. All four major planets have ring systems within the Roche limit. It is named after the French mathematician Édouard Albert Roche (1820–83).

Roche lobe The region surrounding each star in a binary system, within which any material is gravitationally bound to that particular star. The boundary of the Roche lobes is an *equipotential surface, and the lobes touch at the inner *Lagrangian point, L_1, through which *mass transfer may occur if one of the components expands to fill its lobe. It is named after the French mathematician Édouard Albert Roche (1820–83).

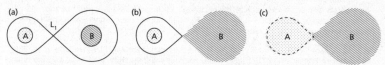

Roche lobe: In a binary system, the Roche lobes of components A and B meet at the L_1 Lagrangian point. (a) In a detached system, neither star fills its Roche lobe. (b) In a semidetached system, one massive component, B, fills its Roche lobe. (c) In a contact binary both components overfill their Roche lobes and share a common envelope.

ROE Abbr. for *Royal Observatory, Edinburgh.

Römer, Ole (or Olaus) Christensen (1644–1710) Danish astronomer. While observing Jupiter's satellites from the Paris Observatory, intending to base an international time standard on their eclipses, he noticed that the times of eclipses disagreed with predictions by G. D. *Cassini, varying by 22 minutes over a six-month period. Römer deduced that light took longer to reach Earth when Jupiter was farther away. In 1676 he calculated the speed of light from his timings, using the contemporary value of the astronomical unit, finding it to be about 200000km/s. This is two-thirds of the true value, but was the first demonstration that light had a finite speed. He also built the first *transit instrument.

Ronchi test A method of testing the shape of a telescope mirror. It is based on the *Foucault test, but instead of a knife edge it uses a fine grating with about 40 dark lines per centimetre. When this apparatus is used, the mirror appears to be crossed by lines. Any curvature of these lines indicates deviations from a spherical figure. It is named after the Italian physicist Vasco Ronchi (1897–1988).

Rood–Sastry type A category in a classification scheme for clusters of galaxies, based on the type and distribution of the ten brightest members. cD clusters contain a *cD galaxy; B clusters have a binary pair of supergiant galaxies less than ten diameters apart; L clusters have their brightest members approximately in line; F clusters are highly flattened; C clusters have a core–halo structure with four or more of the brightest members near the core and fainter members in a halo; I clusters are irregular with no well-defined centre. The classification was introduced in 1971 by the American astronomer Herbert Jesse Rood (1937–2005) and the Indian astronomer Gummuluru Narasimha Sastry (1937–2008).

roof prism A prism with two totally internally reflecting surfaces at 90° to each other, like a roof. An image reflected from one is reflected again from the other, emerging reflected but not laterally inverted as in a normal prism. Roof prisms are used in compact binoculars.

Roque de los Muchachos Observatory (ORM) An observatory owned and operated by the Instituto de Astrofísica de Canarias (IAC) at an altitude of 2396m on La Palma, Canary Islands, founded in 1979. Telescopes belonging to various nations are sited here. The largest is the 10.4-m *Gran Telescopio Canarias (GTC), opened in 2009. Next-largest is the 4.2-m *William Herschel Telescope, part of the *Isaac Newton Group that also includes the *Isaac Newton Telescope and the Jacobus Kapteyn Telescope. Other major telescopes at ORM are the *Telescopio Nazionale Galileo; the *Nordic Optical Telescope; the *Liverpool Telescope; the 1.2-m Belgian Mercator Telescope, opened 2001; the 0.97-m Swedish Solar Telescope (SST), opened 2002; the 0.45-m Dutch Open Telescope (DOT), a solar telescope opened in 1997; and SuperWASP-North, an array of eight wide-angle cameras of 0.1 m aperture for detecting transits of stars by extrasolar planets, which began operation in 2004. The Carlsberg Meridian Telescope (CMT), opened in 1984, is owned and operated by the Real Instituto y Observatorio de la Armada, Spain. ORM also contains two gamma-ray detectors of the *Major Atmospheric Gamma-ray Imaging Cherenkov (MAGIC) Constortium.

((⊕)) SEE WEB LINKS
• Official observatory website.

Rosalind The eighth-closest satellite of Uranus, distance 69930km, orbital period 0.558 days; also known as Uranus XIII. Its diameter is 54km, and it was discovered in 1986 on images from the Voyager 2 spacecraft.

Rosat A joint German–UK–US soft X-ray and extreme ultraviolet (EUV) astronomy satellite; its name is a contraction of Röntgen Satellit and derives from Wilhelm Konrad von Röntgen (1845–1923), the German physicist who discovered X-rays. In the first six months after its launch in 1990 June it carried out an X-ray imaging survey of the entire sky, with a German 0.8-m telescope, and the first-ever survey of the whole sky in the EUV, with the UK's Wide Field Camera (WFC). Subsequently, the mission was devoted to detailed observations of individual objects. It ceased operation in 1999 February.

(⊕) SEE WEB LINKS
• Information page at Goddard Space Flight Center.

Rosetta An ESA probe to rendezvous with and land on a comet. It was launched in 2004 March and underwent three gravity assists from the Earth (in 2005, 2007, and 2009) and one from Mars (2007). It passed the asteroids Šteins, in 2008 September, and Lutetia, in 2010 July, photographing them in close-up, before reaching Comet 67P/Churyumov–Gerasimenko in 2014 May. After going into orbit and mapping the nucleus Rosetta will release a lander named Philae in 2014 November to sample its surface as the comet approaches perihelion.

(⊕) SEE WEB LINKS
• ESA mission website.

Rosette Nebula A diffuse nebula in Monoceros over 1° wide, surrounding a 5th-magnitude star cluster, NGC 2244. The nebula is so named because it resembles a rosette. The brightest parts of the nebula have their own NGC numbers: NGC 2237, 2238, 2239, and 2246. The associated star cluster, consisting of stars of 6th magnitude and fainter, spans about ½°. The nebula and cluster lie 5000 l.y. away.

Ross, Frank Elmore (1874–1960) American astronomer. Working under S. *Newcomb, he calculated an orbit for Phoebe, discovered in 1898. He designed and built the first *photographic zenith tube, completed in 1911. He spent the years 1915–24 developing emulsions and lenses for astrophotography, including coma-correcting lenses for large reflectors. A photographic survey (1924–39) yielded the Ross Catalogue of 869 stars with large proper motion. He photographed Mars and Venus from Mount Wilson, discovering markings in Venus's clouds in the ultraviolet.

Rosse, Third Earl of (William Parsons) (1800–67) Irish astronomer. In 1845 he completed a 72-inch (1.8-m) reflector in the grounds of his family seat at Birr Castle, Parsonstown, in Ireland. This was by far the largest telescope in the world, and its success helped set the trend for large reflectors. With it Rosse detected, in 1845, spiral structure in M51 (the *Whirlpool Galaxy) and embarked on a programme to study the shapes of other 'nebulae'. Some he was able to resolve into star clusters; others, which we now know to be galaxies, remained nebulous. He gave the *Crab Nebula its name, from its appearance in the 72-inch. Among his assistants were his son Laurence Parsons (1840–1908), the fourth earl, and J. L. E. *Dreyer.

Rosseland mean opacity The opacity of a gas of given composition, temperature, and density, averaged over the various wavelengths of the radiation being absorbed and scattered. The radiation is assumed to be in thermal equilibrium with the gas, and hence to have a black-body spectrum. The Rosseland mean opacity is useful for calculating the total amount of energy absorbed over all wavelengths, which is required in calculations of stellar structure. It is named after the Norwegian astrophysicist Svein Rosseland (1894–1985).

Rossi, Bruno Benedetto (1905–93) Italian-American physicist. In 1931 he showed that some cosmic rays have very high energies. After working in particle physics, Rossi seized the opportunities provided by the space age. He showed in 1958 that the interplanetary medium contains ionized gas. In 1961 he mapped the Earth's magnetopause, and in 1962, with

R. *Giacconi, he discovered the X-ray source Scorpius X-1. The *Rossi X-ray Timing Explorer was named in his honour.

Rossiter effect An irregularity in the radial-velocity curve of an eclipsing spectroscopic binary immediately before and after eclipse; also known as the **rotation effect**. In most binaries the stars orbit in the same direction as they rotate on their axes. Immediately before maximum eclipse, the small segment of the star that remains uncovered therefore has a radial velocity that is directed away from the observer. The opposite applies as the star emerges from eclipse. The effect shows up as a short-lived rise and fall on the radial-velocity curve. It is named after the American astronomer Richard Alfred Rossiter (1886–1977).

Rossi X-ray Timing Explorer (RXTE) A NASA satellite launched in 1995 December. It carries a large array of proportional counters, nearly a metre square, to study X-rays of 2–200 keV (0.006–0.6nm). RXTE uses its Proportional Counter Array (PCA) and High Energy X-ray Timing Experiment (HEXTE) to study the variability of the thousand or so brightest X-ray sources at time-scales down to 10μs. Another instrument, the All-Sky Monitor (ASM), scans the sky every 100 min. It is named after B. B. *Rossi.

(⊕) SEE WEB LINKS
• Information page at Goddard Space Flight Center.

rotating variable A star that varies in brightness as it rotates, due to an ellipsoidal shape or non-uniform surface brightness. Irregular surface brightness may result from magnetic effects, as in a *magnetic variable or an *oblique rotator, or the presence of *starspots. *See* ALPHA² CANUM VENATICORUM STAR; BY DRACONIS STAR; ELLIPSOIDAL VARIABLE; FK COMAE BERENICES STAR; SX ARIETIS STAR.

rotation The turning of a body on its axis, such as the daily rotation of the Earth. It is usually measured relative to the stars, and termed the sidereal period of axial rotation.

rotation curve The observed variation in orbital speed of stars and interstellar matter with distance from the centre of a galaxy. The curve is determined by measuring the Doppler shift of spectral lines at optical, infrared, millimetre, or radio wavelengths. The observed 'flatness' of these curves at large radii in many spiral galaxies implies the existence of large amounts of unseen *dark matter beyond the visible regions.

rotation effect Another term for *Rossiter effect.

rotation measure A measure of the change in direction of the polarization of an electromagnetic wave due to the *Faraday effect. It is proportional to the product of the electron density and the radial component of the galactic magnetic field along the line of sight to the source. If the electron density is known, usually from the *dispersion measure of the source, the mean value of the magnetic field can be deduced.

Royal Astronomical Society (RAS) An organization for the promotion of astronomy and geophysics, founded in 1820 as the Astronomical Society of London. It became the Royal Astronomical Society in 1831. Its headquarters are in London. The RAS publishes *Monthly Notices*, *Astronomy & Geophysics*, and *Geophysical Journal International*.

(⊕) SEE WEB LINKS
• Official website.

Royal Greenwich Observatory (RGO) An observatory founded in 1675 at Greenwich, southeast London. After World War II it relocated to Herstmonceux Castle, Sussex, where it operated the *Isaac Newton Telescope from 1967 to 1979. The original building at Greenwich is now maintained as a museum under the name Royal Observatory Greenwich (ROG). In 1990 the RGO moved again, this time to Cambridge, but was closed for good in 1998. The Time Department, which formerly maintained and distributed Greenwich Mean Time, was closed in 1990 before the move to Cambridge.

Royal Observatory, Edinburgh (ROE) An observatory on Blackford Hill, Edinburgh, owned and operated by the UK's Particle Physics and Astronomy Research Council. The ROE was founded in 1818 on Calton Hill, Edinburgh, but moved to its present site in 1894–6. It operated the *United Kingdom Schmidt Telescope in Australia from 1973 to 1988, and still houses its library of photographic plates. ROE now houses the UK Astronomy Technology Centre and the Institute for Astronomy of the University of Edinburgh.

(((⊕))) SEE WEB LINKS
• Official observatory website.

r-process A type of nuclear reaction proposed to explain the formation of elements heavier than bismuth, and certain neutron-rich isotopes of elements heavier than iron. These elements are produced by reactions initiated by the strong flux of neutrons from supernovae. The flux is so strong that unstable nuclei formed by the absorption of a neutron do not have time to decay before another neutron is absorbed, so that chains of reactions involving highly unstable nuclei can take place. The r-process is so named because it is rapid.

RR Lyrae star A type of yellow giant pulsating variable, with a period of 0.2–1.2 days and an amplitude of 0.2–2.0 mag.; abbr. RR. The stars belong to Population II, lying in the galactic halo and globular clusters (hence their earlier name of *cluster variables*). Three subtypes are now recognized: RRAB, which are pulsating in the *fundamental mode; RRC, pulsating in the first overtone; and RR(B) (often abbreviated as RRd), which pulsate in both these modes and hence are known as *double-mode* RR Lyraes. The RRAB type has an asymmetric light-curve, with a steep rise and gentler decline and amplitudes up to 1 mag., whereas light-curves of type RRC are approximately sinusoidal and have amplitudes of about 0.5 mag. Periodic modulation of the periods and amplitudes (the *Blazhko effect) is common. The RRAB type all have approximately the same absolute magnitude (+0.5), which makes them valuable distance indicators. The similarity of the RRAB light-curves to those of Cepheid variables gave rise to the obsolete name of *short-period Cepheid*.

RRs variable *See* DELTA SCUTI STAR.

RR Telescopii star A *symbiotic star that exhibits outbursts similar to those of a nova (up to 10 magnitudes) but with an exceptionally slow decline. There may be Mira-like long-period variations of 1–2 mag. before the outbursts. The secondary is an evolved giant or *Mira star. (This distinguishes the systems from true novae with main-sequence or slightly evolved secondaries.) There appears to be continuous or extremely prolonged mass loss. RR Telescopii and related systems were previously called *very slow novae*, and are now sometimes known as *symbiotic novae*.

RS Canum Venaticorum star A type of close binary star, typically containing solar-type and cool stars in which the light varies due to rotational modulation caused by starspots and chromospheric activity on the cooler component; abbr. RS. The light curve resembles a sine wave, but additional variation of up to 0.2 mag. with periods of years appears to be due to changes in activity akin to the solar cycle. These systems are officially classified as eruptive variable stars but should perhaps be included in the rotating variable star class. They are detached binaries, and many undergo eclipses.

R star *See* CARBON STAR.

Rubin–Ford effect An apparent anisotropy in the expansion of the Universe on a scale of around 100 million l.y., as revealed by a study of the motions of a sample of spiral galaxies by the American astronomers Vera Cooper Rubin (1928–) and William Kent Ford, Jr (1931–). The reason that there is not a purely isotropic *Hubble flow is because the Universe is not homogeneous on these scales.

runaway star A young star moving at high velocity, perhaps hundreds of kilometres per second, suggesting that it has been ejected by some violent event from its birthplace. Such stars were probably once part of a binary, but were ejected either when their companion exploded as a supernova, or else through a close encounter with another binary. Examples of the first type are Zeta Ophiuchi and the pulsar PSR J1932+1059, resulting from a binary disrupted by a supernova about a million years ago. AE Aurigae and Mu Columbae, moving away from the Trapezium region of Orion in opposite directions at around 100 km/s, are thought to result from an encounter between two binaries 2.5 million years ago; their former partners now constitute Iota Orionis, a binary with a highly eccentric orbit.

rupes A scarp on a planetary surface; pl.*rupes*. The name, which means 'cliff', is not a geological term, but is used in the nomenclature of individual features, for example Rupes Recta (the Straight Wall) on the Moon, or the Victoria Rupes on Mercury.

Russell, Henry Norris (1877–1957) American astronomer. By 1910 he had accumulated data on enough stars for him to plot a diagram of their absolute magnitude against their spectral type, finding that red stars fell into two groups, giants and dwarfs, but he was unaware that E. *Hertzsprung had done the same in 1906. This diagram, further developed by Russell in 1913, is now known as the *Hertzsprung–Russell diagram. Russell also evolved a method of calculating the masses of binary stars. In 1928 he established the composition of the Sun's atmosphere from its spectrum, and went on to suggest that all stars contain a high proportion of hydrogen, a conclusion originally reached by C. H. *Payne-Gaposchkin.

Russell–Vogt theorem *See* VOGT–RUSSELL THEOREM.

RV Tauri star A highly luminous, yellow supergiant pulsating variable; abbr. RV. Spectra are F–K at maximum, K–M at minimum. The light-curves have overall amplitudes of 3–4 mag.; all have double maxima (normally with slightly different amplitudes) and an intervening secondary minimum. The amplitudes of the maxima and secondary minimum vary, and the secondary minimum may become as deep as (and replace) the primary minimum. The formal period is the double interval from one deep minimum to the next, and generally lies in the range 30–150 days. In the RVA subtype the mean magnitude remains constant with time. The RVB subtype exhibits a superimposed secondary wave, with amplitudes up to 2 magnitudes and periods of 600–1500 days.

RW Aurigae star A widely used, but now obsolete, term for a young *irregular variable. The designation and its subgroups were based solely on the form of the light-curve, whereas the current system of classification also takes into account the presence of nebulosity, speed of variation, and spectral type. RW Aurigae itself is now classified as of type INT, with rapid variations and a *T Tauri spectrum.

RXTE Abbr. for *Rossi X-ray Timing Explorer.

Ryle, Martin (1918–84) English radio astronomer. After World War II he began to establish Cambridge as a centre for radio astronomy, opting for radio interferometers rather than parabolic receivers. With A. *Hewish he developed the technique of *aperture synthesis and commenced a series of surveys of radio sources that were published as the Cambridge catalogues, the best known of which is the *Third Cambridge Catalogue (objects in it are preceded by the designation 3C). The large number of faint radio sources discovered helped discredit the *steady-state theory. Ryle shared the 1974 Nobel Prize in Physics for his work on aperture synthesis.

Ryle Telescope A former *aperture synthesis array at the *Mullard Radio Astronomy Observatory, near Cambridge, England. It consisted of eight 13-m dishes, four fixed and four movable, on an east–west baseline 4.8 km long. When opened in 1972 it was originally known as the Five-Kilometre Telescope, but was renamed in 1989 in honour of M. *Ryle. It was closed in 2006 and its dishes now form part of the Arcminute MicroKelvin Imager (AMI), an instrument to study the cosmic microwave background.

SAAO Abbr. for *South African Astronomical Observatory.

Sabik The star Eta Ophiuchi. It appears to the naked eye of magnitude 2.4, but it is in fact a close binary consisting of two A-type dwarfs of magnitudes 3.0 and 3.5 that orbit with a period of 88 years. It lies 88 l.y. away.

Sachs–Wolfe effect A phenomenon in which irregularities in the distribution of mass in the early Universe give rise to localized temperature variations in the *cosmic microwave background. It occurs because photons are gravitationally redshifted by regions where the density of matter is higher than average. The ripples detected by the *Cosmic Background Explorer (COBE) satellite in 1992 are thought to be a manifestation of this effect. It is named after the American astrophysicists Rainer Kurt Sachs (1932–) and Arthur Michael Wolfe (1939–).

Sacramento Peak Observatory A solar observatory at an altitude of 2810 m in the Sacramento Mountains, 20 km southeast of Alamogordo, New Mexico. It was founded in 1949 by the US Air Force Cambridge Research Laboratories, which operated it until 1976 when it was taken over by the *Association of Universities for Research in Astronomy (AURA). It became part of the National Solar Observatory when that body was founded in 1984. Its main instruments are a 0.4-m coronagraph (the John W. Evans facility), opened in 1953; the Hilltop Dome, opened in 1963, containing solar patrol cameras; and the Richard B. Dunn Solar Telescope, opened in 1969, a vacuum tower telescope with a 0.76-m entrance window atop a tower 41.5 m high and a 1.6-m main mirror which produces a solar image 0.5 m in diameter.

((⊕)) SEE WEB LINKS
• Official observatory website

Sagan, Carl Edward (1934–96) American astronomer. In the early 1960s he showed that a greenhouse effect should be operating on Venus, and calculated its surface temperature to be 500–800 K. He repeated H. C. *Urey and S. L. Miller's experiment of irradiating a gas mixture resembling the Earth's primordial atmosphere, and found among the products many organic chemicals that are essential to life. He was best known for his espousal of the possibility of extraterrestrial life, and for his popular science writing.

Sagitta (Sge) (*gen.* **Sagittae**) The third-smallest constellation, lying in the northern sky, representing an arrow. Its brightest star is Gamma Sagittae, magnitude 3.5. WZ Sagittae is a recurrent nova seen to erupt in 1913, 1946, 1978, and 2001. M71 is an 8th-magnitude globular cluster.

Sagittarius (Sag) (*gen.* **Sagittarii**) A constellation of the zodiac, representing a centaur with a bow, popularly known as the Archer. The Sun passes through Sagittarius from the third week of December to the third week of January, and hence is in the constellation at the winter solstice. Its brightest star is Epsilon Sagittarii (*Kaus Australis). Beta Sagittarii is a naked-eye double, consisting of unrelated stars of magnitudes 4.0 and 4.3. Sigma Sagittarii is *Nunki. RY Sagittarii is a variable star of the *R Coronae Borealis type, which periodically dips from 6th to 14th magnitude. The constellation contains dense Milky Way starfields that lie towards the centre of our Galaxy; the exact centre is believed to be marked by the radio source *Sagittarius A. M8 is the *Lagoon Nebula; M17 is the *Omega Nebula; M20 is the *Trifid Nebula. M22 is a rich globular cluster of 5th magnitude, the third-best in the heavens.

Sagittarius A A prominent radio source 28000 l.y. away in the constellation Sagittarius, coincident with or close to the centre of our Galaxy. This highly complex region consists of a central core about 50 l.y. in diameter connected to a band of arched, parallel filaments more than 300 l.y. long crossing at right angles to the galactic plane. Sagittarius A*, a compact component at the heart of the central core, is believed to mark the physical centre of the Galaxy. It resembles a small-scale version of the energetic nuclei of active galaxies, and may contain a massive black hole of several million solar masses.

Sagittarius Arm The spiral arm of our Galaxy that is about 5000 l.y. closer to the centre of the Galaxy than the local *Orion Arm, and roughly parallel to it. The *Carina Arm* may be an extension of this arm. According to recent studies of galactic structure, this is one of the two minor arms of our Galaxy, the other being the *Norma Arm. As viewed from Earth, the galactic centre lies behind the Sagittarius Arm and the *Scutum–Centaurus and Norma Arms beyond it, but is highly obscured at visible wavelengths by dust in the Galaxy's disk. The Sagittarius Arm can be traced behind the stars of our local arm from Serpens via Scutum to Scorpius, Sagittarius, Centaurus, and Carina. Among the objects it contains are the *Eagle Nebula, the *Trifid Nebula, the *Lagoon Nebula, the *Jewel Box cluster, and *Eta Carinae.

Sagittarius Dwarf Galaxy A dwarf spheroidal galaxy (type dSph) in the Local Group of galaxies, discovered in 1994 in the constellation of Sagittarius and comparable in size with the Fornax Dwarf Galaxy. It has approximately 2.8×10^7 times the luminosity of the Sun, and is at least 10000 l.y. across. The Sagittarius Dwarf lies about 65000 l.y. from the centre of our Galaxy, and is the Galaxy's closest known satellite. It is on the far side of the Galaxy as viewed from Earth, and is being elongated by strong tidal disruption as it merges with our Galaxy.

Saha ionization equation A formula that relates the fraction of ionized atoms in a star's atmosphere to the temperature of the gas and the electron density (i.e. the number of free electrons per unit volume in the atmosphere). It provides an understanding of the relative prominence of spectral lines of different atoms and ions in the spectra of stars, and enables their temperatures to be derived from their spectral type. The equation was derived by the Indian astrophysicist and nuclear physicist Meghnad Saha (1894–1956).

Sakigake A Japanese space probe to Halley's Comet, launched in 1985 January; its name in Japanese means 'pioneer', and it was Japan's first interplanetary spacecraft. Sakigake passed 7 million km from the sunward side of the comet's nucleus in 1986 March. *See also* SUISEI.

Salpeter function Another name for *initial mass function.

Salpeter process Another name for the *triple-alpha process.

SALT Abbr. for *Southern African Large Telescope.

SAMPEX Abbr. for *Solar, Anomalous, and Magnetospheric Particle Explorer.

Sandage, Allan Rex (1926–2010) American astronomer. In 1950 he became an assistant to E. P. *Hubble, whose work on measuring the distances of galaxies and refining the value of the *Hubble constant he continued, finding a value of around 50 km/s/Mpc. He also calculated an early value for the *deceleration parameter. In 1960 he and the Canadian astronomer Thomas Arnold Matthews (1927–) made the first optical identification of a radio source (3C 48) that would turn out to be a quasar. In 1965 Sandage found the first radio-quiet quasar, and went on to identify many more quasars by their high redshifts.

SAO Abbr. for *Smithsonian Astrophysical Observatory.

SAO Catalog Abbr. for *Smithsonian Astrophysical Observatory Star Catalog*.

saros An interval of 18y 11.3d (or 10.3d, depending on the number of intervening leap years), equivalent to 223 lunations, following which the Sun, Moon, and Earth return to almost the same alignment. Solar eclipses recur one saros apart, but at different geographical locations.

One saros period is almost the same as 19 *eclipse years; the slight discrepancy leads to eclipses in a saros series gradually occurring further north or south until after some 70 eclipses (1262 years) their path no longer falls on the Earth's surface.

SAS Abbr. for *Small Astronomy Satellite.

satellite A small body that orbits a larger one, particularly the natural satellites of the planets. A natural satellite is also known informally as a *moon*. All the planets have at least one natural satellite, except Mercury and Venus. A spacecraft sent into orbit around the Earth or another body is an *artificial satellite*. (See Table 2, Appendix.)

Saturn (♄) The sixth planet from the Sun. Its mean opposition magnitude is between +0.7 and −0.3, depending on the tilt of the rings towards us, the faintest of the five planets known since antiquity. Saturn is the most flattened in shape of all the planets, with an equatorial diameter of 120536 km and a polar diameter of 108728 km. It is also the least dense of all the planets (0.69 cm³) and the only planet less dense than water. The rotation period of the visible surface ranges between about 10h 14m near the equator and 10h 40m at 60° south. The rotation of the planet's interior, derived from observations of Saturn's radio emissions by the Cassini probe, is 10h 47m 6s.

Saturn

Physical data

Diameter (equatorial)	Oblateness	Inclination of equator to orbit	Axial rotation period (sidereal)	
120 536 km	0.098	26°.73	10.233 hours	
Mean density	Mass (Earth = 1)	Volume (Earth = 1)	Mean albedo (geometric)	Escape velocity
0.69 g/cm³	95.16	764	0.47	36.1 km/s

Orbital data

Mean distance from Sun				
10⁶ km	AU	Eccentricity of orbit	Inclination of orbit to ecliptic	Orbital period (sidereal)
1426.725	9.537	0.054	2°.5	29.447 years

Saturn has a thick atmosphere composed of about 96% hydrogen and 3% helium (molecular percentages), with traces of methane and ammonia. The temperature near the top of the atmosphere is around −190°C. Internally, Saturn is thought to possess a rocky high-temperature core, perhaps containing iron, about 20000 km across. This is possibly surrounded by a layer of icy materials 5000 km thick, and a layer of metallic hydrogen and helium over 10 000 km thick. Convection currents within the conductive hydrogen layer are probably responsible for Saturn's magnetic field, which has an equatorial field strength of about 2×10^{-5} tesla, somewhat weaker than the Earth's. Surrounding this layer is normal molecular liquid hydrogen and helium, which gradually merges into a gaseous layer near the visible surface. Like that of the Earth, Saturn's magnetic field is oriented parallel to the planet's spin axis, but with opposite polarity.

Like Jupiter, the visible surface of Saturn is crossed by dark belts or bands of cloud, with bright zones between, although the atmosphere is generally calmer than Jupiter's.

Dark and bright spots occur, but are fainter and far less frequent than on Jupiter. Wisps and festoons suggestive of turbulence in the atmosphere are visible on spacecraft images. There is a 'jet stream' in the equatorial zone, where the rotation period is nearly half an hour faster than elsewhere. There are no long-lived features, but occasional spectacular outbursts of huge white spots occur in the equatorial zone. The first to be well-observed was in 1933 August, soon spreading over much of the equatorial zone. Similar outbreaks occurred in 1960 March and 1990 October.

Saturn's most distinctive feature is its bright rings. They have an albedo of up to 0.60, far higher than any other planetary rings. Through telescopes, three main rings are visible: the outer A Ring, 14600 km wide, extending out to 136800 km from Saturn's centre; the central B Ring, the brightest, 25500 km wide; and the much fainter inner C Ring or *crêpe ring*, 17500 km wide. Darker *spokes are faintly visible on the B Ring. Between the A and B Rings lies a prominent gap, the *Cassini Division, and the A Ring itself is divided by the *Encke Gap and the *Keeler Gap. Space probes have revealed that every ring has dozens of tiny subdivisions.

There are a further four named rings: the D Ring, which lies inside the C Ring; the narrow F Ring, which lies outside the A Ring; the more distant G Ring; and the outermost, the wide and diffuse E Ring. The innermost edge of the D Ring lies 67000 km from the centre of Saturn, while the outer rim of the E Ring is 480000 km from Saturn's centre. Despite their great extent, the rings are extremely thin, a few hundred metres at most, and disappear in all but the largest telescopes when edgewise-on to the Earth, which happens every 15 years or so. Saturn has over 60 known satellites.

Saturn Nebula An 8th-magnitude planetary nebula 3000 l.y. away in Aquarius, also known as NGC 7009. In shape it appears superficially similar to Saturn with its rings.

scale height The vertical distance in an atmosphere over which pressure falls by a factor of e (i.e. 2.718). For example, the scale height in the Earth's *troposphere is 8.5 km.

scattered-disk object (SDO) An object in a high-eccentricity orbit close to the ecliptic plane beyond Neptune; also known as a *scattered Kuiper Belt object* (*SKBO*). SDOs have perihelia close to Neptune's orbit but aphelia extending out to 100AU or more. The first SDO to be discovered was 1996 TL_{66}, now numbered (15874), which has a perihelion of 35.0AU, aphelion 133.5AU, eccentricity 0.584, inclination 24°.0, and a period of 773 years. Such objects may have escaped from the *Kuiper Belt or have been planetesimals scattered from their point of origin between Uranus and Neptune. The dwarf planet *Eris is an SDO.

scatter ellipse The elongated area on the ground within which fragments of a meteorite fall are scattered; also known as a *strewn field*. The long axis of the ellipse is parallel to the meteorite's direction of travel. The sizes of the fragments increase along the ellipse, because air resistance causes smaller fragments to fall to the ground sooner.

scattering The absorption and prompt re-emission of a photon by a particle, with a change of direction but not normally any significant change of energy (wavelength). The type of scattering depends on the particle involved, which can be an electron, a molecule, or a dust speck. If the particle is a free electron, the photon suffers *Thomson scattering; the *Compton effect is the same phenomenon but involving high-energy photons (X-rays and gamma rays) with a change of photon energy. If the scatterer is a bound electron (i.e. within an atom or molecule), the photon undergoes *resonant scattering. If the particle is small compared with the wavelength, the photon suffers *Rayleigh scattering; in this case the particles causing the scattering are molecules and dust, either in the Earth's atmosphere or in space. Alternatively, if the particle is large compared with the wavelength, the photon undergoes *Mie scattering; this can be caused by atmospheric and interstellar dust. The types of scattering differ in their probability of occurrence at different wavelengths, and in the preferred direction of the outgoing photons.

Scheat The star Beta Pegasi. It is an M2 giant 196 l.y. away that varies irregularly between magnitudes 2.3 and 2.7.

Schechter function A mathematical expression that represents the *luminosity function of galaxies. The function correctly reflects the facts that the luminosity function (the number-density of galaxies as function of luminosity) decreases with increasing luminosity and that the decrease is particularly marked at high luminosities. It is named after the American astronomer Paul Leonard Schechter (1948–), who proposed it in 1976.

Schedar The star Alpha Cassiopeiae, magnitude 2.2, a K0 giant 228 l.y. away. The name is also spelt Shedar or Shedir.

Scheiner, Christoph (1573–1650) German scholar and astronomer. In 1611, with the aid of a telescope he built himself, he became one of several independent discoverers of sunspots (the others included *Galileo, who unjustly accused Scheiner of claiming priority). He believed them to be small bodies in orbit round the Sun, and calculated their 'orbital inclination' to the ecliptic as 7° 30'; the present-day value for the Sun's axial inclination is 7° 15'.

Schiaparelli, Giovanni Virginio (1835–1910) Italian astronomer. In the 1860s he discovered that there are similarities between the orbits of some meteor streams and the orbits of particular comets, showing in 1866, for example, that the Perseid meteor shower is associated with Comet *Swift–Tuttle, and postulated that meteors are debris from comets. From 1877 he began a series of careful observations of Mars that resulted in a highly detailed map and a system of nomenclature for its *albedo features. Much speculation and controversy followed his report of straight markings, the *canali*. This term, originally used by P. A. *Secchi, means 'channels' but was translated into English as 'canals'. Schiaparelli also prepared maps of Venus and Mercury, and studied the orbits of binary stars.

schiefspiegler telescope A reflecting telescope that incorporates two spheroidal concave mirrors. The secondary mirror is offset from the primary, and brings light to a focus adjacent to the primary. This gives an all-reflecting system without the central obstruction usually caused by the secondary. However, the focal ratio of the system is high, giving a high magnification and limited field of view. Aberrations limit the aperture to about 100 mm. A variant, called a *trischiefspiegler*, includes a third mirror, which allows the design to be used for larger apertures. The name derives from the German for 'tilted mirror'.

Schmidt, Maarten (1929–) Dutch-American astronomer. In 1963 he obtained the spectrum of the radio source 3C 273. He identified hydrogen lines in its spectrum that were redshifted to an unprecedented degree for an object that appeared to be a star, indicating that it lay far off in the Universe. This was the first *quasar to be discovered. Schmidt continued to study quasars, finding that their number increases with distance, consistent with the Big Bang but not the steady-state theory.

Schmidt camera A wide-field telescopic camera first made in 1930 by the Estonian optician Bernhard Voldemar Schmidt (1879–1935); also called a **Schmidt telescope** or simply a **Schmidt**. It uses a spheroidal primary mirror with a thin *corrector plate at its centre of curvature which eliminates *spherical aberration. The *focal surface is curved and lies within the instrument, which means that photographic plates or films must be curved accordingly. A Schmidt has a very wide field of view free from *astigmatism, *distortion, and coma (*see* COMA, OPTICAL). Large Schmidt cameras are used for sky surveys. The quoted aperture of a Schmidt camera is that of its corrector plate; the mirror is always larger.

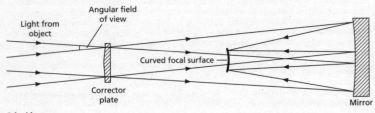

Schmidt camera

Schmidt–Cassegrain telescope (SCT) An adaptation of the *Schmidt camera for visual use. In addition to the concave spheroidal primary mirror it has a convex spheroidal secondary mirror, which reflects the converging beam through a hole in the primary to a *Cassegrain focus. This results in a compact instrument which has become very popular for amateur use. The optical performance is not usually as good as that of a *Newtonian telescope or conventional Cassegrain instrument because of compromises in the design, and the comparatively large central obstruction caused by the secondary mirror, which reduces contrast.

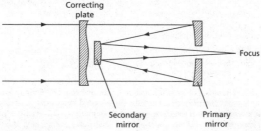

Schmidt–Cassegrain telescope

Schönberg–Chandrasekhar limit *See* CHANDRASEKHAR–SCHÖNBERG LIMIT.

Schröter effect A phenomenon in which the observed phase of Venus appears less than the predicted phase. As a result, *dichotomy* (half phase) occurs early at evening elongations, and late at morning elongations. The time difference between calculated and observed dichotomy is about a week. The effect was discovered by the German astronomer Johann Hieronymus Schröter (1745–1816). The cause may be that the region near the terminator is darker than the rest of the illuminated disk.

Schwabe, (Samuel) Heinrich (1789–1875) German pharmacist and amateur astronomer. From 1826, hoping to detect a planet closer to the Sun than Mercury crossing the Sun's disk, he kept daily records of the number of sunspots visible, and in 1843 announced that the number varied with a period of about 10 years. Schwabe's discovery went largely unnoticed until publicized in 1851 by the German naturalist Friedrich Wilhelm Heinrich Alexander von Humboldt (1769–1859). J. R. *Wolf collated sunspot data and in 1857 announced a period of just over 11 years, whereupon Schwabe received due recognition.

Schwarzschild, Karl (1873–1916) German astronomer. He established a method for determining a star's brightness from photographs, comparing its visual and photographic magnitudes to obtain the *colour index. In 1905 he obtained ultraviolet photographs of a solar eclipse, and went on to study energy transfer in the Sun, deducing that its outer regions had a layered structure. In 1916 he showed that, in the general theory of relativity, a sphere of material (approximating to a star) collapsing under its own gravitational field past its *Schwarzschild radius would cease to radiate energy (i.e. it would become a black hole). His son, Martin Schwarzschild (1912–97), became a naturalized American and studied stellar evolution.

Schwarzschild black hole A non-rotating black hole with no electrical charge. This is the simplest case of a black hole, predicted by K. *Schwarzschild in 1916, but is unlikely to be found in reality. The most likely form for black holes is the rotating *Kerr black hole.

Schwarzschild radius The radius of the event horizon of a black hole. At the Schwarzschild radius the escape velocity becomes equal to the speed of light. The more massive the black hole, the larger the Schwarzschild radius. For a body of mass M the Schwarzschild radius is $2GM/c^2$, where G is the *gravitational constant and c is the speed of light. It was first calculated in 1916 by K. *Schwarzschild.

Schwarzschild telescope A telescope–camera with two concave mirrors producing a flat *focal plane with a wide field of view within the instrument. As both mirrors are *aspheric and the instrument is unsuitable for visual use, it is rarely encountered. It was invented by K. *Schwarzschild.

Schwassmann–Wachmann 1, Comet 29P/ A periodic comet in an unusual near-circular orbit outside that of Jupiter. It was discovered in 1927 November by the German astronomers (Friedrich Carl) Arnold Schwassmann (1870–1964) and Arthur Arno Wachmann (1902–90). Normally faint at around 16th–17th magnitude, the comet is prone to occasional outbursts in brightness of 5 magnitudes (a factor of 100) or more over a few days. A suggested cause is the crystallization of amorphous ice. Its orbit has evolved somewhat since it was first discovered, and currently has a period of 14.7 years, perihelion 5.72 AU, eccentricity 0.05, and inclination $9°.4$.

((⊕)) SEE WEB LINKS

• Information page at Cometography website.

scintillation 1. The rapid fluctuation in brightness of an astronomical object ('twinkling') due to scattering of electromagnetic waves by irregularities in the medium between the source and the observer. It is most pronounced in sources of small angular size, so at optical wavelengths it affects stars more than planets, while at radio wavelengths it affects point sources such as pulsars and quasars. Twinkling of stars at visible wavelengths is caused by convection currents in the atmosphere. Its cause is the same as that of bad *seeing. At radio wavelengths, it occurs in the Earth's ionosphere (*ionospheric scintillation*), in the solar wind (*interplanetary scintillation*), and in interstellar matter (*interstellar scintillation*).
    2. A phenomenon in which high-energy particles emit flashes of visible light (**scintillations**) on collision with atoms in certain materials (**scintillators**). It is the principle of the *scintillation counter used in gamma-ray astronomy.

scintillation counter A device for detecting high-energy X-rays and gamma rays. It makes use of the effect of **scintillation**, in which the incident radiation is converted into a flash of light by interaction with a sensitive material such as caesium iodide or sodium iodide (CsI or NaI). The intensity of the flash is proportional to the energy of the radiation, and it can be converted into an electronic signal by a photomultiplier tube or, in an imaging telescope, by a *microchannel plate detector or *charge-coupled device (CCD).

S-class asteroid A class of asteroid, relatively common in the inner main asteroid belt, having a moderate albedo (0.10–0.28), a reddish spectrum at wavelengths shorter than $0.7 \mu m$, and moderate to non-existent absorption features at longer wavelengths. The proportion of S-class asteroids decreases outwards through the main belt. The S is for silicaceous, since their composition is thought to include silicate minerals such as olivine and pyroxene. They could be the parent bodies of the *stony-iron meteorites. Members of this class include (15) Eunomia, diameter 255 km, and (29) Amphitrite, diameter 212 km.

Sco–Cen Association The nearest *OB association to the Sun, also known as the Sco OB2 association. It stretches along *Gould's Belt from Scorpius to Centaurus and Crux, extending for about 600 l.y. There are three subgroups: Upper Scorpius, Upper Centaurus Lupus, and Lower Centaurus Crux, with estimated ages 5 million years, 17 million years, and 16 million years respectively, and distances 473 l.y., 456 l.y., and 385 l.y. Current estimates place membership of the Sco–Cen association at over 500 stars, mostly of types B and A. *Antares is its brightest member.

scopulus A lobate or irregular scarp on a planetary surface; pl.*scopuli*. The name, which means 'crag' or 'cliff', is not a geological term, but is used in the nomenclature of individual features, for example Tartarus Scopulus on Mars.

Scorpius (Sco) (*gen*. **Scorpii**) A constellation of the zodiac, representing a scorpion. The Sun passes through Scorpius for a week at the end of November. Its brightest star is *Antares. Beta Scorpii (Graffias or Acrab) is a double star of magnitudes 2.6 and 4.9; the stars are unrelated. Another optical double is the naked-eye pair Zeta Scorpii, magnitudes 3.6 and 4.7. Nu Scorpii is a notable quadruple star. M4, a 6th-magnitude globular cluster nearly ½°

wide, is 7000 l.y. away, one of the closest globulars to us. M6 is a 4th-magnitude open cluster ¼°
wide, and M7 is a 3rd-magnitude open cluster more than 1° across. *Scorpius X-1 is the
brightest X-ray source in the sky.

Scorpius X-1 The brightest persistent X-ray source in the sky, and the first known celestial
X-ray source apart from the Sun. It was discovered in 1962 during a sounding-rocket flight.
Scorpius X-1 is a low-mass X-ray binary with an orbital period of 18.9 hours. The optical
counterpart is a 13th-magnitude blue star known as V818 Sco, 9000 l.y. away. The X-rays
arise from the transfer of material from this star on to a companion neutron star via
an *accretion disk. Scorpius X-1 is over three times brighter at X-ray wavelengths than the
second-brightest constant X-ray source, the *Crab Pulsar.

Scotch mount A simple hand-driven camera mount designed to track the stars; also
known as a *barn-door mount*, or as a *Haig mount*, after the Scottish physicist and amateur
astronomer George Youngson Haig (1928–), who invented it in 1972. It consists of two
boards, hinged at one end, with the hinge aligned parallel with the Earth's axis and the lower
board fixed. The camera is carried on the upper board. A bolt is turned to move the upper
board to counteract the Earth's rotation. The apparatus is suited to short exposures with
comparatively wide-angle lenses, its main advantage being cheapness.

SC star *See* S STAR.

SCT Abbr. for *Schmidt–Cassegrain telescope.

Sculptor (Scl) (*gen.* Sculptoris) A faint constellation of the southern sky, representing a
sculptor's workshop. Its brightest star, Alpha Sculptoris, is of magnitude 4.3. The constellation
contains the *Sculptor Dwarf Galaxy, a member of our Local Group, and the south galactic
pole lies in Sculptor.

Sculptor group The nearest group of galaxies to the Local Group, about 10 million l.y. away.
It is a loose association containing the late-type spirals NGC 45, 55, 247, 253, 300, 7793,
and possibly IC 5332. It is centred on the constellation Sculptor, near the south galactic
pole, but its members stray into neighbouring Cetus.

Scutum (Sct) (*gen.* Scuti) The fifth-smallest constellation, lying in the equatorial region of
the sky, representing a shield. Its brightest star, Alpha Scuti, is of magnitude 3.8. Delta Scuti is
the prototype of a class of pulsating variables with small amplitudes, the *Delta Scuti stars. M11
is the *Wild Duck Cluster.

Scutum–Centaurus Arm One of the two major spiral arms of our Galaxy, the other being
the *Perseus Arm. According to current models of the Galaxy, the Scutum–Centaurus Arm
emerges from the end of our Galaxy's central bar on the side closest to the Sun, then
passes between us and the galactic centre at a distance of about 10 000 l.y. The *Sagittarius
Arm, which is one of two minor spiral arms of the Galaxy, is interposed between us and
the Scutum–Centaurus Arm and hence partly obscures it.

SDO Abbr. for *Solar Dynamics Observatory.

S Doradus star A type of extremely luminous, eruptive variable that ejects shells of material;
abbr. SDOR. These stars include some of the most massive (over 30 solar masses) and most
luminous (about 10^6 solar luminosities) stars. Their spectra are Bpeq–Fpeq; the variability is
generally irregular on a timescale of tens of years or even centuries, with amplitudes up to
10 magnitudes. *Eta Carinae is a famous example. Alternative names are *Hubble–Sandage
variable, and *luminous blue variable. The older term *P Cygni star* is now reserved for any star
with spectral lines having a *P Cygni line profile, indicative of emission from an expanding shell.

SDSS photometry A system of photometry used in the *Sloan Digital Sky Survey. It has
five photometric bands with the following central wavelengths and bandwidths: u, 352 and
56 nm; g, 480 and 124 nm; r, 625 and 126 nm; i, 767 and 129 nm; z, 911 and 133 nm.

Search for Extraterrestrial Intelligence (SETI) The attempt to detect artificial transmissions from other civilizations in space. The first such attempt, called Project Ozma, was made in 1960 by the American radio astronomer Frank Donald Drake (1930–), who observed two nearby Sun-like stars, Tau Ceti and Epsilon Eridani, with the 26-m dish at the *National Radio Astronomy Observatory, West Virginia. Since then nearly 200 searches have been made with increasing sensitivity and enlarged coverage.

The two most sensitive searches have been Project SERENDIP (Search for Extraterrestrial Radio Emissions from Nearby Developed Intelligent Populations) and Project Phoenix. SERENDIP is an ongoing all-sky survey that operates on the back of normal astronomical observations at the 305-m Arecibo Telescope in Puerto Rico, while Phoenix was a targeted survey between 1995 and 2004 which observed more than 800 likely stars within about 240 light years. Phoenix used the 64-m radio telescope at Parkes, Australia, the 43-m dish at the National Radio Astronomy Observatory, and the Arecibo Telescope in conjunction with the 76-m Lovell Telescope at Jodrell Bank.

The majority of searches have concentrated on that part of the radio spectrum where the background noise is a minimum, between the radio spectral lines of hydrogen (H) at 21 cm wavelength and hydroxyl (OH) at 18 cm. Since H and OH together make H_2O, this region of the radio spectrum is termed the *water hole*. Radio signals deliberately sent to attract attention over interstellar distances are expected to be of narrow bandwidth, 1 hertz or less, and specialized receivers with millions of channels have been developed to detect them. Project BETA (Billion-channel ExtraTerrestrial Assay) scanned the water hole across all the sky visible from the 25.6-m radio telescope at Harvard University's Oak Ridge Observatory, Massachusetts, in 1995–9.

Another possible mode of communication is by brief (nanosecond) pulses of laser light which could outshine their parent star by 10000 times. Searches for such visible pulses, termed Optical SETI, or OSETI, are now underway, notably an all-sky search using a purpose-built 1.8-m telescope at Oak Ridge Observatory which began in 2006. In 2007 the *Allen Telescope Array, the first radio telescope designed and built specifically for SETI observations, began operations with 42 dishes. When completed it will consist of 350 individual antennas with a collecting area equivalent to that of a 114-m antenna and be capable of observing many stars simultaneously.

SETI can succeed only if there are civilizations elsewhere in the Galaxy transmitting towards us. The *Drake equation attempts to estimate how many such communicative civilizations there might be. Early estimates were optimistic, suggesting that perhaps tens of thousands of them existed in the Galaxy. Most present-day estimates are more conservative, no more than a few hundred, and some astronomers suggest that we might well be the only advanced civilization in the Galaxy at present.

Although simple life may well be widespread throughout the Galaxy, intelligent life on Earth has taken billions of years to evolve, requiring long-term stability of climate. Planets that meet this condition may be very rare, and consequently any advanced civilizations would be few and far between. To have a realistic chance of making radio contact with distant civilizations would require an instrument more sensitive than any yet built. Future radio telescopes such as the *Square Kilometre Array, scheduled for completion in 2024, may be required before there is a realistic chance of contact.

 SEE WEB LINKS
• Official SETI Institute website.

Seashell Galaxy An unusually shaped galaxy about 280 million l.y. away in Centaurus, interacting with the galaxy NGC 5291. The Seashell has no NGC number of its own. Both galaxies are part of a cluster, the brightest member of which is IC 4329.

season Any of the four periods into which the year is divided: spring, summer, autumn, and winter. The seasons arise from the Earth's axis being tilted, so that different latitudes receive varying amounts of sunlight over the course of the year as the Earth orbits the Sun. Astronomically, the seasons are taken to begin at the equinoxes (spring and autumn) and the solstices (summer and winter). Meteorologists regard the seasons as groups of three calendar months, spring in the northern hemisphere being March, April, and May, and so on. The seasons are reversed in the southern hemisphere, southern autumn corresponding to northern spring, and so on.

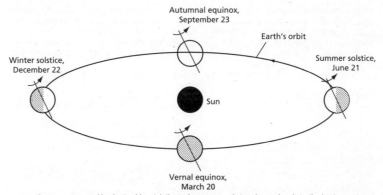

season: Seasons are caused by the Earth's axial tilt. At the summer solstice, the north pole is tilted at its maximum towards the Sun. Northern latitudes then receive the maximum amount of sunlight, and southern latitudes the minimum. The situation is reversed six months later, at the winter solstice. In between, at the equinoxes, everywhere on Earth receives equal amounts of sunlight

Secchi, (Pietro) Angelo (1818–78) Italian astronomer and priest. He was a pioneer of astronomical photography. By photographing a total solar eclipse in 1860 he, and independently the English scientist Warren De la Rue (1815–89), settled the controversy over whether prominences belonged to the Sun or the Moon. Secchi also made the first classification of solar prominences into quiescent and eruptive types. He and W. *Huggins were the first to approach astronomical spectroscopy systematically. Secchi made the first spectroscopic survey of stars, which yielded a catalogue of over 4000 spectra, published 1863. He suggested a spectral classification for stars (*see* SECCHI CLASSIFICATION) from which the modern system of spectral types evolved.

Secchi classification An early attempt to classify the spectra of stars, developed by P. A. *Secchi in the 1860s. Secchi grouped stellar spectra into five types, on the basis of direct visual observation. Type I was the equivalent of today's spectral types B and A; Type II encompassed F, G, and K stars; Type III was the equivalent of the M stars; Type IV encompassed carbon stars (Harvard types R, N); and the rare Type V described spectra with unusual bright lines, such as from Wolf–Rayet stars and planetary nebulae. Secchi's scheme was superseded by the photographic *Harvard classification system.

second (s) The fundamental unit of time in the SI system. It is defined as the duration of 9192631770 cycles of the radiation from the transition between two hyperfine levels in the ground state of caesium-133. This definition is designed to produce 86400 s in the mean solar day. *See also* LEAP SECOND.

secondary cosmic ray One of the shower of atomic particles produced when a *primary cosmic ray enters the Earth's atmosphere. The interaction first produces *pions, which subsequently decay rapidly into muons, some of which may also decay into electrons. The bulk of the secondary cosmic rays present at sea level are muons. *See also* COSMIC-RAY SHOWER.

secondary crater A small crater formed by ejecta thrown out from a larger impact crater. Secondary craters tend to cluster in a ring around the main crater, the greatest number lying slightly more than about one crater diameter away on the Moon, but rather closer on Mercury because of its higher surface gravity. Ejecta falling nearer than about half the crater diameter is travelling too slowly to form craters, and piles up as an *ejecta blanket. Secondary craters may be widely dispersed over the body, if it has no atmosphere. Those far from the primary crater tend to be fairly circular, whereas those nearby may be very irregular in shape, a result of the lower velocities of the nearer ejecta.

secondary minimum **1**. The shallower of the two minima in the light-curve of an eclipsing binary. It occurs when the secondary (the star of lower surface brightness) is eclipsed by the brighter primary. Depending on the relative brightness of the components, the secondary minimum of an *Algol star may be either barely detectable or nearly as deep as *primary minimum. In a *Beta Lyrae star it is shallower, and in a *W Ursae Majoris star it may be nearly as deep as the primary eclipse.
   2. The dip between the double maxima of an *RV Tauri star.

secondary mirror Any mirror that is second in an optical path after the primary mirror, such as the diagonal of a *Newtonian telescope or the convex mirror of a *Cassegrain telescope.

second contact At a total solar eclipse, the moment when the leading (eastern) limb of the Moon just touches the Sun's eastern limb, and totality begins; or, at a total lunar eclipse, the point at which the Moon's trailing limb becomes completely immersed in the Earth's umbra. Second contact in a total solar eclipse may be immediately preceded by the appearance of *Baily's beads or the *diamond ring. At an *annular eclipse, second contact refers to the instant when the Moon's trailing (western) limb leaves the Sun's western limb, so that the whole of the Moon's body is silhouetted against the Sun.

second of arc (arcsec) (″) A very small unit of angular measure, equivalent to one-sixtieth of an arc minute or 1/3600 of a degree. The resolving power of telescopes, the separation of double stars, and the apparent diameter of celestial objects are usually given in seconds of arc.

secular Happening over a long period of time. A secular change in position or brightness, for example, is one that may take several centuries to become apparent.

secular acceleration A perturbation in a body's orbital movement that is cumulative, building up at a rate proportional to the square of the time. Such an effect exists in the orbit of the Moon about the Earth, causing the Moon to recede slowly from the Earth.

secular parallax The angle subtended at a star by a baseline equal to the distance travelled by the Sun in a given time, such as a year or century. It can be used to estimate the average distance of a sample of stars which has a common motion relative to the Sun, such as the stars which define the *local standard of rest. Care must be taken to distinguish other populations, such as RR Lyrae variables, which have different speeds and directions of motion.

secular perturbation A perturbation in a body's orbit that is cumulative, causing an orbital element to grow or decrease. The gravitational attractions of the planets upon each other cause secular changes in the longitude of the ascending node, longitude of perihelion, and times of perihelion passage, as well as causing periodic perturbations.

secular variable A variable star that has supposedly undergone an increase or decrease in brightness on a time-scale of hundreds or thousands of years. Although a number of possible candidates have been advanced, the poor quality of early magnitude estimates means that such changes cannot be confirmed.

seeing The movement or distortion of a telescopic image as a result of turbulence in the Earth's atmosphere. It is thought to be caused by undulations between separate layers of air, which disturb the path of light. The scale of these undulations is often regarded as giving cells of seeing, typically being around 100–150 mm across at sea level, and larger at higher altitude. A small telescope may look through individual cells, whose movements give rise to a sharp but wandering image, while a larger one may look through several at once, producing multiple images. The 'boiling' seen at the limb of the Moon or Sun is another manifestation of bad seeing. The *Antoniadi scale is widely used by amateur astronomers to evaluate seeing.

segmented mirror A telescope mirror that consists of several adjacent components which bring light to a common focus like one larger aperture. Typically the components are hexagonal in shape, and interlock like tiles. Each segment has a separate mounting with position sensors and actuators to maintain its alignment. Each of the two *Keck Telescopes, for example, has 36 components 1.8 m across, forming a combined hexagonal aperture of 9.82 m. Segmented mirrors can be very light for their aperture, and can be made larger than individual mirrors.

selected areas *See* KAPTEYN SELECTED AREAS.

selective absorption Any form of absorption which affects some wavelengths more than others. Examples are *atmospheric extinction and *interstellar absorption, both of which produce a reddening of starlight.

Selene The original name for the Japanese *Kaguya Moon probe.

selenology An obsolete term meaning the study of the geology of the Moon, now referred to as lunar geology.

self-absorption The absorption of *radiation before it can escape from the region from which it is emitted. *Synchrotron emission is prone to self-absorption, causing the intensity of emission at low frequencies to fall off; the spectrum is then said to *turn over*. The term is also used to describe the reduction in intensity at the centre of a spectral line due to a surrounding shell of cool material.

semidetached binary A close binary star in which one component fills its *Roche lobe. *Mass transfer occurs from this star to the other. In some systems the gas stream falls directly on to the surface of the accompanying star, but in most cases it forms an *accretion disk. All active *cataclysmic binary systems are semidetached.

semidiameter The apparent equatorial radius of a celestial body, such as the Sun, Moon, or a planet, expressed as an angle.

semi-forbidden line *See* FORBIDDEN LINE.

semimajor axis (a) Half the longest diameter of an ellipse. The semimajor axis is the average distance of an orbiting object from its primary.

semiregular variable A type of giant or supergiant pulsating variable, with intermediate or late spectra, whose variations show a greater or lesser degree of regularity; abbr. SR. The amplitudes (average 1–2 mag.) are less than those of the *Mira stars, and the mean periods range from 20 days to more than 2000 days. Four subtypes are defined. SRA are giants with reasonably persistent periodicity, although amplitudes (less than 2.5 mag.) and the form of the light-curves vary. SRB are giants with occasional periodicity interspersed with intervals of slow, irregular variations (sometimes with larger amplitude) or even constant brightness. SRC are supergiants with moderate periodicity and low amplitudes (about 1 mag.). These three types have M, C, S, or Me, Ce, Se spectra. SRD are giants and supergiants with F, G, or K spectra, sometimes with emission lines, and amplitudes of 0.1–4 mag.

sensitivity A measure of the weakest signal discernible by a detecting system. The ratio of the amplitude of a signal above the noise level to the amplitude of the noise level itself is known as the *signal-to-noise ratio*. For most purposes, a minimum signal-to-noise ratio of 1 : 1 is required for the signal to be regarded as definitely detected. In the radio and infrared regions, the use of techniques such as integration, chopping, comparison with a stable laboratory source, and phase-sensitive detection can improve the basic sensitivity of the system by several orders of magnitude.

separation The angular distance between two objects on the celestial sphere, particularly the components of a double or multiple star.

Serpens (Ser) (*gen.*** Serpentis)** A constellation of the equatorial region of the sky, representing a large snake. It is unique in being split into two halves, which lie on opposite sides of Ophiuchus: Serpens Caput, the serpent's head, which is the larger half; and Serpens Cauda, the serpent's tail. The two are regarded as a single constellation. The brightest star is Alpha Serpentis (Unukalhai), magnitude 2.6. M5 is a 6th-magnitude globular cluster, and M16 is a 6th-magnitude star cluster in the *Eagle Nebula.

Serrurier truss A tube framework for telescopes. It consists of eight struts arranged in triangles. The struts run between a ring which forms either the top or bottom end of the tube and an open box at the centre of gravity of the assembly. This box is attached to the *declination axis of the mounting (or the altitude axis in an *altazimuth mounting). The triangular framework ensures that strains and loads are carried evenly around the tube, whatever the orientation of the telescope, and that sag is identical at both the top and bottom of the tube. The system was devised by the American engineer Mark Serrurier (1904–88).

Setebos An outer retrograde satellite of Uranus, at a distance of 17 988 000 km; also known as Uranus XIX. It orbits every 2303.1 days at an inclination of 148° to the ecliptic. The orbit is one of the most elliptical of any moon in the Solar System, eccentricity 0.59. Setebos is 48 km in diameter and was discovered in 1999 from ground-based observations.

SETI Abbr. for *Search for Extraterrestrial Intelligence.

setting The moment at which a celestial body disappears below an observer's horizon. For bodies with an observable disk, particularly the Sun and Moon, setting is taken as the moment when their upper limb lies exactly on the observer's horizon. Refraction in the Earth's atmosphere means that an object appears to set later than it actually does, and this effect must be taken into account when calculating observed rising and setting times.

setting circle A graduated scale on each axis of an equatorial mount that indicates the position to which the telescope is pointing. The scale on the *declination axis is fixed, while that on the *polar axis can be rotated to match the particular *right ascension or *hour angle on view. On many modern telescopes, electronic encoders are attached to the shafts of the axes, thus enabling electronic displays and computer control of the pointing.

Seven Sisters A popular name for the *Pleiades star cluster in Taurus.

Sextans (Sex) (*gen.* **Sextantis**) An obscure constellation on the celestial equator, representing a sextant. Its brightest star is Alpha Sextantis, magnitude 4.5. Its only object of note is the *Spindle Galaxy, NGC 3115.

sextant A navigational instrument for measuring the altitude of celestial objects above the horizon. It consists of a graduated 60° arc (one-sixth of a circle) with a movable arm and sighting devices. An index mirror attached to the movable arm reflects the chosen object to another mirror, silvered across half its width, so that its image can be seen alongside the horizon. The object's altitude can then be read off the scale. A telescope and dark filters are provided to improve accuracy and allow observation of the Sun. Because the objects observed are reflected, the range of angles which can be measured is 120°, double that of the graduated scale.

Seyfert, Carl Keenan (1911–60) American astrophysicist. In 1943 he found several spiral galaxies with very faint arms and unusually bright nuclei. Their spectra showed emission lines from hot ionized gases moving at high velocities. These so-called *Seyfert galaxies have since been shown to belong to the wider class of objects with *active galactic nuclei. In 1951 he observed the group of galaxies now known as *Seyfert's Sextet.

Seyfert galaxy A type of galaxy with a small, bright nucleus which shows broad, strong emission lines in its spectrum. The first galaxies of this type were described in 1943 by C. K. *Seyfert. Nearly all known Seyfert galaxies are spirals or barred spirals, and Seyfert-type activity probably occurs in a small percentage of all spiral galaxies. Seyfert galaxies are classified according to the relative widths of the emission lines in their spectra. Type 1 Seyferts (such as NGC 5548) have broad emission lines of hydrogen but narrow forbidden lines of heavier elements. In Type 2 Seyferts (such as NGC 1068) the hydrogen lines and forbidden lines both have the same width, which is broader than the forbidden lines in Type 1 Seyferts but not as broad as the latter's hydrogen lines. Seyferts have *active galactic nuclei that produce strong radiation, probably from an accretion disk around a massive black hole. The radiation excites gas around the central regions, giving rise to the observed emission lines. Seyferts are lower-luminosity examples of quasar activity.

Seyfert's Sextet A group of six galaxies in Serpens; also known as NGC 6027. One of the group is a small spiral with a redshift that indicates a recession velocity some 15000 km/s greater than the rest of the group. This high-redshift member is in fact a background galaxy almost five times as distant as the others, but the sextet has sometimes been cited as evidence for the existence of redshifts not caused by the overall expansion of the Universe.

Shack–Hartmann test *See* HARTMANN TEST.

shadow bands An optical phenomenon seen fleetingly around the onset of totality in a solar eclipse, comprising slow-moving, low-contrast waves of light and dark, perhaps only a few centimetres broad, seen on light-coloured surfaces on the ground. Shadow bands are presumed to originate from differential refraction of the last remaining segment of sunlight in the cooling upper atmosphere around the path of the total eclipse.

shadow transit The passage of the shadow of a satellite across the face of a planet. Shadows of the Galilean satellites of Jupiter, and of Titan on Saturn, are the only ones large enough to be visible through most Earth-based telescopes, although shadow transits of Rhea, and even of Tethys, Dione, and Iapetus, can be seen in large telescopes under good conditions.

Shapiro delay The delay experienced by a radio pulse as it passes through curved spacetime around a massive object such as a star; it is also known as the *gravitational time delay*. In the Solar System, the effect can be detected when bouncing a radar pulse off a planet on the far side of the Sun; in this case the time delay can amount to several hundred microseconds. It arises because the pulse takes a longer path through space on its way to and from the planet than it would have done in the absence of the Sun. A Shapiro delay is also experienced when a pulse emitted by a pulsar in a binary system passes close to its companion star. The effect is an important test of *general relativity and is named after the American astronomer Irwin Ira Shapiro (1929–), who first detected it in 1964.

Shapley, Harlow (1885–1972) American astronomer. From 1911 he studied variable stars, distinguishing Cepheid variables from eclipsing binaries and correctly ascribing their variability to pulsations. He subsequently discovered Cepheids in *globular clusters, whose distances and distribution he was able to estimate using the period–luminosity law discovered by H. S. *Leavitt and a statistical method he devised. His results showed that the Galaxy was much larger than had been supposed (although Shapley initially overestimated its size), with the Sun some way from the centre. He originally sided with the Dutch-American astronomer Adriaan van Maanen (1884–1946) in believing that what were then known as 'spiral nebulae' were relatively small and nearby. In 1920 Shapley propounded this view in the so-called Great Debate with the American astronomer Heber Doust Curtis (1872–1942), who argued (correctly) that spiral nebulae were separate galaxies. The 1932 Shapley–Ames catalogue of 1249 galaxies, compiled with his assistant Adelaide Ames (1900–32), revealed the irregular distribution of galaxies and the existence of clusters of galaxies.

Shapley Concentration *See* SUPERCLUSTER.

shatter cone A ridged, cone-shaped rock fragment produced when strong shock waves, resulting from the impact of a large meteorite, pass through certain types of rock. The shock waves produce weakened zones radiating out from the direction of shock, causing shatter cones, which are later revealed when the shocked layers are exposed by erosion.

Shaula The star Lambda Scorpii, magnitude 1.6, a B1.5 subgiant 571 l.y. away. It is a variable *Beta Cephei star, varying by about 0.05 mag. with a period of 5.1 hours.

Shedir An alternative name for the star *Schedar.

shell burning The nuclear reactions in a shell around a star's core that continue after the fuel in the core itself has been exhausted. As the fuel, whether hydrogen or helium, is progressively exhausted, the shell moves outwards until it enters regions too cool for the reactions to continue.

shell galaxy An elliptical galaxy surrounded by faint arcs or shells of stars, lying at right angles to its major axis. One to twenty shells may be seen, nearly concentric but incomplete. They are 'interleaved' in such a way that successive shells usually occur on opposite sides of the galaxy. About 10% of bright ellipticals show shells, most of them in regions of low density of galaxies. No spirals are known with such shell structure. Shell galaxies may be the result of a giant elliptical digesting a low-mass companion galaxy.

shell star A main-sequence star with a spectral type in the range B–F whose spectrum contains prominent narrow absorption lines originating in surrounding gas. Shell stars are usually *Be stars, in which case both emission and absorption arise in an equatorial disk of ejected gas which we happen to view edge-on from Earth. As well as showing small-amplitude changes in brightness, many shell stars also have variable spectra. Pleione, also known as BU Tauri, a member of the Pleiades cluster, is a well-known example of a shell star, with a range of mag. 4.8–5.5.

shepherd moon A satellite of a ringed planet whose gravitational field has a significant effect on the configuration of planetary rings near it. Shepherd moons were first discovered on Voyager images of Saturn. Saturn's F Ring is kept very narrow by two shepherd satellites, Pandora and Prometheus, which orbit either side of it. Uranus's outer ring, the Epsilon Ring, is kept in place by the shepherd moons Ophelia and Cordelia.

shergottite A very rare type of achondrite meteorite, named after the meteorite that fell at Shergotty, India, in 1865, the first known fall of this type. The shergottites are divided into two subtypes: the *basaltic* or *pyroxene-plagioclase* shergottites, and the *lherzolitic* or *olivine-pyroxene* shergottites. The basaltic shergottites, of which the two earliest recorded shergottite falls, Shergotty (1865) and Zagami (1962), are both examples, consist primarily of pyroxene (pigeonite and augite), together with maskelynite (a plagioclase glass formed by shock metamorphism) and a trace of olivine. The lherzolite shergottites, of which the Antarctic find ALHA 77005 was the first example, contain less pyroxene and maskelynite than the basaltic shergottites, but much more olivine. Shergottites belong to the class of *SNC meteorites, which are thought to come from Mars. Most shergottites share a formation age of 170 million years and an exposure age of about 3 million years.

shield volcano A wide volcano, usually several kilometres in diameter, with gentle slopes, formed from solidified layers of very fluid basaltic lava. Typical examples on Earth are the Hawaiian volcanoes Kilauea and Mauna Loa, which have slopes of 4–10°. Lunar *domes appear to be similar to shield volcanoes, and many shield volcanoes are found on Venus.

shock metamorphism The effects of a high-speed impact on a rock. The extreme temperatures and pressures in a meteorite impact produce a wide range of effects, including fracturing and brecciation, the formation of minerals that are stable only at very high pressures, and heating or extensive melting of the material. Such effects are evident in meteorites, as the results of bombardment of their parent body, and in many lunar rocks. Evidence of shock metamorphism in terrestrial rocks is now regarded as proof of meteoritic impact. On Earth, common forms of shock effects include *shatter cones and the formation of shock-induced minerals such as *coesite and *stishovite.

Shoemaker, Eugene Merle (1928–97) American planetary geologist. In 1952 he began to study *Meteor Crater in Arizona, and became convinced that it and other terrestrial and lunar craters—which many believed to be volcanic—were impact features, especially after his discovery of *coesite at terrestrial craters. In the 1960s he oversaw geological experiments on NASA's lunar missions and helped to train the Apollo astronauts. With the American astronomer Eleanor Kay Helin, née Francis, and later with his wife, Carolyn (Jean) Shoemaker, née Spellmann (1929–), he began in the 1970s to search for *near-Earth asteroids, potential causes of terrestrial impact craters, using the 0.46-m Schmidt telescope at Palomar Observatory. In addition to asteroids, Shoemaker discovered many comets, including *Shoemaker–Levy 9, which demonstrated the effects of an impact on a gaseous body by crashing into Jupiter.

Shoemaker–Levy 9, Comet (D/1993 F2) A comet discovered on 1993 March 25 by the American astronomers E. M. *Shoemaker, his wife Carolyn (Jean) Shoemaker, née Spellmann (1929–), and David Howard Levy (1948–); formerly designated 1994 X. The comet was in a 2-year orbit around Jupiter, having been captured in, or before, 1929. A close approach to Jupiter (21000 km) in 1992 July had disrupted the nucleus into at least 21 fragments. These fragments hit Jupiter over the week of 1994 July 16–22. The impacts produced prominent dark spots, visible even in small amateur telescopes, at latitude 44° S, distributed in longitude; they subsequently merged into a dark belt which persisted for 18 months. 📷

(⊕) SEE WEB LINKS
• Information page at Cometography website.

shooting star Popular name for a *meteor.

short-period comet A comet with an orbital period of less than 30 years. The term was formerly used as a synonym for any *periodic comet.

short-period variable An ill-defined term for a variable that has a short period relative to apparently very similar stars. The suffix *s* is used to designate specific subtypes of variable star with rapid variation.

Sickle Popular name for the asterism formed by the stars Epsilon, Mu, Zeta, Gamma, Eta, and Alpha Leonis. The Sickle makes up the head and chest of Leo, the lion.

side lobe One of several minor lobes to either side of the main lobe in the *antenna pattern of a radio telescope, and analogous to the diffraction rings in an optical telescope. Side lobes are undesirable as they can cause ambiguity in the position of a radio source and also make a radio telescope more open to radio interference.

sidereal Referring to the stars, or to measurements relative to the stars.

sidereal day The interval of time between successive transits of the mean *equinox, equal to 23 h 56 m 04s. Because of the *precession of the equinoxes, the mean equinox is not a completely fixed sidereal point. As a result the sidereal day is 0.0084s shorter than the Earth's rotation period relative to the stars.

sidereal month The mean period of the Moon's orbital revolution about the Earth with reference to the stars, equal to 27.32166 days.

sidereal period The time taken for a planet or satellite to complete one orbit relative to the stars. For example, the Earth's sidereal period is a *sidereal year, and the Moon's sidereal period is a *sidereal month. However, a body's axial spin can also be measured relative to the stars, and this is termed the **sidereal period of axial rotation**; the Earth's sidereal period of axial rotation is the *sidereal day.

sidereal rate The rate of movement of the stars across the sky as the Earth spins, i.e. one rotation in 23 h 56 m 04.091s, a period known as a sidereal day. This is the rate at which a telescope must be driven to follow the stars.

sidereal time Time as measured by reference to the stars; technically, it is the *hour angle of the vernal equinox. The sidereal time is the same as the right ascension of stars currently on the observer's meridian. More generally, the sidereal time is the sum of the right ascension and the hour angle of any celestial object, and hence links these two coordinates. Depending on whether the *true equinox or the *mean equinox is used as the reference point, the resulting form of sidereal time is known as either *apparent sidereal time or *mean sidereal time, respectively. Their difference, which seldom exceeds a second of time, is called the *equation of the equinoxes*. *See also* GREENWICH SIDEREAL TIME; LOCAL SIDEREAL TIME.

sidereal year The period of the Earth's orbital revolution around the Sun with reference to the stars. Its length is 365.256 36 days.

siderite Another name for an *iron meteorite.

siderolite An alternative, now largely obsolete, name for a *stony-iron meteorite.

siderophyre A very rare class of stony-iron meteorite consisting of nickel-iron enclosing bronzite (orthopyroxene) and tridymite (quartz) minerals. The first known siderophyre is the Steinbach meteorite, found in 1724.

siderostat A flat mirror, *equatorially mounted and driven, which directs the light from a celestial object into a fixed telescope. The telescope should be aligned parallel to the Earth's axis and pointing at the mirror, unless a second mirror is used. With a siderostat, the field of view rotates around the object under study.

Siding Spring Observatory An observatory at an altitude of 1150 m in the Warrumbungle mountains near Coonabarabran, New South Wales, owned and operated by the Australian National University. It was founded in 1962 as an outstation of *Mount Stromlo Observatory, its largest telescope at the time being a 1-m reflector. In 1981 Uppsala Observatory, Sweden, moved its 0.5-m Schmidt to Siding Spring from Mount Stromlo. The 2.3-m Advanced Technology Telescope of the Australian National University opened in 1984. Siding Spring is also the site of the *Anglo-Australian Observatory and the *Faulkes Telescope South. The 1.35-m SkyMapper survey telescope came into operation in 2010.

• Official observatory website.

Sigma Octantis The south pole star, being the closest naked-eye star to the south celestial pole. It is an F0 giant of magnitude 5.4, lying 281 l.y. away. It is about 1° from the south celestial pole, and the distance is slowly increasing as a result of precession.

signal-to-noise ratio *See* SENSITIVITY.

Sikhote–Alin meteorite A major shower of iron meteorites that fell in the western part of the Sikhote–Alin mountain range in southeast Siberia on the morning of 1947 February 12. A total of 383 impact sites were found, spread over 1.6 km^2. The largest crater was 27 m in diameter. The parent body broke up at about 5 km altitude, producing about 70 tonnes of metal fragments, containing both *hexahedrite and *octahedrite material. The largest single fragment weighed over 1.7 tonnes. A dark dust train, marking the path of the fireball, remained in the sky for several hours.

silicon burning The set of nuclear reactions in massive stars by which silicon is converted to heavier elements such as iron and nickel.

silicon star A type of *Ap star in which the abundance of silicon is enhanced.

silvering The process of coating a mirror with a layer of reflective silver. It is usually a chemical process, in which metallic silver is precipitated out of solution. The silver will adhere to any perfectly clean surface it encounters. It is normal for the silver coating thus produced to require additional burnishing, yielding a reflectivity of about 93%. Silvering was once the only practical method for coating glass mirrors, but it has now been superseded almost completely by *aluminizing, on grounds of cost and convenience.

SIMBAD Abbr. for Set of Identifications, Measurements, and Bibliography for Astronomical Data, a database created and maintained by the *Centre de Données astronomiques de Strasbourg (CDS). It brings together basic data, cross-identifications, observational measurements, and a bibliography, for celestial objects outside the Solar

System. SIMBAD contains information on some 5 million objects, and can be searched electronically. It first went on line in 1981.

 SEE WEB LINKS

• Online portal to SIMBAD astronomical database.

single-lined binary A spectroscopic binary in which only one set of spectral lines is detectable; sometimes also called an *SB1 system*. The binary nature of the system is indicated by the fact that the spectral lines exhibit periodic Doppler shifts due to orbital motion.

singularity A mathematical point at which certain physical quantities reach infinite values. For example, according to general relativity, the *curvature of spacetime becomes infinite in a black hole. In the *Big Bang theory the Universe was born from a singularity in which the density and temperature of matter were infinite. *See also* NAKED SINGULARITY.

Sinope An outer retrograde satellite of Jupiter, distance 23 848 000 km; also known as Jupiter IX. It orbits Jupiter in 753.2 days at an inclination of 158° to the planet's equator. Its diameter is 28 km, making it the second-largest member of the *Pasiphae irregular group of Jovian satellites. Sinope was discovered in 1914 by the American astronomer Seth Barnes Nicholson (1891–1963).

sinuous rille A long, winding valley, usually with steep sides, on a planetary surface. Sinuous rilles were first seen on the Moon. They have similar characteristics to lava channels and tubes on Earth, but are distinctly different from rivers. For example, they usually start in a crater, become narrower instead of larger as they flow downslope, have abrupt breaks, and sometimes have distributaries (where the main channel splits in two, as in a river delta) but not tributaries. Results from the Apollo 15 mission to Hadley Rille showed that the sinuous rilles are indeed vast lava channels, formed during the eruption of the mare lavas 2.0–3.9 billion years ago. Sinuous rilles have since been identified on Mars and Venus.

sinus An indentation in the edge of high ground on a planetary surface; pl.*sinus*. The name, which means 'bay', is not a geological term, but is used in the nomenclature of individual features, for example Sinus Iridum on the Moon. It is also used for a protrusion from a dark area in the naming of Martian albedo features, for example Sinus Meridiani.

SiO maser A maser source in which the silicon monoxide (SiO) molecule is excited to maser action. SiO has many maser lines, which occur in groups at frequencies near 43, 86, 129 GHz, and so on. SiO masers are commonly found in circumstellar envelopes, close to the surface of the red giant where mass-loss originates. SiO masers are occasionally found in star-forming regions.

Sirius The star Alpha Canis Majoris, magnitude −1.44, the brightest star in the sky. It is popularly known as the Dog Star because it lies in the constellation Canis Major, the Greater Dog. It is an A1 dwarf and lies 8.6 l.y. away, making it the fifth-closest star to the Sun. Sirius has a white-dwarf companion, Sirius B, sometimes known as the Pup, which orbits it in a period of 50 years. Sirius B is of spectral type DA2, absolute magnitude 11.3, visual magnitude 8.4, and can be seen only with large telescopes when at its maximum separation from Sirius, as between the years 2020–2025.

Sirrah An alternative name for the star *Alpheratz.

Sitter, Willem de *See* DE SITTER, WILLEM.

SI units Système International d'Unités, or International System of Units, the generally recognized set of units for scientific and technical measurement. The SI system is based on seven units: the metre (m), kilogram (kg), second (s), ampere (A), kelvin (K), mole (mol), and candela (cd). Other units, such as metres per second for velocity, are derived from these base units.

six-colour system A system of photoelectric photometry that uses six filters with the following central wavelengths and bandwidths: U, 352 and 44 nm; V, 422 and 74 nm; B,

490 and 109 nm; G, 570 and 108 nm; R, 714 and 165 nm; and I, 1027 and 192 nm. The system was introduced primarily for the study of *interstellar absorption, but is now obsolete.

61 Cygni A double star 11.4 l.y. away in Cygnus, consisting of a K5 and a K7 dwarf, magnitudes 5.2 and 6.1, orbiting with a period of 680 years or so. It has the greatest *proper motion of any naked-eye star, 5″.2 per year (first measured by G. *Piazzi, whence the obsolete name Piazzi's Flying Star), and was the first star to have its parallax measured, by F. W. *Bessel in 1838.

SKA Abbr. for *Square Kilometre Array.

sky brightness The brightness of the night sky in the absence of twilight and moonlight, measured in areas of sky devoid of discernible stars. It arises from the *airglow, and from the *zodiacal light and *gegenschein, all of which vary with solar activity. There is a further component from the many stars and galaxies in the sky which are too faint to be seen individually but produce a significant background brightness. Sky brightness is measured in magnitudes per square arc second. The best mountain-top sites have a zenith sky brightness in the V band of around 21.8 magnitudes per square arc second, while 18.1 is typical for light-polluted sites. A lower sky brightness allows the detection of fainter objects. Telescopes in orbit are also affected by sky brightness, but not so greatly.

Skylab A NASA space station, launched on 1973 May 14. The converted upper stage of a Saturn V rocket, it consisted of living quarters, a work section, and the Apollo Telescope Mount (ATM) which contained six telescopes for observing the Sun's chromosphere and corona at X-ray, ultraviolet, and visible wavelengths. Three crews, each of three astronauts, spent respectively 28 days, 59 days, and 84 days aboard the space station. Skylab was then abandoned, and re-entered in 1979 July; fragments of it fell in Western Australia.

Slipher, Earl Carl (1883–1964) American astronomer, brother of V. M. *Slipher. He was a pioneer of planetary photography, achieving a quality rarely surpassed in his lifetime. He obtained unique sequences of photographs of Mars, Jupiter, and Saturn in which changes in visible features can be traced through over 50 years of favourable oppositions. He was one of the first to appreciate the benefits in astrophotography of stacking several negatives together to achieve a highly detailed print.

Slipher, Vesto Melvin (1875–1969) American spectroscopist, brother of E. C. *Slipher. At *Lowell Observatory he used spectroscopy to measure planetary rotation periods and atmospheric compositions. In 1912 he obtained the first radial-velocity measurements from the spectrum of a so-called 'spiral nebula' (the Andromeda Galaxy). By 1925 he had 45 radial velocities of other such nebulae, nearly all of which were receding too fast to belong to the Milky Way. This was the observational foundation for theories of an expanding Universe. In 1912 he discovered the existence of *reflection nebulae from a study of the nebulosity around the Pleiades, proving that there is dust as well as gas in space. He supervised the photographic search that led to C. W. *Tombaugh's discovery of Pluto in 1930.

Sloan Digital Sky Survey (SDSS) A project to investigate the *large-scale structure of the Universe by mapping the three-dimensional distribution of galaxies and quasars over a quarter of the sky, mostly around the north galactic pole but with three strips near the south galactic pole. The survey started in 1998 and used a wide-field 2.5-m telescope at *Apache Point Observatory, New Mexico. The first phase of observations, SDSS-I, was completed in 2005. It was followed by SDSS-II, in three parts: the Sloan Legacy Survey, to fill remaining gaps in the main survey; SEGUE (Sloan Extension for Galactic Understanding and Exploration), to map the structure and stellar makeup of our own Galaxy from the spectra of 240 000 stars; and the Sloan Supernova Survey, searching for distant Type Ia supernovae to measure changes in the rate of expansion of the Universe. SDSS-II was completed in 2008, bringing the total observations to more than 930 000 galaxies and over 120 000 quasars. SDSS-III began at Apache Point Observatory in 2008 and consists of four surveys: the Baryon Oscillation Spectroscopic Survey (BOSS) which maps the walls and voids in the distribution of galaxies resulting from density fluctuations in the early Universe; SEGUE-2, continuing the original SEGUE study with the spectra of an additional 214 000 stars; the Apache Point Observatory Galactic Evolution

Experiment (APOGEE), a spectroscopic survey of 100 000 red giant stars at infrared wavelengths; and the Multi-object APO Radial Velocity Exoplanet Large-area Survey (MARVELS), which monitors the radial velocities of 11 000 bright stars for signs of orbiting planets. SDSS-III will continue until 2014.

(⊕) SEE WEB LINKS
• Official project website with many images.

slow nova A type of nova that exhibits a characteristically slow development and decline, as distinct from a *fast nova or a very slow nova (*see* RR TELESCOPII STAR); abbr. NB. The star takes 150 days or more to fade by 3 magnitudes from maximum, ignoring any minor fluctuations and the deep transition minimum. The initial rise is also slower (2–3 days), and the final rise, following the pre-maximum halt, may take several weeks.

slow pulsator A binary X-ray source in which the emission pulsates with a period of a few minutes rather than the more typical few seconds.

SMA Abbr. for *Submillimeter Array.

Small Astronomy Satellite A NASA series of three small satellites devoted to X-ray and gamma-ray astronomy. SAS-1, also known as Explorer 42, was renamed *Uhuru after launch in 1970. It was the first X-ray satellite, and performed the first X-ray sky survey. SAS-2, otherwise known as Explorer 48, was a gamma-ray mission, launched in 1972 November. The last of the series, SAS-3, or Explorer 53, was launched in 1975 May to study individual X-ray sources.

(⊕) SEE WEB LINKS
• Information page on SAS 1 at Goddard Space Flight Center.
• Information page on SAS 2 at Goddard Space Flight Center.
• Information page on SAS 3 at Goddard Space Flight Center.

small circle A circle on a sphere whose plane does not pass through the centre of the sphere. On the celestial sphere, all lines of declination are small circles except for the celestial equator, which is a *great circle.

Small Magellanic Cloud (SMC) The smaller of the two irregular galaxies that accompany our Galaxy; also known as the Nubecula Minor. It is about 9000 l.y. across and some 200 000 l.y. away, visible to the naked eye as a misty patch about 3° across in Tucana. Its visible mass is less than 2% of our Galaxy's, and it contains relatively more gas and less dust than the Large Magellanic Cloud (LMC), but fewer clusters and nebulae. Its structure may be elongated along the line of sight to Earth. Like the LMC, the SMC shows evidence for star formation early in its history, followed by a lull, and then a more recent burst. The stars and interstellar matter have a lower abundance of heavy elements (from one-quarter to one-tenth) than the stars in our local neighbourhood of the Galaxy. *See also* MAGELLANIC CLOUDS.

SMART-1 An ESA Moon probe, launched in 2003 September; the name is an abbreviation of Small Missions for Advanced Research in Technology 1. It took fourteen months to reach the Moon using a solar-electric propulsion system with xenon as a propellant, and went into a highly elliptical polar orbit in 2004 November. SMART-1 carried a miniaturized high-resolution camera (AMIE) for lunar surface imaging, a near-infrared point-spectrometer (SIR) for lunar mineralogy investigation, and a compact imaging X-ray spectrometer (D-CIXS) for surface composition studies. The mission ended in 2006 September, when it was deliberately commanded to hit the Moon on the southern edge of the Mare Humorum.

(⊕) SEE WEB LINKS
• ESA mission website.

SMC Abbr. for *Small Magellanic Cloud.

Smithsonian Astrophysical Observatory (SAO) A research centre of the Smithsonian Institution, founded in 1890 at Washington, DC, but which moved to Cambridge,

Massachusetts, in 1955. The SAO hosts the International Astronomical Union's *Central Bureau for Astronomical Telegrams and *Minor Planet Center. In 1973 the SAO and the neighbouring *Harvard College Observatory (HCO) established the joint Harvard–Smithsonian Center for Astrophysics. SAO owns and operates the *Fred Lawrence Whipple observatory on Mount Hopkins, Arizona, and jointly owns and operates the *MMT Observatory, also on Mount Hopkins. The SAO, in conjunction with the Academia Sinica of Taiwan, jointly owns and operates the Submillimeter Array (SMA) of eight 6-m dishes on Mauna Kea, Hawaii, opened in 2003.

(⊕) SEE WEB LINKS
• Official observatory website.

Smithsonian Astrophysical Observatory Star Catalog (SAO Catalog) A catalogue of positions and proper motions of 258 997 stars for epoch 1950.0, covering the whole sky down to about 10th magnitude, published by the Smithsonian Astrophysical Observatory in 1966. The original aim was to provide a dense net of stars which could be used as references for deriving positions of artificial satellites observed with wide-angle cameras. It has since been superseded by more accurate catalogues. An associated atlas was published in 1969.

(⊕) SEE WEB LINKS
• Detailed description and full catalogue downloadable from the CDS.

SMM Abbr. for *Solar Maximum Mission.

Smoot, George Fitzgerald III (1945–) American astrophysicist and cosmologist. In the 1970s he began to design and build instruments known as differential microwave radiometers (DMRs) to detect and measure very small variations in the *cosmic microwave background radiation (CMB). Smoot was responsible for the DMR on board the *Cosmic Background Explorer satellite, launched in 1989, which detected variations in the CMB of one part in 10^5, supporting the *inflationary universe variant of the Big Bang theory. For this achievement he shared the 2006 Nobel Prize in Physics with COBE's project scientist, J. C. *Mather.

SNC meteorites A small group of rare and unusual meteorites, consisting of three main subgroups: the *shergottites, *nakhlites, and *chassignites (hence SNC). The SNC meteorites are igneous rocks that solidified from a cooling magma near the surface of their parent body, probably Mars. All but one are relatively young (less than 1.3 billion years old); the single ancient SNC meteorite found to date was formed about 4.5 billion years ago. The proportions and isotopic abundances of noble gases trapped in one shergottite resemble the composition of the Martian atmosphere as analysed by the *Viking landers. SNC meteorites were probably ejected from Mars by impacts, and entered orbits around the Sun before falling to Earth.

S0 galaxy *See* LENTICULAR GALAXY.

SNR Abbr. for *supernova remnant.

SNU Abbr. for *solar neutrino unit.

SOAR Abbr. for *Southern Astrophysical Research telescope.

SOFIA Abbr. for *Stratospheric Observatory for Infrared Astronomy.

soft gamma-ray repeater A source of irregular, repeated bursts of high-energy X-rays and low-energy gamma rays. Such events are believed to occur when the crust of a strongly magnetic neutron star (a *magnetar) suddenly readjusts in a *starquake, due to changes in the star's twisted magnetic field lines. The bursts usually last less than a second and can occur several times a day for a period of months. The source can then remain quiet for decades.

soft X-rays The lowest-energy region of the X-ray spectrum, covering the approximate range 0.1–2.5 keV (0.5–12.4 nm).

SOHO Abbr. for *Solar and Heliospheric Observatory.

Sojourner A six-wheeled roving vehicle which landed on the surface of Mars in 1997 July as part of NASA's *Mars Pathfinder mission. Remotely controlled by an Earth-based operator, it explored the Martian surface around its landing site in Ares Vallis for nearly three months.

Solar, Anomalous, and Magnetospheric Particle Explorer (SAMPEX) A US
satellite launched in 1992 July to study high-energy particles from cosmic rays, solar flares, and the Earth's magnetosphere. In 1994 it detected a belt of cosmic rays trapped by the Earth's magnetic field, within the inner *Van Allen Belt. It ceased operation in 2004 July.

solar activity The collective term for all active phenomena on the Sun, including *sunspots, *faculae, *active regions, *plages, *active prominences, and *flares. Solar activity is strongly associated with magnetic fields, which are thought to arise from a dynamo action within the Sun. Solar activity increases and decreases in a cycle lasting approximately 11 years.

Solar and Heliospheric Observatory (SOHO) A joint ESA–NASA spacecraft launched
in 1995 December to observe the Sun in ultraviolet and visible light, study the solar wind, and measure small oscillations on the Sun's surface. Its twelve instruments included the Large Angle and Spectrometric Coronagraph (LASCO), a set of three coronagraphs that record white-light images of the solar corona out to 30 solar radii. LASCO has returned spectacular views of *coronal mass ejections and led to the discovery of over 2000 comets, including many *Kreutz sungrazers and *sunskirters. Other major instruments are the Extreme Ultraviolet Imaging Telescope (EIT), which images the solar transition region and inner corona at four wavelengths in the EUV, and the Michelson Doppler Imager (MDI), which studies solar oscillations. SOHO is in a *halo orbit around the inner *Lagrangian point (L$_1$) between the Sun and Earth, some 1.5 million km from the Earth in the sunward direction, so that the Sun can be observed without the interruption of eclipses by the Earth.

(⊕) SEE WEB LINKS
• Official mission website.

solar antapex The direction in the sky towards which a sample of stars tends, on average, to be moving because of the motion of the Solar System relative to the sample. This direction cannot be defined precisely because of the random motions of the stars themselves, and it differs according to spectral type because of varying population mixtures, but the approximate position is RA 6h, dec. −30°. *See also* SOLAR APEX.

solar apex The direction towards which the Solar System is moving relative to the local standard of rest. Its position is approximately RA 18h, dec. +30°. As a result of this motion, stars seem to be converging towards a point in the opposite direction, the *solar antapex.

solar atmosphere The region around the Sun including the *photosphere, *chromosphere, *transition region, and *corona. The temperature of the solar atmosphere varies with height, at first decreasing to the *temperature minimum region, and then increasing, very strongly so in the transition region. The cause of the temperature rise is still uncertain, but possible mechanisms include the dissipation of sound waves (in the low chromosphere) and magnetohydrodynamic waves (in the rest of the chromosphere and corona), or heating following magnetic reconnection.

solar constant The amount of solar energy, in all wavelengths, that falls on a given area at the top of the Earth's atmosphere. In fact, the solar 'constant' varies by 0.1% over the eleven-year solar cycle, and perhaps by more over longer periods, so the term *total solar irradiance is now preferred.

solar cycle *See* SUNSPOT CYCLE.

solar day The interval between successive transits of the Sun across the observer's meridian, that is, the rotation of the Earth with respect to the Sun. Strictly this is the *apparent solar day*, which varies slightly during the year because the Earth does not move at a uniform rate in its elliptical orbit about the Sun (*see* EQUATION OF TIME). Its average length, the *mean solar day, is 24 hours or 86400 s. Because of the Earth's orbital motion around the Sun, the solar day is about 4 min longer than the *sidereal day, a discrepancy that adds up to one whole day in the course of a year.

Solar Dynamics Observatory (SDO) A NASA satellite, launched in 2010 February, to improve understanding of solar activity and its effects on Earth. The Helioseismic and Magnetic Imager (HMI) measures solar oscillations and the magnetic fields on the photosphere to determine how processes inside the Sun are related to surface activity; the Atmospheric Imaging Assembly (AIA) monitors the Sun's corona at high resolution; and the Extreme Ultraviolet Variability Experiment (EVE) monitors the Sun's emission at extreme ultraviolet wavelengths. Improved predictions of solar activity should result from the mission.

(⊕) SEE WEB LINKS
• Official mission website.

solar dynamo The action within the Sun whereby the kinetic energy of the hot, highly ionized gas of the solar interior is converted into the magnetic field that gives rise to solar activity. In the picture due to H. W. *Babcock, magnetic field lines under the photosphere running from pole to pole (the *poloidal* field) are twisted parallel to the equator (the *toroidal* field) by the Sun's differential rotation. *Active regions, including sunspots, are thought to be generated as the distorted magnetic field lines rise through the photosphere. Convection currents gradually turn the toroidal field into a poloidal field of the reverse direction for the next cycle.

solar eclipse The passage of the Moon across the Sun's disk. Solar eclipses occur only at new Moon when the Moon lies close to the node of its orbit around the Earth. They do not occur each month because the new Moon is usually either north or south of the node, due to the inclination of the Moon's orbit. Total solar eclipses are rare at any particular place on Earth, since the Moon's shadow falls on only a limited area. The theoretical maximum duration of totality is 7m 32 s, but is usually no longer than 3–4 min. Around totality, a number of interesting phenomena occur, including *shadow bands and *Baily's beads. During totality, solar prominences and the corona can be seen. Solar eclipses around lunar apogee may be *annular eclipses. Either side of the ground track of a total or annular eclipse, a partial eclipse is visible. Partial eclipses produce little obvious diminution in the level of sunlight unless their magnitude exceeds about 0.7 (70%) (*see* MAGNITUDE OF AN ECLIPSE).

(⊕) SEE WEB LINKS
• Solar eclipse predictions, reports, and information.

solar flare *See* FLARE, SOLAR.

solar interior That part of the Sun lying below the deepest visible layer of the photosphere. The solar interior cannot be directly observed, but its structure can be deduced from *standard solar models. Such models can be checked against observations of global oscillations, where the agreement is fairly satisfactory; and against the numbers of solar neutrinos, where the agreement is now excellent (*see* SOLAR NEUTRINO UNIT).

solar irradiance The radiation flux from the Sun received by the Earth at its average distance. The irradiance over all wavelengths is referred to as the *total solar irradiance (TSI). At visible and near-infrared wavelengths the solar irradiance shows very little variation (a fraction of a percent) with time and is the main component of the TSI. The ultraviolet and X-ray irradiances show much larger variations due to *active regions and solar *flares. Although much smaller than the visible and near-infrared irradiance, this high-energy radiation is absorbed by the Earth's atmosphere and is very important for atmospheric chemistry.

solar mass ($M_\odot$) A unit of mass used in stellar and galactic astronomy, equivalent to the mass of the Sun, 1.989×10^{30} kg.

solar maximum The period of time around the peak of the *sunspot cycle when sunspots and other activity are most frequent.

Solar Maximum Mission (SMM) A NASA satellite that observed solar activity at wavelengths from gamma rays to white light. SMM, also known as Solar Max, was launched in 1980 February. A failure of its pointing system after nine months led to a repair mission in 1984 April by Space Shuttle astronauts. SMM continued operations until it re-entered the Earth's atmosphere in 1989 December. The mission led to improved understanding of the energy-release mechanisms of solar flares and related phenomena.

solar minimum The period of time when sunspots and other activity are least frequent, although not usually absent.

solar motion The linear velocity of the Solar System in space relative to the local standard of rest. It is usually given in the form of three components along either galactic or equatorial coordinate axes; alternatively, it may be expressed as a speed towards the *solar apex, about 19.5 km/s or 4 AU/year.

solar nebula The cloud of gas and dust from which our Solar System formed some 5 billion years ago. The cloud is thought to have been shaped like a flattened disk and was dispersed by the *T Tauri wind from the young Sun. Comets, asteroids, and meteorites provide important clues to the composition of the solar nebula. Similar disks of gas and dust have been detected around several nearby young stars, notably Beta Pictoris.

solar neutrino unit (SNU) A measure of the flux of neutrinos from the Sun reaching the Earth. The neutrinos are produced in nuclear reactions at the Sun's centre. The first solar neutrino detector, which began operation in 1968, consisted of a tank of fluid containing chlorine, situated deep underground in the Homestake gold mine, South Dakota. A tiny proportion of the chlorine atoms interact with solar neutrinos to produce argon atoms. The number of neutrino interactions is measured by the solar neutrino unit (SNU), 1 SNU equalling 1 interaction per second per 10^{36} chlorine atoms. The neutrino rate predicted by solar models is about 8 SNU, but measurements from Homestake and from the gallium detectors SAGE (in Russia) and Gallex (near Rome, Italy), indicate a neutrino rate about one-third that expected. However, these instruments detect only so-called electron neutrinos, one of three types of neutrino. More recent results from the Sudbury Neutrino Observatory in Ontario, Canada, which contains heavy water and is sensitive to all three types of neutrino, demonstrate that the right number of neutrinos are produced in the Sun but that many of the electron neutrinos change into muon neutrinos and tau neutrinos on their way to Earth.

Solar Orbiter An ESA solar observatory, planned for launch in 2017 or later, that will orbit the Sun at distances as close as 0.2 AU (45 solar radii), imaging its surface and corona at visible, ultraviolet, and X-ray wavelengths, as well as detecting charged particles and magnetic fields in the solar wind. Its elliptical orbit will bring it close to the Sun every five months, at which time it will be moving at about the same rate as the Sun spins and will temporarily hover over one region like a geostationary satellite to watch the development of solar active regions. Repeated flybys of Venus during the mission will increase the craft's orbital inclination to 35° or more, allowing the probe to examine the Sun's polar regions.

• Official mission website.

solar oscillations *See* HELIOSEISMOLOGY.

solar parallax The angular width of the Earth's equatorial radius as seen from the centre of the Sun, when the Earth is at a distance of 1 AU. Its value is 8″.794148. Historically, the solar parallax was the prime quantity giving the linear scale of the Solar System, and was derived from observations of bodies in the Solar System as seen from widely spaced observatories, for example of transits of Mercury or Venus across the Sun. The solar parallax is now derived from direct measurements of distances in the Solar System by radar and transmissions from spacecraft.

Solar System The collective name for the Sun and all the material that orbits it. It includes the eight major planets, at least five *dwarf planets, and over 170 known satellites, plus countless asteroids, comets, and meteoroids. There is no single measure that defines the boundary of the Solar System. Many *Kuiper Belt objects lie beyond the orbits of the outer planets, the Sun's *heliopause is even more distant (around 100 AU out), and the gravitational influence of the Sun extends halfway to the nearest star. The Solar System has an age of 4.57×10^9 years.

solar telescope A telescope specifically designed for studying the Sun. Lenses or mirrors with very long focal lengths are used so that solar features can be seen in detail. Many large solar telescopes, such as the McMath–Pierce Solar Telescope, are located on high mountains to reduce atmospheric turbulence and cloud cover. Even better conditions are obtained at observatories surrounded by water, such as the Big Bear Observatory in California. *See also* SOLAR TOWER.

solar–terrestrial relations The effects of solar activity on the Earth and its magnetic field. The largest effects originate from magnetic disturbances which travel out from the Sun and produce *geomagnetic storms as they interact with the Earth's magnetosphere. A sharp increase in the strength of the Earth's magnetic field at the start of a storm results from a compression of the field on the sunward side. Disturbances occur world-wide and may continue for a day or so. Magnetic *substorms are more localized, occurring near the geomagnetic poles, and are associated with auroral displays. Links between solar activity and the Earth's weather have been proposed, notably at the time of the *Maunder minimum.

Solar Terrestrial Relations Observatory (STEREO) A pair of NASA spacecraft launched in 2006 October to study coronal mass ejections (CMEs) and the solar wind. One of the pair, STEREO A (Ahead), is in an orbit slightly smaller than that of the Earth and moves ahead of the Earth by just over 20° per year, while the other, STEREO B (Behind), is in a slightly larger orbit than the Earth's and falls behind by a similar amount. Together they provide a three-dimensional view of the Sun and its activity. A set of telescopes called SECCHI on each craft consists of two white-light coronagraphs, Cor1 and Cor2, to observe the inner and outer corona out to 15 solar radii; an Extreme Ultraviolet Imager (EUVI) to observe the chromosphere and inner corona at four different wavelengths; and a Heliospheric Imager (HI) which observes CMEs out to more than 300 solar radii, i.e. beyond Earth's distance. In addition, each craft carries an interplanetary radio burst tracker (SWAVES) to trace radio disturbances in the solar wind, and two experiments to study energetic particles from the Sun and the interplanetary magnetic field (IMPACT and PLASTIC).

((()) SEE WEB LINKS
- Official mission website.
- Mission page at NASA.

solar time Time with respect to the Sun; technically, the *hour angle of the Sun plus 12 hours, added to make the day begin at midnight rather than noon. The true Sun is the basis of *apparent solar time, as shown on a sundial, but this runs irregularly because the Earth's orbit is not circular and the Sun's path (the ecliptic) is inclined to the celestial equator. For accurate timekeeping purposes *mean solar time is used, based on the hour angle of the fictitious *mean sun. Solar time, which is the basis of all civil timekeeping, loses about 4 min a day against *sidereal time, so that a star will rise about 4 min earlier each night.

solar tower A large solar telescope in which the optics are mounted on a high tower to avoid air turbulence near the ground. Sunlight is directed vertically downwards by a movable flat mirror termed a *heliostat into an observing room, usually sited underground. Examples are the two towers at Mount Wilson in California. Tower telescopes can be cooled to reduce convection currents. In a *vacuum solar tower* the tower is evacuated to prevent air turbulence.

solar wind The continuous outflow of ionized gas from the Sun's *corona. The solar wind consists of electrons, protons, and (to a lesser extent) the nuclei of elements such as helium. As the gas expands into interplanetary space it carries with it magnetic field lines that are twisted into a spiral pattern by the Sun's rotation. Three distinct components to the solar wind can be identified: the *fast* and *slow* solar wind streams, and the *transient* wind due to *coronal mass ejections (CMEs). The fast wind arises from *coronal holes where magnetic field lines in the corona directly open out into space. It has generally very stable properties, with speeds at 1 AU of 700–800 km/s. The slow solar wind arises from regions where closed magnetic field structures occur in the Sun's corona; average speeds at 1 AU are 300–400 km/s. It shows more variability in terms of temperature, density, and element composition than the fast solar wind. CMEs can travel outwards from the Sun at velocities up to 3000 km/s, and have enhanced density and magnetic fields over the background solar wind. At solar minimum, when the Sun's global magnetic has its simplest form, the slow solar wind is concentrated in the ecliptic plane around the *heliospheric current sheet. If the current sheet is tilted relative to the ecliptic plane, then the fast wind streams can be detected at Earth on a repeating cycle of 27 days as the Sun rotates. At solar maximum, the structure is more complex, with fast solar wind streams interspersed among the slow solar wind which extends over all latitudes. Both the fast and slow solar wind are accelerated to supersonic speeds within a few solar radii of the Sun's surface. The solar wind interacts with the Earth and other planets with magnetic fields to produce their *magnetospheres. The solar wind extends about 100 AU from the Sun. Its boundary is marked by the *heliopause.

solar year The year in a calendar designed to relate, approximately at least, to the seasonal cycle; alternatively, another name for a *tropical year. Not all calendars use a solar year. For example, the Islamic calendar is closely related to the phases of the Moon, and the calendar year regresses through the seasons.

solid Schmidt telescope A Schmidt camera manufactured from a solid cylinder of glass, with the figuring performed on both ends. A hole is needed at the corrector end to position the film. With this system a focal ratio of f/0.35 is theoretically possible using glass, or even f/0.2 using diamond. However, the design is practicable only for small apertures, and suffers from *chromatic aberration, making it suitable only for use with monochromatic light, such as in spectroscopy.

solstice Either of the two points on the ecliptic at which the Sun reaches its greatest declination north or south of the celestial equator each year; or the dates on which this occurs— on June 21 (*summer solstice* in the northern hemisphere, *winter solstice* in the southern) and December 21 or 22 (vice versa).

solstitial colure The *hour circle that passes through the celestial poles and the summer and winter solstices.

Sombrero Galaxy The galaxy M104 (NGC 4594) in Virgo. It is an 8th-magnitude spiral of type Sa or Sb with a large nucleus and is seen edge-on, giving the appearance of a Mexican hat. A dark lane of dust runs across its centre. It is about 30 million l.y. away, closer than the *Virgo Cluster. ▪

Sothic cycle A cycle of 1460 years in the calendar of ancient Egypt. The Egyptian calendar had a year of fixed length, 365 days, with no leap years. Consequently over one Sothic cycle the calendar regressed by one year with respect to the seasons, since the calendar year was approximately one quarter of a day shorter than the *tropical year. The cycle takes its name from Sothis, the ancient Egyptian name for Sirius, whose *heliacal rising marked the start of their year.

source count A statistical technique used to study the evolutionary history of the Universe. It involves counting the numbers of radio sources, N, above a threshold flux density, S, in a given area of sky. If the Universe is flat and the luminosity and space density of radio sources has not evolved with time, then a plot of N against S on logarithmic axes should be a straight line of slope -1.5. Deviations from this line for the faintest (and hence most distant) sources show that radio sources were both more powerful and more numerous in the early Universe, and are strong evidence that the Universe has evolved with time.

South African Astronomical Observatory (SAAO) An observatory at an altitude of 1770 m near Sutherland, Northern Cape, owned and operated by the South African government's National Research Foundation. It was founded in 1972 by combining the facilities of the Royal Observatory at the Cape of Good Hope (founded in 1820) and the Republic

Observatory, Johannesburg (founded in 1905). A 1-m reflector was moved to Sutherland from the Cape and a 0.5-m reflector from the Republic Observatory, joining a new 0.75-m reflector. The 1.88-m Radcliffe Telescope, originally opened in 1948 at the Radcliffe Observatory, Pretoria, was moved to the SAAO in 1976. The *Southern African Large Telescope (SALT) opened at SAAO in 2005. The SAAO also hosts a station of the *Birmingham Solar Oscillations Network (BiSON), opened in 1990; the 1.4-m Japanese Infrared Survey Facility (IRSF), which began operation in 2000; a 0.5-m telescope of the Korean Yonsei Survey Telescopes for Astronomical Research (YSTAR) project, opened in 2002; a 1.2-m robotic telescope of the German *Monitoring Network of Telescopes (MONET), opened in 2007; and SuperWASP-South, an array of eight wide-angle cameras of 0.1-m aperture for detecting transits of stars by extrasolar planets, which began operation in 2006. The headquarters of the SAAO are at the former Royal Observatory in Cape Town.

(((⊕))) SEE WEB LINKS
• Official observatory website.

South Atlantic Anomaly A region within the inner *Van Allen Belt, which reaches its minimum altitude (250 km) over the Atlantic Ocean off the coast of Brazil. This positioning arises because of the offset between Earth's magnetic and geographical axes. Artificial satellites in low-inclination and low-altitude orbits pass frequently through the South Atlantic Anomaly, with a consequent risk to their electronic components (including degradation of solar cells) from energetic trapped particles.

Southern African Large Telescope (SALT) A reflector at the *South African Astronomical Observatory, opened in 2005, with a hexagonal mirror 11.1 × 9.8 m in diameter, made of 91 hexagonal segments; the maximum effective aperture is 9.8 m. It is a modified version of the *Hobby–Eberly Telescope. SALT has a fixed elevation angle of 37° from the vertical and can rotate only about its azimuth axis. SALT is a joint project of South Africa, Germany, India, New Zealand, Poland, the UK, and the USA.

(((⊕))) SEE WEB LINKS
• Official telescope website.

Southern Astrophysical Research Telescope (SOAR) A 4.1-m reflector jointly owned by the US *National Optical Astronomy Observatory (NOAO), Brazil, Michigan State University, and the University of North Carolina at Chapel Hill. It is sited at an altitude of 2700 m on Cerro Pachón mountain in Chile and is operated by the *Cerro Tololo Inter-American Observatory. It opened in 2004.

(((⊕))) SEE WEB LINKS
• Official telescope website.

Southern Cross Popular name for the constellation *Crux. Its four brightest stars, namely Alpha, Beta, Gamma, and Delta Crucis, form a distinctive cross-shape.

southern lights Popular name for the aurora australis (*see* AURORA).

Southern Reference Stars (SRS) A list of 20488 stars in the southern hemisphere designed to supplement the AGK3R (*see* AGK) in the north, to give a whole-sky coverage of reference stars for calibrating photographic surveys. Observations of stars mostly in the magnitude range 7.5 to 9.5 were made at a dozen observatories between 1961 and 1973, and coordinated at the US Naval Observatory and Pulkovo Observatory. The results were issued in computer-readable form in 1988.

(((⊕))) SEE WEB LINKS
• Detailed description and full catalogue downloadable from the CDS.

Southern Sky Survey A photographic survey of the sky south of declination −17°, produced jointly by the European Southern Observatory's 1-m Schmidt camera and the 1.2-m United Kingdom Schmidt Telescope at Siding Spring. Both telescopes have the same focal length as the 1.2-m Oschin Schmidt, which produced the northern-hemisphere

*Palomar Observatory Sky Survey, so the scale of the plates in each survey is identical. The Southern Sky Survey consists of 606 pairs of red-sensitive and blue-sensitive plates; the red plates were taken in 1978–90 with the ESO Schmidt and the blue plates 1974–87 with the UK Schmidt.

South Pole–Aitken Basin The largest and oldest impact feature on the Moon, and one of the largest in the Solar System. It is centred at −56° lat., 180° E long., on the lunar farside. It takes its name from the fact that the lunar south pole is on its southern rim and the 135-km crater Aitken is on its northern rim. The Basin consists of three main rings with a maximum diameter of some 2400 km and is up to 13 km deep. It is overlain with more recent impacts and contains smoother mare regions resulting from volcanic flooding. The Basin's surface is enriched in mafic minerals (enhanced amounts of iron and magnesium), perhaps derived from the Moon's lower crust. The Basin may have been formed by a glancing blow over 4 billion years ago, before the other lunar basins. 📷

South Pole Telescope (SPT) A 10-m diameter millimetre-wave telescope opened in 2007 at an altitude of 2800 m at Amundsen–Scott station at the south pole in Antarctica where the atmosphere is particularly dry and steady. The SPT operates at wavelengths of 3–1.3 mm, studying the cosmic microwave background and detecting clusters of galaxies via the *Sunyaev–Zel'dovich effect. The SPT is operated by a consortium of US universities and institutes.

(((⊕))) SEE WEB LINKS
• Official telescope website.

space, curvature of *See* CURVATURE OF SPACETIME.

space motion The linear velocity of a star expressed in kilometres per second as components along three axes. Any three axes can be chosen, but most commonly they are oriented towards the galactic centre, in the direction of the Galaxy's rotation, and towards the north galactic pole. Alternatively, axes based on the directions of the equinox and celestial pole can be used.

space probe A spacecraft that is sent away from the Earth to examine another celestial body or the conditions in space. The first space probe was *Luna 1, launched by the Soviet Union on 1959 January 2.

space reddening *See* INTERSTELLAR ABSORPTION.

Space Telescope Science Institute (STScI) A NASA establishment for research with the Hubble Space Telescope. It was founded in 1981 at Johns Hopkins University, Baltimore, Maryland, and is managed for NASA by the *Association of Universities for Research in Astronomy. STScI plans the observing programme for the HST, and receives, analyses, and archives the resulting data.

(((⊕))) SEE WEB LINKS
• Official website.

spacetime A four-dimensional description of the Universe in which the position of an object is specified by three coordinates in space and one in time. According to the theory of special relativity, there is no absolute time which can be measured independently of the observer, so events that are simultaneous as seen by one observer occur at different times when seen from a different place. Time must therefore be measured in a relative manner, as are positions in three-dimensional (Euclidean) space, and this is achieved through the concept of spacetime. The trajectory of an object in spacetime is called its *world line. General relativity relates the *curvature of spacetime to the positions and motions of particles of matter.

space velocity The velocity of a star relative to the Sun. It can be determined from measurements of the star's *radial velocity and *tangential velocity.

space velocity: A star's velocity has two components: its radial velocity and its tangential velocity.

Spacewatch A project to discover and study asteroids and comets, started in 1980 with the 0.9-m reflector at the *Steward Observatory, Arizona. A new 1.8-m reflector was opened in 2000 to extend the survey to fainter objects.

( ) SEE WEB LINKS

• Official project website.

space weather The variable conditions in interplanetary space, usually caused by solar disturbances, which affect human health or technological systems such as satellite navigation (Global Positioning System), radio communications, and spacecraft electronics. Major *geomagnetic storms caused by fast-moving *coronal mass ejections (CMEs) are the principal space weather events. Other effects are caused by X-rays, radio bursts, and energetic particles generated by solar *flares and CMEs. Space weather damage is caused by energetic electrons or ionized atoms penetrating spacecraft electronics or a human body, by ionospheric disturbances that affect radio propagation, and by electrical currents induced in the electrical grid or pipelines on Earth.

( ) SEE WEB LINKS

• Space weather information at ESA.
• Space Weather Prediction Center.

spallation A type of nuclear fission that takes place when matter is bombarded by high-energy atomic particles or cosmic rays. Meteoroids, asteroids, and the Moon are all bombarded by solar particles and cosmic rays, chiefly very high-energy protons. These interact with elements in their surface rocks, producing atoms such as helium-3, neon-21, and argon-38. Spallation also changes the overall composition of high-energy galactic cosmic rays when they collide with stationary atoms of interstellar matter. New atoms of light elements such as lithium, boron, and beryllium are produced. Consequently, these elements are a million times more abundant in cosmic rays than in stars.

spark chamber A device that shows the path of a charged particle. It consists of a series of parallel plates mounted in a gas such as neon. When a particle passes through a detector mounted above the chamber it triggers a high-voltage pulse to the spark chamber. The ions generated by the passage of the particle allow a spark to travel between the plates along the path taken by the particle. The tracks can be recorded using photomultiplier tubes. The device can be used to detect cosmic rays and gamma rays.

spec. Abbr. for speculum, used to denote the mirror of a reflecting telescope. The name derives from the time when telescope mirrors were made of *speculum metal.

Special Astrophysical Observatory An observatory of the Russian Academy of
Sciences, at an altitude of 2070 m on Mount Pastukhova in the Caucasus Mountains of southern
Russia. It is the site of the 6-m Large Altazimuth Telescope (in Russian, Bol'shoi Teleskop
Azimutal'nyi, or BTA), opened in 1975; the mirror was replaced in 1979. Also at the Observatory
are a 1-m reflector, opened in 1990, and a 0.6-m reflector, opened 1994. About 20 km to the
northeast, at an altitude of 970 m, is the *RATAN-600 radio telescope.

(⊕) **SEE WEB LINKS**
Official observatory website.

special theory of relativity A theory proposed by A. *Einstein in 1905, based on the
proposition that the speed of light in a vacuum is constant throughout the Universe, and is
independent of the motion of the observer and the emitting body. A consequence of this
proposition is that three things happen as an object's velocity approaches the speed of light: its
mass goes up, its length shortens in the direction of motion, and time slows down. Hence,
according to special relativity, no object can ever reach the speed of light because its mass
would then become infinite, its length would become zero, and time would stand still. In
addition, Einstein concluded that the mass of a body is a measure of its energy content,
according to the famous equation $E = mc^2$, where c is the speed of light. This equation describes
the conversion of mass into energy in nuclear reactions within stars. *See also* RELATIVITY.

speckle interferometry A technique for detecting detail in celestial objects which are
normally too small to be resolved because of the effects of atmospheric *seeing. Numerous brief
exposures (0.1–0.001 s) are taken at high magnification in a narrow range of wavelengths, each
of which yields an image consisting of fine structure called **speckles**. Each speckle is caused by
cells of seeing. The speckles from a close double star, for instance, will contain double structure
which is detectable even though it is finer than the overall image size. To determine the detail, a
laser beam is passed through the photographic image and focused on to a film. The resulting
image contains fringes which reveal the overall structure of the original image. The technique is
used to measure close double stars and the diameters of red giant stars.

spectral classification The categorization of stars according to the properties of their
spectra. The first attempt to do so was the *Secchi classification in the 1860s, but it was the
*Harvard classification scheme that led to the current system of spectral types. Stars were
classified as type O, B, A, F, G, K, or M in order of decreasing surface temperature, and each type
further subdivided into subclasses from 0 (hottest) to 9 (coolest). The prefixes *d*, *g*, and *c* were
used to signify dwarf stars, giants, and supergiants. To these were added the R and N types (now
known as carbon stars) and S type (heavy metal stars). Spectral types for stars in the Harvard
system were published in the *Henry Draper Catalogue* (published 1918–24). Stars were
originally thought to follow an evolutionary sequence from the 'early' O and B types to the 'late'
K and M types. Although this is now known to be erroneous, the terms 'early type' and 'late
type' are still in use. In the late 1990s, spectral types L and T were added to the sequence to
accommodate the coolest stars and brown dwarfs (with class Y reserved for the coolest brown
dwarfs of all, as yet unobserved).

Astronomers currently use the *Morgan–Keenan classification system introduced in the
1940s. This is a revision of the Harvard system, the key point being the addition of a range
of *luminosity classes indicated by a Roman numeral from I (supergiants) to V (dwarfs, or
main-sequence stars). Luminosity classes VI (subdwarfs) and VII (white dwarfs) were
subsequently added, but are now rarely used. Luminosity subclasses a, b, and c are sometimes
appended, especially for supergiants, while the most luminous hypergiants are assigned
luminosity class Ia-0. The dominant spectral features and other properties of each spectral
type are summarized in the table.

The effective temperatures given are for stars on the main sequence; giants and supergiants
of the same spectral type have slightly different temperatures.

Each spectral type is notionally divided into subtypes 0–9, but not all subtypes are in common
use and some finer divisions have proven necessary, such as types O9.5 and O9.7 for
supergiants. At the extremes, the earliest recognized spectral type is O2, while the coolest brown
dwarfs yet observed are classified T8. Additional suffixes indicate various unusual properties
of the spectrum, as follows:

n	nebulous (diffuse) lines
s	sharp lines
v	variable spectrum
e	emission lines, when not expected (e.g. Be)
ev	variable emission lines
f	certain O-type emission-line stars
p	peculiarity (e.g. anomalous line strengths)
eq	P Cygni emission
k	interstellar lines present
m	metallic-line star

Spectral Classification

Colour	Spectral Type	T_{eff} (K)	M_V	Classification criteria
Very blue	O5	40000	−5.8	Highly ionized atoms He II, Si IV, N III H fairly weak Some emission lines
Blue	B0	28000	−4.1	Lower ionization No He II; He I strong Si III, O II; H stronger
Blue-white	A0	9900	+0.7	No He I; H at maximum (broad) Mg II, Si II strong Fe II, Ti II, Ca II increasing
White	F0	7400	+2.6	H weaker; Ca II strong Ionized metals decreasing Neutral metals increasing
Yellow	G0 (G2 = Sun)	6030	+4.4	Ca II very strong Neutral metals strong H still weaker
Orange	K0	4900	+5.9	Neutral metals strong H very weak Molecular bands increasing
Red	M0	3480	+9.0	TiO bands strong Neutral metals (e.g. Ca I) strong
	R, N	3000		Strong CN, CH, C_2 No TiO; neutral metals
	S	3000		Strong ZrO, YO, LaO Neutral metals
	L0	2200		TiO decreasing. Increasing Na I and K I. Strong H_2O
	T0	1300		Strong CH_4, H_2O

spectral index (S or α) A measure of the way in which the intensity of the continuum emission from a radio source varies with frequency. The intensity usually varies exponentially, and the exponent is called the spectral index. It typically takes positive values from 0 to 2 for thermal emission, while non-thermal emission, such as synchrotron radiation, leads to negative values of the spectral index ranging from about −0.5 to −1.5.

spectral line An emission or absorption line in a spectrum that arises when an electron moves between two energy levels in an atom. A jump to a higher level requires an input of energy, and produces a dark absorption line. A drop to a lower level releases energy, producing a bright emission line. At visible wavelengths, spectral lines from elements such as iron are largely due to jumps involving the outer electrons, but X-rays are produced from the inner electron shells. The precise wavelength (or energy) is a hallmark of the atom involved in producing the line.

spectral type A means of classifying stars according to the details of their spectrum. The spectral type of a star depends largely on its surface temperature, and hence is also a guide to its colour, the hottest stars being the bluest and the coolest ones the reddest. The system of spectral types currently used is the *Morgan–Keenan classification. *See* SPECTRAL CLASSIFICATION.

spectrogram A photographic record of a spectrum.

spectrograph A device for dispersing light into a spectrum so that the intensity at each wavelength can be recorded by a detector. Spectrographs have been designed for use in various regions of the spectrum, with particular emphasis on the ultraviolet, visible, and infrared. Different wavelengths have different technological requirements—for example, many materials that are transparent at optical wavelengths are opaque in the ultraviolet. The main components of a traditional spectrograph include a slit, for selecting a particular object in the telescope field and to limit the instrumental width of spectral features; a collimator for focusing the light into a parallel beam; a disperser (a diffraction grating or prism), which splits the light into a spectrum; and a detector, nowadays usually a CCD, for recording the spectrum. The recorded spectrum is termed a **spectrogram**. The resolving power of the spectrograph is ultimately limited by the size of the disperser.

spectroheliogram A photograph of the Sun taken in the light of a strong Fraunhofer line (or part of the line). Most commonly, the *Hα and calcium K lines are chosen for spectroheliograms, which then show chromospheric features. A spectroheliogram is produced by a *spectroheliograph; if filters are used to select the wavelength, the result is termed a *filtergram.

spectroheliograph An instrument for photographing the Sun at the wavelength of a strong Fraunhofer line. An image of the Sun produced by a telescope is focused on to a primary slit, and light from this slit passes to a grating or prism. The dispersed light so formed is then intercepted by a secondary slit positioned at a desired wavelength (e.g. a part of the *Hα line) and a photographic plate is placed behind this slit. The primary slit is then moved across the image of the Sun and the secondary slit moved to maintain its spectral location, producing a spectroheliogram on the photographic plate or CCD.

spectrohelioscope An instrument with which the Sun can be observed visually at the wavelength of a particular spectral line. The principle is very similar to that of the spectroheliograph, except that the primary and secondary slits are moved rapidly back and forth over the solar image, so rapidly that the eye is unaware of the motion.

spectrometer A *spectrograph in which the output spectrum is scanned by a *photoelectric photometer to produce a record of how the intensity of a spectrum varies with wavelength. A modern example is the *radial velocity spectrometer, in which the positions of the spectral lines are measured to deduce the radial velocity of the star. Otherwise, spectrometers are now rarely used because CCD detectors can record large regions of a spectrum at once, and the device is then known as a *spectrograph.

spectrophotometer An instrument which combines a *spectrograph with a *photometer. In a conventional photometer the wavelengths measured are selected with a filter; in a spectrophotometer, the filter is replaced by a spectrograph to split up the light according to wavelength. A spectrophotometer has the advantage when the wavelength range of interest is 3 nm or less because it is difficult to make filters of such narrow bandwidth. Any modern spectrograph with a *charge-coupled device or similar detector is indistinguishable from a spectrophotometer.

spectrophotometry The science of measuring the brightness of celestial objects at a narrower bandwidth than is possible with a conventional photometer. The instrument used is a *spectrophotometer.

spectropolarimetry The science of measuring the polarization of light from celestial objects at narrower bandwidths than is possible with a conventional *polarimeter. The principles are the same as with a conventional polarimeter, but the wavelengths measured are isolated with a spectrograph rather than with filters. Spectropolarimetry is used to measure the magnetic fields of stars from the *Zeeman effect and to study the origin of spectral lines.

spectroscope An instrument that produces a spectrum for visual observation. Spectroscopes are now used mainly for educational and recreational purposes; research instruments employ CCD detectors to record the spectrum, and hence are *spectrographs.

spectroscopic binary A binary star that cannot be resolved visually or photographically, but whose duplexity is established by periodic Doppler shifts of lines in its spectrum. All close binaries (including *cataclysmic and *interacting binaries) fall into this category, as do most *eclipsing binaries. The information obtainable about a spectroscopic system depends on whether a *radial velocity curve can be obtained for one or both components. With a *single-line binary, it is possible to obtain only an expression known as the *mass function, which is a combination of the masses of each star and the inclination of the orbit. With a *composite-spectrum binary or *double-lined binary, the relative masses may be derived from the *mass ratio. If the orbital inclination can be found (normally only in eclipsing systems), the dimensions and masses of the stars can be calculated. With interacting binaries, there are limitations to the latter procedure because of their complex structure.

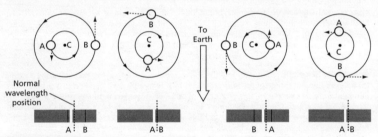

spectroscopic binary: The top diagrams show the positions of two stars, A and B, as they orbit their centre of mass, C. As a result of the Doppler effect, the wavelengths of light from each star are displaced towards the blue or red end of the spectrum, depending on whether the star is approaching or receding from the Earth. Any spectral line in the spectrum of each star is therefore seen to oscillate back and forth across its normal wavelength position, as shown in the row of spectra.

spectroscopic parallax A method of estimating the distance of a star by comparing its apparent magnitude with its absolute magnitude. The star's absolute magnitude is deduced from its *spectral type and *luminosity class, which are found by examining its spectrum. Spectroscopic parallax is the most common method of determining the distances of stars that are too distant to show a reliable trigonometric parallax.

spectroscopy The technique of obtaining and studying the spectra of celestial bodies, from which their compositions, physical properties, and motions can be determined. The instrument used is a spectrograph, a form of spectroscope in which the spectrum is recorded with a detector (a CCD in the case of optical spectroscopy). Three main types of spectrum are observed in astronomical sources: a continuum, or continuous spectrum, consisting of an unbroken band of colours; an emission spectrum, which consists of bright lines; and an absorption spectrum, consisting of dark lines crossing a continuum. However, many classes of objects show both absorption and emission lines superimposed on a continuum, reflecting a range of physical conditions in the source. Study of a star's spectrum allows its temperature, luminosity, and size to be determined (*see* SPECTRAL CLASSIFICATION). The displacement of spectral lines (the *Doppler shift) reveals the object's movement along the line of sight (its *radial velocity).

X-ray spectroscopy concentrates on hot, highly energetic objects. Emission lines are detected from old supernova remnants, coronae of hot stars, and *accretion disks. The field of cosmic X-ray spectroscopy began in earnest with the launch of the *Chandra X-ray Observatory and *XMM-Newton in 1999 and 2000 respectively. Both produce high-resolution X-ray spectra combined with high-sensitivity X-ray CCD detectors.

Ultraviolet spectroscopy was pioneered by the *Copernicus satellite, launched in 1972, and continued by the long-lived *International Ultraviolet Explorer (IUE) satellite (1978–96). This wavelength range is of importance because the *resonance lines of many atoms and ions lie in the ultraviolet. This region is especially useful in studies of the diffuse interstellar medium and of hot stars and their stellar winds. The Hubble Space Telescope has better resolution and much improved sensitivity compared to IUE, while the *Far Ultraviolet Spectroscopic Explorer (FUSE) satellite operates at wavelengths shorter than those accessible to either IUE or HST.

Optical spectroscopy is the key to the widest range of astronomical studies because there are far more telescopes operating in this region. It is also usually possible to obtain much higher spectral resolution than at other wavelengths, making it possible to study line profiles (for temperature and density information) and velocities (e.g. in binaries, from which their masses can be calculated). *Multi-object spectroscopy has become commonplace in the optical region while still not widely applied in other spectral regions.

Infrared spectroscopy is a rapidly developing area. The first attempts at infrared spectroscopy used narrow-band filters with a tunable passband, the detector being a simple *bolometer. The advent of infrared-sensitive arrays, analogous to optical CCDs but constructed of indium antimonide or mercury cadmium telluride, has allowed the development of grating spectrometers which are similar to their visual counterparts, except that the grating is cooled to reduce thermal noise. *Fourier transform spectroscopy is used to achieve the highest resolutions in the infrared. Infrared spectroscopy is making major advances in studies of heavily obscured regions, such as the galactic centre and sites of star formation. Spectroscopy in the far infrared has been conducted successfully from space, most recently with the *Spitzer Space Telescope and the *Herschel Space Observatory. The *James Webb Space Telescope will also be equipped for infrared spectroscopy.

At still longer wavelengths, Fourier transform spectrometers can be used into the millimetre and submillimetre regions, but other techniques, such as *heterodyne spectroscopy*, are also employed here and into the radio region, where wavelengths are too long for conventional interference or diffraction technologies to be feasible. Radio-frequency spectroscopy has great power in studying emission lines from molecular clouds and masers, the 21-cm hydrogen line, and recombination lines in ionized gas. The next generation of radio astronomical instruments, such as the *Atacama Large Millimeter Array, the *Square Kilometre Array, and the *Low Frequency Array, will pursue radio spectroscopy with much increased sensitivity.

spectrum **1.** A range of electromagnetic energies arranged in order of wavelength or frequency (*see* ELECTROMAGNETIC SPECTRUM). The *emission spectrum* of a body or substance is the range of radiations it emits when it is heated, is bombarded by electrons or ions, or absorbs photons. The *absorption spectrum* of a substance consists of dark lines or bands in a continuous spectrum, each line being a wavelength or group of wavelengths at which light is removed from the continuous spectrum by the absorbing medium. These lines and bands are at the same wavelengths as some of the lines and bands in the substance's emission spectrum. Emission and absorption spectra may show a *continuous spectrum* (also called a *continuum*), a *line spectrum*, or a *band spectrum*. A continuous spectrum contains an unbroken sequence of frequencies over a wide range; continuous spectra are produced by incandescent solids, liquids, and compressed gases. Line spectra are discontinuous lines produced by excited atoms and ions as they fall back to a lower energy level. Band spectra (closely grouped bands of lines) are characteristic of molecular gases or chemical compounds.

2. The coloured band produced when visible light is passed through a spectroscope. *See also* SPECTROSCOPY.

spectrum binary A *spectroscopic binary in which the orbits of the stars are viewed almost pole-on, so that there is no *Doppler shift in the stars' light.

Spectrum-Roentgen-Gamma A joint Russian–German X-ray observatory due for launch in 2012 or later into a position at the L_2 *Lagrangian point of the Earth's orbit. It will carry two imaging X-ray telescopes: eROSITA (extended ROentgen Survey with an Imaging Telescope Array) will observe at energies of 0.5–10 keV, while ART-XC (Astronomical Roentgen Telescope X-ray Concentrator) will observe at 6–30 keV.

(((⊕))) **SEE WEB LINKS**
• Official mission website.

spectrum variable *See* ALPHA² CANUM VENATICORUM STAR.

speculum metal An alloy of tin and copper used for telescope mirrors until the mid-19th century, when it was replaced by silvered glass mirrors. Its reflectivity is only approximately 66%, and must be repolished when it tarnishes.

Spencer Jones, Harold *See* JONES, HAROLD SPENCER.

spherical aberration A defect of mirrors and lenses in which rays of light parallel to but far from the optical axis are brought to a different focus from those close to the axis. Spherical aberration is seen in the images formed by spheroidal mirrors and by some lenses and eyepieces. A star image suffering from spherical aberration has no unique focus but instead has a *least circle of confusion* where the image is at its smallest. Beyond this point the image is a bright spot surrounded by a disk; inside it, the image is a ring darkening towards the centre. If rays from the edge of the mirror or lens come to a focus closer than those from its centre, the aberration is called *undercorrected*; in the opposite case it is *overcorrected*.

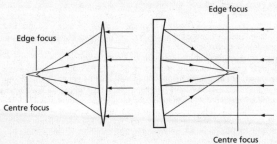

spherical aberration: Light incident at the edge of a lens (left) or spherical mirror (right) is brought to a different focus from light incident closer to the centre.

spherical albedo Another name for *Bond albedo.

spherical astronomy The description of positions and angular displacements of celestial objects by means of coordinates on the celestial sphere. It is one of the oldest branches of astronomy, and is now also concerned with the tracking of artificial satellites and space probes.

spherical coordinates A system for specifying positions in terms of angles on a sphere, such as the celestial sphere or the surface of a planet. Most systems of coordinates used in astronomy are spherical coordinates; examples are *ecliptic coordinates, *equatorial coordinates, and *galactic coordinates.

spheroid A body or surface produced by rotating an ellipse about either its major axis (a *prolate spheroid*) or its minor axis (an *oblate spheroid*). Both are special cases of an *ellipsoid. The Earth's shape is very nearly that of an oblate spheroid, its equatorial radius being 21km greater than the polar radius, while the equator departs from a circle by less than 1km.

Spica The star Alpha Virginis, magnitude 1.0. It is a spectroscopic binary consisting of a pair of B-type stars elongated by gravity so that the received light varies every 4 days as the pair orbit each other, although the amplitude is less than 0.1 mag. In addition, the primary is a variable of the Beta Cephei class, with a period of 0.17 days. Spica lies 250 l.y. away.

spicule A spike-like jet in the upper chromosphere of the Sun, visible especially in *Hα light. Spicules rise upwards into the corona at velocities of about 30 km/s, reaching altitudes of 9000 km in about 90s. They last about 15 min, ending their lives by fading rather than descending. They tend to be clustered, giving the appearance of porcupine quills on the solar limb. Seen against the disk, in Hα light, they seem to be identifiable as fine *mottles, light or dark depending on their height (although they always appear dark when observed in the red *line wing of the Hα line). Fine mottles form clusters with their bases in the coarse mottles which outline the *chromospheric network. At the centre of the Sun's disk, clumps of spicules form radiating patterns called *rosettes*.

spider The support of the central *secondary mirror in reflecting telescopes such as Newtonians and Cassegrains. A spider generally has three or four arms which attach the secondary to the tube.

spin (s) A property of many subatomic particles, which behave as though they are spinning, as they have angular momentum that is additional to any orbital angular momentum. However, since there is no way of marking a particle to observe its rotation, this property should not be taken too literally. The units of spin are $h/2\pi$, where h is the *Planck constant.

spinar A hypothetical type of supermassive star once proposed as the power source of quasars. They were envisaged as having a mass of about 100 million solar masses, to spin rapidly, and to radiate like a scaled-up version of a pulsar. However, even if such an object could form, it would be unstable and would rapidly collapse into a supermassive black hole.

spin casting A technique for casting pieces of glass for mirrors which results in a preformed concave surface, thus considerably reducing the work required to produce the correct figure. Either a kiln containing pieces of glass in a mould is raised to a temperature of 1200°C while being rotated at a few revolutions per minute, or already molten glass is poured into a rotating mould. Centrifugal force causes the surface of the molten glass to take up a parabolic curve. The technique is particularly valuable in preparing large mirrors of small focal ratio.

Spindle Galaxy An elongated 9th-magnitude galaxy in Sextans, also known as NGC 3115. It is classified as either a highly flattened elliptical galaxy, or a spiral with a large bulge and hardly any surrounding disk. It lies about 30 million l.y. away.

spin-down A phenomenon in which the rotation period of a pulsar, P, steadily lengthens as the pulsar loses energy with age. Observed spin-down rates, $\dot{P}$, range from 10^{-12} s per second for the youngest pulsars down to 10^{-19} s per second for *recycled pulsars. The quantity $P/2\dot{P}$ is known as the *characteristic age* or **spin-down age** of the pulsar, and is likely to represent an upper limit on the true age of the pulsar.

spin–orbit coupling A relationship between the orbital period of one body around another and its rotation period on its axis. The relationship has been brought about by tidal forces. For example, the period of rotation of the planet Mercury is two-thirds of its period of revolution. Other examples of the same effect are *co-rotation and *synchronous rotation.

spin temperature A fictitious temperature that describes the average spin of neutral hydrogen (H I) atoms in a gas cloud. H I has two energy states: an upper state in which the spins of the electron and proton are parallel, and a lower state in which the spins are opposite (*antiparallel*). The spin temperature of an H I region is the temperature that would produce the observed ratio of parallel to antiparallel spins if the H I gas were in thermal equilibrium. The transition between the two spin states of the H I atom produces the important radio spectral line at 21 cm wavelength.

spiral arm A curved structure in the disk of spiral (and some irregular) galaxies where young stars, clusters, nebulae (*H II regions), and dust are concentrated. Some galaxies have a well-defined, two-arm spiral pattern, while in others the arms may be three or four in number and sometimes fragmentary. The arms are visible because of the recent formation of bright, massive, short-lived stars within them. This star-forming activity is periodic, corresponding to the movement through the disk of a *density wave.

spiral galaxy A type of galaxy with bright arms of stars, gas, and dust that extend in a spiral pattern from a central hub; Hubble type S. There are usually two arms, which can wrap right around the galaxy, but four-armed and even three-armed examples are known. The arms can also be broken into many short sections. At the centre of a spiral galaxy is a spheroidal bulge of old stars of *Population II. The bulge is large in type Sa spirals, which have tightly coiled arms, but much smaller and less conspicuous in types Sc and Sd, whose arms are coiled more loosely. Types Sa and Sb are, on average, intrinsically brighter and more massive than types Sc and Sd.

The spiral arms are sites of active star formation, and their appearance is dominated by bright, blue, massive young stars of *Population I and gaseous *H II regions. In mass they range from about 10^9 to 5×10^{11} solar masses, and in diameter from about 10000 to over 300000 l.y. Spiral structure can apparently exist only in disk galaxies above a certain size and, although spirals represent 80% of the bright galaxies in regions outside rich clusters, there are no spirals with masses as low as those of many irregulars and dwarf elliptical galaxies. The three brightest members of the *Local Group of galaxies are spirals.

If there is an almost rectangular or cigar-shaped concentration of stars across the central region, the galaxy is termed a *barred spiral. Nearly half of all bright spirals show an obvious bar, but most ordinary spirals probably have some degree of barred structure in their star or gas distribution. 📷

Spirit One of two identical NASA *Mars Exploration Rovers. Spirit was launched in 2003 June and landed in 2004 January inside a crater called Gusev in the southern hemisphere of Mars which is thought to have contained an ancient lake. 📷

Spitzer, Lyman, Jr (1914–97) American astrophysicist. He began to study interstellar matter (ISM) in the late 1930s. Noting that elliptical galaxies by contrast lacked ISM and contained old stars, he suggested that the bright stars seen in the arms of spiral galaxies had recently condensed from interstellar gas and dust. After World War II, he extended a pre-war interest on the effects of magnetic fields on the ISM as a trigger for star formation to techniques for containing a plasma in the laboratory, as a step towards utilizing nuclear fusion as a power source. Spitzer's work on the dynamics of star clusters led to the concept of *relaxation time. His advocacy of an Earth-orbiting telescope in 1946 and his constant lobbying led to the development of the Hubble Space Telescope. The *Spitzer Space Telescope was named in his memory.

Spitzer Space Telescope A NASA spacecraft for infrared astronomy, the last in the series of *Great Observatories. It was launched in 2003 August under the name Space Infrared Telescope Facility but in 2003 December was renamed in honour of Lyman *Spitzer Jr. It carries an infrared telescope with a 0.85-m mirror. The Infrared Array Camera, IRAC, images the sky between 3.6 and 8 µm; the Infrared Spectrograph, IRS, takes spectra at 5–38 µm; and the Multiband Imaging Photometer, MIPS, provides far-infrared imaging and photometry in the 24–160 µm range. The Spitzer Space Telescope trails the Earth in its orbit, away from the heat radiated by our planet.

(⊕) SEE WEB LINKS
• Official mission website.

spokes Dark, approximately radial markings in Saturn's rings, first observed through large telescopes in 1896 and confirmed in 1980–81 by the two Voyager probes. Spokes are transient features—they were not seen between 1999 and late 2005. Their origin is uncertain, but they are thought to be clouds of dust levitated out of the ring plane by electrostatic forces.

sporadic meteor A randomly occurring meteor, not associated with any meteor shower. Sporadic meteors can be seen on any clear night, with rates varying from perhaps 3–4 meteors/h in the spring to 8–10 meteors/h in the autumn. In addition to this seasonal variation there is also a nightly variation, which results in higher sporadic rates before dawn than in the evening hours. The sporadic background represents an end-point in meteor stream evolution, reached when the meteoroids in a stream become so dispersed that a meteor shower is no longer detectable. Because they are random, sporadic meteors are often considered a useful control sample in shower studies.

Spörer, Gustav Friedrich Wilhelm (1822–95) German astronomer. Independently of R. C. *Carrington, he used observations of sunspots to determine the position of the Sun's equator (and hence its axial inclination) and to establish that it has *differential rotation. His finding that the average latitude of spots varies over the course of a solar cycle is known as *Spörer's law. He was the first to draw attention to the lack of sunspots in the period 1645–1715, which is now known as the *Maunder minimum. An earlier period of low solar activity, 1450–1550, is now known as the *Spörer minimum, although he did not discover it.

Spörer minimum A period between about AD 1450 and 1550 when activity on the Sun seems to have been unusually low, as judged from historical records of naked-eye sightings of sunspots and aurorae, and from measurements of carbon-14 in tree rings. The Spörer minimum and the similar *Maunder minimum coincide with a period of lower temperatures on Earth known as the *Little Ice Age*. It was discovered by the American solar physicist John Allen ('Jack') Eddy (1931–2009), who named it after G. F. W. *Spörer.

Spörer's law The drift of the average latitude of sunspots towards the Sun's equator during the 11-year sunspot cycle. Spots of a new cycle are generally located at latitudes 30–40° north or south, but over the course of the cycle the average latitude decreases to 5–10°, as demonstrated by the *butterfly diagram. The 'law' is named after G. F. W. *Spörer, but was actually first noticed by R. C. *Carrington.

spray An explosive solar *prominence, associated with a flare, with very high velocity—up to 2000 km/s, faster than the escape velocity of the Sun. The material of a spray appears fragmented, sometimes in a loop structure.

spring equinox Another name for the *vernal equinox.

Springfield mounting A form of *equatorial mounting that provides a fixed, downwards observing position along the polar axis. The instrument is a *Newtonian telescope, with its *diagonal reflecting the beam through the *declination axis; a *tertiary flat* then reflects the light up along the polar axis. The observing position is particularly comfortable and convenient. However, a large counterbalance is needed to ensure that the declination axis is at the centre of gravity of the tube, and the three reflections result in a dimmed, laterally reversed image. It is named after the town of Springfield, Vermont, where it was invented by American telescope-maker Russell Williams Porter (1871–1949).

spring tide The tide raised in the Earth's oceans at full or new Moon, when the Moon and Sun are pulling on the Earth in line. The amplitude of spring tides (i.e. the difference in water level between high and low tide) is the greatest in the monthly tidal cycle.

s-process A type of nuclear reaction that has been proposed to explain the formation of stable, neutron-rich isotopes of certain heavy elements including tin, cadmium, indium, and antimony. It involves the capture of neutrons by elements such as iron sufficiently slowly (in contrast to the *r-process) for the resulting isotope, if unstable, to have time to decay to a stable isotope before it captures another neutron. This requires a relatively weak flux of neutrons, as found in red giants between shells of nuclear burning.

sputtering A process in which atoms or molecules are removed from the surface of a solid object, such as an interstellar dust grain, by the impact of high-energy atoms and ions; or, a process in which a thin metallic film is deposited on to a surface, such as a telescope mirror, in a vacuum. Sputtering may also occur on asteroid surfaces through the impacts of dust particles.

Square Kilometre Array (SKA) An internationally owned and operated radio telescope that, when completed, will have a collecting area of 1 million square metres, a hundred times greater than the *Very Large Array. At the heart of the SKA will be three core stations each 5 km across, one containing several thousand dish antennas of about 15 m aperture and the other two consisting of *phased arrays for low and medium frequencies; together, these three core stations will comprise about 50% of the SKA's total collecting area. Around this central region smaller stations consisting of a mixture of dishes and phased arrays will be arranged along spiral arms out to about 180 km. The remaining stations, each with about 20 dish antennas, will be positioned across continental distances, out to 3000 km or more. The SKA will operate over a broad spread of wavelengths from about 4 m (70 MHz) to 3 cm (10 GHz) and eventually to at least 1.2 cm (25 GHz). The final site for the SKA was expected to be selected in 2012, with completion by 2024.

(⊕) SEE WEB LINKS
• Official telescope website.

Square of Pegasus A large asterism whose four corners are marked by the stars Alpha, Beta, and Gamma Pegasi, and Alpha Andromedae.

SS Cygni The brightest *dwarf nova, varying from 8th to 12th magnitude, and lying about 500 l.y. away. It is a binary system, consisting of a main-sequence G star and a white dwarf. SS Cygni has an orbital period of 6.5 hours and undergoes an outburst every 50 days on average, but there is a large spread in the intervals.

SS Cygni star A type of *U Geminorum star that exhibits distinct, well-defined outbursts, with an amplitude of 2–6 mag.; abbr. UGSS. The rise time is 1–2 days, with a rather longer decline. The mean interval between outbursts ranges between ten and thousands of days. There are variations in the amplitude, duration, and shape of individual maxima. Although most stars have distinct short and long outbursts, the latter bear little resemblance to the supermaxima of an *SU Ursae Majoris star.

SS433 An X-ray binary about 17000 l.y. away in the constellation Aquila, consisting of a normal star with a compact companion, either a neutron star or black hole. It lies in an old supernova remnant designated W50, which is presumably the debris of the event that produced the compact object. The binary has an orbital period of 13 days and appears optically as a variable star, V1343 Aquilae, of 14th magnitude. The light from an accretion disk around the compact object makes it difficult to identify the spectral type of the normal star, but it is probably of a few solar masses. SS433 is notable for the twin jets of gas being expelled from the face of the system's accretion disk at 78000 km/s (about one-quarter the speed of light). Precession of the accretion disk causes the jets to sweep out a 40° cone which is perpendicular to the face of the disk. The jets are visible at radio, optical, and X-ray wavelengths.

S star A red giant of spectral type K5–M showing bands of zirconium oxide (ZrO) in its spectrum; sometimes also called a *heavy-metal star* or, rarely, a *zirconium star*. Oxides of lanthanum, yttrium, and barium are also found in place of the usual scandium and vanadium. These elements must have been created in the core of the star by nucleosynthesis during its later evolutionary stages. The usual sequence of temperature subtypes is applied, S0–S9, but luminosity classes are difficult to assign to S stars. An alternative classification based on the carbon–oxygen ratio is often used, and given as a number in the range 1–10 following a separator symbol; for example, the long-period variable star Chi Cygni has spectral type S6/1 at maximum brightness. The majority of S stars are irregular or long-period variables. Related subtypes are *MS stars* (*M stars with zirconium oxide in their spectra) and **SC stars**, which are intermediate between S stars and *carbon stars.

ST Abbr. for *standard time.

standard atmosphere A model of the vertical structure of the Earth's atmosphere, approximating to a global, year-long average of the changing conditions of temperature and pressure with altitude. The model takes account of the subdivision of the atmosphere into distinct levels, including the *troposphere, *stratosphere, *mesosphere, and *thermosphere, and the boundaries between them.

standard epoch A date for which star coordinates and other data are calculated in star catalogues; also known as a *fundamental epoch*. The standard place of a star is its right ascension and declination referred to the mean equator and equinox of the chosen epoch. By referring all star positions to a standard date, the effects of proper motion, precession, and nutation are removed. The standard epoch currently used is the *Julian epoch J2000.0, which corresponds to the date 2000 January 1.5 TDB.

standard solar model A mathematical description of the Sun's structure that gives the variation of pressure, temperature, and other quantities with radius. It is obtained by considering how a mass of gas simulating the Sun at its birth evolves to its present state, with radius and luminosity matching measured values. The most recent models give a core temperature of 15.6 million K and a density of $148\,000\,\text{kg/m}^3$.

standard star A star used to calibrate observations of previously unstudied stars, particularly for photometry. In *spectrophotometry, standard stars are compared with a black-body source located close to the telescope so that the amount of radiation emitted at each wavelength is known. In conventional photometry, standard stars have accurately known magnitudes and colours, against which stars under study are compared. Each system of photometry (e.g. *Johnson, *Kron–Cousins RI, or *Strömgren photometry) has its own set of standards which have been carefully intercompared. Standard stars must be bright enough to be easily observable with small telescopes, but not so bright as to saturate photometers on large telescopes.

standard time The civil time within a country, or within a time zone on Earth. It usually corresponds to the *mean solar time on some meridian of longitude within the country. Standard time usually differs from Universal Time by an exact number of hours, although in some countries half-hour differences are used.

standstill An interval during which a *Z Camelopardalis star, a subtype of dwarf nova, remains at approximately constant intermediate brightness, and normal outbursts are suppressed. Standstills are of unpredictable occurrence and duration, but normally begin on the decline from an outburst.

star A luminous ball of gas that, at some stage of its life, produces energy by the nuclear fusion of hydrogen to form helium. The term thus not only includes stars such as the Sun, which is currently burning hydrogen, but also protostars, not yet hot enough for such burning to have begun, and various evolved objects such as giant and supergiant stars, which are burning other nuclear fuels, or white dwarfs and neutron stars, which consist of spent nuclear fuel.

The maximum mass of a star is about 120 solar masses, above which it would be blown apart by its own radiation. The minimum mass is 0.08 solar masses; below this, objects never become hot enough at their cores for hydrogen burning to begin, and instead become *brown dwarfs. Star luminosities range from about half a million times the Sun's luminosity for the hottest to under one-thousandth of the Sun's for the faintest dwarfs. Although the most prominent stars visible to the naked eye are more luminous than the Sun, most stars are actually fainter than the Sun and hence insignificant to the naked eye.

Stars shine as a result of the conversion of mass into energy through nuclear reactions, of which those involving hydrogen are the most important. For every kilogram of hydrogen thus burnt, about 7 grams of mass is converted to energy. According to the famous equation $E = mc^2$, the 7 grams provides an energy of 6.3×10^{14} joules. Nuclear reactions not only provide the stars' heat and light, they also produce elements heavier than hydrogen and helium. These heavy elements have been distributed throughout the Universe either by supernova explosions or via planetary nebulae and stellar winds.

Stars are classified in a number of ways. One way is by their evolutionary stage: into pre-main-sequence, main-sequence, giant, supergiant, white-dwarf, or neutron stars. Another classification is in terms of their spectra, which indicate their surface temperature (*see* MORGAN–KEENAN CLASSIFICATION). Another way is into Populations I, II, and III, which represent stars with progressively lower abundances of heavy elements, indicating progressively greater age. *See also* STELLAR EVOLUTION; Tables 4 and 5, Appendix.

starburst galaxy A galaxy whose energy output is dominated by radiation from recently formed stars. The implied rate of star formation is much greater than could be sustained for the whole of the galaxy's life, so the burst of activity must be transient. Radiation is emitted mainly in the ultraviolet region from hot, young massive stars, but this radiation is absorbed and re-emitted by dust in interstellar space to give very high luminosities at far infrared wavelengths. The *Infrared Astronomical Satellite, IRAS, discovered the first starburst galaxies, some of which are the most luminous galaxies known, radiating at 10^{14} solar luminosities in the far infrared. Prominent examples of starburst galaxies are M82, which has undergone an encounter with its larger neighbour M81, and the pair of colliding galaxies known as the *Antennae. 📷

star cluster A collection of stars formed together from the same gas cloud and bound together by mutual gravitational attraction. There are two main types of star cluster. *Globular clusters contain from tens of thousands to millions of old stars packed into a near-spherical ball, and are found in the haloes of galaxies. *Open clusters are much more sparse and irregular in shape, consisting of no more than a few hundred relatively young stars in the spiral arms of galaxies. A group looser than an open cluster is known as a stellar *association.

star count The number of stars in a specific area of sky and within a given range of magnitudes. Star counts are used extensively in statistical studies of the distribution of stars in the Galaxy.

Stardust A NASA space probe launched in 1999 February to collect dust from Comet 81P/Wild 2 and return it to Earth. The dust particles were collected on the surface of a paddle coated with a low-density glass foam called *aerogel*. On its way to the comet, Stardust collected samples of interstellar dust passing through the Solar System and in 2002 November flew past asteroid (5535) Annefrank. In 2004 January Stardust passed within 240 km of the nucleus of Comet Wild 2, photographing the nucleus and catching dust particles. A return capsule carrying the dust samples parachuted to Earth in 2006 January. Following this successful encounter, the spacecraft was retargeted to fly past Comet 9P/Tempel 1 in 2011 February, and was renamed Stardust-NExT (short for New Exploration of Tempel 1).

(⊕) SEE WEB LINKS
- Official Stardust website.
- Stardust-NEXT website.

Stark broadening The broadening or splitting of lines in a spectrum by the electric field of electrons and ionized atoms. The effect is much more pronounced in main-sequence stars, which have higher densities of ions and electrons in their atmospheres (hence raising the frequency of collisions between atoms), than in the rarefied atmospheres of red giants. The hydrogen lines are sensitive to Stark broadening, and the resulting linewidths can be used to assign a luminosity class to A stars. The effect is named after the German physicist Johannes Stark (1874–1957).

starquake A proposed explanation for the abrupt disturbances (*glitches) observed in the regular trains of pulses from certain pulsars. According to the starquake theory, a rapidly spinning neutron star will be slightly oblate, and as it slows down it will become more spherical. As it does so, the solid crust will undergo sudden fracturing and readjustments called starquakes, similar to earthquakes.

starspot The stellar equivalent of a sunspot: a region of enhanced magnetic field strength and low surface brightness. The existence of starspots explains certain types of variability, as occurs for example in a *BY Draconis or *RS Canum Venaticorum star.

star streaming A feature of the statistical properties of stellar motions. Instead of being randomly distributed, the peculiar motions of stars in the Sun's neighbourhood appear to favour two opposite directions, or **star streams**, roughly towards and away from the centre of our Galaxy. This arises because the orbits of stars around our Galaxy are not perfectly circular. The phenomenon of star streaming was first noticed by J. C. *Kapteyn in 1904, and was early evidence that the Galaxy is rotating, although it was not recognized as such at the time.

static limit The region around the event horizon of a rotating black hole (a *Kerr black hole) within which it is impossible for any object to remain stationary; also known as the **stationary limit**. Within the static limit, an object would be forced to spin with the black hole, as though in a whirlpool. The static limit touches the event horizon at the black hole's poles, where there are no rotational forces, but becomes larger than the event horizon towards the black hole's equator. The static limit is the outer boundary of the *ergosphere. In a non-rotating black hole, the static limit coincides with the event horizon.

static universe A universe that is neither expanding nor contracting. A. *Einstein constructed such a model by adding the *cosmological constant to his equations of general relativity. Such a universe is incompatible with the observed *Hubble law. *See also* EXPANDING UNIVERSE.

stationary point 1. A position at which a planet temporarily does not move in right ascension on the celestial sphere. Planets exterior to the orbit of the Earth reach a stationary point when changing from direct motion to retrograde motion and back again.
    2. A point in space at which a body can remain in equilibrium with other bodies, maintaining a constant distance from them, for example a *Lagrangian point.

statistical parallax The mean parallax of a group of stars, found by analysing their peculiar motions (i.e. their proper motions with the effect of the Sun's motion removed).

steady-state theory A cosmological model of a universe which is expanding but has the same density at all times due to the *continuous creation of matter. The steady-state theory is based on the *perfect cosmological principle, which requires the universe to be the same at all times, as well as in all places. The mathematical solution of the equations of general relativity that results from this principle is the *de Sitter universe. For many years the steady-state theory was a rival to the Big Bang theory, but it has fallen out of favour because it is inconsistent with the observed properties of the *cosmic microwave background and cannot explain the evolution in the appearance of our Universe with time. Unlike the Big Bang theory, the steady-state theory has no initial *singularity and does not require the existence of a *primordial fireball phase. It was put forward in 1948 by H. *Bondi, T. *Gold, and F. *Hoyle.

Stefan–Boltzmann constant (σ) A constant (appearing in the *Stefan–Boltzmann law) that relates the luminosity of a black body to its thermodynamic temperature in kelvin. It has the value $5.670\,373 \times 10^{-8}$ W/m^2/K^4. Also called the **Stefan constant**.

Stefan–Boltzmann law A law relating the energy emitted by a black body (such as a star) to its temperature; also known as **Stefan's law**. According to the law, the total energy radiated in watts per square metre is proportional to the fourth power of its thermodynamic temperature in kelvin; hence a doubling of temperature leads to a sixteen-fold increase in energy output. Expressed mathematically, $E = \sigma T^4$, where σ is the *Stefan–Boltzmann constant. The total power per square metre can vary from 3 μW for the microwave background radiation, to 75 MW for the Sun, and thousands of gigawatts for hot stars such as white dwarfs. The law was discovered by Joseph Stefan (1835–93) and derived theoretically by Ludwig Edward Boltzmann (1844–1906).

Stefan's law An alternative name for the *Stefan–Boltzmann law.

stellar association *See* ASSOCIATION, STELLAR.

stellar atmosphere The low-density outer region of a star that is mostly transparent to light. Most stellar atmospheres consist of a *photosphere, *chromosphere, *transition region, and *corona analogous to those of the solar atmosphere, although a number of evolved stars, as well as B and A main-sequence stars, do not have transition regions or coronas. Atoms and ions in the photosphere and chromosphere absorb radiation at specific wavelengths, giving rise to dark *absorption lines in the stars' spectra, notably the *Fraunhofer lines. Absorption lines provide information about the chemical composition, pressure, temperature, rotation, and magnetic field strength in the stars' atmospheres.

stellar evolution The series of changes that stars undergo during their lifetimes, the time-scale of which depends strongly on the star's mass and also, to some extent, on its initial composition. The progress of a star during its evolution can be followed on a graph called the *Hertzsprung–Russell (HR) diagram.

A star is born when a dense region of a cloud of gas collapses under its own gravity. A star first shines because the gravitational potential energy lost in this collapse is released as heat and light. Eventually the temperature at the centre of the *protostar reaches 1 million K, igniting nuclear reactions involving deuterium (an isotope of hydrogen), and for some time the energy from this is sufficient to prevent further collapse. Once the deuterium has been exhausted, the collapse continues, and the star is classified as a pre-main-sequence object, following a characteristic path on the HR diagram (see HAYASHI TRACK; HENYEY TRACK). For a star the mass of the Sun, this phase lasts several million years.

Eventually the core of the star reaches temperatures of around 10 million K, hot enough to initiate the nuclear reactions that convert hydrogen to helium, and the star joins the *main sequence on the HR diagram. This hydrogen-burning phase will last from a few million years, in the most massive stars, to (potentially) more than the present age of the Universe for low-mass stars. Once the hydrogen in the core has been exhausted, the core contracts under its own gravity until, in stars of more than 0.4 solar masses, the core temperature reaches 100 million K, initiating further reactions which transform helium into carbon (the *triple-alpha process).

Subsequent evolution depends on the star's mass. In stars of similar mass to the Sun and greater, while helium burning proceeds hydrogen burning may continue in a shell outside the core. In this post-main-sequence phase the star is cooler, larger, and brighter than it was on the main sequence, and is classified as a *giant or, for the most massive stars, a *supergiant. Once the helium in the core is exhausted, the process of core contraction, followed by the initiation of a new set of nuclear reactions, may be repeated several times. Thus the more massive giants and supergiants can develop a layered structure, with the heaviest fuel burning in the centre and overlying layers containing lighter fuels from previous burning cycles. Throughout these processes the stars become larger and brighter. Eventually, however, either the contraction of the core fails to bring about a high enough temperature for further nuclear reactions or, in supergiants, the point is reached at which the core consists of iron, which cannot be used as a nuclear fuel. At this point, with no more energy being produced at the star's centre, the core collapses. The collapsing core becomes a neutron star, or possibly a black hole, while the outer layers are ejected explosively in a Type II *supernova explosion.

In less massive stars, evolution proceeds rather differently, in part because their cores are dense enough for *degeneracy effects to be important. When helium ignites in a degenerate core it does so explosively in a *helium flash, causing the core to expand. Thereafter, with the star on the *horizontal branch of the HR diagram, helium continues to burn non-explosively in the core while hydrogen burns in a surrounding shell. Once helium is exhausted in the core, it continues to burn in a shell during the *asymptotic giant branch phase. Details of later evolutionary phases are uncertain. However, it is thought that the outer layers of the red giant are puffed off to form a *planetary nebula, leaving the core of the star exposed as a *white dwarf. Hence the end-point of stellar evolution, in both high- and low-mass stars, is that much of the star is dispersed into interstellar space, leaving a collapsed remnant of spent nuclear fuel.

stellar interferometer An apparatus devised by A. A. *Michelson which, in its original form, consisted of two pairs of flat mirrors on arms mounted across the top of the 100-inch (2.5-m) Hooker Telescope at Mount Wilson so as to increase its effective aperture. Interference fringes produced when the light was combined in the telescope allowed the diameters of stars to be estimated. A modern variant uses telescopes separated by many metres as collectors, with their light brought to a central point where techniques similar to those of aperture synthesis in radio astronomy are used to detect fine detail on stars.

stellar population See POPULATION, STELLAR.

stellar structure The calculated properties of the interior of a star. Although only the properties of a star's surface can be observed directly, these can be used to calculate how temperature, pressure, density, and so on vary throughout the star. The equations used assume that the star is in equilibrium: that at each point within it, the outward pressure of the star's radiation balances the inward pull of gravity, and that the rate at which energy is generated matches the rate at which energy is transported outwards.

stellar wind The outflow of gas from the surface of a star. The Sun itself loses about 10^{-14} of its mass through the *solar wind each year, but winds are of much greater importance in pre-main-sequence stars (e.g. *T Tauri stars) and in giants and supergiants. Extreme stellar winds are found in X-ray binaries, in which stars of spectral type O or B can lose around 10^{-6} of a solar mass per year to a compact companion (a neutron star or black hole). Stars of intermediate mass can lose up to 10^{-4} solar masses per year through stellar winds as they evolve from the *asymptotic giant branch stage into a *planetary nebula.

Stephano An outer retrograde satellite of Uranus, at a distance of 7 942 000 km; also known as Uranus XX. It orbits every 675.7 days at an inclination of $141°.5$ to the ecliptic. It is about 32 km in diameter and was discovered in 1999 from ground-based observations.

Stephan's Quintet A group of five apparently interacting galaxies, one of which shows a very different redshift from the other four. These four (NGC 7317, 7318A and B, and 7319) are probably physically associated, with redshifts implying a recessional velocity of 5700 to 6700 km/s, while NGC 7320 has a redshift implying a recessional velocity of 800 km/s and is now recognized to be a foreground object superimposed by chance. Hence the group should perhaps be renamed **Stephan's Quartet**. The group, which lies about 250 million l.y. away in Pegasus, is named after the French astronomer Édouard Jean Marie Stephan (1837–1923), who discovered it in 1876. 📷

step method *See* ARGELANDER STEP METHOD; POGSON STEP METHOD.

steradian (sr) The unit of solid angle: an angle in three dimensions, as for example subtended by the base of a cone. One steradian encloses an area on the surface of a sphere that is equal to the square of the sphere's radius. The solid angle subtended at the centre of a sphere by its entire surface is 4π steradians.

STEREO Abbr. for *Solar Terrestrial Relations Observatory.

stereo comparator A device in which each of two photographic plates to be compared is viewed through separate eyepieces arranged side by side as in binoculars. Any object which has changed position slightly on the plates will have a stereoscopic appearance. This instrument is particularly suited to locating asteroids.

Steward Observatory The observatory of the University of Arizona, founded at Tucson in 1922 and named after Lavinia Steward, its benefactor. It currently operates telescopes at five main sites in Arizona. Steward Observatory's site on Kitt Peak, at an altitude of 2070 m, was founded in 1963. It has three telescopes: the 2.3-m Bok reflector opened in 1969, named after former Steward Observatory director B. J. *Bok; a 0.9-m reflector opened in Tucson in 1922 and moved to Kitt Peak in 1963; and a 1.8-m, opened in 2000; these latter two telescopes are used in the *Spacewatch project. Also on Kitt Peak, Steward Observatory operates a 12-m millimetre-wave dish, on loan from the National Radio Astronomy Observatory; this dish was opened in 1967 and upgraded in 1984, and transferred to Steward Observatory in 2000. Steward Observatory's Mount Lemmon Station in the Santa Catalina Mountains, altitude 2780 m, lies 38 km northeast of Tucson. Here, Steward Observatory operates a 1.5-m and a 1-m reflector, both opened in 1970; the University of Minnesota operates a 1.5-m reflector at the same site, also opened 1970. At the Mount Bigelow Station in the Santa Catalina Mountains, altitude 2510 m, Steward Observatory operates the 1.55-m Kuiper Telescope, opened in 1965, and a 0.68-m Schmidt, opened 1965 and upgraded in 2003. Steward Observatory jointly owns and operates the *MMT Observatory on Mount Hopkins; it also owns and operates the *Mount Graham International Observatory and is a partner in the *Magellan Telescopes.

(🌐) **SEE WEB LINKS**
• Official observatory website.

stishovite A rare mineral that is produced when a particularly high-pressure shock wave meteorite impact passes through rock containing quartz. It is a high-density form of silica produced at a pressure of 130000 bar. Stishovite-bearing fragments in the vicinity of large craters are taken as evidence that the crater is of meteoritic origin. It is named after the Russian mineralogist Sergei Mikhailovich Stishov (1937–) who helped synthesize it in 1962. *See also* COESITE.

Stokes parameters Four quantities used to describe the polarization properties of an electromagnetic wave: the total intensity, I; the circular polarization, V; and the components of linear polarization, Q and U. In radio astronomy, the Stokes parameters are normally measured in janskys. They are named after the British mathematician and physicist George Gabriel Stokes (1819–1903).

Stonyhurst heliographic coordinates One of two *heliographic coordinate systems used for identifying the position of features on the Sun's surface. In the Stonyhurst system the zero point is set at the intersection of the Sun's equator and central meridian as seen from the Earth. Longitude increases towards the Sun's western limb. A solar feature will have a fixed latitude as it rotates across the solar disk, but its longitude will increase. This is in contrast to the *Carrington heliographic coordinate system, where the longitude remains approximately fixed in time. Stonyhurst coordinates are named after Stonyhurst College in Lancashire, England, where they were devised in the 19th century.

stony-iron meteorite A meteorite composed of nickel–iron and silicate minerals in roughly equal proportions; also known as a *siderolite*. Such meteorites were apparently formed within asteroid-sized parent bodies. Stony-iron meteorites comprise only about 1% of observed falls. Four groups of stony-iron meteorites have been defined, the two main ones being the *pallasites and *mesosiderites. The other two, the *lodranites and *siderophyres, are extremely rare.

stony meteorite A meteorite composed primarily of silicate minerals, usually with some nickel–iron; also known as an *aerolite*. Around 95% of all meteorites observed to fall are stones. However, they are difficult to distinguish from terrestrial rocks, and many go unnoticed. Their fusion crust is thicker than that of irons and is often black in colour, either dull or shiny. The interior of stones is usually grey or dark grey, and granular in texture. Stony meteorites are divided into two main subtypes: *chondrites, which comprise 82% of all observed falls, and *achondrites.

stratigraphy The branch of geology concerned with the formation, chronology, and correlation of strata deposited on the surface of a planet or other body. Through the use of stratigraphy, the geological history of the body can be worked out.

stratopause The upper boundary of the *stratosphere, immediately underlying the *mesosphere at an altitude of 50 km above the Earth's surface. Stratospheric temperatures reach a maximum of about 0°C at the stratopause.

stratosphere The layer of the Earth's atmosphere immediately above the *troposphere, extending to the *mesosphere (i.e. between altitudes of 10–15 km and 50 km). In contrast with those in the troposphere, temperatures in the stratosphere rise with increasing altitude, thus suppressing convection. The concentration of water vapour in the stratosphere is low, and clouds are rare. During the winter months, *nacreous* or *polar stratospheric clouds* may occasionally form at high latitudes, around altitudes of 15–30 km. The *ozonosphere, including the ozone layer, lies within the stratosphere, and there is also some overlap with the lower part of the *chemosphere.

Stratospheric Observatory for Infrared Astronomy (SOFIA) A NASA Boeing 747 SP aircraft modified to carry a 2.5-m reflector built by the German Aerospace Center, DLR. It began operation in 2010. SOFIA flies at altitudes of about 13 km to make infrared observations at wavelengths from 0.3 to 1600 μm. SOFIA is based at NASA's Dryden Aircraft Operations Facility in Palmdale, California, and is operated jointly by NASA and DLR.

(⊕) SEE WEB LINKS

• Official telescope website.

streamer A large-scale structure seen in white-light images of the Sun's corona. Streamers are brighter than the surrounding corona because they are of higher density, and are often bottle-shaped, narrowing away from the Sun. They are also referred to as *helmet streamers* because of their resemblance to a helmet with a spike on top. The broad base of the streamer is where coronal plasma is trapped along closed magnetic field lines, but at larger distances the magnetic field weakens and the plasma is able to break free from the Sun, producing the narrow spike. Streamers are principally found above active regions and prominences, but also occur at solar minimum when there are few or no active regions. In the latter case, two streamers are seen on opposite sides of the Sun, close to the equator. They are a projection of a **streamer belt** that encircles the Sun. At the poles of the Sun during solar minimum the magnetic field opens out directly into space, and narrow *coronal plumes are found to lie along these field lines.

Strehl ratio A measure of the quality of an optical system. It is the ratio of the amount of light actually delivered by an optical system into the *Airy disk compared with the theoretical maximum. Perfect optics have a Strehl ratio of 1.0. Optics that meet the *Rayleigh criterion have a Strehl ratio of 0.82. The ratio is named after the German physicist Karl Wilhelm Andreas Strehl (1864–1940).

strewn field 1. An area on Earth where tektites are found. The largest strewn field covers the whole of southern Australia and Tasmania (the *australites). Other major strewn fields are found in the Czech Republic (the *moldavites), Africa (the Libyan Desert glass and Ivory Coast tektites), the USA (the *bediasites, *georgiaites, and Martha's Vineyard tektites), across Southeast Asia (the *billitonites, *indochinites, javanites, malaysianites, philippinites, and rizalites), and in central Russia (the irghizites).
2. The area over which pieces of a fragmented meteorite are scattered, also known as a *scatter ellipse.

string *See* COSMIC STRING.

Strömgren, Bengt Georg Daniel (1908–87) Swedish astronomer. After working with his father, Svante Elis Strömgren (1870–1947), he studied stellar atmospheres and interstellar gas clouds, in particular H II regions. From this work evolved the idea of the *Strömgren sphere—a region of ionized gas surrounding a hot star. His later studies were concerned with the changes in chemical composition undergone by stars during their evolution. He developed *Strömgren photometry.

Strömgren photometry A system of *intermediate-band photometry in which the names, wavelengths, and bandwidths of the filters are: u, 352 and 31 nm; v, 410 and 17 nm; b, 469 and 18 nm; and y, 548 and 23 nm. Strömgren photometry is often supplemented by two interference filters to measure the strength of the Hβ line (*see* HYDROGEN SPECTRUM). They share the same central wavelength of 486 nm, but have different bandwidths of 15 and 3 nm. The difference in the two magnitudes is the Hβ index, symbol β. The different colours can be combined to determine the temperature, luminosity, and reddening of stars of spectral types O and B. The metallicity can also be determined for stars of spectral types A and F. The system is named after its originator, B. G. D. *Strömgren.

Strömgren sphere The region of ionized gas surrounding a hot star. B. G. D. *Strömgren first derived the relationship between the density of the interstellar gas, the temperature of the star, and the radius of the region which it ionizes. The ionized gas is usually called an *H II region; its size is termed the **Strömgren radius**.

strontium star A form of *Ap star with enhanced strontium lines in its spectrum.

Struve Russian–German family which produced four generations of astronomers. The most important were: F. G. W. von *Struve; his son Otto W. *Struve (Otto I); and Otto I's grandson Otto *Struve (Otto II). Karl Hermann Struve (1854–1920) was the elder son of Otto I. He refined the orbits of several satellites and studied Saturn's rings. Gustav Wilhelm Ludwig Struve (1858–1920), the younger son of Otto I and the father of Otto II, continued his father's work on the motion of the Solar System and studied lunar occultations. Georg Struve (1886–1933), Hermann's son, also studied the Solar System, notably the satellites of Saturn and Uranus.

Struve, Friedrich Georg Wilhelm von (1793–1864) German astronomer, father of O. W. *Struve. In 1819 he began to observe double stars, aiming to continue where F. W. *Herschel left off. The next year he published the first of many double-star catalogues, the last of which (1852) compared separations and position angles with historical data (his catalogue numbers for doubles are still in use). Struve showed that double and multiple stars are more common than had been suspected. In 1833 Struve moved to Russia to set up the Pulkovo Observatory, near St Petersburg. He made the third reliable measurement of a stellar parallax, that of Vega, in 1840. He also refined values for astronomical constants, including the constant of aberration (*see* ANNUAL ABERRATION).

Struve, Otto (1897–1963) Russian-born German-American astronomer, grandson of O. W. *Struve. His early spectroscopic studies were on rapidly rotating stars. In the late 1930s, with the American astronomer Christian Thomas Elvey (1899–1970), he built a spectrograph for studying nebulae. They discovered interstellar hydrogen concentrated near the galactic plane, and also detected ionized hydrogen (*see* H II REGION). Struve also studied binary stars (as had his predecessors), stellar evolution, and the origin of planetary systems.

Struve, Otto Wilhelm (1819–1905) Russian-born German astronomer, son of F. G. W. *Struve. He worked at Pulkovo Observatory, where he collaborated with his father in the search for double stars, publishing catalogues of his own. He accurately determined the rate of change of precession, and estimated the velocity of the Sun in the Galaxy from the proper motions of nearby stars.

STScI Abbr. for *Space Telescope Science Institute.

style The shadow-forming edge of a horizontal or vertical plate sundial. In such a sundial the style is a narrow edge aligned at the same angle to the horizontal as the dial's latitude. Occasionally, where there is additional information on the dial's face, such as to show the length of the day, there may be a small notch or *nodus* in the style. The shadow of this notch indicates the correct reading.

Subaru Telescope A Japanese 8.2-m reflector for optical and infrared astronomy, opened in 1999 at an altitude of 4139 m on *Mauna Kea, Hawaii. It is owned and operated by the National Astronomical Observatory of Japan. Subaru is the Japanese name for the Pleiades.

(((( ))) SEE WEB LINKS
• Official telescope website.

subatomic particle Another name for an *elementary particle.

subdwarf A star that lies appreciably below the main sequence on the Hertzsprung–Russell diagram, and is somewhat bluer (i.e. of earlier spectral type) than a main-sequence star of the same luminosity; abbr. sd. Subdwarfs belong to Population II, being stars from the galactic halo which are passing through the solar neighbourhood. Their spectral types range from F to M and they differ from normal dwarfs in having a low abundance of heavy elements.

subgiant A star that has exhausted the hydrogen at its centre and is evolving into a giant. They are of *luminosity class IV. The subgiants we see are usually less massive than the Sun, because more massive stars move very quickly through this stage into giants. Low-mass stars take many billions of years to evolve this far, so low-mass subgiants are very old. A **subgiant branch** on the Hertzsprung–Russell diagram, linking the main sequence to the giant branch, is therefore found only for old clusters such as globular clusters.

subluminous star A star that is less luminous than a main-sequence star of the same mass. Subluminous stars include *subdwarfs, *white dwarfs, and *brown dwarfs.

sublunar point The point on Earth at which the Moon is directly overhead at a particular time.

Submillimeter Array (SMA) A radio interferometer jointly owned by the Smithsonian Astrophysical Observatory and the Academia Sinica Institute of Astronomy and Astrophysics of Taiwan, situated at an altitude of 4080 m on Mauna Kea, Hawaii. It consists of eight dishes each of 6 m aperture, arranged in an approximately triangular shape with baselines up to 509 m. The SMA was opened in 2003 and operates at wavelengths of 0.4 to 1.7 mm (700–180 GHz). It can also work with the nearby *James Clerk Maxwell Telescope and the *Caltech Submillimeter Observatory to form an extended array (eSMA) with higher resolution and more than twice the collecting area of the SMA alone.

((⊕)) SEE WEB LINKS
• Official telescope website.

Submillimeter Wave Astronomy Satellite (SWAS) A NASA satellite launched in 1998 December to study regions of star formation. It carried an elliptical radio telescope 0.55 × 0.71 m which collected submillimetre waves emitted by water, oxygen, carbon, and carbon monoxide in *molecular clouds. SWAS ceased operations in 2005.

((⊕)) SEE WEB LINKS
• Official mission website.

submillimetre astronomy The study of the sky at radio wavelengths shorter than 1 mm, specifically 0.3–1.0 mm. As with *millimetre-wave astronomy, this part of the spectrum is rich in lines emitted by interstellar molecules. This region was previously considered part of the far infrared, but is now regarded as an extension of the radio spectrum since instrumental techniques developed to investigate it have more in common with radio astronomy than infrared astronomy. Major submillimetre telescopes include the 15-m *James Clerk Maxwell Telescope and the *Atacama Large Millimeter Array.

sub-pulse One of several distinct bursts of radiation making up a single pulse from a pulsar. Sub-pulses are themselves composed of many brief flashes (*micropulses*), which are believed to be the basic components of the pulsed emission. Occasionally a sub-pulse is seen to change its position gradually within the pulse, giving the appearance of drifting (a *drifting sub-pulse*), but the cause is poorly understood.

subsolar point The point on the Earth, or other body, at which the Sun is directly overhead at a particular time.

substellar object An object with a mass too small to ignite hydrogen and thus become a true star, such as a *brown dwarf.

substellar point The point on the Earth, or other body, at which a particular star is directly overhead at a given time.

substorm An intensification of activity along the *auroral oval at high latitudes, resulting from a relatively minor geomagnetic disturbance. Substorm activity initially takes the form of a bright arc on the night-side of the oval, then spreads polewards and westwards before subsiding. The associated ground-level magnetic effects can be detected with sensitive magnetometers. There may be as many as five substorms per day, each lasting 1–3 hours.

sudden ionospheric disturbance A rapid enhancement in the ionization of the dayside *D layer of the ionosphere following a solar flare, resulting in disruption of radio communications. Strong X-ray emission from the flare causes increased ionization which absorbs radio waves, leading to signal fade-outs. Sudden ionospheric disturbances are most common around sunspot maximum, when there may be as many as 25 solar flares per day.

al-Ṣūfī, 'Abd al-Raḥmān (903–86) Arab astronomer (Latinized name Azophi), born in modern Iran. His *Book of the Fixed Stars* (*c.*964) contained a star catalogue and atlas based on Ptolemy's *Almagest*, revised in the spirit of Arab astronomy and with Arab star names that are still in use, albeit in corrupted form. It incorporated al-Ṣūfī's own observations, and contains the earliest known reference to the Andromeda Galaxy.

Suisei A Japanese space probe to Halley's Comet, launched in 1985 August; its name means 'comet' in Japanese. It passed 150000 km from the sunward side of the comet's nucleus in 1986 March, observing the comet's hydrogen halo. *See also* SAKIGAKE.

sulcus A complex network of parallel linear depressions and ridges on a planetary surface; pl.*sulci*. The name, which means 'furrow', is not a geological term, but is used in the nomenclature of individual features, for example the Sulci Gordii on Mars, or the Harran Sulci on Enceladus.

summer solstice *See* SOLSTICE.

Summer Time The version of *civil time that is applied in the summer months to allow better use of the daylight hours; also known as Daylight Saving Time. In most countries, clocks are put forward one hour relative to standard time in the spring and back again to standard time in the autumn.

Summer Triangle A large triangle in the northern sky formed by the first-magnitude stars *Vega, *Altair, and *Deneb. It is best seen during northern summer and autumn evenings.

Sun (☉) The central body of the Solar System, and by far the nearest star—the only one that can be studied in great detail. It is classified as a G2V star: a yellowish star with an *effective temperature of 5770 K (spectral type G2) and a *main-sequence dwarf (luminosity class V). Its apparent visual magnitude is −26.7, but its absolute magnitude is only +4.82. The Sun is largely hydrogen (71% by mass), with some helium (27%) and heavier elements (total 2%). Its age is estimated to be about 4.6 billion years. The energy produced by nuclear reactions at its core is transferred to the surface layer, or *photosphere, by radiation through the inner two-thirds of its radius and thence by convection for the outer one-third. The transfer of energy from the core to the surface takes 10 million years. At its centre the temperature is calculated to be 15.6 million K and the density 148 000kg/m³. The vast majority of its energy escapes to space from the photosphere. The photosphere is marked by *granulation and *supergranulation, both of which are relatively small-scale convection currents, and features related to *solar activity, such as *sunspots and faculae, which are associated with regions of strong magnetic field. The Sun's sidereal rotation period is about 25 days at the equator, and 27–28 days at latitude 40°; near the poles the period is 33.5 days. The adopted mean value, corresponding to latitude 17°, is 25.38 days.

Sun: Physical Data			
Diameter	Inclination of equator to ecliptic	Mean axial rotation period (sidereal)	Mean density
1 392 530 km	7°.25	25.38 d	1.41 g/cm³
Mass	Luminosity	Volume (Earth = 1)	Escape velocity
1.989 × 10³⁰ kg	3.85 × 10²⁶ W	1.3 × 10⁶	617.3 km/s

The photosphere is only a few hundred kilometres thick, and its temperature steadily decreases with height to about 4400 K at the *temperature minimum. Above this is the *chromosphere, where the temperature is between that of the temperature minimum region and about 20000 K. There is a rapid rise of temperature with height—in the *transition region—towards the *corona, where the temperature is 2 million K or more.

The number of active regions on the Sun follows an 11-year cycle. The magnetic polarities of pairs of sunspots, which are opposite in the leading and following parts, are reversed in each successive cycle, so that there is a 22-year magnetic cycle. A continuous stream of particles, the *solar wind, flows out into interplanetary space at 300–750 km/s, with high-speed streams emanating from *coronal holes.

sundial A device for telling *solar time from the position of the Sun's shadow thrown by a *gnomon. If the dial is a horizontal or vertical plate, the hour lines are at unequal angles to each other. A vertical dial which faces one of the cardinal points of the compass is called a *direct dial*; one which does not is a *declining dial*. An alternative design of sundial is the *armillary sphere.

sundog Popular name for a *parhelion.

Sunflower Galaxy The galaxy M63 (NGC 5055) in Canes Venatici. It is a 9th-magnitude spiral midway between types Sb and Sc. The estimated distance is 30–35 million l.y.

sungrazer *See* KREUTZ SUNGRAZER.

sunrise The instant when the upper limb of the Sun first appears above the horizon. At this instant the true zenith distance of the centre of the Sun's disk is about 90° 50′, since the centre of the Sun is 16′ below its upper limb and atmospheric refraction on the horizon is about 34′.

sunset The moment when the upper limb of the Sun disappears below the horizon. *See also* SUNRISE.

sunskirter A comet with a perihelion distance between 6 and 15 solar radii and which is a member of one of several comet families having similar orbital elements. Three major groups so far identified are the *Kracht group, the *Marsden group, and the *Meyer group. Sunskirting comets have been seen only with the coronagraphs aboard the *SOHO spacecraft. They differ from *Kreutz sungrazers in not passing quite so close to the solar surface.

sunspot A dark area on the Sun's photosphere that is cooler than its surroundings, associated with very strong (0.4 tesla) magnetic fields. Spots generally appear in pairs or groups, the leading and following spots having opposite magnetic polarities. Spot sizes vary from small *pores* about 300 km across to groups spanning more than 100000 km. The largest spots usually last longest, up to 6 months; some small spots may last for less than an hour. Sunspots are mostly confined to belts either side of the equator between about 40° and 5° latitude north and south, appearing at higher latitudes at the start of the *sunspot cycle and moving towards the Sun's equator as the cycle develops. Well-developed spots have a darker interior, the *umbra (2), about 1600 K cooler than the photosphere, and a lighter outer *penumbra (2), which accounts for up to 70% of the spot's area and is about 500 K cooler than the photosphere. All spots start their lives as tiny dark pores, and may then develop into small penumbra-less spots arranged in pairs. In a developing group, the spots become much larger and more separated in the first two days, attaining their maximum area and complexity by the tenth day. Broad categories of spot groups can be defined. In the *McIntosh scheme*, which has replaced the formerly used Zürich scheme, a three-letter code describes the class of sunspot group (single, pair, complex), penumbral development of the largest spot, and compactness of the group. The *Mount Wilson scheme* is used to describe the magnetic field structure, which may be simple (bipolar or, if a single spot, unipolar) or complex. Spots that give rise to *flares tend to be highly complex in appearance and magnetic field structure. Gas flows outwards from spots at low altitudes (the *Evershed effect), and inwards at coronal altitudes. ◙

sunspot cycle The variation in the number of sunspots and other forms of solar activity with an average period of about 11 years. In each successive cycle the north and south magnetic polarities of the Sun are reversed, so that there is a 22-year magnetic cycle. The monthly averages of *relative sunspot numbers are around 6 in years of sunspot minimum and 116 at sunspot maximum. Not only sunspot number varies, but also the numbers of *active regions and *flares, and the level of ultraviolet, radio, and other emission associated with such regions. The 11-year periodicity is thought to arise through the action of the *solar dynamo. The sunspot cycle may not always have operated in its present observed form, as suggested by the existence of the *Maunder minimum.

sunspot number *See* RELATIVE SUNSPOT NUMBER.

Sunyaev–Zel'dovich effect A phenomenon in which the apparent temperature of the cosmic microwave background is reduced when the radiation passes through hot, ionized gas in a cluster of galaxies. Paradoxically, the effect is caused by an increase of the mean photon energy due to *inverse Compton scattering, which appears as a reduction in brightness temperature of no more than 0.001 K at centimetre and millimetre wavelengths. At submillimetre wavelengths the brightness temperature increases. Measurement of the effect is difficult, but it has now been observed in more than ten clusters of galaxies. It may allow an independent determination of the Hubble constant. The effect is named after the Russian astrophysicists Rashid Alievich Sunyaev (1943–) and Yakov Borisovich Zel'dovich (1914–87), who first drew attention to it.

supercluster A grouping of clusters of galaxies. Many superclusters are known, containing anything from around 10 to more than 50 rich clusters. The most prominent known supercluster is the *Shapley Concentration*, about 450 million l.y. away in the region of Centaurus, while the nearest is the *local supercluster. Superclustering is known to exist on scales up to 300 million l.y., and possibly more. *See also* LARGE-SCALE STRUCTURE.

supergalactic plane A great circle around the sky that represents the equator of the *local supercluster of galaxies. The supergalactic plane is nearly at right angles to the disk of our Galaxy, with its north pole in the constellation Aquila.

supergiant The largest and brightest type of star, with a luminosity of about 10000–100000 times the Sun's, and a diameter from 20 to several hundred times the Sun's. Stars of at least 10 solar masses become supergiants when they swell up and leave the main sequence towards the end of their lives. They occupy a band at the top of the Hertzsprung–Russell diagram and are of *luminosity class I. Rigel and Betelgeuse are well-known supergiants.

supergiant elliptical A very luminous *elliptical galaxy in a cluster; also known as a *cD galaxy. Supergiant ellipticals may be over 10 times brighter than the next-brightest cluster member, and are often radio sources.

supergranulation A convection pattern in the Sun's photosphere, consisting of large (30000 km) convection cells within which there is a mainly horizontal flow outwards to the cell's edge, with a slight upward flow at the cell's centre and downward flow at the cell's edges. The velocity of the horizontal flow is only 0.4 km/s. A typical lifetime of a supergranule cell is a day. Small areas of strong magnetic field are transported outwards from cell centres to the edges, where the field is mostly concentrated. This appears to give rise to the *chromospheric network, consisting of clumps of *spicules.

superheterodyne receiver A receiver widely used in radio astronomy in which the incoming radio-frequency (RF) signal is mixed with a signal of lower frequency generated by a local oscillator, thereby reducing its frequency to a value (the *intermediate frequency*, IF) which can be more readily amplified and filtered. Receivers designed for high RFs may have more than one such stage. The receiver is tuned by varying the frequency of the local oscillator so that the IF is kept constant.

superhump *See* SU URSAE MAJORIS STAR.

superior conjunction The moment when an inferior planet (i.e. Mercury or Venus) lies directly behind the Sun as seen from Earth. *See also* CONJUNCTION.

superior planet Any planet whose orbit is farther from the Sun (and hence larger in radius) than that of the Earth, namely Mars, Jupiter, Saturn, Uranus, and Neptune.

superluminal velocity A velocity that apparently exceeds that of light. In some double radio sources or quasars, the components seem to be moving away from each other many times faster than the speed of light. The theory of special relativity does not allow objects to accelerate to speeds faster than that of light, but the observed superluminal velocities are thought to be illusory, caused by the two components moving in a direction very close to the observer's line of sight.

supermassive black hole A black hole formed from the collapse of an object, such as a massive gas cloud, significantly more massive than typical stars. By 'supermassive' astronomers usually mean greater than 10^5 solar masses. Such objects can increase their mass by accreting material from their surroundings, and the energy released in this process may be responsible for the activity seen in galaxies and quasars.

supermassive star A hypothetical star of over about 120 solar masses, so luminous that it would be expected to disintegrate under the outward pressure of its own radiation. Supermassive stars were proposed as an explanation for very bright objects in the Large Magellanic Cloud, but these are now known to be clusters of ordinary O stars.

supermaximum *See* SU URSAE MAJORIS STAR.

supernova A violent explosion in which certain stars end their lives; given the variable-star type designation SN. In a supernova explosion, the star may become over a billion times brighter than the Sun, and for weeks may outshine the entire galaxy in which it lies. However, the optical luminosity represents only 0.01% of the energy released in the explosion. Most of the energy emerges in the form of neutrinos, and 1% goes into the kinetic energy of the gas expelled. The last supernova seen in our Galaxy was in 1604 (*Kepler's Star), although *Supernova 1987A in the Large Magellanic Cloud reached naked-eye brightness. However, two or three supernovae are expected to occur every century in a typical spiral galaxy like ours, which suggests that many have been missed due to absorption of light by dust in the galactic plane.

Supernovae are classified into Types I and II, the latter type showing hydrogen in its spectrum, whereas the spectrum of a Type I supernova shows no hydrogen. There is further subclassification into Types Ia, Ib, and Ic according to other details of the spectrum. Type Ia supernovae reach a maximum magnitude of about −19, while Types Ib and Ic are about 1.5 magnitudes fainter. Type II have a wide range of peak magnitudes, but on average are similar to Types Ib and Ic. Types II, Ib, and Ic occur in young stars of Population I, and thus are concentrated in the disks of spiral galaxies. Type Ia supernovae occur among old stars of Population II, as are found in elliptical galaxies and the halos of spirals.

Type Ia supernovae are believed to be due to the explosion of a white dwarf in a binary as a result of matter falling on to it from its companion star. When the mass of the white dwarf eventually exceeds the *Chandrasekhar limit, it undergoes runaway carbon burning and explodes, ejecting about 1 solar mass.

Type Ib and Ic supernovae are thought to result from the collapse of the cores of massive stars which have lost their hydrogen envelopes, either through a stellar wind or by transfer of matter to a companion in a binary. Types Ib and Ic show minor differences in spectra, indicating different compositions of the progenitor stars, which are probably *Wolf–Rayet stars stripped of different amounts of their outer layers. Despite their classification, Type Ib and Ic supernovae are more closely related to Type II than to Type Ia.

Type II supernovae arise from the explosion of stars of more than 8 solar masses. Nuclear reactions cease once the star's core consists of iron and heavier elements, because these elements cannot be burnt to produce energy. The stars then collapse under their own gravity, reaching densities so high that protons and electrons combine to form neutrons, producing a neutron star or even a black hole. The formation of a neutron star causes the overlying layers of material to rebound violently. In this process much of the envelope of the original star, amounting to many solar masses, is ejected at speeds of 2000–20 000 km/s.

Type II supernovae can be subdivided into II-P, II-L, and IIb. Type II-P (for *plateau*) remain at near-constant brightness for 2–3 months after the outburst before fading. The rarer II-L (for *linear*) type fades more rapidly from an initial peak. Type IIb show a double maximum in brightness. Type IIb are thought to result from massive stars that have lost most, but not all, of their hydrogen envelope before exploding. The second peak, a few weeks after the initial outburst, is caused by the decay of radioactive nickel and cobalt in the supernova debris. *See also* SUPERNOVA REMNANT.

(⊕) SEE WEB LINKS
• Database of all known supernovae, regularly updated.

Supernova 1987A A Type II supernova that erupted in the Large Magellanic Cloud on 1987 February 24, and which reached a maximum apparent magnitude of 2.8 in mid-May of that year. It was the first supernova to reach naked-eye brightness since 1604, and remained visible to the naked eye until the end of 1987. Its progenitor was Sanduleak −69° 202, a blue supergiant, contrary to predictions that a Type II supernova arises from a massive, red supergiant. A burst of neutrinos from the supernova was detected on Earth. Gamma-ray lines from the radioactive decay of ^{56}Co to ^{56}Fe were observed, confirming that heavy elements are produced in supernovae. A ring of gas about 1 l.y. in diameter observed around the star is believed to have been shed in an earlier red supergiant phase.

supernova remnant (SNR) A diffuse nebula consisting of the remains of the outer layers of a star that have been blown into space by a supernova explosion, together with swept-up interstellar matter. The expansion velocity can approach a few per cent of the speed of light. SNRs are strong sources of X-ray emission and radio waves (*synchrotron radiation). SNRs can be classed as either **shell remnants**, such as Cassiopeia A, or the rare *filled-centre remnants* (*plerions), such as the Crab Nebula. Plerions are believed to be energized by a central pulsar, the collapsed core of the exploded star.

super-rotation A phenomenon in which the atmosphere of a planet rotates independently of the surface. It is notable on Venus, where the top of the atmosphere rotates in 4 days, while the surface beneath rotates in 243 days. The cause of super-rotation is not well understood. Several interacting processes may be at work which transport momentum from the lower to the upper atmosphere.

super-Schmidt telescope A development of the *Schmidt camera which has a very wide field of view and a *focal ratio as fast as $f/1$. This is achieved by placing a *meniscus lens on either side of the *corrector plate. It was designed for meteor photography.

superstring theory An example of a general class of theories of high-energy physics known as **string theories**. In string theories, the fundamental objects are not point-like objects (particles) but one-dimensional objects known as strings. Superstrings are strings which also have the property of **supersymmetry** (i.e. every particle which is a *boson has a partner which is a *fermion). The superstring theory tries to describe all classes of observed elementary particle in terms of properties of such strings. The theory is entirely speculative, but may have some implications for the physics of the very early Universe.

supersynthesis A variant of the *aperture synthesis technique in radio astronomy, in which the rotation of the Earth is employed to sweep out the aperture of a radio telescope. It is also known as *Earth-rotation aperture synthesis*.

surface brightness The brightness of an extended astronomical object, such as a planet, nebula, galaxy, or the sky background, expressed as magnitudes per unit area. Surface brightness is calculated by dividing the object's magnitude by its dimensions. For example, the average surface brightness of the planetary nebula M57 is given as magnitude 17.6 per square arc second, compared with 5.2 for Jupiter and 23.0 for the darkest night sky. Contours of surface brightness are termed *isophotes*.

surface gravity (g) The value of the acceleration produced in a freely falling object near the surface of a celestial body. If the body is rotating, the effect of centrifugal force on the object must be taken into account. *See also* ACCELERATION DUE TO GRAVITY.

surface temperature The temperature of the surface of a celestial body, as opposed to its underlying layers (for example, the central temperatures of stars are vastly greater than their surface temperatures). The surface temperature a star would have if it radiated exactly like a black body of the same size and energy output is called its *effective temperature. The *colour temperature of a star is found by comparing its energy distribution at different wavelengths with that of a black body; this usually differs from the effective temperature because stars are not perfect black bodies. In practice, a star's surface temperature is an average of the temperatures of a series of layers at different heights in its photosphere. A planet's surface temperature depends on the relative amounts of solar radiation absorbed by the planet and the radiation emitted by its surface.

surge prominence A type of solar *active prominence in which material ascends, then descends, almost vertically. Maximum velocities are about 200 km/s, and the surge prominences reach heights of 100000 km. They often recur in identical locations, and the more active surges occur at the onset of flares. The brightest surges last for half an hour or more.

Surveyor A series of NASA probes which soft-landed on the Moon, acting as pathfinders for the Apollo manned landings. All the probes carried TV cameras; in addition, Surveyors 3 and 7 had mechanical scoops to test the strength of the soil, and Surveyors 5, 6, and 7 carried equipment to analyse the surface composition. 📷

Successful Surveyor Probes[a]		
Probe	Launch date	Results
Surveyor 1	1966 May 30	Landed June 2 in Oceanus Procellarum
Surveyor 3	1967 April 17	Landed April 20 in Oceanus Procellarum
Surveyor 5	1967 September 8	Landed September 11 in Mare Tranquillitatis
Surveyor 6	1967 November 7	Landed November 10 in Sinus Medii
Surveyor 7	1968 January 7	Landed January 10 near crater Tycho
[a] Surveyor 2 crash-landed; radio contact with Surveyor 4 was lost shortly before it landed.		

SU Ursae Majoris star A type of *U Geminorum star that occasionally exhibits a *supermaximum* (also called a *superoutburst*), i.e. an outburst that is about 2 magnitudes brighter and five times as long as a normal maximum; abbr. UGSU. The intervals and properties of normal maxima are similar to those found in an *SS Cygni star, and are three to four times as frequent as supermaxima. Supermaxima show a pronounced periodicity, unlike all other outbursts of dwarf novae. During a supermaximum, a brightening known as a *superhump* is visible in the light-curve. This feature has an amplitude of about 0.2–0.3 mag., and a period that is a few per cent longer than the binary's orbital period (which is less than 0.1 day). The interval between supermaxima is typically a few hundred days. However, some systems, the *ER Ursae Majoris stars, have much shorter intervals (20–50 days) while others, the *WZ Sagittae stars, have very long intervals, the most extreme being WZ Sagittae itself, with an interval of around 12 000 days.

Suzaku A joint Japanese–US X-ray astronomy satellite, known as Astro-E2 before its launch in 2005 July. It observes at a wide range of energetic wavelengths, ranging from soft X-rays to gamma rays (0.2–700 keV), with two instruments: an X-ray Imaging Spectrometer (XIS), consisting of four X-ray CCDs, and a hard X-ray detector (HXD). Four grazing-incidence X-ray telescopes (XRTs) focus X-rays onto each of the four XISs. A third instrument, an X-ray spectrometer (XRS), failed shortly after launch.

(((●))) SEE WEB LINKS

- Official mission website.
- Information page at Goddard Space Flight Center.

S Vulpeculae star A yellow giant or supergiant semiregular variable (previously type SRD, but now classified DCEP) that shows intervals of reasonable periodicity, interspersed with unpredictable episodes of irregular variations.

Swan bands Features of the spectrum of the carbon molecule C_2 (diatomic carbon), first investigated by the Scottish scientist William Swan (1818–94). These bands are prominent in *carbon stars and cometary spectra. Numerous lines are present throughout the optical and red region of the spectrum, with strong features at 438, 474, and 516.5 nm.

Swan Nebula An alternative name for the *Omega Nebula.

SWAS Abbr. for *Submillimeter Wave Astronomy Satellite.

Swift A NASA satellite, with UK and Italian involvement, launched in 2004 November to detect and study *gamma-ray bursts (GRB). Swift gets its name from its ability to detect a GRB with its wide-angle Burst Alert Telescope (BAT), and then to turn rapidly towards the outburst to search for *afterglow radiation with its X-ray Telescope (XRT) and Ultraviolet/Optical Telescope (UVOT). Redshifts from the XRT and UVOT help determine the burst's distance.

(((●))) SEE WEB LINKS

- Official mission website.
- Information page at Goddard Space Flight Center.

Swift–Tuttle, Comet 109P/ A periodic comet, the parent of the *Perseid meteor shower, independently discovered in 1862 July by the American astronomers Lewis Swift (1820–1913) and Horace Parnell Tuttle (1837–1923). It reached magnitude +2 and showed a tail 25–30° long in 1862 August. Swift–Tuttle's nucleus was particularly active, with many jets or fountains of material. Such activity is the source of *non-gravitational forces which influence the comet's orbital period. It was recovered in 1992 September, reaching perihelion (0.96 AU) on December 12 and peaking at 5th magnitude. The comet has an average orbital period of about 130 years. It is now known to be identical with Comet Kegler of 1737, and appearances back to 69 BC have been identified. Comet Swift–Tuttle will make a close passage to Earth on its next return in 2126. Its orbit has an eccentricity of 0.963 and inclination 113°.5. The nucleus has an estimated diameter of around 25 km.

(((●))) SEE WEB LINKS

- Information page at Cometography website.

Sword Hand of Perseus Part of the constellation of Perseus, marked by the *Double Cluster, which appears to the naked eye as a brighter part of the Milky Way.

Sword of Orion A region of sky around the *Orion Nebula. As well as the Orion Nebula and the stars within it, the Sword of Orion includes the open cluster NGC 1981, the diffuse nebula NGC 1977, and the double star Iota Orionis.

SX Arietis star A low-amplitude (less than 0.1 mag.), *extrinsic, main-sequence variable with a B0p–B9p spectrum; abbr. SXARI. The spectral lines of He I and Si III are of variable intensity. The variations in brightness, spectrum, and magnetic field are caused by the star's rotation, with a typical period of about 1 day. Many features resemble those of an *Alpha^2 Canum Venaticorum star, but the spectrum is earlier and the period generally shorter.

SX Phoenicis star A type of pulsating variable that closely resembles a *Delta Scuti star; abbr. SXPHE. These stars are subdwarfs and appear to belong to the spherical and older disk components of the Galaxy (Population II), as well as being present in globular clusters. The spectra are A2–F5, and multiple periods are present (range 0.04–0.08 days), with variable amplitudes up to 0.7 mag.

Sycorax An outer retrograde moon of Uranus, distance 12 216 000 km; also known as Uranus XVII. Sycorax orbits every 1289.0 days at an inclination of 152°.7 to the ecliptic. It is about 150 km in diameter and was discovered in 1997 from ground-based observations.

symbiotic star A star (in many cases a *cataclysmic variable) that exhibits spectral lines at highly contrasting temperatures, such as those typical of a late-type red giant or supergiant (3000 K), and a dwarf B star (20000 K). Such characteristics indicate that the star is an *interacting binary. The group is very heterogeneous, but includes the relatively well-defined *Z Andromedae stars, the **symbiotic novae** or *RR Telescopii stars, and objects such as R Aquarii that appear to have an *accretion disk and jets emerging approximately perpendicular to the plane of the disk. Mass transfer occurs either through mass streaming or as a stellar wind from the giant secondary.

synchronous orbit An orbit in which a satellite moves around a planet in the same time as the planet takes to rotate on its axis. In the Earth's case, a body in synchronous orbit has a period of one sidereal day and an average distance of 42162 km from the Earth's centre. If the orbit is circular and in the Earth's equatorial plane, the satellite remains above a particular point on the equator. Such an orbit is called a *geostationary orbit* and is used for communication satellites. If the orbit is synchronous but in a plane inclined to the equator, the satellite appears to trace out a figure-of-eight each day, as seen from Earth. The centre of the figure-of-eight lies on the equator, and the loops of the figure-of-eight lie between latitudes equal to the orbit's inclination to the equator.

synchronous rotation The circumstance in which a satellite spins on its axis in the same time as it takes to orbit a planet, thereby keeping the same face turned towards the planet at all times; also known as *captured rotation*. Most major satellites in the Solar System, including our own Moon, are in synchronous rotation, as a consequence of tidal action. In addition, in some short-period binary star systems the tidal interaction between the stars is sufficient to make one or both stars rotate synchronously. Examples are *contact binaries, in which both stars rotate synchronously, and *cataclysmic binaries, where the mass-losing star rotates synchronously but the white dwarf does so only in the magnetic *AM Herculis systems.

synchrotron radiation Electromagnetic radiation emitted by a charged particle (normally an electron) moving in a magnetic field at a velocity very close to that of light. It is similar to *gyrosynchrotron radiation except that, because of the extreme velocity of the particle, the emission appears in a continuous spectrum extending over a very wide range of wavelengths. Examples of synchrotron sources are pulsars, supernova remnants, our Galaxy, and radio galaxies. It is sometimes also known as *magnetobremsstrahlung.

syndyname The curve connecting regions in a comet's dust tail which contain particles of identical size; also known as the **syndyne**. Particles with diameters of around 1 μm are subject to the influence of *radiation pressure, which leads to sorting on the basis of size; smaller particles are carried further from the nucleus by radiation pressure in a given time and therefore have a syndyname farther from the Sun than do large particles.

synodic Relating to conjunctions of planets or satellites.

synodic month The mean period between successive occurrences of identical lunar phases, e.g. from new Moon to new Moon; also known as a *lunation*. The synodic month is over two days longer than the Moon's orbital period (the *sidereal month), because of the Earth's orbital motion around the Sun. The number of synodic months in the year is exactly one fewer than the annual number of revolutions of the Moon around the Earth. The mean length of the synodic month is 29.53059 days.

synodic period The average time taken for a planet to return to the same position with respect to the Sun as seen from Earth, for example from one opposition to the next; or the average time taken for a satellite to return to the same position relative to the Sun as seen from the planet it is orbiting, for example from one full Moon to the next.

Syrtis Major The most prominent dark marking (albedo feature) on Mars, roughly triangular, about 1000 km wide and 1200 km long, centred at +10 ° lat., 290 ° W long. It is the Greek name for the Gulf of Sirte on the north African coast, whose shape it resembles, and means 'the great sand-bank'. It is an area of dark basalt rock sloping gently downwards from west to east with a variable covering of wind-blown dust. It contains regions with hydrated minerals, evidence for surface water in the past. It was first sighted by C. *Huygens in 1659. The official name of this surface region is now Syrtis Major Planum. 📷

Systems I, II, and III Three rotation schemes for different parts of Jupiter. System I refers to the equatorial region of the planet's visible atmosphere, and System II to the rest of its atmosphere. The period of System I is 9 h 50 m 30.003 s, and that of System II is 9 h 55 m 40.632 s. These figures are derived from angular rotations of 877 °.90 and 870 °.27 in 24 hours, the latter representing the motion of the Great Red Spot in 1890–91. The systems are used for reference by visual observers, particularly when making longitude plots of visible surface markings. System III, derived from radio observations, refers to the rotation of the solid interior. Its period is 9 h 55 m 29.711 s.

syzygy The configuration in which the celestial longitude of a planet or the Moon is the same as the Sun's, or differs from it by 180 °; the Earth, Sun, and the third body are then in line as seen from above the plane of the ecliptic. Hence, for the Moon, syzygy is the time of new or full Moon, and for a planet it is the time of conjunction or opposition.

S

tachocline A layer within the Sun where the internal plasma movements change from the rigid rotation of the inner radiative zone to the *differential rotation of the outer convective zone. This occurs about 70% of the way from the Sun's core to the surface, although this varies with latitude, being slightly more distant at higher latitudes. The thickness of the tachocline is believed to be around a few per cent of the Sun's radius.

tachyon A hypothetical particle that can travel faster than the speed of light. According to the theory of special relativity, it is impossible to accelerate a particle to the speed of light since its mass would then be infinite. However, the theory does not exclude particles that travel faster than light. No evidence for such particles has yet been found.

TAI Abbr. for Temps Atomique International, the French name for *International Atomic Time.

tail, cometary The part of a comet containing dust and gas released from the comet's head. Many comets fail to develop a tail but, when present, tails are always directed away from the Sun, so that comets move tail-first after perihelion. Tails do not normally develop until a comet is within about 2 AU of the Sun, and are usually most impressive shortly after perihelion. Comet tails have two main components: the Type I or gas tail (also known as the *ion tail* or *plasma tail*), and the Type II or *dust tail*. Some comets may have a third form of tail, known as Type III, of neutral sodium, which lies between the gas and dust tails. This was first seen in 1997 in Comet *Hale–Bopp. Gas tails consist of ionized gas carried away from the coma by the solar wind, and is more or less straight. They can reach 10^8 km or more in length. Gas tails appear bluish or greenish and are dominated by emission from singly ionized carbon monoxide (CO^+) at 420 nm wavelength, resulting from excitation by solar ultraviolet radiation. They are subject to *disconnection events. The dust tail, by contrast, appears yellowish because it shines by reflected sunlight, and often appears markedly curved. Dust tails are usually shorter than gas tails, but can still reach 10^7 km. They consist of micrometre-sized solid particles which are pushed away from the head by *radiation pressure on parabolic trajectories. Larger particles of dust (millimetre- to centimetre-sized) shed by comets give rise to *meteor streams. *See also* ANTITAIL.

tangential velocity The component of a star's velocity that is perpendicular to our line of sight (i.e. in the tangent plane); also known as **transverse velocity**. A star's tangential velocity can be computed from its observed proper motion and measured distance.

Tarantula Nebula The largest and brightest nebula in the Large Magellanic Cloud, also known as 30 Doradus or NGC 2070. It has a diameter of over 800 l.y., with faint extensions to 6000 l.y., and contains half a million solar masses of ionized gas. The ionization is produced by several clusters of O and B stars, including the very powerful and compact cluster R136 near its centre. The nebula's name comes from its spider-like shape. 🔳

Tarazed The star Gamma Aquilae, magnitude 2.7. It is a K3 bright giant 395 l.y. away.

T association A loose grouping of very young, low-mass stars of the *T Tauri type, still associated with the nebula from which they formed, and often partly obscured by it. They have relative velocities of several kilometres per second and are not bound together by gravity, so the association is likely to break up after only about 10 million years. The closest T association to the Sun is the Taurus–Auriga association, at a distance of 460 l.y.

Taurid meteors A moderate meteor shower, producing steady activity from October 1 to November 25 from two principal radiants. The parent body is Comet *Encke. The maximum is broad and flat, with the Northern Taurid radiant peaking around November 12 from RA 3h 52m, dec. +22°, close to the Pleiades, while the Southern component peaks around November 5 from RA 3h 28m, dec. +15°. Peak activity typically reaches a ZHR around 10. The shower's long duration and comparatively low activity indicates that the meteor stream is ancient and well-dispersed. Taurid meteors are slow (geocentric velocity 28 km/s). Combined with a reasonable proportion of negative-magnitude events, often of long duration, this has led to the shower acquiring a reputation as a rich fireball source. *See also* BETA TAURIDS.

Taurus (Tau) (*gen.* **Tauri**) A constellation of the zodiac, representing a bull. The Sun passes through Taurus from mid-May to the third week of June. Its brightest star is *Aldebaran (Alpha Tauri). Beta Tauri is *Elnath. The constellation contains two large, bright open clusters, the *Hyades and the *Pleiades. Theta Tauri is a wide double in the Hyades, magnitudes 3.4 and 3.8. Within *Hind's Variable Nebula lies T Tauri, the prototype of a class of young variable stars. M1 is the famous *Crab Nebula, a supernova remnant. The *Taurid meteors radiate from the constellation from late October to the end of November every year.

Taurus A A strong radio source in the constellation Taurus, identified with the *Crab Nebula. The radio and optical emissions from the nebula are produced by *synchrotron radiation. At the centre of the nebula lies the *Crab Pulsar, but this is a much weaker radio source than the nebula itself.

Taurus Molecular Clouds A large group of clouds of gas and dust 400 l.y. away in the constellation Taurus. The whole group of clouds contains some 30 000 solar masses of material. One particular cloud in the group, known as Taurus Molecular Cloud 1 (TMC-1), contains only 1 solar mass of material yet is one of the coldest molecular clouds known, with a temperature of only 10 K. It contains some of the most complex interstellar molecules yet identified. The complex chemistry is thought to have developed because the cloud has remained stable for many years without forming stars.

Taurus Moving Cluster A large cluster, containing stars scattered over the constellation Taurus, sharing the same space motion as the Hyades cluster, which forms its nucleus.

Taurus X-1 The second X-ray source to be discovered, during a 1963 sounding-rocket flight, and the first to be identified optically. The X-ray emission originates from the pulsar at the heart of the *Crab Nebula supernova remnant.

TCB Abbr. for *Barycentric Coordinate Time.

T-class asteroid A class of asteroid distinguished by a low albedo (0.04–0.11), with a moderate absorption feature at wavelengths shorter than 0.85 μm and a generally flat reflectance spectrum in the near-infrared. Members of this class include (96) Aegle, diameter 170 km, (308) Polyxo, diameter 141 km, and (114) Kassandra, diameter 100 km.

TDB Abbr. for Temps Dynamique Barycentrique, the French name for *Barycentric Dynamical Time.

T dwarf A spectral classification applied to *brown dwarfs at the very bottom end of the *main sequence, cooler and fainter even than the *L dwarfs. They have temperatures from about 1300 to 600 K or even cooler, and their spectra show strong infrared absorption bands due to methane and water. Being so cool, they are detectable only at infrared wavelengths.

Teapot An asterism in the constellation Sagittarius. The stars Gamma, Delta, and Epsilon Sagittarii form the spout of the teapot; Zeta, Sigma, Tau, and Phi Sagittarii are its handle; and Lambda Sagittarii marks the lid.

Tebbutt, Comet (C/1861 J1) A long-period comet discovered on 1861 May 13 by the Australian amateur astronomer John Tebbutt (1834–1916); also called the Great Comet, formerly designated 1861 II. It reached perihelion (0.82 AU) on June 12 and was 0.13 AU from Earth on June 30, when the Earth passed through its tail. Comet Tebbutt reached a peak magnitude of 0, and had a tail of 100° or more. The comet's orbit has a period of 409 years, eccentricity 0.985, and inclination 85°.4.

(⊕)) SEE WEB LINKS
• Information page at Cometography website.

technetium star A *carbon star with technetium lines in its spectrum. Since the longest-lived isotope of technetium that can be created by stellar nucleosynthesis has a half-life of only 210 000 years, this material must have been created recently inside the star and then brought to its surface.

tectonics The study of the structural features of a planet that result from crustal movement or deformation, or the processes associated with such movement. *Faults, *graben, and *wrinkle ridges are all tectonic features, and give clues to past movements of a planetary crust. *See also* PLATE TECTONICS.

Teide Observatory (OT) An observatory of the Instituto de Astrofísica de Canarias (IAC) on Tenerife, Canary Islands, at an altitude of 2390 m, founded in 1959. Its largest telescope is the 1.52-m Carlos Sánchez infrared reflector, opened in 1972. Other telescopes include the 1.5-m GREGOR solar telescope, owned by a consortium of German institutes, due to begin operation in 2011; STELLA I and II, a pair of 1.2-m robotic telescopes owned and operated by the Astrophysical Institute of Potsdam, Germany, opened in 2007; the 0.9-m Franco-Italian Heliographic Telescope for the Study of Solar Magnetism (THEMIS), opened in 1996; the 0.8-m reflector of the IAC (IAC-80), opened in 1991; the 0.7-m Vacuum Tower Telescope (VTT) of the Kiepenheuer Institute of Solar Physics, Germany, opened in 1989; the 0.5-m Mons reflector of the University of Mons, Belgium, opened in 1972; a Solar Laboratory, which began operation in 1976 and now contains six instruments (including a station of the GONG network) devoted to helioseismology, asteroseismology, and planetary transits; and the *Very Small Array.

(⊕)) SEE WEB LINKS
• Official observatory website.

tektite A small, glassy object 1–30 mm in size, usually dark green, brown, or black. Tektites are scattered over certain areas of the Earth's surface, often in vast *strewn fields*. They are 0.75–65 million years old and have a variety of shapes, including spheroids, dumbbells, droplets, lenses, and disks. The word 'tektite' is derived from the Greek word **tektos**, meaning 'melted'. They were probably formed from molten material splashed into space by meteorite impacts on Earth. This material later condensed into droplets and re-entered the atmosphere, undergoing *ablation. Tektites are named according to the area in which they are found, for example *australites (Australia and Tasmania), *bediasites (Texas, USA), *billitonites (Billiton Island, Indonesia), *georgiaites (Georgia, USA), *indochinites (former Indo-China), malaysianites (Malaysia), *moldavites (Czech Republic), and philippinites (Philippine Islands).

telecompressor A converging lens placed between a telescope's objective and the *focal point to reduce the *effective focal length. A telecompressor provides a smaller *focal ratio and, ideally, a wider field of view. Telecompressors are designed particularly for photography with *Schmidt–Cassegrain telescopes, typically changing the usual $f/10$ to $f/5$, but have a similar effect on refractors.

tele-extender A tube that fits into a telescope's drawtube, designed to hold an eyepiece which will increase the *effective focal length of a telescope for astrophotography. An adapter allows a single-lens reflex (SLR) camera to be attached. A tele-extender with eyepiece gives larger images of small, bright objects such as planets, and also makes it possible to use an

SLR camera which otherwise could not be located at the focal plane of telescopes that have a limited focus range.

telescope An optical instrument that collects light from faint and distant objects and magnifies their images. Telescopes have an objective lens or mirror which gathers light and focuses it at a *focal plane. The resulting image may be observed with an eyepiece or recorded by a photographic emulsion, a *charge-coupled device, or some other detector. In a *refracting telescope the objective is a lens, and in a *reflecting telescope it is a mirror. A telescope that uses both lenses and mirrors is called a *catadioptric system. The larger the aperture of a telescope the more light it collects, making it possible to observe fainter stars and to see finer details. The highest magnification that can be used on a telescope of given aperture is restricted by the effects of *diffraction. In practice, the finest detail visible with a telescope is set by the *Dawes limit. Telescope apertures range from about 25mm for the smallest refractors up to many metres for the largest reflectors.

telescope drive See DRIVE.

telescope mounting See MOUNTING.

Telescopio Nazionale Galileo (TNG) An Italian 3.58-m reflector at the *Roque de los Muchachos Observatory, opened in 1998. It is operated by the Italian National Institute for Astrophysics. The telescope has a thin mirror similar to that of the *New Technology Telescope, utilizing *active optics and *adaptive optics.

(((⊕))) SEE WEB LINKS
• Official telescope website.

Telescopium (Tel) (*gen.* **Telescopii**) An obscure constellation of the southern sky, representing a telescope. It contains little of interest. Its brightest star is Alpha Telescopii, magnitude 3.5.

Telesto A satellite of Saturn, occupying the leading *Lagrangian point (L_4) in the orbit of *Tethys; also known as Saturn XIII. Its distance from the planet is 294710km, and its orbital period 1.888 days. Telesto is $33 \times 24 \times 20$km in size, with a smooth and sparsely cratered surface. It was discovered in 1980 on images from Voyager 1.

telluric lines Absorption and emission lines seen in astronomical spectra arising in the Earth's atmosphere. In the optical region the most prominent features are the Fraunhofer A band and B band, at 760nm and 687nm respectively, both due to molecular oxygen; other features are due to water vapour, carbon dioxide, and various metals.

tellurium A mechanical model showing the Earth and Moon in orbit around the Sun, used to demonstrate the occurrence of seasons and eclipses; also known as a **tellurian**. *See also* ORRERY.

Tempel–Tuttle, Comet 55P/ A periodic comet, discovered independently in 1865 December by the German astronomer (Ernst) Wilhelm (Leberecht) Tempel (1821–89) and in 1866 January by the American astronomer Horace Parnell Tuttle (1837–1923). The comet's orbital period is close to 33 years, and past apparitions back to AD 1366 have been identified. Comet Tempel–Tuttle is notable as the parent of the *Leonid meteor stream, from which outbursts of high activity are sometimes seen when the comet returns to the inner Solar System. The comet's orbit has a perihelion of 0.976AU, eccentricity 0.906, and inclination 162°.5.

(((⊕))) SEE WEB LINKS
• Information page at Cometography website.

temperature A measurement of the heat energy possessed by a body. Temperature can be measured in various ways, such as the total energy emitted (*effective temperature), its emission at a single frequency (*colour temperature), the mean velocity of the particles (*kinetic temperature), the levels of excitation of electrons in atoms (*excitation temperature), or degree of ionization (*ionization temperature). In thermal equilibrium all measures of temperature will give the same value, but under other conditions they may give quite different temperatures. The *kelvin scale is used for scientific measurements of temperature.

temperature minimum   A region in the Sun's atmosphere that marks the boundary between the photosphere and the chromosphere. It lies about 550 km above the base of the photosphere, where the temperature reaches a minimum of about 4400 K. Above the temperature minimum, the temperature rises steadily to about 6000 K at a height of about 1000 km, and then more rapidly higher up.

Tenma   A Japanese X-ray astronomy satellite, known as Astro-B before its launch in 1983 February; Tenma is Japanese for 'Pegasus'. It studied the variability and spectra of relatively bright X-ray sources with a gas scintillation proportional counter of large area (800 cm²). It ceased operation in 1985 November.

((⊕)) SEE WEB LINKS
• Information page at Goddard Space Flight Center.

termination shock   A boundary within the *heliosphere, situated 70–90 AU from the Sun, where the solar wind slows down to subsonic speeds. The termination shock lies within the *heliopause, and the intervening region is referred to as the *heliosheath.

terminator   The dividing line between the sunlit and night portions of a planet or satellite. It is the line of sunrise or sunset.

terra   An extensive upland area on a planetary surface; pl.*terrae*. The name, which means 'ground' or 'land', is not a geological term, but is used in the nomenclature of individual features, for example Lada Terra on Venus, or Roncevaux Terra on Iapetus.

terracing   Stepped terrain found around the inside slopes of larger impact craters. Terraces are usually found in impact craters bigger than about 20 km. They may also occur within volcanic craters, particularly collapse craters and calderas. Terracing is caused by the slumping and sliding of large blocks of land down the steep inner crater slopes.

terrestrial age   The time that has elapsed since a meteorite fell to Earth. It can be estimated from the decay of certain fairly short-lived radioactive isotopes, such as argon-39, carbon-14, chlorine-36, and beryllium-10, which were formed by cosmic-ray bombardment of the meteorite while it was in space.

Terrestrial Dynamical Time (TDT)   The original name for *Terrestrial Time, introduced in 1984. In 1991 the word 'dynamical' was dropped; the two time-scales are equivalent in all but name.

terrestrial planet   A small, high-density planet with a solid, rocky surface. In the Solar System, the terrestrial planets are Mercury, Venus, Earth, and Mars.

Terrestrial Time (TT)   A time-scale used for calculating geocentric positions of Solar System bodies as an approximation to the relativistic time-scale Barycentric Dynamical Time (TDB). It is a continuation of *Ephemeris Time. TT was introduced in 1984 under the name Terrestrial Dynamical Time (TDT) but was renamed in 1991. The fundamental unit of TT is the day of 86 400 SI *seconds. TT is related to *International Atomic Time (TAI) through the definition that 1977 January 1.0 TAI corresponds to 1977 January 1.0003725 TT. This means that TT runs permanently 32.184 seconds ahead of TAI, but behind UTC, which moves further ahead of TT each time a leap second is introduced.

tesla   (T) The unit of magnetic flux density. It is defined as one weber of magnetic flux per square metre. It is named after the Croatian-American physicist Nikola Tesla (1856–1943). One tesla is equal to 10⁴ gauss.

tessera   A complex, polygonally patterned arrangement of troughs on a planetary surface; pl.*tesserae*. The name, which means 'tile' or 'mosaic', is not a geological term, but is used in the nomenclature of individual features, for example Tellus Tessera on Venus.

Tethys   The fifth-largest satellite of Saturn, average diameter 1062 km, sharing the same orbit as the much smaller moons *Calypso and *Telesto; also known as Saturn III. Its orbital distance is 294 990 km, and its orbital period 1.888 days, the same as its period of axial rotation.

Tethys was discovered in 1684 by G. D. *Cassini. It is densely covered with impact craters; one of these, Odysseus, is 440 km across, nearly half the satellite's diameter. Another prominent feature is Ithaca Chasma, a valley over 1000 km long, up to 100 km wide, and up to 3 km deep. Tethys has a density close to that of water, suggesting that it is composed largely of ice.

Thalassa The second-closest satellite of Neptune, orbiting between the Galle and Le Verrier Rings in 0.311 days at a distance of 50075 km; also known as Neptune IV. It is about 80 km in diameter, and was discovered in 1989 on images taken by the Voyager 2 spacecraft.

Thales of Miletus (c.625–c.547 BC) Greek philosopher, born in modern Turkey. He was one of the first Greeks to adopt a scientific approach, seeking natural rather than divine causes for natural events. However, his reputed astronomical achievements are known only via the writings of others. He was said to have used the Babylonian *saros to predict a solar eclipse that took place on May 28, 585 BC, but there is little evidence for the prediction or even for the Babylonian saros. According to Aristotle, Thales believed that everything was created from a primordial mass of water.

Tharsis Montes A large upland region on Mars over 1800 km across, centred at +1° lat., 113°W long.; also known as the Tharsis ridge. It includes the three giant volcanoes Ascraeus Mons, Pavonis Mons, and Arsia Mons, each of which rises to about 20 km high. Tharsis is over 9 km high at the bases of these volcanoes and reaches over 11 km high at Noctis Labyrinthus, a complex network of fractures at the western end of the *Valles Marineris.

Thebe The fourth-closest satellite of Jupiter, distance 221900 km, orbital period 0.675 days; also known as Jupiter XIV. Its axial rotation period is the same as its orbital period. Thebe's size is $116 \times 98 \times 84$ km. It was discovered in 1979 on images from Voyager 1.

Themis family A *Hirayama family of asteroids in the outer part of the main asteroid belt at a mean distance of 3.13 AU from the Sun. It is one of the most populous and best-defined families, consisting of a core of large objects surrounded by a cloud of predominantly smaller ones. The core includes (24) Themis, (62) Erato, (90) Antiope, (468) Lina, (526) Jena, and (846) Lipperta. Most family members have low albedos and are of the carbonaceous C class. Themis itself is a C-class asteroid, diameter 198 km, discovered in 1853 by the Italian astronomer Annibale de Gasparis (1819–92). It is the first asteroid discovered to have water ice on its surface. The orbit of Themis has a semimajor axis of 3.129 AU, period 5.54 years, perihelion 2.72 AU, aphelion 3.54 AU, and inclination 0°.8.

Themisto A small satellite of Jupiter, distance 7450000 km; also known as Jupiter XVIII. It orbits every 130.0 days at an inclination of 46° to the planet's equator and is about 8 km in diameter. Themisto was first seen in 1975 but not confirmed until 2000, in both cases from ground-based observations. It gives its name to the Themisto prograde irregular group of Jovian satellites, of which it is so far the sole member.

theory of everything *See* GRAND UNIFIED THEORY.

thermal bremsstrahlung Another name for emission produced by the process of *free–free transition.

thermal equilibrium 1. A state in which two objects, or an object and its surroundings, have the same temperature so that there is no exchange of heat energy between them. For example, a telescope mirror should ideally be in thermal equilibrium with its supports and with the atmosphere to prevent distortion of the mirror or the creation of air currents within the telescope's tube.
2. A state in which the available energy of an object is distributed uniformly among all the possible forms of energy; also known as **thermodynamic equilibrium**. For example, deep inside a star the radiation field, the kinetic energy, the excitation energy, and the ionization levels will all have equal amounts of energy. Furthermore, all processes are in balance so that, for example, there will be as many ionizations of helium per second as there are recombinations of free electrons and helium ions. The condition of *local thermodynamic equilibrium is often taken as an approximation when modelling stellar atmospheres.

thermal radiation Electromagnetic radiation that arises from the thermal energy of an object. It takes two main forms: *black-body radiation, and thermal *free–free radiation. All objects emit thermal radiation by virtue of their temperature; for many astronomical objects, such thermal radiation will be very similar to black-body radiation. In free–free radiation, the thermal energy of electrons is converted into radiation as they are accelerated past ions. This accounts for much of the radio emission from hot nebulae. *See also* NON-THERMAL RADIATION.

thermocouple An instrument that turns heat into an electric current. It consists of two wires of different metals joined together at their ends. When heat falls on one of the junctions while the other junction is kept cold, an electric current is produced. A thermocouple thus acts as a thermometer for radiant heat. Thermocouples are of low sensitivity but can be used from the microwave region to the ultraviolet. They are used in astronomy mainly to calibrate more sensitive detectors operating in different parts of the spectrum.

thermodynamic equilibrium *See* THERMAL EQUILIBRIUM (2).

thermodynamic temperature A temperature scale, measured in kelvin (K), that is related to the energy possessed by matter; it was formerly known as *absolute temperature*. The zero point on the scale (0 K) is *absolute zero. Thermodynamic temperature can be converted to temperature on the Celsius scale by subtracting 273.15.

thermopile An instrument for measuring the heat radiated by an object. It consists of a series of *thermocouples connected together for greater sensitivity, and produces an electric current from which the intensity of the radiation can be measured.

thermosphere An outer layer of the Earth's atmosphere, which includes the *ionosphere, extending from an altitude of 85 km (above the *mesopause) to the lower level of the *exosphere at 500 km. Temperatures increase with altitude in the thermosphere, reaching 1500 °C at 500 km above the Earth's surface. Phenomena which occur in the thermosphere include *meteors and the *aurora.

thinned chip A *charge-coupled device (CCD) that has been reduced in thickness to improve its *quantum efficiency and sensitivity to ultraviolet and infrared light. This is necessary because the electrodes connected to each *pixel of a CCD are deposited over the top of the light-sensitive surface and have a restricted light transmission. To overcome this, the silicon chip in which the pixels are incorporated is thinned from its back, by acid etching, until it is typically only 10 μm thick. Light entering from the back of the thinned chip can then reach the pixels more readily. Many CCDs in professional use are thinned.

Third Cambridge Catalogue (3C) The first reliable radio survey of the northern sky, made at the *Mullard Radio Astronomy Observatory, Cambridge. Objects in the catalogue are given the prefix 3C. The Catalogue was published in 1959, with a revised version in 1962 which contained 328 sources north of dec. −5°. Many famous objects appear in the 3C catalogue, including the brightest quasar (3C 273); the brightest radio source outside the Solar System, the supernova remnant *Cassiopeia A (3C 461); and the brightest extragalactic radio source, *Cygnus A (3C 405).

third contact In a total solar eclipse, the moment when the trailing (western) limb of the Moon coincides with the Sun's western limb, at which time totality ends; or, in a total lunar eclipse, the point when the Moon's leading limb reaches the eastern edge of the Earth's umbra, ending totality. Immediately after third contact in a solar eclipse, the *diamond ring or *Baily's beads may become visible. At an *annular eclipse, third contact refers to the instant when the Moon's leading (eastern) limb leaves the Sun's eastern limb, and the annular phase ends.

third quarter Another name for *last quarter.

30 Doradus *See* TARANTULA NEBULA.

Thirty Meter Telescope (TMT) A large reflector being built jointly by the US and Canada on Mauna Kea, Hawaii, that will operate at visible to mid-infrared wavelengths (0.3–30 μm). It will have a mirror of 30 m diameter consisting of 492 hexagonal segments each 1.44 m wide and is planned to come into operation in 2018. Partners in the venture are the California Institute of Technology, the University of California, and the Association of Canadian Universities for Research in Astronomy (ACURA). The National Astronomical Observatory of Japan is a collaborating institution.

 SEE WEB LINKS
• Official telescope website.

tholus A small domed hill on a planetary surface; pl.*tholi*. The name, which means 'dome', is not a geological term, but is used in the nomenclature of individual hills, for example Ceraunius Tholus on Mars, and Inachus Tholus on Io.

Thomson, William *See* KELVIN, LORD.

Thomson scattering The scattering of photons by free electrons. Such scattering is independent of wavelength and equal numbers of photons are scattered forwards and backwards. Thomson scattering occurs in stellar atmospheres. It is named after the British physicist Joseph John Thomson (1856–1940).

three-body problem The mathematical problem of finding the positions and velocities of three massive bodies that attract each other gravitationally for any time in the past or future, given their positions, masses, and velocities at a starting time. No general mathematical solution has been found, but certain general results are known, namely that the system's centre of mass travels with constant velocity; the total energy of the system is constant; and the total angular momentum of the system also does not alter. *See also* N-BODY PROBLEM; RESTRICTED THREE-BODY PROBLEM.

three-colour photography *See* ASTROPHOTOGRAPHY.

three-colour photometry *See* JOHNSON PHOTOMETRY.

three-kiloparsec arm An inner spiral arm in our Galaxy about 9000 l.y. from the galactic centre (i.e. at a radius of three kiloparsecs), just over a third of the way out to the Sun's orbit. It is detected by the 21-cm radiation from interstellar hydrogen gas. The three-kiloparsec arm appears to be moving outwards at a speed of about 50 km/s as well as orbiting the centre of the Galaxy.

Thuban The star Alpha Draconis, magnitude 3.7. It is an A0 giant 303 l.y. away, and was the north pole star about 4800 years ago.

Thule Asteroid 279, discovered in 1888 by the Austrian astronomer Johann Palisa (1848–1925). Thule is a D-class asteroid of diameter 127 km. With its semimajor axis of 4.270 AU, near the 3 : 4 commensurability of period with Jupiter, it is regarded as marking the outer edge of the main asteroid belt. Its orbital period is 8.82 years, perihelion 4.23 AU, aphelion 4.31 AU, and inclination 2°.3.

tidal bulge The distortion of a planet, satellite, or star caused by the gravitational attraction of other bodies on it. For example, the Moon and Earth produce tidal bulges on each other, the solid bodies being distorted as well as the Earth's oceans and atmosphere. The components of a close binary star grossly distort each other along the line joining their centres.

tidal evolution The changes in the orbital and axial periods of two bodies in mutual orbit caused by the tides that each raises on the other. The bodies can have their axial rotation periods altered until one or both of the rotation periods equals their orbital period, a situation known as *synchronous rotation. In the Solar System most of the major satellites have synchronous rotation periods as a result of tidal evolution. *See also* TIDAL FRICTION.

tidal force The ability of a massive body to raise tides on another body. Tidal force depends on the mass of the tide-raising body and the distance between the two bodies.

Tidal forces can bring about the destruction of a satellite orbiting a planet or a comet approaching too close to the Sun or a planet. When the orbiting body crosses the *Roche limit, tidal forces across the body are stronger than the cohesive forces holding the body together. *See also* TIDAL FRICTION; TIDAL HEATING.

tidal friction A force between the oceans of Earth and the ocean floors caused by the gravitational attraction of the Moon. The Earth tries to carry the ocean waters round with it, while the Moon tries to keep them heaped up under it and on the far side of the Earth (*see* TIDES). Over long periods of time tidal friction decreases the Earth's rate of spin, so lengthening the day. In turn, the Moon has angular momentum added to it in its orbit and gradually spirals away from Earth. Ultimately, when the day equals the Moon's orbital period (each being about 40 times the length of the present day) the process will cease. A new process will then begin in which the Sun's tide-raising power takes angular momentum from the Earth–Moon system. The Moon will then spiral in closer to the Earth until it is torn to pieces when it enters the Earth's *Roche limit.

tidal heating Heating caused by tidal action on a planet or satellite. The strongest example of tidal heating in the Solar System is Jupiter's effect on its satellite Io, in which the tidal effects produce such high temperatures that the interior of the satellite melts, producing volcanism.

tides The rise and fall of the Earth's oceans, caused by the gravitational attractions of the Moon and Sun. At the Earth's centre, centrifugal force due to the motion of the Earth about the barycentre of the Earth–Moon system is balanced by gravitational force. On the side of the Earth nearest the Moon, the Moon's gravitational pull is greater than centrifugal force, while on the opposite side the centrifugal force is greater. In both cases a tidal bulge is produced in the oceans. The Sun's tide-raising power is about one-third that of the Moon, as a result of its greater distance from the Earth. The Sun and Moon act together at new or full Moon to produce extra-high and extra-low tides; these are the so-called *spring tides*. When the Moon is at first or third quarter the Sun and Moon are acting in different directions. The high tides are then at their lowest and the low tides are at their highest; these are the so-called *neap tides*. The Moon and Sun also raise measurable tides in the solid Earth and its atmosphere. The Earth raises tides in the Moon, its force being a maximum when the Moon is at perigee.

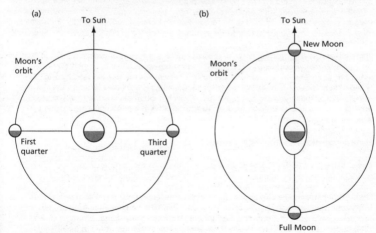

tides: (a) neap tide; (b) spring tide

time The dimension that allows events occurring at the same place to be distinguished. In the classical physics of Galileo and Newton, time had an absolute significance and a time-scale could in principle be adopted so that all observers would agree on the time at which any event occurred. Different observers would see the event occurring at different times, but these differences were explained by the travel time of light from the event to the observer. Moreover, in classical physics this common time-scale was in step with each observer's own local measure of time, the *proper time. All observers would agree that the time of any event is simply the time recorded by the local clock.

Newton's theory of gravity gives a very accurate description of orbital motion within the Solar System, allowing the calculation of each body's position at any given instant of time. *Ephemeris Time (ET) was intended to represent this concept of time, and for many years the positions of planets and other bodies were tabulated at fixed intervals of ET in *The Astronomical Almanac* and elsewhere. ET contrasts with *Universal Time (UT) which, like *sidereal time, is based on the rotation of the Earth and hence is susceptible to irregularities in the Earth's rotation. ET could be checked only from the detailed reduction of many astronomical observations, whereas UT has a strict link to sidereal time, which can be obtained directly from observations of stars. Departures of UT from uniformity are detected by comparison with atomic clocks. Atomic clocks provide *atomic time, which depends on the constants of atomic physics but is independent of the constant of gravitation. Unlike ET, atomic time is not a form of *dynamical time but, if all the constants of physics are genuinely in fixed mutual ratios, the two time-scales should be strictly and uniformly related. Atomic time is the most accurate time-scale available today.

In the theory of relativity, time appears in the equations in much the same way as the space dimensions. Even in Newtonian physics, these space dimensions are relative and have different meanings for different observers. In relativity, time is relative as well, so different observers measure their own proper time and time loses its absolute significance. It is still necessary, however, to have a global version of time as a means of labelling events throughout space and time. This is provided by *coordinate time, which can be thought of as the proper time for one specially selected observer.

To allow consistently for relativistic effects such as gravitational redshift, ET was replaced in 1984 by two new dynamical time-scales. The first of these is *Terrestrial Time (TT, originally known as Terrestrial Dynamical Time). It is used for calculating geocentric positions of Solar System bodies, as published in *The Astronomical Almanac.* It is effectively the proper time for any observer at sea level. For calculating the orbits of Solar System bodies, ET has been replaced by *Barycentric Dynamical Time (TDB). This is a form of coordinate time free from the influences on terrestrial timekeeping resulting from the Earth's motion and the masses of the Sun and planets, but rescaled so that it has only periodic differences from TT. Both TDB and TT use the same fundamental unit as International Atomic Time (TAI), namely the SI second.

time dilation The slowing down of time that occurs at speeds approaching that of light, as predicted by the special theory of relativity. A clock moving relative to an observer will appear to slow down by a factor of $\sqrt{[1 - (v^2/c^2)]}$, where v is the relative velocity and c the speed of light. At everyday speeds, as on Earth, the effect is not noticeable, but the slowing increases rapidly as v approaches c. Not just clocks but all processes slow down, so that an astronaut would appear to have aged far less during a high-speed flight than a person who remained on Earth.

time of perihelion passage *See* PERIHELION PASSAGE, TIME OF.

time zone One of the 24 longitudinal bands into which the Earth is divided; or, any region of the Earth that keeps the same civil time. The 24 standard time zones are each 15° wide in longitude, and each is assigned a zone time that differs by 1 hour from the adjacent time zone, being 1 hour later to the east and 1 hour earlier to the west, until the *International Date Line is reached when a whole day changes. The Earth's prime time zone is centred on the meridian of Greenwich and its standard time is *Universal Time. Large countries such as the USA, Russia, and Australia are divided into several time zones according to local geographical or political considerations rather than strict parallels of longitude.

(())) SEE WEB LINKS
• Map of current time zones at HM Nautical Almanac Office.

TiO bands Absorption bands due to the molecule titanium oxide that are prominent in the spectra of K and M stars. Only in such cool stars can molecules such as TiO survive.

tip–tilt mirror A mirror used in *adaptive optics, which overcomes overall movements of the incoming wavefront of light caused by atmospheric turbulence. This is done by tipping and tilting the mirror rapidly in response to overall changes in position of a reference star, using fast-acting motors.

Tisserand parameter (T) A measure of the orbital motion of a comet or asteroid with respect to a planet, usually Jupiter (T_J). The Tisserand parameter takes into account the semimajor axis, eccentricity, and inclination of the small body's orbit, and remains broadly constant during the small body's lifetime. It is a form of the *restricted three-body problem and is useful in identifying small bodies observed before and after encounters with planets, as its numerical value remains largely unchanged by the encounter. The Tisserand parameter can also be used to classify planet-crossing bodies. For example, members of the *Jupiter comet family have T_J between 2 and 3, whereas for most asteroids T_J is larger than 3. If T_J is just less than 3, a comet can experience very strong encounters with Jupiter. The parameter is named after the French astronomer François Félix Tisserand (1845–96).

Titan The largest satellite of Saturn, and the second-largest satellite in the Solar System, diameter 5150 km; also known as Saturn VI. Titan orbits Saturn in 15.945 days at a distance of 1 221 800 km. Its axial rotation period is the same as its orbital period. At opposition it is magnitude 8.3. It was discovered in 1655 by C. *Huygens. Titan is probably composed of rock and ice in about equal parts. It is the only satellite in the Solar System to have a substantial atmosphere, the surface atmospheric pressure being higher than on Earth. The atmosphere is mainly nitrogen, with 2–10% methane, 0.2% hydrogen (molecular percentages), and traces of ethane, propane, ethyne, hydrogen cyanide, and carbon monoxide. The atmospheric pressure at Titan's surface is 1.5 bar and the temperature about $-180°$C. Photochemical reactions have generated dense orange clouds of hydrocarbons at a height of 200 km, while layers of atmospheric haze exist up to 500 km. The Voyager probes showed a north polar cap in the clouds of Titan with a slightly darker collar around it. Also, the northern hemisphere was distinctly darker than the southern. These are probably both seasonal effects. The satellite was closely studied by the *Cassini–Huygens mission. Radar and spectroscopy revealed a complex surface of hilly terrain and extensive darker fields of dunes. There are a few raised regions, one called Xanadu which appears to show weathering and layered deposits, perhaps of organic matter. The *Huygens probe successfully landed on Titan in 2005 in a region of high ground known as Adiri, east of which is a dark lower-lying area called Shangri-la. Adiri is etched by river valleys, fed perhaps by sporadic liquid methane rain. Weathering on Titan is not restricted to fluid erosion. Radar observations of Titan's surface by the Cassini craft show what appear to be thousand-kilometre-long dune fields of an unknown material, as well as lakes of liquid methane and other hydrocarbons in the polar regions. 📷

Titania The largest satellite of Uranus, 1578 km in diameter; also known as Uranus III. It orbits in 8.706 days at a distance of 436 300 km. Its axial rotation period is the same as its orbital period. It was discovered in 1787 by F. W. *Herschel, soon after the discovery of Uranus itself. Titania has an icy surface with a non-uniform distribution of impact craters; the largest crater, Gertrude, is 326 km in diameter. A number of steep-sided fractures with stepped faults suggests different eras of surface reworking. The longest fracture, Messina Chasmata, is 1500 km long and up to 100 km wide.

Titius–Bode law *See* BODE'S LAW.

TLP *See* LUNAR TRANSIENT PHENOMENON.

TMT Abbr. for *Thirty Meter Telescope.

TNG Abbr. (in Italian) for *Galileo National Telescope.

TNO Abbr. for *trans-Neptunian object.

Toby Jug Nebula A reflection nebula in the constellation Carina, also known as IC 2220, shaped like a tankard. It is a *bipolar nebula surrounding the red giant HD 65750, about 300 l.y. away. The red giant, also known as V341 Carinae, is an irregular variable, ranging between 6th and 7th magnitudes.

Tolman test A test of the assumption that the redshifts of distant galaxies imply that the Universe is expanding. If the Universe is expanding, the surface brightness of galaxies should decrease rapidly with redshift. In fact, the surface brightness of galaxies does not decrease as rapidly as the Tolman test predicts, but most astronomers assume that this is because galaxies were brighter in the past because of *galaxy evolution. The test is named after the American physicist Richard Chace Tolman (1881–1948), who first proposed it in 1930.

Tombaugh, Clyde William (1906–97) American astronomer. In 1929 he became an assistant at the *Lowell Observatory, and began a search for the trans-Neptunian planet that had been predicted by P. *Lowell. To do this, he took pairs of photographs a week apart and compared them using a *blink comparator. Pluto was discovered on 1930 February 18, on a pair of plates exposed the previous month. Tombaugh continued the search for another ten years, finding star clusters, clusters of galaxies, and nearly 800 asteroids, but no further planets. He later made a similar search for small moonlets of the Earth, but found none.

topocentric As seen from the surface of the Earth, as in for example *topocentric coordinates.

topocentric coordinates A system of coordinates with their origin at a point on the Earth's surface. For objects in the Solar System there are slight differences between topocentric and *geocentric coordinates that must be allowed for in precise measurements, but for more distant objects such as stars and galaxies there is no detectable difference. *Horizontal coordinates are topocentric.

Torino scale A scale for estimating the potential danger of an Earth-approaching asteroid. The levels of danger are rated from 0–10, with corresponding colour coding: 0, coloured white, means the asteroid presents no hazard; 1 (green zone) means the asteroid will pass near the Earth but poses no unusual level of danger; 2–4 (yellow zone) mean the discovery merits attention by astronomers; 5–7 (orange zone) mean the asteroid is rated as threatening; 8–10 (red zone) refer to a certain collision, the exact number assigned indicating the anticipated level of destruction. Most discoveries start at 1 or 2 on the scale (green or yellow) but rapidly drop to 0 (white) as further observations are made. No objects have yet been assigned higher numbers than this.

Toro Asteroid 1685, a member of the *Apollo group, discovered in 1948 by the American astronomer Carl Alvar Wirtanen (1910–90). It passed within 0.14 AU (20.9 million km) of Earth in 1972. Toro is an S-class asteroid about 3–4 km in diameter. Its orbit has a semimajor axis of 1.367 AU, period 1.60 years, perihelion 0.77 AU, aphelion 1.96 AU, and inclination $9°.4$; and is remarkable for a resonance involving both Earth and Venus.

torquetum A medieval instrument for measuring the position of an astronomical object in either equatorial, ecliptic, or terrestrial coordinate systems. It consisted of a series of circular plates with *alidades, marked in degrees, hinged one upon the other at the appropriate angles.

torus An object shaped like the inner tube of a car tyre or a ring doughnut: a cylinder bent into a circle and the ends joined together leaving a hole in the middle; pl.*tori*. The cross-section through the cylinder is often circular, but may take other shapes. The *Van Allen Belts around the Earth are toroidal in shape. Other examples include *accretion disks and the ionized material ejected from Jupiter's volcanic satellite Io, which spreads along Io's orbit to form the *Io plasma torus*.

total eclipse A solar eclipse during which the Moon completely covers the Sun's visible disk; or, a lunar eclipse during which the Moon is completely immersed within the Earth's umbra. *See also* TOTALITY.

totality In a total solar eclipse, the period during which the Sun's disk is completely obscured by the Moon; or, in a total lunar eclipse, the period when the Moon is completely immersed in the Earth's *umbra (1). Totality at a solar eclipse may last from only a few seconds to a theoretical maximum of 7m 32s, depending on the Moon's distance from the Earth; the Moon appears larger, and hence takes longest to traverse the Sun's disk, around perigee. Totality at a lunar eclipse may last up to 1h 47m, depending both on the Moon's distance from the Earth and on its path through the umbra.

total magnitude *See* INTEGRATED MAGNITUDE.

total solar irradiance (TSI) The amount of solar radiation at all wavelengths reaching the Earth at its mean distance from the Sun; also known as the *solar constant*. Its average value is 1368W/m^2. The total solar irradiance varies by about 0.1% between solar maximum and minimum, and larger variations are found on a daily basis. The highest values of the TSI are found at solar maximum, because the brightening effect of increased numbers of *faculae on the photosphere more than compensates for the darkening effect of sunspots.

Toutatis Asteroid 4179, a member of the *Apollo group, discovered in 1989 by the French astronomer Christian Pollas (1947–). At present, Toutatis can approach to within 0.006 AU (900000km) of the Earth's orbit. Of the largest near-Earth asteroids, Toutatis is perhaps the most potentially dangerous to the Earth in the foreseeable future, because of its frequent approaches plus a 3 : 1 resonance with Jupiter which makes the orbit chaotic. Radar observations show that it consists of two irregular, cratered bodies in close contact, with maximum widths of about 4.2km and 2.0km, respectively. Its orbit has a semimajor axis of 2.531 AU, period 4.03 years, perihelion 0.94 AU, aphelion 4.12 AU, and inclination $0°.4$.

TRACE Abbr. for *Transition Region and Coronal Explorer.

train Anything remaining along the trajectory of a meteor or fireball after the head of the object has passed. The train may consist of light, dust, vapour, or ionized gas. The faintly glowing train of ionization left along the path of a bright meteor is generally termed a *wake* if it glows for a second or less, or a *persistent train* if the duration is longer. A train may be bent and broken up by winds in the upper atmosphere. Meteorite-dropping fireballs often leave a train of fine dust.

transfer function A measure of the difference in contrast between an object and its image as formed by an optical system. The transfer function is used to evaluate the system's overall performance.

transfer lens A lens used to transfer a beam of light from one place to another without reimaging the beam; also known as a *relay lens*. An example is often found in a *Dall–Kirkham or *Maksutov telescope, where it makes the *focal point, which would otherwise be within the instrument, accessible.

transfer orbit The path followed by a spacecraft when moving from one orbit (the *departure orbit*) to another (the *destination orbit*). For example, the transfer orbit of a spacecraft from Earth to Mars intersects the Earth's orbit and the orbit of Mars, and involves two firings of the spacecraft's engines, one to put it into the transfer orbit and the other to put it into the destination orbit when it arrives. Such a procedure is also used for transferring communications satellites from low Earth orbit to *geostationary orbit. The special case of a transfer orbit that involves minimum expenditure of rocket fuel is known as a *Hohmann orbit.

transient lunar phenomenon *See* LUNAR TRANSIENT PHENOMENON.

transit, meridian The instant at which a celestial object crosses the observer's meridian. When in transit, an object is at its highest altitude above the horizon and has an hour angle of 0h. Meridian transit is also known as *culmination or *meridian passage*.

transit, planetary **1**. The passage of one object across another of larger apparent diameter, such as Mercury or Venus in front of the Sun, or a satellite or its shadow across the face of a planet (*see also* SHADOW TRANSIT). Mercury and Venus transit the Sun only when they are close to a node of their orbits at inferior conjunction. With Mercury this occurs in early November (ascending node) or early May (descending node), and with Venus in early December (ascending node) or early June (descending node). Transits of Mercury are more common than those of Venus. Forthcoming transits of Mercury are on 2016 May 9, 2019 November 11, 2032 November 13, and 2039 November 7. The next transit of Venus is on 2012 June 5/6. 📷

(⊕) SEE WEB LINKS
• NASA transits website.
　　2. The passage of a feature across the *central meridian of a planet.

transit circle The combination of a transit instrument and meridian circle, for simultaneously measuring time of meridian transit and zenith distance at the same passage across the meridian.

transit instrument A telescope constrained to rotate in the plane of the meridian about a horizontal axis mounted east–west. It is used for measuring the times of meridian transit of stars and hence deriving clock error or longitude. *See also* TRANSIT CIRCLE.

transition probability A measure of the likelihood of an electron making a jump (**transition**) between a particular pair of energy levels in an atom or ion. The atom can either absorb energy as the electron jumps to a higher level (the likelihood of which is the *absorption probability*), or emit energy as it drops to a lower level (the likelihood of which is the *emission probability*). There is also the probability of stimulated emission taking place, in which an atom is in a higher level but the probability of it making a transition to the ground state is low. However, if the atom is struck by a photon of just the right frequency (i.e. having the same energy as the difference in energy between the higher level and the ground state), the electron is stimulated into making the transition.

transition region A region of the Sun's atmosphere between the upper chromosphere and corona, where temperatures rise rapidly, from about 20 000 to 2 million K. Its altitude depends on the local magnetic field strength, but is about 2500 km above the base of the photosphere.

Transition Region and Coronal Explorer (TRACE) A NASA satellite launched in 1998 April to study the Sun's corona, and the *transition region where the temperature of the solar atmosphere rises sharply. It carried a 0.3-m telescope and imaged gas on the Sun at a range of temperatures in ultraviolet and far ultraviolet wavelengths. TRACE ceased operation in 2010 June.

(⊕) SEE WEB LINKS
• Official mission website.

transmission grating A *diffraction grating consisting of a transparent material such as glass or plastic on which lines are ruled. A beam of light passed through the grating will be partly split into sets, or *orders*, of spectra on either side of it, the blue light being diffracted the least and the red the most in each order. The orders of spectra increase in dispersion and faintness with distance from the direct beam.

trans-Neptunian object (TNO) Any small, icy body orbiting the Sun beyond Neptune. There are probably tens of thousands of such objects with diameters exceeding 100 km in a ring-shaped zone extending outwards from the orbit of Neptune (at 30 AU). They include objects in the main *Kuiper Belt, *scattered-disk objects, and objects in resonant orbits with Neptune. Most of the latter are in a 3:2 orbital resonance and are known as *Plutinos, but there are smaller populations in other resonances, such as 2 : 1 and 4 : 3.

transparency, atmospheric The extent to which the Earth's atmosphere transmits light from celestial objects. The atmosphere's clarity varies widely, depending on the amount of absorbing material such as water vapour, dust, aerosols (suspended fine particles), and polluting gases. In general, transparency improves with altitude, particularly above *inversion layers*, which occur where cold surface conditions produce a layer of cold air close to the ground which traps much of the absorbing material. In temperate latitudes, transparent air often follows a cold front. *See also* ATMOSPHERIC EXTINCTION.

transverse velocity Another name for *tangential velocity.

Trapezium A multiple star 1350 l.y. away at the heart of the Orion Nebula, also known as Theta[1] Orionis. The Trapezium consists of four main stars, magnitudes 5.1, 6.7, 6.7, and 8.0. There are additional fainter stars, forming a loose cluster born from the gas of the Orion Nebula.

Triangulum (Tri) (*gen.* **Trianguli**) A small but distinctive constellation of the northern sky, representing a triangle. Its brightest star is Beta Trianguli, magnitude 3.0. It contains M33, the *Triangulum Galaxy, a member of the *Local Group.

Triangulum Australe (TrA) (*gen.* **Trianguli Australis**) The sixth-smallest constellation, lying in the southern sky, and popularly known as the Southern Triangle. Its brightest star, Alpha Trianguli Australis, is *Atria. The triangle is completed by Beta and Gamma Trianguli Australis, of magnitude 2.8 and 2.9 respectively.

Triangulum Galaxy The galaxy M33 (NGC 598) in Triangulum, a member of the *Local Group. It is a type Sc spiral seen nearly face-on, up to 1° wide and of 6th magnitude. It lies 2.8 million l.y. away and has a true diameter of 52 000 l.y.

Trifid Nebula The nebula M20 in Sagittarius, also known as NGC 6514. It is a combined emission and reflection nebula, extending for nearly ½°. The Trifid takes its name from three dark lanes of dust that trisect it. It is centred on the multiple star HN 40, the brightest members of which evidently illuminate it. The Trifid lies 5000 l.y. away. 📷

trigonometric parallax The angle subtended at a star by a standard distance of one astronomical unit, the radius of the Earth's orbit. It is usually expressed in seconds of arc. The reciprocal of trigonometric parallax is the star's distance in parsecs. From Earth, accurate parallaxes can be obtained only out to about 50 l.y.; the *Hipparcos satellite extended this to several hundred light years and future satellites such as *Gaia will extend this even farther. The first star to have its parallax measured was *61 Cygni. *See also* ANNUAL PARALLAX.

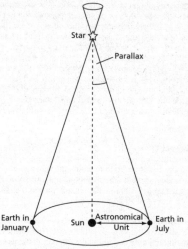

parallax: The apparent change in position of a star as seen from diametrically opposite points of the Earth's orbit

triple-alpha process The chain of nuclear reactions by which carbon is synthesized in the interiors of stars through the fusion of three helium nuclei (alpha particles); also known as the *Salpeter process*, after the American astrophysicist Edwin Ernest Salpeter (1924–2008). It requires a temperature of at least 100 million K, and operates inside stars of at least 0.4 solar masses when all available hydrogen has been converted to helium in their cores. It is the main source of energy production in red giants.

triplet A lens consisting of three individual optical *elements.

triquetrum A medieval angle-measuring instrument consisting of an *alidade supported between a horizontal and a vertical spar. The horizontal spar was calibrated in degrees.

trischiefspiegler *See* SCHIEFSPIEGLER TELESCOPE.

Triton The largest satellite of Neptune, 2705 km in diameter; also known as Neptune I. It lies 354 760 km from Neptune's centre, and orbits in a retrograde direction every 5.877 days at an inclination of 156°.3 to Neptune's equator. Its period of axial rotation is the same as its orbital period. Triton was discovered by W. *Lassell in 1846, a few weeks after the discovery of Neptune itself. Triton has a tenuous nitrogen atmosphere with some methane. The surface pressure, however, is only about 16 μbar, little more than 10^{-5} of the Earth's atmospheric pressure, and its surface temperature is −235°C, the coldest surface so far measured in the Solar System. Triton has a young icy surface with an average albedo of 0.75, one of the brightest surfaces of any planetary satellite. Few impact craters are seen on Triton, but it has many different types of terrain including smooth plains, hummocky plains, *cantaloupe terrain, long linear features, and also a large, probably seasonal, polar cap of nitrogen ice. The Voyager 2 space probe photographed dark plumes rising from dark spots on the surface to an altitude of 8 km before drifting downwind. These plumes may be geyser-like eruptions from subsurface pockets of liquid nitrogen or methane.

Trojan asteroid A member of one of two groups of asteroids, at a mean distance of 5.2 AU from the Sun, which share the orbit of Jupiter. They lie clustered around the leading (L_4) and following (L_5) *Lagrangian points, 60° ahead of and behind Jupiter. Perturbations by other planets cause the Trojans to oscillate along the orbit of Jupiter about 45–80° from the planet, taking 150–200 years per cycle. The first Trojan to be discovered was *Achilles in 1906 in the leading (L_4) group, followed that same year by (617) Patroclus in the trailing (L_5) position. As of the end of 2010 over 4700 Trojan asteroids were known, 64% in the leading (L_4) group. In 1990 the first Martian Trojan, (5261) Eureka, was discovered in the L_5 region of Mars's orbit, and the first at the L_4 point was (121514) in 1999. In 2001 the first Neptunian Trojan (2001 QR$_{322}$) was discovered at Neptune's leading Lagrangian point; the first Neptunian L_5 Trojan was 2008 LC$_{18}$ in 2008.

(((( )))) SEE WEB LINKS
• Database of Trojan minor planets.

tropic Either of the two latitudes on Earth at which the Sun appears overhead at the summer and winter solstices each year. The tropics lie at latitudes of about 23½° north and south, an angle defined by the Earth's axial tilt. *See* TROPIC OF CANCER; TROPIC OF CAPRICORN.

tropical month The mean orbital period of the Moon about the Earth with respect to the vernal equinox. Its length is 27.321 58 days. This definition differs from that of the *sidereal month only through the inclusion of the precession of the equinoxes. The difference is only 7 s of time, but this very small difference will build up to one complete month over the 25 800-year cycle of precession.

tropical period Any orbital period of a planet or moon that is defined by the interval between successive passages through the equinox, for example the *tropical month and the *tropical year.

tropical year The interval of time between successive passages of the Sun through the *mean equinox. This differs slightly from the period of the Earth's revolution around the Sun, since the equinox itself moves continually due to precession. The length of the tropical year is 365.242 19 days. This is the exact period over which the seasons recur, so it is the length of the year that must be matched by a solar calendar.

Tropic of Cancer The latitude on Earth at which the Sun lies directly overhead at noon on the June solstice, when it reaches its most northerly declination. The Tropic of Cancer lies at a latitude of about 23½° north.

Tropic of Capricorn The latitude on Earth at which the Sun lies directly overhead at noon on the December solstice, when it reaches its most southerly declination. The Tropic of Capricorn lies at a latitude of about 23½° south.

tropopause The uppermost limit of the *troposphere, at its boundary with the *stratosphere. The precise altitude of the tropopause depends on latitude, being around 10 km in temperate regions, but closer to 15 km at the equator. Seasonal variations also occur, the height of the tropopause being greater in summer. Temperatures in the troposphere reach their minimum at the tropopause, typically around −40°C in the summer.

troposphere The lowest distinct layer within the atmosphere, extending from sea-level to the *tropopause at an altitude of 10–15 km. Weather systems are confined principally to the troposphere, which accounts for about 75% of the Earth's total atmospheric mass. Temperatures fall with increasing height in the troposphere, reaching a minimum at the tropopause.

true anomaly (ν) The angle between the periapsis of an elliptical orbit and the body's position in orbit, as seen from the body being orbited. The angle is measured in the direction of orbital motion.

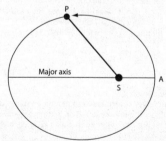

true anomaly: The true anomaly of a planet's position is the angle between the perihelion (A), Sun (S), and planet (P).

true equator The great circle perpendicular to the direction of the *true pole.

true equinox The intersection of the ecliptic with the true equator. Its path on the sky consists of oscillations due to nutation superimposed on the progressive motion due to precession.

true pole The direction towards which the Earth's axis points at a given time. Its path on the sky consists of oscillations due to nutation superimposed on the precessional motion about the pole of the ecliptic.

Trumpler classification A classification of open clusters according to three criteria: degree of central concentration (from I, the most concentrated, to IV, little different from the surrounding stars); range of brightness of individual stars (from 1, small range, to 3, large range); and the total number of stars in the cluster (p meaning poor, less than 50 stars; m, moderately rich, 50–100 stars; r, rich, more than 100 stars). The suffix 'n' indicates that nebulosity is involved with the cluster, as in the Pleiades which is classified as I 3rn. The system was introduced by the Swiss-born American astronomer Robert Julius Trumpler (1886–1956).

Trumpler star The name given to certain highly luminous stars which were once erroneously believed to be hundreds of times the mass of the Sun (*see* SUPERMASSIVE STAR).

TT Abbr. for *Terrestrial Time.

T Tauri star A very young eruptive variable star, less than 10 million years old, with a mass similar to or somewhat less than that of the Sun; abbr. INT (or IT if no nebulosity is present). T Tauri stars have diameters several times the Sun's, and are still contracting. Their spectral types are F–M. On the Hertzsprung–Russell diagram they lie on *Hayashi tracks above the main sequence. T Tauri stars are classified according to their spectra as either *classical*, *weak-line*, or *naked*. Classical T Tauri stars have strong emission lines, and are much brighter in the infrared than other stars of similar temperature. This is thought to be due to surrounding disks of warm dust, similar to the early solar nebula, which may be sites for planet formation. Weak-line objects show less evidence of surrounding material, while naked T Tauri stars show none at all. T Tauri stars exhibit irregular variability ranging from ultraviolet flares on a time-scale of minutes to optical variations on time-scales of days, months, or years. The prototype, T Tauri itself, lies within *Hind's Variable Nebula and varies irregularly between 8th and 13th magnitudes. *See also* FU ORIONIS STAR.

T Tauri wind The vigorous outflow of gas from a *T Tauri star at the rate of about 10^{-8} of a solar mass per year. T Tauri winds are believed to drive the large-scale molecular outflow (a *bipolar outflow) observed near some young stars.

TÜBİTAK National Observatory (TUG) An observatory founded in 1996 at an altitude of 2500 m on Mount Bakirlitepe about 50 km west of the city of Antalya in southern Turkey. Its main telescope is the 1.5-m Russian–Turkish Telescope (RTT150), opened in 1999. The observatory is owned and operated by the Scientific and Technological Research Council of Turkey (TÜBİTAK).

(((⊕))) SEE WEB LINKS
- Official observatory website.
- Information page on RTT150 telescope.

Tucana (Tuc) (gen. Tucanae) A constellation of the southern sky, representing a toucan. Its brightest star is Alpha Tucanae, magnitude 2.9. Beta Tucanae is an optical triple, magnitudes 4.4, 4.5, and 5.1. *47 Tucanae is a major globular cluster. The constellation's other notable feature is the *Small Magellanic Cloud.

Tully–Fisher relation An observed correlation between the width of the 21-cm hydrogen line from spiral galaxies, caused by their rotation, and their absolute magnitude. This correlation can be exploited to measure the relative distances of spiral galaxies. It was first published in 1977 by the Canadian astronomer R(ichard) Brent Tully (1943–) and the American astronomer J(ames) Richard Fisher (1943–).

Tunguska event An explosion above the Podkamennaya Tunguska (Stony Tunguska) River region of central Siberia at about 7.30 a.m. local time on 1908 June 30, caused by a large stony meteorite or cometary fragment. A fireball as bright as the Sun detonated in mid-air, producing dust which caused abnormally bright nights across Europe for days afterwards. Expeditions to the site found a vast (2200 km²) area of devastation, with trees knocked over and scorched up to 40 km away, but no craters were found. The incoming body is thought to have been 50–100 m in diameter, and to have exploded 8–9 km above the ground with the energy of 15–20 megatonnes of TNT.

tuning-fork diagram A diagram showing the different types of galaxy in the *Hubble classification, so named because its shape resembles a tuning fork. The 'handle' consists of the elliptical galaxies (in order of increasing flattening) followed by lenticular galaxies. Along the two parallel 'prongs' of the fork are the spiral and barred spirals respectively, from type a (near the handle) to type d (tip of the prongs).

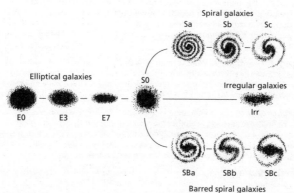

tuning-fork diagram

turbulence, atmospheric Unsteadiness of the Earth's atmosphere, the result of differential heating of the surface and variations of air density within an air stream. It gives rise to bad *seeing and to the twinkling of stars.

turnoff point The point on the Hertzsprung–Russell diagram at which members of a star cluster leave the main sequence. Since all the stars of a particular cluster were formed more or less simultaneously, the most massive of them will leave the main sequence first. The location of the turnoff points allows the age of the cluster to be calculated.

al-Ṭūsī, Naṣīr al-Dīn (in older texts **Nasir** or **Nasser Eddin**) (1201–74) Arab scholar, born in modern Iran. In 1259 he persuaded the Mongol ruler Hūlāgū to build an observatory at Marāgha in Persia equipped with large quadrants. From observations with these instruments, al-Ṭūsī compiled accurate tables of planetary motion and the positions of stars. He refined the *Ptolemaic system by replacing the epicycle with a small circle rolling around inside the circumference of another circle of twice the radius. Any point on the smaller circle then describes a straight line. The *Tusi couple*, as this device is called, was later used by N. *Copernicus.

Tuttle–Giacobini–Kresák, Comet 41P/ A periodic comet, first seen in 1858 by the American astronomer Horace Parnell Tuttle (1837–1923) and rediscovered in 1907 by the French astronomer Michel Giacobini (1873–1938). No reliable orbit could be computed until a second rediscovery was made in 1951, by the Slovakian astronomer Lubor Kresák (1927–94). Observations finally yielded an orbit with a period of 5.5 years. At its 1973 return, the comet showed two surges of about 9 magnitudes in brightness, which remain unexplained and have not recurred. Its orbit has a perihelion of 1.05 AU, aphelion 5.12 AU, eccentricity 0.66, and inclination 9°.2 (which has reduced from nearly 19° when it was first discovered).

(⊕) SEE WEB LINKS
• Information page at Cometography website.

21-centimetre line An emission or absorption line from neutral hydrogen atoms (H I) in free space at a frequency of 1420 MHz (21.1 cm); also known as the *H I line* or *neutral hydrogen line*. It was the first radio spectral line to be discovered, in 1951, and has proved invaluable in mapping the distribution and motions of hydrogen clouds in our Galaxy and others. The 21-cm line is caused by electrons in the ground state reversing their direction of spin (*spin-flip transition*).

twilight The period before sunrise and after sunset during which the sky is not completely dark. Three different periods of twilight are defined according to how far the Sun is below the horizon. During *civil twilight*, the centre of the Sun is less than 6° below the horizon; during *nautical twilight*, the centre of the Sun is between 6° and 12° below the horizon; and during *astronomical twilight* the centre of the Sun is between 12° and 18° below the horizon.

twinkling The rapid fluctuations in brightness of stars, more properly known as *scintillation. Twinkling is caused by atmospheric turbulence distorting the incoming wavefront from a star. Twinkling is less noticeable in planets, which have an extended area. Excessive twinkling is a sign of bad *seeing.

two-body problem The mathematical problem of finding the positions and velocities of two massive bodies attracting each other gravitationally, given their positions, masses, and velocities for a starting time. The problem was first solved by I. *Newton, who showed mathematically that the orbit of one body about the other was either an ellipse, a parabola, or a hyperbola, and that the centre of mass of the system moved with constant velocity.

two-colour diagram The plot of a star's U – B colour against B – V, on the system of *Johnson photometry. Stars of normal chemical composition lie on a tightly defined sequence on the diagram, if they are not reddened by interstellar absorption. Deviations from the sequence can arise for two reasons. For hot stars of spectral type O or B they are caused by interstellar reddening, which can be measured from the observed deviation. Alternatively, they indicate the *ultraviolet excess in Population II stars which is related to their metal abundance. The diagram thus provides a method for measuring the metal abundances of large numbers of stars, which is useful for studying the distribution of different stellar populations in our Galaxy.

two-colour photometry *See* BV PHOTOMETRY.

Two-Micron All-Sky Survey (2MASS) A survey of the entire sky in three near-infrared wave bands (J, H, and K) carried out jointly by the University of Massachusetts and the Infrared Processing and Analysis Center (IPAC) of the California Institute of Technology (Caltech). Two identical 1.3-m reflectors were used, one at the *Fred Lawrence Whipple Observatory in Arizona and the other at *Cerro Tololo Inter-American Observatory in Chile. The northern survey started in 1997 and the southern in 1998; both were completed in 2001, and the 2MASS All-Sky Catalog was released in 2003, containing positions and photometry for 470992970 objects.

(⊕) SEE WEB LINKS
• Official project website with numerous images.

two-spectrum binary An alternative term for a *composite-spectrum binary.

Tycho Brahe *See* BRAHE, TYCHO.

Tycho Catalogue A catalogue published in 1997 containing positions and two-colour photometry of 1 058 332 stars, compiled from observations made by the star-mapping instrument on the *Hipparcos satellite. In 2000 the improved Tycho-2 Catalogue was issued, based on a re-analysis of the original star-mapper data, giving positions, motions, brightness, and colours of 2539913 stars down to magnitude 12.5.

((⊕)) SEE WEB LINKS
• Detailed description and full catalogue downloadable from the CDS.

Tychonic system A geocentric model of the Solar System proposed by Tycho *Brahe in which the Sun and Moon revolve around the Earth but the planets orbit the Sun.

Tycho's Star A supernova in Cassiopeia that appeared in 1572 November, and was extensively studied by Tycho *Brahe. The maximum magnitude was −4, and the star remained visible until 1574 April–May. The light-curve shows that it was a Type Ia *supernova. The radio source 3C 10, faint optical nebulosity, and a weak X-ray source comprise the supernova remnant, about 8000 l.y. away.

Type I, Type II supernova *See* SUPERNOVA.

UBV photometry *See* JOHNSON PHOTOMETRY.

UBVRIJKL photométry A subset of *Johnson photometry.

U-class asteroid A class of asteroids included in older classification schemes, where the 'U' stood for unclassifiable, for asteroids which could not be allocated to one of the five compositional types C, S, M, E, and R. However, more recent schemes group the asteroids into 9, 11, or 14 different classes, depending on the technique used for analysis. In the 14-class scheme the original U class has been subdivided into R, A, M, P, Q, E, V, and T classes, in addition to the C, G, B, F, D, and S classes used. In the new scheme, the letter U is appended to the classification of objects that possess an unusual spectrum for their class (e.g. CU, MU).

U Geminorum star A type of cataclysmic binary showing abrupt outbursts at unpredictable intervals, brightening within a day with typical amplitudes of 3–8mag.; abbr. UG. They are also known as *dwarf novae*. A U Geminorum star is an *interacting binary, with *mass transfer from the main-sequence or slightly evolved secondary to an *accretion disk around the white-dwarf primary. Eclipses are observed in some systems, together with a hump caused by a *hot spot (1), located either where the gas stream hits the disk or at the inner edge of the disk. Orbital periods range from 80 minutes to 14 hours, and both components have masses similar to that of the Sun. Specific subtypes are *SS Cygni stars, *SU Ursae Majoris stars, and *Z Camelopardalis stars.

Uhuru The SAS-1 satellite, launched from a platform called San Marco off the coast of Kenya in 1970 December and named Uhuru after the Swahili for Freedom; also known as Explorer 42. It carried out the first survey of the X-ray sky. In all it discovered 339 sources before it ceased working in 1973 April.

(((●))) **SEE WEB LINKS**
• Information page at Goddard Space Flight Center.

UKIDSS Abbr. for *UKIRT Infrared Deep-Sky Survey.

UKIDSS photometry A system of photometry used by the *UKIRT Infrared Deep-Sky Survey. It is carried out in five broad bands and two narrow bands. The broad bands are J, H, and K (*see* J MAGNITUDE; H MAGNITUDE; K MAGNITUDE), together with a Z band centred on 880nm with bandwidth 95nm, and a Y band centred at 1020nm with bandwidth 100nm. This wavelength region is nearly free from lines due to the Earth's atmosphere and lies between the traditional optical and infrared regions.

UKIRT Abbr. for *United Kingdom Infrared Telescope.

UKIRT Infrared Deep-Sky Survey (UKIDSS) A near-infrared survey being undertaken with the *United Kingdom Infrared Telescope (UKIRT) on Hawaii. It consists of five parts: the Large Area Survey (LAS), covering 4000 square degrees of sky away from the galactic plane down to a K magnitude of 18.4; the Galactic Plane Survey (GPS) of 1800 square degrees along the galactic plane down to a K magnitude of 19.0; the Galactic Clusters Survey (GCS), imaging ten open star clusters and associations to a K magnitude of 18.7; the Deep Extragalactic Survey (DXS), targeting four areas of sky at high galactic latitudes to a K magnitude of 21.0; and the Ultra Deep Survey (UDS), imaging the same field as surveyed by the Subaru optical telescope and XMM-Newton satellite, to a K magnitude of 23.0. The LAS will be repeated after an interval of at least two years to obtain proper motions. The total area of sky covered by all surveys is about 7500 square degrees. UKIDSS began in 2005 May and is expected to be completed in 2012.

(⊕) SEE WEB LINKS
• Official project website.

UKST Abbr. for *United Kingdom Schmidt Telescope.

ultra-compact dwarf galaxy (UCD) A small, faint type of galaxy with a diameter between about 60 and 200 l.y. and containing some 10 million stars, midway in size between a globular cluster and a dwarf elliptical galaxy. Discovered in 2000 in the Fornax cluster, UCDs are thought to be the nuclei of dwarf elliptical galaxies whose outer layers have been stripped away by the tidal effects of much larger neighbour galaxies. Some ultra-luminous globular clusters around giant elliptical galaxies may in fact be UCDs.

ultra-luminous infrared galaxy (ULIRG) A luminous galaxy which emits most of its radiation in the far-infrared. ULIRGs were discovered by the *Infrared Astronomical Satellite in the 1980s. The energy source that is heating the dust cannot generally be seen with optical telescopes because the dust hides it, but observations in other wavebands generally show that the dust is being heated either by an *active galactic nucleus or by a *starburst, a large number of recently formed stars.

ultraviolet astronomy The study of the Universe in the ultraviolet region of the electromagnetic spectrum, approximately 91.2–350 nm. These wavelengths are largely blocked by the Earth's atmosphere, so observations became possible only with the use of rockets after World War II. Balloons were also used, but the altitude they could achieve allowed observations only in the near ultraviolet, longer than 200 nm.

The *Orbiting Astronomical Observatory series of ultraviolet missions commenced in 1968. In 1972 OAO-3, also known as *Copernicus, revealed some of the detailed structure of interstellar matter, in particular its patchiness. TD-1A, a European satellite, made an ultraviolet survey from 1972 to 1974 at 135–290 nm. ANS, the *Astronomical Netherlands Satellite, made photometric observations of a large number of stars at 155–330 nm, also in the 1970s. Ultraviolet observations were carried out from the Skylab space station and the Voyager interplanetary probes, the latter covering the range 50–170 nm.

Ultraviolet astronomy entered a new era in 1978 with the launch of the *International Ultraviolet Explorer (IUE), which obtained tens of thousands of spectra of various objects. Highlights include the discovery of hot haloes of gas surrounding our own and many other galaxies; the monitoring of mass loss by stellar winds in many different types of stars; and the study of the processes operating in novae and X-ray binaries. IUE also observed Halley's Comet and contributed substantially to the understanding of Supernova 1987A.

The *Hubble Space Telescope (HST) has extended the work of IUE, obtaining higher spectral resolution and observing significantly fainter objects. In addition, various ultraviolet telescopes have been carried in the cargo bay of the Space Shuttle. Ultraviolet astronomy has been carried into the *extreme ultraviolet by *Rosat and the *Extreme Ultraviolet Explorer (EUVE). In 1999 the *Far Ultraviolet Spectroscopic Explorer (FUSE) was launched to make high-resolution spectroscopic measurements in the 90–120 nm range. ▣

ultraviolet excess star A star that shows an excess of ultraviolet radiation compared to normal stars. An ultraviolet excess can be used to identify hot O and B stars, white dwarfs, and objects surrounded by an *accretion disk, such as neutron stars and black holes.

ultraviolet photometry Photometry at wavelengths shorter than 300 nm, to which the Earth's atmosphere is totally opaque. The observations therefore have to be made from space. Many ultraviolet *spectrophotometry observations have been made with the *International Ultraviolet Explorer. The Hubble Space Telescope has ultraviolet filters well suited to photometry, and can reach much fainter objects than earlier satellites. Hot stars emit most of their radiation at wavelengths shorter than 300 nm, as do accretion disks in cataclysmic variables. *See also* U MAGNITUDE.

ultraviolet radiation The region of the electromagnetic spectrum spanning the wavelength range from the *Lyman limit at 91.2 nm to 350 nm. It can be split into the shorter-wavelength *far ultraviolet* and the longer-wavelength *near ultraviolet*, the boundary between the two lying at approximately 200 nm. Overlapping slightly with the far ultraviolet is the *extreme ultraviolet waveband, spanning the range 10–100 nm.

Ulugh Beg Title assumed by Muhammad Taragi (1394–1449), Mongol ruler and astronomer, born in modern Iran. In 1420 he established an observatory at Samarkand, in modern Uzbekistan. He equipped it with a huge (radius 40 m) sextant of stone which allowed observations accurate to a few seconds of arc to be made. The result was the first original star catalogue since those of *Ptolemy and al-*Ṣūfī, which it surpassed in precision.

Ulysses A joint ESA–NASA spacecraft launched in 1990 October to study the solar wind, in particular from the unexplored regions around the Sun's poles. The spacecraft passed Jupiter in 1992 February, which swung it on to a trajectory that took it over the Sun's poles. Ulysses passed through the solar wind dominated by the south polar coronal hole, measuring wind speeds in the region of 700 km/s, in 1994 June–November, and the corresponding region over the Sun's north pole in 1995 June–September. Ulysses then embarked on a second orbit of the Sun, overflying the south pole for a second time from 2000 September to 2001 January and the north pole in 2001 September–December. Ulysses ceased operation in 2009 June.

(((∰))) SEE WEB LINKS
• ESA mission website.

U magnitude The magnitude of a star in ultraviolet light on the *Johnson photometry system. The U filter has a central wavelength of 366 nm and a bandwidth of 65 nm. It is that part of the older *photographic magnitude which contains the ultraviolet light. There are also U filters in *Geneva, *RGU, *Vilnius, and *Walraven photometry, and in the *six-colour system, but the context should always make it clear when these are intended.

umbra 1. The conical inner region of shadow cast by a planet or satellite, from within which the Sun's disk is completely obscured. Passage of a body through this shadow results in its *eclipse as, for example, when the Moon traverses Earth's umbra. The dark umbra is surrounded by a broader *penumbra (1) in which the Sun's disk is only partially obscured.
   2. The central, darkest part of a sunspot. It is also the coolest part, with a temperature of about 4200 K. Umbrae are not uniformly dark but contain light **umbral dots**. These are smaller versions of photospheric granules, only about 300 km across, and last somewhat longer than granules, up to 25 min. They seem to be penumbral grains that have just entered the umbra.

Umbriel The third-largest satellite of Uranus, 1169 km in diameter; also known as Uranus II. It orbits at a distance of 266000 km in 4.144 days. Its axial rotation period is the same as its orbital period. Umbriel was discovered in 1851 by W. *Lassell. Its surface shows several subdued impact craters, the largest being Wokolo, 208 km across.

undae Dunes on a planetary surface, used in the plural form; sing.*unda*. The name, which means 'wave', is not a geological term, but is used in the nomenclature of individual features, such as Menat Undae on Venus and Hyperboreae Undae on Mars.

undersampling The circumstance in which a detector's resolution elements, or *pixels*, are too big to resolve the detail in the spectrum or image under study. Undersampling of astronomical images can occur when the atmospheric seeing conditions are exceptionally good. If almost all the light in a star's image lies within one pixel of a CCD, then the image appears square. The image is said to be **undersampled**, and it is not then possible to obtain accurate estimates of the star's image size or its position. *See also* ALIASING.

Undina family A small family of asteroids in the outer part of the main asteroid belt, at a mean distance of about 3.2 AU from the Sun. The largest member is (94) Aurora, diameter 205 km. The family is named after the M-class (92) Undina, diameter 126 km, discovered in 1867 by the Danish-American astronomer Christian Henry Frederick Peters

(1813–90). Its orbit has a semimajor axis of 3.190 AU, period 5.70 years, perihelion 2.86 AU, aphelion 3.52 AU, and inclination 9°.9.

unfilled aperture An aperture synthesis array in which the individual antennas do not sweep out the entire synthesized aperture, usually because they are immobile and a long distance apart. Examples include the *Multi-Element Radio-Linked Interferometer Network, the *Very Large Array, and all *very long baseline arrays.

unit distance A distance taken as a convenient unit of length. Within the Solar System the unit distance is the astronomical unit (AU), the average distance of the Earth from the Sun. For stellar distances the light year (l.y.) or the parsec (pc) is used. Larger units such as the kiloparsec (1000pc) or the megaparsec (1000000pc) are used for distances within our Galaxy or to other galaxies.

United Kingdom Infrared Telescope (UKIRT) A 3.8-m infrared telescope opened in 1979 at an altitude of 4194 m at the *Mauna Kea Observatories, Hawaii, owned by the UK and operated by the *Joint Astronomy Centre in Hawaii. It observes in the 1–30 μm near-infrared band. UKIRT was the first large instrument to use a thin mirror to achieve a lightweight and compact design.

(⊕) SEE WEB LINKS
• Official telescope website.

United Kingdom Schmidt Telescope (UKST) A 1.2-m *Schmidt camera at Siding Spring, New South Wales, opened in 1973. It is part of the *Australian Astronomical Observatory (formerly the Anglo-Australian Observatory); before 1988 it was owned and operated by the Royal Observatory, Edinburgh.

(⊕) SEE WEB LINKS
• Official telescope website.

Universal Time (UT) A world-wide standard time-scale, the same as Greenwich Mean Time. Universal Time is the mean solar time on the meridian of Greenwich. It is defined as the *Greenwich hour angle of the *mean sun plus 12 hours, so that the day begins at midnight rather than noon. It is closely linked to *Greenwich Mean Sidereal Time (GMST), since the mean sidereal day is a precisely known fraction of the mean solar day. In practice, UT is determined by a formula from GMST, which in turn is derived from the observations of the meridian transits of stars. The version of UT derived directly from such observations is designated *UT0*, which is slightly dependent on the observing site. When UT0 is corrected for the variation in longitude due to the *Chandler wobble, a version of Universal Time, *UT1*, is derived which has genuine world-wide application. When UT1 is compared with *International Atomic Time (TAI), it is found to be losing approximately a second a year against TAI. Broadcast time signals use the time-scale known as *Coordinated Universal Time (UTC). This is TAI with an offset of a whole number of seconds. The offset is adjusted when necessary by the introduction of a *leap second, and UTC is always kept within 0.9s of UT1.

Universe Everything that exists, including space, time, and matter. The study of the Universe is known as *cosmology. Cosmologists distinguish between the Universe with a capital 'U', meaning the cosmos and all its contents, and universe with a small 'u', which is usually a mathematical model derived from some physical theory. The real Universe consists mostly of apparently empty space, with matter concentrated into galaxies consisting of stars and gas. The Universe is expanding, so the space between galaxies is gradually stretching, causing a *cosmological redshift in the light from distant objects. There is now strong evidence that space is filled with unseen *dark matter that may have many times the total mass of the visible galaxies; and even more mass may be accounted for by a still-mysterious *dark energy. The most favoured concept of the origin of the Universe is the *Big Bang theory, according to which the Universe came into being in a hot, dense fireball 13.7 billion years ago.

unsharp masking A photographic and image-processing technique for reducing overall brightness variations in an image. A negative mask is made from the original, of low density and slightly out of focus. This is then sandwiched with the original, thus reducing the brightness range of the original while retaining fine detail. The result enables small brightness variations to be seen across an image with little reduction in their contrast.

upper culmination The point on an observer's meridian at which a celestial body reaches its greatest altitude in the sky. *See* CULMINATION.

Upsilon Sagittarii star A form of interacting binary in which an evolved supergiant secondary is transferring hydrogen-deficient material on to a main-sequence primary. The type star is also classified as a low-amplitude (0.1 mag.) *Beta Lyrae star, with a period of 138 days.

Uranometria The first star atlas to cover the entire sky, published in 1603 by J. *Bayer. Each constellation was depicted on a page of its own, with star positions from the catalogues of *Ptolemy and Tycho *Brahe, with the exception of the 12 newly described southern constellations invisible from Europe, which were combined on a single page. This atlas introduced the system of labelling the brightest stars by Greek letters, known as *Bayer letters*.

( ) SEE WEB LINKS
• Scans of the original atlas.

Uranus (♅ or ⛢) The seventh planet from the Sun. It is blue-green in colour due to the absorption of red light by methane in the upper atmosphere. Its mean opposition magnitude is +5.5, making it just visible to the naked eye under favourable conditions, but it was unknown until 1781 when discovered telescopically by F. W. *Herschel. Uranus is distinctly ellipsoidal in shape (equatorial diameter 51 118 km, polar diameter 49 946 km). Its density, 1.3 g/cm^3, is the lowest of the planets except Saturn. Its rotation axis is tilted at over 90° to its orbital plane, so that its rotation is retrograde, and it presents its poles and its equator alternately towards the Sun as it orbits. The rotation period of the visible surface ranges from about 16 hours at 70° south to about 18 hours near the equator, but radio observations indicate that the core rotates in 17 h 14 m.

Uranus has a thick atmosphere composed of 83% hydrogen, 15% helium, and 2% methane (molecular percentages). Thicker clouds of methane at the 1 bar pressure level overlie deeper opaque clouds that probably consist of hydrogen sulphide. The temperature near the top of the atmosphere is around −220°C. Internally, Uranus is thought to have a small rocky core at a high temperature, probably surrounded by a layer of icy materials, topped by a layer of hydrogen and helium. Unlike the other gas giants, Uranus does not emit more heat than it receives. The interior is probably as hot as that of Neptune, but the heat may be prevented from escaping as effectively as a result of some unknown process. Its magnetic field has a strength of about 2.5×10^{-5} tesla at the equator, similar to the Earth's. Uranus's magnetic axis is not centred in the core, but one-third of the way to the surface, and is tilted at nearly 60° to its spin axis.

The atmosphere of Uranus shows few visible features. Computer-enhanced images have revealed extremely faint dark belts or bands with bright zones between them. Dark and bright spots occur, but these are also very faint. Each latitude has its own characteristic rotation period, with the shortest rotation periods at around 70° south. This may be a result of the high axial inclination of Uranus; since each pole of Uranus spends long periods facing the Sun, convection cells originating at the pole could produce winds that blow against the rotation of the planet near the equator, due to the *Coriolis force.

Uranus has thirteen known rings, all far fainter and less extensive than those of Saturn. The innermost ring, known as Zeta, is 35–40 000 km from the centre of Uranus; the brightest ring, called Epsilon, is 51 100 km from the centre and 20–100 km wide. The rings differ from those of Jupiter and Saturn in that they are slightly eccentric, and do not perfectly lie in the equatorial plane of the planet. They are comparatively dust-free and composed of much bigger particles, generally over a metre in size. The Epsilon Ring is composed mainly of large icy boulders, but is dark grey in colour with an albedo of a few per cent. Uranus has 27 known satellites. 📷

Uranus				
Physical data				
Diameter (equatorial)	Oblateness	Inclination of equator to orbit	Axial rotation period (sidereal)	
51118km	0.023	97°.77	17.24 hours	
Mean density	Mass (Earth = 1)	Volume (Earth = 1)	Mean albedo (geometric)	Escape velocity
1.27 g/cm³	14.5	63	0.51	21.4km/s
Orbital data				
Mean distance from Sun				
10⁶ km	AU	Eccentricity of orbit	Inclination of orbit to ecliptic	Orbital period (sidereal)
2870.972	19.19	0.047	0°.8	84.02 years

Urca process A cycle of nuclear reactions in which an electron is absorbed by a nucleus and is subsequently re-emitted as a beta particle (a fast electron) with the generation of a neutrino–antineutrino pair. The process makes no change to the composition of the nucleus, but removes energy from it in the form of the neutrino and antineutrino. While insignificant in ordinary stars, it is important at very high densities and temperatures such as those which give rise to supernovae. The process is named after the Urca Casino in Rio de Janeiro, where it was said that money disappeared as rapidly as energy from the stars.

ureilite A class of calcium-poor achondrite meteorite; also known as *olivine–pigeonite achondrites*. They consist mainly of millimetre-sized grains of olivine and pigeonite minerals enclosed in a matrix of carbonaceous veins, with some metal also present. They are named after the meteorite that fell at Novo Urei, Russia, on 1886 September 10. The ureilites are noted for their carbon content (1% in Novo Urei), much of which is in the form of diamond, possibly produced by shocks from impacts while in space. The ureilites probably originated within the solar nebula, which would make them the most primitive of the achondrites.

Urey, Harold Clayton (1893–1981) American chemist. He studied the Earth's origin from a chemical viewpoint, deducing its constituents by supposing that final accretion occurred at about 0°C. This would have given it a primitive atmosphere of mainly water vapour, methane, ammonia, and hydrogen. In 1953 his student Stanley Lloyd Miller (1930–2007) subjected a mixture of these gases to an electrical discharge to simulate lightning, in what has come to be called the Miller–Urey experiment. After several days there had formed organic molecules important to life, in particular amino acids. Similar experiments were later carried out by others, including C. E. *Sagan. Urey also studied the abundance of elements on the cosmic scale.

Ursa Major (UMa) (*gen.* **Ursae Majoris**) The third-largest constellation, lying in the northern sky and better known as the Great Bear. Its seven main stars make up the familiar shape known as the Plough or Big Dipper. Its two brightest stars are *Dubhe (Alpha Ursae Majoris) and *Alioth (Epsilon Ursae Majoris), both magnitude 1.8. *Mizar (Zeta Ursae Majoris) is a famous naked-eye double. Xi Ursae Majoris is a binary of magnitudes 4.3 and 4.8 and a period of 60 years; it was the first binary to have its orbit computed. *Lalande 21185 is the fourth-closest star to the Sun. M81 and M101 are spiral galaxies of 7th and 8th magnitudes. M82 is an 8th-magnitude peculiar galaxy, now thought to be a spiral seen edge-on that is undergoing a burst of star formation. M97 is the *Owl Nebula.

Ursa Major Moving Cluster A widespread group of stars with similar space velocities, about 14 km/s, which includes five of the seven main members of the Plough (Beta, Gamma, Delta, Epsilon, and Zeta Ursae Majoris). It is the nearest star cluster to us, about half the distance of the Hyades. There is evidence that stars over a much wider area of sky, including Sirius, have similar motions to those in the core of the cluster.

Ursa Minor (UMi) (*gen.* **Ursae Minoris**) The constellation that contains the north celestial pole, popularly known as the Little Bear. Its brightest star is *Polaris (Alpha Ursae Minoris), the north pole star. The stars *Kochab (Beta Ursae Minoris) and *Pherkad (Gamma Ursae Minoris) are popularly known as the Guardians of the Pole. The constellation itself is sometimes referred to as the Little Dipper.

Ursid meteors A rather poorly observed meteor shower, generally regarded as producing low rates (peak ZHR 5–10) around maximum on December 23. The radiant, at RA 14h 28m, dec. +76°, is circumpolar from northern temperate latitudes, lying near the Guardians in Ursa Minor. Higher activity (ZHR perhaps as high as 50) has occasionally been reported, as in 1945, 1982, and 1986. The parent comet is 8P/Tuttle. Ursid meteors are fairly swift, the brighter ones appearing yellow.

US Naval Observatory (USNO) The US government observatory in northwest Washington, DC. It was founded in 1844, taking over the astronomical work of the US Navy's Depot of Charts and Instruments. It moved to its present site in 1893, absorbing the US Navy's office of The *Nautical Almanac. Its main instrument is a 26-inch (0.66-m) refractor, opened in 1873. Since 1955 the USNO has operated an observing station near Flagstaff, Arizona, at an altitude of 2315 m. The instruments here include the 1.55-m Kaj Strand astrometric reflector, opened in 1964; a 1-m reflector moved from Washington in 1955 and given a new mirror in 1969; and a 1.3-m wide-field telescope opened in 1999. USNO also jointly owns and operates the Navy Prototype Optical Interferometer (NPOI) at *Lowell Observatory.

(⊕) SEE WEB LINKS
• Official observatory website.

USNO CCD Astrograph Catalog (UCAC) An all-sky catalogue of positions and proper motions of stars between 8th and 16th magnitudes produced by the US Naval Observatory; proper motions are derived by reference to older catalogues. Observations of the southern hemisphere were made at Cerro Tololo Inter-American Observatory in Chile with a 0.2-m astrograph in 1998–2001. The telescope was then moved to the US Naval Observatory's Flagstaff Station in Arizona to observe the northern sky. Intermediate data releases, UCAC1 and UCAC2, were made in 2000 and 2003. Observations ended in 2004; the final catalogue (UCAC3), containing over 100 million stars, was published in 2009.

(⊕) SEE WEB LINKS
• Official catalogue web page.

UT Abbr. for *Universal Time.

UTC Abbr. for Universel Temps Coordonné, the French name for *Coordinated Universal Time.

Utopia Planitia A vast sloping plain on Mars some 3200 km across, centred at +50° lat., 242° W long. Eastern Utopia was the landing site for Viking 2 in 1976, nearly 200 km west of the crater Mie.

UU Herculis star A yellow giant or supergiant semiregular variable (type SRD) of spectral type F–K. Amplitudes range from 0.1 to 4 mag. and the periods range from 30 to 1100 days.

UV Abbr. for ultraviolet.

uvby system *See* STRÖMGREN PHOTOMETRY.

UV Ceti star A type of eruptive variable (often known as a *flare star) exhibiting sudden flares with a rise time of seconds to tens of seconds, and decay times of minutes to tens of minutes; abbr. UV. Amplitudes of up to 6 magnitudes have been recorded. X-ray and radio flares also occur, sometimes coincident with the optical flares. The stars are K–M emission-line dwarfs with very strong magnetic fields, and those in *associations are rapidly rotating, young stars. Tidal effects in a binary system may create the strong magnetic fields of other UV Ceti stars. The flare activity is similar to that found on the Sun, but appears more prominent because of the intrinsic faintness of these cool red dwarfs. *See also* BY DRACONIS STAR.

UXor *See* UX ORIONIS STAR.

UX Orionis star A subclass of the intermediate-mass, pre-main-sequence *Herbig Ae/Be stars; also known as **UXors**. The light curves show distinctive Algol-like fadings of 1 mag. on a period of days to weeks. These random fadings are accompanied by changes of colour during minima and have been interpreted as due to non-uniform dust structure around the star.

UX Ursae Majoris star A term formerly used as a synonym for a *nova-like variable, but now restricted mainly to those nova-like variables that always show broad absorption lines in their spectra.

u

vacuum energy The minimum possible energy of a field. In contrast to classical physics, in which the minimum energy of a field is zero, in quantum physics the minimum energy is not zero. This **vacuum energy** has been proposed as an explanation of the discovery that the expansion of the Universe is accelerating (*see* DARK ENERGY).

Valhalla A multi-ringed impact basin on Callisto, Jupiter's second-largest moon. The centre of the basin is marked by a bright region 600 km across, surrounded by at least 15 concentric ridges. The total diameter is about 3000 km, making it the largest multi-ringed basin in the Solar System.

Valles Marineris A large system of canyons on Mars, centred at about –11° lat., 60° W long. It is the largest canyon in the Solar System, 3800 km long, up to 500 km wide, and over 4 km deep in places. It is really a complex of several parallel troughs, the central ones being the Ius, Melas, Coprates, and Eos Chasmata, with Tithonium, Ophir, Candor, and Gangis Chasmata to the north. Hebes and Juventae Chasmata are separate from the main complex, lying further north. They are all thought to have been caused by tensional faulting within Mars, and all have been modified and enlarged by the action of ancient water channels running down their walls. Valles Marineris, named after the Mariner 9 spacecraft that discovered it in 1971, is visible from Earth as the dark streak *Coprates. Its darkness may be due to dust covering part of the canyon floor. 📷

vallis A valley on a planetary surface; pl.*valles*. The name is not a geological term, but is used in the nomenclature of particularly long or large valleys, such as *Valles Marineris on Mars, or Leprechaun Vallis on Ariel. The term was first used for features on the Moon in the 17th century.

Van Allen, James Alfred (1914–2006) American space scientist. In 1945 he began high-altitude rocket research, initially using captured German V2 missiles. He was responsible for instruments aboard the first American satellites, the Explorer series, with which the radiation zones now called *Van Allen Belts were discovered. In all, Van Allen worked on 24 missions, including Pioneers 10 and 11, studying planetary magnetospheres (Saturn's in particular), solar X-ray emission, and the solar wind.

Van Allen Belts Two *radiation belts surrounding the Earth, containing trapped charged particles. The inner Van Allen Belt lies around 1.5 Earth radii (9400 km) above the equator, and contains protons and electrons of both solar and ionospheric origin. The outer Van Allen Belt, at an equatorial distance of 4.5 Earth radii (28000 km), contains mainly electrons from the solar wind. The belts were discovered in 1958 by J. A. *Van Allen from measurements obtained by the Explorer 1 satellite.

Van Biesbroeck's Star One of the intrinsically faintest stars known; also designated GJ 752B and LHS 474. It is an M8 dwarf less than 1/350000 as luminous as the Sun (absolute magnitude 18.7), about 19 l.y. away in Aquila; visually it appears of 18th magnitude. It is a companion to another red dwarf, BD +4° 4048, and is named after its discoverer, the Belgian astronomer Georges Achille Van Biesbroeck (1880–1974).

van de Hulst, Hendrik *See* HULST, HENDRIK VAN DE.

van Maanen's Star A white dwarf 14.1 l.y. away in Pisces, the nearest white dwarf after the companions of Sirius and Procyon; also designated GJ35 and LHS 7. It appears of 12th magnitude and has the relatively large proper motion of $2''.99$ per year. It is named after the Dutch-American astronomer Adriaan van Maanen (1884–1946), who discovered it in 1917.

variable star Any star that varies in brightness. Two broad categories are recognized: *extrinsic variables, which vary for a mechanical reason (e.g. rotation); and *intrinsic variables, which undergo a real change in luminosity of either an individual star or some element in a binary system. Certain stars may combine both forms of variation. The standard reference for the classification of types of variable stars and their nomenclature is the *General Catalogue of Variable Stars (GCVS). The current total of designated variables in our Galaxy is over 43 500.

The classification of variable stars was originally based upon the form of their *light-curve, its *amplitude (1), and periodicity (or lack of it). Increasingly, however, the physical mechanisms that underlie the different forms of variation, or the physical structure of the stars (and binaries) are used to define six groups of variables, each with specific features. The groups are subdivided into individual types of variation, often named after specific stars, and commonly referred to by capital-letter abbreviations, as listed in Table 8, Appendix. Some stars exhibit more than one form of variability, in which case a combined abbreviation is used, such as E+UG, BY+UV, or EA+UV+BY. Eclipsing binaries are classified in three ways: by their light curve, their physical characteristics, and the evolutionary state of their components. For these, combined abbreviations of the form E/DM, EA/DS/RS, EB/AR, EW/KW are used.

*Eruptive variables exhibit unpredictable changes in the form of flares or fades, most of which originate in chromospheric or coronal activity. The group includes *flare stars, *Gamma Cassiopeiae stars, *Orion variables, *R Coronae Borealis stars, and *T Tauri stars.

*Pulsating variables expand and contract or experience wave-like motion of the surface because of fluctuations in the flow of energy from their interiors. Notable examples are *Cepheid variables, *Mira stars, *RR Lyrae stars, *RV Tauri stars, *semiregular variables, and *ZZ Ceti stars.

*Rotating variables are a small group whose variations arise from non-uniform surface brightness or ellipsoidal shape.

*Cataclysmic variables generally exhibit powerful outbursts with a sudden release of energy. The group includes *novae, *dwarf novae, and *supernovae, and should not be confused with eruptive variables.

Eclipsing variables are binary stars which exhibit partial or total eclipses of one or both components. They are subdivided, on the basis of the shape of the light-curve, into *Algol stars, *Beta Lyrae stars, and *W Ursae Majoris stars. This group also includes some systems that do not eclipse, but where the distorted shape of the components produce fluctuations in the light-curve.

Optically variable X-ray sources bear many similarities to cataclysmic variables (specifically to *cataclysmic binaries). The optical variations are often induced by the X-ray variability. Two examples are the *AM Herculis stars (or *polars*) and the HZ Herculis stars (or *X-ray pulsars).

variation 1. A periodic disturbance in the Moon's celestial longitude caused by changes in the Sun's gravitational attraction as the Moon orbits the Earth. It has an amplitude of $40'$ and a period of half a synodic month.

2. The yearly change in the coordinates of a star due to the effects of precession and proper motion; also known as *annual variation*.

variation of latitude *See* LATITUDE VARIATION.

vastitas An extensive lowland plain on a planetary surface; pl.*vastitates*. The name, which means 'desolation' or 'vastness', is not a geological term, but is used in the nomenclature of individual features, for example Vastitatis Borealis on Mars.

Vatican Advanced Technology Telescope (VATT) A 1.8-m reflector, opened in 1993 at *Mount Graham International Observatory, Arizona. It is owned by the Vatican Observatory, and jointly operated by them and *Steward Observatory. It consists of the Alice P. Lennon Telescope and the adjoining Thomas J. Bannan Astrophysics Facility, named after their benefactors. The telescope's mirror has the unusually short focal ratio of $f/1.0$, making the tube very compact. The Gregorian secondary employs adaptive optics to improve image quality.

(((⊕))) SEE WEB LINKS

• Official telescope website.

VBLUW photometry *See* WALRAVEN PHOTOMETRY.

V-class asteroid A rare class of asteroid distinguished by a moderately high albedo and a reflectance spectrum displaying a strong absorption feature at wavelengths shorter than 0.7 μm. There is also a strong absorption feature at wavelengths near 0.95 μm, in the near-infrared. This dip is characteristic of the silicate mineral pyroxene. The 'V' is for Vesta, the first asteroid to be assigned to the class.

Vega The star Alpha Lyrae, magnitude 0.03, the fifth-brightest star in the sky. It is an A0 dwarf 25 l.y. away. The Infrared Astronomical Satellite (IRAS) discovered that it is surrounded by a disk of dust and gas that may be the site of a forming planetary system.

Vega probes Two identical space probes launched by the former Soviet Union in 1984 December, each consisting of a Venus lander and a comet Halley fly-by probe. The name 'Vega' was a contraction of the Russian words *Venera* ('Venus') and *Gallei* ('Halley'). In 1985 June the spacecraft released balloons into the atmosphere of Venus and landers on to the planet's surface before flying on to comet Halley, which they reached in 1986 March. They passed 8900 km and 8000 km respectively from Halley's nucleus, photographing it and analysing the comet's dust and gas.

Veil Nebula A beautiful filamentary nebula which is part of the *Cygnus Loop, a large supernova remnant. The brightest part of the Veil is designated NGC 6992-5; NGC 6960 and NGC 6979, to the west, are fainter parts of the same structure. The filaments are gas which has cooled from a supernova explosion some 5000 years ago. It is 1400 l.y. away. ◙

Vela (Vel) (*gen.* Velorum) A constellation of the southern sky, representing the sails of the ship *Argo Navis. Gamma Velorum is a wide double star of magnitudes 1.8 and 4.3, the brighter of which is the brightest known *Wolf–Rayet star. The constellation contains the *Gum Nebula and the *Vela pulsar.

Vela pulsar A pulsar in the *Vela Supernova Remnant; also known as PSR 0833–45. It has a period of 89 ms and is slowing down at a rate of 10.7 ns per day. The Vela pulsar is one of the youngest pulsars known, with a maximum age (*spin-down age) of about 11 000 years, and one of the few to have been positively identified with a supernova remnant. In 1977 it was discovered to be flashing at visible wavelengths, making it the second known optical pulsar.

Vela Supernova Remnant An extensive nebulosity over 5° wide in the constellation Vela, the remains of a supernova that exploded about 11 000 years ago. It lies about 1600 l.y. away, within the more extensive and much older *Gum Nebula, and contains the *Vela pulsar, which is the collapsed core of the exploded star.

velocity curve A plot of the radial velocity of an object against time, derived from the Doppler shift of spectral lines. With a *spectroscopic binary, it may be possible to derive a curve for each component and thence obtain the relative orbit and the *mass ratio. With a *pulsating star, a radial-velocity curve provides details of the star's expansion and contraction, and thus of the phase relationship between the variations in radius and the changes in brightness, surface temperature, and spectrum.

velocity dispersion The spread of velocities of stars or galaxies in a cluster. Galaxies in a cluster have individual orbital velocities about the cluster's centre of mass, which depend on the total cluster mass. By measuring the radial velocities of selected members, the velocity dispersion of the cluster can be estimated, and used to derive the cluster's mass from the *virial theorem. The velocity dispersion of individual stars in a star cluster can be measured in the same way, but if the stars are unresolved (as in a distant galaxy) the velocity dispersion can be determined from Doppler broadening in the spectrum of the integrated starlight.

velocity–distance relation The relationship between the recession velocity of an extragalactic object and its distance from the observer. In 1929 E. P. *Hubble showed that for nearby galaxies this relationship is a straight line—the *Hubble law—which is evidence that we live in an expanding homogeneous Universe. This relationship is predicted to deviate from a straight line for more distant galaxies (see HUBBLE DIAGRAM).

Venera A series of space probes to the planet Venus, launched by the former Soviet Union. Veneras 4–6 ejected capsules which sent back information on the atmosphere of Venus before being crushed by the intense pressure. Venera 7 made the first successful landing on Venus; the Venera 9 lander took the first pictures of the planet's surface, while its orbiter section was the first craft to orbit Venus. Venera 13 made the first analysis of the planet's soil.

Successful Venera Probes[a]

Probe	Launch date	Results
Venera 4	1967 June 12	Ejected capsule into atmosphere of Venus October 18
Venera 5	1969 January 5	Ejected capsule into atmosphere of Venus May 16
Venera 6	1969 January 10	Ejected capsule into atmosphere of Venus May 17
Venera 7	1970 August 17	Ejected capsule which landed on Venus December 15
Venera 8	1972 March 27	Ejected capsule which landed on Venus July 22
Venera 9	1975 June 8	Combined orbiter and lander, arrived at Venus October 22
Venera 10	1975 June 14	Combined orbiter and lander, arrived at Venus October 25
Venera 11	1978 September 9	Ejected lander which reached surface of Venus December 25
Venera 12	1978 September 14	Ejected lander which reached surface of Venus December 21
Venera 13	1981 October 30	Ejected lander which reached surface of Venus 1982 March 1
Venera 14	1981 November 4	Ejected lander which reached surface of Venus 1982 March 5
Venera 15	1983 June 2	Went into orbit around Venus October 10, mapped surface by radar
Venera 16	1983 June 7	Went into orbit around Venus October 14, mapped surface by radar

[a] Veneras 1–3 were failures

Venus (♀) The second planet from the Sun. It has the most circular orbit of all the planets. Its mean geometric albedo, 0.65, is the highest of all the planets, a result of its unbroken cover of white clouds. At its brightest it reaches magnitude −4.7, far brighter than any other planet. Its rotational axis is tilted at nearly 180° to the upright, so that its rotation is retrograde. It rotates on its axis every 243 days, and so presents the same face to the Earth every time the two planets are at their closest.

Venus has a dense atmosphere composed (by volume) of about 96.5% carbon dioxide and 3.5% nitrogen, with traces of sulphur dioxide, water vapour, argon, carbon monoxide, and helium. The pressure of the atmosphere at the surface is about 92 bar (i.e. 92 times the Earth's sea-level pressure). Surface temperatures average about 460°C due to the *greenhouse effect in the planet's atmosphere. There is inconclusive evidence for lightning activity. A thick cloud layer at 45–65 km altitude, composed of sulphuric acid and water droplets, permanently obscures the surface. Through an optical telescope Venus appears virtually featureless, but at ultraviolet wavelengths streamers of cloud can be seen extending directly from the equator to the poles, indicating that there is a single *Hadley circulation cell in each hemisphere carrying hot air directly from the equator to the poles, where atmospheric vortices are seen. The temperature difference in the atmosphere between equator and pole is no more than about 10°. At the equator the upper atmosphere rotates in only four Earth days in a retrograde direction, a phenomenon known as *super-rotation*. At the surface, wind speeds are only around 1 m/s, as measured by lander craft.

Radar maps of the surface have been made, both from the Earth and from space probes, which indicated the presence of highland areas such as *Aphrodite Terra and *Ishtar Terra and extensive rolling plains. In the early 1990s the *Magellan spacecraft revealed that much of Venus is intensely fractured, and there are vast assemblages of graben and wrinkle ridges, suggesting both crustal expansion and contraction. Volcanism is an important surface process, and eruptions may be going on at present. There are many circular structures hundreds of kilometres across, called *coronae*, which appear to be super-volcanoes that have undergone phases of uplift, faulting, and degradation; the largest, Artemis Corona, is 2600 km across. Volcanic features more like those on Earth are also found, such as steep-sided circular volcanic domes, volcanic cones, large shield volcanoes, calderas, and lava flows. There are also volcanic features more like those on the Moon, including vast plains-forming lavas of low viscosity, and many sinuous rilles, one of which, Baltis Vallis, at 6000 km long, is the longest lava channel in the Solar System.

The dense atmosphere protects the surface from all but the largest meteoroids, so impact craters are scarce compared with Mercury or Mars, although more frequent than on Earth. About one thousand impact craters have so far been identified on Venus, with diameters greater than 1 km; the largest, Mead, is 270 km across. Many appear remarkably fresh, with central peaks or inner rings, but most are flooded or partially covered by subsequent lavas. All the best-preserved ones have ejecta outflows somewhat similar to those associated with *rampart craters on Mars, but the flows are longer, sometimes winding away from the crater for hundreds of kilometres like lava. Greater impact melt due to the higher temperatures of Venus, and also debris flow due to the thick atmosphere, may both be responsible.

Internally, Venus is probably similar to the Earth, with a thin lithosphere perhaps 50 km thick, a rocky asthenosphere, and a metallic core about half the planet's diameter. Venus has no significant magnetic field, and no natural satellites. 📷

Venus

Physical data

Diameter	Oblateness	Inclination of equator to orbit	Axial rotation period (sidereal)	
12 104 km	0.0	177°.36	243.02 days	
Mean density	Mass (Earth = 1)	Volume (Earth = 1)	Mean albedo (geometric)	Escape velocity
5.24 g/cm³	0.82	0.86	0.65	10.36 km/s

Orbital data

Mean distance from the Sun				
10⁶ km	AU	Eccentricity of orbit	Inclination of orbit to ecliptic	Orbital period (sidereal)
108.209	0.723	0.007	3°.4	224.701 days

Venus Express An ESA space probe, launched in 2005 November, which went into a highly elliptical polar orbit around Venus in 2006 April to study the atmosphere and surface of the planet. The Venus Monitoring Camera (VMC) images the clouds at ultraviolet, visible, and near-infrared wavelengths; three spectrometers (VIRTIS, PFS, and SPICAV) observe at ultraviolet, visible, and infrared wavelengths to plumb the composition, temperature, and structure of the atmosphere down to the surface; and a particle detector (ASPERA) and a magnetometer study the planet's magnetic field and interaction of the solar wind with its atmosphere.

(⊕) SEE WEB LINKS
• ESA mission website.

VERITAS Abbr. for *Very Energetic Radiation Imaging Telescope Array System.

vernal equinox (♈) The point on the celestial sphere at which the Sun passes from south to north of the celestial equator each year. It is also known as the *spring equinox* or the *first point of Aries. It currently occurs on March 20. At the vernal equinox, as with the autumnal equinox six months later, night and day are equal in length the world over. *Right ascension is measured from the vernal equinox, which by definition has zero right ascension and declination.

vertical circle A form of meridian circle which can be rotated about a vertical axis so that observations may be made in two orientations 180° apart. Zenith distance is measured by combining the circle readings on a star in the two orientations; the change in the instrumental zero point between orientations is measured by a spirit level.

Very Energetic Radiation Imaging Telescope Array System (VERITAS) An array of four 12-m optical reflectors for gamma-ray astronomy in the 50 GeV–50 TeV range at the *Fred Lawrence Whipple Observatory, Arizona. Each dish consists of 350 individual hexagonal mirrors. The array, opened in 2007, is owned and operated by a consortium of institutions in the US, the UK, Ireland, and Canada, led by the Smithsonian Astrophysical Observatory.

(⊕) SEE WEB LINKS
• Official telescope website.

Very Large Array (VLA) An aperture-synthesis radio telescope sited at an altitude of 2124 m on the Plains of San Agustin, west of Socorro, New Mexico, and operated by the US National Radio Astronomy Observatory. It consists of 27 movable dishes of 25-m aperture, mounted along a Y-shaped railway track with arms 19, 21, and 21 km long. The telescopes can be arranged in four different configurations with maximum baselines of 1, 3.6, 10, and 36 km. The data from the antennas is combined electronically to give the resolution of an antenna 36 km across with the sensitivity of a dish 130 m in diameter. The VLA was completed in 1980, and has its headquarters in Socorro, New Mexico. The array is now being upgraded to produce the Expanded Very Large Array (EVLA) with new electronics and up to eight new dishes giving baselines to 350 km.

(⊕) SEE WEB LINKS
• Official telescope website.

Very Large Telescope (VLT) An instrument consisting of four 8.2-m reflectors, with the combined collecting area of a 16-m mirror, owned and operated by the *European Southern Observatory at an altitude of 2635 m at the *Paranal Observatory in Chile. The four individual reflectors are known as the Unit Telescopes (UT). Unit Telescope 1, named Antu, was opened in 1998, followed by UT2 (Kueyen) in 1999, with UT3 (Melipal) and UT4 (Yepun) in 2000. Next to the Unit Telescopes are four movable 1.8-m Auxiliary Telescopes (AT). Light from the Unit Telescopes and the Auxiliary Telescopes can be combined in the VLT Interferometer (VLTI) for high-resolution imaging. The Auxiliary Telescopes can be placed in 30 different positions, providing interferometer baselines up to 200 m. The first AT was installed in 2004 and the last in

v

2006. Adjacent to the VLT is the 2.61-m VLT Survey Telescope (VST), a wide-field survey instrument built by ESO and Italy, which began observations in 2011. ◙

(⊕) SEE WEB LINKS
• Official telescope website.

Very Long Baseline Array (VLBA) A purpose-built VLBI array of ten radio telescopes spread across the USA with a maximum baseline of 8600 km, giving a resolution better than $0''.001$ at its shortest operating wavelength, 3.5 mm. Eight of the 25-m telescopes are in the continental USA, with the other two on Hawaii and the Virgin Islands. The VLBA was completed in 1993 and shares headquarters in Socorro, New Mexico, with the *Very Large Array. It is operated by the US National Radio Astronomy Observatory.

(⊕) SEE WEB LINKS
• Official VBA website.

very long baseline interferometry (VLBI) A technique in radio interferometry where telescopes, often many thousands of kilometres apart, may be operated as an interferometer to achieve angular resolutions better than $0''.001$. Each telescope works independently, recording the signals together with timing information on arrays of hard disks. Later the data are replayed and the signals combined. In the new technique of *e-VLBI*, the data are transferred in real time to the central processor over the Internet. A side-benefit of VLBI work is the location of the observatories to accuracies of a few centimetres, useful in the study of continental drift. Examples of VLBI arrays are the *European VLBI Network (EVN), the *Very Long Baseline Array, and the *Australia Telescope. The advent of space-borne radio telescopes will extend VLBI baselines to hundreds of thousands of kilometres.

very slow nova *See* RR TELESCOPII STAR.

Very Small Array (VSA) A radio telescope array at *Teide Observatory in the Canary Islands, jointly owned and operated by the UK and Spain. The array consists of 14 radio dishes making observations of the *cosmic microwave background. It began operation in 2000 and has been used in three configurations: the Compact Array, with antennae of 0.14-m diameter; the Extended Array, with antennae of 0.32-m diameter; and, from 2006, the Super-Extended Configuration with antennae of 0.61-m diameter giving resolutions down to 0.1 degree.

(⊕) SEE WEB LINKS
• Official telescope website.

Vesta Asteroid 4, the fourth asteroid to be discovered, by H. W. M. *Olbers in 1807. Its mean diameter is 530 km, similar to that of *Pallas. Vesta's mass is 3×10^{20} kg and its mean density is 3.9 g/cm^3. Its orbit has a semimajor axis of 2.362 AU, period 3.63 years, perihelion 2.15 AU, aphelion 2.57 AU, and inclination of $7°.1$. Vesta's colour changes slightly as it rotates every 5.34 hours, indicating that its surface is not uniform in composition. The Hubble Space Telescope has shown bright and dark features as small as 80 km across on Vesta, suggesting a geologically diverse terrain with an exposed mantle, ancient lava flows, and impact basins. The average albedo is 0.42. Vesta's spectrum is characteristic of the mineral pyroxene, common in lava flows. Vesta may be a source of the *eucrite meteorites and some other small asteroids. It is sufficiently different from other asteroids to be assigned a special class, V (for Vesta). Its apparent mean magnitude at opposition is 6.5, but at particularly favourable oppositions it can reach magnitude 5.5, the only asteroid bright enough to be visible with the naked eye.

Viking Two US space probes to Mars, each of which consisted of an orbiter and a lander (see table). The orbiters studied the planet and its satellites, while the landers photographed the surface around them, took weather readings, and analysed the composition of the soil. Of particular interest was the possible presence of living organisms on Mars, but none were found.

Viking Probes

Probe	Launch date	Results
Viking 1	1975 August 20	Entered orbit around Mars 1976 June 19; lander descended in Chryse Planitia 1976 July 20
Viking 2	1975 September 9	Entered orbit around Mars 1976 August 7; lander descended in Utopia Planitia 1976 September 3

VILGEN photometry A combination of seven selected filters from the *Vilnius and *Geneva systems of photometry (hence VILGEN), used to distinguish peculiar and metal-poor stars. The filters used are U from the Geneva system; P from the Vilnius system; B_1 (Geneva); Y (Vilnius); Z (Vilnius); V (Geneva or UBV); and S (Vilnius).

Vilnius photometry A system of *intermediate-band photometry devised at the Vilnius observatory, Lithuania. The eight filters with their wavelengths and bandwidths are U, 345 and 40 nm; P, 374 and 26 nm; X, 405 and 22 nm; Y, 466 and 26 nm; Z, 516 and 21 nm; V, 544 and 26 nm; T, 625 and 20 nm; and S, 656 and 20 nm. These filters are designed to enable temperatures, luminosities, and peculiarities in composition and reddening to be derived. *See also* VILGEN PHOTOMETRY.

violent relaxation The rapid evolution of a star cluster or galaxy that has formed in a configuration far from equilibrium. During violent relaxation, the orbits of individual stars change dramatically because of changes in the gravitational potential of the system. It is thought to play an important part in shaping elliptical galaxies during the first billion years of their evolution.

virga A coloured streak or patch on the surface of Titan; pl.*virgae*. The name is not a geological term, and the origin of such features is not yet known. An example is Hobal Virga, over 1000 km long.

Virginid meteors A meteor shower, producing low rates (ZHR around 5) from various radiants around the ecliptic in Virgo throughout March and April. Several maxima occur, with the main one perhaps around April 12, from radiants at RA 14h 04m, dec. $-9°$ (east of the Virgo bowl) and RA 13h 36m, dec. $-11°$ (close to Spica). Virginid meteors are slow, often long, and sometimes bright. The Virginids are part of a continuous flux of activity from around the ecliptic known as the *anthelion radiant, extending to the *Ophiuchids and *Alpha Scorpiids in May and June.

Virgo (Vir) (*gen.* **Virginis**) The largest constellation of the zodiac, and the second-largest constellation of all, representing the goddess of justice. It lies on the celestial equator. The Sun passes through Virgo from the third week of September to the end of October, and hence lies within the constellation at the *autumnal equinox. Its brightest star is *Spica (Alpha Virginis). Gamma Virginis is a double star known as *Porrima. The constellation contains numerous members of the *Virgo Cluster of galaxies. M104, the *Sombrero Galaxy, is not part of the cluster. Virgo also contains the brightest quasar, 3C 273, magnitude 12.9, about 3 billion l.y. away.

Virgo A A strong radio source in the constellation Virgo, identified with the 9th-magnitude giant elliptical galaxy M87 (also known as NGC 4486), lying near the centre of the *Virgo Cluster of galaxies about 55 million l.y. away. It is a classical radio galaxy, with two lobes, the brighter one fed by a prominent jet 4000 l.y. long which is also visible at optical wavelengths.

Virgo Cluster The nearest large cluster of galaxies, and the centre of the *local supercluster. It is an irregular, roughly elliptical cluster of over 2000 galaxies, the brightest of which are within the reach of amateur telescopes. Its distance is approximately 55 million l.y., and its diameter 9 million l.y. The cluster spans $12°$ of sky, spilling northwards over the border of Virgo into Coma Berenices; hence it is sometimes also known as the *Coma–Virgo Cluster*, but should not be confused with the separate *Coma Cluster, which is considerably more distant. The Virgo Cluster has a comparatively high proportion of spirals, many dwarf galaxies, and a mass of 2 to 5×10^{14} solar masses. The four brightest members, all ellipticals, are M49 (NGC 4472), M87 (NGC 4486, a giant elliptical galaxy with an unusual jet emanating from its nucleus), M60 (NGC 4649), and M86 (NGC 4406). There is evidence of a slight asymmetry in redshifts between the spirals and the ellipticals, suggesting that the cluster is in fact two clusters superimposed along the line of sight.

Virgo Supercluster An alternative name for the *local supercluster.

virial theorem A way of estimating the total mass of an object such as a galaxy or a cluster of galaxies from the movement of its constituent members. The theorem states that the average gravitational potential energy of the constituent objects is twice their average kinetic energy. Calculations with the virial theorem show that galaxies and clusters contain up to ten times as much mass as can be seen telescopically, providing strong evidence for the existence of large quantities of *dark matter. A modified version of this theorem, called the *cosmic virial theorem*, applies on cosmological scales. It relates the statistics of galaxy motions and the *correlation function (which describes the way galaxies cluster in space) to the average density of the Universe. Since the first two quantities are measurable, the *density parameter can thus be estimated. The usual result obtained is around 0.2, indicating that there is dark matter on cosmological scales, but not enough to reach the *critical density.

Virtual Observatory A large astronomical database and associated software, which is accessed remotely. The downloading and analysis of the huge volumes of data currently produced by modern telescopes is becoming increasingly impractical. In the Virtual Observatory, the data analysis is performed remotely at the same location as the data are stored, with just the final results being transmitted to a user. Numerous nations are collaborating on an International Virtual Observatory that will be usable by anyone in the world. The European Virtual Observatory (EURO-VO) aims to establish an operational Virtual Observatory involving all European astronomical data centres.

(∰) SEE WEB LINKS
- International Virtual Observatory Alliance website.
- European Virtual Observatory website.
- Astrogrid website.

virtual particle A particle–antiparticle pair which comes into being out of nothing and then rapidly annihilates without releasing energy. Virtual particles populate the whole of space in large numbers but cannot be observed directly. The principle of conservation of mass and energy is not broken provided the virtual particles appear and disappear so quickly that the change in mass or energy cannot be detected. However, if the members of a virtual particle pair move too far apart to rejoin they may become real particles, as happens in *Hawking radiation from a black hole; the energy to make the particles real is extracted from the black hole. The lifetime of a virtual particle pair increases as the mass or energy involved decreases. Thus an electron and positron can exist for only up to about 4×10^{-21} s, but a pair of radio photons with wavelengths of 600000km could last for up to one second.

Visible and Infrared Survey Telescope for Astronomy (VISTA) A United Kingdom 4.1-m wide-field reflector at an altitude of 2635m at *Paranal Observatory, Chile. VISTA began observations in 2009, surveying the sky at near-infrared wavelengths of 0.84–2.5 µm. The VISTA Hemisphere Survey (VHS) covers the entire southern sky while other surveys will look at smaller regions in greater detail.

(∰) SEE WEB LINKS
- Official telescope website.

VISTA Abbr. for *Visible and Infrared Survey Telescope for Astronomy.

visual binary A binary star in which the two components are sufficiently well separated to be individually detectable visually or photographically. The brighter star is known as the *primary*, and the fainter as the *secondary, companion*, or *comes*. Even when orbital motion is extremely slow, visual binaries may sometimes be identified from the *common proper motion of their components.

visual magnitude (m_v) The apparent magnitude of a star as estimated by the human eye. The eye is capable of ranking stars in order of brightness and estimating equality between two stars, or a star and an artificial source. These were the only ways to measure visual magnitudes until the advent of the *photovisual magnitude technique and *photoelectric photometry. Now that magnitudes are measured precisely by photometers, it is more usual to use the apparent *V magnitude.

VLA Abbr. for *Very Large Array.

VLBA Abbr. for *Very Long Baseline Array.

VLBI Abbr. for *very long baseline interferometry.

VLBI Space Observatory Programme (VSOP) A project for extremely high-resolution radio astronomy, extending the principle of *very long baseline interferometry by putting an antenna into space. From 1997 to 2003 an 8-m antenna on board the Japanese satellite *HALCA observed in conjunction with radio telescopes on Earth. The combination of signals from the spacecraft and Earth-based telescopes synthesized a radio telescope equivalent in diameter to three times the Earth's radius. A follow-on mission, VSOP-2, using a satellite with a 9-m antenna and working at shorter frequencies, is planned for launch in 2013 or later.

(((●))) SEE WEB LINKS
• Official project website.

VLT Abbr. for *Very Large Telescope.

V magnitude The magnitude of a star in yellow-green light, the centre of the range of wavelengths to which the eye is sensitive, on the system of *Johnson photometry. The V filter has a central wavelength of 545 nm and a bandwidth of 84 nm. It is the photoelectric equivalent of the older *photovisual and *visual magnitudes. There are also V filters in *Geneva photometry, *Vilnius photometry, *Walraven photometry, and the *six-colour system, but the context should always make it clear when these are intended.

Vogel, Hermann Carl (1841–1907) German astronomer. In the 1870s he began an ambitious survey to amass stellar spectra, from which he hoped to establish a *spectral classification that would reveal something about stellar evolution. In this objective he failed, but in 1888 he detected Doppler shifts in stellar spectra which enabled him to measure reliable radial velocities for the first time. The next year he discovered that Algol and Spica were spectroscopic binary stars, and derived their masses and orbits.

Vogt–Russell theorem The theorem, found valid except in rare circumstances, that there is only one internal structure possible for a star of given mass and chemical composition. The calculation of that structure depends on knowing how quantities such as pressure, rate of energy production, and opacity depend on local gas properties such as temperature and chemical composition. The *mass–radius and *mass–luminosity relations in main sequence stars are among the theorem's consequences. It is named after the German astronomer Heinrich Vogt (1890–1968) and H. N. *Russell.

void A region of space containing far fewer galaxies than average, or even no galaxies at all; also known as a *cosmic void*. Voids with less than one-tenth the average density of the Universe on scales of up to 200 million l.y. have been detected in large-scale surveys. These regions are often (but not always) approximately spherical. The first large void was detected in Boötes in 1981; it has a radius of about 180 million l.y. and its centre is approximately 500 million l.y. from the Milky Way. The existence of large voids is not surprising, given the existence of *clusters of galaxies and *superclusters on very large scales.

V

Voigt profile The profile of a strong absorption line in a stellar atmosphere which combines the broadening effects of thermal (Doppler) motions of the atoms in the atmosphere with those of atomic collisions (*pressure broadening) which dominate the *line wings. It is named after the German physicist Woldemar Voigt (1850–1919).

Volans (Vol) (*gen.* **Volantis**) An unremarkable constellation of the southern sky, representing a flying fish. Its brightest star is Beta Volantis, magnitude 3.8. Gamma and Epsilon Volantis are attractive double stars.

 volatile A term describing an element or compound that melts or boils (vaporizes) at a relatively low temperature, or (equivalently) condenses from a gas at a low temperature. Hydrogen, helium, carbon dioxide, and water are examples. A volatile substance is the opposite of a *refractory* one. The terrestrial planets and meteorites are depleted in a wide range of elements that are volatile below about 1000 K. These volatile elements are, however, widespread in the outer Solar System, beyond 4 AU from the Sun.

von Zeipel theorem The mathematical expression of the relationship between stellar rotation and the circulation that is induced within the body of the star. It was originally advanced in 1924 by the Swedish astrophysicist Edvard Hugo von Zeipel (1873–1961), and still forms the basis of all studies of the circulation of material within a stellar interior.

Voyager Two US probes to the outer planets. Voyager 1, which was launched after Voyager 2 but on a faster trajectory, visited Jupiter and then Saturn, passing close to Saturn's largest satellite, Titan. Its trajectory was swung out of the ecliptic by the Saturn encounter, so that it could not reach other planets. Voyager 2's course took it on to Uranus and then Neptune, giving the first close-up views of these remote worlds, their satellites, and rings. Both Voyagers are now on trajectories that will take them out of the Solar System. On the way, it is hoped that they will determine the location of the *heliopause, the boundary of the region within which the solar wind flows.

(((🌐))) SEE WEB LINKS
• Official mission website.

Voyager Probes		
Probe	Launch date	Results
Voyager 1	1977 September 5	Jupiter fly-by 1979 March 5; Saturn fly-by 1980 November 12
Voyager 2	1977 August 20	Jupiter fly-by 1979 July 9; Saturn fly-by 1981 August 26; Uranus fly-by 1986 January 24; Neptune fly-by 1989 August 25.

VSOP Abbr. for *VLBI Space Observatory Programme.

Vulcan A hypothetical planet supposedly lying within the orbit of Mercury, invoked in 1859 by U. J. J. *Le Verrier to explain the advance of Mercury's perihelion (later accounted for by S. *Newcomb's improved calculation of perturbations by the other planets, and A. *Einstein's general theory of relativity). It is now known not to exist.

Vulpecula (Vul) (*gen.* **Vulpeculae**) A faint constellation of the northern sky, representing a fox. Its brightest star, Alpha Vulpeculae, is of magnitude 4.4. It contains the striking star cluster known popularly as the *Coathanger. M27 is the *Dumbbell Nebula, a large planetary nebula. The first *pulsar was discovered in Vulpecula.

VV Cephei star *See* ZETA AURIGAE STAR.

V

VY Sculptoris star A nova-like variable that occasionally shows a sudden drop in brightness, reminiscent of a *Z Camelopardalis star with an exceptionally long standstill; however, these stars never show the rise above standstill that is seen in Z Camelopardalis stars after their drop to quiescence, nor any other form of outburst activity. Because their light-curves look like an inverted version of dwarf nova light-curves, VY Sculptoris stars are sometimes known as *anti-dwarf novae*.

walled plain A now largely obsolete name for a large lunar crater with a flat floor.

Walraven photometry Photometry with a unique form of spectrophotometer devised by the Dutch astronomer Theodore Walraven (1916–2008) and his wife Johanna Helena Walraven, née Terlinden (1920–89). There are no filters; instead, the wavelength ranges are defined by quartz lenses and prisms. The names of the wavebands with their wavelengths and bandwidths are W, 325 and 14 nm; U, 363 and 24 nm; L, 384 and 23 nm; B, 432 and 45 nm; and V, 547 and 72 nm. All five bands are observed simultaneously.

waning The circumstance when the phase of the Moon is decreasing from full to new.

Washington Double Star Catalog (WDS) The world's principal database of astrometric information on double and multiple stars, maintained at the *US Naval Observatory, Washington. The *WDS Catalog* includes positions, magnitudes, discoverers, position angles, separations, spectral types, and proper motions for the components of over 100 000 systems covering the entire sky. It was first published in 1984, with updated editions in 1996 and 2001. The online version is updated daily.

(⊕) SEE WEB LINKS
• Full catalogue online, updated daily.

Water Jar A Y-shaped asterism in the constellation Aquarius consisting of the stars Gamma, Zeta, Eta, and Pi Aquarii.

water maser *See* H_2O MASER.

water of hydration Water embedded within the mineral structure of a rock or other substance, or water chemically combined with a substance to form a hydrate. About two-thirds of the main-belt C-class asteroids, including Ceres, show spectral evidence for water of hydration.

wavefront An imaginary surface connecting points at which the electromagnetic waves from a source are in phase. The wavefront is perpendicular to the direction of travel.

waveguide A hollow metal pipe, normally of rectangular cross-section, designed to carry radio waves with little attenuation. A waveguide is commonly used to carry waves from a horn antenna at the focus of a radio telescope to the receiver.

wavelength (λ) The distance between successive peaks or troughs of a wave. Wavelength is equal to the speed of the wave divided by its frequency, f. For electromagnetic radiation, which travels at c, the speed of light, $\lambda = c/f$. Wavelengths of electromagnetic radiation range from hundreds of metres for radio waves to 10^{-16} m for gamma rays.

wavenumber (σ) The inverse of wavelength. The SI unit is the reciprocal metre. For a spectral line it is the number of waves in a given distance; for example, a line of wavelength 500 nm has 2 000 000 wavelengths in 1 m.

wave plate A transparent plate which retards the phase of one plane of vibration of light relative to the plane at right angles; also known as a *retardation plate*. The retardation is usually a quarter of a wavelength (a *quarter-wave plate*) or half a wavelength (a *half-wave* *plate*). A quarter-wave plate converts elliptically or circularly polarized light into linear polarized light, or vice versa, while a half-wave plate rotates the direction of the plane of polarization of linearly polarized light. Wave plates are used in *polarimeters.

waxing The circumstance when the phase of the Moon is increasing from new to full.

WC star *See* WOLF–RAYET STAR.

WDS Abbr. for *Washington Double Star Catalog*.

weber (Wb) A unit of magnetic flux. It is named after the German physicist Wilhelm Eduard Weber (1804–91).

weird terrain An informal name for hilly and lineated terrain found on Mercury and the Moon. On Mercury, weird terrain is found on the opposite side of the planet to the large *Caloris Basin, and is thought to have formed when the seismic waves from the Caloris impact converged, a process known as *seismic focusing*. Weird terrain on the Moon, not so clearly defined as on Mercury, occurs in two places, one directly opposite the centre of the *Imbrium Basin, the other opposite the centre of the *Orientale Basin.

Weizsäcker, Carl Friedrich von (1912–2007) German theoretical physicist and philosopher. In 1938, independently of H. A. *Bethe, he proposed that stars generate their energy via the *carbon–nitrogen cycle, converting hydrogen into helium by nuclear fusion. In 1944 he set out a modern version of the nebular hypothesis proposed in the 18th century, first by I. *Kant and later by P. S. de *Laplace, to account for the origin of the Solar System.

Werner lines A sequence of absorption or emission lines due to molecular hydrogen. They lie at ultraviolet wavelengths in the range 100–123 nm, the same region of the spectrum as the *Lyman series of atomic hydrogen. They are named after their discoverer, the Danish physicist Sven Theodor Werner (1898–1984).

West, Comet (C/1975 V1) A long-period comet discovered in 1975 November by the Danish astronomer Richard Martin West (1941–); formerly designated 1976 VI. Comet West reached perihelion (0.20 AU) on 1976 February 25. In early March, when closest to Earth at 0.8 AU, the comet reached a maximum brightness of magnitude −1. The broad fan-shaped dust tail was 30–35° long. A few days after perihelion, the nucleus split into four fragments. The comet's orbit has a period of about 500 000 years, eccentricity 0.99997, and inclination 43°.1.

⊕ SEE WEB LINKS
• Information page at Cometography website.

Westerbork Synthesis Radio Telescope (WSRT) An aperture-synthesis radio telescope at Westerbork, about 40 km southeast of Groningen, in the Netherlands, owned and operated by the Netherlands Institute for Radio Astronomy (ASTRON). It consists of ten fixed and four movable dishes, each 25 m in diameter. The fixed dishes are in a 1.2-km east–west line, with two movable dishes on a 300-m length of track at the eastern end; this array was opened in 1970. In 1980 a separate 200-m track with two more movable dishes was opened 1.5 km further east, thereby increasing the maximum baseline to 3 km.

⊕ SEE WEB LINKS
• Official telescope website.

Wezen The star Delta Canis Majoris, magnitude 1.84. It is an F8 supergiant about 1600 l.y. away.

Whipple, Fred Lawrence (1906–2004) American astronomer. In the 1930s and 1940s he conducted photographic meteor patrols, and found that the Taurid meteor stream has the same orbit as Comet *Encke. In 1949 he proposed an 'icy conglomerate' theory of comets (later dubbed the 'dirty snowball' model), according to which a comet's nucleus consists of frozen gases mixed with dust; this was borne out in 1986 when the *Giotto space probe flew past the nucleus of Halley's Comet. Whipple discovered six comets, and studied planetary nebulae, flare stars, and stellar evolution.

Whipple Observatory *See* FRED LAWRENCE WHIPPLE OBSERVATORY.

Whirlpool Galaxy The 8th-magnitude galaxy M51 in Canes Venatici, also known as NGC 5194. It is an Sc-type spiral seen nearly face-on, and was the first galaxy in which spiral structure was clearly seen, by Lord *Rosse. It is interacting with a small irregular galaxy, NGC 5195, which appears to lie at the end of one of its spiral arms but is actually slightly behind it. The Whirlpool Galaxy lies about 25 million l.y. away. 📷

whistler A phenomenon in which audible whistles of falling pitch are occasionally received by radio telescopes. Whistlers are caused by radio waves from distant lightning flashes suffering dispersion in the ionosphere and being reflected back to the ground.

white dwarf A small, dense star that is the end-result of the evolution of all but the most massive stars. White dwarfs are thought to form from the collapse of stellar cores once nuclear burning has ceased there. The core is exposed to view when the outer parts of the star are driven off to form a planetary nebula. Such a core contracts under its own gravity until, having reached a size similar to that of the Earth, it has become so dense ($5 \times 10^8 \, \text{kg/m}^3$) that it is supported against further collapse by the pressure of *electron degeneracy. White dwarfs are formed with high surface temperatures (above $10\,000 \, \text{K}$) because of the heat trapped within them, released both by previous nuclear burning and through gravitational contraction. They gradually cool, becoming fainter and redder. White dwarfs may constitute 30% of the stars in the solar neighbourhood, but because of their low luminosity (typically 10^{-3} to 10^{-4} of the Sun's) they are very inconspicuous. The maximum possible mass for a white dwarf is 1.44 solar masses, the *Chandrasekhar limit. An object of greater mass would contract further and become either a neutron star or a black hole. *See also* D STAR.

white hole The time-reverse of the collapse of an object into a black hole. The equations of general relativity which describe such a collapse are time-symmetric, so there is no theoretical reason why they should not run in reverse. A white hole would therefore be a place where matter spontaneously appeared into our Universe. However, no such object has been observed.

white-light corona The Sun's corona as seen at visible wavelengths during total solar eclipses and with *coronagraphs. The white-light emission arises from light from the Sun's photosphere that is scattered by free electrons (the *K corona) and dust (the *F corona). A small amount of visible light is from emission lines (the *E corona).

WHT Abbr. for *William Herschel Telescope.

Wide Field Infrared Survey Explorer (WISE) A NASA satellite launched 2009 December into polar orbit around the Earth to perform an imaging survey of the sky at mid-infrared wavelengths from 3.4 to 22 μm with a 0.4-m telescope. WISE is expected to return over a million images containing information on objects ranging from asteroids to nearby brown dwarfs, planet-forming disks around other stars, and ultra-luminous galaxies. It will produce a reference catalogue of objects for the *James Webb Space Telescope. WISE ceased observations in 2011 February.

(⊕) SEE WEB LINKS
• Official mission website.

Widmanstätten pattern A cross-hatched pattern appearing on the surface of *octahedrite iron meteorites which have been polished and etched with acid. It was discovered in 1804 by the Austrian mineralogist Aloys Joseph (Beck Edler) von Widmanstätten (1754–1849). It results from an intergrowth of the minerals kamacite and taenite as the meteorite parent body cooled in space.

Wielen dip A feature of the *luminosity function of stars. Instead of rising steadily towards fainter absolute magnitudes, the number of stars increases little between absolute magnitudes 6 and 9. This feature may provide a clue to the formation of stars of different masses, but its full significance is not yet understood. It is named after its discoverer, the German astronomer Roland Wielen (1938–).

Wien's displacement law The relation between the wavelength of peak emission from a black body and its temperature. At low temperatures *black-body radiation is confined mainly to the infrared region of the spectrum, but at progressively higher temperatures the peak of the emission is displaced to progressively shorter wavelengths. According to the law, the wavelength of peak emission, λ_{max}, multiplied by T, the thermodynamic temperature of the body, is a constant. Although celestial bodies are not perfect black bodies, the displacement law is still useful for predicting the wavelengths near which most of their radiation is emitted. For example, the cosmic background radiation has $T = 2.7$ K and $\lambda_{max} = 1$ mm; a cool, red star has $T = 3000$ K and $\lambda_{max} = 1$ μm (in the infrared); the Sun, $T = 6000$ K and $\lambda_{max} = 500$ nm (visible); the hottest normal stars, $T = 30000$ K and $\lambda_{max} = 100$ nm (ultraviolet); planetary nebula nuclei $T = 100000$ K and $\lambda_{max} = 30$ nm (extreme ultraviolet). The law is named after the German physicist Wilhelm Wien (1864–1928).

Wild Duck Cluster The 6th-magnitude open cluster M11 in Scutum, also known as NGC 6705. Visually it appears somewhat fan-shaped, like a flight of ducks, covering nearly ¼°. It lies 6500 l.y. away.

Wild's Triplet A group of three disturbed barred spiral galaxies with luminous connecting bridges, about 200 million l.y. away in the constellation of Virgo. The bridges are probably the result of gravitational tidal interactions between the galaxies. The trio is named after the British-born Australian astronomer (John) Paul Wild (1923–2008), who investigated it in the early 1950s.

Wildt, Rupert (1905–76) German-American astronomer. In the 1930s he showed that prominent absorption bands in the spectra of the giant planets indicate the presence of methane and ammonia in their atmospheres. He also proposed models for the giant planets in which an iron–silicate core is surrounded by a deep ice mantle and a deep atmosphere. Wildt also worked on solar and astrophysical problems.

Wilkinson Microwave Anisotropy Probe (WMAP) A NASA satellite launched in 2001 June to study small fluctuations in the *cosmic microwave background left over from the Big Bang, in succession to the *Cosmic Background Explorer. It was positioned at the L_2 *Lagrangian point of the Earth's orbit, 1.5 million km from Earth in the direction opposite the Sun, from where it could observe the Universe without interference from the Sun, Earth, or Moon. WMAP determined the age of the Universe to be 13.73 billion years to within 1% (0.12 billion years); found that the *curvature of spacetime is within 1% of flat (Euclidean); determined that ordinary atoms (baryons) make up 4.6% of the Universe, to within 0.1%; *dark matter (not made up of atoms) makes up 23.3%, to within 1.3%; and that dark energy makes up the remaining 72.1%, to within 1.5%. WMAP was named in honour of the American cosmologist David Todd Wilkinson (1935–2002). It ceased observations in 2010 August.

(()) SEE WEB LINKS
• Official mission website.

William Herschel Telescope (WHT) A 4.2-m reflector at the *Roque de los Muchachos Observatory in the Canary Islands, opened in 1987. It is jointly owned and operated by the UK, the Netherlands, and Spain.

(()) SEE WEB LINKS
• Official telescope website.

w

Wilson, Robert Woodrow (1936–) American physicist. Using a sensitive horn antenna at Bell Laboratories in Holmdel, New Jersey, originally developed for satellite communications, he and A. A. *Penzias found a faint background noise emanating from all parts of the sky. They had discovered the *cosmic microwave background, interpreted as energy left over from the *Big Bang. Their discovery, published in 1965, earned Penzias and Wilson the 1978 Nobel Prize in Physics.

Wilson–Bappu effect A relationship between a property of the calcium *K line in late-type stars and the luminosity of the star. When the strong calcium absorption line is observed at high resolution it is often found to have a weak emission at its centre. The strength of this line has been found to correlate with the star's brightness. The effect is named after the American astronomer Olin Chaddock Wilson (1909–94) and the Indian astronomer (Manali Kallat) Vainu Bappu (1927–82).

Wilson effect The apparent displacement of a sunspot's *umbra towards the Sun's centre as the spot approaches the limb, named after the Scottish astronomer Alexander Wilson (1714–86), who first observed it in 1769. The effect gave rise to the belief that sunspots were saucer-shaped depressions. However, good examples of the effect are few. Sometimes, displacement of the umbra *away* from the Sun's centre is seen (the *reverse Wilson effect*). The effect could be explained by the fact that spots rarely have a circular outline.

Wilson–Harrington, Comet 107P/ An object classified as a periodic comet in 1949 when discovered by the American astronomers Albert George Wilson (1918–) and Robert George Harrington (1904–87), but subsequently found to be identical to asteroid 4015 discovered in 1979. This was the first evidence of a comet evolving into an asteroid-like body. Asteroid 4015 is a member of the *Apollo group, and has a diameter of about 4 km. It can approach to within 0.049 AU (7.3 million km) of the Earth's orbit. Its orbit has a semimajor axis of 2.639 AU, period 4.29 years, perihelion 0.991 AU, aphelion 4.29 AU, and inclination $2°.8$.

((∰)) SEE WEB LINKS
• Information page at Cometography website.

WIMP Abbr. for weakly interacting massive particle, a hypothetical particle of non-baryonic *dark matter. In some theories, WIMPs are assumed to pervade the cosmos. Examples could be *cold dark matter, such as axions or photinos, or *hot dark matter, such as neutrinos. *See also* MACHO.

Wind A NASA satellite launched in 1994 November to study the incoming solar wind and the Earth's magnetosphere. Initially it was placed in a highly elliptical orbit with apogee from 80 to 250 Earth radii over the Earth's day side, then in 1997 moved to a halo orbit around the L_1 *Lagrangian point. In 1998, it returned to an orbit around the Earth, this time at an inclination of 60°, allowing it to study conditions far from the ecliptic. In 2004 it was returned to the L_1 point.

((∰)) SEE WEB LINKS
• Official mission website.
• Mission page at NASA.

winter solstice *See* SOLSTICE.

WISE Abbr. for *Wide Field Infrared Survey Explorer.

WIYN Telescope A 3.5-m reflector jointly operated by Wisconsin, Indiana, and Yale Universities, plus the National Optical Astronomy Observatory (hence WIYN). It is situated at *Kitt Peak National Observatory and began operation in 1994. It is designed for wide-field spectroscopic observations. In 2001 the WIYN consortium took over operation of the NOAO 0.9-m reflector, originally opened on Kitt Peak in 1960.

((∰)) SEE WEB LINKS
• Official telescope website.

WMAP Abbr. for *Wilkinson Microwave Anisotropy Probe.

W. M. Keck Observatory (WMKO) An observatory at an altitude of 4146 m on *Mauna Kea, Hawaii, the site of the twin 10-m Keck Telescopes which have main mirrors consisting of 36 hexagonal segments. Keck I was completed in 1992 and Keck II in 1996. The two telescopes are sited 85 m apart and can be used as an optical interferometer. The observatory and telescopes are owned and operated by the California Association for Research in Astronomy (CARA), a partnership between the University of California, the California Institute of Technology, and NASA. Construction of the observatory and telescopes was funded by the W. M. Keck Foundation, a charitable organization, after which they are named. ◙

((⊕)) SEE WEB LINKS
• Official observatory website.

WN star *See* WOLF–RAYET STAR.

Wolf, Maximilian ('Max') Franz Joseph Cornelius (1863–1932) German astronomer. In 1891 he instituted a programme of wide-field photography for the discovery of asteroids, which appeared on the developed plates as trails among the stars. The first to be found in this way was (323) Brucia, in 1891. In all Wolf discovered over 200 asteroids; they include *Achilles, the first *Trojan asteroid to be found; in 1906. Independently of E. E. *Barnard, Wolf proposed that 'voids' in the Milky Way are dark nebulae. He also discovered the North America Nebula (and gave it that name), Comet 14P/Wolf (1884), and various clusters of galaxies, including the *Coma Cluster.

Wolf, (Johann) Rudolf (1816–93) Swiss astronomer. Following the discovery of the solar cycle by S. H. *Schwabe, Wolf set about collecting all available data on sunspot numbers. He established the dates of all sunspot maxima and minima going back to the year 1610, and calculated the length of the solar cycle to be 11.1 years. He devised the system of measuring solar activity by counting sunspots (*see* RELATIVE SUNSPOT NUMBER). Wolf was one of several astronomers to note that variations in terrestrial magnetism and auroral activity mirrored the solar cycle.

Wolf diagram A graph of star counts at various apparent magnitudes, from which the distance to a dark nebula can be determined. Max *Wolf first plotted star counts for a dark nebula and a comparison field on the same graph, and used the magnitude at which the two counts started to diverge to estimate the distance at which the obscuration began.

Wolf–Lundmark–Melotte system (WLM system) An 11th-magnitude dwarf Magellanic-type irregular galaxy, a member of the Local Group; also known as DDO 221. It was discovered in 1909 by Max *Wolf and later rediscovered by K. E. *Lundmark and the English astronomer Philibert Jacques Melotte (1880–1961). It lies about 3 million l.y. away in Cetus.

Wolf–Rayet star (WR star) A type of very luminous star, believed to be a late stage in the evolution of stars born with spectral type O which have lost their hydrogen envelopes through strong stellar winds, exposing the helium core. Wolf–Rayet stars are of over 10 solar masses and have surface temperatures of 20 000–40 000 K. They have strong stellar winds with mass-loss rates of order 10^5 solar masses per year, and outflow velocities of 2000–3000 km/s. Their spectra show strong emission lines, and they are classified as WN or WC types according to whether nitrogen or carbon dominates. They may be the progenitors of supernovae of Types Ib and Ic, and of some *gamma-ray bursters. They are named after the French astronomers Charles Joseph Étienne Wolf (1827–1918) and Georges Antoine Pons Rayet (1839–1906), who discovered them in 1867. The brightest example is 2nd-mag. Gamma Velorum.

Wolf sunspot number An obsolete name for *relative sunspot number.

Wolf 359 The third-nearest star to the Sun, 7.8 l.y. away in the constellation Leo. It is an M6 dwarf of magnitude 13.5, and has a luminosity 1/50 000 of the Sun's. It is a flare star, with the variable-star designation CN Leonis.

Wollaston prism A device used for analysing and producing polarized light, used in some designs of *polarimeter. It consists of two prisms of either quartz or calcite cemented together. The Wollaston prism divides incoming light into two linearly polarized beams with perpendicular planes. It is named after the British scientist William Hyde Wollaston (1766–1828).

Wolter telescope Any of a series of designs for *grazing-incidence telescopes by the German physicist Hans Karl Herman Wolter (1911–78). To obtain a true image over an extended field of view requires photons to undergo two successive reflections from combinations of paraboloid–hyperboloid or paraboloid–ellipsoid surfaces. For imaging the commonly used design is the Wolter I telescope; for spectroscopy, where a smaller field of view but very high spatial resolution is required, the Wolter II design is appropriate.

Woolley, Richard van der Riet (1906–86) English astronomer. His work covered solar studies, globular clusters, and stellar dynamics. His analyses of the radial velocities and proper motions of large numbers of stars clarified the structure of our Galaxy and identified groups of stars that formed at different stages in its evolution. This led to the modern view that a galaxy such as ours is born from a collapsing cloud of gas that gradually flattens out into a disk, with the oldest stars being in a surrounding halo and the youngest stars in the disk. Woolley was the eleventh Astronomer Royal (1956–71).

world line The trajectory followed by an object in *spacetime. The fact that spacetime is four-dimensional makes world lines difficult to visualize but, if the Universe had only one dimension in space and one in time, the world line could be drawn on a graph with time plotted vertically and distance plotted horizontally. A particle at rest with respect to the coordinate system would have a world line which runs along the vertical axis, while moving particles would have world lines which are curves or straight lines sloping upwards. In the real Universe, the path of a moving particle is a curved line in spacetime.

wormhole A hypothetical hole or tunnel in the fabric of *spacetime. Standard cosmological theories are based on the assumption that spacetime is smooth and simply connected. To give a three-dimensional analogy, spacetime is assumed to be like a sphere rather than a *torus; a sphere is said to be simply connected, whereas a torus is not. In *quantum cosmology it is thought that, on scales of the order of 10^{-35} m, spacetime has a very complicated, multiply connected structure in which 'tunnels' and 'handles' supply short cuts between apparently distant points. In principle, sufficiently large wormholes might allow one to travel to a distant part of the Universe much more quickly than light and, in some situations, to travel in time. The existence of wormholes is, however, highly speculative.

Wright telescope A variant on the *Schmidt camera proposed in 1935 by the American Franklin B. Wright (1891–1967). It uses an oblate spheroidal mirror and a *corrector plate which has a greater corrective effect than in a conventional Schmidt camera, reducing the tube length by half. The *focal surface is flat, and the instrument can also be adapted for visual use by inserting a secondary mirror.

wrinkle ridge A meandering ridge on a planetary surface that resembles a wrinkle in skin or cloth. Wrinkle ridges were first named on the Moon, but are found on all the terrestrial planets and many large satellites. They are thought to arise from some combination of subsurface faulting and compression.

WR star Abbr. for *Wolf–Rayet star.

W Serpentis star See ALGOL STAR.

W Ursae Majoris star A type of eclipsing binary with a period of less than 1 day; abbr. EW. The system is a *contact binary, with ellipsoidal components, generally of spectral type F–G (or later). Light variation is continuous throughout the orbit, such that the beginning and end of eclipses cannot be defined. *Primary and *secondary minima are usually of similar amplitude, which is normally below 0.8 mag.

W Virginis star A type of Population II Cepheid variable; abbr. CW. They are giants that undergo *radial pulsation in the *fundamental mode, with periods from 0.8 to 35 days and amplitudes of 0.3–1.2 mag. Their light-curves generally resemble those of the *classical Cepheids, particularly if they have a period of 3–10 days, but outside this range the curves have different amplitudes or shapes. In addition, W Virginis stars have lower masses (0.4–0.6 solar masses) than classical Cepheids and obey a different *period–luminosity relation, with absolute magnitudes that are fainter by 0.7–2 mag. There are two subtypes: CWA, with periods of more than 8 days, which includes W Virginis itself; and CWB, with periods of less than 8 days. The latter are sometimes known as *BL Herculis stars*.

Wyoming Infrared Observatory (WIRO) An observatory at an altitude of 2943 m on Jelm Mountain near Laramie, Wyoming, operated by the University of Wyoming. Its main instrument is the 2.3-m Wyoming Infrared Telescope, opened in 1977.

(⊕) SEE WEB LINKS
• Official observatory website.

WZ Sagittae star A form of dwarf nova with an extremely long outburst period, comparable with that of a *recurrent nova (i.e. decades). WZ Sagittae itself erupted in 1913, 1946, 1978, and 2001. Small-amplitude variations (less than 0.3 mag.) occur between the major eruptions, and *Algol-like eclipses are sometimes visible. The outbursts resemble those of an *SU Ursae Majoris star with superhumps, and these systems are now regarded as SU Ursae Majoris stars with an exceptionally long interval between superoutbursts; no normal outbursts are observed between them. The mass transfer rate is considerably lower than in typical SU Ursae Majoris stars. Observations suggest that an *excretion disk* of matter is being ejected from the system.

w

Xinglong Observatory An observatory of the Chinese Academy of Sciences founded in 1968 at an altitude of 950 m some 170 km north of Beijing, China. Its main telescope is a 2.16-m reflector opened in 1989. Other instruments include a 1.26-m infrared telescope.

(⊕) SEE WEB LINKS
• Official observatory website.

XMM-Newton An ESA satellite for X-ray astronomy, launched in 1999 December, named from its X-ray multi-mirror design and to honour Isaac Newton. It carries three grazing-incidence telescopes with CCD detectors for imaging and spectroscopy in the energy range 0.1–12keV (120–1 nm), plus a 0.3-m telescope that checks for visual variability. XMM-Newton is in a highly elliptical 48-hour orbit to allow extended observations of sources.

(⊕) SEE WEB LINKS
• ESA mission website.

X-ray astronomy The observation of X-ray emission from celestial objects, which allows us to study violent and energetic processes in the Universe. The subject was born in 1949, when X-rays from the Sun were discovered by a rocket-borne experiment. The first celestial X-ray source, *Scorpius X-1, was detected in 1962, also by a sounding rocket. Rockets and high-altitude balloons during the 1960s revealed other individual sources, the existence of a diffuse X-ray background, and the variability of several objects.

The first dedicated X-ray astronomy satellite was *Uhuru, launched in 1970, which carried out the first X-ray survey of the sky. Other satellites with X-ray instruments included the *Copernicus satellite and the *Astronomical Netherlands Satellite. Balloon and rocket experiments also continued throughout this period. The UK's Ariel 5 and the US SAS-3 satellites launched in 1974 and 1975 extended the catalogues of known X-ray sources and monitored their variability.

In 1977 NASA launched the first of its *High Energy Astrophysical Observatories (HEAO). HEAO-1, considerably larger than any previous mission, surveyed the sky with unprecedented sensitivity across a large energy range, from 0.1keV to 10MeV. HEAO-2, later renamed the *Einstein Observatory, was the first satellite to carry a grazing-incidence telescope and record true X-ray images of the sky. This imaging capability, coupled with the ability to study faint sources, revolutionized X-ray astronomy, putting it on an equal footing with other wavelengths.

Progress continued with European, Japanese, and Russian missions. ESA's *Exosat carried imaging and non-imaging experiments in a unique 96-hour elliptical orbit in 1983. This allowed long unbroken viewing of sources. Beginning in 1979, Japan launched three missions—*Hakucho, *Tenma, and *Ginga—carrying experiments of increasing size. In particular, the large array of proportional counters on Ginga improved spectral studies of X-ray sources. In 1987 the Roentgen Observatory was docked to the Mir space station. It was followed by the *Granat satellite in 1989.

X-ray astronomy has continued to develop rapidly. *Rosat, launched in 1990, performed the first imaging sky survey, detecting more than 60000 sources, a huge increase on the numbers previously catalogued. The Broad Band X-ray Telescope on the Space Shuttle Astro-1 mission in 1990 and the launch of a fourth Japanese mission, *ASCA, improved spectral resolution with the use of CCD detectors. *BeppoSAX was launched in 1996. Two major X-ray observatories were launched in 1999: *Chandra, concentrating on high-resolution imaging and spectroscopy, and *XMM-Newton, which has high-throughput spectroscopy. The Japanese *Suzaku observatory was launched in 2005.

X-ray background X-ray emission covering the entire sky, discovered in the 1960s. For 35 years it was uncertain whether this background consists of a large number of point sources, or emission from hot intergalactic gas. The *ROSAT and *Chandra observatories have shown that over 75% of this background can be resolved into distinct sources. About half of these individual sources are *active galactic nuclei and *quasars. The other half are *X-ray binaries in other galaxies.

X-ray binary An X-ray emitting binary system consisting of a neutron star (or in a few cases a black hole) and a normal stellar companion. They are divided into two groups according to the mass of the companion. In a low-mass X-ray binary (LMXB), for example *Scorpius X-1, the normal star is less than about 2 solar masses; it is greater than about 10 solar masses in a high-mass X-ray binary (HMXB), such as Cygnus X-1. Binaries with intermediate-mass companions are very rare. The X-ray emission is a result of accretion of material from the companion on to the neutron star.

X-ray bright point *See* CORONAL BRIGHT POINT.

X-ray burst A strong surge of emission from an X-ray binary. Bursts are of short duration, typically less than 1 min. So-called type I bursts are a result of thermonuclear fusion of material accreted on to the surface of a neutron star. Type II bursts are only found in one object, the Rapid Burster (MXB 1730–335), and are caused by spasmodic accretion.

X-ray calorimeter A device used to detect X-rays by converting the absorbed energy from an X-ray photon into heat. It consists essentially of an absorber, cooled to a very low temperature (about 0.1 K), coupled to one or more heat detectors (thermistors), and gives good energy resolution. The first X-ray calorimeter was carried on the Japanese–US *Suzaku mission launched in 2005 July.

X-ray nova A nova-like outburst at X-ray wavelengths, which may also have an optical counterpart. The main difference between X-ray and optical novae may be that the compact object is a neutron star or black hole rather than a white dwarf. *See also* X-RAY TRANSIENT.

X-ray pulsar A regularly variable X-ray binary, in which the pulsation is associated with the spin period of the compact companion, a magnetized neutron star; abbr. XP. Periods range from a few seconds to a few minutes. These pulsations are thought to be caused by the magnetic field channelling the accreting gas on to the poles of the star, producing localized 'hot spots' which move in and out of view as the star spins. An example of such a system is *Hercules X-1.

X-rays Electromagnetic radiation with wavelengths between those of ultraviolet and gamma rays, approximately 0.01–10 nm. At these short wavelengths it is more usual to talk in terms of photon energies. The energy range for X-rays is approximately 0.1–100 keV.

X-ray source A celestial object in which high-energy processes such as accretion on to a compact object, the shock wave from a supernova, a stellar wind, or the hot gas in stellar coronae give rise to detectable X-ray emission. The first X-ray sources to be discovered were mostly objects such as *X-ray binaries. Subsequently, supernova remnants, active galactic nuclei, and hot white dwarfs were also found to be X-ray sources. Higher-sensitivity missions have revealed that most objects are X-ray sources at some level. Currently, around 200 000 X-ray sources are known, the majority discovered by *XMM-Newton.

X-ray telescope An instrument used to focus X-rays into an image. Most X-ray telescopes are *grazing-incidence telescopes, based on a technique first developed in the 1940s and 1950s. A complete unit may combine a number of individual mirrors mounted concentrically, inside one another. The most commonly used types are *Wolter telescopes. Since the mid-1980s normal-incidence X-ray telescopes have been developed, exploiting the reflecting properties of multi-layer coatings on conventional mirrors. However, their efficiency is restricted to only a very narrow wavelength range, determined by the particular coating used. More recent research has led to the development of *micropore optics, which consist of a large array of small holes in a silicon wafer, with one wall of each hole being an X-ray reflecting surface. In all designs of telescope, the X-rays are focused on to an X-ray detector, such as a *gas scintillation proportional counter, a *proportional counter, a *CCD spectrometer, or a *microchannel plate detector.

X-ray Timing Explorer (XTE) *See* ROSSI X-RAY TIMING EXPLORER.

X-ray transient A burst of X-ray emission that, after rising to a maximum brightness, fades until no longer detectable. Examples are novae, supernovae, cataclysmic variables, stellar flares, and active galactic nuclei. Many sources originally detected as transients have subsequently been found to be weak quiescent X-ray sources.

XTE Abbr. for *Rossi X-ray Timing Explorer.

XUV Abbr. for *extreme ultraviolet.

X

Yagi antenna A directional aerial consisting of a dipole and a number of parallel rods arranged perpendicular to the line of sight. The rods in front of the dipole are termed *directors*, while the one behind is the *reflector*. The design is commonly used in television aerials, but in radio astronomy Yagi antennas are used as elements in interferometers. The design is named after the Japanese electrical engineer Hidetsugu Yagi (1886–1976).

Yale Bright Star Catalogue An alternative name for the *Bright Star Catalogue*.

Yarkovsky effect A change in orbital momentum experienced by small rotating particles as a result of absorption and re-emission of solar radiation, similar to the *Poynting–Robertson effect. Whereas the Poynting–Robertson effect is equal in all directions, the re-emission which causes the Yarkovsky effect is anisotropic. The object may be accelerated or decelerated by the momentum carried away by the radiated photons, depending on the spin direction. The effect is named after the Russian scientist Ivan Osipovich Yarkovsky (1844–1902).

Y dwarf A spectral classification reserved for as-yet unobserved brown dwarfs with surface temperatures below about 500 K, cooler and fainter even than the T dwarfs. Such stars would be so cool that water features would disappear from the spectrum, as H_2O condenses into water vapour at temperatures around 150 K, the same temperature as the atmosphere of Jupiter.

year The Earth's period of orbital revolution around the Sun, and hence by extension the orbital period of any planet. Astronomically the Earth's year can be defined in several ways. The actual orbital period with reference to the fixed stars is the *sidereal year, 365.25636 days. However, the positions of the stars are gradually changing because of precession. The interval between successive passages through the *mean equinox is termed the *tropical year, 365.24219 days, which takes precession into account. This is the most commonly adopted definition of the year as it is the one that relates directly to seasonal changes. A further form of year is the *anomalistic year, which is the average interval between successive passages of the Earth through the perihelion point of its elliptical orbit, 365.25964 days. This differs slightly from the sidereal year due to small perturbations of the Earth's orbit by the gravitational influence of the other planets. Another definition of the year in astronomical use is the *eclipse year of 346.62003 days. This is the average interval between successive passages of the Sun through a node of the Moon's orbit. *See also* CALENDAR YEAR.

Yepun The fourth 8.2-m Unit Telescope (UT4) of the European Southern Observatory's *Very Large Telescope in Chile, opened in 2000. Its name means 'Venus' as the evening star in the local Mapudungun language.

Yerkes Observatory The observatory of the University of Chicago at Williams Bay, Wisconsin, on the shores of Lake Geneva, at an altitude of 330 m. It was founded in 1897 by G. E. *Hale with a grant from the American businessman Charles Tyson Yerkes (1837–1905). Its main instrument is the world's largest refractor, aperture 40 inches (1.02 m), opened in 1897. It also has a 1-m reflector, opened in 1967, with adaptive optics added in 1994.

(⊕) SEE WEB LINKS
• Official observatory website.

Yerkes system 1. Another name for the *Morgan–Keenan classification system for stellar spectra.
 2. An alternative name for *Morgan's classification for galaxies.

Yinghuo-1 A Chinese Mars orbiter to be launched with the Russian *Phobos-Grunt mission in 2011 or 2012. Once Phobos-Grunt reaches Mars, Yinghuo-1 will be released into a highly elliptical equatorial orbit around Mars to study the planet's atmosphere and magnetic field.

Y magnitude An infrared magnitude used in the *UKIDSS system of photometry.

Yohkoh A Japanese satellite, with US and UK involvement, launched in 1991 August to study flares and other solar activity at X-ray and gamma-ray wavelengths. The name means 'sunbeam' in Japanese. Yohkoh carried four instruments: a Soft X-ray Telescope (SXT), a Hard X-ray Telescope (HXT), a Bragg Crystal Spectrometer (BCS), and a Wide Band Spectrometer (WBS). It ceased observations in 2001 December and re-entered the Earth's atmosphere in 2005 September.

 SEE WEB LINKS
• Mission page at Japan Aerospace Exploration Agency.

yoke mounting A form of *English mounting in which the telescope is held within the yoke, rather than in the cross-axis variant.

YY Orionis star A subtype of the irregular *Orion variables, related to the *T Tauri stars; abbr. IN(YY). Their spectrum shows absorption on the long-wave side of the emission lines, indicating an infall of material.

y

z Symbol for *redshift.

ZAMS Abbr. for *zero-age main sequence.

Z Andromedae star A type of cataclysmic variable with irregular brightness fluctuations, exhibiting the spectra typical of a hot star (20000 K), a cool star (3000 K), and extended nebulosity excited by radiation from the hot component; abbr. ZAND. Interaction occurs either through mass transfer or a stellar wind from the cool giant secondary. Currently, a heterogeneous collection of objects is classified by this type (often known as *symbiotic stars), including the *symbiotic novae* (*RR Telescopii stars), and stars with an *accretion disk and jets, such as R Aquarii.

Zanstra method A process for estimating the temperature of the central star in a planetary nebula. The basic assumption is that the nebula absorbs all the ultraviolet light from the star which can cause ionization. For each ultraviolet photon absorbed a Hα photon is emitted when the ionized hydrogen subsequently recombines with an electron. Thus the strength of the Hα line is related to the ultraviolet magnitude of the star. The strength of the red continuum (broad-band) emission underlying the Hα line gives the red magnitude and hence a colour temperature for the star. This method is not affected by interstellar absorption, since both colours are effectively measured at the same wavelength. It is named after the Dutch astrophysicist Herman Zanstra (1894–1972), who first published it in 1927.

ZC Abbr. for *Zodiacal Catalogue.

Z Camelopardalis star A type of *U Geminorum star that occasionally exhibits a standstill at intermediate light, but which otherwise behaves like an *SS Cygni star; abbr. UGZ. A standstill normally begins on the decline from an outburst and may last for a few days or persist for years. During this state there is increased *mass transfer between the components. The amplitude of normal outbursts is similar to those of the UGSS type (2–5 mag.), but the mean interval tends to be shorter (10–40 days).

ZD Abbr. for *zenith distance.

Zeeman effect The splitting of a spectral line due to a magnetic field. It is named after the Dutch physicist Pieter Zeeman (1865–1943). The Zeeman effect is widely used for determining magnetic fields in astronomical objects, notably the Sun. However, field strengths are often insufficient to cause actual line splitting (sunspot fields, of about 0.4 tesla, are an exception); instead, line broadening occurs. The strengths of weak fields can be deduced by measuring the polarization of each *line wing, which is the principle of the solar *magnetograph*, an instrument for measuring magnetic fields.

Zelenchukskaya The location in the Caucasus Mountains, Russia, of the *Special Astrophysical Observatory of the Russian Academy of Sciences.

zenith The point on the celestial sphere directly above an observer. A line to the zenith is at right angles to the plane of the horizon. The zenith is in the opposite direction from the *nadir. Strictly speaking, this definition refers only to the *astronomical zenith*; two other, slightly different forms of zenith may also be defined. The *geocentric zenith* is the direction indicated by a line from the centre of the Earth through the observer. The *geodetic zenith* is at right angles to the *geoid at the observer's location. All three forms of zenith differ slightly because of the non-spherical shape of the Earth. If used without qualification, 'zenith' means the astronomical zenith.

zenithal attraction An effect in which the observed radiants for meteors appear closer to the zenith than their true position. It occurs because meteoroids impacting on the upper atmosphere are subject to the Earth's gravitational pull, which increases their geocentric velocities. The change in apparent orbital velocity is most marked for slow meteoroids. The effect must be corrected for when analysing positional observations to determine meteor shower radiants and heliocentric orbits.

zenithal hourly rate (ZHR) An index of the activity of a meteor shower. The ZHR gives the activity in meteors per hour which would be expected by an experienced visual observer under a perfectly clear sky (limiting magnitude +6.5) with the radiant overhead. In practice, observed rates are almost always lower than the derived ZHR due both to sky conditions and the fact that the radiant is seldom directly overhead. Meteor showers with ZHRs lower than about 5 are usually difficult to distinguish from the background level of *sporadic meteors.

zenith distance (ZD) (z) The angular distance of a celestial body from the zenith. The zenith distance is 90° minus the body's altitude above the horizon (i.e. the complement of the altitude) and hence is also known as *coaltitude*.

zenith telescope An instrument for measuring the zenith distances, and/or times of meridian transit, of stars close to the zenith. The telescope is mounted so that it points vertically, and so does not distort under its own weight, as it would if it were not vertical. In addition, the effects of atmospheric refraction are smallest at the zenith. Hence star positions can be measured at the zenith with maximum accuracy.

zero-age horizontal branch That part of the Hertzsprung–Russell diagram occupied by low-mass stars that have just undergone the *helium flash. The scatter in properties of stars on the zero-age horizontal branch is due to the range of stellar masses resulting from mass loss on the giant branch. *See also* HORIZONTAL BRANCH.

zero-age main sequence (ZAMS) The diagonal strip on the Hertzsprung–Russell diagram occupied by stars that have just begun to convert hydrogen to helium in their cores. As this conversion proceeds, stars become slightly redder and more luminous, so they move upwards and to the right on the HR diagram. As a result, the main sequence is a broad band that is displaced slightly from this zero-age strip.

Zeta Aurigae star An eclipsing binary with a *composite spectrum, in which the components are a late-type bright giant or supergiant and an early-type star. Immediately before and after primary eclipse the hot star is visible through the outermost layers of the secondary. If the spectrum shows features which indicate the presence of an M-type supergiant, emission, or mass transfer (but not necessarily eclipses), the star is sometimes known as a *VV Cephei star*. In Zeta Aurigae and VV Cephei themselves the orbital periods are 972 and 7430 days, respectively. The secondary in VV Cep is estimated to have a diameter over 1000 times that of the Sun, making it one of the largest stars known.

ZHR Abbr. for *zenithal hourly rate.

zirconium star *See* S STAR.

Z magnitude An infrared magnitude used in the *UKIDSS system of photometry.

zodiac The strip of sky up to 8° either side of the ecliptic against which the Sun, Moon, and major planets appear to move. The strip is divided into twelve *signs of the zodiac*, each 30° long. These signs were named by the ancient Greeks after the constellations of the zodiac that occupied the same positions some 2000 years ago. As a result of *precession, the constellations have moved eastwards by over 30° and no longer coincide with the signs.

Zodiacal Catalogue (ZC) Popular name for the *Catalog of 3539 Zodiacal Stars* published in 1940 by the Nautical Almanac Office of the US Naval Observatory. It lists the positions, magnitudes, spectral types, and proper motions of stars within 8° of the ecliptic, mostly brighter than magnitude 8.5, for epoch 1950.0. ZC numbers are still used for identifying stars that are occulted by the Moon.

zodiacal dust The population of small (1–300 μm) particles of cometary and asteroidal origin lying roughly in the ecliptic plane out to about 5 AU from the Sun; measurements from the Pioneer and Voyager probes have indicated much lower dust concentrations beyond the orbit of Jupiter. The distribution of the zodiacal dust is controlled mainly by Jupiter's gravitational influence. Reflection of sunlight from the zodiacal dust gives rise to the *gegenschein and *zodiacal light. The total mass of zodiacal dust is estimated to be similar to that of an average comet nucleus.

zodiacal light A faint diffuse glow, comparable in intensity to the fainter parts of the Milky Way, produced by the reflection of sunlight from *zodiacal dust particles in the ecliptic plane. From temperate latitudes, the zodiacal light is best seen on spring evenings about 90 min after sunset, or in the autumn about 90 min before sunrise; at these times, the ecliptic, along which the light appears to extend for some 60° to 90°, lies at a steep angle relative to the horizon. An observing site free from light pollution is essential. The intensity of the zodiacal light may vary with the solar cycle, being greatest around sunspot minimum when the interplanetary solar wind is dominated by fast-flowing particle streams from *coronal holes.

Zond A series of space probes launched by the former Soviet Union for testing spacecraft equipment and flight techniques. Zonds 1 and 2 in 1964 were launched towards Venus and Mars to test long-distance communications, but contact was lost with both. Zond 3 in 1965 July took photographs of the parts of the Moon's farside not covered by Luna 3. Zonds 4–8 in 1968–70 were rehearsals for a planned manned Soyuz mission around the Moon and back that was cancelled after the success of Apollo.

zone catalogue A catalogue that gives positions of stars within a restricted band of declination but extending over a large range of right ascension. Historically, zone catalogues were compiled from meridian observations; they are now usually compiled from measurements of a series of wide-angle photographs centred on a particular declination.

zone of avoidance The band around the sky in which very few galaxies are seen. It is a result of obscuration by interstellar dust in the plane of the Milky Way, although the zone is actually inclined at 2° to the galactic equator. The detailed structure of the zone is irregular, varying in width between 38° (towards the galactic centre) and 12°, with some fairly transparent regions known as *galactic windows.

zone time The civil time that is kept within one of the 24 time zones into which the Earth's surface is divided. It differs from Universal Time by a whole number of hours, being ahead of UT in time zones to the east of Greenwich and behind UT in time zones to the west.

Zubenelgenubi The star Alpha Librae, magnitude 2.7. It is an A3 giant or subgiant 75 l.y. away. It has a wide companion of magnitude 5.2, an F3 dwarf.

Zürich relative sunspot number An obsolete name for *relative sunspot number.

Zwicky, Fritz (1898–1974) Swiss astronomer, born in Bulgaria, who worked in America. In 1933 he found that the mass–luminosity ratio for clusters of galaxies was fifty times that for lone galaxies—an early indication of *dark matter, which he attributed to intergalactic matter and dwarf galaxies. In 1934 he and W. *Baade coined the term 'supernova' to distinguish such enormous eruptions from the less powerful novae, and proposed that a supernova explosion leaves behind a neutron star, confirmed when the *Crab Pulsar was identified in 1968. Zwicky observed supernovae in other galaxies, and began a long study of galaxy clusters (*see* ZWICKY CATALOGUE). He also drew attention to intergalactic matter, in particular to bridges of stars between neighbouring galaxies.

Z

Zwicky Catalogue Popular name for the *Catalogue of Galaxies and of Clusters of Galaxies* published in six volumes in 1961–8 by F. *Zwicky and his collaborators. It contains 31 350 galaxies and 9700 clusters recorded on the *Palomar Observatory Sky Survey photographs. Its system of classifying clusters as compact, medium compact, or open is still used.

(⊕) SEE WEB LINKS

• Detailed description and full catalogue downloadable from the CDS.

• Updated Zwicky Catalogue downloadable from the CDS.

ZZ Ceti star A type of variable white dwarf that undergoes *non-radial pulsation, with minimal variation in size but significant temperature changes; abbr. ZZ. Periods range from 30s to 25min, and generally several periods exist simultaneously. Amplitudes are 0.001–0.2 mag. There are three subtypes: ZZA, with DA (hydrogen absorption) spectra; ZZB, with DB (helium absorption) spectra; and ZZO, sometimes called *GW Virginis stars*, with extremely hot, DO spectra. Some ZZ Ceti stars exhibit flares, thought to be caused by a companion *UV Ceti star.

Appendix

Table 1. Apollo lunar landing missions

Crew	Mission duration	Results
Apollo 11 Neil A. Armstrong Michael Collins Edwin E. Aldrin	1969 July 16–24	Armstrong and Aldrin made first manned lunar landing in Sea of Tranquility, July 20
Apollo 12 Charles Conrad Richard F. Gordon Alan L. Bean	1969 November 14–24	Conrad and Bean landed in Ocean of Storms, near unmanned robot probe Surveyor 3, November 19
Apollo 13 James A. Lovell John L. Swigert Fred W. Haise	1970 April 11–17	Landing attempt cancelled after explosion in Service Module
Apollo 14 Alan B. Shepard Stuart A. Roosa Edgar D. Mitchell	1971 January 31 – February 9	Shepard and Mitchell landed near crater Fra Mauro on February 5
Apollo 15 David R. Scott Alfred M. Worden James B. Irwin	1971 July 26 – August 7	Scott and Irwin landed near Hadley Rille at foot of Apennine mountains on July 30. First use of lunar rover
Apollo 16 John W. Young Thomas K. Mattingly Charles M. Duke	1972 April 16–27	Young and Duke landed near crater Descartes in lunar highlands on April 21
Apollo 17 Eugene A. Cernan Ronald E. Evans Harrison H. Schmitt	1972 December 7–19	Cernan and Schmitt landed near crater Littrow, on edge of Sea of Serenity, December 11

Table 2. Main satellites of the planets

Planet and satellite		Diameter (km)	Mean distance from centre of planet (10^3 km)	Orbital period (days)[a]	Orbital inclination (degrees)[b]	Orbital eccentricity	Year discovered
Earth							
	Moon	3475	384.4	27.322	18.28–28.58[c]	0.055	
Mars							
I	Phobos	26 × 23 × 18	9.38	0.319	1.08	0.015	1877
II	Deimos	16 × 12 × 10	23.46	1.262	1.79	0.000	1877
Jupiter							
XVI	Metis	60 × 40 × 34	128.0	0.295	0.02	0.001	1979
XV	Adrastea	20 × 16 × 14	129.0	0.298	0.03	0.002	1979
V	Amalthea	250 × 146 × 128	181.2	0.498	0.39	0.003	1892
XIV	Thebe	116 × 98 × 84	221.9	0.675	1.07	0.015	1979
I	Io	3643	422.0	1.769	0.04	0.004	1610
II	Europa	3122	671.0	3.551	0.47	0.009	1610
III	Ganymede	5262	1070.0	7.155	0.17	0.002	1610
IV	Callisto	4821	1883.0	16.689	0.31	0.007	1610
XIII	Leda	10	11 150	240.5	27.46	0.164	1974
VI	Himalia	170	11 443	250.1	27.50	0.162	1904
X	Lysithea	24	11 700	258.5	28.30	0.112	1938
VII	Elara	80	11 716	259.1	26.63	0.217	1905
XII	Ananke	20	21 048	624.1(R)	148.9	0.244	1951
XI	Carme	30	23 280	726.3(R)	164.9	0.253	1938
VIII	Pasiphae	36	23 658	744.2(R)	151.4	0.409	1908
IX	Sinope	28	23 848	753.2(R)	158.1	0.250	1914
Saturn							
XVIII	Pan	34 × 31 × 21	133.58	0.575	0.00	0.000	1990
XV	Atlas	41 × 35 × 19	137.67	0.602	0.3	0.002	1980
XVI	Prometheus	136 × 79 × 59	139.38	0.613	0.00	0.002	1980
XVII	Pandora	104 × 81 × 64	141.72	0.629	0.00	0.004	1980
XI	Epimetheus	130 × 114 × 106	151.41	0.694	0.34	0.009	1980
X	Janus	203 × 185 × 153	151.46	0.695	0.14	0.007	1966[d]
I	Mimas	416 × 393 × 381	185.54	0.942	1.56	0.019	1789
II	Enceladus	513 × 503 × 497	238.20	1.370	0.03	0.005	1789
III	Tethys	1077 × 1057 × 1053	294.99	1.888	1.10	0.000	1684
XIII	Telesto	33 × 24 × 20	294.71	1.888	1.16	0.001	1980
XIV	Calypso	30 × 23 × 14	294.71	1.888	1.47	0.001	1980
IV	Dione	1123	377.65	2.737	0.01	0.002	1684
XII	Helene	43 × 38 × 26	377.42	2.737	0.21	0.000	1980
V	Rhea	1527	527.37	4.518	0.35	0.000	1672
VI	Titan	5150[e]	1221.8	15.945	0.30	0.029	1655
VII	Hyperion	360 × 266 × 205	1481.1	21.277	0.64	0.103	1848
VIII	Iapetus	1469	3561.8	79.330	18.5	0.028	1671
IX	Phoebe	219 × 217 × 204	12 893	548.2(R)	173.73[c]	0.176	1898
Uranus							
VI	Cordelia	26	49.75	0.335	0.14	0.000	1986
VII	Ophelia	30	53.76	0.376	0.09	0.010	1986
VIII	Bianca	42	59.17	0.435	0.16	0.001	1986
IX	Cressida	62	61.77	0.464	0.04	0.000	1986
X	Desdemona	54	62.66	0.474	0.16	0.000	1986
XI	Juliet	84	64.36	0.493	0.06	0.001	1986
XII	Portia	108	66.10	0.513	0.09	0.000	1986
XIII	Rosalind	54	69.93	0.558	0.28	0.000	1986
XIV	Belinda	66	75.26	0.624	0.03	0.000	1986
XV	Puck	154	86.00	0.762	0.31	0.000	1985
V	Miranda	481 × 468 × 466	129.87	1.413	4.34	0.001	1948
I	Ariel	1162 × 1156 × 1155	190.95	2.520	0.04	0.001	1851
II	Umbriel	1169	266.00	4.144	0.13	0.004	1851
III	Titania	1578	436.30	8.706	0.08	0.001	1787

Planet and satellite		Diameter (km)	Mean distance from centre of planet (10³ km)	Orbital period (days)[a]	Orbital inclination (degrees)[b]	Orbital eccentricity	Year discovered
IV	Oberon	1523	583.52	13.463	0.07	0.002	1787
XVI	Caliban	72	7170.0	579.6(R)	139.8[c]	0.159	1997
XX	Stephano	32	7942.0	675.7(R)	141.5[c]	0.229	1999
XVII	Sycorax	150	12 216	1289.0(R)	152.7[c]	0.522	1997
XVIII	Prospero	50	16 089	1948.1(R)	146.3[c]	0.445	1999
XIX	Setebos	48	17 988	2303.1(R)	148.3[c]	0.591	1999
Neptune							
III	Naiad	58	48.227	0.294	4.74	0.000	1989
IV	Thalassa	80	50.075	0.311	0.21	0.000	1989
V	Despina	148	52.526	0.335	0.07	0.000	1989
VI	Galatea	158	61.953	0.429	0.05	0.000	1989
VII	Larissa	208 × 178	73.548	0.555	0.20	0.001	1989
VIII	Proteus	436 × 416 × 402	117.647	1.122	0.04	0.000	1989
I	Triton	2705	354.76	5.877(R)	156.3	0.000	1846
II	Nereid	340	5513.4	360.13	6.68	0.751	1949

[a] (R) = retrograde orbit
[b] Orbital inclinations are relative to the planet's equator, except where noted.
[c] Relative to the ecliptic plane.
[d] Confirmed 1980.
[e] Diameter of solid body; diameter at cloud top is 5550 km.

Orbital data from *The Astronomical Almanac*. Diameters from the IAU/IAG Working Group on Cartographic Coordinates and Rotational Elements (2009) and *The Astronomical Almanac*.

(⊕) SEE WEB LINKS

• Orbital and physical data for the satellites of the planets from *The Astronomical Almanac*.

(⊕) SEE WEB LINKS

• Up-to-date listings of all satellites of the planets, with orbital and size data.

(⊕) SEE WEB LINKS

• NASA's Jet Propulsion Laboratory pages on the satellites of the planets.

(⊕) SEE WEB LINKS

• Database of planetary moons at the Natural Satellites Data Centre, Paris.

Table 3. The constellations

Name	Genitive	Abbr.	Area (square degrees)	Order of size	Origin (see notes)
Andromeda	Andromedae	And	722	19	1
Antlia	Antliae	Ant	239	62	6
Apus	Apodis	Aps	206	67	3
Aquarius	Aquarii	Aqr	980	10	1
Aquila	Aquilae	Aql	652	22	1
Ara	Arae	Ara	237	63	1
Aries	Arietis	Ari	441	39	1
Auriga	Aurigae	Aur	657	21	1
Boötes	Boötis	Boo	907	13	1
Caelum	Caeli	Cae	125	81	6
Camelopardalis	Camelopardalis	Cam	757	18	4
Cancer	Cancri	Cnc	506	31	1
Canes Venatici	Canum Venaticorum	CVn	465	38	5
Canis Major	Canis Majoris	CMa	380	43	1
Canis Minor	Canis Minoris	CMi	183	71	1
Capricornus	Capricorni	Cap	414	40	1
Carina	Carinae	Car	494	34	6
Cassiopeia	Cassiopeiae	Cas	598	25	1
Centaurus	Centauri	Cen	1060	9	1
Cepheus	Cephei	Cep	588	27	1
Cetus	Ceti	Cet	1231	4	1
Chamaeleon	Chamaeleontis	Cha	132	79	3
Circinus	Circini	Cir	93	85	6
Columba	Columbae	Col	270	54	4
Coma Berenices	Comae Berenices	Com	386	42	2
Corona Australis	Coronae Australis	CrA	128	80	1
Corona Borealis	Coronae Borealis	CrB	179	73	1
Corvus	Corvi	Crv	184	70	1
Crater	Crateris	Crt	282	53	1
Crux	Crucis	Cru	68	88	4
Cygnus	Cygni	Cyg	804	16	1
Delphinus	Delphini	Del	189	69	1
Dorado	Doradus	Dor	179	72	3
Draco	Draconis	Dra	1083	8	1
Equuleus	Equulei	Equ	72	87	1
Eridanus	Eridani	Eri	1138	6	1
Fornax	Fornacis	For	398	41	6
Gemini	Geminorum	Gem	514	30	1
Grus	Gruis	Gru	366	45	3
Hercules	Herculis	Her	1225	5	1
Horologium	Horologii	Hor	249	58	6
Hydra	Hydrae	Hya	1303	1	1
Hydrus	Hydri	Hyi	243	61	3
Indus	Indi	Ind	294	49	3
Lacerta	Lacertae	Lac	201	68	5
Leo	Leonis	Leo	947	12	1
Leo Minor	Leonis Minoris	LMi	232	64	5
Lepus	Leporis	Lep	290	51	1
Libra	Librae	Lib	538	29	1
Lupus	Lupi	Lup	334	46	1
Lynx	Lyncis	Lyn	545	28	5
Lyra	Lyrae	Lyr	286	52	1
Mensa	Mensae	Men	153	75	6
Microscopium	Microscopii	Mic	210	66	6
Monoceros	Monocerotis	Mon	482	35	4
Musca	Muscae	Mus	138	77	3
Norma	Normae	Nor	165	74	6
Octans	Octantis	Oct	291	50	6
Ophiuchus	Ophiuchi	Oph	948	11	1
Orion	Orionis	Ori	594	26	1
Pavo	Pavonis	Pav	378	44	3

Name	Genitive	Abbr.	Area (square degrees)	Order of size	Origin (see notes)
Pegasus	Pegasi	Peg	1121	7	1
Perseus	Persei	Per	615	24	1
Phoenix	Phoenicis	Phe	469	37	3
Pictor	Pictoris	Pic	247	59	6
Pisces	Piscium	Psc	889	14	1
Piscis Austrinus	Piscis Austrini	PsA	245	60	1
Puppis	Puppis	Pup	673	20	6
Pyxis	Pyxidis	Pyx	221	65	6
Reticulum	Reticuli	Ret	114	82	6
Sagitta	Sagittae	Sge	80	86	1
Sagittarius	Sagittarii	Sgr	867	15	1
Scorpius	Scorpii	Sco	497	33	1
Sculptor	Sculptoris	Scl	475	36	6
Scutum	Scuti	Sct	109	84	5
Serpens	Serpentis	Ser	637	23	1
Sextans	Sextantis	Sex	314	47	5
Taurus	Tauri	Tau	797	17	1
Telescopium	Telescopii	Tel	252	57	6
Triangulum	Trianguli	Tri	132	78	1
Triangulum Australe	Trianguli Australis	TrA	110	83	3
Tucana	Tucanae	Tuc	295	48	3
Ursa Major	Ursae Majoris	UMa	1280	3	1
Ursa Minor	Ursae Minoris	UMi	256	56	1
Vela	Velorum	Vel	500	32	6
Virgo	Virginis	Vir	1294	2	1
Volans	Volantis	Vol	141	76	3
Vulpecula	Vulpeculae	Vul	268	55	5

Origin

1. Original Greek constellations listed by Ptolemy.
2. Considered by the Greeks as part of Leo, made separate by Caspar Vopel in 1536.
3. The 12 southern constellations of Keyser and de Houtman.
4. Constellations added by Plancius.
5. Seven constellations of Hevelius.
6. The southern constellations of Lacaille.

Table 4. The brightest stars

Common name	Star	RA h	RA (2000.0) m	dec °	dec ′	Apparent magnitude	Spectral type	Distance (l.y.)	Absolute magnitude
Sirius	α Canis Majoris	06	45.2	−16	43	−1.44	A1V	8.60	1.45
Canopus	α Carinae	06	24.0	−52	42	−0.62	F0II	309	−5.50
Rigil Kentaurus	α Centauri	14	39.6	−60	50	−0.28[a]	G2V + K1V	4.32	4.11[a]
Arcturus	α Boötis	14	15.7	+19	11	−0.05[b]	K1.5III	36.7	−0.35
Vega	α Lyrae	18	36.9	+38	47	+0.03[b]	A0V	25.0	0.60
Capella	α Aurigae	05	16.7	+46	00	0.08[b]	G8III + G0III	42.8	−0.51
Rigel	β Orionis	05	14.5	−08	12	0.18[b]	B8Iab	863	−6.93
Procyon	α Canis Minoris	07	39.3	+05	14	0.40	F5IV–V	11.5	2.67
Achernar	α Eridani	01	37.7	−57	14	0.45[b]	B3Vpe	139	−2.70
Betelgeuse	α Orionis	05	55.2	+07	24	0.45[b]	M2Iab	498	−5.47
Hadar	β Centauri	14	03.8	−60	22	0.61[b]	B1III	392	−4.79
Altair	α Aquilae	19	50.8	+08	52	0.76[b]	A7V	16.7	2.21
Acrux	α Crucis	12	26.6	−63	06	0.77[a]	B0.5IV + B1V	322	−4.20[a]
Aldebaran	α Tauri	04	35.9	+16	31	0.87	K5III	66.6	−0.68
Spica	α Virginis	13	25.2	−11	10	0.98[b]	B1III–IV + B2V	250	−3.44
Antares	α Scorpii	16	29.4	−26	26	1.06[b]	M1.5Iab–b	554	−5.09
Pollux	β Geminorum	07	45.3	+28	02	1.16	K0IIIb	33.8	1.08
Fomalhaut	α Piscis Austrini	22	57.7	−29	37	1.17	A3V	25.1	1.74
Becrux	β Crucis	12	47.7	−59	41	1.25[b]	B0.5III	279	−3.41
Deneb	α Cygni	20	41.4	+45	17	1.25[b]	A2Iae	1412	−6.93
Regulus	α Leonis	10	08.4	+11	58	1.36	B7V	79.3	−0.57
Adhara	ε Canis Majoris	06	58.6	−28	58	1.50	B2Iab	405	−3.97

[a] Combined magnitude of double star.
[b] Variable. For the individual components of α Centauri see Nearest Stars table. Data from *The Hipparcos Catalogue New Reduction*. Spectral types from the SIMBAD database.

Table 5. The nearest stars

Star	RA (2000.0) h	m	dec °	′	Apparent magnitude	Spectral type	Parallax ″	Distance (l.y.)	Absolute magnitude
Sun	—	—	—	—	−26.75	G2V	—	—	4.82
Proxima (V645 Cen)	14	30	−62	41	11.01[a]	M6Ve	0.7716	4.23	15.45
α Centauri A	14	40	−60	50	−0.01	G2V	0.7548	4.32	4.38
B					1.35	K1V	0.7548	4.32	5.74
Barnard's Star (V2500 Ooh)	17	58	+04	42	9.54	M4Ve	0.5483	5.95	13.24
CN Leonis (Wolf 359)	10	56	+07	01	13.54[a]	M6.5Ve	0.419	7.78	16.65
Lalande 21185 (HD 95735)	11	03	+35	58	7.49	M2V	0.3926	8.31	10.46
Sirius A	06	45	−16	43	−1.44	A1V	0.3792	8.60	1.45
B					8.44	DA2	0.3792	8.60	11.33
UV Ceti A (BL Ceti)	01	39	−17	57	12.57[a]	M5.5Ve	0.3737	8.73	15.43
B (UV Ceti)					11.99[a]	M5.5Ve	0.3737	8.73	14.85
V1216 Sgr (Ross 154)	18	50	−23	50	10.37	M3Ve	0.3367	9.69	13.01
HH And (Ross 248)	23	42	+44	11	12.28	M6Ve	0.316	10.32	14.78
ε Eridani	03	33	−09	27	3.72	K2V	0.3109	10.49	6.18
Lacaille 9352 (HD 217987)	23	06	−35	51	7.35	M2V	0.3053	10.68	9.77
FI Vir (Ross 128)	11	48	+00	48	11.08	M4V	0.2980	10.94	13.45
EZ Aqr	22	39	−15	18	12.18	M6Ve	0.290	11.27	14.49
HD 173740 (BD +59° 1915 B)	18	43	+59	38	9.70	M3.5V	0.2895	11.27	12.01
61 Cygni A (V1803 Cyg)	21	07	+38	45	5.20[a]	K5V	0.2868	11.37	7.49
B					6.05[a]	K7V	0.2859	11.41	8.33
Procyon A	07	39	+05	14	0.40	F5IV-V	0.2846	11.46	2.67
B					10.92	DA	0.2846	11.46	13.19
HD 173739 (BD +59° 1915 A)	18	43	+59	38	8.94	M3V	0.2802	11.64	11.18
GQ And (BD +43° 44 B)	00	18	+44	02	11.04	M3.5V	0.2803	11.64	13.28
GX And (BD +43° 44 A)	00	18	+44	01	8.09[a]	M2V	0.2788	11.70	10.32

[a] Variable. Data from *The Hipparcos Catalogue New Reduction*, the RECONS database, and the SIMBAD database.

[handwritten annotations: "Catalog of French astronomer, Charles Mess..." with arrows pointing to the table title]

Table 6. The Messier objects

Number M	NGC	Constellation	Size (')	Mag.	Type
1	1952	Tau	6 × 4	8.4[a]	Supernova remnant
2	7089	Aqr	13	6.5	Globular cluster
3	5272	CVn	16	6.4	Globular cluster
4	6121	Sco	26	5.9	Globular cluster
5	5904	Ser	17	5.8	Globular cluster
6	6405	Sco	15	4.2	Open cluster
7	6475	Sco	80	3.3	Open cluster
8	6523	Sgr	90 × 40	5.8[a]	Diffuse nebula
9	6333	Oph	9	7.9[a]	Globular cluster
10	6254	Oph	15	6.6	Globular cluster
11	6705	Sct	14	5.8	Open cluster
12	6218	Oph	14	6.6	Globular cluster
13	6205	Her	17	5.9	Globular cluster
14	6402	Oph	12	7.6	Globular cluster
15	7078	Peg	12	6.4	Globular cluster
16	6611	Ser	7	6.0	Open cluster
17	6618	Sgr	46 × 37	7[a]	Diffuse nebula
18	6613	Sgr	9	6.9	Open cluster
19	6273	Oph	14	7.2	Globular cluster
20	6514	Sgr	29 × 27	8.5[a]	Diffuse nebula
21	6531	Sgr	13	5.9	Open cluster
22	6656	Sgr	24	5.1	Globular cluster
23	6494	Sgr	27	5.5	Open cluster
24		Sgr	90	4.5[a]	Starfield in Sagittarius
25	IC 4725	Sgr	32	4.6	Open cluster
26	6694	Sct	15	8.0	Open cluster
27	6853	Vul	8 × 4	8.1[a]	Planetary nebula
28	6626	Sgr	11	6.9[a]	Globular cluster
29	6913	Cyg	7	6.6	Open cluster
30	7099	Cap	11	7.5	Globular cluster
31	224	And	178 × 63	3.4	Spiral galaxy
32	221	And	8 × 6	8.2	Elliptical galaxy
33	598	Tri	62 × 39	5.7	Spiral galaxy
34	1039	Per	35	5.2	Open cluster
35	2168	Gem	28	5.1	Open cluster
36	1960	Aur	12	6.0	Open cluster
37	2099	Aur	24	5.6	Open cluster
38	1912	Aur	21	6.4	Open cluster
39	7092	Cyg	32	4.6	Open cluster
40		UMa	—	8[a]	Faint double star
41	2287	CMa	38	4.5	Open cluster
42	1976	Ori	66 × 60	4[a]	Diffuse nebula
43	1982	Ori	20 × 15	9[a]	Diffuse nebula
44	2632	Cnc	95	3.1	Open cluster
45		Tau	10	1.2	Open cluster
46	2437	Pup	27	6.1	Open cluster
47	2422	Pup	30	4.4	Open cluster
48	2548	Hya	54	5.8	Open cluster
49	4472	Vir	9 × 7	8.4	Elliptical galaxy
50	2323	Mon	16	5.9	Open cluster
51	5194–5	CVn	11 × 8	8.1	Spiral galaxy
52	7654	Cas	13	6.9	Open cluster
53	5024	Com	13	7.7	Globular cluster
54	6715	Sgr	9	7.7	Globular cluster
55	6809	Sgr	19	7.0	Globular cluster
56	6779	Lyr	7	8.2	Globular cluster
57	6720	Lyr	1	9.0[a]	Planetary nebula
58	4579	Vir	5 × 4	9.8	Spiral galaxy
59	4621	Vir	5 × 3	9.8	Elliptical galaxy
60	4649	Vir	7 × 6	8.8	Elliptical galaxy
61	4303	Vir	6 × 5	9.7	Spiral galaxy
62	6266	Oph	14	6.6	Globular cluster
63	5055	CVn	12 × 8	8.6	Spiral galaxy
64	4826	Com	9 × 5	8.5	Spiral galaxy
65	3623	Leo	10 × 3	9.3	Spiral galaxy

Number M	NGC	Constellation	Size (')	Mag.	Type
66	3627	Leo	9 × 4	9.0	Spiral galaxy
67	2682	Cnc	30	6.9	Open cluster
68	4590	Hya	12	8.2	Globular cluster
69	6637	Sgr	7	7.7	Globular cluster
70	6681	Sgr	8	8.1	Globular cluster
71	6838	Sge	7	8.3	Globular cluster
72	6981	Aqr	6	9.4	Globular cluster
73	6994	Aqr	—		Group of four stars
74	628	Psc	10 × 9	9.2	Spiral galaxy
75	6864	Sgr	6	8.6	Globular cluster
76	650–1	Per	2 × 1	11.5[a]	Planetary nebula
77	1068	Cet	7 × 6	8.8	Spiral galaxy
78	2068	Ori	8 × 6	8[a]	Diffuse nebula
79	1904	Lep	9	8.0	Globular cluster
80	6093	Sco	9	7.2	Globular cluster
81	3031	UMa	26 × 14	6.8	Spiral galaxy
82	3034	UMa	11 × 5	8.4	Irregular galaxy
83	5236	Hya	11 × 10	7.6[a]	Spiral galaxy
84	4374	Vir	5 × 4	9.3	Elliptical galaxy
85	4382	Com	7 × 5	9.2	Elliptical galaxy
86	4406	Vir	7 × 6	9.2	Elliptical galaxy
87	4486	Vir	7	8.6	Elliptical galaxy
88	4501	Com	7 × 4	9.5	Spiral galaxy
89	4552	Vir	4	9.8	Elliptical galaxy
90	4569	Vir	10 × 5	9.5	Spiral galaxy
91	4548	Com	5 × 4	10.2	Spiral galaxy
92	6341	Her	11	6.5	Globular cluster
93	2447	Pup	22	6.2[a]	Open cluster
94	4736	CVn	11 × 9	8.1	Spiral galaxy
95	3351	Leo	7 × 5	9.7	Spiral galaxy
96	3368	Leo	7 × 5	9.2	Spiral galaxy
97	3587	UMa	3	11.2[a]	Planetary nebula
98	4192	Com	10 × 3	10.1	Spiral galaxy
99	4254	Com	5	9.8	Spiral galaxy
100	4321	Com	7 × 6	9.4	Spiral galaxy
101	5457	UMa	27 × 26	7.7	Spiral galaxy
102					Duplicate of M101
103	581	Cas	6	7.4[a]	Open cluster
104	4594	Vir	9 × 4	8.3	Spiral galaxy
105	3379	Leo	4 × 4	9.3	Elliptical galaxy
106	4258	CVn	18 × 8	8.3	Spiral galaxy
107	6171	Oph	10	8.1	Globular cluster
108	3556	UMa	8 × 2	10.0	Spiral galaxy
109	3992	UMa	8 × 5	9.8	Spiral galaxy
110	205	And	17 × 10	8.0	Elliptical galaxy

[a]Approximate value.

Adapted from a table in A. Hirshfeld and R. Sinnott (eds), *Sky Catalogue 2000.0*, Vol. 2 (Sky Publishing Corp.).

Notes

M1	Crab Nebula
M8	Lagoon Nebula
M11	Wild Duck Cluster
M17	Omega Nebula
M20	Trifid Nebula
M24	Contains open cluster NGC 6603
M27	Dumbbell Nebula
M31	Andromeda Galaxy
M42	Orion Nebula
M44	Praesepe, or Beehive Cluster
M45	Pleiades
M51	Whirlpool Galaxy
M57	Ring Nebula
M64	Black Eye Galaxy
M97	Owl Nebula
M104	Sombrero Galaxy

Table 7. Main members of the Local Group of galaxies

Galaxy	RA (2000.0) h	m	Dec. °	'	Type[a]	Absolute magnitude	Visual magnitude	Distance from Sun (kpc)
M31 (NGC 224)	00	42.7	+41	16	Sb I–II	−21.2	3.4	770
Milky Way	—		—		S(B)bc I–II	−20.9	—	—
M33 (NGC 598)	01	33.9	+30	39	Sc II–III	−18.9	5.9	850
Large Magellanic Cloud[M]	05	24	−69	45	Ir III–IV	−18.5	0.2	50
Small Magellanic Cloud[M]	00	53	−72	50	Ir IV/IV–V	−17.1	2.0	63
M32 (NGC 221)[A]	00	42.7	+40	52	dE2	−16.5	8.1	770
M110 (NGC 205)[A]	00	40.4	+41	41	dE5 pec	−16.4	8.4	830
IC 10[A]	00	20.3	+59	18	dIr IV	−16.0	10.3	660
NGC 6822 (Barnard's Galaxy)[M]	19	44.9	−14	52	dIr IV–V	−16.0	8.3	500
NGC 185[A]	00	39.0	+48	20	dE3 pec	−15.6	9.0	620
IC 1613[A]	01	04.8	+02	07	dIr V	−15.3	9.1	730
NGC 147[A]	00	33.2	+48	30	dE5	−15.1	9.9	755
Wolf–Lundmark–Melotte (DDO 221)	00	02.0	−15	28	dIr IV–V	−14.4	10.6	920
Fornax Dwarf[M]	02	40.0	−34	27	dSph	−13.1	7.7	138
Pegasus Dwarf (DDO 216, Peg DIG)[A]	23	28.6	+14	45	dIr/dSph	−12.9	12.3	760
Sagittarius I Dwarf[M]	18	55.0	−30	29	dSph	−12.7	7.7	26
Sagittarius (SagDIG)	19	30.0	−17	41	dIr V	−12.0	13.8	1040
Andromeda I[A]	00	45.7	+38	00	dSph	−11.8	12.9	810
Andromeda II[A]	01	16.5	+33	25	dSph	−11.8	12.6	680
Cassiopeia Dwarf (Andromeda VII)[A]	23	26.5	+50	42	dSph	−11.7	13.0	790
Leo A (Leo III, DDO 69)	09	59.4	+30	45	dIr V	−11.7	12.8	690
Andromeda VI (Pegasus II)[A]	23	51.8	+24	35	dSph	−11.3	13.4	820
Leo I (DDO 74)[M]	10	08.4	+12	18	dSph	−11.0	10.4	250
Aquarius Dwarf (DDO 210)	20	46.9	−12	51	dIr/dSph	−10.9	14.2	940
Andromeda III[A]	00	35.6	+36	30	dSph	−10.2	14.4	760
Cetus Dwarf	00	26.2	−11	03	dSph	−10.1	14.4	775
Leo B (Leo II, DDO 93)[M]	11	13.5	+22	09	dSph	−10.1	11.5	205

[a] dE = dwarf elliptical; dIr = dwarf irregular; dSph = dwarf spheroidal.
[A] = companion of the Andromeda Galaxy;
[M] = companion of the Milky Way.
There are numerous additional faint dwarfs, most of them companions to the Andromeda Galaxy or the Milky Way.
Based on a table by Eva K. Grebel, John S. Gallagher III, and Daniel Harbeck, *Astronomical Journal*, vol. 125, p. 1928 (April 2003).

Table 8. An A to Z of variable star types

ACV	Alpha^2 Canum Venaticorum type
ACVO	rapidly oscillating Alpha^2 Canum Venaticorum type
ACYG	Alpha Cygni type
AM	AM Herculis type
AR	AR Lacertae type
BCEP	Beta Cephei type
BCEPS	Beta Cephei type of short period
BE	variable Be star not of GCAS type
BLBOO	anomalous Cepheid
BY	BY Draconis type
CEP	Cepheid type
CEP(B)	beat Cepheid
CST	constant, although previously suspected of variability
CW	W Virginis type
CWA	W Virginis type with period longer than 8 days
CWB	W Virginis type with period shorter than 8 days
D	detached binary
DCEP	classical Cepheid
DCEPS	classical Cepheid with near-symmetrical light curve
DSCT	Delta Scuti type
DSCTC	Delta Scuti type of low amplitude
DM	detached binary with two main-sequence stars
DS	detached binary with a subgiant
DW	detached binary similar to W Ursae Majoris type but not in contact
ELL	ellipsoidal type
E	eclipsing binary
EA	eclipsing binary of Algol type
EB	eclipsing binary of Beta Lyrae type
EP	star eclipsed by its own planet
EW	eclipsing binary of W Ursae Majoris type
FKCOM	FK Comae Berenices type
FU	FU Orionis type
GCAS	Gamma Cassiopeiae type
GS	eclipsing binary with a giant or supergiant
I	irregular variable with uncertain features
IA	irregular variable of early (O–A) spectral type
IB	irregular variable of F–M spectral type
IN	Orion variable, also known as a nebular variable
INA	Orion variable of early spectral type (B–A)
INB	Orion variable of F–M spectral type
INT	Orion variable of T Tauri type
IN(YY)	Orion variable of YY Orionis type
IS	irregular variable of short period
ISA	irregular variable of short period and early spectral type (B–A)
ISB	irregular variable of short period and F–M spectral type
IT	Orion variable of T Tauri type but without nebulosity
K	contact binary
KE	contact binary of early spectral type (O–A)
KW	contact binary of W Ursae Majoris type
L	irregular variable of long period
LB	irregular variable of long period and late spectral type (K, M, C, S)
LC	irregular variable supergiant
LPB(LBV)	long-period pulsating B star
M	Mira type
N	nova
NA	fast nova
NB	slow nova
NC	very slow nova; RR Telescopii type
NL	nova-like variable
NR	recurrent nova
PN	eclipsing binary with one component the nucleus of a planetary nebula
PSR	optically variable pulsar

(continued)

Table 8. Continued

PVTEL	PV Telescopii type
R	close binary with strong reflection effect
RCB	R Coronae Borealis type
RR	RR Lyrae type
RR(B)	RR Lyrae type with two pulsation modes, i.e. beats
RRAB	RR Lyrae type with asymmetric light curve
RRC	RR Lyrae type with near-symmetric or symmetric light curve
RS	RS Canum Venaticorum type
RV	RV Tauri type
RVA	RV Tauri type with no variation in mean magnitude
RVB	RV Tauri type with periodic variation in mean magnitude
SD	semidetached binary
SDOR	S Doradus type
SN	supernova
SNI	supernova Type I
SNII	supernova Type II
SR	semiregular variable
SRA	semiregular giant with persistent periodicity
SRB	semiregular giant with occasional periodicity
SRC	semiregular supergiant with moderate periodicity
SRD	semiregular giant or supergiant with F–K spectral type
SRS	semiregular red giant of short period
SXARI	SX Arietis type
SXPHE	SX Phoenicis type
UG	U Geminorum type
UGSS	SS Cygni type
UGSU	SU Ursae Majoris type
UGZ	Z Camelopardalis type
UV	UV Ceti type
UVN	Orion variables of Ke-Me spectral type with flares
WD	eclipsing binary with white dwarf components
WR	Wolf–Rayet type
X	X-ray binary
XB	X-ray burster
XF	X-ray binary with rapid fluctuations
XI	X-ray binary with irregular variations
XJ	X-ray binary with jets
XND	X-ray transient with a dwarf or subgiant of G–M spectral type
XNG	X-ray transient with an early-type supergiant or giant
XP	X-ray pulsar
XPR	X-ray pulsar with reflection effect
XPRM, XM	X-ray binary with a late-type dwarf and a pulsar with strong magnetic field
ZAND	Z Andromedae type
ZZ	ZZ Ceti type
ZZA	ZZ Ceti type with DA spectrum
ZZB	ZZ Ceti type with DB spectrum
ZZO	ZZ Ceti type with DO spectrum

Table 9. The planets: orbital and physical data

	Mercury	Venus	Earth	Mars	Jupiter	Saturn	Uranus	Neptune
Mean distance from Sun (AU)[a]	0.387	0.723	1.000	1.524	5.203	9.537	19.189	30.069
Min. distance from Sun (AU)[a]	0.307	0.718	0.983	1.381	4.952	9.021	18.286	29.811
Max. distance from Sun (AU)[a]	0.467	0.728	1.017	1.666	5.455	10.054	20.096	30.327
Eccentricity[a]	0.206	0.007	0.017	0.093	0.048	0.054	0.047	0.009
Inclination to ecliptic (°)[a]	7.00	3.39	0.00	1.85	1.31	2.48	0.8	1.77
Sidereal period	87.969d	224.701d	365.26d	686.980d	11.863y	29.447y	84.02y	164.79y
Mean synodic period (d)	115.88	583.92	–	779.94	398.88	378.09	369.66	367.49
Axial rotation (sidereal)[b]	58.646d	243.019d	23.934h	24.623h	9.842h	10.233h	17.240h	16.110h
Inclination of equator to orbit (°)[c]	0.01	177.36	23.44	25.19	3.13	26.73	97.77	28.32
Equatorial diameter (km)[c]	4879	12 104	12 756	6792	142 984	120 536	51 118	49 528
Oblateness	0	0	0.0034	0.0059	0.065	0.098	0.023	0.017
Mass (Earth = 1)	0.06	0.82	1.00	0.11	317.83	95.16	14.5	17.15
Volume (Earth = 1)	0.06	0.86	1.00	0.15	1321	764	63	58
Mean density	5.43	5.24	5.52	3.94	1.33	0.69	1.27	1.64
Geometrical albedo	0.11	0.65	0.37	0.15	0.52	0.47	0.51	0.41
Mean visual magnitude[d]	0.0	–4.4	—	2.0	–2.7	+0.7[e]	+5.5	+7.8

[a] Keplerian elements for epoch 2000.0.

[b] Venus and Uranus are retrograde. The rotation period for Jupiter is at its equator (System I). The rotation periods for Uranus and Neptune are those of their magnetic fields.

[c] At the 1-bar level in the atmosphere for Jupiter, Saturn, Uranus, and Neptune.

[d] Greatest elongation for Mercury and Venus, mean opposition distance for Mars to Neptune.

[e] With rings closed; –0.3 with rings open.